D1205563

INTRODUCTION TO TELECOMMUNICATIONS

SECOND EDITION

M. A. Rosengrant

Rochester Institute of Technology

PEARSON

Prentice
Hall

Upper Saddle River, New Jersey
Columbus, Ohio

Library of Congress Cataloging-in-Publication Data

Rosengrant, M. A. (Martha A.)
  Introduction to telecommunications/M.A. Rosengrant.—2nd ed.
    p.cm.
  Includes index.
  ISBN 0-13-112615-6
  1. Telecommunication—Textbooks. I. Title.
TK5101.R595 2007
384—dc22   2006046002

Editor in Chief: Vernon Anthony
Acquisitions Editor: Jeff Riley
Production Editor: Rex Davidson
Design Coordinator: Diane Ernsberger
Editorial Assistant: Lara Dimmick
Cover Designer: Candace Rowley
Cover Art: Digital Vision
Production Manager: Matt Ottenweller
Director of Marketing: David Gesell
Senior Marketing Manager: Ben Leonard
Marketing Assistant: Les Roberts
Senior Marketing Coordinator: Liz Farrell

This book was set in Times Roman by Carlisle Publishing Services and was printed and bound by Courier Kendallville, Inc.
The cover was printed by Coral Graphic Services, Inc.

Copyright © 2007, 2002 by Pearson Education, Inc., Upper Saddle River, NJ 07458. Pearson Prentice Hall. All rights
reserved. Printed in the United States of America. This publication is protected by copyright and permission should be obtained
from the publisher prior to any prohibited reproduction, storage in a retrieval system, or transmission in any form or by any
means, electronic, mechanical, photocopying, recording, or likewise. For information regarding permission(s), write to: Rights
and Permissions Department.

Pearson Prentice Hall™ is a trademark of Pearson Education, Inc.
Pearson® is a registered trademark of Pearson plc.
Prentice Hall® is a registered trademark of Pearson Education, Inc.

Pearson Education Ltd.   Pearson Education Canada, Ltd.
Pearson Education Australia Pty. Limited   Pearson Educatión de Mexico, S.A. de C.V.
Pearson Education Singapore Pte. Ltd.   Pearson Education—Japan
Pearson Education North Asia Ltd.   Pearson Education Malaysia Pte. Ltd.

10 9 8 7 6 5 4 3 2
ISBN: 0-13-112615-6

To my family

THE TELECOMMUNICATIONS INDUSTRY

The information age is upon us. The raw materials are ones and zeros. The technology used to transport the ones and zeros provides numerous opportunities for entrepreneurs, scientists, and engineers. Telecommunications is dynamically changing the way we work, learn, communicate, and view society. At no other time in history have so many people been given the ability to exchange ideas, sell their products, or learn through research and study. E-commerce, business-to-business information exchange, surfing the Web, chatting on the Internet, calling anywhere in the world, low long distance and local telephone rates, and hundreds of calling features are just some of the services being carried across the information superhighway.

The dilemma we all face is that we must adapt to the information age and the constant changes it presents. We all need to be versed in the way information is managed in order to function in the new digital society. Think of telecommunications as a huge, jagged jigsaw puzzle with pieces that change continuously. In order to become part of the new digital technology, you must understand how the pieces fit together.

PURPOSE

The purpose of *Introduction to Telecommunications*, Second Edition, is to provide a comprehensive guide for telecommunications students and specialists alike. The text is written and organized in a manner that is conducive to information retention. Numerous examples and analogies, designed to help simplify technical concepts and improve students' assimilation of key topics, are used throughout. Instead of simply stating facts, the text presents entire scenarios that help students understand the "big picture," thus encouraging them to think critically instead of simply memorizing information.

STRUCTURE AND CONTENT

This text has been structured so that key facts are linked to the fundamental concepts necessary to develop true mastery of the discipline of telecommunications.

Each chapter begins with a list of *Objectives* and an *Outline,* both of which help point out the most important topics to be covered in that chapter. Having read through these first two sections, students will be prepared to pay extra attention to each objective as they read through the balance of the chapter. *Key terms* and *definitions* placed in the margins help readers more fully understand each discussion point. Chapters contain many informative *photos* and *line drawings,* which are used to illustrate important aspects of telecommunications.

All chapters have *Review Questions,* and many chapters have a *Troubleshooting* section. The review questions are designed to improve information retention, while the troubleshooting sections require critical thinking and application of theory. A *Case Study* is provided in most chapters. Beginning with Chapter 4, students are asked to make decisions about new services being offered to a small, fictitious town in Wyoming. Throughout the remaining chapters of the book, students will be presented with new challenges. They will need to decide where to place telephone poles, cables, and switching devices and will have to determine which types of applications to offer. This very involved case study helps students put together the telecommunications "puzzle." Upon finishing the last case study, students will have designed a complete communications network for Green Grass, Wyoming. Finally, chapters conclude with a list of the *Key Terms*—the words defined in the margins of the chapter.

The foreword, "The Evolution of the Communications Network: A Historical Overview of Telecommunications," contains two sections. "Part I: 1845 to 1920" delves into the origins of telecommunications. "Part II: 1920 to the Present" outlines the development of the telecommunications network as we now know it. Taken as a whole, the foreword gives students the background needed to fully understand the modern networks of telephone companies and the technical aspects of their services.

"Part One: High-Level Overview" includes "Chapter 1: The Basics—Sound, Electrical Signal, Electromagnetic Spectrum"; "Chapter 2: The Telephone and the Telephone Line"; and "Chapter 3: Connecting the Dots—Transporting Information across the Superhighway." The first few chapters present an overview of the telecommunications industry and address technical components necessary for the telephone network to function.

"Part Two: Networking Fundamentals" includes Chapter 4 through Chapter 8. The goal of Part Two is to explain at a high level the fundamental concepts associated with communications systems. The student will gain insight into how communications systems use common methods and protocols to prepare, package, transport, and interpret the message traveling across the medium. The section starts by presenting core concepts related to handling and manipulating information into a standard format that is acceptable to the medium being used to carry the signal end to end. This may be a voice information signal riding on a copper wire or a high-speed digital signal riding on a fiber optic strand. The goal of "Chapter 4: Manipulating Information for Transmission," the first chapter in the section, is to explain the nuts and bolts of the communications signal from the modulation technique to the one and zero pulse shape. "Chapter 5: Open System Interconnection" covers the OSI model as it relates to communications networks. The chapter first defines each layer of the seven-layer model, then associates each layer to a function in the end-to-end communications flow. The main goal of Chapter 5 is to present the OSI model in a format that relates the abstract to the actual application. "Chapter 6: Layer 3 and Layer 4 Networking and Transport Protocols" covers the most common Layer 3 and Layer 4 protocols—IP, TCP, UDP, and RTP. "Chapter 7: Layer 2 Protocols: Ethernet, Frame Relay, ATM" focuses on Layer 2 protocols, specifically Ethernet, frame relay, and ATM. "Chapter 8: Layer 1 Networking Protocols: TDM Protocols—DS-1, DS-3, SONET" focuses on Layer 1 protocols, specifically the North American digital standard. In summary, the primary purpose of Part Two is to present fundamental principles related to transporting communications signals across many different types of networks. Later chapters build on the concepts and information presented in Part Two.

"Part Three: Switching and Routing" includes Chapter 9 through Chapter 12. Part Three of the text covers three principal areas in communications: how voice information is switched across the network, how data information is routed across the network, and how signaling protocols are used to direct and navigate the information signals. The part starts with an overview of the traditional centralized switching architecture. This includes a section on the standard switch hierarchy, class 5, class 4, and so on, in addition to a closer view of the typical structure of a circuit switch in "Chapter 9: The Digital Circuit Switch." "Chapter 10: Signaling" looks at the signaling methods used to transport traditional voice services across the PSTN. It focuses on SS-7 networks and other analog switching techniques deployed throughout North America. "Chapter 11: Distributed Switching Architecture—Voice Over IP" covers the new distributed switching architecture used to carry VoIP, also called digital voice traffic. "Chapter 12: IP Routing Fundamentals" focuses on how information is routed across the Internet or across any network that uses the standard Internet Protocol. Chapter 12 takes a high-level view of IP addressing, routing protocols, and so forth. The goal of Part Three is to show how information is parsed and moved through the network by very smart devices such as circuit switches or gateway routers.

"Part Four: Telecommunications Networks" includes Chapter 13 through Chapter 17. At this point in the text, the student should be comfortable with the fundamental rules of the road—the OSI model—and how the network handles switching and routing of the information. Part Four focuses on the networks that connect all of the switching and routing devices—simply, the structure of the highway used to carry the information between the switching and routing devices. "Chapter 13: Transmission Media: Copper, Fiber, Wireless" provides an overview of the physical medium. Its goal is to describe the physical structure of the medium, the types of signals that ride across the medium, how much information can be carried, and how far the signal can travel on the medium. "Chapter 14: Telecommunications Networks' Physical Infrastructure" covers how the physical communications plant is constructed. The chapter provides a view of the physical structure of the communications plant, including poles, underground systems, and so forth. Chapter 15, Chapter 16, and Chapter 17 provide an overview of the three primary classes of networks used by all communications providers today: core networks, metropolitan networks, and access networks. All service providers—cellular, telephone company, wireless, ISPs, and CATV—utilize the core, metropolitan, and access networks described in this section. The primary reason for presenting the information in this sequence is to show the overall architecture of the telecommunications infrastructure and how it doesn't matter whether it is carrying a voice over IP signal or a peer-to-peer gaming session.

"Part Five: Communications Services Providers" includes Chapter 18 through Chapter 24. Part Five of the text focuses on the three primary service providers in telecommunications: the telephone company, the cable television provider, and the wireless providers. "Chapter 18: The Public Switched Telephone Network Central Office" provides an overview of central office services. Chapter 19 looks at the PSTN's topology, and Chapter 20 looks at the customer premises. Chapter 21 and Chapter 22 present the CATV provider's headend and network architecture. Chapter 23 looks at the cellular telephone network and switching center. Chapter 24 presents wireless broadband providers' networks and equipment overview. The purpose of Part Five is to expand on the previous chapters' discussion related to fundamental concepts, information switching/routing, and networking. This part shows how the previously discussed concepts are used by telecommunications providers to offer information services to the end subscriber. The service providers discussed in this section provide a service to the end user. The goals of Chapter 18 to Chapter 24 are to show how each provides those services.

"Part Six: Telecommunications Applications" includes Chapter 25 through Chapter 27. Part Six covers the various types of communications applications sold to end subscribers by the different service providers. Chapter 25 discusses the communications services sold to residential customers; Chapter 26 focuses on services sold to business customers; and Chapter 27 discusses how the Internet functions as a service to all end users. Having completed Part Five of the text, the student should be comfortable explaining the various piece parts of a telecommunications network. The purpose for ending the text with a description of the different applications available to the end user is twofold. First, it provides the student with a comprehensive view of communications systems from the fundamental concepts as presented in Chapter 4 to how the networks are used by end subscribers. Second, the information presented in Part Six completes the story of how information flows end to end. At this point in the text, students should feel comfortable discussing key terms relevant to communications networks. In addition, they should have a good feel for what telecommunications professionals deal with day to day and the overall structure of the current telecommunications industry.

"Part Seven: Emerging Technologies" includes Chapter 28. Chapter 28 focuses on several areas related to new communications systems, equipment, and networks.

REVIEWERS

Rodney A. Crater, Northwest Iowa Community College

Phillip Davis, Del Mar College

Ed Margraff, Marion Technical College

Robert Robertson, Southern Utah University

Carl Shinn, University of Denver

Joseph Sloan, Lander University

Timothy Staley, DeVry University—Irving

David Whitmore, Champlain College

Jianchao Zeng, Howard University

SUPPLEMENTARY MATERIAL

To access supplementary material online, instructors need to request an instructor access code. Go to *http://www.prenhall.com/*, click the Instructor Resource Center link, and then click Register Today for an instructor access code. Within forty-eight hours of registering, you will received a confirming e-mail including an instructor access code. Once you have received your code, go to the site and log on for full instructions on downloading the materials you wish to use.

An *Instructor's Manual* with PowerPoints® is available.

PH TestGen®, an electronic test bank with questions that can be used to develop customized quizzes, tests, or final exams, is available.

BRIEF CONTENTS

CONTENTS

The Evolution of the Communications Network: A Historical Overview of Telecommunications

Part I: 1845 to 1920

Part II: 1920 to the Present

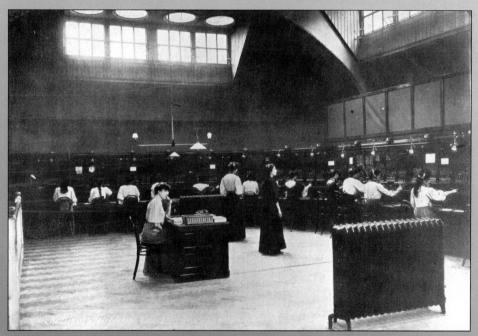

Photo courtesy of the Library of Congress

1845 to 1920

Objectives

After reading this section, you should be able to

- ■ Describe the communications network from 1844 to 1875.
- ■ Describe the telephone industry from 1876 to 1893.
- ■ Describe the growth of the telephone industry from 1894 to 1918.

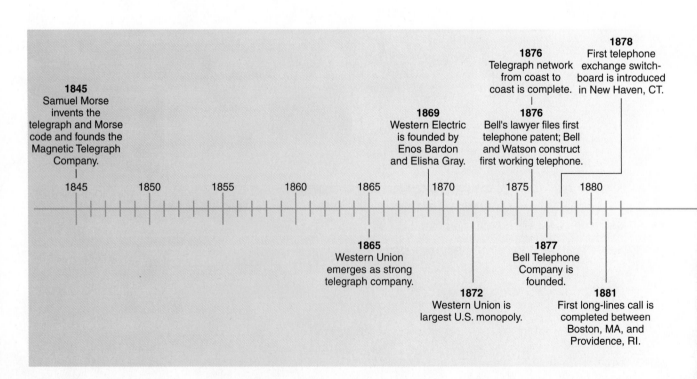

1845
Samuel Morse invents the telegraph and Morse code and founds the Magnetic Telegraph Company.

1869
Western Electric is founded by Enos Bardon and Elisha Gray.

1876
Telegraph network from coast to coast is complete.

1876
Bell's lawyer files first telephone patent; Bell and Watson construct first working telephone.

1878
First telephone exchange switchboard is introduced in New Haven, CT.

1845 1850 1855 1860 1865 1870 1875 1880

1865
Western Union emerges as strong telegraph company.

1872
Western Union is largest U.S. monopoly.

1877
Bell Telephone Company is founded.

1881
First long-lines call is completed between Boston, MA, and Providence, RI.

Outline

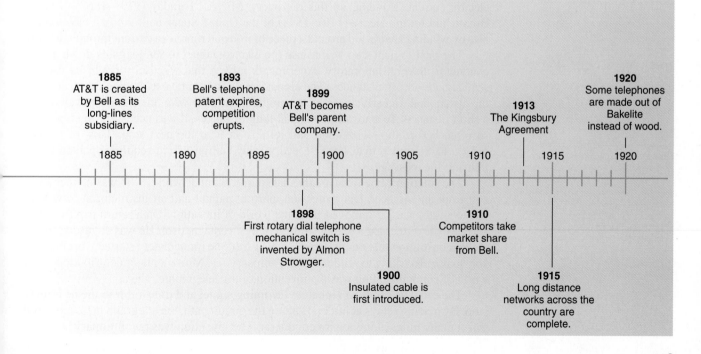

INTRODUCTION

Try to imagine life without a telephone. How would you invite a friend to dinner, ask your significant other to bring home a loaf of bread, or summon an ambulance to take you to the hospital? Try to imagine life without a computer. How would you e-mail an overseas colleague when researching a paper, find information about the latest fad diet on the Internet, or chat online with a celebrity? The telephone network allows instantaneous communication among millions of people. Whether the information being sent is in the form of data from the computer or voice from the telephone, today's communication networks provide reliable connections anywhere in the world.

The purpose of presenting historical information before Chapter 1 through Chapter 20 is to explore the evolution of the telephone industry and, in doing so, provide you with a foundation in telecommunications. This foundation will help explain why the telecommunications network evolved into what it is today. In order to understand the modern network, it is critical to explain the history of the telephone industry and the technological advances of the past century. Therefore, the information presented in Part I and Part II focuses on the foundation of the telephone network, the telephone industry, and the regulatory decisions that shaped the telecommunications landscape.

Part I covers the years between 1845 and 1920. The material has been organized into three main topics: the birth of the communications network, the telegraph, and the telephone; the regulatory and business climate; and technological innovations. The discussion begins with an introduction to the birth of the communications network in the United States. From there, the discussion focuses on the regulatory and business aspects of the early communications industry. The final portion of Part I concentrates on the evolution of the network, including topics such as telephone switching and the type of cable strung between poles.

■ I.1 THE COMMUNICATIONS INDUSTRY

I.1.1 The Telegraph: 1845 to 1876

The mid-1800s were a time of great discoveries. The idea that communication could take place across long distances using wire and electricity was just being imagined. Several discoveries during this time period helped evolve the idea of sending signals across wire. One of the most significant came from Hans Oersted (1777–1851), a Danish physicist who discovered electromagnetism—the relationship between magnetism and an electric current. Building on this discovery, Michael Faraday (1791–1867) of Great Britain and Joseph Henry (1797–1878) of the United States built the first electromagnets by winding a wire coil around a piece of iron and running a current through the coil.

The next logical step was to use the electromagnet to send signals down a wire conductor. Several laboratory experiments proved that electricity could be used to send signals across distances. The signals could in turn be used to communicate with the distant end. In early attempts to communicate over wire, the signals traveled only a short distance. To transmit the signal long distances, it was necessary to stop every so often before the signal died and retransmit it to the next stop. This, of course, proved to be impractical because every relay station would require a person to retransmit the message.

The solution came from an unusual source, Samuel Morse, who was neither a scientist nor an engineer. His profession, portrait painter and art instructor at New York Art Institute, was far removed from the world of invention. On a return trip from Europe, Morse witnessed a demonstration of electromagnetism. He was so fascinated by the demonstration that on his arrival in New York he immediately started working on a way to use electricity to send messages across wires. Morse's major contribution to the world of communication was the invention of the telegraphic relay.

The relay was able to repeat an incoming signal and transmit it to the next relay stop. By placing relays at intervals along the circuit path, the telegraph message could travel many miles along a wire conductor. The invention was revolutionary.

In addition to the relay, Morse invented Morse code. The code was constructed of dots and dashes that were used to represent letters of the alphabet. For instance, the letter E is represented by a dot-dash-dot. The dot created a short interruption in the current flow; the dash created a longer interruption in the current flow. Therefore, the telegraph operator receiving the message could interpret the dots and dashes as letters of the alphabet. For a complete list of Morse code characters, see Table PI–1.

The telegraph consisted of a transmitter connected to a battery that placed voltage on a line. The transmitter key interrupted the current that flowed as a result of the battery voltage. The transmitter could be compared to a water faucet and the wire conductor to a water pipe. Water flows when the faucet is in the open position and stops flowing when the faucet is in the closed position. Imagine that the letter Y is sent whenever the water stops flowing for one second and the letter N is sent whenever the water stops flowing for a half second. The person standing at the end of the water pipe watches and times the water flow. When the water stops flowing for one second, the person jots down a Y; when it stops for a half second, an N; and so on.

A telegraph transmitter functions like a faucet. The current is the water, and the receiver is the end of the water pipe where the water starts and stops flowing. When the transmitter is open, the current flows steadily, like water through the pipe. The receiver does not click because there is no change in the current to force it into a different position. When the transmitter is closed, similar to the faucet turning off, the current temporarily stops flowing onto the line. In turn, the receiver clicks shut. The dots and dashes stop the current for a specific amount of time forcing the receiver to click according to the number of times the transmitter is open and closed. Dots and dashes define letters. As in the Y and N analogy, the telegraph operator uses the number of clicks and the time period between clicks to determine what letters are being received from the distant end.

Put simply, when communicating across a wire medium using electricity the transmitter affects the current and the current then affects the receiver. Morse specifically chose short interruptions for the most frequently used letters, thereby reducing the amount of time needed to send messages. Morse code is still a standard communication code used in the telecommunications industry in the United States.

The outside plant portion of the telegraph network consisted of telegraph poles, wire conductors, glass insulators, and the telegraph relay. The wire conductors (wires) were strung between the poles. To isolate the wire from the pole, the wire was wrapped around glass insulators mounted on the cross arms or horizontal timber on top of the telegraph pole. Today the glass insulators can be found at flea markets or garage sales. They are often green, purple, or cobalt blue. The relays were placed on

Table PI–1
Sample of Morse Code.

Letter	Code	Letter	Code
A	.-	N	-.
B	-...	O	---
C	-.-.	P	.--.
D	-..	Q	--.-
E	.	R	.-.
F	..-.	S	...
G	--.	T	-
H	....	U	..-
I	..	V	...-
J	.---	W	.--
K	-.-	X	-..-
L	.-..	Y	-.--
M	--	Z	--..

Figure PI–1
The telegraph network consisted of telegraph stations where telegraph operators entered a message. The network, strung between towns, was made up of telegraph poles, glass insulators, open copper wire, and repeaters.

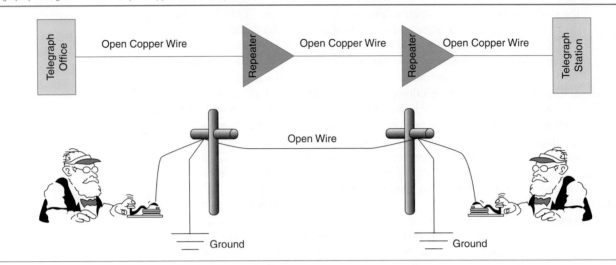

poles at intervals with the wire conductor connected to them. The relay regenerated the incoming signal and passed it on either to the next relay or to the telegraph station depending on its location on the route. Refer to Figure PI–1 for an illustration of a typical telegraph network.

On May 24, 1843, Morse demonstrated his invention to the postmaster general of the United States. Morse had constructed a telegraph line between Washington, D.C., and Baltimore, Maryland, using a $30,000 grant from the government. The first message Morse sent across the line was, "What hath God wrought?" Morse was sure the government would want to build and operate the telegraph network in the same way it operated the postal service, but the postmaster general dismissed the invention as interesting but not of any practical use.

Because the postal service refused to back the telegraph, Morse had to look for capital in the private sector. One of the greatest differences between the United States' telephone system and those of other countries is that the U.S. network was built and operated by the private sector. In most countries, as with Great Britain's Postal Telegraph and Telephone (PTT), telephone networks were started by the government. If the postmaster general had realized the value of the telegraph, AT&T might not exist today and telephone company employees might be government employees.

Morse received financial backing from the private sector and started the first telegraph company in the United States—Magnetic Telegraph Company—in 1845. Other telegraph companies were formed over the next decade offering telegraph services in different regions. The greatest problem facing the industry was that of interconnection between regions. Telegraph companies needed to connect to adjoining companies to transmit messages from one region to another; however, competition and squabbling resulted in a disjointed telegraph network.

The evolution of the telegraph network was accepted enthusiastically by many industries. The railroad granted right-of-way to the telegraph companies in exchange for reduced rates. It used the telegraph to communicate the whereabouts of each train, to notify stations of any delays, and to let them know when there was trouble. Newspapers loved the telegraph. The New York Associated Press was created as the hub for news reporters. Reporters could transmit a story immediately from the telegraph station instead of rushing back to the office.

The Civil War proved to be a turning point for the telegraph network. It was used to relay movements of troops and supplies, and many historians credit the North's victory to the sophisticated telegraph network in the northern states.

After the Civil War, Western Union emerged as one of the strongest telegraph companies. It used the wealth it had acquired during the war years to purchase other telegraph companies throughout the United States. By the 1870s, Western Union was offering telegraph services in almost every region of the country; and, by 1872, it was the largest monopoly in the United States. After the Civil War, the government realized that the interconnection of regions was critical for national security and the government became convinced that the communications network should be run by one company—Western Union.

In 1876, the telegraph network reached from coast to coast with relays from New York to California. Even in a small western town with the train rumbling through, horses stirring up dust, and citizens going about their business, the telegraph operator clicked away sending and receiving messages from as far away as New York City. People who used the telegraph to communicate with family members and friends no longer viewed the transmission of messages across wire as magical. The stage was set for the next revolutionary invention in communications—the telephone.

I.1.2 The Telephone: 1875 to 1920

The year 1875 was a time of invention; among new inventions were the lightbulb, the refrigerator, the camera, and the telegraph. Inventors were making money then like stockbrokers at the millennium. Any invention that enhanced the telegraph network was guaranteed to bring prosperity to its inventor; consequently, many inventors were busy working on devices that would improve the new communications network.

In 1874, Alexander Graham Bell was in his late twenties. He was not a rich man and, if he continued at his current profession—teaching the deaf—he would never be rich. This would not have mattered to Bell except that he had fallen in love with Mabel Hubbard, the daughter of one of Boston's noted attorneys, Gardiner Hubbard. Hubbard refused Bell's request to marry his daughter, insisting that Bell make his fortune before receiving his consent.

Bell was inspired to succeed and started working on a device (the harmonic telegraph) that would allow multiple telegraph calls to traverse one wire. Bell soon employed a young machinist named Thomas Watson to assist in building the device. The match was perfect. Bell was clumsy and not mechanically inclined, but he understood the physical nature of sound and had an irrepressible enthusiasm. Watson was meticulous, good with his hands, and very mechanically inclined but lacked theoretical knowledge and education.

The two started working on their invention with the backing of Gardiner Hubbard and Thomas Sanders. Though Hubbard and Sanders expected the harmonic telegraph, Bell and Watson instead invented the telephone, a result, in large part, of happenstance. Bell and Watson had placed a wire conductor between Bell's bedroom and the study. They placed their new device, the harmonic telegraph, on each end of the wire—one in the study and one in the bedroom. One evening Watson tightened down one of the screws on the harmonic telegraph, accidentally creating a complete connection. A twang crossed the wires and traveled from the study into the bedroom where Bell was working. Because of his knowledge of sound's properties, Bell immediately understood the significance of the twang. If the wire could carry a twang, it could carry sound or, more specifically, the human voice. The event changed the direction of Bell's invention, and he excitedly explained his findings to his financial backers, expecting them to be as intrigued as he. They were skeptical about whether the invention would work and were not convinced that people would want to talk to each other across a wire, therefore demanded that Bell work on the harmonic telegraph and put aside his impractical idea. Bell tried very hard to honor their demand, but he could not give up on his idea and continued to work on the new invention.

Bell designed the first telephone on paper. His attorney filed the patent for the telephone on May 7, 1876. A few hours later, an engineer named Elisha Gray filed a caveat to file a patent for a similar device. Over the years, Bell fought legal battles with Gray and others over the patent.

Figure PI–2
The box telephone, the first functional phone installed at a customer location. (Photo courtesy of AT&T, Warren, NJ)

Three days after filing the patent, Bell and Watson constructed the first working telephone (see Figure PI–2). They used sulfuric acid in the transmitter and one cone for the speaker and receiver. Again, Bell placed one of the devices in his bedroom and one in the study. This time, as the story has been told, Bell spilled sulfuric acid on his pants and yelled, "Watson, come quickly, I need you." Watson heard Bell's voice come across the wire and into the device in the study. History was made—the telephone worked. Bell and Watson had built a device that transformed sound waves into electrical waves, then transported the electrical waves across a wire conductor to an end device that converted the electrical waves back into sound waves.

Within a year, Bell and Watson had revised the design of the telephone and filed an addendum to the first patent. The box telephone was the first functional phone installed at a customer location. Today's telephones are built around the same design parameters as the first two telephones described in the patents filed by Alexander Graham Bell. The telephone proved to be one of the most significant and longest lasting patents ever filed.

I.1.3 The Birth of the Bell Telephone Company

Imagine what it was like during the first years of Microsoft®, Apple™, or Cisco®. Apple started in a garage, Cisco in the inventor's living room. Bell, in 1877, was no different. He and Watson lacked money, experience, and support; the chances that they would succeed were slim. In the 1870s, people could not conceive of a device that allowed people to talk to each other over distances. Bell and Watson realized that they not only needed to acquire capital, they also needed to persuade the American public that a telephone was a practical, real device. They soon found a way to do both. They began touring and giving performances during which Bell stood on stage while Watson spoke into the telephone from a distant location, usually blocks away. The audiences loved the performance, and the result was perfect. The people in the audience learned what a telephone was, and Bell and Watson received all ticket proceeds.

Using money from the lecture circuit, in addition to financial backing from Hubbard and Sanders, the Bell Telephone Company was formed in 1877. The company's trustees included Gardiner Hubbard, Thomas Sanders, Thomas Watson, and Alexander

Graham Bell. The earnings from the lecture circuit helped carry the new company initially; but, because Hubbard and Sanders were convinced the telephone was not going to be profitable, they persuaded Bell to try to sell the telephone patent to Western Union for $100,000. Western Union officials, however, were not interested in Bell's talking box. The CEO was not convinced that the American public needed to talk to each other over distances, and officials believed the telegraph provided a suitable transmission method for communicating across distances.

The Bell Company decided to continue on its own and started offering telephone service in the New England area. Western Union officials realized that they had made a major blunder by refusing to purchase the Bell patent. Thanks to weeks on the road demonstrating the telephone, the request for the new-fangled service was exploding.

Western Union decided to correct its initial mistake by offering telephone service using Thomas Edison's and Elisha Gray's designs. Because Western Union owned the extensive telegraph network infrastructure and possessed enormous amounts of capital, it was able to rapidly deploy telephone services. The Bell Company counteracted by suing Western Union for patent infringement. Western Union's legal team realized that Bell would easily win the suit and agreed to settle out of court.

Bell was given Western Union's customer base in exchange for a monthly fee. During the seventeen-year agreement, Bell paid Western Union seven million dollars. In addition, Western Union agreed not to offer telephone service and Bell agreed not to offer telegraph services. The playing field, therefore, was defined. Western Union continued as the telegraph monopoly, and the Bell Company worked at becoming the telephone monopoly.

Over the next few years, Bell filed numerous suits against companies that tried to break into the telephone market. He won every suit. Bell's plan was to build infrastructure in large urban areas; but, because the company lacked capital, it resorted to selling franchise licenses that allowed the franchisees to use the Bell patent. The franchise companies later became subsidiaries of Bell, known as Bell Operating Companies or BOCs. After divestiture in 1984, they became known as RBOCs or Regional Bell Operating Companies, which are today called incumbent local exchange carriers (ILECs).

The Bell Company prospered between the years of 1877 and 1893, facing minimal competition because it owned the telephone patent. The building of the telephone network happened almost overnight; the technology that evolved continued to improve the service and, in turn, increased the demand for telephones. When Alexander Bell again asked Gardiner Hubbard for permission to marry Mabel, Hubbard gave his approval.

■ I.2 REGULATIONS AND BUSINESS

I.2.1 Bell's Patent Expires

From 1845 until 1918, the government placed few regulations on the communications industry. The only restrictions were the result of the patents filed by Bell and the agreements made between Western Union and Bell. From 1876 until 1893, the Bell Company controlled the telephone network and grew into a large, wealthy company. Starting in 1893, the year the Bell patents expired, the landscape changed dramatically. Competition in the local telephone network sprang up and stole a large portion of Bell's customer base. By the early 1900s, Bell served only 50% of the telephone subscribers in the United States and the new independent telephone companies served the other 50%. General Telephone and Electric (GTE) emerged as the leading independent telephone company. Other hometown companies survived by promoting their hometown image and comparing it with the big, bad Bell outsider.

One of the greatest problems facing the telephone network of the late 1800s was Bell's refusal to interconnect with other independent telephone companies. Many business customers were forced to subscribe to multiple telephone companies' services in order to serve all of their customers. They found it increasingly difficult to keep up with the burgeoning numbers of phone companies. Both telephone customers and independent phone companies complained to Congress and demanded reform. They argued that Bell's refusal to allow them to interconnect with his long-lines division gave Bell an unfair advantage and therefore stifled competition.

This battle between the independents and the Bell Company resulted in the formation of the first independent telephone trade organization. The USTA, or United States Telephone Association, as it is known today, represented the many independent telephone companies' interests. The USTA provided a consolidated, powerful lobbying organization for the independent phone companies' interests. Today, the USTA represents independent telephone companies along with the new competitive local exchange carriers (CLECs) and the RBOCs. The organization provides support to the telecommunications industry in Congress by providing lobbying power. In addition, it provides vital information on market trends, governmental updates, and other information relevant to the telephone industry.

American Telephone and Telegraph (AT&T) was formed as the long-lines subsidiary of Bell Telephone in the late 1800s. Tax and regulatory issues forced the Bell Company in the early 1900s to make AT&T the controlling company for the entire Bell system. In the early 1900s, when AT&T began to have financial problems, J. P. Morgan stepped in to provide assistance and managerial direction and immediately began buying independent telephone companies on behalf of the company. Morgan also purchased controlling stock in Western Union. The significance of this purchase did not go unnoticed by the government; the creation of one communications monopoly for both the telegraph and the telephone concerned regulators.

I.2.2 The Kingsbury Agreement

The Interstate Commerce Commission (ICC) was the government agency responsible for regulating the communications network. Between 1876 and 1913 little regulation existed; but outcries from customers, independents, and antimonopoly advocates forced the U.S. attorney general to react. Nathan Kingsbury, vice president of AT&T, quickly formulated a compromise and presented the agreement to the attorney general in 1913. He accepted the compromise, making the Kingsbury Agreement of 1913 the first set of regulatory guidelines to be imposed on the telephone companies. The agreement stated the following:

1. AT&T would allow independent telephone companies to interconnect with its AT&T long-lines network.
2. AT&T would divest itself of any Western Union stock.
3. AT&T would stop buying independent telephone companies.

AT&T was given, in return, the right to be the only telephone provider in the geographic regions it was presently serving. The areas where AT&T did not have a presence or had a very small presence were given to one of the independent telephone companies. AT&T was to be the only long-lines company, but it was required to interconnect with all other telephone companies. The BOCs, owned by AT&T, became the exclusive local telephone providers for most urban areas, while an independent telephone company served the rural and smaller cities. GTE was the largest of these independents. Many small "mom and pop" telephone companies emerged to serve rural America, and AT&T was required to interconnect with each of them.

The Kingsbury Agreement of 1913 established the era of the telephone monopoly; independent companies and Bell each controlled their own regions. The monopoly status of the telephone industry remained intact for the next fifty years.

In 1920, AT&T serviced 80% of the U.S. telephone subscribers' local calls and 100% of long distance calls. Independent telephone companies such as GTE serviced many rural areas and small cities, and many independent telephone companies owned by individual families provided telephone service to small towns throughout the United States.

■ I.3 EVOLUTION OF THE NETWORK

Throughout this text, technical descriptions of the telephone network have been organized into specific areas—the central office, the transmission network, the outside plant, customer equipment, the data communications network, and the long distance network. The following sections detail the technological components used to build the telecommunications infrastructure.

I.3.1 The Central Exchange Office (Central Office)

One of the most important developments in the telephone network was the establishment of the first telephone exchange in New Haven, Connecticut, in 1878. Until then, customers were connected wire for wire. To talk to a person three blocks away, a dedicated wire connecting the two telephones was needed. To talk to a next-door neighbor, another point-to-point wire was needed to connect those telephones. Obviously, this topology is both impractical and unsuitable; the number of wires needed to connect telephone subscribers was overwhelming.

One of the greatest differences between the telephone network and the telegraph network was that telephone wires had to extend all the way to a customer's home or business. Telegraph wires connected only to the telegraph office, and customers went there when they needed to send a telegram. In addition, telegraph offices were normally located in a train station or near train tracks where the wire passed.

The introduction of the wire center, also called telephone exchange, solved the problem of wiring every home and business to every other home and business. Inside each wire exchange, a large metal structure, the main distribution frame (MDF), was used to terminate at a defined point all of the incoming wire loops from the customers' homes. (*Local loop* refers to the wire that extends from the central office to the customer's home or business.)

The MDF was wired into a new device called the switchboard, which was used to connect an incoming caller's line to the line of a person being called—calling party to called party. The switchboard was essentially the first central office switch. Figure PI–3 shows an early switchboard.

The term *central office* is used today to identify the building that telephone lines feed into from houses and businesses. The term *wire exchange* is still used to describe what we know as a central office. There are some technical differences between the two that we will not discuss at this time. The main point to remember is that the central office was one of the first innovations in the growing telephone network and that today these same central offices perform the same function—connecting customer to customer from a central hub and connecting central hub to central hub.

Figure PI–3
This switchboard was used to connect subscriber lines together for the duration of a call. A light or drop-down door was used to signal an incoming call, at which point the switchboard operator connected one port to the second using a plug cord inserted into each of the two jacks. Upon completion of a call, the operator removed the plug cord. (Photo courtesy of M. A. Rosengrant)

The Switchboard

The switchboard was a terrific breakthrough for young telephone companies. No longer did the idea of everyone owning a telephone seem bizarre. Every subscriber's line was terminated on the frame and also on the rear of the switchboard. Each had a wire connection point on the back of the switchboard associated with a loop plug hole on the front. When the loop cord plug was inserted into the loop plug hole, the subscriber was connected to the operator. The operator then asked the subscriber whom they wished to reach. He or she would find that person's associated loop plug hole and, using a loop cord with loop plugs on each end, connect the two subscribers (see Figure PI–4).

The functions of the first switchboard and the functions of today's digital switches are similar. The main difference is that the digital switch uses computers for brains while the switchboard depended on the brains of the operator.

The First Mechanical Switch

The independent telephone companies were the first to introduce mechanical switches in the late 1800s and early 1900s. Like the inventors of the telegraph and the telephone, the inventor of the first switch was not an engineer or a scientist. Almon Strowger was an undertaker from Kansas City, Missouri. Strowger's motivation to build the telephone switch evolved from his belief that the local telephone operator was causing him to lose business. Strowger believed that the operator, who was dating his competitor, automatically connected callers to her relative's funeral parlor instead of his. Strowger, ignorant of the complexity of telephone switching, built a durable mechanical switch in the carriage house behind his home. Along with the switch, Strowger introduced the first rotary dial telephone set. The introduction of the dial on the telephone set and the mechanical switch revolutionized the way local telephone calls would be handled. Customers dialed the telephone number of the person they wished to call and the Strowger step-by-step switch mechanically connected the lines. Subscribers no longer had to ask the operator to connect them to the person they were calling—as long as the person they were calling lived in the same area.

The independent telephone companies were the first to install the Strowger switches, beginning with the first switch in Laporte, Indiana. The Bell Telephone company eventually followed suit and replaced local switchboards with step-by-step switches. The step-by-step switch remained a part of the telephone network for many years. Though they began to be phased out in the late 1970s, some people still use them. Figure PI–5 illustrates the Strowger step-by-step switch.

Today's digital switches perform the same function as the step-by-step switch invented by Strowger. The switch receives the dialed number, looks it up in memory to

Figure PI–4
The switchboard was the first switch used to connect two subscribers' telephone conversations. The switchboards shown are tied together with trunk connections. The operator connects the first subscriber's port to one of the trunk connections, then connects the second subscriber to the same trunk creating a complete connection between the two subscribers.

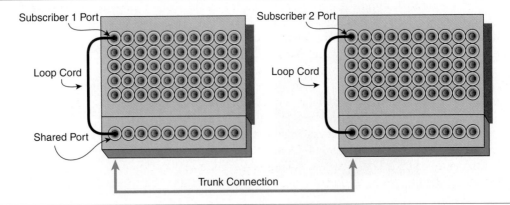

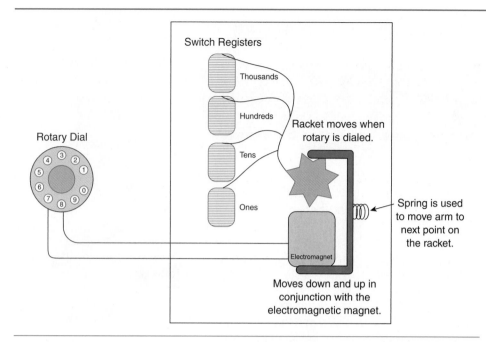

Switch Registers

Thousands

Hundreds

Racket moves when rotary is dialed.

Tens

Rotary Dial

Ones

Spring is used to move arm to next point on the racket.

Electromagnet

Moves down and up in conjunction with the electromagnetic magnet.

Figure PI–5
The Strowger step-by-step switch was the first central office exchange switch; it replaced the switchboard.

determine where it is located, and connects the two lines, thus creating an end-to-end connection between subscribers. Even cellular switches perform the same function. The only difference is there are not any copper wires connecting the subscriber to the switch—just air.

Trunking Circuits

New telephone companies have central buildings that house termination points for all the wires coming in from customers' homes and a switchboard or switch to handle switching of incoming calls. The next problem that telephone engineers had to deal with was the increasing number of customers who needed to be terminated on the switchboard at the central office. The term *trunking* is universal throughout the tele-communications industry. It first evolved in the late 1800s as multiple switchboards were placed in central offices to serve the increasing number of customers. There was a limit to the number of ports that could physically fit on a switchboard or a Strowger switch. Therefore, as the number of subscribers increased, the need for multiple switch-boards in the central office also increased.

Connecting the lines of a customer terminated on one switchboard to a customer terminated on a different switchboard was almost impossible. The operator needed a very long loop cord and had to run from one switchboard to the other switchboard, which took extra time. Some operators wore roller skates when connecting two cus-tomer lines together, as shown in Figure PI–6.

The solution put forth by early telephone engineers was to connect all the switch-boards together with shared connections. A one-for-one connection between switch-boards was not needed and, in fact, was not feasible because it would require hundreds of wires and switchboard ports. Instead, the switchboards were equipped with shared ports and shared wires that were used to connect the switchboards. The shared wire connections were called *trunks*. The operator could now use the shared trunk ports to establish a call between two switchboards.

The operator who sat at the switchboard and received the request to make a call notified the operator sitting at the switchboard where the call needed to terminate. The customer's voice then traveled from his or her house into the switchboard, across the trunk to the terminating switchboard (see Figure PI–7), and finally out onto the line that was connected to the party being called. Once the calling party disconnected, the op-erator would disconnect the line from the shared trunk and the shared trunk could be used for a different call.

Figure PI–6
Some switchboard operators wore roller skates to help them move quickly up and down the rows of switchboards.

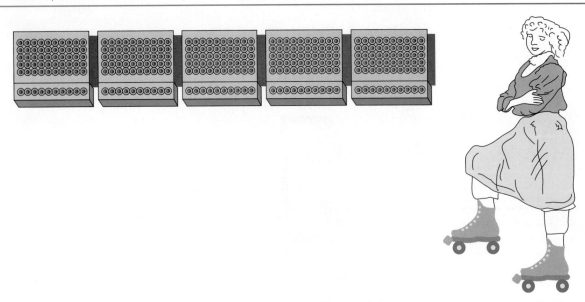

The beauty of this innovation was the efficiency of using shared trunks. The engineers realized they could base the number of trunks needed between switchboards on how busy the office was at the busiest time of day. They performed statistical calculations to determine the number of trunks required to handle all of the calls coming in.

As the need to connect central offices to central offices increased, so did the use of shared trunks. Like the first trunking circuits that were used to connect switchboards, the use of shared trunks between central offices became a standard configuration. Trunking circuits also connected the Strowger mechanical switches.

Figure PI–7
Termination boxes like this one were constructed of wood, metal conductors, and doors and had an appearance that was similar to a small medicine cabinet. The boxes provided a point where wires could be terminated and connected to a second cable pair. (Photo courtesy of M. A. Rosengrant)

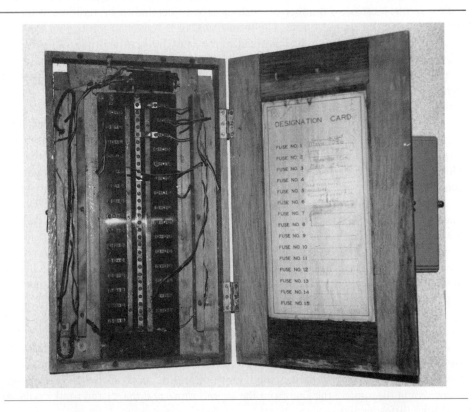

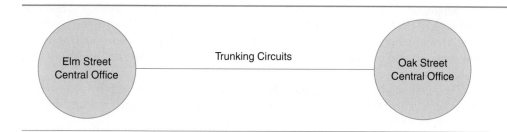

Figure PI–8
Trunking circuits established between central offices helped reduce the number of copper wires between offices.

Today, trunking circuits perform the same function. They are used to connect local central offices to other local central offices, local central offices to long distance companies' switching centers, and long distance switching centers to other long distance switching centers. The cellular industry also uses trunking cables to connect into the local telephone and long distance networks and to connect their switching hubs together. Telephone engineers continue to use the same statistical analysis when determining the number of trunking circuits needed between switching centers. Figure PI–8 depicts trunking circuits between switching centers.

The Internet has caused telephone traffic engineers to look again at the busy-hour calculations they have used for years. The accepted call time has been about three minutes per call per subscriber. The Internet caller stays online for multiple hours, skewing the calculation used to build switch port size and trunking capacity into and out of the switch. A subscriber receives a fast busy notification if the switch ports are being used by other customers or if the trunking circuits are filled to capacity. Because of the Internet, fast busy signals are quite common today.

Signaling

Signaling is one of the functions required in every telephone network. Chapter 8 is devoted to the telephone signaling network, but for now it is important to understand its purpose. *Signaling*, as we know it, is the ringing of the telephone to notify us that a call is coming in, the digits dialed that tell the switch where the call needs to go, the dial tone, and busy signals. Signaling is used to set up a call, supervise what is going on, and tear down a call.

Early customers were notified of an incoming call by the operator yelling down the telephone line that a call was waiting and that they needed to answer the phone. It became obvious that this method was crude and ineffective. Thomas Watson solved the problem by placing bells in the telephone that rang whenever a ringing voltage was induced onto the line. The operator induced the voltage by cranking a magneto on the switchboard producing an alternating current that traveled from the central office to the telephone ringers. Today, telephone switches induce ringing voltage on the telephone line that turns on an electronic ringer in the telephone.

To make a call, the customer cranked a magneto on the telephone set to induce a current that would light a light on the switchboard thus notifying the operator that the customer wanted to make a call. The first switchboards had small doors that dropped down above the customer port signaling the operator that the customer wished to place a call. The introduction of the mechanical switch changed the way the switch was notified. When the customer picked up the handset, the circuit completed, allowing current to flow. The switch sensed the flow of current and immediately placed a tone on the line, notifying the customer that the switch was ready to accept digits. Today, dial tone is still used as the notification method that the switch is ready to accept digits.

When calls terminated on the switchboard, the customer asked the operator to be connected to a particular central exchange. The operator would then connect the customer to the correct line or route him or her to the exchange where a second operator would connect to the terminating line. Initially, names were used to identify customers. The change to using numbers was the result of an outbreak of measles in a small Massachusetts town. The local doctor was concerned that the operator was the only person who knew how everyone's lines terminated on the switchboard and that if she took ill the town would be left without telephone service. To solve the problem, he set up a numbering plan to identify all telephone subscribers. Telephone numbers soon replaced names; and, eventually, each telephone office received an identifying number.

The numbers dialed on the rotary dial were interpreted by the step-by-step switch. The switch stepped through a maze of mechanical connections until it reached the correct location on the switch, and the call was connected.

The light above the switchboard port turned off once the customer hung up the handset. The operator would then disconnect the line and continue to monitor the switchboard for additional requests. Today, we receive dial tone and punch in the telephone number of the person we wish to speak to and the switch connects our two lines by using routing tables. Routing of calls is discussed in Chapter 9. When the handset is placed back on the telephone cradle, the switch sees the on-hook condition on the line. The switch instantaneously disconnects the two lines and continues to scan the line for a request for dial tone.

Operators

The first switchboard operators were teenage boys. The job required only that the operator connect the incoming call to the person being called, but the boys were vulgar and rude and played endless practical jokes. To eliminate the problem, the Bell Company decided to change its hiring practices and hired young women instead of young men. The women were required to follow a strict set of rules. They were not allowed to talk to customers other than to request the connecting party name or number; each followed a strict dress code, used correct posture and proper etiquette, and behaved like a lady in all situations—both at work and outside of work. The women proved to be efficient, polite, and reliable. Unlike the young men, the women had very few employment alternatives and therefore tolerated the strict climate of the telephone company. The section opener on page 1 shows operators at work in a switching center in the late 1800s.

The chief operator at the telephone company was often referred to as the queen. Her employees were responsible for connecting and routing all calls. It was the chief operator's responsibility to make sure calls were completed, customers were happy, and outages were kept to a minimum. To do that, she ruled with an iron fist.

Today, many switch translation engineers are women. Because building translations is the process of defining how calls are routed through the switch, it was a natural transition for a woman to move from operator to translation switch engineer.

Operators are now being replaced by automatic voice systems but still provide telephone number information. The roles and responsibilities of operators have changed dramatically due to the technological innovations of the past twenty years.

Centralized Power

Central office battery is the term used to identify the power plant located in every central office. The telephone circuit requires power to energize the flow of current and, in turn, to allow the voice signal to travel down the telephone wire. Rather than use local power (the power source at each customer's house), the telephone company decided to provide centralized power from each central office. Using central power instead of local power is a decision that we still live with today.

The central office places voltage on each telephone line. The electrical characteristics of a telephone line and the power parameters are covered in more detail later in this text. At this time, it is important to realize that the telephone company provides the power needed to make the telephone work and to make the telephone signal strong enough to travel from a house to the central office.

To understand how this affects us today, answer the following question. When commercial power goes out, the lights, television, and refrigerator go out. Does the phone still work? Yes, of course it does. But why? For the reason just mentioned. The central office supplies power on the telephone line. When commercial power goes out, the central office switches to a reserve power source—first batteries and then a petroleum-fed generator. Therefore, you have telephone service even during power outages—another legacy attributed to the early telephone engineers.

Carrier Serving Area

The central office is the hub for a particular area in the telephone company. Each home is fed with wires that connect to a particular central office. For instance, if I live on Elm Street, I know that I am fed out of the Elm Street central office. The early telephone engineers designed a central office network that would serve each area of the city. Large cities had multiple central offices strategically placed to serve the most subscribers. Normally, a small town had one central office that fed everyone in town and everyone within a certain radius of the town. The largest central offices served as many as 70,000 subscribers. Figure PI–9 shows a typical carrier serving area (CSA) map.

Determining where to place the central offices and how many offices were needed within a given region was one of the tasks of the first telephone engineers. They had to worry not only about the distance limitations of the telephone lines, referred to as the local loop, but also about the cost associated with copper wires, telephone poles, and other needed equipment. Their mission was to build as few central offices as possible because of the expense of building and then maintaining them. A design rule was formulated. Because the voice signal was able to travel only so far before it was no longer discernable, distance became the most critical parameter used when placing central offices. A second parameter was the availability and cost of building space. The radius drawn around the central office was referred to as the carrier serving area (CSA). Each office served anyone living within its radius. Thus, every subscriber line had a serving central office and every central office had a CSA associated with it.

Today's local telephone map is quite similar to what it was 100 years ago. Every central office has a CSA associated with it, and subscribers are fed out of their serving central office.

I.3.2 The Outside Plant—Telephone Poles, Telephone Cables, Telephone Problems

Everyone knows what a telephone pole looks like. It is conical, tall, and made of wood. Citizens in the late 1800s were very aware of what telephone poles looked like. In fact, in some areas of the city, about all they could see were telephone poles and telephone wires strung from block to block to block. The first wires used in the telephone network were made of steel, which was soon replaced by copper. Today's telephone networks continue to use copper wire conductors to connect homes and businesses to the serving central office.

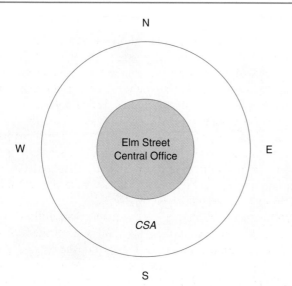

Figure PI–9
The carrier serving area (CSA) is the region surrounding the central office. The perimeter of the CSA defines the area that the central office serves. All subscribers living inside the perimeter are served by the Elm Street central office.

The early telephone industry copied the telegraph industry's wire scheme, using one wire to connect a customer to the central office. Similar to the telegraph, they used ground to complete the circuit. The electrical characteristics of the telephone line are discussed in Chapter 1. For now, we will discuss why two wires instead of one wire connect our telephones to the central office.

During a routine turnup of a circuit, a telephone technician accidentally connected two wires together at each end of the circuit. The result was a clearer, less noisy, louder voice signal. From that discovery, the two-wire circuit was born and since that time the telephone network has been built around a one-pair per subscriber wiring scheme. One pair refers to two wires that work together to form a telephone line. This can be seen by pulling the cover off the jack that a telephone is plugged into. The two wires wrapped under two screws are the wires that connect a home to the serving central office.

The telephone network of the late 1800s consisted of open wire pairs strung from pole to pole with each wire wrapped around glass insulators. Imagine the mess created by the multitude of wires, especially in the urban areas. As telephone poles became taller and taller, the cross arms teetered from the weight of the wires. The urban landscape was becoming inundated by telephone wires and telephone poles. The expiration of the Bell patents in 1893 only increased the congestion as competitive telephone companies began stringing their own wire and placing their own poles. Figure PI–10a shows the congestion of the 1800s outside plant.

Open wire cable was not only unsightly and an inefficient use of pole space, it also reduced the clarity of the telephone conversation. Streetcars, other telephone conversations, and the weather all affected the telephone signal riding on the open wire. The problem was resolved in the late 1800s when insulated cable was developed. Many farmers claimed they could hear a storm rolling in by listening to the telephone. Insulated cable was a technique that used paper to separate wires from each other. After wrapping each wire with paper, the wire was bundled together and encapsulated in an insulating outer jacket made of lead. The bundled cable with the protective jacket was then placed in ducts underground. This solved the problem of having too many poles, too many open wires, and not enough space for growth. It also helped improve the quality of the voice signal because it insulated the wires from one another and from the outside environment.

Beginning around the turn of the century, underground insulated cable was installed in all urban areas in the United States. Soon the outside plant telephone network consisted of buried cable in cities and open wire cable in rural areas. Telephone poles no longer had to be as tall or as large to carry the growing number of telephone lines, as shown in Figure PI–10b.

Figure PI–10
(a) Due to the increase in telephone lines, more cross arms were added and poles were built taller and taller. The weight of the wires sometimes stressed the poles and cross arms to the point of breaking. (b) Paper insulated cable was soon found to alleviate telephone pole and wire congestion in downtown cities. Insulated wires were buried underground, as shown in this photo, therefore lessening the need for aerial builds. (Photos courtesy of the National Archive)

(a)   (b)

The outside plant network consisted of multitudes of local telephone lines connecting into their serving central office and all the other central offices. Trunking cable was similar to regular subscriber cable in the late 1800s. Both were made from copper, and both carried only one voice signal at a time.

Today the outside plant consists of many, many copper local loops connecting telephone subscribers to their serving central office. However, trunking cable has changed from copper to fiber optic cable in most instances. In addition, millions of calls ride on one fiber optic strand, compared with one call on the old copper trunking cables of the late 1800s and early 1900s.

I.4 CUSTOMER PREMISES EQUIPMENT

One of the first and most significant decisions made by the Bell Company was to lease telephone sets to their subscribers for a monthly fee. Gardiner Hubbard's plan was to provide the Bell Company with a continuous flow of cash by collecting on this lease. It worked. The amount of money paid to the Bell Company over the years for a telephone set far exceeded the cost of the set. Figure PI–11 shows one of the early telephones.

In the beginning, one of Bell's major problems was the poorly made telephone sets produced by multiple manufacturing companies. In 1869, Elisha Gray (Bell's former competitor) and Enos Barton founded a manufacturing company called Western Electric, which was known for producing superior telephone equipment. The Bell Telephone Company purchased Western Electric in 1881 and made it its sole telephone equipment manufacturer, thereby resolving the problem of poorly constructed telephone sets from other manufacturers. The most significant result of the purchase of Western Electric was that Bell could now dictate its own standards. From 1881 until the beginning of deregulation in the early 1960s, every telephone set and piece of telephone equipment was built around strict guidelines defined by Western Electric engineers.

Many small independents, however, bought their equipment from GTE's manufacturing arm. These companies proved to be more innovative and more willing to try new inventions than the big Bell companies. For example, the first rotary dial telephone, the first telephone mechanical switch, and the first telephone using a combined handset were all introduced by independent telephone companies.

The second greatest change in the telephone set, after the introduction of the rotary dial, was combining the transmitter and receiver into one unit called the handset. By 1920, telephones were made of Bakelite instead of wood; they had rotary dials and a handset that rested on a switch hook, as shown in Figure PI–12a.

Figure PI–11
Telephones consisted of a receiver, a separate transmitter, and batteries for power. A bell on the phone rang when the operator cranked a magneto at the far end. When a subscriber wished to place a call, she turned a crank, which signaled the operator. (Photo courtesy of M. A. Rosengrant)

Figure PI–12
(a) A traditional black Bakelite telephone, as shown in this photo, merged transmitter and receiver into one unit called a handset. Rotary dial improved the call connection time and allowed telephone companies to switch calls mechanically. (b) Long distance calls were connected to switchboards that were used to connect the local network to the long distance network. Long distance operators used special cost books to log and charge the customer for long distance calls. (Photos courtesy of M. A. Rosengrant)

(a)

(b)

I.5 LONG LINES

Long lines, as they were first called, used trunking cable to connect switchboards in distant cities, similar to the method used to connect calls from one central office exchange to another within a city. The first long-lines call was completed between Boston and Providence in 1881.

The central office housed a switchboard used exclusively for long-lines calls, where long-lines cable terminated on the back of the switchboard. The cable used to connect two cities was normally open wire cable strung between telephone poles and wrapped around glass insulators. Often, railroad right-of-ways were used for long distance telephone line runs. The cable entered a central office called the tandem central office or toll office, which was connected to the serving central offices located throughout the region. The toll office became the hub for connecting local subscribers to the long distance network. The number of toll offices in a city depended upon the geographic area. Figure PI–12b shows a long distance switchboard.

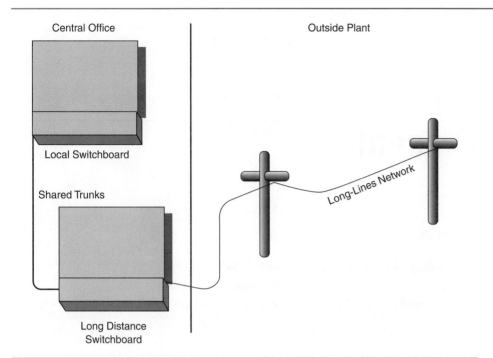

Central Office   Outside Plant

Local Switchboard

Shared Trunks

Long-Lines Network

Long Distance
Switchboard

Figure PI–13
Long-lines service was made possible by connecting long distance switchboards directly to the long-lines network. The long distance switchboard connected to local switchboards via shared trunks.

All subscribers shared the lines connecting the distant cities. Like the trunking cable placed between switchboards within the central office, telephone traffic studies determined the number of trunking cables needed to connect two cities. The local and long distance switchboards were connected using looping cords and shared trunks whenever a local caller wished to talk to someone outside the local area. The same happened on the terminating end of the call. The step-by-step switch was used as a local switch and had to connect to the long distance switchboard whenever a long distance call was made, as shown in Figure PI–13.

The long distance operator timed the call, used a long distance rate book to determine the cost of the call, and then applied the cost to the customer's record. Today, the switch turns on a timer to determine the length of the call. This information is passed on to an external billing system that calculates the cost and applies the cost to the customer's account. Detailed records in each month's phone bill show when calls were made, to what number they were made, and the charges for the calls.

In 1885, AT&T was created by Bell Telephone Company as its long-lines subsidiary, and in 1899 AT&T became Bell's parent company. By 1915, a person could place a call from New York to Los Angeles. Unlike the local telephone network that was comprised of independent telephone companies and the Bell Company, the long distance network was exclusively owned and operated by AT&T.

SUMMARY

The telegraph network was born in 1845, the telephone followed, and by 1920 the telephone and telegraph network spanned the globe. The communications revolution had begun. This chapter looked at the emergence of the telecommunications network, discussed the business and regulatory climate, and explained the impact of the technological innovations that were introduced between 1845 and 1920. It specifically highlighted the technological evolution of the network, including the introduction of telephone cable, telephone amplifiers, the switch, the central office exchange, and the two-wire circuit. These technological innovations helped change the way people communicated all over the world.

PART II

1920 to the Present

Objectives

After reading this section, you should be able to

- Describe the regulatory events between 1920 and the present.
- Describe the technical evolution of the network between 1920 and the present.

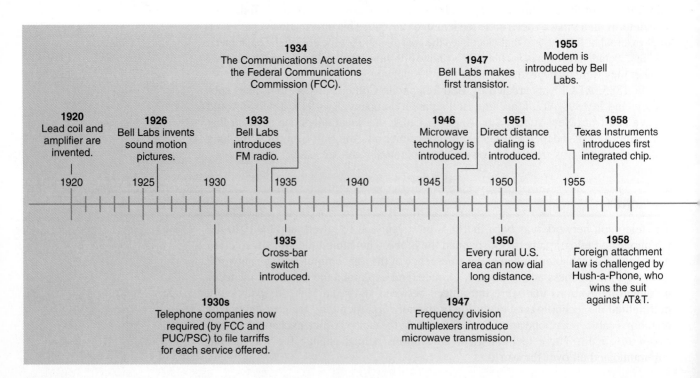

1920
Lead coil and amplifier are invented.

1926
Bell Labs invents sound motion pictures.

1933
Bell Labs introduces FM radio.

1934
The Communications Act creates the Federal Communications Commission (FCC).

1946
Microwave technology is introduced.

1947
Bell Labs makes first transistor.

1951
Direct distance dialing is introduced.

1955
Modem is introduced by Bell Labs.

1958
Texas Instruments introduces first integrated chip.

1935
Cross-bar switch introduced.

1930s
Telephone companies now required (by FCC and PUC/PSC) to file tarriffs for each service offered.

1947
Frequency division multiplexers introduce microwave transmission.

1950
Every rural U.S. area can now dial long distance.

1958
Foreign attachment law is challenged by Hush-a-Phone, who wins the suit against AT&T.

Outline

Introduction

Summary

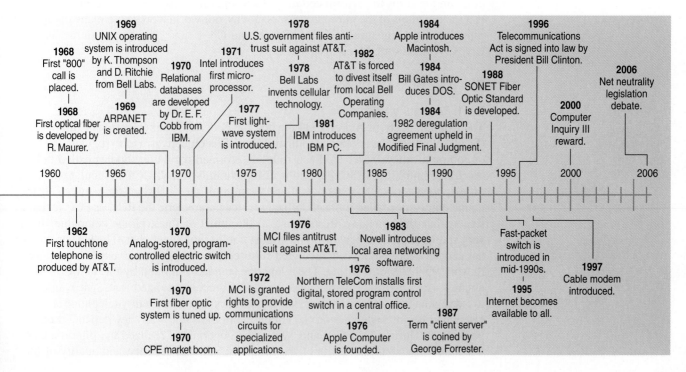

INTRODUCTION

The years between 1920 and the present have been marked by numerous technical break-throughs in the telecommunications industry along with significant regulatory mandates. Part II looks at these years, focusing on the technological innovations, the business climate, and the regulatory landscape of the time period.

The material has been organized into five time periods: 1920 to 1950, 1950 to 1970, 1970 to 1990, 1990 to 2000, and 2000 to the Present. Within each time period are seven topics: regulatory milestones, the central office, the transmission network, the outside plant, customer premises equipment, the long distance network, and the data network. The technical descriptions are not as complete as those in Part I because topics discussed herein are covered in detail in later chapters.

■ II.1 IMPORTANT TECHNICAL BREAKTHROUGHS

Today's communications network has been developing for the past 100 years. No single technological breakthrough is responsible for today's sophisticated communications network, but each invention and discovery helped the network evolve and grow at an amazing pace. Following are some of the most significant inventions and their roles in the development of the modern communications network.

During the early 1900s, AT&T created a research and development division staffed by Western Electric and Bell engineers. The new division, Bell Labs, was responsible for many, many scientific innovations. For example, in 1926, Bell Labs invented sound motion pictures; in 1933, it introduced FM radio, transporting the signal across telephone lines; and in 1947, three scientists from Bell Labs made the first transistor. The modem was introduced by Bell Labs in 1955; in 1978 Bell Labs invented cellular technology; and the list goes on. Today Bell Labs is Lucent Technologies, a separate publicly traded company that is no longer part of AT&T.

The invention of the transistor in 1947 by three scientists—John Barden, Walter Brattain, and William Shockley—from Bell Labs played an important role in the birth of the electronic revolution. In 1958, building on their success, Jack S. Kilby from Texas Instruments introduced the first integrated chip; and, by 1971, Ted Hoff from Intel had developed the microprocessor. The microprocessor was the first programmable device consisting of millions of transistors and logic circuitry that enabled scientists to program the chips to perform certain tasks. Program languages were developed to direct the functions of the microprocessors. Today, our world works with many computer programming languages, including Pascal and C++, and operating systems such as UNIX, Windows 98, Macintosh, and Linx.

■ II.2 1920 TO 1950

II.2.1 Regulatory Milestones

The Communications Act of 1934 created the Federal Communications Commission (FCC) to oversee the telephone industry. One of its objectives was to regulate the cost of communications services. Each state also formed regulatory bodies called public service commissions (PSCs) or public utility commissions (PUCs) to regulate the cost of telephone service at a state level and to monitor the quality of the network. If the phone companies failed to meet defined service parameters, the PUC or FCC levied a fine against the company. Responding to the threat of fines, phone companies built very stable, expensive networks and almost bombproof telephones. The legacy, which has carried over to today's deregulated culture, is that the customer expects the phone to work 99.999% of the time. The telephone standards defined by the Bell Company assured users that they would; and the Bell Company realized that, if it wanted to avoid regulatory action, it would need to provide consistent, quality telephone service. Users had begun to depend on the telephone in the same way they depended on the post office. During the 1930s and 1940s, the government realized the importance of telephone service to the health of the economy, national security, and quality of life.

The government believed each citizen living in the United States should have access to the telephone network and that the quality of the telephone service would not vary between regions. Universal service was born.

At the same time, AT&T understood the need to provide universal service. It, too, realized that if it refused or failed at offering reasonably priced telephone service, the government would intervene and force complete government control. The dilemma created by the universal service philosophy was the cost associated with offering telephone service to residential areas and small populations of rural users. The population density per square mile was not large enough to pay for such things as the copper cable, telephone poles, and central offices needed to provide quality telephone service. In addition, residential users were not able to pay high prices for dial tone.

AT&T, along with the independent telephone companies, knew that it would have to find some way to cover the costs of providing residential services. The solution was obvious. Higher fees would be charged for long distance calls and business customers to subsidize the residential phone subscriber. Even though providing service to the rural and residential user was often more costly than service to the business user, the residential subscriber was allowed to pay less for phone service. Today, most residential customers do not pay a per usage or per minute charge for local calls; some business customers do. This policy can be attributed to the universal service philosophy.

The method used by the regulatory agencies to police the phone companies was first introduced in the 1930s. The FCC and PUCs/PSCs required the companies to file tariffs for each service they offered. A tariff is a document that defines a specific service, such as a residential telephone line, and then states the cost of the service, the quality parameters the telephone company must meet, and the responsibilities of the customer. Today, telephone companies file state and federal tariffs whenever they wish to add a new type of service or change the pricing or definition of an existing service. One of the best ways to understand a specific telephone service is to read the tariff. Filed tariffs are public documents.

To summarize, the most significant regulatory events between 1920 and 1950 were the concept of universal service, the formation of the FCC and PUCs/PSCs by the government, and the introduction of the tariff as a contract between the government and the telephone company.

II.2.2 Technological Milestones
Central Office

The central office changed little between 1920 and 1950. The greatest change was the conversion from the Strowger step-by-step switch introduced in the late 1900s to the cross-bar mechanical switch. The cross-bar switch could handle a greater number of incoming calls than the step-by-step could handle switch-to-switch calls, eliminating the need to hand the call to a manual switchboard. A significant breakthrough in 1951 was the introduction of direct distance dialing (DDD) for all long distance calls. The deployment of DDD took several years after the initial introduction. Now customers could dial a long distance number using the 1 prefix followed by the telephone number. Direct distance dialing reduced the telephone company's dependence on operators, and the phrase "long distance, please" became a thing of the past.

Transmission Network

Transmission network and *transmission equipment* are common terms still used to refer to special pieces of equipment that help transport information from location to location. Initially, transmission equipment was used exclusively for circuits between central offices. As the demand for communications links increased, transmission equipment moved out into the local-loop network to service large customer buildings or business complexes. Today, transmission refers to fiber optic equipment placed on the ends of fiber optic lines, fast-packet concentrators, and many other unique devices that aid in the transport of information.

Figure PII–1
The frequency division multiplexer was the first equipment used to aggregate multiple signals onto one connection.

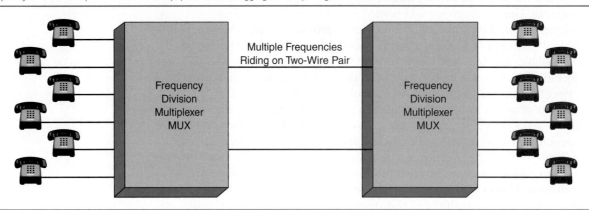

The first piece of transmission equipment introduced into the telephone network was the frequency division multiplexer (FDM) shown in Figure PII–1. The FDMs, which aggregated multiple circuits onto one link, were installed in central offices and connected by copper links. Amplifiers were placed at intervals along the circuit route to boost the strength of the signal, performing the same function as the telegraph relay. The amplifier increased the signal strength, allowing the signal to travel longer distances. The multiplexers' main purpose was to carry multiple trunks or calls on fewer copper pair wires.

For instance, prior to the introduction of the multiplexer, a central office with twenty-four trunking circuits required twenty-four copper pairs to be dedicated between the two central offices. Once a multiplexer was placed on both ends, only two copper pairs were needed to carry twenty-four trunking circuits. The multiplexer, whose purpose was to reduce the amount of copper needed in the outside plant, gained popularity due to the increased need for trunking circuits between switches. The objective of most outside plant engineers and transmission engineers was to reduce the amount of copper that was used in the network. Engineers today face a similar objective—to reduce the volume of network fiber. The amount of capital saved by using multiplexers to reduce copper in the network was enormous. They were installed in most central offices and became as common as the main distribution frame, as shown in Figure PII–2.

Figure PII–2
Using FDM provided a way to carry multiple voice signals on fewer copper pairs between central offices.

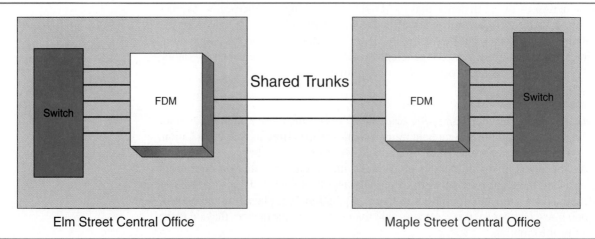

Long distance networks benefited especially from the use of multiplexed circuits. The long-lines connections had required miles of copper to be strung between cities, so aggregating the traffic onto fewer links allowed for less expensive long distance loops. The term *carrier*, given to the circuits that carried multiple calls, is still used to define the digital T1 circuits that connect all switch sites, both local and long distance.

By 1950, most switch-to-switch connections and all long distance circuits used multiplexed analog carrier circuits, or FDM technology. A special group of technicians trained to install and maintain the new class of circuits spent hours tuning the amplifiers and multiplexers of the carrier circuits. When the weather changed from hot to cold or cold to hot, the circuits fluctuated away from the calibrated settings. The technicians then had to work their way through the circuits again tuning one amplifier after the other. Many retired communications technicians fondly remember the overtime paychecks received every spring and fall.

The introduction of multiplexers in the network dramatically changed the telecommunications network. The transmission plant continued to evolve and has become the backbone of today's information superhighway.

Outside Plant

The outside plant network changed little between 1920 and 1950. Improvements were made in the cable terminating equipment, the insulation methods, and the way the cable traveled through the network during this time period. The standards for running cable were clarified and refined. Workers became more skilled, and an apprenticeship was created to ensure skills were transferred from one generation to the next. Overall, the outside plant continued to grow at a very rapid pace with changes marked as improvements to the total network. Figure PII–3 shows a wire termination panel used during this period.

Customer Premises Equipment

Little changed at the residential customer site during the thirty-year period between 1920 and 1950. Telephones were still owned by AT&T or the independents. They did not vary in style, color, or hardware: The style was either desk set or wall mount; the hardware still contained a handset, switch hook, and rotary dial; and the color was always black.

Business customers continued to demand special services such as alarm circuits, automatic switchboards, and better multiline telephone sets. In response to these demands, a special group of telephone technicians (special services technicians) was formed to service the business customer. These technicians work on computer circuits,

Figure PII–3
Wire termination panel used between 1920 and 1950. (Photo courtesy of M. A. Rosengrant)

Figure PII–4
Customer premises switchboards used from 1920 to 1950. Small switchboards, as shown in the photos, were purchased by business customers to help switch calls at the customer's location. An operator routed incoming and outgoing calls similar to operators at the telephone company. (Photos courtesy of M. A. Rosengrant)

(a)

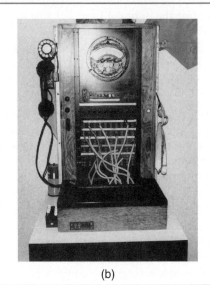

(b)

special signaling circuits, and other sophisticated business customer circuits. Customer premises switchboards illustrated in Figure PII–4 were common.

Long Lines

The most significant change in the long-lines service was the extensive network that was built across the country. By 1950, every rural area in the United States had the ability to call long distance to anyone in the country.

The long distance network fed into each local telephone company network via a toll center. The toll center was normally located in one of the downtown central offices. The toll switchboards were fed by trunking circuits to tandem central offices in the region. The toll switch handed its call off to a local tandem switch that determined which local central office the call was destined for, as shown in Figure PII–5. Chapter 8 describes the switching hierarchy between the local and long distance network in detail.

Figure PII–5
End-office switches in the local telephone network provide telephone service to subscribers living near the central office. The central offices connect to one another to carry calls across the region, while the tandem switch office is a central switch site with the ability to switch calls to any office in the network. Some end offices are connected directly to accommodate traffic volumes.

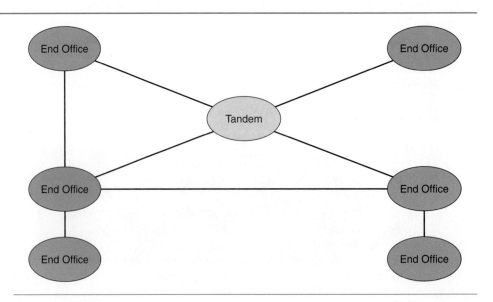

Summary

The most notable technological breakthroughs during this time period were the cross-bar switch, the frequency division multiplexer, and direct distance dialing. Many other improvements continued to expand and improve the telephone network. Customers were beginning to demand special services from the telephone company, and telephone workers were becoming more specialized and skilled.

■ II.3 1950 TO 1970

II.3.1 Regulatory Milestones

During the years after the Kingsbury Agreement, AT&T controlled the use of any device placed on the telephone network. Customers were allowed to place only AT&T equipment on the end of their phone lines. In fact, if they violated this law, AT&T had the right to deny them phone services. The term *foreign attachment* was used to define non-AT&T standard equipment.

The foreign attachment law was challenged in 1958 by a small company called Hush-a-Phone that produced a passive device that fit on the telephone transmitter to ensure privacy. AT&T would not allow customers to attach the Hush-a-Phone device to its telephones. Hush-a-Phone claimed that, because the device did not affect the telephone network, AT&T had no right to deny the attachment on the telephone. Hush-a-Phone won the suit.

Within a few years, the Carterfone Company filed a similar suit against AT&T. Carterfone produced a device that was used to link the telephone to a mobile radio connection. It argued in court that the device did not adversely affect the telephone network and therefore should be allowed to connect to the telephone line. Carterfone won the suit against AT&T; as a result, interconnection companies started building customer premises equipment, such as PBXs, key systems, and telephone sets. Northern Telecom, Mitel, and Rolm were some of the first interconnect companies to provide customer equipment. Today, a long list of interconnection companies produce everything from sophisticated PBXs to $10 telephones.

II.3.2 Technological Milestones
Central Office

Few changes occurred in the central office during this time. The most significant change was the introduction of the first analog-stored, program-controlled electronic switch in 1970. The advantage of the electronic switch was the reduction in mechanical relays. The relays used by the cross-bar switches required constant maintenance, and they limited the total call-handling capacity of the switch. In addition, the electronic switch was much smaller, required less power, and had fewer mechanical connections than the cross bar. The call-handling capacity of the electronic switch was far greater than that of the cross bar. The electronics revolution was just beginning.

The main distribution frame and the power plant changed little between 1950 and 1970, but new ways of connecting the wire on the frame did improve the quality of the connection. The power plant also became more sophisticated and robust, ensuring constant telephone service even during the worst disasters.

Transmission Network

The most significant change here was the introduction of microwave technology, which radically changed the way information was transmitted in the long distance network. The military first used microwave technology during World War II. Soon after the war ended, the long distance telephone network started to replace sections of copper wire with microwave transmission links. By 1970, the majority of circuits in the long-haul network traveled on microwave links.

Microwave technology consisted of microwave antennas spaced about thirty miles apart. The antennas were connected to multiplexing equipment and microwave radio electronics. Most microwave sites were placed at the highest point in the area, and, in

Figure PII–6
Microwave transmission helped eliminate the need to run large copper cables between two sites. DS-3s carrying T1 circuits are commonly transported on the microwave systems. Antennas are spaced thirty to thirty-five miles apart.

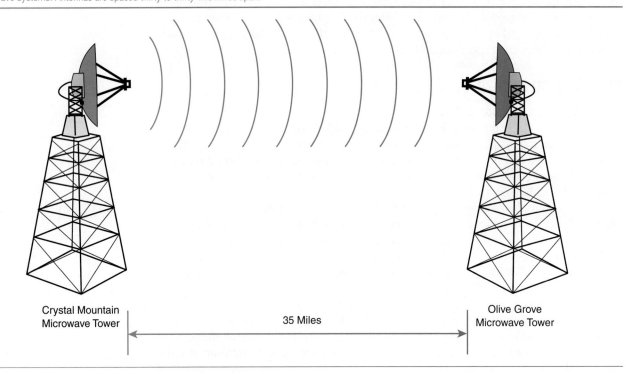

Crystal Mountain
Microwave Tower

35 Miles

Olive Grove
Microwave Tower

metropolitan areas, the antennas were placed on the roofs of the tallest buildings. Microwave technology transmitted information signals through the air from antenna to antenna as shown in Figure PII–6.

Local networks used microwave technology to transmit television signals and special events after connecting to the long distance links at the toll office. Multiplexers were used to aggregate the incoming lines onto one link. The only difference between analog carrier circuits and microwave links was that microwave links used air as the transport medium instead of land line copper.

Outside Plant

The most significant development in the outside plant network was the introduction of polyurethane-insulated cable. Paper insulation cable, introduced in the late 1800s, though better than open wire, posed several problems in the outside plant network. If the outer sheath of the cable cracked or was punctured, water seeped in and dissolved the paper insulation protecting the bare copper wire. Ultimately, copper wire touching copper wire shorted out the circuit and caused the line to go dead. The polyurethane-insulated cable was more resistant to water and therefore more resistant to cable troubles. Beyond that, one of the greatest advantages of the polyurethane cable was the color-coding scheme that was developed to help technicians locate particular pairs.

When a cable is cut in the outside plant network, technicians are called in to splice the two ends of the cable back together. If they chose the wrong pair, the customer would be connected to the wrong line unit in the switch and consequently would receive calls meant for someone else. Paper cable did not provide an identifying method to determine which wire belonged to which.

The polyethylenel cable solved this problem by defining a color code, so technicians simply matched wire colors when splicing two ends together. For example, a blue/white pair in the blue binder group had to connect with the blue/white pair of the blue binder group.

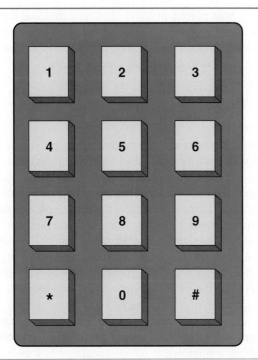

Figure PII–7
Touchtone telephone service was introduced as a value-added service to the subscriber, and for several years consumers paid extra for the convenience. Today, phone companies prefer that subscribers use touchtone telephones.

Customer Premises Equipment

In 1962, AT&T introduced the first touchtone telephone, which used frequencies to define each number and special character, as shown in Figure PII–7. Before this time, the Strowger rotary dial was the only dialing method available. Today, a rotary dial telephone is an anomaly.

After the Hush-a-Phone and Carterfone decisions, the "foreign attachment" clause was revised to allow non-AT&T vendors into the interconnection market. Customer premises switches called private branch exchanges (PBXs) were introduced and companies such as Rolm, Northern Telecom, and Mitel found a niche in building business systems such as the key system, PBXs, and multiline telephone sets. (PBXs, key systems, and other pieces of customer premises equipment are discussed later.) The new equipment companies were referred to as *interconnect companies*—companies that built interconnection equipment. For the first time in fifty years, AT&T was facing competition.

Residential phone subscribers also benefited from the Carterfone and Hush-a-Phone agreements as non-AT&T telephones were now allowed on the telephone network. Competition also forced AT&T to add to its current selection of telephone sets. The popular Princess phone was introduced by AT&T as a way to hedge competition. Soon, department stores stocked telephones of every color, style, and shape.

Today, of course, the selection of styles and manufacturers is enormous, and phones can be purchased from electronics stores, department stores, and even the home shopping network.

Long Distance

Important milestones for the long distance network ranged from the first communications satellite launch to the first 800 call. Satellite communications provided a way to connect the telephone networks of distant countries and was used mainly for international calling and television broadcasts. Today, satellite technology links every country in the world. Also, a new low-orbitingearth / (LEO) satellite technology called Iridium (to be discussed later) allows telephone conversations from any location in the world.

The first 800 call was placed in 1968. Today, a national database allocates 800 numbers for long distance companies. The database allows customers to keep their 800 numbers even when they switch long distance providers. Many businesses rely on 800 numbers to route their callers to the appropriate database.

■ II.4 1970 TO 1990

II.4.1 Regulatory Milestones

There were several reasons the telephone monopoly started to erode between 1970 and 1990, one of which was the introduction of new technologies. During the preceding fifty years, switches had required a great deal of space within the central offices and copper cable had been the only means of transporting information. At the time, the Kingsbury Agreement was correct in mandating one telephone company per given region. The infrastructure that was required, along with the lack of electronic equipment, forced the industry to accept this mandate. However, by 1970, with the introduction of digital transmission equipment, electronic switches, and fiber optic cable, the industry was ready for a change. In addition, the government was wary of allowing one company to supply and control the growing communications network.

By 1970, the customer premises market was inundated by multiple equipment manufacturers. The next area to be divested was the long distance network. It all started when a small company called MCI filed suit against AT&T claiming unfair practices. MCI wished to establish a microwave communications link between Chicago and St. Louis to carry radio traffic for trucking companies. AT&T claimed MCI had no right to carry voice traffic because AT&T was sanctioned by the government as the sole long distance provider.

In 1972, thanks to MCI's persistence, the company was granted the right to provide communications circuits for specialized applications. The specialized common carrier was born. During the next ten years, MCI continued to fight AT&T for the right to become an official long distance carrier. In 1976, MCI filed an antitrust suit against AT&T; two years later, the Justice Department also filed an antitrust suit against AT&T.

The result of the suits was a compromise agreement issued in 1982 by the FCC. The agreement forced AT&T to divest itself from its local Bell Operating Companies. The local "baby bells" would consolidate and become Regional Bell Operating Companies (RBOCs). AT&T was allowed to keep the long distance network, the Western Electric manufacturing company, and the Bell Labs research division. In addition, AT&T was allowed to offer computer services. Two years passed before the ruling took effect.

Judge Harold Green agreed with the FCC's 1982 ruling by handing down one of the most significant pieces of legislation in telecommunications. The 1984 Modified Final Judgment ruled that AT&T must divest itself of all twenty-two Bell Operating Companies—the local service arm of AT&T. The operating companies would be organized into regional holding companies to be known as Regional Bell Operating Companies (RBOCs). The RBOCs would provide local telephone services within their regions. The regions were defined by metropolitan statistical areas (MSAs), which took into account population studies and the structure of the current operating area. The country was divided into seven RBOCs to be controlled by the following:

1. Bell Atlantic
2. Nynex
3. BellSouth
4. Pacific Bell
5. Ameritech
6. Southwestern Bell
7. U.S. West

Local access and transport areas (LATAs) were formed to define the local calling regions across the United States. Figure PII–8 shows the LATA regions for New York State.

LATA Map for New York State

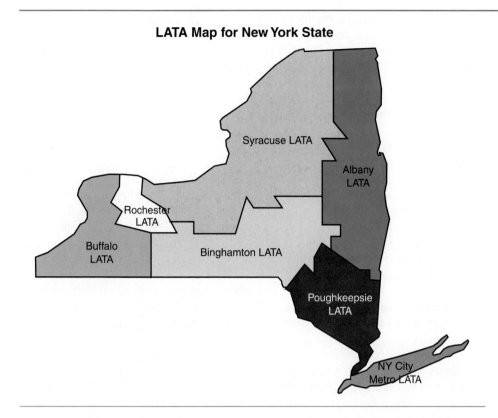

Figure PII–8
The United States is divided into regional areas referred to as LATAs for local access and transport areas. New York State has seven LATAs. A call placed from Albany, New York, to Rochester, New York, is an InterLATA call.

AT&T was allowed to keep the AT&T long distance network, Cincinnati Bell, Bell Labs, Western Electric, and its computer division. It was not allowed to offer local services, and the RBOCs were not allowed to offer long distance services. AT&T believed it had won the war because the area it coveted was the computer networking portion. The RBOCs felt they had won because they no longer had to be ruled by AT&T. Finally, the new long distance carriers, such as MCI, believed they were the winners because they realized the enormous profits available in the long distance market.

The year 1984 was one of the most significant in the telecommunications industry in the twentieth century. The Modified Final Judgment (MFJ) changed the U.S. telephone network forever. The philosophy of the 1980s was deregulation. A noticeable effect of the judgment was the introduction of television advertisements for long distance companies. Sprint, MCI, AT&T, and many others purchased a great deal of air time to promote their long distance prices. Suppertime telephone calls inundated the American public as persuasive salespeople sold the new long distance services by claiming lower prices and complicated calling plans.

The result for the consumer was lower long distance rates. The result for the local telephone companies was that they needed to determine how to connect to each of the new long distance carriers. The term *equal access* was used to describe the interconnection with all long distance companies. Local telephone companies were given a specific time frame in which to comply with the equal access ruling.

The government realized that providing residential service was more costly than providing long distance service. The cost of the outside plant and the maintenance of millions of residential telephones placed an unfair burden on the local telephone companies. One outcome of the MFJ was that the residential subscriber was no longer subsidized by long distance calls. The FCC mandated that an access fee be paid by every long distance carrier for every call terminated into the local telephone companies' networks. Access charges are still a point of contention between local telephone companies and the long distance carriers. One of the most controversial issues today is whether or not Internet providers have the right to terminate voice calls over the

Internet without paying access fees to the local telephone company. So far, the FCC
has ruled in favor of the Internet companies.

One last result regarding the MFJ is the status of the independent telephone companies. GTE, Rochester Telephone, and Alltel were given the right to become long distance carriers; and all three did so. The independent companies were required to
provide equal access to all long distance carriers; and, in turn, they benefited from the
access fees paid for terminating long distance calls.

II.4.2 Technological Milestones
Central Office

The first digital, stored-program control switch, built by Northern Telecom, was installed in a central office in 1976. In the next fifteen years, the majority of central offices were converted to digital switching technology. AT&T produced the 4ESS and
5ESS; Northern Telecom's digital switches were the DMS-10, DMS-100, and DMS-
250; and other switch manufacturers produced similar products. The digital switches
required less space and less maintenance, could handle millions of calls per second, and
provided new value-added features.

Structurally, the digital switch is a giant computer with about three million
lines of code telling it what to do next. Caller ID, call forwarding, call waiting, and
call return are all features designed by switch engineers to provide added benefits
to customers.

Many central offices were built to accommodate large mechanical switches. Several floors were needed to house these switches. Today's digital central office requires
a fraction of that space, so vacant floor space is common. Walking through a central office today is very different from walking through a central office in 1970.

The introduction of operating systems, such as UNIX in 1969 by Ken Thompson
and Dennis Ritchie from Bell Labs, and the development of relational databases in 1970
by Dr. E. F. Cobb at IBM helped lay the groundwork for the first digital switch. For the
first time, telephony and information technology worked together to produce a superior
product for the telecommunications network. At that point, they became synergistic
partners in building the voice/data network.

Transmission Network

International Telephone and Telegraph (ITT) scientist Alec Reeves developed a way to
convert the voice signal into a digital signal called *pulse code modulation* (PCM). A
digital signal is very similar to the dots and dashes used to send telegraph messages. A
one is represented by an on voltage and a zero is represented by no voltage. How a voice
signal is transformed into ones and zeros will be explained later. For now, what is important to remember is that during the 1970s the transmission network—the network
that connects switches together—became a network of digital circuits. Digital multiplexers called *time division multiplexers* (TDMs) replaced frequency division multiplexers. Repeaters replaced the amplifiers. T1 circuits or T-Carrier circuits were the
names given to the new digital transmission lines, as shown in Figure PII–9.

Digital multiplexers were placed on each end of the circuit, and the repeaters were
placed every mile along the circuit path. Unlike old analog amplifiers, repeaters did not
require tuning. In fact, new digital circuits were much more robust and less susceptible
to noise, they were not affected by the change in climate, they could travel greater distances, and they required less maintenance. The introduction of the digital circuit was
the beginning of the digital revolution in the telephone network. Today, T1 circuits form
the backbone of the telecommunications infrastructure.

A second revolutionary invention, introduced in 1977, was the first lightwave
system. Robert Maurer developed the first optical fiber in 1968. Within ten years, the
technology progressed into the communications system and was used to carry information signals between locations. Fiber optic cable is truly one of the most important
inventions in communications. Not only does it allow millions and millions of calls
to travel on one hair-sized fiber strand, the cable itself is very resistant to outside

Figure PII–9
Time division multiplexing replaced FDM carrier systems. The circuits between two locations carried digital signals consisting of ones and zeros. The T1 circuit was the first digital signal to be deployed in the network.

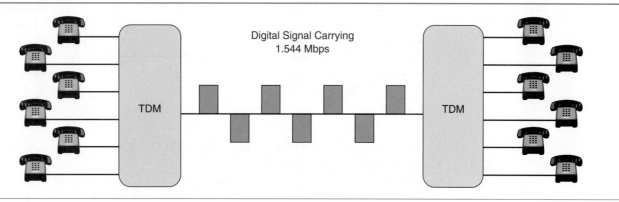

influences such as noise from power stations, water, and unauthorized phone taps, as shown in Figure PII–10.

The first fiber optic systems consisted of a fiber optic multiplexer that used the digital TDM scheme. Laser technology was used to transmit the on and off pulses over two strands, one for transmit and one for receive. A fiber optic multiplexer was placed on each end of the fiber optic link. T1 circuits fed into the multiplexer and were aggregated onto the link. The reverse happened at the distant end, as shown in Figure PII–11. By 1980, fiber optic light-wave systems were replacing the microwave and copper transport technologies in long distance and local trunking circuits.

Outside Plant

The outside plant was transformed between 1970 and 1990. Fiber optic splicers were in great demand. New splicing techniques improved the quality of the fiber optic link and saved enormous amounts of time during the splicing and repair processes. Better insulating sheaths were developed to protect the delicate fiber strands, and a color code similar to that of the copper PICs cable was standardized to help organize the fiber layout and improve repair and installation time.

Even though fiber optic cable was being laid throughout urban, long distance, and international routes, copper circuits were still the mainstay in most local communications networks.

A third noteworthy introduction was subscriber line carrier equipment, also called digital loop carriers (DLCs), which were placed strategically in the area around the central office. (Refer to Figure PII–12.) The purpose of the DLCs was to reduce the amount of copper cable feeding residential and business complexes. DLCs were placed on cement pads along the road or in fields, or they were buried in controlled environmental vaults (CEVs). The units were connected to the central office with T1 circuits. For example, an AT&T SLC-96 required four T1 circuits to be connected to the central office in order to serve ninety-six customers. The customer's local loop fed directly into the electronic unit in the field where it was multiplexed onto the T1 circuits that fed back into the switch.

The advantage of placing a DLC was the reduction in the number of copper pairs needed to feed a particular area. The ninety-six telephone lines connected directly to

Figure PII–10
Fiber cable was introduced in the 1970s as a communications medium. The glass strands were not affected by water or electromagnetic influence such as in power lines, intrusive taps, or temperature; and one strand of fiber could carry many more signals much farther than could copper cable.

Figure PII–11
Fiber optic multiplexers gathered many slower signals such as T1 1.544 Mbps into one 45 Mbps signal, which was transported on fiber optic strands.

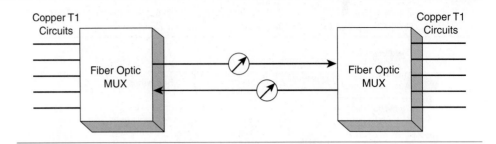

the DLC while the four T1 lines connected the DLC to the central office. See Figure PII–13 for an example of a DLC.

The most popular DLCs were Northern Telecom's Urban DLC and AT&T's SLC-96. The DLCs extended the CSA of the central office and helped reduce the amount of copper cable needed to feed a given area around the central office. Many of the DLCs are now fed by fiber optic links and are often placed in the basement of large office buildings.

Computer

The computer revolution started in the late 1950s, and true data networking began in the early 1970s. Microcomputers began to be linked together in the 1970s; thus began the introduction of data communications circuits provided by the telephone company. It is important to understand that we can no longer talk about telephony without discussing the invention of the computer and all of the computing systems. Many

Figure PII–12
Digital loop carrier systems are deployed to help reduce the amount of copper cable needed to feed specific areas in a region. The DLC connects to the central office via T1 circuits that carry the many voice channels from the individual subscribers.

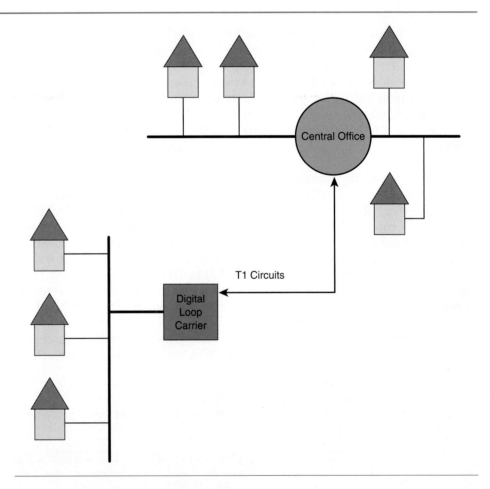

Figure PII–13
A digital loop carrier (DLC) showing a typical site with a DLC, a cross box and a cable termination pedestal. The DLC is placed on a cement pad or below ground in a vault and is used to terminate individual POTS lines in the area, then multiplex them on shared trunks back to the central office. (Photo courtesy of M. A. Rosengrant)

inventions and developments shaped today's data communications network. Following are descriptions of a few that contributed to the design of the data communications network.

The U.S. Defense Department created the Advanced Research Projects Agency Network (ARPANET), the organization that is credited with creating the Internet, in 1969. In 1973, Vinton Cerf defined the routing topologies and global addressing schemes that defined the Internet. The Ethernet white paper detailing the protocol was published in 1976 by Dr. Robert N. Metcalfe and Dr. David Boggs of Xerox Corporation. Apple Computer was founded in 1976 by Steven Jobs and Steve Wozniak; in 1984, Apple introduced the Macintosh and Bill Gates introduced the Disk Operating System (DOS). IBM introduced the IBM PC in 1981. The age of personal computing was under way. Novell introduced its local area networking software in 1983, and the term *client server* was first used in 1987 by George Forrester. Each invention and discovery formed a piece of the puzzle that we know today as the data communications network.

The computer operating systems, the databases, the Ethernet standard, the Internet routing structure, the personal computer, the modem, and LAN interfaces have allowed us to communicate with people all over the world. We pull down information from university research databases, we order products off the Internet, and we chat with friends and colleagues. The developments of the 1970s and 1980s formed the cornerstone of today's information age.

Customer Premises

The customer premises network was transformed between 1970 and 1990 due to three significant events. The first was the steady growth of interconnect companies, the second was the introduction of data communications circuits, and the third was the beginning of the cellular telephone industry.

Large mainframe computers were commonplace during this time. The mainframes were connected locally to computer terminals where data entry workers typed in information. The next logical progression was to connect distant terminals to the mainframe computers. The Sabre® airline reservation system was one of the first data communications networks in the United States. The telephone company provided special data circuits that interfaced into the airline computer terminal and the airline's mainframe computer. Airline reservation clerks were then able to assign seats, cancel reservations, and reserve space by typing the necessary information into their "dumb" terminals and waiting for a response from the mainframe. The ability to provide real-time communication between the mainframe computer and distant terminals was revolutionary.

The telephone companies played a critical role in building the infrastructure to handle data traffic between terminals and the distantly located mainframe computer.

They designed a network of private analog point-to-point circuits capable of handling data rates of 2.4 kbps, or 2400 bps (bits per second). The Sabre system was a huge success. Research universities, financial institutions, and the military soon built networks connecting distant locations to their mainframe computers via data circuits.

The phone company placed equipment at the customer site that would interface with the data circuit, but the customer owned the terminating equipment that interfaced with his or her database. Special service technicians were trained to handle the new data circuits, and special service transmission engineers were trained to design the circuits and guarantee that transmission standards were followed.

By 1990, data communications networks were becoming as common as telephone lines. During the next few years, new digital circuits quickly replaced old analog circuits. AT&T copyrighted the circuit name DDS for digital data service. The DDS circuit provided a faster data speed—up to 56,000 bps—and a more reliable circuit. Large customers started purchasing full T1 circuits—1,500,000 bps—to connect their offices.

A phenomenal growth rate in business telephone systems was seen between 1970 and 1990. New technologies, spurred on by the electronics revolution, provided business customers with new solutions to fit their needs. The era of business solutions was born.

The final technological introduction of the era was that of cellular telephone technology. Bell Labs invented cellular technology in 1978. It was made available in 1982 after the FCC granted its first radio license. By 1987, there were more than one million cellular telephone subscribers.

Long Distance

The greatest transformation in the long distance network was the replacement of microwave transmission links with fiber optic light-wave systems. The fiber optic systems transport information, either voice or data, for long distances on single strands of fiber optic cable. The capacity of the cable is enormous compared with microwave systems and copper long-haul circuits. The fiber does require regeneration sites about every thirty to forty miles, but the capacity of the fiber outweighs the cost of the repeaters.

Another advantage of fiber over copper or microwave is its reliability. The only maintenance nightmare associated with fiber optic systems is backhoe fade. This happens when a backhoe or other destructive device accidentally cuts the fiber cable. By 1990, the long distance network was comprised of miles of fiber optic cable and fiber optic equipment.

Long distance networks participated in the new data communications arena as they carried data information from location to location. The local phone lines then handed the point-to-point data circuits to AT&T to carry to the distant end. The long distance company had become an integral part of the data network.

■ II.5 1990 TO 2000

II.5.1 Regulatory Milestones

During the first half of the 1990s, few regulatory changes occurred in the telecommunications industry. In 1993, the U.S. government auctioned off additional frequencies for cellular networks. Large numbers of investors partnered to purchase lucrative chunks of the limited spectrum. Before 1993, only two cellular providers had been allowed per region, but the additional frequencies opened up the cellular market for multiple cellular telephone providers per area.

President Bill Clinton signed the second most important regulatory decision of the twentieth century into law in 1996. The purpose of the 1996 Telecommunications Act was to extend deregulation to the local telephone industry. The law listed specific mandates that local telephone companies would have to abide by. The incumbent telephone company would have to open up its local network to competitive service providers called competitive local exchange carriers (CLECs.) The existing local telephone companies were now referred to as incumbent local exchange carriers (ILECs).

CLECs quickly emerged in the marketplace. Many were started by long distance companies such as MCI, AT&T, and Frontier; but small entrepreneurial players who found funding from venture capitalists also formed start-up CLECs, hoping to cash in on the billion-dollar local telephone market. The ILECs were required to provide access to their facilities, and the RBOCs were only allowed to offer long distance service in their region once they opened their network to competitive service providers.

By 1999, the central offices owned by the ILECs were housing CLEC transmission equipment. Colocation cages were constructed to hold the new competitive companies' equipment. The local copper loops owned by the ILEC were now available for the CLEC's customers. The large space left when digital switches replaced mechanical switches were then filled with cages containing CLEC network equipment. The local network was going through what the long distance network experienced in 1984—competition, competition, and more competition.

II.5.2 Technological Milestones
Central Office
The changes that occurred in the central office in the 1990s were dramatic. Many central offices were converted to digital circuit switches. The reduction in workforce directly correlated to the introduction of the switch, which could be programmed remotely, lacked moving mechanical parts, and provided excellent redundancy or robustness.

A new type of network was also evolving in response to the increased need to combine data traffic with voice traffic. The fast-packet switch was introduced in the mid-1990s. Fast-packet technology includes several protocol types, such as frame relay, ATM, and IP, that depend on statistical multiplexing methodologies to help improve the efficiency of transporting information across the network.

Data Communications
The fastest and most phenomenal growth in the computer industry was the introduction of the local area network (LAN). A LAN is simply a network of PCs—file servers that hold shared software programs, printers, and other PC equipment. The LAN evolved as a result of the PC. Twenty years earlier, mainframes required huge rooms with special cooling plants and dumb terminals depended on the mainframe for processing power. In the 1990s, the PC carried the same computing power as the old mainframes. It was no longer necessary to depend on the mainframe to perform tasks, so the PC became the storage, processing, and presentation unit in the network. The next logical step was to connect all the PCs so that the users could communicate.

With the advent of the LAN, e-mail became a ubiquitous term in the office place and file sharing became quite simple. The information age had reached almost every office in the country. Manufacturing plants built LANs to link their workstations to their manufacturing lines. Once the LAN was perfected, the next step was to link the LANs together, creating wide area networks (WANs). The WAN used high-speed digital links to tie distant office LANs together so that employees at different sites could communicate with one another. Exchanging e-mail, transferring files, sharing software programs, and more were accomplished by connecting LANs.

From the day the telephone was invented to the late 1990s, we can pinpoint the technological inventions that dramatically changed the way human beings live their lives. One, of course, was the telephone; the second was the Internet. The Internet burst onto the scene in 1995, offering a way to browse for information and to communicate with distant users. Until then, only research scientists and computer gurus and the like had used the Internet. With the advent of browsing software and the introduction of the World Wide Web, the American public, along with the rest of the world, became connected.

Internet service providers (ISPs) popped up like dandelions. Large corporate ISPs fought for customer shares, and most areas had a choice of servers. AOL, CompuServ, and many of the long distance and local providers created Internet branches to serve the millions of customers who wanted to browse the World Wide Web.

By 2000, the Internet had become a way for companies to communicate with one another, for vendors to sell products, and for advertisers to advertise. Chat rooms were used by anyone who wanted to chat about any subject imaginable. The Internet boom seemed a little like the gold rush in California in 1849 and in Alaska at the end of the century: speculators wanted instant gold. The times were unpredictable but exciting.

Transmission Network

The most important development in the transmission area was that of the SONET fiber optic standard. In 1988, the SONET or Synchronous Optical Network was developed by industry experts—vendors, carriers, and research engineers—to improve the interoperability of fiber optic networks. In simple terms, the SONET standard provided rules that allowed Vendor A's fiber optic equipment to connect to Vendor B's fiber optic equipment. In addition to being vendor compatible, the SONET was much more reliable due to its ring architecture. A SONET ring topology provided a resilient network structure that reduced the chance of a complete outage. If the fiber were cut, the traffic would reverse itself and travel in the opposite direction, as shown in Figure PII–14.

Improved lasers, improved fiber optic cable, and improved splicing methods helped increase the capacity on one strand of fiber to 192×672 calls or 129,024 calls traveling simultaneously on one strand of fiber. The SONET standard was accepted by the telecommunications industry and soon began to replace the older point-to-point fiber optic networks. By 2000, SONETs connected every corner of the United States. The ring architecture improved the resiliency of the network and provided a better method of transporting information around the world. However, the introduction of wavelength switching dense wave division multiplexing (DWDM) and packet protocols were threatening the popularity of the SONET standard.

Outside Plant

The outside plant continued to grow as the demand for additional lines for business and residential customers increased. The outside plant technician was still responsible for building and maintaining the network. Tools that assisted the technicians in their jobs became more and more intelligent. Test equipment scanned lines to determine where a break in a line was located, reducing the number of hours necessary to troubleshoot an

Figure PII–14
SONET multiplexers, introduced in the early 1990s, changed the topological layout of networks. The ring carries traffic in both directions. If one section is cut, the traffic between the two sites is routed the opposite way around the ring. The reliability of the network increased dramatically with the introduction of ring architecture.

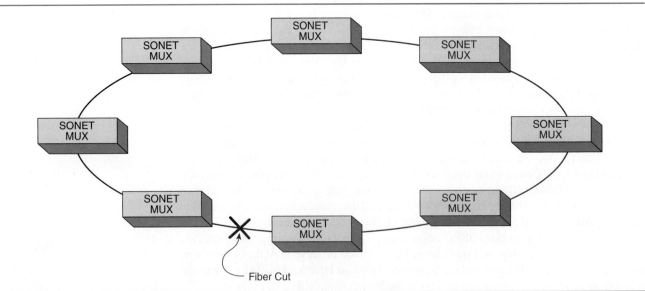

outage. Putting in poles required less effort because huge augers were used to dig the holes for the poles. The outside plant was continually improving.

Customer Premises

The integrated services digital network (ISDN) was introduced as a new way to provide telephone service to the end customer. Just as the transmission plant converted from analog carrier lines to digital carrier lines, and the switch changed from an analog electronic switch to a digital switch, the last mile to the customer site was converted from an analog loop to a digital loop. The standard that performed this conversion was the ISDN.

From its inception, the ISDN was expected to revolutionize the local loop. The clean, efficient digital network reached the very last termination point—the end customer. Unfortunately, the ISDN was not received the way all of the experts predicted. The deployment was complicated; ISDN telephone sets were expensive (around $600); unlike a regular analog set, the ISDN depended on commercial power, and the world in the early 1990s was simply not ready for the benefits provided by a digital line.

By 2000, the ISDN was more prevalent due to the persistence of the telephone companies and the boom of the Internet. An ISDN line gave the customer two full channels that could be used for voice or data and an additional slower speed data channel all on one telephone line.

A rival technology that some claimed would eliminate ISDN completely was digital subscriber line (DSL) technology. DSL provided a high-speed data link to the customer site. By 2000, the fastest modem available was the 56,000 bps modem. DSL could carry as many as six million bps. The beauty of DSL technology was that it could run on the same copper pair as the regular telephone service. DSL was already being deployed in many areas, and analysts predicted that it would grow at exponential rates over the next few years.

Long Distance

The long distance companies continued to face stiff competition throughout the 1990s, but the greatest change came from the fast-packet technologies, the SONET fiber optic rings, and the Internet. Long distance companies carried voice, data, and video over sophisticated networks. The greatest change during the last few years of the twentieth century was the interconnection with international networks. No longer could a long distance company survive if it only supplied domestic telephone service; therefore, most long distance companies continued to build out their networks to include links throughout the world.

■ II.6 2000 TO THE PRESENT

II.6.1 Regulatory Milestones

The most dramatic regulatory mandate since the 1996 Telecom Act is the elimination of the Computer Inquiry regulation that was enacted thirty years ago. ILECs are no longer required to separate services as "basic" and "enhanced." The CLEC or ISP will not have access to the broadband portion of the bundle—bandwidth greater than 200 kbps in one direction. Internet broadband services no longer fall under the Computer Inquiry mandate. Though this may sound insignificant, it has far-reaching consequences—particularly if you are a non-facilities-based ISP that has based your business plan on purchasing portions of a broadband pipe from a facilities-based carrier. On the flip side, the mandate helps free up the ILECs that offer Internet broadband services competitively with other broadband providers such as cable modem and wireless offerings—neither subject to regulatory restrictions.

CLECs still have access to unbundled network elements (UNEs), as defined by the 1996 Telecom Act, Section 251, meaning that a CLEC will be able to purchase the copper loop extending from the CO to the customer premises. Unfortunately for the CLEC, this does not include access to the fiber to the home pipe owned by the ILEC.

The reasons the FCC decided to eliminate the "basic" and "enhanced" separation are the improvements in network, switching, and data processing technologies; the new entrants offering broadband services; and the realization that the services have converged blurring the distinction between what is what. The FCC felt comfortable that CATV's cable modem and the wireless providers' new EV-DO broadband wireless product provide the needed competitive incentive to keep the facilities-based common carriers—ILECs—in check.

ILECs have aggressively started to deploy fiber to the home in order to compete with CATV providers. The pipe to the home, called the *access pipe*, is becoming the defining factor in who can compete in local networks. The FCC sees the reduction in laser costs, the reduction in fiber costs, and the improvement in aggregation equipment as signs that it no longer needs to regulate the industry to the degree it did in the beginning.

A second regulatory mandate that has caused heated discussions around the water cooler is the FCC's ruling on E911 functionality requirement for VoIP. Initially, many of the non-network-based VoIP carriers provided N911 services, meaning that the 911 call was routed to a call center that would then direct the call to the correct E911 center. Due to the routing of the call, the E911 address was not displayed on the E911 operator's screen. If the caller was unable to talk or if the call was placed by a child, the 911 operator would not be able to determine the caller's location. In 2005, the FCC placed an E911 requirement mandating all VoIP providers to comply to E911 capability. The ruling forces VoIP providers to build trunks into each market's public safety answering point (PSAP) or to interface with a CLEC or other provider that has established trunks to the center. The VoIP provider must also submit customer addresses to the Master Street Address Guide (MSAG), the E911 database holder, in order to guarantee address verification at the time of turnup. A third portion of the mandate states that the carrier must give customers the ability to change their address location when they move even when the move is temporary. This ruling is significant because the IP device—called an ATA, analog telephone adapter—is portable. It can connect to any broadband connection in the world and provide dial tone. A person who wants to take his or her adapter on vacation needs to be able to enter his or her new address into a graphical user interface (GUI), which adds the address information into the local MSAG database.

The E911 ruling is the FCC's first regulatory mandate directed toward the emerging VoIP carriers. Carriers such as CLECs, Tier 1 ISPs that have built substantial E911 networks across the country, and E911 validation centers are benefiting from this ruling. Carriers that built their networks on N911 have to redesign their infrastructure and back office systems to meet the FCC requirements.

One other note that should be made regarding the emerging packet-based voice providers is that one of the major advantages VoIP offers a customer is its low price. Up to this point, the regulatory and tax agencies have not instituted extensive taxes or fees on VoIP services. If the regulatory and taxing agencies continue to see VoIP as data services, the price advantage offered by VoIP services will remain stable. If, however, carriers, regulators, and taxing agencies begin to require the same fees and taxes as those placed on traditional voice services, the cost advantage VoIP offers will disappear.

II.6.2 Technological Milestones

The most significant technological milestone from 2000 to 2005 was the emergence of the converged network. During the telecom boom of the late 1990s, *converged networks* was one of the most overused terms in the industry. Though the idea was enticing at that time, the technology and the infrastructure were not ready to provide the cost savings anticipated by "non-network"-savvy analysts. The result was a hodgepodge of network types, each trying to pretend to be converged in order to make the investors happy. By the early 2000s, the telecom bust left a glut of network bandwidth available to service providers. In addition, improved, highly resilient routing and switching equipment entered the marketplace at very low prices—thanks to the telecom crash.

The large amounts of fiber optic cable crisscrossing the country and world reduced the cost to ship information to almost every area. Network planners were able to build substantial packet networks that required highly reliable links such as those offered by fiber providers. In addition, distributed switching architectures emerged with carrier class switch functionality allowing new providers to enter the market and provide VoIP and other services with fairly low entrance costs.

The technological advances since 2000 are not as dramatic as those of the earlier decades, fiber and digital switching in particular. The most significant changes in today's networks are improvements in efficiency and cost savings, not in new inventions such as fiber and the transitor. Improved wave division multiplexing; new wireless schemes; smaller, cheaper, and more efficient chips; and so on are driving the industry in the 2000s.

II.6.3 Business Milestones

A new category that has been added to the final section of this part is "Business Milestones." After the 1996 Telecom Act, hundreds or possibly even thousands of new CLECs—competitive local carriers—entered the market to offer telecommunications services. Auxiliary companies sprang up to provide services and equipment to the new carriers. Long distance switch companies purchased and ran miles of fiber optic cable around the world opening up areas to the "information society." The telecom boom followed the .com boom through the end of the 1990s, amazing analysts, investors, and everyone else watching the nightly news. Unfortunately, by the end of the 1990s, the sector's bubble started to burst. CLECs, equipment vendors, and large, well-known carriers started to file for bankruptcy. Jobs were lost, investment houses stopped handing out money, and the entire industry went bust. The telecom depression was severe.

Today, the industry continues to suffer from too much investment too fast. Competition is still very stiff, and carriers continue to file for Chapter 11, though less frequently than before. Telecom services are now more of a commodity forcing companies to streamline and consolidate. Luckily, the outlook for the industry started to improve slightly in late 2004. The improvement was due primarily to wireless technologies such as broadband cellular and Wi-Fi. An expansion of fiber deployments in the access network also increased in the early 2000s. As the century changed from the 1990s to the 2000s, the industry looked bleak, similar to the years after Bell's patent expired. Today, though, it is unlikely telecom will ever experience a boom like that of the 1990s; the field continues to grow and emerge, making it an exciting area in which to work.

SUMMARY

The years between 1920 to today produced many technological and regulatory changes in the Telecommunications industry. The 1934 Communications Act defined the industry for over 60 years. In the 1970s AT&T lost its fight with MCI producing the Modified Final Judgment allowing the emergence of long distance telephone companies in the United States. In 1996 the Telecom Act opened the local side to competition further straining the existing local telephone monopoly. Though the mergers after the Telecom Act of 1996 produced substantial changes in the industry, the crash that occurred in 2000 was much more significant. AT&T decided to sell off its TCG subsidiary, Worldcom decided to change its name to MCI and many of the RBOCs now called ILECs decided to hold off on mergers and acquisitions. Luckily, over the past two or three years the industry has slowly started to recover from the crash. ILECs have started to build fiber to home, cellular providers continue to build 3G cellular networks and the cable company's continue to expand their all in one full service offering. In the year 2000 the Computer Inquiry Acts were remanded eliminating the restrictions on the ILECs related to basic and enhanced services.

As of 2006 the most significant changes in the industry are mergers between large cellular companies, ILECs and surviving CLECs. The long distance providers are experiencing demand for bandwidth from the new VoIP industry as CATV and others build out full service networks. Equipment vendors are also seeing their bottom line turn black as fiber to home and VoIP expand. On the regulatory side, the battle between Internet "freedom" advocates and the large carriers over network neutrality or pay per what you use rages.

Though the industry has experienced ups and downs over the past century, it is unlikely we will see another telecom boom like that experienced in the late 1990s. Similar to the boom experienced in 1900, it is possible the telecom boom cycle is tied to 100 year intervals. In summary, over the past 85 years the regulatory climate and the technological advances such as the transistor, fiber optic cable and digital multiplexing schemes produced the communications infrastructure we all use and depend on today.

High-Level Overview

Photo courtesy of the Library of Congress

The Basics—Sound, Electrical Signal, Electromagnetic Spectrum

Objectives

After reading this chapter, you should be able to

- Describe the characteristics of sound.
- Describe the frequencies human beings use to communicate.
- Describe how sound waves are converted into electrical analog waves.
- Describe basic electricity as it relates to circuits.
- Describe the electromagnetic spectrum.
- Describe telecommunications standards organizations.

Outline

Introduction

1.1 The Characteristics of Sound

1.2 Frequencies Humans Use in Communication

1.3 Sound Waves versus Electrical Waves

1.4 Basic Electricity As It Pertains to Circuits

1.5 Defining the Electromagnetic Spectrum

1.6 Standards Organizations/Industry Forums and Professional Groups

■ INTRODUCTION

Every time you speak, sound waves are generated from your mouth and are carried across a medium, such as air, to a receiver. Communication in its most primitive form is nothing more than transferring information between transmitter and receiver across a medium. As your voice travels through the air, several characteristics such as the loudness of the signal, the variations in frequencies, and the distance between the sender and receiver determine whether the signal will arrive and make sense to the recipient. Our study of telecommunications begins with an overview of the physical characteristics that comprise the nature of communications.

Chapter 1 discusses what makes up a sound wave and how that sound wave is converted into an electrical wave. It defines certain terms that are used frequently in telecommunications, such as *frequency* and *wavelength*. The chapter also explores the electrical properties associated with communications circuits and how voltage, current, power, and impedance affect a signal as it travels down a copper medium. In addition, the electromagnetic spectrum is defined, as well as how it relates to the way information travels across the network. Finally, descriptions of each of the standards bodies responsible for developing the rules and protocols for the telecommunications industry are presented.

■ 1.1 THE CHARACTERISTICS OF SOUND

1.1.1 Sound

Have you ever thought about *sound?* Sound is all around us—music on the radio, a crying baby, a whistle, a door slamming, or a whisper. Before beginning the journey into the world of telecommunications, we first need to define the physical properties that make up sound and the associated electrical equivalents. To understand how sound travels through our atmosphere, we discuss a few concepts from basic physics.

Pretend you are standing on a street corner talking to your friend Henry. In your mind's eye, create a picture of your words floating through the air and entering Henry's ear. If you say, "Henry, are you going to the midnight movie marathon?," the word "Henry" comes out of your mouth, travels through the air, and enters Henry's ear. This brings up several questions. How are sounds that make up the word "Henry" created? How does the air carry the sounds? And how does Henry's ear distinguish and interpret the sound?

The first question can be answered by defining sound and how it travels. Sound is a periodic variation in air pressure. As we speak, vibrations created in vocal cords travel out of the mouth and disturb the air around us. Sound creates a disturbance in the media it travels across, whether that medium is air or a copper wire. An analogy that may help you visualize sound disturbing air is that of comparing air to water and sound waves to water waves. When water is undisturbed, the surface is flat and the water is still. When a boat speeds past, waves are produced that travel outward from the boat. Figure 1–1 shows waves projecting from their source.

Air can also be disturbed. In our case, instead of a boat speeding through the water, the word "Henry" is disturbing the air by pushing through it and creating waves. We now know that air is a *medium* like water and that *sound waves* change or disturb the air as they push outward from their source, the mouth. So, sound travels through air in the form of sound waves, but how do different sounds or different words distinguish themselves from one another? To understand how one sound is distinguished from another, we need to understand what components make up a sound wave. The three main components of a

sound
Audible frequencies created by a source such as a person talking. A periodic variation in air pressure.

medium
Information carrier such as copper wire, the air, or fiber optic strands that carries information signals from location to location.

sound waves
Pressure waves that vibrate molecules in a medium such as the air.

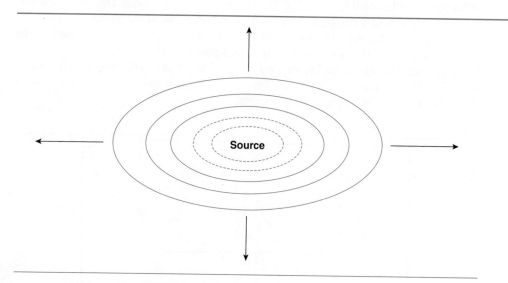

Figure 1–1
Waves projecting from the source.

sound wave are the frequency, the wavelength, and the amplitude of the waveform. We begin by defining terms and explaining how they relate to transmitting a voice signal across a telephone line.

1.1.2 Frequency

Frequency is a term used every day to explain how often something occurs. Again, an analogy might help explain the concept. Imagine that it is a very hot day in July and you are standing waist deep in the Atlantic Ocean off the shore of Delaware. You pull out your new waterproof watch, set the precision timer, and press "start" every time a wave crashes into you and knocks you down. You soon discover that a wave hits you once every minute; therefore, the frequency of the waves is one wave per minute.

Frequency in nature is defined as the number of occurrences during a specific period of time—the frequency of the seasons, the frequency of a full moon, the frequency of a full day, the frequency of a year, the frequency of a weekly television program.

The term *frequency*, as it relates to communications, is used to define the different sounds we are able to speak and hear. We can distinguish different sounds by measuring the number of waves that are completed each second in the same way that we can define one year by counting 365 days.

The time period used to measure the frequency of sound waves is one second. The event that is measured is the number of waves that are completed. Therefore, the frequency of a sound is the number of times the waveform repeats itself each second.

Frequency is a critical concept in the world of communications. Every sound has its own frequency, although the volume may be so low that we are not able to hear it. For example, the number of *cycles per second* (cps) for the musical note middle C never changes. The term used to define frequency when it relates to sound waves or electromagnetic waves is *cycles per second. Hertz* (Hz) is the unit of measurement for cycles per second. Therefore a frequency measurement for sound or electromagnetic waves is given in hertz rather than cycles per second.

If we were to display a 5 Hz (5 cps) waveform onto a screen, we would see a waveform that repeated itself five times per second, as shown in Figure 1–2. Stated simply, you would be able to count five complete waveforms every second. As we observed the 5 Hz waveform on the wall, we could also listen to the sound it produces. If the frequency of the tone increased to 7 cps (7 Hz), the waveform on the wall would change and so would the sound of the tone. You could then count seven complete waveforms every second.

1.1.3 Wavelength

What are the differences between the 5 cps tone and the 7 cps tone? The obvious difference is the length of a complete waveform. Remember that in the world of sound, one second is the constant reference time used to define sound's frequency. In Figure 1–3, the wavelength of the 5 Hz tone is obviously longer than that of the 7 Hz tone. To fit 7 cps into a one-second period, the cycles must repeat more often, so the length of the waveform is shorter, as more waveforms are forced into the same space. The term *wavelength* is used to describe the length of the waveform.

---

cycles per second
The number of times a waveform cycles every second, also referred to as the frequency of the signal.

hertz
Measurement used to define the number of cycles per second, or frequency, of a signal. Ten Hz equals 10 cps.

wavelength
The length of one complete waveform. The higher the frequency, the shorter the wavelength; and the lower the frequency, the longer the wavelength.

ppening over a ___ period. Frequency us___ communications is measured in hertz, which represents the number of cycles per second.

47

Figure 1–2
Five cps analog waveform projected onto a white screen.

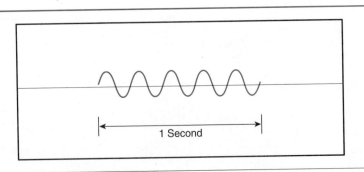

1 Second

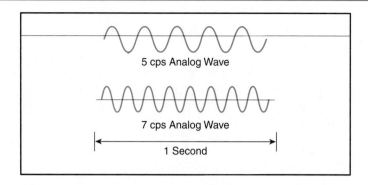

Figure 1–3
Comparison between a
5 cps analog signal and a
7 cps analog signal.

What sounds are you able to hear? Or rather, what frequencies are you able to hear? Human beings should be able to hear frequencies between 20 and 20,000 Hz. This means that humans can distinguish waveforms that range from 20 cycles every second up to waveforms of 20,000 cycles every second. The middle C key on a piano has a frequency or sound wave of about 250 cps. Since we cannot draw all 250 cycles, we will project one cycle, as shown in Figure 1–4. The distance between the start of the cycle and the end of the cycle is referred to as *wavelength*. The length of the waveform is determined by frequency. The higher the frequency, the shorter the wavelength. The lower the frequency, the longer the wavelength.

Now, visualize a frequency of 30,000 Hz or 30,000 cps. Again, we can project the waveform onto a screen and measure the distance between the beginning and the end of the cycle—10,000 meters (m). Will a 10,000 m wave fit into a human ear? Let's hope not.

Now imagine that you own an elephant and a dog. The dog is very jealous of the elephant. You want to play with the elephant without being pestered by the dog, but every time you call the elephant the dog jumps up and runs to you as well. What would solve this problem?

The size of the ear determines the frequencies that can be heard. A dog's ear is constructed specifically to hear very high frequencies; the eardrum is sensitive to very short wavelengths. A small ear like the dog's can hear very high frequencies. A large ear like the elephant's can hear very low frequencies. To solve your problem, you could buy an elephant whistle that projected low frequencies the dog could not hear; consequently, he would miss out on special treats reserved for our pet elephant.

Using the same concept, we now understand how Henry can hear the sound made by saying his name. His ear, like all human ears, is able to accept frequencies between 20 and 20,000 Hz. The wavelength of these frequencies fits into our ears and vibrates against our eardrums to be picked up by the brain and interpreted as the word "Henry,"

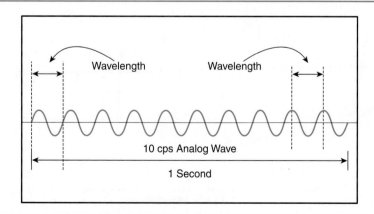

Figure 1–4
One wave of a 10 cps waveform
measured from the beginning of the
cycle to the end of the cycle or from
the peak of one wave to the peak of
the next wave.

Figure 1–5
Abstract depiction of a typical
waveform generated by human
speech.

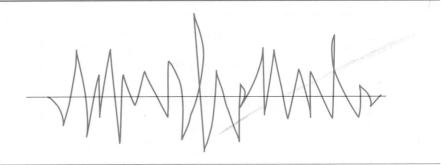

which has many frequencies. If we were to project the sound onto our blank wall, it would look something like the diagram in Figure 1–5.

1.1.4 Amplitude

Amplitude is the term used to define the power or strength of a signal. The amplitude of a sound wave determines the loudness of the sound. If a radio is turned up, the amplitude or the height of the sound wave increases. If it is turned down, the amplitude, or height, decreases, as shown in Figure 1–6. Again, mentally project the frequency of middle C onto the white wall, and listen to the sound it produces. What happens to the waveform if we turn up the volume? The height of the waveform increases in both the positive and negative directions. The amplitude of the waveform is what determines the volume of the sound as it travels through the air. The louder you yell, the higher the waveform and the stronger the signal.

The frequency of the waveform does not change—a middle C is still a 250 Hz middle C. The sound projects farther away from the source as the amplitude, or volume, is increased. Conversely, when we whisper, sound does not travel as far from the source. The softness or loudness of the sound is determined by the strength of the signal. A shout produces a stronger signal than a whisper. The strength of the signal relates to its amplitude. The greater the amplitude, the stronger the signal.

amplitude
The amplitude of the sine wave is the measurement of the waveform above/below the zero reference. For example, +3 Volt (V) refers to a 3 V amplitude above the zero reference.

Figure 1–6
Voltage is used to measure the signal strength of various amplitudes. The greater the amplitude, the louder the sound and the stronger the signal.

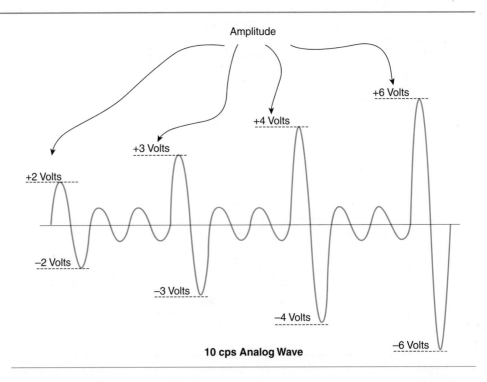

10 cps Analog Wave

Amplitude can be thought of as the power of the signal. The stronger the amplitude, the more power the signal possesses—hence, the farther the sound will travel. The signal reach of a radio station depends on the power of the radio transmitter. A high-powered transmitter can project a *radio signal* a great distance from the antenna by sending out very strong waveforms that cover a large geographic area. The FCC dictates how far a radio station's signal is allowed to travel. Radio stations that must buy air space to cover a specific region compete for radio spectrum.

radio signal
Signal projected from a radio transmitter's oscillator. Projects the frequencies out into the air, which serves as the transport medium.

1.2 FREQUENCIES HUMANS USE IN COMMUNICATION

Sound is made up of multiple frequencies, and each frequency within the range of 20 to 20,000 Hz has a specific sound that can be heard by humans. Normal human speech falls within the *frequency range* of 300 to 4000 Hz. Therefore, if we were to place filters on Henry's ears that blocked out all frequencies above 3000 Hz and below 300 Hz, Henry would still be able to hear me ask, "Henry, are you going to the midnight movie marathon?" However, he would not enjoy listening to his favorite rock band if we filtered out the low and high frequencies. Music generally ranges from 20 to 20,000 Hz.

frequency range
Range of frequencies a signal covers; 300 to 3000 Hz is the frequency range of a voice signal.

Early telephone engineers used the 300 to 4000 Hz frequency range when designing the telephone network and they built the telephone to handle these frequencies. These frequencies were used to define the boundaries of the *carrier serving area* (CSA) that surrounded every central office (CO). Telephone engineers realized they would need to use a specific gauge of wire and limit the length of the wire in order to carry voice frequencies to distant telephone subscribers.

carrier serving area
Serving area surrounding a central office that the CO will provide service to.

1.3 SOUND WAVES VERSUS ELECTRICAL WAVES

The electrical waves produced by a telephone transmitter and traveling down telephone wires are referred to as analog signals. Figure 1–7a compares a sound wave to an electrical voice wave. An analog waveform is analogous to the sound waves coming from

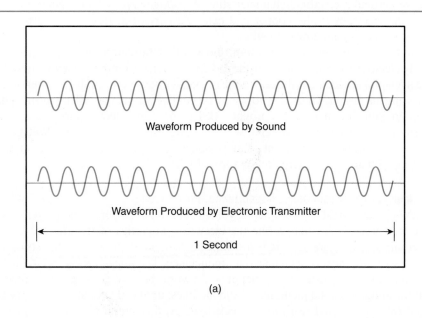

(a)

(b)

Figure 1–7
(a) A waveform of a defined frequency produced by sound compared with a waveform of the same frequency produced by an electronic transmitter. (b) DC voltage waveform.

human vocal cords. The telephone transmitter takes pressure waves and varies the current flow on telephone wire. When sound waves are converted into electrical signals by using the telephone, the signal created by the sound of the voice mirrors the analog signal coming out of the telephone transmitter. If we were to project the sound waves coming from our vocal cords to the analog electrical signal leaving the telephone transmitter onto the wall, they would look identical. In essence, we can transmit conversations across copper wires because the waveforms created by vocal cords can be converted to electrical analog signals that mirror sound waves. The medium used to transmit these signals is irrelevant, other than to determine the distance the signal is able to travel. Cellular phones perform the same function as the standard land line telephone performs.

Now suppose that Mary, who lives in New York City, places a call to her friend Mabel in Los Angeles. Mabel says "hello," and Mary responds with "Mabel, how are you? This is Mary." The response is instantaneous even though the signal is traveling over 3000 miles.

Electrical analog signals travel through a conductor much faster than sound waves travel through air. Analog signals are able to travel so quickly because of the *shock wave effect* of powerful electrons.

The voltage induced onto a conductor creates a current flow by exciting the electrons, which then push against one another all the way down the wire. *Shock wave effect* is the term used to describe the process, which is similar to the way dominos create a chain reaction when someone gives the first one in line a push. This effect allows the signal to travel at an extremely fast rate—millionths of a second—to its final destination, which, in this case, is the end telephone.

Older satellite connections produced a noticeable delay between the speaker and the listener. Today, special electronics have all but eliminated that delay.

■ 1.4 BASIC ELECTRICITY AS IT PERTAINS TO CIRCUITS

1.4.1 Voltage and Current

The copper wires that connect your telephone to the telephone company's central office have particular electrical characteristics. A *voltage* applied to the copper line energizes the electrons in the copper and causes a *current* to flow. That current flow carries the voice signal down the wire.

Find an old electric bill and examine the way that the electric company charges for monthly electricity use. It charges by the number of watts per month that you used. Now locate a telephone bill. Does the telephone company charge you for the number of watts used during the month? Of course not, but it could. Every time you talk on the telephone or dial into your Internet service provider (ISP), you cause electrons to flow and voltage to drop, which in turn causes energy to be exerted in the form of watts. Voltage multiplied by electron flow (current) equals power, and power is measured in watts. Therefore, you are using energy in the form of watts.

1.4.2 Impedance—Resistance, Capacitance, Inductance

As with other electrical measurements, voltage, current in amperes, and power in watts are three characteristics that define the electrical properties of the circuit. A fourth and very important characteristic is the impedance of a circuit, which consists of three properties—resistance, capacitance, and inductance. The *impedance* of a circuit must be considered when designing a telephone network, because it adversely affects the signal traveling down the medium. Telephone lines, unlike electrical wires, carry both *AC* and *DC* signals, and *resistance* affects both. At the central office, a negative DC voltage is applied to a line to energize the electrons and cause a current to flow, which in turn allows a voice signal to travel down a copper medium. The voltage varies from −48 to −52 V. Figure 1–7b shows a typical DC voltage. Direct current travels in one direction as evidenced by its name, while alternating current travels back and forth, as shown in Figure 1–8a. The AC source on the line is the information signal. The

shock wave effect
Electrons in a copper medium pushing each other down a line. The shock wave effect allows signals to cover great distances almost instantaneously.

voltage
Energy source of electricity that, when applied to a line, causes current to flow.

current
Flow of electrons in a conductor caused by an electrical source being applied.

impedance
Combination of resistance, capacitance, and inductance.

AC voltage
Alternating current.

DC voltage
Direct current.

resistance
Electrical characteristic of a medium that inhibits the flow of electrons. Copper wire has a resistive value based on the gauge and length of the wire. High resistance impedes the signal flow.

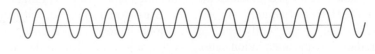

(a) Alternating Current

0 V ——————————————————————————————

−48 V ——————————————————————————————

(b) Direct Current

Figure 1–8
(a) Alternating current (AC) waveform is the electrical power received from an AC outlet in a home. (b) Direct current (DC) is used to feed telephone equipment, telephone lines, and other telecommunications devices.

analog voice signal shown in Figure 1–9 is an alternating sine wave, in which electrons alternate just as they do in electrical circuits. Therefore, the voice signal can be thought of as an alternating current.

Inductance and *capacitance* affect AC signals; resistance affects both AC and DC signals. All three variables must be considered when determining the gauge of wire to use, the distance the circuit can extend from the central office, and whether or not special conditioning will be required.

Resistance, as the name implies, resists the forces of electrons in the circuit. Imagine yourself walking through water. The water slows your pace and causes you to lose energy faster than if you were walking on land. Resistance in a telephone line is similar to the water pushing against your stride. Every copper line has a certain amount of resistance associated with it. The resistance impedes the flow of electrons, thus causing the information signal to lose power as it travels down the path. You would also lose strength hiking through hip-deep water. The same is true with a copper line. Therefore, when designers determine how far a copper line can be extended from a central office, they need to take into account the amount of resistance per foot associated with that line. Resistance is measured in *ohms* and is necessary to determine the voltage drop of a circuit. Use Ohm's law to determine the voltage drop:

$$V \text{ (voltage)} = I \text{ (current)} \times R \text{ (resistance in ohms)}$$
or
$$I \text{ (current)} = V \text{ (voltage)}/R \text{ (resistance in ohms)}$$
or
$$R \text{ (resistance in ohms)} = V \text{ (voltage)}/I \text{ (current)}$$

Resistance is a common measurement taken by technicians when they need to troubleshoot a circuit. A high-resistance short tells the technician the line has a problem, while a zero-resistance reading tells the technician that the circuit is open somewhere along the route.

Capacitance and inductance can be viewed as siblings with completely opposite personalities. Capacitors consist of two electrical plates separated by an insulator. In electronics, capacitors are placed strategically to store charges. In telephony, capacitance happens naturally when two conductors run long distances side by side with air acting as the insulator. Capacitance, like resistance, affects the rate at which the signal attenuates. The longer the line runs, the greater the loss of the AC signal, especially at higher frequencies. The distance the signal can travel is reduced due to the amount of ohms that are formed from the capacitive reactance of the telephone line. Stated

inductance
A measurement called henrys that defines an electrical field around a conductor.

capacitance
A measurement called farads that defines the electrical storage capacity of the medium.

ohms
Measurement of resistance. For example, 100 ohms is an amount of resistance measured.

Figure 1–9
Analog waveform.

simply, capacitive reactance increases the signal attenuation, or loss of higher frequencies on the line. Higher-frequency AC waveforms are stored by the capacitor, thus heating up the line and consequently using up additional energy. The distance a signal can travel is reduced by the attenuation caused by the capacitive effect of the circuit.

A solution was developed in the early 1900s that is still used today when designing circuits longer than 18,000 feet. Inductors are placed every 6000 feet to eliminate frequencies higher than 3000 Hz. An inductor is formed by wrapping wire around a magnetic core. Inductance cancels out capacitance, and vice versa. A circuit with 100 ohms of capacitance and 100 ohms of inductance would equal 0 ohms of impedance; however, we would still have impedance from resistance. Telephone engineers understand this relationship and use it to extend the distance a signal can travel down a copper wire. A device called a *load coil* is placed in long circuits to cancel out the effect of capacitive reactance. A load coil is an inductor. It does not allow frequencies higher than 3000 Hz to pass through a circuit; thus, it reduces the number of stored charges in a line and, in turn, decreases the total attenuation of the information signal. The result is that the signal can travel greater distances before it fades away. Capacitance and inductance only affect AC signals.

1.4.3 Power—Watts

Similar to the electrical line going into a house, a telephone line also carries a certain amount of energy, referred to as *watts*. A watt (W) is the unit used to measure the electrical power extended at a given time. When the telephone handset is taken off the cradle, the circuit is complete, and voltage applied to that line causes electrons to flow through the copper wire, also known as causing current. Multiplying the voltage and current values, as shown in the following equation, produces the power value, or watts. Watts may be viewed as the value that defines the strength of the signal. For example, if we measure 48 V and 0.6 mA (0.006 A), the energy extended is 0.28 W.

$$W \text{ (watts)} = V \text{ (voltage)} \times I \text{ (current)}$$

1.4.4 Decibels—Measuring Signal Strength

Have you ever had trouble hearing the person at the other end of the phone? Or experienced a high-pitched squeal on the telephone line? Both are the result of an improperly tuned circuit. Low volume and overdriving the signal producing squeal on the line are a result of faulty equipment, improper circuit termination, or an unbalanced circuit. The electrical parameters of a circuit must be tuned properly for a quality signal—your voice, computer data, or a stream of video or audio—to be delivered. If the voltage is too low or the resistance of the circuit increases due to a faulty cable, the signal strength may be too weak to be heard. If the amplitude of the signal is too low, the result is low volume or corrupted data. If the amplitude is too high, squealing or humming on the line causes a distorted voice signal or corrupted data. Therefore, the ability to measure and define the information signal at any point in the circuit is critical when designing, installing, or repairing circuits. The range between the softest sounds humans can hear and the loudest their eardrums can tolerate is one-thousand-trillionth (0.000000000000001) of a watt to one-thousandth (0.001) of a watt. Understandably, measuring a change in watts within this range is impossible and thus requires an alternative way to determine signal strength. The unit used in telecommunications to measure signal strength is the *decibel (dB)*. The decibel was devised to relate the increase of power to a logarithmic value instead of adding and subtracting exponents. Figure 1–10 lists several common sounds and their related signal level in *dBm*. Figure 1–11 shows milliwatts with corresponding dBm values. Figure 1–12 depicts the relationship between signal loss and frequency. As shown in the graph, the higher the frequency, the greater the loss. The telecommunications industry depends on decibel measurements to ensure that the circuits, equipment, and network perform as designed.

Early scientists determined that the human ear could perceive a change in volume when a signal was either doubled or reduced by half and the decibel gain or loss was

load coil
Inductor placed in the telephone cable to offset the effect of capacitance. Load coils are placed about a mile apart along the cable run.

watts
Measurement term used to define power. For example, a lightbulb is rated as 60 W.

decibel
Logarithmic value used to measure the strength of a signal. The decibel was named after Alexander Graham Bell and has been used to define the strength of communications signals from the early days of telecommunications.

dBm
Decibel measurement of 1 milliwatt.

Signal Levels of Common Sounds	
0 dB	Lowest level for hearing
10 dB	Sound of crumpling paper
20 dB	Sound of a whisper
65 dB	Sound of normal conversation
110 dB	Sound of a tractor trailer
120 dB	Sound that causes pain

Figure 1–10
Measuring any of these sounds using a dB meter would result in values similar to those listed.

three. For example, when you double the power output of your radio by turning up the volume knob, the decibel level increases. Specifically, if the power was 10 milliwatts (mW) initially and it increases to 20 mW, you have just added 3 dB of gain to the signal. The same holds true if you increase from 20 to 40 mW—you add 3 dB to the output. On the other hand, when you turn the radio down from 40 to 20 mW, you decrease the signal by half or take away 3 dB from the signal.

The decibel is used throughout telecommunications as the measurement unit for *signal levels*. It is nothing more than a value that represents a change in a signal, whether it is an increase or a decrease in strength. We know that humans hear sounds ranging between one-thousand-trillionth W and 1 mW. Therefore, the loudest sound humans can tolerate has a signal strength of 1 mW. Scientists realized early on that if they used 1 mW as a base reference point, they could easily develop ways to measure the strength of an electrical signal. The scale developed around the 1 mW reference point is the dBm scale. As shown in Figure 1–11, 0 dBm has the same value as 1 mW.

Looking at how a circuit is tested end to end should further simplify this concept. A test box is attached to a telephone circuit to measure dBm levels. Algorithms are built into the test box to convert watt readings to dBm values. When a technician takes measurements along a circuit path, the value tells the technician whether or not the circuit

signal level
Strength of a signal measured in decibels.

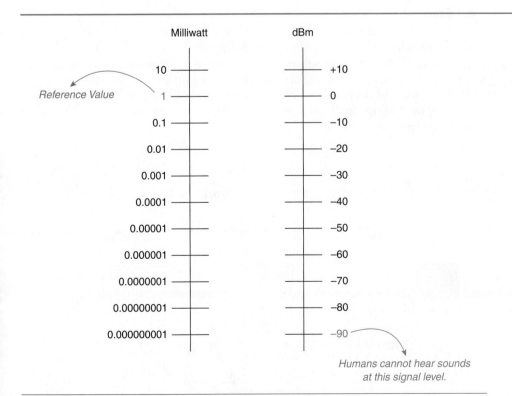

Figure 1–11
Milliwatt values compared with dBm values. For example 0.000001 mW and −60 dBm represent the same value. Using −60 dBm to identify the strength of a signal is much simpler than using 0.000001 mW.

Humans cannot hear sounds at this signal level.

Figure 1–12
Correlation between loss of dBm and frequency. The greater the frequency, the greater the loss. Other variables also affect loss of circuit, such as the gauge of the cable.

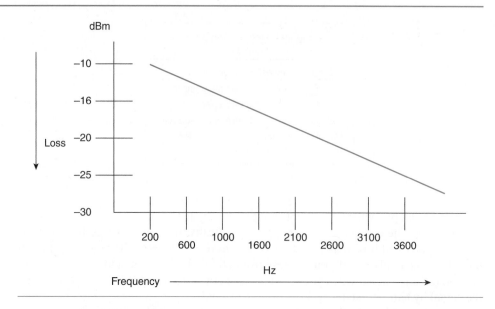

is too low. This is indicated by a signal level that is too low, or too hot, shown by a signal level that is overdriven or has too much power. The dB levels are noted and checked against the original circuit design record to make sure the signal has not moved too far from its original value. Different types of circuits have different level parameters and are designed accordingly. As shown in Figure 1–13, the signal strength changes along the circuit route depending on the type of equipment it travels through, the distance the signal travels, the gauge or type of medium used, and any other imperfections that may cause the signal to attenuate. The decibel is the measurement unit used to design, test, and maintain communications signals whether they be analog, digital, voice, data, video, or something else.

1.4.5 Exponential Notation

Several mathematical values are used extensively in telecommunications. The terms *nano*, *micro*, *milli*, *kilo*, *mega*, and *giga* are defined in Table 1–1. Each is used to simplify very small or very large values. For example, if a circuit carries 45 million bits every second, it can be written as 45 Mbps (megabytes per second) instead of 45,000,000. Each of the listed values is used repeatedly throughout the text. The most commonly used values are listed in the table, but many others found in most technical math books have been left out.

Figure 1–13
End-to-end circuit loss using dBm as the measurement value. The total end-to-end loss of the circuit on this one-way termination is −8.5 dBm. At position C, the signal is regenerated, thus canceling out any loss up to that point.

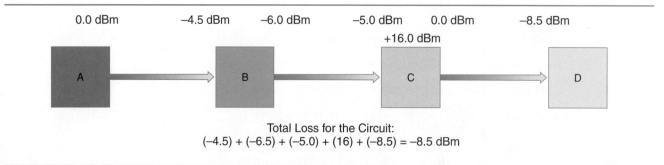

Table 1–1
Summary of Commonly Used Values.

Term	Abbreviation	Value	
Nano	n	10^{-9} or	0.000000001
Micro	μ	10^{-6} or	0.000001
Milli	m	10^{-3} or	0.001
Kilo	K	10^{3} or	1000
Mega	M	10^{6} or	1,000,000
Giga	G	10^{9} or	1,000,000,000

■ 1.5 DEFINING THE ELECTROMAGNETIC SPECTRUM

1.5.1 Electromagnetic Waves

The discovery of electromagnetic waves was one of the most important discoveries of the last millennium. We cannot see them, touch them, or talk to them but we can use them to transmit information around the world.

The electromagnetic spectrum was first defined in the 1800s. The best way to understand it is to visualize a magnet in the shape of a ball with a defined top and bottom—north and south. Imagine that invisible lines of force similar to rivers tie the north and south points of the ball together. See Figure 1–14 for an illustration of this concept. Thanks to Gauss's law, Faraday's law, and Maxwell's formulas, it was determined that when an electrical charge vibrates in space, a magnetic field is induced and when the magnetic field changes, an electrical field is induced. The changing magnetic and electrical fields create a wave projecting out from the source of the charge. A transmitted signal travels through space, riding on the fluid lines, or rivers, of force that surround the earth. The waves traveling along the river are identical to the waves of the signal being transmitted.

1.5.2 Frequencies of the Electromagnetic Spectrum

The *electromagnetic spectrum* has been grouped into specific bands of frequencies. For instance, the frequency band used to transmit *AM radio* is 530 to 1710 kilohertz (kHz). The frequency band set aside for FM radio is 87.7 to 107.9 kHz. Hertz is a measure of cycles per second. One cycle is one full waveform, so it would be very difficult to project an AM radio signal onto our blank white wall. There would be from 530,000 to 1,710,000 cps.

Sound waves are not the only frequencies defined within the electromagnetic spectrum. Light waves, X-rays, infrared waves, and gamma waves are all defined in the spectrum. For instance, black ink on white paper has a specific frequency that

electromagnetic spectrum
The magnetic field surrounding the earth capable of carrying multiple frequencies of varying wavelengths. Infrared, radio frequencies, ultrasound, and others make up the different spectra.

AM radio
Amplitude modulated radio. A source signal that impresses itself onto a carrier wave by manipulating the amplitude of the signal.

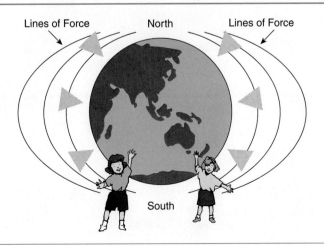

Figure 1–14
Lines of force surround the earth and are tied together at the North and South poles.

makes up our definition of the color black. Like middle C, black has its very own frequency. Each color is made up of a specific frequency within the visible light frequency range.

A superhero's (or superheroine's) X-ray vision allows him (or her) to decipher frequencies that are higher than those a normal human being could see. Remote controls used to change channels on the television or to open and close the garage door are often built to transmit frequencies within the infrared spectrum. Every time you change a channel, lower the volume, or turn off the television, small signals travel from the remote control unit to a sensor on the television that interprets the message. Heaters, toasters, and ovens also emit infrared frequencies. We may not see them, but we can feel the heat that they produce. Communications carriers designed to transmit billions of bits of information on a single fiber strand use a laser. Lasers emit frequencies that can be used in surgery and in communications. The human eye cannot see the laser light, yet the eye can be severely damaged by the laser's frequencies. Just because we do not see it does not mean it is not there.

Radio stations use frequencies allocated by the FCC. When you turn the dial on your AM radio to 1370, you are tuning into the radio frequency 1370 kHz for radio station WXXI. Each radio station is given a specific transmitting frequency. Your radio is the signal receiver. It has to be tuned to 1370 to pick up WXXI's AM signal, which is riding on the 1370 kHz frequency. When you travel across the country, you will notice that radio stations located in separate geographic areas use the same frequency ranges. For instance, 1320 in Los Angeles may be a pop station, while 1320 in Nashville is a country station. The signal projected from the radio transmitter in Nashville is not powerful enough to reach the Los Angeles radio spectrum, so it does not interfere with the 1320 kHz pop station.

When you place a call from your cellular phone, you are actually projecting your voice signal out into a frequency range of between 30 million and 300 million Hz. Your voice is riding one of those "rivers" of force surrounding the earth. The U.S. government reduced the national debt between 1993 and 1999 by auctioning off additional spectrums for new frequencies. The new blocks of frequencies allowed additional wireless providers to enter into the cellular marketplace. The cost for one slice of spectrum in the larger cities was in the billions of dollars, so many consortiums formed to vie for the most lucrative areas, such as New York City, Los Angeles, and Atlanta. Table 1–2 defines the frequency spectrum in more detail.

Table 1–2
Frequency Spectrum.

Frequency Type	Application	Frequency Range
Low Frequency—LF	Telephone signal	0 to 4000 Hz
Medium Frequency—MF	AM radio signal	300,000 to 3,000,000 Hz
High Frequency—HF	Shortwave radio	3 million to 30 million Hz
Very High Frequency—VHF	FM radio signal, cell phones, television, aeronautical and maritime transmission	30 million to 300 million Hz
Ultra High Frequency—UHF	Television, landmobile communications	300 million to 3 billion Hz
Super High Frequencies	Microwave, satellite	3 billion to 30 billion Hz
Infrared	TV remote, inside LAN networks	Tera Hz—10^{11} to 10^{14} Hz
Visible Light	Colors, shadows	Tera Hz—10^{15}
Lasers	Optical transmission, medical applications	Tera Hz—10^{14}
Ultraviolet	Sun	Tera Hz—10^{15} to 10^{17}
X-rays	Medical applications	Tera Hz—10^{17} to 10^{19}
Gamma rays	Radioactive medicine	Tera Hz—10^{20} and up

■ 1.6 STANDARDS ORGANIZATIONS/INDUSTRY FORUMS AND PROFESSIONAL GROUPS

What would happen if I thought the word "yes" meant "no"? For example, when a server asked whether or not I would like gravy on my potatoes, I would answer. Expecting that "no" meant "yes," I would be very disappointed when my potatoes arrived without gravy. There would be complete confusion between the server and myself, and communication would break down due to duplicate meanings for one word. The world of communications functions similarly. If one vendor builds equipment that understands certain proprietary codes that other vendors' equipment does not, the two will not communicate. Standards have been, and continue to be, initiated to ensure that the hundreds of communications carrier networks can interconnect and carry user information transparently around the network.

Today's standards organizations are not the same as those established before divestiture. At that time, AT&T was responsible for all communications standards in the United States. After divestiture, Bell Communications Research (Bellcore) was formed to provide assistance to the newly established RBOCs and to help provide a source for industry initiatives. Bellcore has since changed its name to Telcordia and has become an equipment manufacturer and research company that provides special services to all carriers and equipment manufacturers. To ensure that the communications industry continued to strive for open communications standards, the International Telecommunications Union (ITU) and the American National Standards Institute (ANSI) were given the job of setting the course for the industry. These two organizations, along with several other standards bodies, contribute to the vast amount of data that define how technologies will work together.

In addition to the traditional standards organizations, newly formed industry forums have evolved to help define and initiate new communications protocols such as SONET, ATM, frame relay, and the Internet. The forums are composed of equipment vendors, software providers, and *service providers*. A service provider is a company that provides communications services, such as telephone companies, Internet service providers, and other companies that offer services to the end users. For example, the ATM forum developed standard quality of service parameters that all equipment vendors used when it developed its ATM switch architectures. Service providers implement services using the standard quality of service parameters agreed upon within the forum.

Both the technology-dependent forums and the various standards organizations are comprised of industry experts, equipment manufacturers, software companies, service providers, and governmental representatives. The standards organizations are responsible for implementing guidelines for new protocols, physical terminations, interconnections, and much more.

Both the state and the federal governments have organizations that deal with communications issues. The federal government established the Federal Communications Commission (FCC) in 1934 and granted the agency power over all communications at a federal level, which includes all telecommunications. In addition, the FCC mandated that the 1996 Telecom Act ensures that both *ILECs* and *CLECs* adhere to it. Each state also has a communications regulatory agency known as a public utility commission (*PUC*) or a public service commission (*PSC*). Both perform the same function of regulating the telecommunications providers.

Following are the main organizations that are setting the rules for the communications industry today.

1.6.1 Standards Organizations

- ITU—International Telecommunications Union—The ITU receives its authority from the United Nations. It is the standards organization responsible for ensuring that each country will be able to exchange information with other countries. The Consultative Committee on International Telegraphy and Teleplay Committee (CCITT) has been absorbed by the ITU, making it the world-recognized standards

service provider
Communications provider such as a long distance carrier or local carrier.

ILEC
Incumbent local exchange carrier. Term born from the 1996 Telecom Act, which defines the existing local telephone company (RBOCs).

CLEC
Competitive local exchange carrier. Term born from the 1996 Telecom Act, which defines the newly formed competitive local telephone companies offering local telephone service.

PUC or PSC
Public utility commission or public service commission. A state's regulatory commission that oversees telecommunications policy and rates.

organization for telecommunications. The ITU consists of several branches; the ITU-T represents telecommunications and the ITU-R represents radio. Within the ITU-T, subcommittees called study groups review and formulate policy. The study groups present their work to a plenary committee that reviews the information and determines the final composition of the standard. A country such as the United States whose telecommunications are privatized does not have to comply to ITU standards. The prestige of the ITU has grown in the past few years due to the need for fully compatible networks throughout the world.

- ANSI X.3T9.5—American National Standards Institute X.3T9.5—The X.3T9.5 is the telecommunications standards body within the ANSI standards group. There are hundreds of such subgroups. ANSI is a nonprofit organization that sets telecommunications standards in the United States; and, similar to the ITU, ANSI is comprised of industry experts, vendors, service providers, and government agents. The significance of ANSI to the industry can be seen in equipment specs in which technical specifications are referenced to particular ANSI standards.

- ISO—International Standards Organization—The ISO is composed of several standards organizations, including ANSI X.3T9.5. An ISO representative acts as a liaison between the U.S. government and the ITU committee.

1.6.2 Industry Forums and Professional Groups

- IEEE—Institute of Electrical and Electronic Engineers—The IEEE is a professional organization founded in 1884 for the purpose of establishing standards for electrical, electronic, and computing systems. The IEEE is responsible for setting standards around several computing protocols, such as the IEEE 802.3, known as the Ethernet protocol standard. IEEE study groups have also standardized several other LAN protocols.

- IETF—Internet Engineering Task Force—The IETF sets standards for the Internet and, in doing so, is responsible for much of the work done on TCP/IP. The IETF is comprised completely of volunteers and is the offshoot of ARPANET, the Defense Department agency that was responsible for the birth of the Internet. The significance of IETF's work has increased over the past two to three years due to the increased need to converge networking protocols such as SONET, ATM, frame relay, and IP. The IETF is gaining the respect and attention of communications professionals around the world.

- NFOEC—The National Fiber Optic Engineering Consortium—The NFOEC, like the IETF, has increased in size and clout over the past few years. The NFOEC pulls together organizations from around the world to discuss issues pertaining to the role of fiber optic communications in a network. The organization holds annual meetings that include the presentation of white papers and discussions about the changing role of the fiber optic industry as electronics and software improve and evolve. The NFOEC is becoming an important professional organization in the industry.

- OEC—Optical Engineering Consortium—The OEC is similar to the NFOEC in that it also deals with issues of fiber optic deployment in the industry. The OEC is more concerned with the transportation of pure protocols over glass.

- ATM Forum—Asynchronous Transfer Mode Forum—The ATM forum is, as it sounds, a forum that deals with issues surrounding the ATM protocol.

- Frame Relay Forum—Similar to the ATM forum, the frame relay forum's purpose is to work on issues relating to frame relay protocols.

- SONET Forum—Synchronous Optical Network Forum—The SONET forum consists of a group of industry experts who deal with issues concerning SONET protocol.

- DSL Forum—Digital Subscriber Line Forum—The DSL forum is a newer forum that has established standards within the DSL arena.

- OBF—Ordering and Billing Forum—The OBF's purpose is to establish ordering standards between service providers. For example, the order form used to order a circuit between carriers has been standardized by the OBF working group and adopted by the industry as the standard. When ordering a circuit from a carrier, the established order form and the agreed-upon codes must be used to specify the circuit parameters.

- NANC—North American Numbering Council—The NANC is a group of thirty-four industry experts appointed by the FCC to oversee the distribution of telephone numbers in North America. The NANC does not distribute the numbers directly but appoints a third party to administer the North American Numbering Plan (NANP). In the past, Bellcore was responsible for overseeing the distribution of telephone numbers in the United States. Today, Lockheed Martin is the NANP administrator and, as such, is responsible, along with the local municipality, for establishing new area codes, initiating number pooling, overseeing NPA splits, and policing the use of the limited supply of telephone numbers.

- NPA—Numbering Planning Area—NPA splits happen when an area code is exhausted. The local telephone companies, municipalities, and residents work with NANP to establish a new area code in the region.

- IANA—Internet Assigned Numbers Authority—The IANA watches over domain names and Internet addresses in addition to the Internet protocol.

- ARIN—American Registry for Internet Numbers—The ARIN is responsible for IP number assignments in North America, South America, the Caribbean, and Sub-Saharan Africa. Two other organizations oversee IP address allocation in the remaining parts of the world. They are RIPE NCC—Regional Internet Registry Europe, which oversees Europe, the Middle East, and parts of Africa; and APNIC—Asia Pacific Internet Registry, which oversees all of Asia.

  In addition, ARIN manages Autonomous System Numbers (ASNs) and oversees INADDR, ARDA, or IP6 inverse mapping standards. ARIN's main objective is to work with the Internet community to conserve IP addresses and to help guarantee an Internet that functions efficiently.

- DARPA—Defense Advanced Research Projects Agency—The DARPA was originally called ARPANET, which was the organization that founded the Internet. DARPA is a U.S. agency comprised of industry experts who work on advanced research projects in communications.

SUMMARY

Sound causes vibrations in the air much like a boat traveling through water creates waves. Each sound or tone has a specific frequency associated with it, which is measured in cycles per second called hertz (Hz). Human beings are able to hear frequencies ranging from 20 to 20,000 Hz. Human speech frequencies range from 300 to 4000 Hz. The wavelength is what determines the frequencies that humans, dogs, elephants, and other creatures can hear. If the wavelength is too long, it will not fit inside the human ear. The eardrum vibrates from the sound wave; but, if the waveform is too short, the eardrum is not sensitive enough to vibrate.

The telephone network was designed around 300 to 4000 Hz in order to support human speech. A sound wave is converted into an electrical wave by a telephone transmitter whenever the telephone handset is picked up and the subscriber speaks into the transmitter. The electrical wave produced by the telephone transmitter is referred to as an analog signal, which mirrors that of the sound wave. A telephone conversation between two people is instantaneous due to the shock wave effect caused by electrons pushing electrons. Amplitude relates to the strength of a signal. The louder the signal, the higher the amplitude; the softer the signal, the lower the amplitude. Turning up the volume on a radio or television increases the amplitude of the waveform.

The electromagnetic spectrum surrounds all Earth inhabitants. The earth is similar to a huge round magnet with lines of force tying the North and South poles together. The lines of force are similar to rivers that flow between two points. When an electrical force is generated, the rivers of force carry the charges away from the source, such as when a radio transmitter vibrates an antenna and causes charged electrons to flow away from the transmitter. The waves carrying the information signal are received by the receiving equipment and converted back into the original signal format.

Devices that take advantage of the electromagnetic highway include the television remote control, the laser, the telephone, the X-ray, and the FM radio.

Communication depends on sound, an electrical signal, and the electromagnetic spectrum. Communication, both voice signals or electrical analog signals, has specific parameters such as frequency, amplitude, and wavelength. The electromagnetic spectrum is what allows all communications to happen.

REVIEW QUESTIONS

1. How is sound communicated from one person to another?

2. Explain *frequency* as it pertains to human communications.

3. When the frequency of a waveform increases, does the wavelength increase or decrease?

4. Draw a 3 Hz waveform with an amplitude of 3 V and note the beginning and end of the first wavelength.

5. What is the minimum frequency a human's eardrum is able to perceive?

6. What is the maximum frequency a human's eardrum is able to perceive?

7. What happens when voltage is applied to a telephone line?

8. What are three components that make up impedance?

9. Explain the term *resistance* and how it affects the voice signal as it travels through a circuit.

10. Explain what current does on a telephone line.

11. Explain the effect of capacitance on a telephone circuit.

12. What solutions are available to help eliminate the problems associated with capacitance?

13. Because capacitance and impedance only affect AC signals, why do they affect the voice signal on an electrical circuit?

14. Why is the decibel used in telecommunications to measure signal strength?

15. Draw a dBm scale and note the zero reference point.

16. My test set measures a 6 dBm signal difference since the last time I measured the signal at that point. Has the signal increased or decreased in strength since my first reading? By how much?

17. Write out in long form the number that represents the following values:

    a. 6 Mbps   d. 0.6 mA
    b. 1310 nm   e. 0.054 μA
    c. 2.5 Gbps   f. 450 kbps

18. What is the electromagnetic spectrum?

19. To what frequencies within the spectrum do voice signals belong? Microwave signals? Lasers?

20. Why are standards so important to the telecommunications industry?

21. What is the difference between the ITU and the ANSI X.3T9.5?

KEY TERMS

sound (47)	frequency range (51)	capacitance (53)	electromagnetic spectrum (57)
medium (47)	carrier serving area (51)	inductance (53)	AM radio (57)
sound waves (47)	shock wave effect (52)	ohms (53)	service provider (59)
frequency (48)	voltage (52)	load coil (54)	ILEC (59)
cycles per second (48)	current (52)	watts (54)	CLEC (59)
hertz (48)	impedance (52)	decibel (54)	PUC (59)
wavelength (48)	AC voltage (52)	dBm (54)	PSC (59)
amplitude (50)	DC voltage (52)	signal level (55)	
radio signal (51)	resistance (52)		

2

The Telephone and the Telephone Line

Objectives

After reading this chapter, you should be able to
- Discuss the components of a telephone.
- Discuss the parts of a telephone line.

Outline

■ INTRODUCTION

Chapter 2 explains how the telephone works and how telephone lines connect to telephone switches. It goes through the call flow process, from the time the handset is picked up until it is placed back on the telephone cradle. Terms such as *called party*, *calling party*, *ringing voltage*, and *RJ-11* are defined in this chapter; and *dialing*, *ringing*, *circuit closure*, and *routing of the call* are explained in detail. The call that Chapter 2 discusses is one that travels between two telephones that are served by the same central office switch.

■ 2.1 OVERVIEW OF A CALL

A telephone is a small and unassuming device. Colors and styles vary dramatically, yet the internal components conform to universal standards. Five major pieces of hardware make up a telephone set: the *transmitter,* the receiver, the dial pad, the switch hook, and the ringer. All of these parts work together to produce a device that is able to convert a person's voice into electrical signals, and vice versa.

A transmitter is a device that converts pressure waves—human voice traveling through the air—into electrical waves. The electrical signal produced by the telephone transmitter is projected onto the communications medium, which varies from a pair of copper wires, to a fiber optic cable, or to the air. The receiver is the device that receives the electrical voice signal and converts it into pressure waves—the human voice. Both can be viewed as *transducers*—devices that convert one form of energy into another. A telephone's transmitter converts sound waves into electrical waves. The receiver converts electrical waves into sound waves.

A dial pad is used to dial the called party digits and send them to the serving central office. The ringer notifies the called party that a call is coming in and should be answered. The term *called party* refers to the person receiving a call. The term *calling party* refers to the person originating a call, and a telephone subscriber is any telephone customer.

A *switch hook* opens or closes the circuit that connects the telephone to the central office. When the telephone handset is picked up from the telephone cradle, the switch hook releases and closes the circuit, allowing current to flow from the central office to the telephone.

The telephone circuit that connects a home to the central office consists of two wires that form a complete circuit whenever the handset of the telephone is picked up. One end of the pair of wires terminates at the telephone; the other end terminates at the switch in the central office.

In order to explain the parts of a telephone set and the circuit that connects it to the switch, we will follow a call from the time the handset is picked up to the time the call is completed. At this point, we look at a call traveling through one *central office* switch, often referred to as an intraoffice call. In later chapters, we discuss calls that travel from switch to switch within the local network and outside onto the long distance network.

■ 2.2 COMPLETING THE CIRCUIT

Andy decides to call his good friend Daisy and ask her to dinner. He picks up the *handset* of his telephone. When the telephone handset is picked up, the telephone circuit instantly closes. This completes the circuit and current is allowed to flow between the switch and the telephone.

Pick up the *handset* (the unit you talk into) from your own telephone and look for the switch hook. The switch hook is the "on" and "off" button that you depress when you wish to hang up the telephone. When the button is depressed, the phone is on-hook. When it is released, the telephone is off-hook. "On-hook" and "off-hook" are common terms used in telephony to describe the status of the telephone line. When the switch hook is depressed, the circuit is idle. If you listen on the line without allowing the switch hook button to release, you do not hear a dial tone. An *idle circuit* is an open circuit.

Today, the switch hook may be in the handset, as it is with portable and cellular phones. It does not matter whether the switch hook resides on the telephone cradle or in the handset; its purpose is to close the circuit when the handset is picked up or the switch hook button is pressed.

When the circuit is closed, the switch places between –48 and –52 V onto the line and sends a *dial tone* to Andy's telephone. For current to flow on a telephone line, there must be a power source to stimulate the electrons in the copper wire. The switch places power on the line, and current flow allows the voice signal to travel across the wires.

transmitter
Source that produces a signal, such as a telephone transmitter or an optical laser.

transducer
Converts one form of energy into a different form of energy. The standard telephone is a transducer that converts sound waves to electrical waves.

called party
Person receiving a call.

calling party
Person originating a call.

switch hook
Part of the telephone that is used to open the circuit when the handset is returned to the cradle or to close the circuit when the handset is picked up off the cradle.

central office
Physical building that houses all of the communications equipment, including switches, transmission equipment, and central office. It also terminates all of the outside plant cables that feed the telephone subscribers.

handset
Portion of the telephone that contains the receiver and transmitter. The receiver is placed against the ear and a person speaks into the telephone transmitter.

idle circuit
Telephone line not in use.

dial tone
A tone sent by the switch when the telephone switch hook is released.

Figure 2–1
Closing an electrical circuit by releasing a switch hook is similar to lowering a drawbridge across a river.

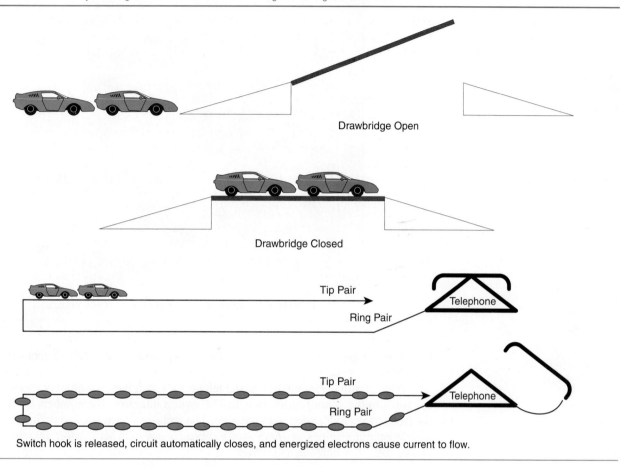

Drawbridge Open

Drawbridge Closed

Tip Pair

Telephone

Ring Pair

Tip Pair

Telephone

Ring Pair

Switch hook is released, circuit automatically closes, and energized electrons cause current to flow.

Therefore, there is a direct correlation between the voltage on the line and the current flow across the line. When the switch hook is released, the circuit closes, the switch energizes the line, a current flow begins, and a dial tone is provided to the subscriber.

A common analogy is that of a drawbridge being lowered across a river. (Refer to Figure 2–1 for an illustration of this concept.) When the handset is on-hook, the drawbridge is open. Nothing can cross, and no current can flow. As soon as the handset is picked up, the drawbridge is closed and current can flow across the line. The telephone is off-hook.

Once the handset is picked up (the drawbridge is closed), the circuit is completed (the two wires are connected) and the switch immediately knows the telephone is waiting for a dial tone. In fact, the switch continuously scans every single line attached to its line cards to see if a circuit has been completed.

■ 2.3 DIALING THE TELEPHONE

2.3.1 Touchtone Dialing

When Andy hears the dial tone coming out of a telephone handset, he punches in Daisy's telephone number on the dial pad and waits for a response. The *touchtone* pad has twelve buttons, each of which generates a specific tone when pressed. If Andy presses the number 3, frequencies of 697 and 1477 Hz are generated. The number 4 generates 770 and 1209 Hz. Pick up your telephone handset, press each number from 1 to 10, and listen to the difference in each number's sound. Figure 2–2 identifies the frequency associated with each digit. There are a total of twelve buttons on each

touchtone
Dialing method using tones to signify digits.

Figure 2–2
Frequencies associated with each
number on a touchtone telephone.

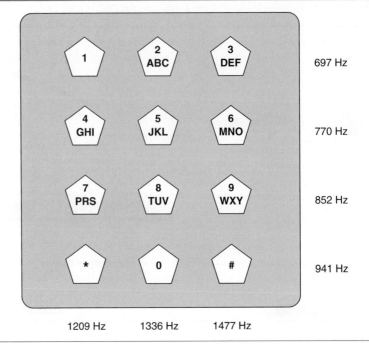

touchtone telephone set; ten are number keys and two are special characters—the # and the * keys.

When a number is entered on the telephone touch pad, the number travels across the wires to the central office switch. The switch interprets the numbers by detecting the frequencies associated with the tone. The technical term used to define touchtone in telecommunications is *dual-tone multifrequency* (DTMF). Most telephony-wise people refer to touchtone as DTMF.

2.3.2 Rotary Dialing

rotary dial
Dialing method using pulse
to signify digits.

Some people still have *rotary dial* telephones, like the one in Figure 2–3. In the late 1800s, Armond Strowger invented the rotary dial, which uses pulses of specific lengths rather than frequencies to identify each digit. The pulses interrupt the current flow of the circuit, and the switch counts the pulses to determine what number is being dialed. For instance, dialing a 5 produces five pulses or, in technical terms, five interruptions in the line current. The 0 travels the farthest when dialed; the 1 travels the shortest distance. You can hear the pulses tick off when you dial a 0 or a 1. Many phones offer the option of choosing pulse or dial tone.

Figure 2–3
Rotary dial.

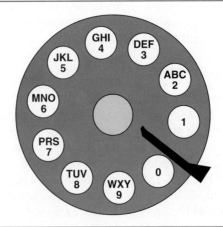

When touchtone telephones were introduced into the telephone network around 1972, many telephone companies charged an extra fee for the service. Some still do. In the past, many telephone manufacturers provided the option for touchtone or pulse because some switches were not equipped for touchtone, and some people refused to pay the extra fee for touchtone service.

Touchtone telephones can be dialed much faster than rotary dial phones and the network has now transitioned to the digital switch, which finds touchtone service much easier to interpret than pulse. Today, touchtone is as common as the telephone itself and phones are very rarely set up for pulse dialing.

■ 2.4 ROUTING THROUGH THE SWITCH

Andy has finished dialing Daisy's telephone number and is now waiting for his serving central office switch to ring her telephone. The switch interprets the incoming digits and uses its routing tables to decide where to route the call. Each routing table holds the numbers of all the telephone subscribers who are served out of that switch. The routing tables also hold the *exchange codes* of all the switches in the local serving area. The exchange code is the first three digits of a telephone number. The exchange code for the phone number 523-2345 is 523. In addition to the local telephone numbers and the exchange codes of all of the local switches, the local serving central office switch also holds routing tables for all long distance telephone companies. Each customer's record indicates which long distance telephone company he or she selected for his or her long distance provider.

exchange code
The three-digit code in front of the subscriber number.

When a switch receives the call, it first determines whether or not the telephone number belongs to one of the customers served by the switch or if it needs to travel to a different local or long distance switch. With this information, the switch is able to route the call. There are three possible routes: to a line within the switch, onto a trunking circuit that connects to another local central office switch, or onto a trunking circuit that connects either to the toll switch or directly to the long distance telephone company's switch.

Our example is fairly simple. Daisy's telephone line is served out of the same central office switch as Andy's telephone line. Therefore, the call does not leave the switch or ride on a shared trunking circuit. Instead, it will travel in on Andy's line and out on Daisy's line. This type of call is referred to as a line-to-line connection.

■ 2.5 RINGING THE TELEPHONE

2.5.1 Ring Back

Andy has finished dialing Daisy's number and listens to the *ring back* tone generated by the serving central office switch. *Ring back* is a term used to define the ringing you hear when you call someone and wait for them to answer. You might think that you are listening to the ringers of the called party's phone, but you are not. You are listening to a ringing tone generated by the switch.

ring back
The ringing signal the calling party hears while waiting for the called party to pick up the telephone.

2.5.2 Ringing

Daisy's telephone rings when her serving central office applies *ring voltage* on the line. Ring voltage is a varying AC voltage that activates the ringer in a telephone. Today, most phones have electronic ringers that simulate the sound of ringing. Many of these phones allow you to change the sound of the ring. Ringing voltage is still needed to turn on electronic ringers just as it was for the older style bells. Ringing voltage is comprised of a 20 Hz frequency that rides on a 70 V AC signal.

ring voltage
AC voltage sent down the line to ring the telephone.

Daisy is busy repotting some plants when the phone rings. She wipes off her hands and runs to pick up the phone. Once she picks up the handset, the circuit between Daisy's phone and her serving central office is complete. (The drawbridge closes.) When the switch senses the circuit is closed, it knows that it needs to complete the connection between the two lines. The circuit physically connects Daisy's line to Andy's line.

At this point, we need to look at the physical connection between the two telephones. To understand the end-to-end signal flow across the circuit from telephone to telephone, imagine that we are very small people who can fit into the telephone and travel down the telephone wire. We will walk the entire path of the circuit between Daisy's telephone and Andy's telephone.

■ 2.6 TELEPHONE TRANSMITTERS

Beginning our journey at Andy's telephone, we jump onto the sound wave coming from his vocal cords. The ride is similar to being on a roller coaster. The waveform is composed of multiple frequencies that move up and down at varying rates. Sometimes the wave is high and sometimes it is low, depending on the loudness or softness of the sound. The frequencies within the sound wave vary between 300 and 4000 Hz. Figure 2–4 illustrates the "roller coaster ride" of the sound wave.

We now enter the telephone through the microphone on the handset—the unit we speak into on the telephone (see Figure 2–5a). It is also referred to as the telephone transmitter. The transmitter is the device that converts your voice's sound waves into electrical waves. It consists of a metal cup filled with carbon granules, a metal *diaphragm* covering the cup, a cap that holds the lid over the carbon granules, and a metal lining that helps complete the circuit to the outside world. Figure 2–5b shows a carbon-filled transmitter. We enter through the porous structure of the microphone and press against the metal diaphragm. The diaphragm vibrates in proportion to sound waves that press against the diaphragm, which in turn pushes in on the cup that holds the carbon granules. The pressure fluctuating against the diaphragm causes the granules to compress and expand in relation to the sound waves made by Andy's vocal cords. As the granules compress and expand, the resistance of the carbon granules varies. A change in resistance causes a change in the electrical current flowing across the closed telephone circuit. Figure 2–5c shows a typical telephone receiver.

So, when the sound waves vibrate the diaphragm, the carbon granules vary the resistance of the entire circuit and the current flow. The varying current flow resembles the sound wave that we jumped on before we entered the microphone. The transmitter converted the pressure waves into electrical waves. We are now riding on an electrical waveform that, just like the sound wave, has numerous frequencies ranging from 300 to 4000 Hz. We continue to ride up and down, as shown in Figure 2–6.

■ 2.7 THE TELEPHONE CIRCUIT

The term *conductor* refers to a medium, such as copper, that is able to conduct electricity. In the telephone business, the word *conductor* refers to a copper wire. The telephone circuit is made up of two copper conductors. The wire conductors used to connect the telephone subscriber to the serving central office are referred to as the *tip and ring* pair. The term *tip and ring* is derived from the plug cords used by operators working on a switchboard and is one of the most common terms in telephony. Figure 2–7 shows a plug cord and illustrates the tip contact at the very end

diaphragm
Flexible device found in the telephone that moves as electrical or sound waves push against it.

conductor
Physical medium used to carry a signal—copper, fiber, air.

tip and ring
Wire pair.

Figure 2–4
The "roller coaster ride" of a sound wave produced by a voice signal.

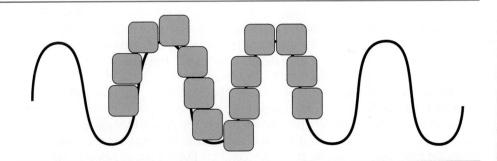

Figure 2–5
(a) Front view of a telephone microphone; (b) side view of a carbon-filled telephone transmitter; (c) the coil, the receiver, and the wire conductor in a telephone receiver.

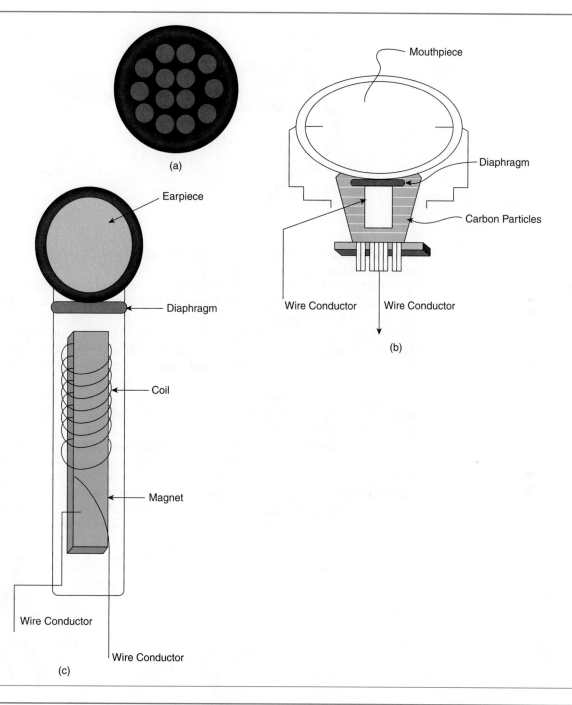

(a)

Earpiece

Diaphragm

Coil

Magnet

Wire Conductor

Wire Conductor

(c)

Mouthpiece

Diaphragm

Carbon Particles

Wire Conductor   Wire Conductor

(b)

Figure 2–6
Waveform produced by human speech.

Figure 2–7
Plug cord used to connect a tip and
ring pair into a jack panel.

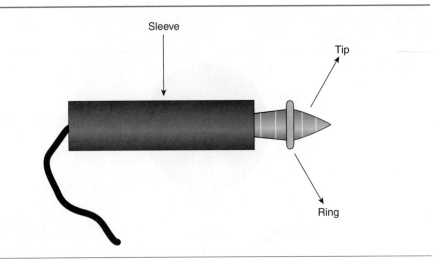

of the plug and the ring contact, a metal ring sandwiched between two sections of insulation behind the tip of the plug.

plug cord
Cord used to connect two
circuits together.

In Figure 2–8, the outer insulation of the *plug cord* has been stripped back. Inside the insulation are two separate insulated wire conductors. One of the conductors is physically connected to the tip contact on the inside of the plug. The other conductor is physically connected to the ring section on the inside of the plug. Figure 2–9 illustrates the tip and ring pair that terminates most telephones in the United States. Notice that the wires terminated on an RJ-11 jack, the jack you plug your telephone into, are defined as tip and ring.

Today, the tip and ring pair are wire wrapped separately onto a block on the main distribution frame (MDF) in the central office. The only time a plug cord is used is while troubleshooting a circuit.

In most cases, the two strands of wire that travel from your house to the telephone central office are dedicated to your house, although there are now devices that allow people to share the same wires. Each telephone subscriber has a tip and ring pair that travels from the home or business to the serving central office.

The switch has to make sure that all of the wires have a –48 to –52 DC voltage on the ring pair of the two wires. Using a multimeter—a meter that can measure DC

Figure 2–8
Stripped-back view of a telephone
plug that would fit into a jack board in
a central office.

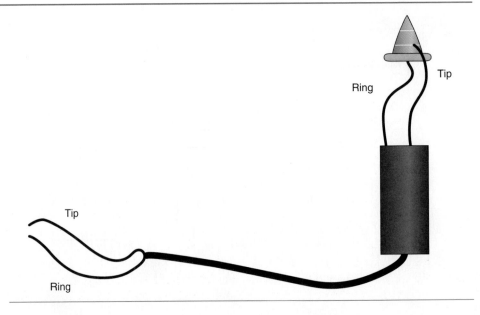

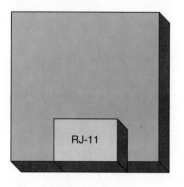

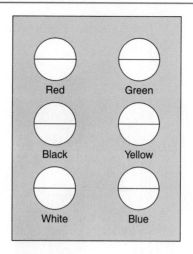

Line 1
Red = Tip
Green = Ring

Line 2
Black = Tip
Yellow = Ring

Figure 2–9
The wiring color code associated with a standard RJ-11 two-line telephone jack. Each line requires a two-wire pair to be terminated onto a modular RJ-11 jack with four active conductors.

voltage—we could place one meter lead on the ring pair and the other to a ground. The needle on the meter would deflect to a –48 to –52 V reading. With one of the leads on the tip pair and the other to the ground, there would be a minute amount of DC voltage—about –5 V. In other words, the switch places this voltage on the telephone circuit's ring pair. Therefore, when the phone is on-hook and idle, voltage is still induced on one of the pair. It takes the off-hook condition to complete the circuit allowing voltage to be induced on both pairs. If one multimeter lead is placed on the tip and one on the ring, the voltage will be about –50.

Ringing travels on the ring path. If we were walking on the ring path and the switch needed to ring the station, we would crash right into the AC waveform—20 Hz generated by the ringing voltage. Ringing voltage routinely zaps telephone technicians while they are splicing wire pairs back together.

■ 2.8 TELEPHONE-TO-TELEPHONE COMPANY PROTECTORS

We work our way through the telephone's circuitry and venture into the outside world. The door we use to leave the telephone is called an *RJ-11 modular jack*. The terms *RJ* and *modular* will be defined later. For now, simply remember that it is the jack you plug the phone cord into to connect the jack to the telephone (see Figure 2–10). We travel a short distance on this cord and then enter another RJ-11 modular jack on the wall. If we popped the cover off the jack, we would see two wires tightened beneath two screws. The screws make a contact between the house wires and the modular jack's pins that connect to the telephone. The jack connects the telephone cord to the inside house wiring. If you removed the wires from underneath the screws, you would immediately lose dial tone. Your line would be disconnected from the central office switch.

Normally, house wire is hidden within the walls of the house, so we are now traveling through the walls of Andy's house. In this case, the wire extends from the kitchen through the walls to the basement and out a hole in the concrete to a *protector block* mounted on the outside of the house. The phone company has mounted the block and terminated the outside line coming from the telephone pole onto lugs in the protector box. Both the inside wire coming from Andy's kitchen and the wires coming from the central office terminate on the lugs in the protector block. The termination forms a bridge between the house wiring and the outside plant telephone wire. Figure 2–11 shows a protector block and the *drop wire* that connects the block to the nearest telephone pole termination point. Later, we will look at every component in the outside plant

RJ-11 modular jack
Registered six-pin jack used to connect telephones to the outside network.

protector block
Block placed at the subscriber premises by the local telephone company to terminate the telephone wire pair.

drop wire
Wire used to connect the outside plant cable to the subscriber's premises.

Figure 2–10
Telephone wall jack with two modular
RJ-11 cables and an enlarged view of
an RJ-11 modular jack. Each of the
stripes is a pin used to terminate the
tip or ring wire conductor. The RJ-11
modular connector has six pin
conductors with four functional pins.

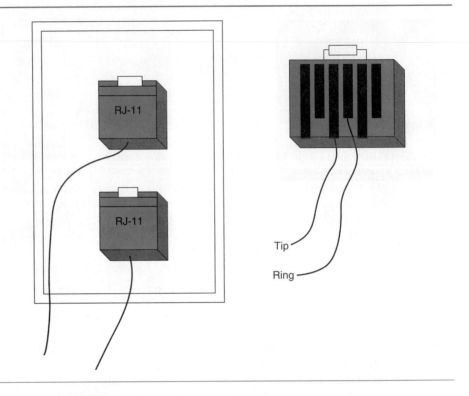

network, but for now we are interested only in the view of the path between the subscriber and the central office.

Once we leave the telephone protector, we start riding across the telephone network. We may journey from telephone pole to telephone pole or venture underground through cable laid in ductwork. The path the cable travels does not affect our ride, however. We are still traveling up and down and up and down within the wire conductor.

Figure 2–11
Protector block mounted on a
building and connected to a terminal
mounted on a telephone pole.

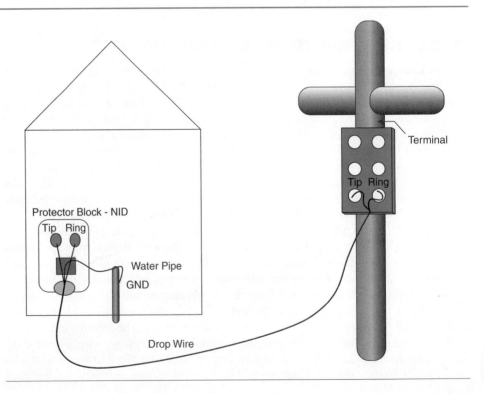

■ 2.9 THE CENTRAL OFFICE

2.9.1 The Main Distribution Frame

Finally, we enter the serving central office after leaving the last section of the outside plant network—the entrance cable. The first object we see is a huge frame-type structure called the main distribution frame *(MDF)*. The MDF is used in every local central office to connect a subscriber's line to the switches. We enter the frame on the subscriber side, or vertical side, of the frame. Every local central office also has a vertical side of the MDF. Figure 2–12 illustrates a typical MDF.

The opposite side of the MDF is the horizontal side. The termination blocks mounted there have connecting cables that run from the MDF to the central office switch. The switch terminates on the blocks that are mounted on the horizontal side of the MDF, and the subscriber lines terminate on wire terminating points on the vertical side of the MDF. A frame wire is used to connect the two points and complete the connection between the subscriber line and the switch. A frame wire connects each subscriber to the central office local switch.

We ride across the frame wire and into the switch block on the horizontal side of the frame, then leave the block and ride on the wire that connects the frame block to the switch. The switch line card is the entry point into the switch.

MDF
Main distribution frame.

2.9.2 Switches

A switch has built a temporary path between Andy's line and Daisy's line. Daisy's telephone number was the information used to build this connection within the switch. The method the switch used to build the link between the two lines was to look at the number Andy dialed, find the line card associated with that number, and establish an electronic connection between Andy's line card and the line card of the number he dialed. Each subscriber served by the switch has an individual wire port where the copper line is terminated but the switch does not provide an electronic port per subscriber. Because the port is needed only for the duration of the call, subscribers unknowingly share electronic switch ports with their neighbors. The call path through the switch between the two electronic ports is held in place until one of the callers hangs up the telephone and the switch senses the loss of current, resulting in an idle port. The switch relies on the telephone number to route the call to the correct line card and also to know which port to tear down at the end of the call. Once Andy or Daisy goes on-hook, the switch immediately realizes the port has gone idle and disconnects the link between the two telephone lines.

The telephone company determines how many people will need to be connected at one time by performing traffic studies. Normally, they use Mother's Day as a benchmark to determine the size of the switch. The number of lines needed for subscribers varies depending on the telephone company, but the value commonly used is six customers for every one switch connection. Therefore a 6:1 ratio of customers to switch connections is used to design switch capacity. This is called *concentration ratio*.

Today, the Internet is causing headaches for many telephone traffic engineers. Engineers assume that subscribers will stay online for a short period of time, so when Internet users stay online for hours, it skews the traffic engineers' call-time assumptions. The problem that has arisen throughout the industry is overloaded switches that are unable to provide a dial tone on demand. Instead, a fast busy signal indicates the switch does not have a path available. This system, which worked for almost 100 years, is now being reevaluated.

concentration ratio
The ratio of subscribers to circuits.

We continue on our trip through the switch from Andy's line card, that is, the circuit pack that resides in the switch chassis, through the internal back plane connections that connect to Daisy's line card. We leave the switch and travel out onto physical wire that links the switch to the block on the horizontal side of the MDF. We ride from there to the vertical side on the frame wire and exit the central office from the point on the vertical side of the MDF where Daisy's outside telephone line terminates. We then leave the central office and head into the outside plant network. Again, we travel along

Figure 2–12
Main distribution frame—vertical side connects to the horizontal side via a frame wire.

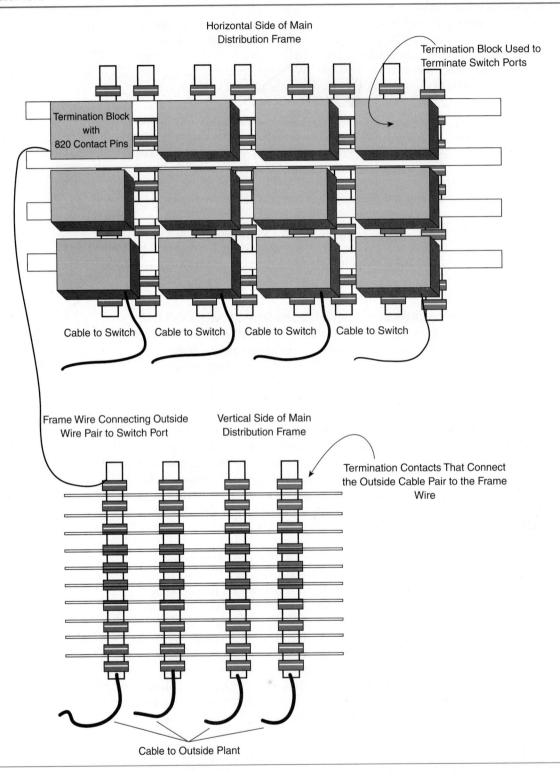

Horizontal Side of Main
Distribution Frame

Termination Block Used to
Terminate Switch Ports

Termination Block
with
820 Contact Pins

Cable to Switch   Cable to Switch   Cable to Switch   Cable to Switch

Frame Wire Connecting Outside
Wire Pair to Switch Port

Vertical Side of Main
Distribution Frame

Termination Contacts That Connect
the Outside Cable Pair to the Frame
Wire

Cable to Outside Plant

aerial cable from pole to pole or underground through a labyrinth of ductwork to reach Daisy's protector block on the side of her house. The outside plant wires coming from the telephone pole meet the inside house wires at the protector block on Daisy's house. Traveling through the house on wiring, we find our way to the RJ-11 jack on the wall in Daisy's dining room. From there, we step onto the telephone cord that connects the jack to the telephone and finally reach the RJ-11 jack, the hole in the back of the telephone, and walk inside. This time, we follow the circuitry inside the telephone that leads us to the *receiver* located in the handset instead of the transmitter.

receiver
Device that receives the signal.

2.10 THE TELEPHONE RECEIVER

The four physical components that make up the receiver are the voice coil, the magnet, the diaphragm, and the signal cone. The voice signal enters the receiver on the tip and ring wires connected to the receiver's voice coil (which is the coil wrapped around the permanent magnet). Traveling through the voice coil around the magnet we cause the permanent magnet to be transformed into an electromagnet. The varying current from the electrical voice signal causes the current around the magnet to vary proportionally. The varying current causes the diaphragm, positioned between the face of the electromagnet and the speaker cone, to vibrate. The vibration in the diaphragm produces pressure waves in the speaker cone—the part held next to our ear when we are talking or listening on the telephone.

Moving from the electromagnet in the receiver to the diaphragm, the electrical waves hit the diaphragm, which in turn disturbs the air in the same way our voice disturbs the air, or causes vibrations, when we speak. The vibrations travel out of the receiver speaker cone and into Daisy's ear as sound waves that are identical to the electrical voice signal.

The line current flowing across the closed circuit is also fed to the receiver to provide a side tone for the user. Side tone eliminates the hollow dead sound that is inherent on the telephone receiver.

Our journey is over. At this point in the call walk-through process, Daisy and Andy are chatting away discussing their plans. Strong, clear voice signals move back and forth along the path we just traveled.

2.11 GOING ON-HOOK

Now we need to find out what happens when the telephone handset goes on-hook. Does it matter which party hangs up first? Not to the telephone subscribers, but it does matter to the telephone switch. Daisy and Andy enjoy their conversation, she accepts his dinner invitation, and they say good-bye and place the handsets back onto the telephone cradles.

As soon as Daisy places the handset on the cradle, the circuit connection between tip and ring opens. Current can no longer flow between the switch and the telephone. The switch sees the open circuit condition and tears down the talk path it built in the switch between Daisy's line and Andy's line. As soon as Andy hangs up his phone, the current ceases to flow due to the break in the circuit. The switch recognizes both Andy and Daisy's line as idle or on-hook.

SUMMARY

At this point, you should be familiar with the way a call flows between two telephones served out of the same central office switch. We learned that the telephone is made up of five main components: the *transmitter*, the *receiver*, the *ringer*, the *dial pad*, and the *switch hook*. The transmitter and receiver are devices that convert sound waves into electrical waves or electrical waves into sound waves. Each telephone circuit has two wire conductors referred to as tip and ring. A complete circuit is comparable to a drawbridge. Current flows whenever the switch hook is released, which is similar to closing a drawbridge.

The main distribution frame (MDF), located in the central office, has a vertical side and a horizontal side. The subscriber termination from the outside plant is on the vertical side, and the switch is connected to the horizontal side.

Ringing voltage must be applied to the ring side of the pair whenever the switch needs to ring a telephone. The telephone protector block connects the inside house wire to the outside plant wire. Finally, once the telephone is placed on-hook, the switch disconnects the lines.

REVIEW QUESTIONS

1. Explain how a transmitter converts a sound wave into an electrical signal in a standard telephone set.

2. Explain how the receiver converts an electrical wave into a sound wave in a standard telephone set.

3. List and define the five components of the standard telephone.

4. Describe the path that the signal travels from the time it leaves the telephone to the time it reaches the switch.

5. Explain how the switch handles incoming calls.

6. Where is an RJ-11 jack found in the network?

KEY TERMS

transmitter (64)

transducer (64)

called party (64)

calling party (64)

switch hook (64)

central office (64)

handset (64)

idle circuit (64)

dial tone (64)

touchtone (65)

rotary dial (66)

exchange code (67)

ring back (67)

ring voltage (67)

diaphragm (68)

conductor (68)

tip and ring (68)

plug cord (71)

RJ-11 modular jack (71)

protector block (71)

drop wire (71)

MDF (73)

concentration ratio (73)

receiver (75)

3

Connecting the Dots—Transporting Information across the Superhighway

Objectives

After reading this chapter, you should be able to

■ Describe the local telephone network infrastructure.

■ Describe the long distance telephone network infrastructure.

■ Describe the type of circuits used to connect switches.

■ Describe the data network infrastructure.

■ Define new telephone terms.

Outline

■ INTRODUCTION

One of the best ways to learn how information travels across the telecommunications network is to visualize our system of roads. The multitude of interstates, state highways, and rural one-lane roads can be compared with the multitude of transport media found in the telecommunications network—the local loop, the local switch-to-switch connection, and long distance connections. Interstates are able to accommodate large numbers of vehicles while the local one-lane roads usually only carry local traffic. The same holds true in telecommunications. The fiber optic systems—the superhighways—connect switching centers, large office complexes, and other high-volume areas. The local loop services one customer and provides an ample amount of bandwidth in the same way the local road carries residential traffic without too much congestion.

Telecommunications also mimics the railroad network, with its hubs or switching centers where trains come in, drop off railroad cars, pick up other railroad cars, and continue on to their destinations. The telecommunications network also has switching centers where traffic is dropped off or added to the information stream. We do not need separate telephone lines to talk to Aunt Sally in California and Uncle Art in Florida, and the phone company does not need separate telephone networks. The switching center distributes the information instantaneously.

The goal of this chapter is to "connect the dots" of the network. We demonstrate how the local network connects to other switches within the local area, how the local network connects with the long distance network, and how the long distance network connects to the international network. We then discuss specific pieces of the network and the types of circuit media used to carry the zillions of bits of information. The chapter concludes with a simple high-level explanation of the Internet.

■ 3.1 THE LOCAL TELEPHONE NETWORK

3.1.1 Connecting Local Telephone Company Switches

In Chapter 2, we traced a call from Andy's house to Daisy's house. The call traveled on the copper loop that extended from the serving central office to Andy's house. We now trace a call as it leaves the serving central office and travels to a central office that serves a different area across town.

Andy decides to make dinner reservations at Alfonso's. He picks up the phone, listens for dial tone, and dials Alfonso's number. The digits travel down the wire into the Wilmount Street central office (the *serving CO*). (All central offices are referred to by a street or location name.) The switch receives the digits 232-5522 and determines that the number belongs to a switch located elsewhere in the local network. In other words, Alfonso's restaurant is not physically connected to the Wilmount Street switch.

The Wilmount Street switch, through the magic of the *SS-7 network* (to be discussed later) determines that the number 232-5522 resides in the Maple Street central office. The Wilmount Street switch routes Andy's call onto a trunking circuit that connects the two central offices. Figure 3–1 illustrates how the two switches connect using shared trunking circuits.

What would happen if the Maple Street CO and the Wilmount Street CO were not connected directly? An intermediary, or tandem, switch is directly connected to both the Wilmount CO and the Maple Street CO. The tandem switch would route the incoming call from the Wilmount CO onto a *trunk* riding on a circuit to the Maple Street CO. Figure 3–2 illustrates how the call would travel through the tandem switch. Both scenarios—direct connection between the two *end-office* switches and connection through a tandem—are possible. Both are very common in the local telephone network.

The Maple Street CO happens to be the serving central office for Alfonso's restaurant; that is, Alfonso's telephone line is connected to the Maple Street CO switch. Therefore, when Andy dials Alfonso's telephone number, the call travels into Andy's serving central office switch, where it is routed onto shared trunks that terminate at Alfonso's serving central office switch located on Maple Street.

The Maple Street CO examines the number, knows that it belongs to Alfonso's business, and routes the call onto the switching line card attached to Alfonso's telephone.

serving CO
The central office that provides telephone service to the subscribers within the CSA.

SS-7 network
Signaling system 7—The out-of-band signaling network used to set up calls.

trunk
Circuit shared by multiple users that connects two switches together. A trunk carries calls between switches, such as a long distance switch and a local switch. Once the call is complete, the trunk becomes available for a different call.

end office
The serving central office providing dial tone to the end subscriber.

Figure 3–1
Shared trucks between two switches, which pass calls via the shared trunks.

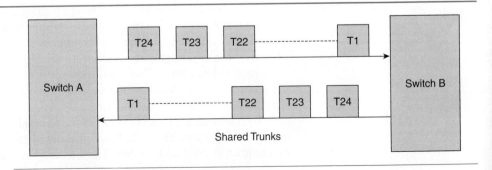

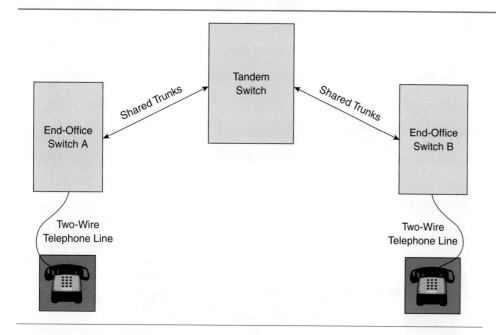

Figure 3–2
ILEC local network with two end offices connecting into a tandem switch office. Traffic passes from end-office switch A to the tandem, and from the tandem to end-office switch B. Shared trunks are used to pass traffic between the end-office switch and the tandem.

The term *trunk* is used to define the channel connection between two central office switches. Andy's call to Alfonso's traveled on Andy's line into the Wilmount CO switch, then onto a trunk that connected the Wilmount CO to the Maple Street CO, and finally onto Alfonso's line from the Maple Street switch. In telephony, that is referred to as a line/trunk/line call.

It is also an intraoffice trunk connection, which brings us to a discussion of intraLATA and interLATA. The telephone network is divided into 161 *LATAs* (local access and transport areas). LATAs represent geographical regions as defined by the Modified Final Judgment in 1982, and every area within the United States resides within a specific LATA.

LATA
Local access and transport area. Geographic areas defined after the Modified Final Judgment of 1982.

Calls traveling within a LATA are charged differently than calls traveling across LATAs. InterLATA calls travel between LATAs and are carried by long distance telephone carriers. Calls traveling within a LATA are intraLATA calls, which may have long distance charges.

3.1.2 Connecting Two Local Telephone Companies' Switches

The second type of call we trace is one that travels between two different local telephone companies, or service providers. This is an intercompany/intraLATA call. Thanks to the 1996 Telecommunications Act, intercompany/intraLATA calls are very common. New competitive local exchange carriers (*CLECs*) are offering telephone services in local areas. The incumbent local exchange carrier (*ILEC*) must allow CLECs to connect to their end-office switches or to their access tandem; therefore, trunking circuits must be established between the ILEC's tandem and the CLEC's local switch.

CLEC
Competitive local exchange carrier.

ILEC
Incumbent local exchange carrier.

If Alfonso's dial tone was provided by one of the many new CLECs, the call would be routed as shown in Figure 3–3. The digits Andy dialed would travel into the Wilmount CO switch as before. The switch would then determine that the call should go to the access tandem to be routed to the destination switch. The call would then travel onto trunking circuits connected to the local ILEC's tandem. The access tandem would examine the incoming number and determine that it belonged to CLEC Spidercom. The access tandem would switch the call onto a trunk on a circuit traveling to Spidercom's class 5 switch located in its State Street regional switch center. The call would arrive at Spidercom's switch, which would route it to Alfonso's telephone line.

Connecting the Spidercom switch directly to the ILEC's end-office switch that serves Andy's telephone is a second option often used in the local network. The call

Figure 3–3
An ILEC local network connecting into a Spidercom (CLEC) local network via shared trunks at the tandem switch office. Traffic passes from end-office switch A to the tandem, from the tandem to Spidercom's office switch B, and then to a digital loop carrier located at the ILEC end office serving Alfonso's restaurant. Shared trunks are used to pass traffic between the two companies' networks. The end connection between Spidercom's switch and Alfonso's telephone requires a special piece of equipment called a UNE (unbundled network element), which is located in a cage in the ILECs central office. The cable pair serving Alfonso's home is owned by the ILEC and leased to Spidercom.

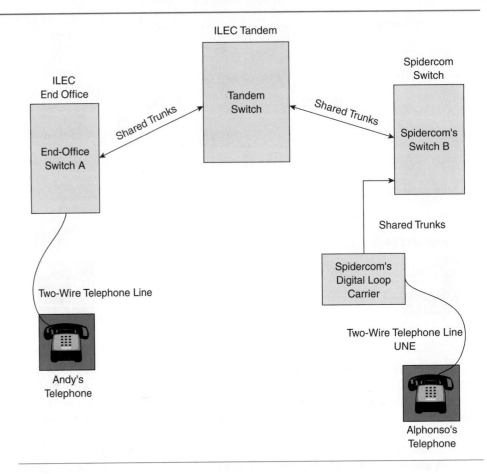

would travel from Andy's CO switch to the CLEC's CO switch, completely avoiding the ILEC's tandem switch, as shown in Figure 3–4.

3.1.3 Connecting the Local Telephone Switch to a Toll Tandem Switch

A *toll switch* owned by the local telephone company switches calls between the local telephone network and the long distance network. A toll tandem switch interfaces with long distance switches. All 1+ calls are routed to the toll tandem switch that connects to long distance companies. The toll *tandem switch* contains routing software that looks at the digits to determine which trunk group to send the call out on. The toll tandem switch is the gateway between the local telephone network and the long distance telephone network. The toll tandem switch is concerned with the first three digits (area code) for long distance calls and the exchange code (second set of three digits) for local switch calls. The toll tandem switch routes calls between trunks, not between lines.

■ 3.2 THE LONG DISTANCE NETWORK

Since divestiture in 1982, local telephone networks have had to connect with more than one long distance telephone company. Initially, to connect two networks, callers had to input special access codes in order to directly dial any of the many long distance carriers. Thanks to equal access, which is discussed in detail in Chapter 6, we now just dial 1 before the ten-digit number. The local switch knows which long distance carrier the call should be routed to. The only stipulation is that the long distance company must have a point of presence within the LATA where it connects with the local phone company or other CLECs offering local telephone service.

How does the local network connect to the many long distance networks? The answer is fairly simple, having just discussed intraswitch trunking. The difference is the way

toll switch
A switch that is capable of switching long distance traffic. The local class 5 switches may function as a toll switch if long distance software has been installed. Therefore, a toll switch is a switch that has the ability to switch long distance calls.

tandem switch
Switch that has connections to other switches—end office, toll, cellular, and others. A tandem switch is used to switch traffic between switches. Today, access tandems are used to interconnect ILEC networks to CLECs and other providers.

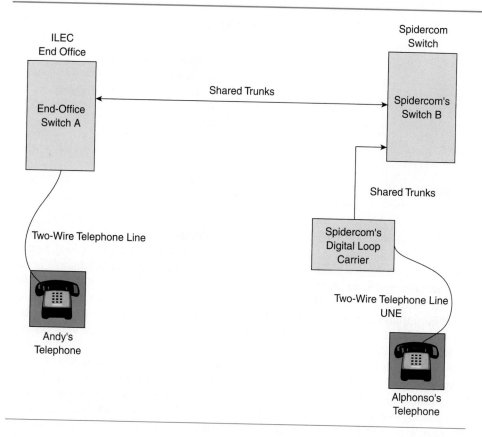

ILEC
End Office

End-Office
Switch A

Shared Trunks

Spidercom
Switch

Spidercom's
Switch B

Two-Wire Telephone Line

Shared Trunks

Spidercom's
Digital Loop
Carrier

Andy's
Telephone

Two-Wire Telephone Line
UNE

Alphonso's
Telephone

Figure 3–4
An ILEC local network connecting into the Spidercom (CLEC) local network via shared trunks between the ILEC end office serving Andy's home and the Spidercom switch. Traffic passes from end-office switch A to Spidercom's class 5 switch, then from Spidercom's office switch B to a digital loop carrier located at the ILEC end office serving Alfonso's restaurant. Shared trunks are used to carry traffic between the two companies' networks. The end connection between Spidercom's switch and Alfonso's telephone requires a special piece of equipment called a UNE (unbundled network element) which is located in a cage in the ILECs central office. The cable pair serving Alfonso's home is owned by the ILEC and leased to Spidercom.

the trunks are defined. The trunks used to connect long distance telephone companies to local telephone companies are called *Feature Group* trunks. The FCC's National Exchange Carrier Association defined the Feature Groups used to connect the many long distance carriers with the local telephone companies.

The purpose of the standard is to ensure that all long distance carriers are treated equally by local telephone companies. For instance, if you chose MCI as your long distance company in 1985, you had to dial 10322 1 + the number to place a long distance call. The FCC was concerned that consumers would hesitate to leave AT&T if they had to dial additional numbers every time they wished to make a long distance call.

The solution was *equal access*. Equal access trunks are Feature Group D trunks. A switch's database holds the customer record that shows which long distance carrier (primary interexchange carrier [PIC]) the customer has chosen and in turn eliminates the need for the customer to dial the special access code. If you could climb into the switch, you would actually see the special access code in front of the number the customer dials. Today, the most common type of trunk used to connect a long distance company switch to a local switch—tandem or end office—is the Feature Group D trunk. Figure 3–5 shows Feature Group D trunks connecting a local tandem switch to a long distance carrier's switch. Figure 3–6 shows Feature Group D trunks connecting a local end-office switch to a long distance carrier's switch.

We finish connecting the dots between the local and long distance network by walking through a long distance call scenario. The call will travel from Andy's home in Portland, Maine, to his friend Sam's home in Philadelphia, Pennsylvania. Andy picks up his telephone and dials Sam's number—1-818-555-7788. The Wilmount Street CO receives the digits and looks up Andy's profile in the customer information record in the switch's database to determine which long distance telephone company Andy has chosen. It shows that Andy has chosen ABC Long Distance Company. The switch attaches the access code—343—to the beginning of 818-555-7788 and ships the number out on an intraoffice trunk that connects Andy's switch to the local toll tandem switch.

equal access
All carriers have equal access to the local telephone network. Allows any long distance provider to terminate trunks into the local telephone network tandem switch or end-office switch.

Figure 3–5
Long distance switches connect to local telephone switches via Feature Group D trunks, which provide special signaling options that allow the switches to communicate and carry traffic correctly.

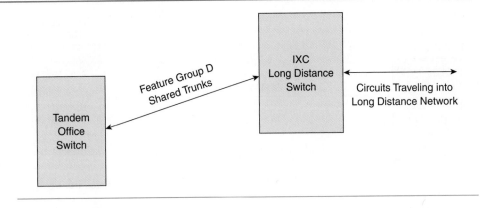

POP
Long distance company point of presence in a region. The long distance provider must have a presence in an area in order to route calls onto its network.

The tandem switch receives digits from the Wilmount Street switch, then examines the access code and sends the call to ABC's point of presence (*POP*). Each long distance carrier must establish a point of presence in each LATA it wishes to terminate/originate long distance traffic in. A POP may consist of anywhere from one relay rack holding transmission equipment to a full-blown switch site with a long distance switch. The POP must have trunks connecting it to the local telephone company in order to pass the traffic to the local network. Many POPs also have connections to other long distance providers, ISPs, and CLECs in the area. The call is switched onto the trunk that connects the local telephone company's tandem to ABC's long distance switch. ABC's switch may be in the same city as the local toll tandem switch or it may be miles away in another city. If they are in the same city, the trunks that connect the two switches travel a few blocks, much like the local end-office switches. If ABC's switch is not located in the same city, long-haul circuits are used to connect the two switches. The trunks connect the local tandem switch into ABC's long distance switch. Figure 3–7 illustrates the connection between ABC's long distance switch and the local telephone company's toll tandem switch.

The route that the call takes from ABC's long distance switch in Portland to ABC's Philadelphia long distance switch may vary. In one scenario, shown in

Figure 3–6
Long distance switches connect to local telephone switches via Feature Group D trunks and provide special signaling options that allow the switches to communicate and carry traffic correctly.

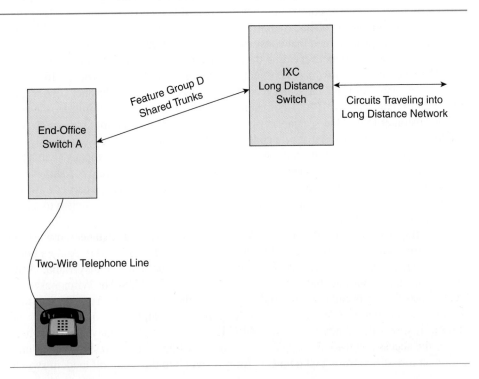

Figure 3–7
Long distance connection between two ILECs through ABC's long distance network.

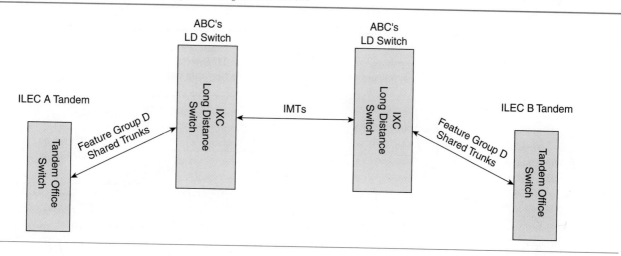

Figure 3–8, the call travels entirely on ABC's long distance network. In the other scenario, illustrated in Figure 3–9, the call travels on ABC's network from Portland to New York City and then on a wholesale long distance carrier's network from New York City to Philadelphia.

In the first scenario, the call leaves the ABC long distance POP in Portland on a trunk that connects directly into ABC's Philadelphia long distance switch. The trunks connecting the two long distance switches are *intermachine trunks* (IMTs). These are similar to Feature Group D trunks and are discussed in detail in Chapter 6.

Figure 3–8
Long distance connection between two ILECs through ABC's long distance network. ABC has switches in Portland, Boston, New York City, and Philadelphia.

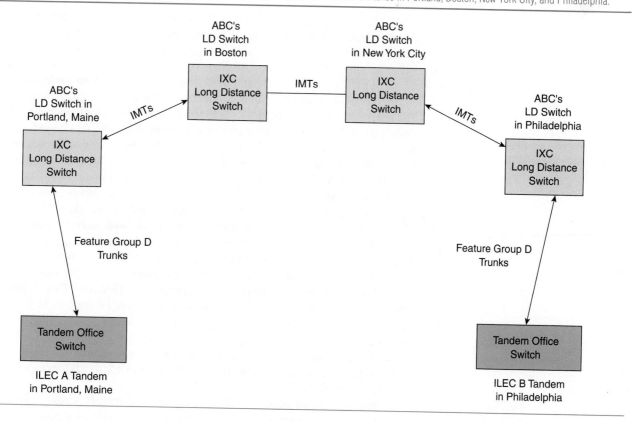

Figure 3–9

Connection between two ILECs through ABC's long distance network. The long distance portion of the network contains two LD carriers' networks. ABC leases the network between New York City and Philadelphia from a wholesale LD carrier. Most LD calls travel across multiple networks, but customers are not usually aware of this.

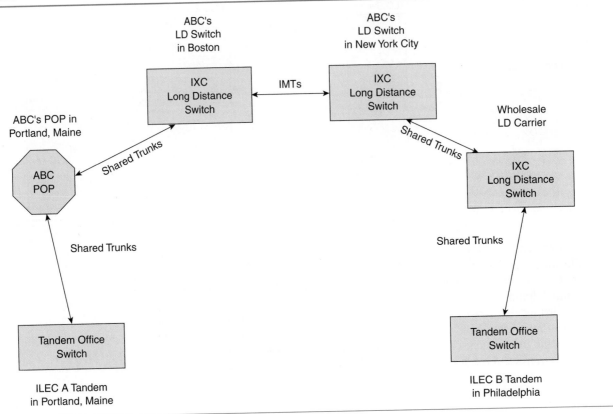

From ABC's long distance switch in Philadelphia, the call is routed onto a trunk that connects the local telephone network's tandem to ABC's long distance switch, probably via a Feature Group D trunk. The local tandem, as before, directs the call to the correct end office serving Sam's telephone line. The serving CO points the call to the line pair connecting to Sam's telephone and sends a ringing voltage. Sam's telephone rings, he picks up the handset, and the call is connected all the way through the network.

The other route the call could take is to travel onto ABC's long distance network, then onto a wholesale long distance carrier's network that ABC has leased, and then into the ABC's long distance telephone switch in Philadelphia. The difference between the two scenarios is the additional trunks that connect from the ABC long distance switch in Portland to XYZ's wholesale carriers network. Figure 3–10 shows how ABC's long distance networks and XYZ's network connect. This is a very common method of transporting calls. Many smaller long distance carriers do not have end-to-end networks across the country and therefore must use third-party networks to complete connections.

A third scenario, and the most complicated, is one in which ABC long distance company does not have a switch in either Philadelphia or Portland. Instead, the closest switch to Portland is Boston, and the closest switch to Philadelphia is New York City. Figure 3–11 shows the path of the call as it travels from the local tandem in Portland to ABC's POP in Portland and from there on circuits that connect into ABC's long distance switch in Boston. From the switch in Boston, the call travels on IMT trunks that connect into ABC's long distance switch in New York City, and from there on Feature Group D trunks that connect into the local tandem switch in Philadelphia. Again, the tandem switch routes the call to the local end office that serves Sam's home.

Even though the call hops off ABC's network and travels across a third-party wholesale network, the telephone still rings within thousandths of a second once the

Figure 3–10
Two ILEC networks are connected through ABC's long distance network. The long distance portion of the network contains two LD carriers' networks. ABC does not own the network from Portland to Philadelphia but leases it from XYZ who transports all of ABC's traffic across the country through special shared trunks between the switches. Most LD calls travel across multiple networks, but customers are not usually aware of this.

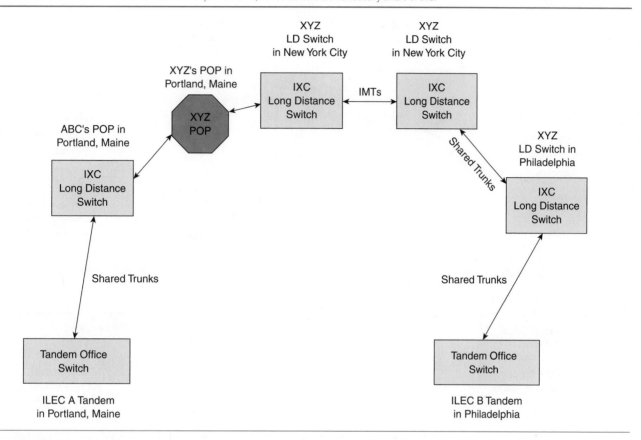

digits are dialed. The digital switching network using special SS-7 signaling links is responsible for the instantaneous switch times.

■ 3.3 THE INTERNATIONAL NETWORK

Andy now decides to call his good friend Shawn who lives in Ireland. Andy looks up Shawn's number and dials the digits—011-353-555-229292.The numbering plan for international calls requires a country code, a city code, and the subscriber's number. In addition, the 011 is needed in the same way a 1+ is required for a long distance call. Therefore, an international call to Ireland from the United States always starts with the 011 *international code*, and is followed by the 353 country code, a city code, and the subscriber number. A list of country codes can be found in Appendix B.

The method used to route a call from the United States to Ireland is very similar to that of routing a long distance call. The international number Andy dials is accepted by his local serving switch on Wilmount Street. The switch looks in the customer profile record to see which long distance carrier Andy uses and routes the call to that carrier. The difference occurs when the call arrives at the ABC long distance switch. The switch sees the 011 prefix and realizes that the call must be routed to an international *gateway switch*. A gateway switch is one that interfaces with the far-end international gateway switches located at the international gateway POP.

ABC has a gateway switch in New York City. The call travels from ABC's Portland, Maine, long distance switch to the long distance switch in New York City. The long distance switch in New York City has IMT that connect it to ABC's international gateway switch located at the same site. The gateway switch strips off the international 011 prefix and determines where to route the call by analyzing the 353 country code. The

international code
The 011 prefix used to tell the switch that an international call is being placed.

gateway switch
Telephone switch capable of switching international calls and converting the SS-7 signaling protocol into protocols used elsewhere in the world such as C7—the European Standard.

Figure 3–11
Long distance connection between two ILECs through ABC's long distance network. ABC has switches in New York City and Boston. Portland, Maine, and Philadelphia are POP sites. POP sites are not capable of switching calls; instead, they transport traffic to a switch site where calls are routed to the correct destinations. The diagram represents an architecture commonly used by long distance companies.

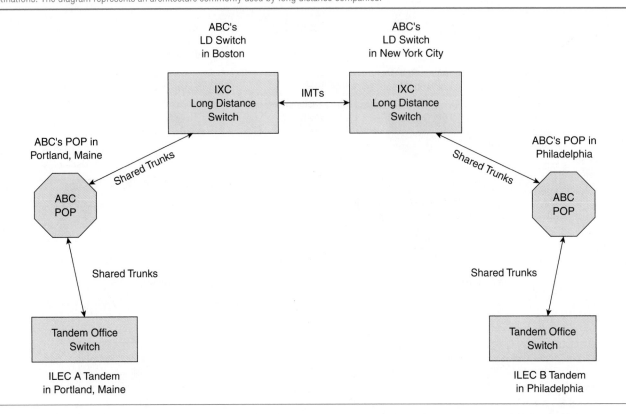

gateway switch sees that the country is Ireland and that all calls going to Ireland must be routed onto the trunk group that connects to a gateway switch located in Dublin.

The gateway switch in New York directs the call onto the IMT trunk riding on a circuit that connects into the Dublin gateway switch. The call arrives in Dublin where the gateway switch directs the call to a long distance switch that then directs the call to a local switch that finally directs the call to the telephone sitting in Shawn's kitchen. Figure 3–12 provides a comprehensive description of how the call travels between countries. There are several scenarios that are actually more common than directly connecting into a country gateway switch. In fact, a small long distance company such as ABC would not be allowed to connect directly into Ireland's telephone network. Due to regulatory

Figure 3–12
For calls between countries, a long distance network connects to an international gateway switch that performs two critical functions. It converts T1 to E1 format or it converts signaling information.

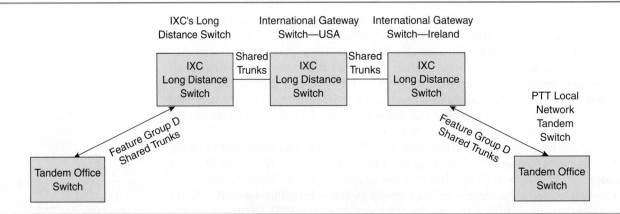

Figure 3–13
A long distance network that does not have direct connections to a country connects to an international gateway switch owned by another carrier.

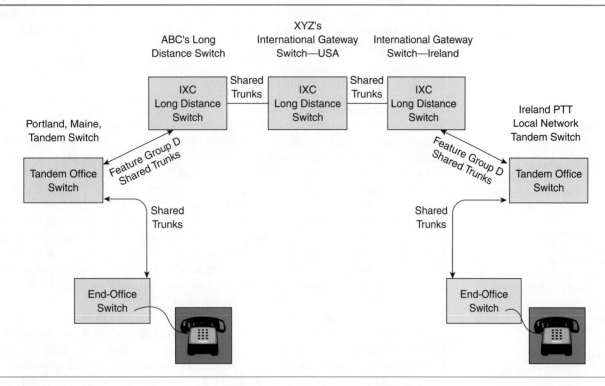

restrictions, special *bypass agreements* are required for a long distance carrier to directly connect into a country's phone network. Today, many European countries have opened their networks, just like the United States divested the long distance and local telephone companies' networks. Initially, the European Postal, Telephone, and Telegraph (PTT) networks were controlled exclusively by the individual governments. Unlike the United States, whose network was privately owned and operated by AT&T, the rest of the world depended on governments to operate the phone network—much like our post office. Third-world countries are just beginning to open up their telephone networks to competition, thereby allowing long distance companies to directly interconnect.

Therefore, a company like ABC long distance would connect its gateway switch with a large older carrier such as AT&T or MCI for international calls. Figure 3–13 shows a common method of connecting a long distance company to a bypass carrier going into a country's PTT network.

bypass agreement
Agreement established between a country and a carrier allowing the carrier to legally terminate international calls from its network into the country's network.

■ 3.4 THE NETWORK CIRCUIT

A *network* is a type of circuit used to connect two switches. The circuits may transport intraswitch trunking, interswitch trunking, and even international IDDD trunking. In the United States, telephone networks use T1 circuits.

The medium used to connect the two switches may be copper, radio (microwave), or fiber optic cable. The information traveling on the medium behaves the same, looks the same, and sounds the same whether it moves on fiber, copper, or air. The connection between the two switches is shared by all of the residents terminating on that switch. The physical medium is like a highway built to carry ones and zeros, with the makeup of the medium determining how fast the ones and zeros are able to travel.

Andy's call shares the same medium as it travels between his serving central office on Wilmount Street and Alfonso's on Maple Street. A T1 circuit-carrying trunk has twenty-four information channels. Andy's voice signal is placed in one of the twenty-four channels and carried to the terminating switch. T1 circuits are also used to connect long distance switches. The circuits connecting the switches are similar to a train with

network
Communications devices connected by communication circuits.

twenty-four boxcars; the switch is comparable to a railroad yard. The boxcars carry voice signals from point *A* to point *B*. The switch manages the flow of traffic in and out of the boxcars knowing which are empty and full. Sharing the medium between switches enables communication carriers to better utilize their network and consequently reduce prices to the end subscriber.

■ 3.5 DEFINING THE DATA NETWORK

3.5.1 Dial-Up Data Network

dial-up data
Establishing data communications using the public switched telephone network. The dial-up data circuit requires a modem to establish the call.

LAN
Local area network—Multiple computing devices such as PCs, servers, and printers are connected to a LAN. Each device on the LAN is able to exchange information.

modem
Modulator/demodulator. Device that converts the digital signal coming from the PC into an analog signal that can be transported onto a voice grade POTS line.

The *dial-up data* circuit was one of the first methods used to transport data information across the telephone network, and it can still be used to connect one PC to another PC. A simple example is using a dial-up connection to connect a home PC to a file server that is connected to a local area network (*LAN*) at a remote business site. Many people today dial into their company's LAN network from home. To establish this connection, a dial-up line and an access port into the company's LAN are required.

Andy remembers that he has to respond to some work-related e-mail. He turns on his PC and clicks on the local dial-up icon on his desktop. The PC's *modem* dials out the telephone number of the LAN at Andy's place of business. The switch receives the digits in the same way it would receive and handle the digits for a normal telephone call. It sees that the number 234-1000 is a local number and that it resides in the Elm Street CO. The switch directs the call onto a trunk group connected to the Elm Street CO. The Elm Street switch sends the call onto the line attached to Boxes, Incorporated, and rings the end station. In this case, the end device is a modem pool attached to a router. The router is a piece of communication equipment that routes data packets using the source and destination addresses of the devices on the network such as the LAN. Once the call travels through the modem pool and into the LAN, Andy is able to access his e-mail just as he would if he were attached to the LAN at work. The dial-up connection has established a point-to-point connection that will last as long as the PC and end station remain off-hook. Once Andy logs off the LAN and disconnects his modem from the line, the point-to-point connection is dropped and the telephone line becomes idle waiting for an off-hook condition.

3.5.2 Dedicated Data Circuits

dedicated data circuit
Point-to-point circuit established for data communications only. Dedicated data circuits carry information between two locations, such as between a bank and its branches. The network is closed to anyone other than the bank.

A *dedicated data circuit* is one that is permanently open. A good example is the connection between a bank's ATM (automated teller machine) and the bank's main office. Andy suddenly realizes he is low on cash. He rushes down to the corner ATM and slides in his ATM card. As Andy punches his personal identification number (PIN) into the machine, the bank receives the digits and determines if Andy is allowed to access the account defined on his ATM card. The information packet travels between the ATM and the main hub on the data communication circuit. The ATM transfers Andy's request for $200 to the mainframe where his account information is stored. The mainframe responds to the request from the ATM by replying with a "yes, approved" transaction. The ATM discharges $200 and a receipt detailing the transaction. Andy picks up his money, asks for his card back, and leaves. The communication circuit between the ATM and the bank's mainframe is a dedicated data connection used only by the two end devices. The circuit, unlike a dial-up line, is always up. The ATM or the mainframe initiates sessions by sending special wake-up characters to the receiving device. During idle times, a keep-alive signal is transmitted to ensure the two ends remain synchronized. The purpose of a dedicated data line is twofold. One is to ensure the security of the information; the second is to improve the speed of the transmission.

When Andy punches in his PIN, the circuit sends the digits out on the copper line that is connected to the Wilmount CO. The difference between this circuit and Andy's telephone line is that the bank/ATM circuit does not go through the switch. Instead, it travels into special service equipment connecting it to the Walnut Street CO, which serves the bank. The copper line travels from one point at the ATM to the Wilmount Street CO, then to the Walnut Street CO, and finally to the bank office. Dedicated point-to-point lines are very common in telecommunications.

3.5.3 Packet Networks—The Internet

The last network we discuss is the *packet network*. Andy wants to order some flowers for Daisy. He decides to order the flowers by logging onto the Internet and visiting a flower shop, which will require the use of a dial-up modem to establish a connection through the network.

Andy double-clicks on the Internet icon on his desktop and waits as the modem dials the Internet service provider's (ISP's) telephone number. Again, the digits are received by the Wilmount Street switch and routed to the tandem switch, which directs the call to a trunk connecting into his ISP, Surf the NET. Surf the NET's modem pool synchronizes with Andy's modem and establishes a connection. Andy types in the flower shop's Web address and waits for a response. Surf the NET's modem is connected to a gateway router that sends the Web address out across Surf the Net's Internet backbone. The Internet backbone is connected to several peering points that continue to route the request through the World Wide Web to the destination Web site. The response from the flower shop Web site travels back across the network to Andy's PC, where it "paints" its Web page across his screen.

The information traveling between Andy's PC and the flower shop's Web site is in packets, something like envelopes, containing information. The shop's Web site graphics traversed the phone lines as bits of data (ones and zeros). These packets do not have to travel in sequence on the same link. The first packet of information may travel down one path and the second down a completely separate path. The local loop between the serving central office, Andy's home, and the flower shop are the only dedicated portions of the network. The information travels through the network across multiple routes depending on traffic load and circuit availability. The address attached to each of the packets directs it through the network.

packet network
Statistically multiplexed network that carries information in the form of variable-length packets. The information is typically all data and is used to help increase the amount of information that can be transferred between multiple locations using the same links.

■ 3.6 NEW TELEPHONY TERMS

The telecommunications industry has gone through dramatic changes during the past 100 years. A change in lingo accompanied the technological changes. Following are some of the terms used to describe the new telecommunications players and where they fit in the industry.

- ILEC—Incumbent Local Exchange Carrier—An ILEC is a local telephone company that served an area before the 1996 Telecom Act. For example, Southwestern Bell was the ILEC for Dallas, Texas; Bell Atlantic (now Verizon) for Pittsburgh; and Frontier for Rochester, New York.
- CLEC—Competitive Local Exchange Carrier—CLECs evolved as a result of the 1996 Telecom Act. They compete with ILECs as local exchange services, building switches and offering dial tone but using the ILECs' wire facilities to provide services to subscribers served by the ILECs' COs. A facilities-based CLEC places its own equipment in the ILEC's CO and only relies on the ILEC for the last mile of copper. A non-facilities-based CLEC buys a port on the ILEC switch but does not place equipment in the ILEC's CO. It simply resells dial tone using the ILEC's equipment and facilities.
- DLEC—Data Local Exchange Carrier—DLECs are similar to CLECs except that they offer only data services in the local telephone network. They colocate in the ILECs' COs and offer data services, such as DSL, to subscribers fed out of those offices.
- BLEC—Building Local Exchange Carrier—A BLEC also provides services in the local area but does so by placing equipment in the basement of high-rise buildings or in strip malls. The BLEC offers local services to building tenants and bypasses the ILEC's network.
- ICP—Interexchange Communications Provider—An ICP, though the term has not caught on, is a company that offers data in the local market, voice, data services such as Web hosting, and long distance.
- ASP—Application Service Provider—An ASP offers application services such as Web hosting and storage units to hold content. They are popping up all around the country and are normally strategically placed to service entire regions.

- IXC—Interexchange Carrier—An IXC is a long distance provider such as MCI, Sprint, Global Crossing, Qwest, or AT&T.
- CAP—Competitive Access Provider—A CAP offers access in local markets. For example, an IXC or a CLEC might lease facilities from a CAP to gain access to the local telephone network. The CAP provides an alternative to the ILEC's local network, helping reduce the cost to the CLEC and the IXC.

SUMMARY

Chapter 3 provided an overview of the telephone network. Calls were traced through the local network and the long distance network. T1 circuits were introduced as the backbone of the network used to carry voice and data traffic. The chapter concluded with a discussion of the data network and how information is transported between two PCs.

REVIEW QUESTIONS

1. Make a sketch of how a local telephone network interfaces with a long distance telephone network.

2. Make a sketch of how a long distance telephone network interfaces with an international telephone network.

3. Explain what is meant by *interoffice trunk connection*.

4. I've decided to call my friend Jane who lives next door. Jane's telephone line and my telephone line are both served out of the Oak Street CO. Draw the call path between Jane's telephone and my telephone.

5. I now need to call my friend Ernie who lives on the other side of town and is served out of the Monroe Street CO. Draw the call path between Ernie's telephone and my telephone.

6. I decide that I need to call my friend José who lives two states away. Trace the call path between my telephone set hanging off the Oak Street CO in Burlington, Vermont, and José's telephone set hanging off the Clinton Street CO in Muncie, Indiana.

7. After talking with José, I realize I haven't talked with my friend Karl who lives in Denmark for some time and decide to call him. Trace the call path between my telephone hanging off the Oak Street CO in Burlington, Vermont, U.S.A., and Karl's telephone hanging off the Boulevard Street CO in Copenhagen, Denmark.

8. Connections are built between switches, whether the switches are used to switch local calls or long distance calls. The circuits connecting the switches are of different types of media. List the three common types of media used to connect two switches together.

9. What is a T1 circuit used for?

10. Define the term *LATA*.

11. Define the term *intraLATA*.

12. Define the term *interLATA*.

13. List the three types of data networks described in this chapter.

14. Draw a dial-up data connection.

15. What is the difference between an ILEC and a CLEC?

16. What is the difference between a BLEC and a DLEC?

17. What is the difference between a CAP and an IXC?

18. What is the difference between an ASP and an ISP?

19. What is an ICP?

KEY TERMS

serving CO (78)	ILEC (79)	international code (85)	LAN (88)
SS-7 network (78)	toll switch (80)	gateway switch (85)	modem (88)
trunk (78)	tandem switch (80)	bypass agreement (87)	dedicated data circuit (88)
end office (78)	equal access (81)	network (87)	packet network (89)
LATA (79)	POP (82)	dial-up data (88)	
CLEC (79)			

Networking Fundamentals

Photo courtesy of the Library of Congress

4

Manipulating Information for Transmission

Objectives

After reading this chapter, you should be able to

- Define modulation schemes.
- Describe analog-to-digital conversion.
- Define multiplexing.
- Define data communications basics.

Outline

Introduction

4.1 Defining Modulation Schemes

4.2 Defining Analog-to-Digital Conversion

4.3 Multiplexing

4.4 Structure of Data Communications

■ INTRODUCTION

Chapter 4 focuses on the various ways information is transformed, packaged, and readied for transport. Analog-to-digital and digital-to-analog conversion is detailed, as is the method used to format digital-to-digital signals for transport. Multiplexing methods, including frequency division multiplexing, time division multiplexing, statistical multiplexing, and wave division multiplexing are also covered. Methods used to code and format data such as binary encoding schemes are also presented. The concepts presented in this chapter lay the groundwork for information presented in the rest of the book.

modulation
Superimposing the information signal into a steady state signal.

steady state signal
Transmitter produces a signal at a set "steady" frequency that is used to carry the information signal across the medium.

■ 4.1 DEFINING MODULATION SCHEMES

Modulation techniques are core to all communications systems whether they have a digital or analog circuit, a wire-line or wireless scheme, or a fiber or copper medium. At a high level, modulation techniques can be categorized as digital, analog, or a combination of digital and analog. Regardless of the type, the term *modulation* in communications refers to manipulating a *steady state signal* by superimposing the *message signal* onto it. The steady state signal is referred to as the *carrier signal*. The message signal may be an analog radio signal, a one-zero bit pattern coming from a modem, a voice signal coming from a telephone transmitter, and so forth. The method used to modulate a signal depends on the system used to carry the signal and the type of signal being transmitted.

In all systems, the term *modulator* is used to define the transmitting device that modulates or superimposes the message signal onto the carrier wave. The term *demodulator* is used to define the receiving device that pulls the message signal off the carrier wave and reformats it into the original communications signal. Therefore, modulation consists of a modulator, a demodulator, and a carrier wave used to "carry" the communications from one point to another. Smoke signals, one of the earliest forms of communications, used a modulator, a demodulator, and a carrier medium. The modulator was the blanket or device that when lifted allowed a puff of smoke to rise into the air. The air was the carrier, and the person reading and interpreting the smoke signals from three mountaintops away was the demodulator.

4.1.1 Analog Modulation Techniques

It would be nice if modern communications methods were as simple to explain as smoke signal communications systems. Today, in order to explain modulation, we need to segment the various techniques by analog versus digital modulation schemes. We start with analog modulation methods. The best example of analog modulation is AM and FM radio transmission. A carrier wave is transmitted by the radio system out the radio antenna at a set frequency. The radio signal coming from the modulator is superimposed onto the carrier signal and pushed out across the airwaves. In the case of AM radio, the amplitude of the carrier wave is manipulated and used to convey the message to the far end. FM (frequency modulation) manipulates the frequency of the carrier wave in order to send the communications signal. AM radio transmits at a frequency range of 530 to 1710 kHz, and FM radio transmits at a frequency range of 87.7 to 107.9 kHz. Figure 4–1 illustrates how a signal is transmitted from a radio station into the atmosphere and picked up by a radio. In AM or FM transmission, the radio station controls the modulator that outputs the message signal into the atmosphere. The standard radio, car radio, or radio in the house, for example, contains the demodulator. Every time you listen to your favorite radio talk show host while driving down the highway, think of the host's voice signal riding through the air on an invisible carrier wave generated by the radio transmitter.

message signal
The information signal such as voice, data, and video.

carrier signal
Signal produced by the transmittal equipment used to carry the message signal.

modulator
Component in the transmitter that superimposes the message signed onto the carrier wave.

demodulator
Component in the receiver that pulls the message signal off the carrier waves.

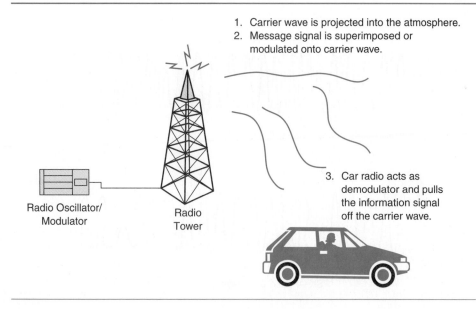

1. Carrier wave is projected into the atmosphere.
2. Message signal is superimposed or modulated onto carrier wave.
3. Car radio acts as demodulator and pulls the information signal off the carrier wave.

Radio Oscillator/ Modulator

Radio Tower

Figure 4–1
Amplitude and frequency modulation are used to push message signals such as radio signals into the atmosphere. A car radio is a good example of a demodulator of AM and FM radio signals.

4.1.2 Digital Modulation Techniques

Numerous digital modulation techniques are used in communications systems. Our goal is to present a few of the most common methods deployed in the network and explain the basic concepts relevant to all types of digital modulation. Digital modulation refers to superimposing a digital signal—a one-zero pattern—onto a carrier wave. Unlike analog modulation, which superimposes a continuous analog signal onto a carrier wave, digital modulation uses symbols consisting of one-zero patterns that superimpose finite message signals onto a carrier wave. The carrier wave's amplitude, frequency, and phase (or a combination frequency and phase) are manipulated to represent the binary pattern.

The simplest form of digital modulation is amplitude shift keying (ASK). For example, when a modem sends a one, the carrier wave's amplitude is shifted. In frequency modulation referred to as frequency shift keying (FSK), the frequency of the wave is adjusted to represent a one or a pattern of ones and zeros. Finally, phase shift keying (PSK) refers to adjusting the phase of the sinusoidal carrier wave to represent the message signal. Figure 4–2 depicts all three methods—ASK, FSK, and PSK.

Real-life systems depend on a combination of the three basic modulation techniques—amplitude, frequency, and phase—in order to increase the number of ones and zeros being shipped per signaling state or baud rate. Before discussing how the modulation technique works, we need to define a few terms used in the process.

The first is the signaling rate or baud rate. A *baud rate* refers to the number of times a signal changes per second. For example, 1200 baud refers to a signaling rate of 1200 signaling changes every second. Baud does not have to match frequency, and, most important, baud does not have to match bit-per-second (bps) rate. A 1200-baud signal may carry 4800 bps or four bits per baud.

baud rate
The number of times the signal changes or cycles per second.

Figure 4–2
Message signals are superimposed onto carrier waves by manipulating the amplitude, frequency, or phase of the sine wave. Modem modulation techniques typically use a combination of methods; for example, QAM uses PSK and ASK.

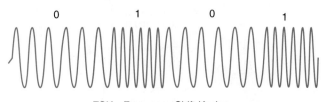

ASK—Amplitude Shift Keying

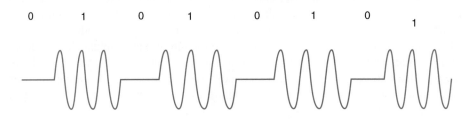

FSK—Frequency Shift Keying

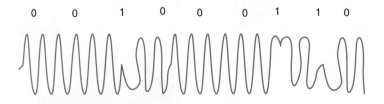

PSK—Phase Shift Keying

Modulation techniques are developed around a graphical plane referred to as a *constellation*, as shown in Figure 4–3. The constellation consists of two axes—the real *I* axis and the imaginary *Q* axis. Though relevant if you are a design engineer, at this time it is nice-to-know information but not critical to what we are explaining. What you do need to understand is that the signal can be manipulated according to the number of symbol states available within the constellation. This brings us to our next term, *symbol*. A symbol is the one-zero bit pattern that is assigned to a quadrant or portion of the quadrant in the constellation, also shown in the figure. As with quadrature amplitude modulation (QAM), the carrier wave is manipulated according to phase and amplitude providing multiple symbols and bits per symbol.

The symbols defined by the modulation scheme, such as QAM, are referred to as the *modulation alphabet*. As is obvious, the larger the alphabet, the higher the bandwidth. For example, 16 QAM sends four bits per signaling state change by defining four symbols per quadrant or sixteen symbols per constellation. This is a very difficult concept to comprehend as it relates to how the sine wave rides on an imaginary *x, y* axis, the constellation. In very simple terms, the sine wave, as shown in Figure 4–3, resides in one of the four quadrants of the constellation. Each quadrant has been assigned a specific one-zero pattern or symbol, also shown in the figure. When the sine wave is manipulated to fall within the quadrant, the demodulator is smart enough to know that it represents that particular symbol. As stated earlier, all digital transmission is made up of patterns that are translated into information whether that be letters of the alphabet or an analog waveform. The different modulation techniques—QAM, PSK, CAP, DMT—define the constellation according to the method they use to manipulate the carrier wave and the number of ones and zeros they ship per signaling state. The underlying concept to grasp at this point regardless of the digital modulation technique is that the number of ones and zeros transmitted per second relates directly to the number of symbols defined and the number of bits per symbol. For example, 8 PSK has eight

constellation
Term used to describe the abstract graphical plane of the modulation type of constellation, is divided into four quadrants.

symbol
One-zero bit pattern that is used to represent a code that in turn represents a pattern of bits.

modulation alphabet
In digital modulation, the changes in the signal area chosen from a fixed list (the modulation alphabet) each entry of which conveys a different possible piece of information (a symbol).

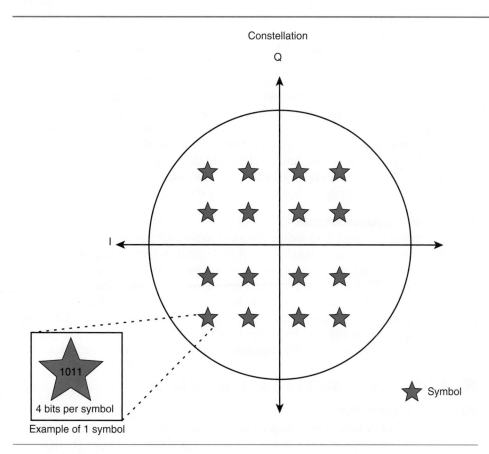

Figure 4–3
Though technically simplistic, the goal of this diagram is to show how a signal falls into different quadrants according to the phase and amplitude of the signal. The diagram also shows an abstract view of a constellation and the symbols within the constellation. The diagram provides an abstract view of a 16 QAM.

symbols—also called the modulation alphabet—and three bits per symbol. Likewise, 16 QAM refers to sixteen symbols containing four bits per symbol. The greater the number of symbols and bits per symbol or modulation alphabet, the higher the transmission rate. Following is a brief summary of the most common digital modulation techniques.

- ASK—Amplitude Shift Keying—ASK refers to a modulation scheme that manipulates the amplitude of the carrier sine wave. ASK is very susceptible to noise on the line, in addition to providing a limited transmission rate. Consequently, ASK is used in combination with other modulation schemes such as PSK.
- PSK—Phase Shift Keying—PSK uses the phase of the carrier sine wave to manipulate data. Many types of PSK are available—4 PSK, 8 PSK, and so forth. Similar to ASK, PSK is seldom deployed in today's networks due to limited transmission rates.
- QAM—Quadrature Amplitude Modulation—QAM combines ASK and PSK methods to produce a very robust, high bit rate transmission system used to carry data across analog telephone lines. QAM, developed by Bell Labs, is the modulation technique deployed in most modems today—such as V.22bis, V.32bis, V.34. QAM was also the precursor to other modulation techniques such as carrierless amplitude and phase (CAP) used in DSL broadband systems.
- QPSK—Quaternary Phase Shift Keying—QPSK is a modulation scheme commonly deployed by cable television (CATV) systems utilizing the Data Over Cable Systems Interface Specification (DOCSIS). QPSK works with QAM schemes to produce broadband data connections similar to DSL.
- CAP—Carrierless Amplitude and Phase—CAP modulation was the first modulation scheme deployed in DSL systems. CAP, an offspring of QAM, deployed by Globspan Semiconductor is used in ADSL systems to provide broadband asymmetrical transmission speeds over the traditional POTS copper pair. The bit rates related to CAP DSL are dependent on the manufacturer's modulation techniques and the loop characteristics. Typically, most service providers offer speed ranging from 256 kbps up to 4 Mbps.
- DMT—Discrete Multitone—DMT modulation, also used as a DSL modulation technique, uses multiple subcarriers to carry the symbol alphabet. DMT modulation uses QAM techniques to modulate symbols onto each subcarrier. DMT modulation has 256 subcarrier channels divided into either upstream or downstream. The message signal superimposes itself on the subcarriers depending on the quality of the link. If, for example, the link is noisy, fewer carriers are available to use for transmission. The result is lower bandwidth or simply fewer bits per second. The ability to adjust to the line is referred to as *rate adaptive* or RADSL (Rate-Adaptive Digital Subscriber Line). CAP modulation schemes also provide a means for rate adapting the signal depending on the quality of the line.

rate adaptive
Electronics have the ability to automatically adjust the bandwidth of a circuit depending on line conditions.

4.1.3 Hybrid Modulation Schemes

The most common hybrid modulation scheme is pulse code modulation (PCM), which is used to convert an analog signal into a digital signal. A section later in the chapter explains PCM in detail, though it is important to note that PCM is being used today in V.90 modems in order to increase the data rates across traditional dial-up lines. PCM schemes used in V.90 modems use sixty-four levels or a sixty-four-level scale with six bits per every 8000 Hz sample. (Traditional G.711 voice encoding using PCM uses a 256-level scale with an eight-bit PAM word for every 8000 Hz sample.) The data rate obtained can equal as much as 53.3 kbps depending on the quality of the line.

■ 4.2 DEFINING ANALOG-TO-DIGITAL CONVERSION

4.2.1 Digital Signals Defined

The analog voice or data signal must be manipulated into a digital one and zero bit stream before transporting the information onto a digital circuit. In communications,

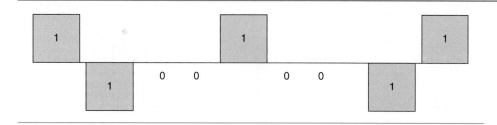

Figure 4–4
Digital waveform showing 110010011
bipolar return to zero format.

the term *digital* refers to information being transmitted using a one state or a zero state. Voltage being turned on is a one and no voltage is a zero. Figure 4–4 illustrates a typical digital waveform showing a 110010011 word.

Similar to the analog waveform, digital waveforms can be defined by the way the bit pattern is outpulsed onto the transmission medium. The three common digital patterns are *unipolar*, bipolar nonreturn to zero (*BPNRZ*), and bipolar return to zero (*BPRZ*), as illustrated in Figure 4–5. A unipolar digital signal consists of positive *pulses* moving in one (uni) direction. A unipolar waveform is never transmitted farther than a few feet. Signals traveling between circuit packs inside the equipment often use unipolar digital waveforms. Unipolar signals are difficult to synchronize and hard to monitor with error-checking schemes because there are no opposite polarity pulses to count.

A BPNRZ signal uses opposite polarity states for each bit transmitted. If the first state is transmitted as a positive voltage, the second bit or state sent will be a negative voltage. A pulse represents a zero and a pulse represents a one. The signal is comprised of voltage pulses of opposite polarity; a zero or no voltage state is transmitted. BPNRZ is less common in telephone networking equipment, due to the increased complexity of sending all pulses. Each pulse is the polar opposite of the preceding pulse. If pulse 1 is sent as a +3, pulse 2 will be sent as a −3 thus representing a bipolar bit stream.

The most commonly used digital bit pattern and the one most commonly used in the communications network is the BPRZ waveform. Again, the bipolar pattern means that each consecutive pulse is of the opposite polarity as the previous one, forming a + voltage state and a − voltage state. Return to zero tells us that a zero is represented by a zero state. Simply, a voltage state is transmitted whenever a one is sent. A zero voltage state is sent when a zero is transmitted, thus the term *bipolar return to zero*.

In the digital world, everything is built around patterns and defined standards. We only have two states—a one and a zero—to represent letters of the alphabet, numbers, voice signals, video signals, and so forth. To do this, digital signal standards have been defined and are followed within the industry. A pulse has a specific voltage, normally +3 or −3 V, and it lasts a specific amount of time. Each one pulse has a predefined standard height and width. The end equipment is expecting a +3 or −3 *voltage state* and a specific pulse width, as shown in Figure 4–6. The zero state must conform to a standard pulse length. The term *pulse mask* is used to represent the standard pulse shape for that bit stream. Test equipment uses the pulse mask to determine whether a digital word is corrupted. As shown in Figure 4–7, it is a little like a template that is placed against the pulse to see if it fits within its border. When the pulse width and height varies beyond pulse mask borders, it means that errors are occurring and that the information is viewed as corrupted. The end equipment defines the pulse shape and pulse width. Computer equipment transmits a pulse with a pulse height of +3/−3 volts. Therefore if you were to measure a one's pulse, you would need a +3 or −3 volts depending on the polarity of the pulse.

Bipolar return to zero provides the best method for transmitting digital signals longer distances. It also allows for an error-checking scheme in which bits are counted based on their *polarity*. The number is then tallied to determine whether a bipolar error occurred, that is, two consecutive pulses of the same polarity. The BPRZ bit stream is less cumbersome to deploy than the nonreturn to zero and thus provides the best method

digital signal
Signal that has limited, discrete number of values such as zero and one.

unipolar
All pulses are of the same polarity. Every state change produces a pulse of the same polarity such as +,+,+.

pulses
Electrical occurrences of a defined length and amplitude. The pulse created to represent a one state is defined by the amplitude and the length of time it lasts.

BPNRZ
Bipolar nonreturn to zero— Pulse states do not return to zero but do reverse polarity for each state change.

BPRZ
Bipolar return to zero—Zero pulse is represented by the zero state and the state reverses polarity for each state change.

voltage state
The voltage level of the pulse.

pulse mask
Shape the pulse should fall within in order to be considered a pulse. Similar to a standard outline the pulse should match.

polarity
Positive and negative values represent the polarity of the signal.

Figure 4–5

(a) A state equaling a pulse at +3 V and a state equaling 0 voltage. (b) A state equaling a pulse at +3 V and a zero state equaling a pulse at −3 V. (c) A pulse equaling a one state alternating between a +3 V pulse and a −3 V pulse. Zero is represented by no voltage.

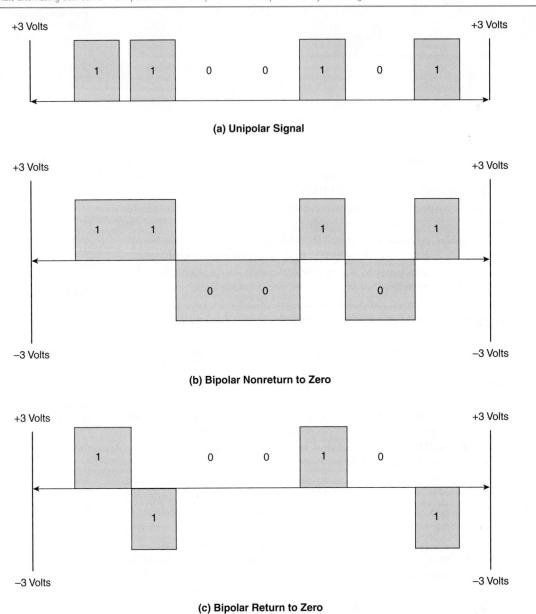

(a) Unipolar Signal

(b) Bipolar Nonreturn to Zero

(c) Bipolar Return to Zero

Figure 4–6

Six-volt peak-to-peak digital waveform.

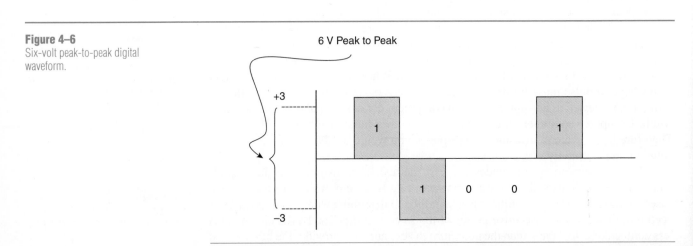

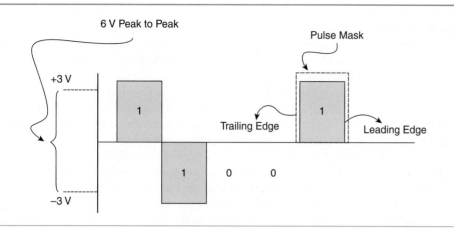

Figure 4–7
Pulse mask placed around a one pulse. The pulse mask represents the correct size of a pulse. When the pulse falls outside the mask or does not fill the mask, the end equipment has difficulty distinguishing where the pulse starts, where it stops, and whether it exists at all. Both the leading and the trailing edge of the pulse must meet the pulse mask parameters.

for transmitting information across a digital signal. The pulse shape is also used to determine whether the signal is clean or corrupted.

Beyond pulse shape and voltage states, the FCC has mandated specific rules regarding the transmission of multiple zeros. Digital equipment uses synchronized clocks to align the bits being received. If too many zeros are sent, the receiving equipment becomes confused and doesn't know what portion of the bit stream it is looking at. The equipment depends on the opposite polarity pulse states to synchronize the incoming bit stream and determine the demarcation point for each channel. The rule states that there can be no more than fifteen consecutive zeros, or 12.5% of the pulses can be zeros. Two methods have been devised to solve the all-zeros problem—*B8ZS* and *AMI*.

To discuss analog-to-digital conversion, you must first understand circuit direction. There are three modes of transmission—*simplex*, *half duplex*, and *full duplex*, as illustrated in Figure 4–8. An example of a simplex circuit is a loudspeaker in a school. The principal speaks into the intercom, which projects his voice out to a classroom. The signal travels in one direction—from principal to student. The circuit is not capable of carrying the student's response back to the principal. Because the circuit carries information in only one direction, it is a one-way transmission circuit.

A half-duplex circuit can carry information in both directions, but not simultaneously. A good example would be a two-way radio. You can talk to your friend by pressing the transmit button, but your friend can't talk to you until you release the button. The circuit carries information in only one direction at a time.

The full-duplex circuit can carry information in both directions simultaneously. Most digital circuits are full-duplex circuits. Telephone lines are technically full duplex but are used as if they are half duplex because it is difficult to carry on a simultaneous conversation.

B8ZS
Zero suppression code used to replace eight consecutive zeros in order to meet the one's density rule defined by the FCC.

AMI
Alternate mark inversion—Line code. Each alternate pulse is the opposite polarity.

simplex
Communication occurs in one direction only—source to receiver.

half duplex
Communication can occur in both directions, but in only one direction at a time.

full duplex
Communications occur in both directions at the same time.

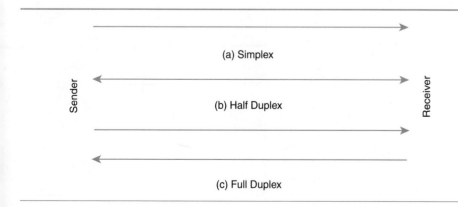

Figure 4–8
(a) Simplex transmission mode allows information to travel in only one direction—from sender to receiver. (b) Half-duplex mode allows information to travel between sender and receiver in both directions but in only one direction at a time. (c) Full-duplex mode allows simultaneous transmission between sender and receiver.

4.2.2 Building the PCM Word
Analog-to-Digital Conversion

Converting an analog waveform into a digital waveform was first accomplished in the 1960s when the advent of microprocessors and digital electronics spilled over into the world of telecommunications. Scientists and engineers were quick to recognize the simplicity of sending a signal composed of ones and zeros, compared to an analog signal with multiple frequencies. Analog signals required much more time to turn up and maintain due to the complex nature of dealing with many frequencies. For example, every time the weather changed, the analog carrier circuits had to be tuned to line up the equipment on both ends and the amplifiers located along the circuit path between them. When temperature increases, so does the resistance on a link. When the equipment on one end of the circuit varied because of the increased resistance, the receiving equipment was unable to decipher the content of the message.

In addition, the analog signal was limited in the range or number of frequencies allowed over distances, thus reducing the total bandwidth of the circuit. The digital signal bandwidth far exceeds that of the analog signal because the transmission of one and zero states is much less complicated. Analog signals have multiple frequencies to deal with, increased noise on the line, greater signal degradation (especially on the higher frequencies), and complex waveforms. Converting the analog telephone network into a digital network produced an extremely reliable network capable of carrying huge amounts of information, thus making way for the "information superhighway."

Converting signals from analog to a digital is one of the fundamental processes of telecommunications networks. The only analog portion of the network left today is that found between the serving central office and the subscriber's residence. Large businesses use digital circuits for their voice and data networks. When a signal enters the switch, or arrives at the ISP's modem pools or a company's PBX, the signal is converted into a digital signal, multiplexed, and transported digitally to its destination, as illustrated in Figure 4–9. Pundits often speak of the new digital society. As you can see, it is an accurate description.

Moving to the analog-to-digital process, we will revisit Andy and Daisy's conversation as it travels across the telephone circuit and enters the class 5 switch. Andy's voice signal, "Hi Daisy, how ya doing?" enters the line card in the switch. The line card converts the analog signal into a digital signal and, conversely, the digital signal into an analog signal. A coder/decoder (Codec) that is located on the line card does the analog-to-digital and digital-to-analog conversion. In this example, the Codec

Figure 4–9

Analog data signals terminate into a modem pool. The modems connect to a TDM multiplexer that converts the analog signal into a digital signal, which then feeds into a telephone switch that routes the calls accordingly.

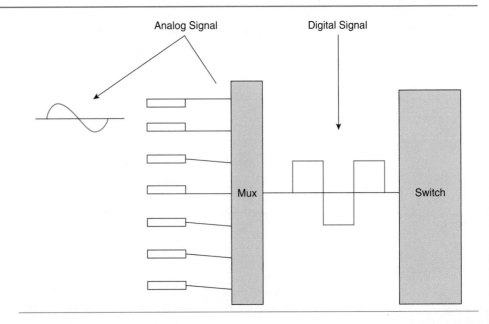

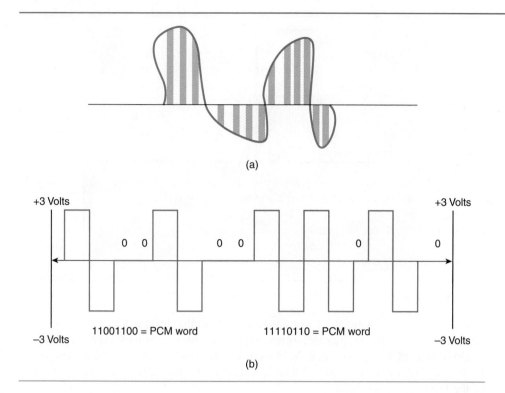

(a)

(b)

Figure 4–10
Comparison of an analog voice waveform and its digital equivalent. The analog waveform goes through an analog-to-digital conversion that produces multiple eight-bit words consisting of ones and zeros or voltage pulses and no voltage pulses. (a) Analog waveform showing pulse amplitude modulated samples from the first step of the digital-to-analog conversion. (b) Two eight-bit PCM words representing the analog frequencies shown in (a).

looks at the analog waveform created by Andy's voice frequencies saying "Hi," as depicted in Figure 4–10. A three-stage process occurs during the analog-to-digital conversion: sampling, quantizing, and encoding.

The Codec samples the analog waveform 8000 times every second. Think of the Codec as a photographer who takes 8000 pictures of an erratic waveform speeding past every second. The samples taken are pulse amplitude modulated (*PAM*) samples. They move through the microprocessor chip where they are converted into binary ones and zeros to make a digital waveform. This is called *quantizing*. During the quantizing process, the sample is placed against a 255-level quantizing scale—similar to a yardstick. Each of the 255 levels has an eight-bit one and zero word associated with it as shown in Figure 4–11. The sample, which is simply a voltage level, is matched to one

PAM
Pulse amplitude modulation—Part of the analog-to-digital conversion process.

quantizing
Step that converts a voltage value of the analog waveform to a digital word in analog-to-digital conversion.

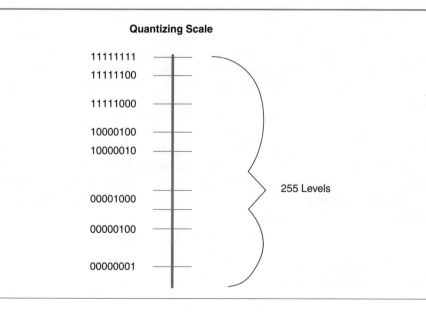

Figure 4–11
The 255 levels that are used to define eight-bit words from voltage samples.

Figure 4–12
A digital signal can be fed into a piece of test equipment called an oscilloscope, which can be slowed down to show the actual one and zero pulses of the signal.

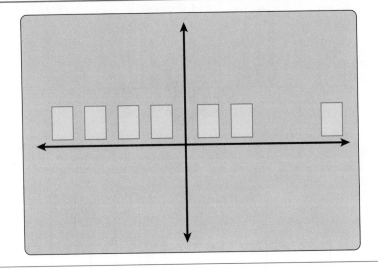

PCM
Pulse code modulation digital word.

encoding
Step in which digital words are encoded for transport in analog-to-digital conversion.

of the 255 levels. The eight-bit word associated with that level is the pulse code modulation (*PCM*) word. Technicians often refer to the bit pulses they view on their test equipment as PCM words. Each PCM word refers to one sample of the analog waveform. Early scientists determined that 255 eight-bit words sufficiently represented all of the levels of human speech.

The final step in the process is *encoding*. Encoding formats the PCM words and ships them out onto the physical transmission medium, whether it is a copper wire or a fiber optic strand.

Figure 4–12 shows a PCM word as it would appear on an oscilloscope screen. Notice the unipolar structure of the bit stream and the similarity in the height and width of each pulse. Bit format does not vary. It might be viewed as an army of similarly clad soldiers marching at the same pace in the same direction, following the same order and routine, but the entire process happens within millionths of a second. In fact, it takes only 125 microseconds (μs) to take a sample from the analog waveform, quantize it, and encode it onto the line.

Now that we understand how analog-to-digital conversion works, we can discuss how the reverse—digital-to-analog conversion—is performed. The eight-bit PCM word is received by the receiving end codec. Using the quantizing scale, it translates the eight-bit one and zero word back into one sample or voltage level. The word "Hi" becomes multiple voltage samples, as shown in Figure 4–13. The samples are then connected to form the original analog waveform. Figure 4–14 illustrates the decoding process of the digital signal. The eventual analog waveform is not an exact mirror image of the original "Hi Daisy, how ya doing?" because the waveform is a compilation of the samples of the original. However, it is so close that the ear can't tell the difference.

The Nyquist theorem was used when determining the sampling rate for PCM words. The theorem states that a waveform should be sampled at two times its highest frequency in order to obtain a toll-quality signal. Because the frequency range for a voice signal is 300 to 4000 Hz, the sample must be 2×4000, or 8000 samples every second to create a digital waveform that will adequately represent the source voice waveform. Consequently, a toll-quality voice signal requires 64 kbps of bandwidth, as shown in the following equation.

8-bit PCM word × 8000 samples per second = 64,000 bps or 64 kbps.

toll quality
Voice signal meeting specs provided by standards organizations.

The term *toll quality* refers to a voice signal that meets the toll voice specifications, referred to as mean opinion scores (MOSs), by standards organizations such as the ITU and ANSI.

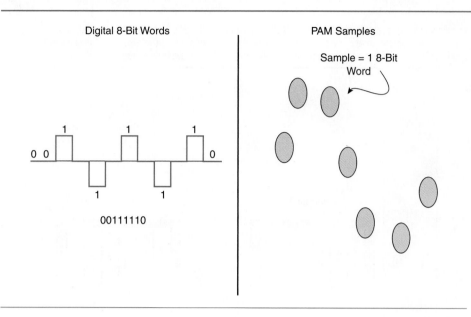

Digital 8-Bit Words

PAM Samples

Sample = 1 8-Bit Word

00111110

Figure 4–13
In digital-to-analog conversion, the digital word is converted into voltage samples. Each of the samples is shown as a circle representing one eight-bit word.

With the advent of the transistor in the 1960s, Alec Reeves from ITT patented a PCM word process based on the Nyquist theorem. The transistor allowed for the fast processing required for analog-to-digital conversion.

Microprocessor speed and buffering capability (temporary information storage) have made it possible to decrease the number of samples needed in order to transmit toll-quality voice. Today, sampling rates have been halved, thus reducing the amount of bandwidth required to ship an analog signal. For example, when an analog signal is sampled 2000 times every second instead of 4000 times, a 32 kbps word is formed instead of a 64 kbps word. Adaptive pulse code modulation (*ADPCM*) refers to a process that uses 2000 samples per second to form the PCM word. Other more involved techniques such as LD-CELP have also been created to compress the voice signal and reduce the number of ones and zeros needed for each PCM word. The bandwidth required per voice channel has been reduced to levels as low as 16 kbps and even 8 kbps while still providing toll-quality voice.

Once analog-to-digital conversion is complete, the signal leaves the switch line card and is routed to the correct trunk port, or trunk group. Normally, the trunk group interfaces with the outside world via a T1 circuit. The multiple 64 kbps words are multiplexed onto the T1 circuit and carried to the terminating end.

ADPCM
Adaptive pulse code modulation—Compressed digital word.

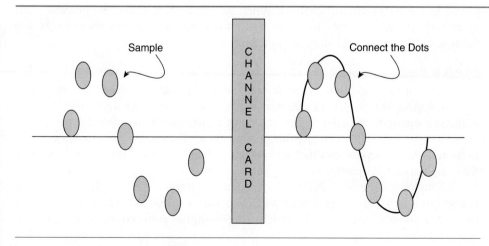

Sample

CHANNEL CARD

Connect the Dots

Figure 4–14
An abstract view of how the PAM samples are converted back into an analog waveform.

Figure 4–15
To multiplex individual signals into channels on a digital circuit, a sample is taken from signal 1, an analog-to-digital conversion is performed, and the resulting eight-bit word is placed in the first channel. The process is repeated on the second signal where the resulting eight-bit word is also placed in channel 1. The process continues through the twenty-fourth signal and then begins again. The process continues until all signals have been converted.

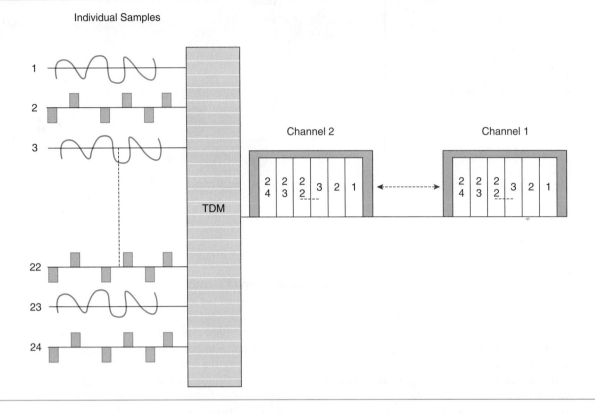

T1 circuits form the backbone of our communications network. A T1 circuit has twenty-four time slots capable of carrying twenty-four digital signals or twenty-four customers' signals on a transmit-and-receive pair. The method used by the switch is to sample each line in sequence starting with the first customer circuit and working down to the twenty-fourth, as shown in Figure 4–15. For example, when Andy's voice is sampled, the sample becomes a PCM word that is placed in the first channel (time slot) of the T1. The second sample is taken from the second subscriber's line and is converted into an eight-bit PCM word that is placed in the second channel of the T1. The same is done with the third customer's line and the process continues until the twenty-fourth customer's analog waveform is sampled, converted into an eight-bit PCM word, and placed in the twenty-fourth channel. When that channel is filled, the process starts again with the next frame. To better understand this concept, review Figures 4–9 through 4–15. Figure 4–16 illustrates the entire process.

Digital-to-Digital Conversion
The source of the analog waveform may be a voice, a data call placed through a modem, or a video signal. The T1 circuit may also carry digital data traffic that does not require an analog-to-digital or digital-to-analog conversion. Remember that the type of signal leaving your PC is a binary one or zero signal. The modem interfacing your PC to the telephone line converts the one and zero signals into an analog waveform using the same frequencies produced by the human voice.

Though regular analog POTS lines feed most homes, some have digital lines that connect to a digital telephone and/or a PC. When a digital line interfaces with our PC, the signal doesn't have to go through analog-to-digital conversion. Instead, special

Figure 4–16
(a) Analog signal converted into a digital signal. The signal is sampled at a rate of 8000 times every second, then compared to a predefined scale called quantizing, converted into one and zero eight-bit words, and encoded into the digital bit stream. (b) Digital signal converted into an analog signal. The digital eight-bit PCM words are decoded into PAM samples, which are connected much like connecting the dots to form the analog signal.

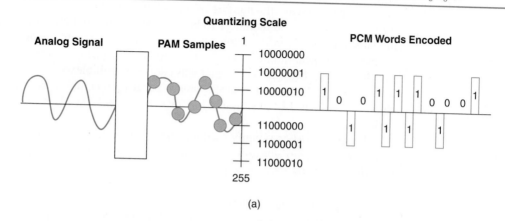

(a)

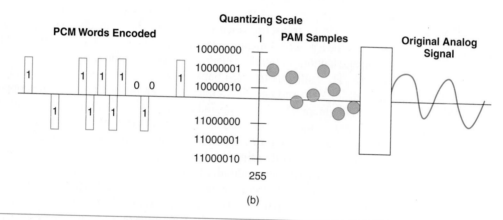

(b)

interface equipment formats the digital signal leaving the PC and prepares the signal for transport across the communication link. The one and zero bits coming from the PC are normally in a unipolar bit format and need to be converted into a bipolar return to zero format. Devices placed between the PCs and the outside telephone circuit perform the conversion process and prepare the signal in much the same way as the encoding process discussed earlier. The ones and zeros are pushed into the predefined channels and then sent out across the network. If you could somehow open a signal midstream, you would not be able to decipher which of the ones and zeros belonged to an analog signal and which to a digital signal from a PC. The end pieces of equipment are able to count the bits and determine the borders of each channel and frame. From this segmentation of the bit stream, the end equipment is able to send the correct eight-bit words to the appropriate *channel* cards. There are cases in which a T1 circuit or ISDN circuit will carry digital data traffic; and, in order to accept the digital data traffic, the data stream must be formatted to fit into the 64 kbps channels. Formatting digital signals directly onto a digital circuit is simpler than formatting a voice. The information is segmented into groups of eight ones and zeros. The ones and zeros are pushed into the time slot (or channel) of the circuit. If the signal does not have enough bits to fill the channel, stuffing bits are added to fill up the space. The stuffing bits are then stripped out at the termination end. The digital-to-digital conversion is more formatting of information than converting it. The interface equipment formats the digital signal into a BPRZ signal before transmitting it out onto the network.

channel
Term used to define the individual signal riding within the multiplexed signal.

■ 4.3 MULTIPLEXING

multiplexing
Process of combining
multiple signals into one
larger signal.

The dilemma facing engineers since the first communications network began was the cost of handling hundreds of calls between two points. The options for the early engineers were to increase the number of wires between two points or have customers vie for space on the shared facilities. Today, engineers have multiple options for transporting large amounts of information between two points. The introduction of *multiplexing* technology helped eliminate the need to add new facilities or force customers to compete for space every time a customer was added. A multiplexer combines numerous information signals onto one transmission pipe. It may combine as few as twenty-four channels onto one T1 transmission link or as many as 129,024 channels onto one fiber optic strand.

The four most commonly used multiplexing methods in telecommunications networks are *frequency division multiplexing*, *time division multiplexing*, *statistical multiplexing*, and *wave division multiplexing*.

4.3.1 Frequency Division Multiplexing

FDM
Frequency division
multiplexing—Method that
multiplexes different
frequencies together into one
signal.

In the early 1900s, frequency division multiplexing (*FDM*) was added to the telephone network. The purpose of the first FDM systems was to reduce the number of copper wires needed to transport voice calls across the long distance network. Once FDM proved to be a practical means to connect two sites, the technology was introduced into the local telephone network as central offices were connected to one another using FDM systems. The calls that traveled between switch sites were fed into an FDM that combined the signals into one signal that was then pushed out onto the copper link. At the far end, the incoming signal was fed into a second multiplexer that pulled out the individual signals and fed them into the switch. These multiplexed links were called *shared trunks* and were similar to the trunking connections used to connect switchboards.

FDM works by dividing the frequency spectrum into subchannels, or frequency chunks, and transporting individual voice signals across each channel. For example, a twelve-channel FDM system using the 60 to 108 kilohertz (kHz) frequency spectrum divides the bandwidth into 3 kHz chunks, plus guard bands placed before and after each subchannel. As the signals arrive, they are multiplexed into one of the twelve subchannels illustrated in Figure 4–17. For example, signal 1 is multiplexed into the 60 kHz band and signal 12 is multiplexed into the 108 kHz band. The example shown illustrates twelve voice lines feeding into the FDM, which places the channels into the correct frequency subchannel, then transports the frequencies out onto a four-wire copper circuit. The multiplexer is also capable of demultiplexing the incoming frequencies back to the original 3 kHz signal and sending them out onto the copper line interfacing the multiplexer and the customer's site as shown in Figure 4–18.

Obviously, FDM reduces the number of copper wires needed in the network. Early systems carried twelve signals (voice calls) on two pairs of wires. Running two wire

Figure 4–17
Frequency division multiplexing aggregates multiple channels onto one transmission circuit.

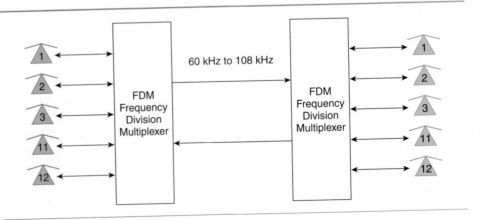

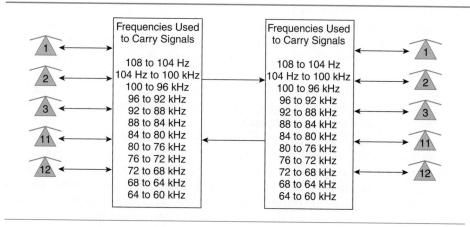

Figure 4–18
A typical frequency division multiplexing scenario. The example shows twelve telephone lines being multiplexed onto one four-wire pair. The four-wire pair carries the twelve signals by shoving each signal into a separate frequency within the range shown between the multiplexers.

pairs instead of twelve wire pairs produced an immediate financial incentive to place multiplexers in every central office and long distance switch site. When FDM was invented, it solved many problems and reduced costs for outside plant facilities for telephone companies. Today, FDM is seldom used because of several technical and logistical problems associated with it. The three major ones are the power required to energize the circuits, the increased noise in the line, and the task of tuning the circuits.

In every FDM system, the signal must be amplified at regular intervals. Each time the signal passes through an amplifier, both the signal and the line noise are amplified. After the signal has passed through multiple amplifiers, the noise on the line becomes so loud that the information signal is drowned out and rendered useless. Because of the noise on the line, the distance between multiplexers was limited. The engineers constantly measured and calculated when they designed trunking FDM circuits between sites. If the signal passed through too many amplifiers, the end equipment found it hard to distinguish between the noise and the good signal. Therefore, the distance a signal could travel on an FDM circuit was limited and the engineer had to invest in either additional multiplexers or more copper.

A second problem with FDM amplified systems was the need to tune the systems whenever seasonal temperature change occurred. Analog signals must arrive at a certain signal level and continue to mimic the original signal. When the weather changes in the spring and fall, the physical characteristics of the in-line amplifiers and the resistance value of the copper cable change. The higher the temperature, the higher the resistance values. The change in the resistance of the circuit causes the signal to change as it travels down the wire. Because the signal is a representation of the original, any change is unacceptable. If changes do occur, deciphering the signal at the terminating end becomes impossible. To alleviate this problem, technicians must tune the circuit between each amplifier. The procedure requires two technicians—one at each amplifier—adjusting the levels of the circuit to maintain signal quality.

The third problem with FDM systems was the need to line power the circuit. Because the amplifiers placed in the circuit required power, the only solution was to place power on the line at the central office. The amount of power required was much higher than for a POTS line and longer circuits needed more power to energize the amplifiers. Therefore, power requirements also limited the distance an FDM system could send a signal.

FDM systems are beginning to be used again today. New DSL line codes that were accepted as standard in 1998 use an FDM scheme called DMT. DMT divides the 10 megahertz (MHz) frequency spectrum that is available on a copper twisted pair line into 256 4000 Hz segments. Information is multiplexed into these 4 kHz bands. FDM is still a viable multiplexing method for certain applications such as DSL, which is used for short distances between central offices and subscribers' locations. Also, newer microprocessors are able to automatically align the signal levels, relieving the technicians from manually adjusting the line cards.

4.3.2 Time Division Multiplexing

Time division multiplexing (*TDM*) is the most common method of multiplexing used today. TDM was first introduced into the network in the late 1970s when the first T1 circuit turned up. TDM, like FDM, combines multiple signals onto one pipe. TDM rides on all types of media—copper, radio frequencies, and fiber. TDM is the most widely used transport protocol in the worldwide communications network. ILECs, CLECs, IXCs, enterprise customers, global communication providers, and ICPs all have built networks using the TDM protocol. The two most significant differences between the FDM and TDM technologies is that TDM uses time slots to carry signals instead of frequencies and multiplexes digital signals into the time slots, not analog signals, as with FDM. The TDM time slot technique requires that the signals are clocked at a defined rate and that the bits are interleaved serially onto the link. The bits are fed into the time slots, also referred to as channels, according to the framing structure of the circuit. For example, the time required to encode and place one eight-bit PCM word into a channel is 125µs. The two ends of the circuit are able to synchronize because both use the same time calculations.

There are two distinct steps in the TDM process, *conversion* and multiplexing. The conversion process involves either an analog-to-digital conversion or formatting the digital words. The next step in the process is to multiplex the digital words into the time slots (channels). The number of channels is dependent on the type of multiplexer and the medium being used. TDM systems vary in size from as small as two 64 kbps channels multiplexed onto one copper line, as in an ISDN BRI circuit, to as large as an OC-192 signal that carries 129,024 channels on one single-fiber strand. The method, however, remains constant. The signal is first converted into digital binary words and then multiplexed into individual time slots on the circuit, using the rules defined by the TDM protocol standard.

A standard T1 multiplexer accepts information signals on a per-port basis. For a voice signal, the multiplexer runs the signal through a codec chip where it is converted from analog to digital. Newly converted ones and zeros are packaged into eight-bit words and placed in specific channels on the T1 bit stream. If the signal is already in digital format, the signal is packaged into eight-bit words and is then also placed into specific channels in the T1 bit stream. The multiplexer we are discussing has twenty-four input ports that are receiving twelve voice signals and twelve data signals. The multiplexer uses a round-robin approach when determining which signal to grab and convert. The first input channel is selected, as shown in Figure 4–19. The channel card in the multiplexer converts the analog signal into an eight-bit digital word, then places the word into the first channel of the digital signal. The second channel is then selected, and an eight-bit word is produced and placed. The process is repeated until the twenty-fourth signal is complete, and then it starts all over with channel 1.

Multiple eight-bit words form full sentences or full streams of data, depending on the input signal. Piecing all of the eight-bit words together in a round-robin fashion produces smooth coherent speech and uncorrupted data because the process happens so fast that there is no delay between words. Each sample takes 125 µs—a microsecond (µs) is one millionth of a second. We learned earlier that one fiber optic strand can carry as many as 129,024 signals or channels and that one microwave link can carry multiple T1 signals. These systems are able to handle such large quantities of information because of TDM.

4.3.3 North American Digital Hierarchy and SONET

There are two standard digital hierarchies used in North America. The older of the two is the North American Digital Hierarchy; the newer is the synchronous optical network (*SONET*) standard. The North American Digital Hierarchy has been around since the beginning of digital circuits in the late 1970s. The hierarchy defines the circuit bit-per-second rate. For instance, twenty-four 64 kbps DS-0 channels represent a T1 circuit. A digital signal 0 (DS-0) is a channel that can hold 64 kbps. In other words, one time slot can carry 64,000 ones and zeros each second. The T1 can carry twenty-four 64 kbps time

TDM
Time division multiplexing. Method that multiplexes digital words into time slots together in one signal.

conversion
First step in TDM process. Analog-to-digital words.

SONET
Synchronous optical network—Optical standard that defines specific bit-per-second rates.

Figure 4–19
Signals entering the channel bank may be either analog or digital. Analog signals go through an analog-to-digital conversion, which requires a Codec chip. A digital signal must be encoded and buffered or stuffed, if necessary.

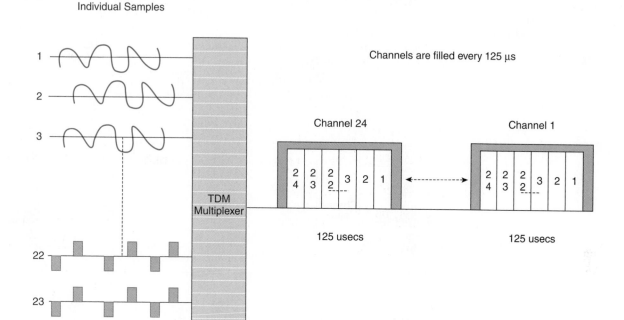

slots and, therefore, has a bit-per-second rate of 1.544 Mbps or 1,544,000 bps. The hierarchy's purpose is to allow equipment vendors to produce standard equipment interfaces. If an equipment vendor produced a multiplexer with a T1 card that allocated thirty time slots instead of twenty-four, the circuit would not work with other vendors' equipment. The North American Digital Hierarchy is shown in Table 4–1. The most common rates are the *DS-3*, also called T3; the *DS-1,* also called T1; and the *DS-0.* Even within the SONET structure, the DS-3, DS-1, and DS-0 are used to define the subnetwork.

TDM is used to multiplex lower-rate signals into higher-rate signals. Figure 4–20 illustrates how a TDM system can aggregate thousands of lower-speed channels into one high-speed pipe. As shown in the diagram, twenty-four 64 kbps DS-0s are time division multiplexed into one 1.544 Mbps T1 signal. The second step illustrates how twenty-eight 1.544 Mbps signals are time division multiplexed into one 45 Mbps DS-3. The third step shows how forty-eight 45 Mbps DS-3s are time division multiplexed into one OC-48 (2.5 gigabits per second [Gbps]) signal.

DS-3
Digital signal level 3—
45 Mbps.
DS-1
Digital signal level 1—
1.544 Mbps.
DS-0
Digital signal level 0—
64 kbps.

Digital Signal Level	Bit-per-Second Rate	No. of DS-0s	No. of DS-1s	No. of DS-3s
DS-0	64 kbps	1	NA	NA
DS-1	1.544 Mbps	24	1	NA
DS-2	6.44 Mbps	96	4	NA
DS-3	45 Mbps	672	28	1
560 Systems	560 Mbps	8064	336	12

Table 4–1
Digital Signal Levels as Defined by the North American Digital Hierarchy.

Figure 4–20
Multiplexing lower-rate signals into higher-rate signals. Individual DS-0s are multiplexed into a T1 circuit in the Channel Bank. Twenty-eight DS-1s are multiplexed in an M13 multiplexer into a DS-3. The DS-3 is multiplexed into an OC-48 in the Optical SONET multiplexer.

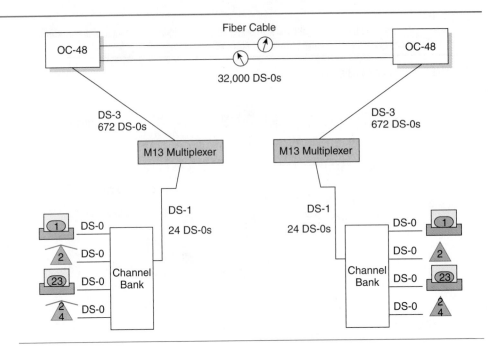

4.3.4 Connecting the Dots

The following is an example of using the North American Digital Hierarchy and SONET in an end-to-end network scenario. Top Spot telephone company requires trunks between two of the towns it serves—Crystal Falls and Olive Grove. The transmission engineer decides to use an existing microwave link between the two towns. The microwave site in Crystal Falls sits on top of Crystal Mountain. Olive Grove's microwave antenna sits on top of a 3000-foot tower in the middle of farmer Dempsey's cow pasture. In order for the residents in the two towns to call each other, one T1 circuit is needed to carry the shared trunks between the two towns' central office switches.

First a link must be established between the central office in Olive Grove and the microwave site in the cow pasture. Fortunately, a fiber optic cable with spare capacity exists between the switch site and the tower in the pasture. The Olive Grove central office switch multiplexes twenty-four DS-0s into one T1 bit stream. It is then fed into a DS-3 multiplexer where the 1.544 Mbps T1 signal is multiplexed into a 45 Mbps DS-3. The DS-3 from the DS-3 multiplexer is fed into the existing fiber optic multiplexer. It is then multiplexed into a high-speed OC-12 signal that rides across the fiber optic strand to the microwave tower site in farmer Dempsey's field. Once at the microwave tower site, the OC-12 signal is demultiplexed into twelve DS-3s, one of which is carrying the new T1 circuit from the central office switch. One of the DS-3s is then fed up the microwave waveguide and transmitted out across the airways toward the Crystal Mountain microwave antenna. The T1 carrying the shared trunks is in the fifth time slot of the DS-3.

A copper T1 circuit connects the Crystal Springs central office switch to the Crystal Springs microwave tower. Once at the tower, the T1 is fed into a DS-3 multiplexer where it is multiplexed into channel 5 on the 45 Mbps DS-3. The T1 is multiplexed together with twenty-seven other T1 circuits into a DS-3. The DS-3 that travels between the two microwave towers carries twenty-eight T1 circuits that can carry information between the two towns. The Top Spot engineer has used the fifth time slot to carry the T1 signal for shared trunks.

From this example, we can see that different digital signal rates can be multiplexed together, demultiplexed, and multiplexed again. Figure 4–21 documents the example.

Figure 4–21

Connection between the Olive Grove local switch and the Crystal Mountain local switch. A DS-3 connects the two towns via a microwave transmission link. Local switches are connected via land line. T1s connect the Crystal Mountain microwave tower to the local switch, while the Olive Grove microwave tower is connected to the local switch with fiber optic cable.

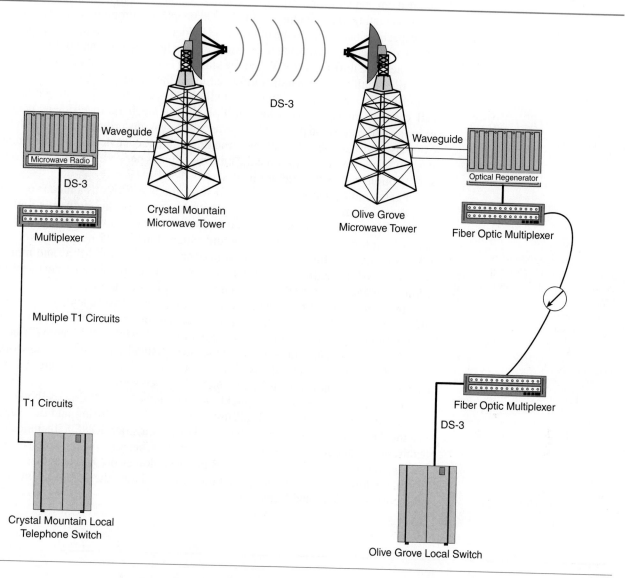

4.3.5 Statistical (Fast-Packet) Multiplexing

Fast-packet multiplexing is a method used to transmit information. It is much more efficient than TDM or FDM. Statistical multiplexing does not reserve time slots or frequencies. Instead, it dynamically allocates bandwidth only when it is needed. For instance, information being transported using TDM requires an assigned time slot or channel, similar to box cars on a train. FDM reserves specific frequencies per user channel. The consequence is that bandwidth is unused when the user is not transmitting information. For example, if the subscribers in Olive Grove and Crystal Falls decided never to call one another again, the twenty-four trunks connecting the two towns would never be used. The circuit would sit idle. Statistical (fast-packet) multiplexing is similar to the post office. The route a letter will take through the postal system depends on many factors. It might take a certain route depending on the day, the time, and the letter's priority. It is very possible that two letters going to the same address would take completely different routes. Many packet protocols handle packet transmission in a

fast packet
Method using packets, cells, or frames to statistically transmit information. IP, frame relay, and ATM are common fast packets.

similar way. Packets are statistically multiplexed and routed according to the availability, time, and number of hops along the route.

In packet multiplexing, the pipe is similar to free space. A user's information is shipped out and vies for available bandwidth—similar to buying concert tickets without assigned seating. If the stadium holds 10,000 people, the promoter may sell 15,000 tickets, expecting that a percentage of the ticket holders will not show up. The promoter may prioritize tickets, guaranteeing high-priority ticket holders a seat and only allowing others if there are unfilled seats. When the stadium is full, those outside are turned away. Statistical multiplexing handles information in the same way. Circuits may be oversubscribed in the same way that the stadium seats are oversold. Being able to oversell capacity makes fast-packet multiplexing a much more efficient way to use expensive transmission pipes. The engineer can oversubscribe the bandwidth because every user will not be online at the same time. Statistical calculations are made when designing networks to determine the number of users per pipe. A concentration (oversubscription) rate is then determined.

Packet multiplexing has been used since the late 1960s as a means of transmitting data information. The difference between early packet networks and today's fast-packet networks is the increased transmission rate and the ability to ship time-sensitive information such as voice and video. Early *X.25* networks were able to transmit only 2.4 kbps. Today's ATM networks transmit 155 Mbps and more. Digital transmission facilities and fiber optic pipes have increased transmission rates while reducing transmission errors encountered in the facility.

In order to send time-sensitive information, such as voice and video, fast packet uses very fast DSPs (digital signal processors) to prioritize time-sensitive information by stamping a high-priority bit on the frame. By using cell-based *ATM* networks, voice, video, and data can all share the same statistically multiplexed network without producing choppy and delayed voice or video streams. Even Internet protocol (*IP*) networks are now able to carry voice and video signals.

Fast-packet mulitplexing is quickly becoming the multiplexing method of choice. ATM, frame relay (*FR*), and IP all use fast-packet multiplexing and are quickly replacing the standard TDM scheme while still using the fiber SONET. Technically, ATM and frame are not always called packet technologies, because ATM is a cell-based technology and frame relay ships frames. Both technologies do, however, depend on statistical multiplexing methodologies. Figure 4–22 shows the statistical multiplexing protocols—IP, ATM, and frame relay.

X.25
Early packet transmission protocol used to transmit slower-speed data rates.

ATM
Asynchronous transfer mode—Fast-packet transmission protocol that statistically multiplexes fixed-length cells.

IP
Internet protocol—Transmission protocol that statistically multiplexes variable-length packets.

FR
Frame relay—Fast-packet transmission protocol that statistically multiplexes variable-length frames.

Figure 4–22
(a) IP multiplexing. (b) ATM multiplexing. (c) Frame relay multiplexing.

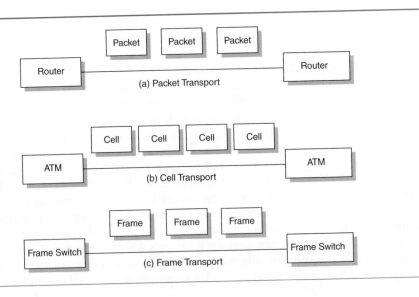

(a) Packet Transport

(b) Cell Transport

(c) Frame Transport

4.3.6 Dense Wave Division Multiplexing

DWDM is the newest multiplexing technology in the communications network. DWDM systems multiplex different colors of light onto fiber strands, thus increasing the bandwidth available on fiber optic systems. The rapid development of DWDM systems has helped increase the amount of bandwidth available throughout the world.

To understand these concepts, imagine three flashlights emitting different colors of light—red, blue, and purple—each shining into a fiber strand. Space is left between each light beam to ensure that the colors do not mix together. The red beam carries a signal from an OC-48 multiplexer, the blue beam carries a signal from an OC-192 multiplexer, and the purple beam carries a signal from an OC-12 multiplexer. The fiber strand carrying the three colors (wavelengths) of light is transporting 169,344 DS-0s of traffic, as shown in the summation following.

$$OC\text{-}48 = 48 \times 672 = 132,256 \text{ DS-0s}$$
$$OC\text{-}192 = 192 \times 672 = 129,024 \text{ DS-0s}$$
$$OC\text{-}12 = 12 \times 672 = \underline{\quad 8,064 \text{ DS-0s}}$$
$$169,344\text{-0 DS} = 0$$

or

$$169,344 \text{ DS-0s} \times 64 \text{ kbps} = 10,838,016,000 \text{ bps}$$

Systems deployed today carry 80 to 120 wavelengths (colors) of light on one strand of fiber, and each of these wavelengths could carry an OC-48 or OC-192 signal. The amount of bits per second being transported over one strand of fiber the size of a human hair is phenomenal. Lab trials have proven as many as 700 wavelengths carried by one strand of fiber.

Wavelengths used in DWDM systems vary, depending on a manufacturer's lasers. The ITU has defined wavelength standards and segmented the spectrum into six windows, as shown here.

Window 1 = 850 nanometer (nm) wavelength

Window 2 = 1300 nm wavelength = S-band

Window 3 = 1550 nm wavelength = C-band

Window 4 = 1600 nm wavelength = L-band

Window 5 = 1350 to 1530 nm wavelength

Window 6 = 1260 to 1650 nm wavelength

The ITU has designated specific wavelengths within each of the windows by segmenting the spectrum into grids. The most common grid deployed today is the C-band as shown in Table 4–2. Wavelengths are often referred to as *channels* or *lambdas*, the Greek word for *wavelength*, and are capable of carrying a high-speed signal such as an OC-48 (see Figure 4–23). Figure 4–24 shows a typical DWDM system using the 1550 nm C-band grid.

An additional parameter that determines the number of wavelengths per fiber is the amount of spacing placed between each channel; 50 and 100 gigahertz (GHz) are the most commonly used spacing arrangements. If the spacing is too narrow, adjacent channels could cross over and corrupt their neighbor's signal. The term *coarse wavelength division multiplexing (CWDM)* refers to systems with very wide spacing fields between each wavelength that limit the number of channels per system. The purpose for deploying CWDM is to reduce the cost of the system. For example, an eight-lambda or wavelength system will work with large gaps between the channels, thus allowing less precise lasers and consequently reducing the cost of the system. Eighty-wavelength systems depend on very expensive, finely tuned lasers to maintain signal credibility and eliminate any crossover effect (as cross talk). Spacing as small as 25 GHz is being promoted as a way to increase the number of channels per system. The cost, of

DWDM
Dense wave division multiplexing. The newest multiplexing technology in the communications networks.

CWDM
Systems with very wide spacing fields between each wavelength.

Table 4–2
ITU Wavelength Grid—C-Band.

Wavelength	Frequency	Wavelength	Frequency
1528.77 nm	196.10 THz	1544.53 nm	194.10 THz
1529.16	196.05	1544.92	194.05
1529.55	196.00	1545.32	194.00
1529.94	195.95	1545.72	193.95
1530.33	195.90	1546.12	193.90
1530.72	195.85	1546.52	193.85
1531.12	195.80	1546.92	193.80
1531.51	195.75	1546.32	193.75
1531.90	195.70	1547.72	193.70
1532.29	195.65	1548.11	193.65
1532.68	195.60	1548.51	193.60
1533.07	195.55	1548.91	193.55
1533.47	195.50	1549.32	193.50
1533.86	195.45	1549.72	193.45
1534.25	195.40	1550.12	193.40
1534.64	195.35	1550.52	193.35
1535.04	195.30	1550.92	193.30
1535.43	195.25	1551.32	193.25
1535.82	195.20	1551.72	193.20
1536.22	195.15	1552.12	193.15
1536.61	195.10	1552.52	193.10
1537.00	195.05	1552.93	193.05
1537.40	195.00	1553.33	193.00
1537.79	194.95	1553.73	192.95
1538.19	194.90	1554.13	192.90
1538.58	194.85	1554.54	192.85
1538.98	194.80	1554.94	192.80
1539.37	194.75	1555.34	192.75
1539.77	194.70	1555.75	192.70
1540.16	194.65	1556.15	192.65
1540.56	194.60	1556.55	192.60
1540.95	194.55	1556.96	192.55
1541.35	194.50	1557.36	192.45
1541.75	194.45	1557.77	192.40
1542.14	194.40	1558.17	192.35
1542.54	194.35	1558.98	192.30
1542.94	194.30	1559.39	192.25
1543.33	194.25	1559.79	192.20
1543.73	194.20	1560.20	192.15
1544.13	194.15	1560.61	192.10

course, is extremely high. Figure 4–25 shows the difference between narrower and wider spacing methods.

■ 4.4 STRUCTURE OF DATA COMMUNICATIONS

4.4.1 Defining Data Communications

Data communications, quite simply, is the use of a physical medium to transmit data between two points. The medium can be copper wire, fiber optic cable, or air. The information may be a batch file carrying the bank balance for everyone from Tin Cup, Indiana, or a full unabridged encyclopedia of the world. Data communications is used by the majority of businesses today from the largest corporations to the smallest establishments. According to analysts, data transmission is growing exponentially and will soon exceed voice traffic on the PSTN (Public Switched Telephone Network).

Figure 4–23

A nine-wavelength DWDM multiplexing system. Nine OC-48s are fed into a DWDM, where they are multiplexed onto different wavelengths before being transported out onto the fiber. Two strands of fiber, one for transmit and one for receive, carry all nine OC-48 signals.

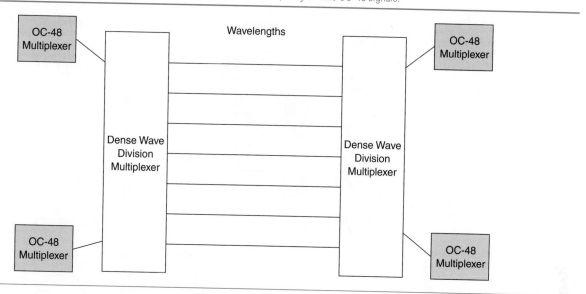

4.4.2 Binary Information Encoding—ASCII and EBCDIC

Every letter in the alphabet must have an equivalent binary code unique to that letter. For example, when a *C* is typed, a *C* appears on the monitor, but the computer sees 1000011. That is the binary word for *C* in the seven-bit *ASCII* binary code. Every computer must adhere to the same code to know what the transmitting computer is saying. In our example, the seven-bit binary code is from the ASCII standard developed by the American National Standards Institute (ANSI). As shown in Table 4–3, each letter, number, and special character has an equivalent seven-bit one and zero number.

There are other types of encoding schemes, but ASCII is the most popular for personal computing devices. A second type of ASCII is eight-bit ASCII. The additional bit is added to increase the number of characters the code can represent. For example, 128 letters, numbers, and special characters are possible with a seven-bit code. To determine the number of possible combinations of ones and zeros, take 2 to the power of the number of bits. In our example, that would be 2 to the power of 7 or 2 multiplied by itself 7 times. The letters in the alphabet must each have a character for lowercase and a character for uppercase. The numbers from 0 to 9, punctuation, and special numeric symbols must all be represented. Eight-bit ASCII can represent 256 characters, or 2 to the power of 8. The additional characters are used for foreign language letters. If one computer is using eight-bit ASCII and the second computer is using seven-bit ASCII, the first 128 characters of both codes are the same. So even if one computer is using seven-bit ASCII and the other is using eight-bit ASCII, the two computers will be able to communicate.

The second most popular code in North America is the extended binary coded decimal interchange code (*EBCDIC*). It is an eight-bit code that IBM developed specifically for its machines. If I were using EBCDIC and my neighbor was using either seven- or eight-bit ASCII, we would not be able to communicate until a translation between the codes was completed. Converting ASCII to EBCDIC, and vice versa, is handled by the computer and is fairly simple.

4.4.3 Transmission Modes—Serial and Parallel

Transmission mode refers to the way data bits travel across the physical medium and defines the pulse sequence of the bits as they are transported. In the world of data transmission, there are two common types of physical interface modes—*serial* and *parallel*.

ASCII
American Standard Code for Information Interchange.

EBCDIC
Extended binary coded decimal interexchange code.

serial
Serial interface carries information one bit after the other.

parallel
Parallel interface carries bits across multiple conductors at one time.

Figure 4–24
A DWDM system using the ITU/
C-band grid. Each wavelength may be
viewed as a channel that carries a
signal. The system shown is carrying
sixteen wavelengths.

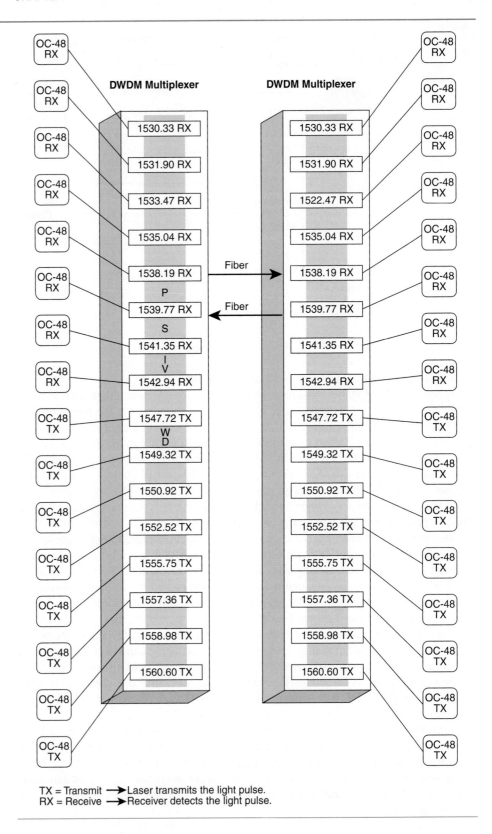

Figure 4–24
A DWDM system using the ITU/
C-band grid. Each wavelength may be
viewed as a channel that carries a
signal. The system shown is carrying
sixteen wavelengths.

TX = Transmit →Laser transmits the light pulse.
RX = Receive →Receiver detects the light pulse.

When information travels across the network in serial fashion, bits are shipped out
onto the medium one at a time. When Henry hit the send key on the computer in Katmandu, his message left the computer by way of the modem interface jack connecting
to the telephone line on the back of the PC. A serial interface means that each signaling change is sent out one after the other. Multiple changes in the signal cannot be sent

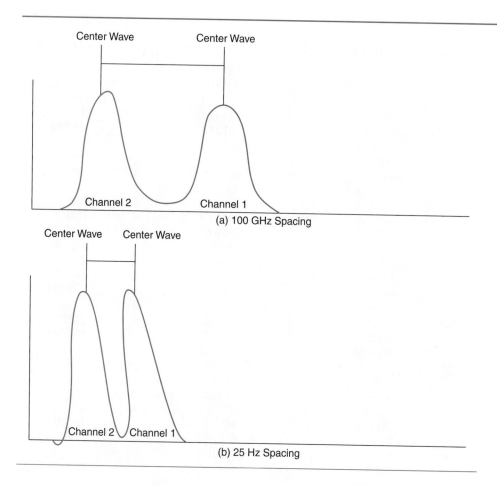

Figure 4–25
(a) Wider-spaced DWDM system.
(b) Narrower-spaced DWDM system.

simultaneously on the same wire. It seems logical to say that only one bit is transmitted on a serial interface at one time, but because of sophisticated modulation schemes that allow for multiple bits to be transmitted every time a pulse is sent, this is not the case. Rather, serial transmission is able to handle one pulse change at a time because there is only one physical connection between the two devices. Figure 4–26 illustrates two devices connected to one another with a serial interface.

The other mode used in data communications is parallel transmission mode. As it sounds, it is a way to send multiple signals over parallel facilities. The best example of a parallel interface is the connection between the printer and the PC. If you were to cut in half the cable that connects the two, you would find eight individual wires. Because the distance between your computer and the printer is relatively short, placing eight individual wires is cost-effective. The eight wires each carry information from the PC to the printer. When we hit the print key, the information flows out of the computer onto the wire eight bits at a time. The bits that make up one character are shipped simultaneously. Figure 4–27 illustrates parallel transmission mode between Henry's PC and the printer. Because parallel transmission can carry eight signals at once instead of only one (as in serial mode), you might wonder why the network doesn't use parallel transmission. The answer is that eight lines would be needed between each person's home and the central office, which would not be cost-effective.

4.4.4 Asynchronous Data Transmission

Asynchronous transmission is used to tell receiving equipment which bit is the first, which is the second, and so on when receiving information. It delineates each of the characters that are shipped across the network. For example, like every other character, the character *H* is preceded by a start bit and followed by a stop bit. The end equipment synchronizes the bit stream according to start and stop bits. In a sense, it is similar to saying "over" during shortwave radio conversations to signal the far end that

asynchronous transmission
Data transmission using stop and start bits to define frames.

Table 4–3
ASCII Code Values.

Symbol	One-Zero Code	Letter	One-Zero Code	Letter	One-Zero Code
!	1000010	A	1000001	a	1000011
"	0100010	B	0100001	b	0100011
#	1100010	C	1100001	c	1100011
$	0010010	D	0010001	d	0010011
%	1010010	E	1010001	e	1010011
&	0110010	F	0110001	f	0110011
'	1110010	G	1110001	g	1110011
)	0001010	H	0001001	h	0001011
(	1001010	I	1001001	I	1001011
*	0101010	J	0101001	j	0101011
+	0011010	K	1101001	k	1101011
,	0011010	L	0011001	l	0011011
-	1011010	M	1011011	m	1011011
.	0111010	N	0111001	n	0111011
/	1111010	O	1111001	o	1111011
:	0101110	P	0000101	p	0000111
;	1101110	Q	1000101	q	1000111
<	0011110	R	0100101	r	0100111
=	1011110	S	1100101	s	1100111
>	0111110	T	0010101	t	0010111
?	1111110	U	1010101	u	1010111
[	1101101	V	0110101	v	0110111
\	0011101	W	1110101	w	1110111
]	1011101	X	0001101	x	0001111
∧	0111101	Y	1001101	y	1001111
—	1111101	Z	0101101	z	0101111
{	1101111	0	0000110		
|	0011111	1	1000110		
}	1011111	2	0100110		
~	0111111	3	1100110		
@	0000001	4	0010110		
'	0000011	5	1010110		
		6	0110110		
		7	1110110		
		8	0001110		
		9	1001110		

Figure 4–26
(a) Serial interface between the laptop and a modem. The serial interface is a two-wire connection that has data transmitted one bit after the other bit across the wires.
(b) Section of a serial interface.

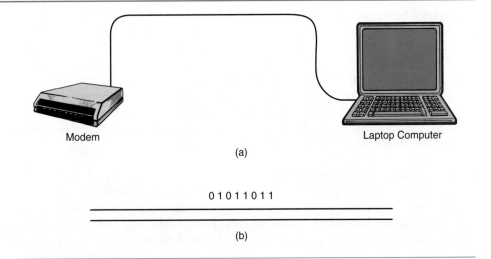

Modem

Laptop Computer

(a)

0 1 0 1 1 0 1 1

(b)

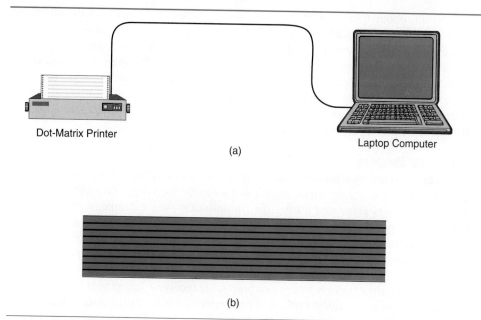

Figure 4–27
(a) Parallel interface between the laptop and the dot-matrix printer.
(b) Section of ribbon cable that provides a parallel interface.

Dot-Matrix Printer

Laptop Computer

(a)

(b)

the message is over. In the case of asynchronous transmission, the stop bit serves the same purpose.

The problem associated with using start and stop bits is the amount of bandwidth used every time a character is sent. In fact, the asynchronous transmission format contains 20% overhead due to the use of stop and start bits. The need to synchronize the bit stream between transmitting and receiving devices is essential. Asynchronous transmission is the most common format used for slower-speed terminals. The maximum bit-per-second rate handled by asynchronous transmission is 9.6 kbps. See Figure 4–28 for an illustration of asynchronous transmission.

4.4.5 Synchronous Data Transmission

Synchronous transmission, like asynchronous transmission, is responsible for synchronizing, transmitting, and receiving equipment to ensure that the information is received in the same order that it is sent. Synchronous transmission does not use stop and start bits to align the bit stream. Instead, synchronous transmission uses a clock source as a reference to align the bit stream. The end equipment uses the same clock as a reference. The use of one clock to reference the start and stop of the bit stream allows the equipment to send characters without individual stop and start bits.

synchronous transmission
Data transmission scheme using clock source to sync up the frames.

Start	Information Bits	Stop	Information Bits	Start	Information Bits	Stop Bit

(a)

Figure 4–28
(a) Asynchronous transmission uses a start bit before each character to organize the bit stream.
(b) Synchronous transmission allows a variable amount of information to be shipped. The signal uses a clock source to keep the end equipment in sync.

Flag	Address	Control Bits	Information	Frame Check	Flag

(b)

Synchronous transmission, as illustrated in Figure 4–28, formats the bit stream with sync bits and flags placed around blocks of data. The block becomes a frame that is shipped intact ready to be read by the receiving equipment. The receiving equipment opens the frame and reads the information inside using the same clock signal as the transmitting equipment. The clock is either embedded in the data stream or it rides on a separate clock circuit. Digital ones and zeros require a timing source in order to move to the same reference. Synchronous systems require a more advanced timing system than asynchronous data systems.

4.4.6 Error Checking

Asynchronous and synchronous transmission formats ensure that receiving equipment is able to synchronize the incoming bit stream to the bit stream being transmitted. If one of the bits is corrupted during the transmission—for example, a one pulse becomes a zero pulse—error checking tells the receiving equipment that the zero should be a one. Error checking is performed continuously on data communications circuits to ensure that the number of ones and zeros being sent is the same as the number being received.

As information flows between two points, the receiving equipment looks at the incoming bit stream, counts the bits, and determines whether the information is correct or corrupted. The equipment receiving the bit stream runs the ones and zeros through an algorithm that calculates the number of odd or even ones in the bit stream. When the bit stream is corrupted, the end equipment sends a *no acknowledgment* (*NAK*) to the transmitting equipment telling it the information was corrupted.

Error checking happens at different levels of the OSI stack, including the physical level that watches the digital bit stream, the data link level looking for errored frames, and higher levels looking for applications errors.

Parity checking is an error-checking scheme that counts the positive and negative pulses to determine whether or not there is a polarity violation. A parity violation is when a positive one pulse follows a positive one pulse or a negative one pulse follows a negative one pulse. Bit streams are formatted as alternate mark inversions (AMIs), meaning that every other pulse is the opposite in polarity. A positive one follows a negative one. Parity checks are looking for polarity violations; when one is found, it means that the information being sent is incorrect and must be retransmitted. There are two types of parity checking: *odd parity* and *even parity*. Odd parity wants to see an odd number of either positive or negative one pulses; even parity wants to see an even number of either positive or negative one pulses. The equipment on the ends are optioned for either odd or even parity, but both parity checking schemes perform the same function.

Cyclic redundancy check (CRC) is another error-checking scheme used to monitor the data stream. CRC characters are appended to a block of data leaving the transmitting equipment. The receiving equipment reads the block and uses the CRC algorithm to calculate whether or not the information has been corrupted. CRC error checking is a robust method that guarantees the circuit is free of errors. A 99.99% validity score is possible when using CRC error checking. This means that 99.99% of what CRC claims is true. If there are no CRC errors, then you can feel confident there are no errors.

4.4.7 Analog Data Transmission

Analog data transmission occurs every time you log onto the Internet using your PC, your modem, and your telephone line. The information from your PC is converted into an analog signal within the modem, similar to a voice signal, and shipped out onto the telephone line. This section deals with the way the analog signal carries data information. A typical telephone line is designed to carry frequencies between 300 and 4000 Hz. Frequencies of 0 to 300 Hz and 3300 to 4000 Hz are reserved as guardbands. The guardbands ensure that the data transmission will not interfere with other signals. The typical data sent through a modem out onto the PSTN are restricted to the frequency range of

odd parity
Error-checking method that counts the bits to make sure they equal an odd number. If they do not, the assumption is that an error has occurred.

even parity
Error-checking method that counts the bits to make sure they equal an even number. If they do not, the assumption is that an error has occurred.

cyclic redundancy check
Error-checking algorithm used to ensure data transmission.

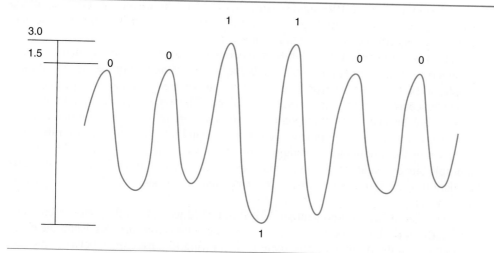

Figure 4–29
Data may be transmitted across an analog signal by manipulating the amplitude of the waveform. This waveform represents a one when the amplitude is 3 V or zero when the amplitude is 1.5 V.

300 to 3300 Hz, so the digital data arriving from the computer must be modulated onto frequencies within this range.

Therefore, the digital signal leaving the computer is converted into an analog signal format before being transmitted out onto the telephone line. The modem converts the digital signal into an analog waveform, and the analog signal carries the one and zero bits across the network. The modem uses various modulation techniques to convert the digital signal into an analog signal. The three most common methods are *amplitude modulation (AM)*, *frequency modulation (FM)*, and *phase modulation (PM)*.

AM is the simplest technique. The amplitude of the waveform is varied to represent either a one or a zero bit. For example, in Figure 4–29 the higher amplitude represents a one and the lower amplitude represents a zero. The modem increases the amplitude of the carrier wave to adjust the voltage value when a one is transmitted, and reduces the amplitude of the waveform when a zero is transmitted. This process is called modulation. The amplitude of the waveform is being modulated.

FM is a second modulation method. In this case, the waveforms' frequencies are modulated when a one or a zero pulse is transmitted. A one pulse is represented by 2400 cycles every second and a zero is represented by 1200 cycles every second. The frequency of the carrier wave is modulated, as illustrated in Figure 4–30. "Frequency shift keying" is the phrase used to describe FM.

amplitude modulation (AM)
Modulation technique that manipulates the amplitude of the signal.

frequency modulation (FM)
Modulation technique that manipulates the frequency of the signal.

phase modulation (PM)
Modulation technique that manipulates the phase of the signal.

Figure 4–30
Data may be transmitted across an analog signal by manipulating the frequency of the waveform.

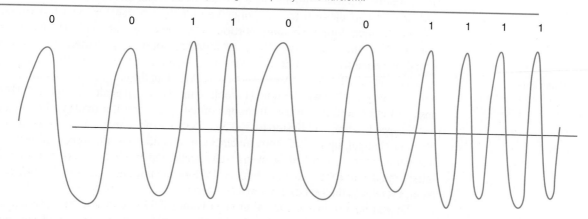

Phase modulation (PM) manipulates the phase of the waveform. For example, a one is represented by a 180-degree phase shift in the waveform. The phase of the carrier wave is modulated to represent the digital bit.

To determine the number of bits per second carried by an analog signal, we need to understand the term *baud*. Baud stands for the signaling rate of the circuit. For example, a 2400-baud-per-second circuit means there are 2400 signaling changes or changes occurring in the waveform every second. The two terms are often used interchangeably, but they are not the same. The type of circuit, the physical limitations of the circuit, and the end equipment determine the baud rate of a link. The bit rate is determined by the modulation technique performed by the equipment. The ability to send multiple bits per baud is how a normal analog line is able to carry a 56 kbps signal. The modem modulates multiple bits per baud rate and in so doing increases the bits-per-second rate.

The first modems used straight AM, FM, or PM techniques to modulate bits onto a waveform. Each method was limited to no more than a four-bits-per-baud rate every second due to the difficulty encountered when trying to differentiate between the two states. When two modulation techniques are combined, the number of bits per baud increases dramatically. *Quadrature amplitude modulation* (QAM) is an example of combining modulation techniques to increase the number of bits per baud. QAM combines PM and AM to produce a four-bit-per-baud signal. *Trellis-coded modulation* (TCM) allows up to eight bits per baud. An analog data line is not 2400 bps or 56 kbps because the frequency range of the circuit has been increased or changed. Instead, modulation techniques have been perfected to allow multiple bits to be transmitted on every baud.

The amount of data the analog data circuit is able to carry depends on the type of modem on each end, the length of the circuit from the subscriber to the central office, the cable gauge, and the total impedance of the line. Subscribers often purchase 56 kbps modems and then complain when they are only able to transmit 33 kbps or less. Unfortunately, there is little a telephone company can do to increase the speed of the modem.

Early data communications circuits used specially designed, dedicated, point-to-point analog circuits that, at the time, provided a more reliable, faster means to transmit data. These dedicated data lines are now obsolete, thanks to new digital line technologies. Business customers still use the existing analog data lines, mainly because of the cost to upgrade the end equipment. Unfortunately for the telephone company, the circuits are difficult to maintain.

4.4.8 Digital Data Transmission

We live in a digital age. The telecommunications network has evolved into a digital network, changing the way we all communicate. First, trunking circuits were converted from analog carrier lines into T1 digital circuits. Next, analog switches were converted into digital switches while the size and price of the digital computer reduced, allowing for mass distribution of data to business and residential customers. Following a voice call or data call through the network, we quickly see that for most residential customers, the local loop is the only portion of the network that continues to use an analog signal. To understand why the local loop has not been converted to carry a digital signal, we need to find out how a digital data signal is transmitted across the network.

The first digital circuit was deployed by AT&T in the early 1980s. The *digital dataphone service* (DDS) was designed to replace the dedicated analog circuits used by business customers to transmit data from one office to another. It provided a link that could handle speeds from 2.3 kbps up to 56 kbps. Because older analog circuits were only capable of carrying up to 19.2 kbps under the best conditions, the trend was established. Business owners soon asked for even faster speeds, and T1 circuits were deployed to help satisfy the desire for bandwidth. Finally, ISDN was introduced as a POTS line replacement. ISDN was the first digital-switched service.

Digital data transmission, whether it is an ISDN line or a DDS, requires special terminating equipment to format the bit stream before it is transmitted out onto the

quadrature amplitude modulation
Modulation technique that manipulates the amplitude of a signal to increase the number of bits shipped per signal change.

trellis-coded modulation
Modulation technique that manipulates the phase and amplitude of the signal to increase the number of bits shipped per signal change.

digital dataphone service
An AT&T copyrighted term to define a digital data service.

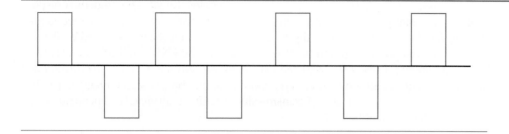

Figure 4–31
Alternate mark inversion describes a waveform that alternates between a positive pulse and a negative pulse. This waveform depicts an AMI digital signal.

line. In place of a modem, the dedicated digital line requires a CSU/DSU and the ISDN line requires an NT-1. A dedicated DDS digital circuit's CSU accepts the one and zero bit stream from the computing device. The signal is converted from a unipolar signal, in which all of the pulses are of the same polarity, to an alternate mark inversion (AMI) format, as shown in Figure 4–31. The alternating one and zero signal is transmitted out onto the network at the speed set by the end equipment. For instance, if a PC is set to transmit 56 kbps, the DSU receives the signal and outpulses it onto the copper pair. If the PC ships a 9.6 kbps signal, the CSU receives the signal and outpulses the bits onto the digital circuit.

An ISDN circuit is similar to a DDS circuit. The significant differences are that an ISDN circuit is able to carry two 64 kbps signals and a 16 kbps signal and interface into the class 5 switch. The unit used to interface an ISDN line is called an *NT-1*. The NT-1, like the CSU, is responsible for the bit stream format as it is transmitted onto the physical circuit. ISDN was built upon a *2B1Q* line-code format that allows two bits to be transmitted for each pulse shape, thereby producing the ability to send more bits every second. The 2B1Q line-code format carries two bits per signal change compared with the AMI line-code format that carries one bit per signal change. The difference between the two line-coding schemes is shown in Figure 4–32.

NT-1
ISDN termination device.

2B1Q
Line-code format used to encode ISDN circuits and DSL circuits.

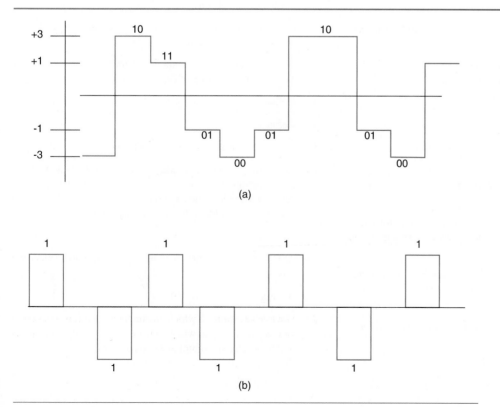

Figure 4–32
(a) The ISDN BRI signal is a quarternary signal that increases the number of bits per pulse to two showing the 2B1Q line code.
(b) Illustrates one pulse per signal change and each alternate pulse occurs at the opposite polarity.

To review, an analog data circuit carries an analog signal. The modem is responsible for modulating the signal onto the circuit and determining the bit-per-second rate. The digital data circuit carries a digital data signal that has been formatted by the end device. The two most common types of devices are the CSU, used for dedicated circuits, and the NT-1, used for switched ISDN circuits. Both the analog data circuit and the digital data circuit are able to carry more than one bit per signal change. The bit-per-second rate for both types of transmission circuits is ultimately dependent on the quality of the physical link.

SUMMARY

Chapter 4 covers fundamental concepts in communications. The primary purpose of the chapter is to explain how information is manipulated into a format that can be placed onto a medium and carried to the end destination. Simply, the chapter discusses how the signal, whether it be data or voice, is prepared and packaged according to standard communications methodologies. The key topics presented in the chapter are modulation techniques, signal conversion—analog to digital and digital to analog—and multiplexing methods. In addition, the chapter highlights data communications techniques used to establish and carry information between computing devices such as two computers, a computer and a printer, and so forth.

REVIEW QUESTIONS

1. Why is it necessary to convert an analog signal into a digital signal?

2. List and explain each of the steps involved in converting an analog signal to a digital signal.

3. List and explain each of the steps involved in converting a digital signal to an analog signal.

4. When a voice signal is converted into a digital signal, do the ones and zeros of the digital signal represent letters?

5. Explain what is meant by *PAM*.

6. Explain what is meant by *PCM*.

7. Explain why analog waveforms are sampled 8000 times every second.

8. What portion of the telephone network still depends on analog signals?

9. Draw and define the three encoding schemes used to format a digital signal—unipolar, bipolar nonreturn to zero, and bipolar return to zero.

10. Explain the ones density rule defined by the FCC and define the two encoding schemes used to solve the all-zeros dilemma.

11. Draw a $+6/-6$ Vpeak-to-peak digital waveform using a bipolar return to zero format.

12. Why do digital signals require timing sources?

13. What is the purpose of placing a multiplexer at the end of a circuit?

14. Explain how a frequency division multiplexer multiplexes lower-rate signals onto one link.

15. Explain how a time division multiplexer multiplexes lower-rate signals onto one link.

16. What advantage does a statistical multiplexer have over a TDM multiplexer or an FDM multiplexer?

17. Define *modulation*.

18. Define *carrier wave*.

19. Why is it necessary to modulate a signal onto a carrier wave?

20. The term *asynchronous data* indicates that the bandwidth in one direction (bandwidth from the Internet to the customer versus bandwidth from the customer to the Internet) is greater than the bandwidth in the opposite direction.

    a. true   b. false

21. ASCII and EBCDIC are binary information encoding schemes.

    a. true   b. false

22. List two situations in which you might use a serial connection and two situations in which you would use a parallel connection between two communications devices.

TROUBLESHOOTING

A channel card in a channel bank in the Ontario central office has stopped working. The circuit interfacing the card has gone into alarm. What are the steps you should take to correct the problem?

ABC uses Spidercom's data network to transport intracompany information between New York City and Los Angeles. Lately, the network manager has noticed that numerous retransmissions are occurring on that link. List five possible reasons why the information has to be retransmitted between the two sites.

KEY TERMS

modulation (92)
steady state signal (92)
message signal (93)
carrier signal (93)
modulator (93)
demodulator (93)
baud rate (94)
constellation (95)
symbol (95)
modulation alphabet (95)
rate adaptive (96)
digital signal (97)
BPNRZ (97)
BPRZ (97)
pulses (97)
voltage state (97)
pulse mask (97)
polarity (97)

B8ZS (99)
AMI (99)
simplex (99)
half duplex (99)
full duplex (99)
PAM (101)
quantizing (101)
PCM (102)
encoding (102)
toll quality (102)
ADPCM (103)
channel (105)
multiplexing (106)
FDM (106)
TDM (108)
conversion (108)
SONET (108)
DS-3 (109)

DS-1 (109)
DS-0 (109)
fast packet (111)
X.25 (112)
ATM (112)
IP (112)
FR (112)
DWDM (113)
CWDM (113)
ASCII (115)
EBCDIC (115)
serial (115)
parallel (115)
asynchronous
  transmission (117)
synchronous
  transmission (119)

odd parity (120)
even parity (120)
cyclic redundancy
  check (120)
amplitude modulation (AM)
  (121)
frequency modulation (FM)
  (121)
phase modulation (PM) (121)
quadrature amplitude
  modulation (122)
trellis-coded
  modulation (122)
digital dataphone
  service (122)
NT-1 (123)
2B1Q (123)

5

Open System Interconnection

Objectives

After reading this chapter, you should be able to

- Define the OSI—Open System Interconnection—model.
- Associate the OSI layers with networking protocols.
- Relate the OSI layers to end-to-end communications.

Outline

Introduction

5.1 Open System Interconnection Architecture: View of the Entire Stack

5.2 Linking the Layers through Real-Life Applications

■ INTRODUCTION

The Open System Interconnection (OSI) model is the communications industry protocol standard defining how data should be packaged and transported across a circuit. The seven-layer OSI stack is used by equipment vendors, Internet service providers (ISPs), telephone service providers (TSPs), cellular service providers, and cable television (CATV) service providers when designing and implementing telecommunications networks. The goal of Chapter 5 is to relate the OSI layers to the network function and explain how one layer depends on the rest to provide end-to-end communications services.

■ 5.1 OPEN SYSTEM INTERCONNECTION ARCHITECTURE: VIEW OF THE ENTIRE STACK

As shown in Figure 5–1, the OSI model consists of seven layers stacked one on top of the other, similar to a seven-layer cake. The seven-layer architecture defines how end-to-end communication processes function starting at the physical medium and ending with the presentation of the application. In real-world terms, this relates sending one and zero pulses on a copper line to receiving an e-mail from a friend. The goal of this section is to associate the OSI model to common communications protocols used in the industry. Normally, as soon as the term *OSI* is uttered, eyelids start to droop. Because it is difficult to avoid, it is important to understand the significance of the OSI model in communications. Industry experts commonly refer to the seven-layer stack during network design discussions, troubleshooting exercises, and so forth. Therefore, drink a glass of your favorite caffeinated beverage and read on.

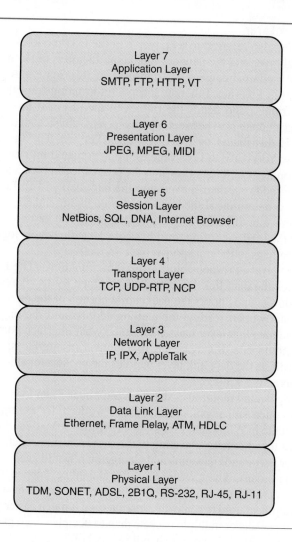

Figure 5-1
Open System Interconnection protocol stack.

5.1.1 The Physical Layer

Before we explain each layer of the model, it is important to define the term *encapsulation*. Each of the layers manages the flow of information by packaging the one-zero bit pattern into organized groups also called *frames, packets,* or *cells.* Each layer encapsulates the previous layer's groups into its own package. The process compares to placing letters inside a box, then "encapsulating" that box inside a second box, and so forth. Therefore, when the term *encapsulation* is used in communications, don't panic. It simply refers to packaging information together while keeping the previous layer's protocol intact. The same is true in the reverse: The box is opened and the contents spilled out onto the floor for inspection.

The physical layer's job is to format the signal according to mechanical, electrical, and functional parameters associated with the type of signal and medium deployed. For example, the amount of voltage required to represent a one pulse on the line is defined by the physical layer. The mechanical structure of an RJ-45 modular jack and the termination pins are defined by the physical layer. The modulation scheme, error-correction schemes, and the alarming method are tied back to the physical layer standards. DS-1, DS-3, and SONET frame formats reside at the physical layer, as do the DSL, Wi-Fi, OFDM, and DOCSIS modulation schemes. The physical layer defines the highway, the speed limit, and the physical shape of the vehicle.

5.1.2 The Data Link Layer

The data link layer, also referred to as Layer 2 in the OSI model, is the interface between the physical layer and the network link layer. Its job is to tie the network protocol to the physical medium being used to transport the information. A good example of a Layer 2 protocol is Ethernet. Ethernet accepts information from the network layer, packages it into frames, and hands the frames to the physical layer to place on the medium. The intermediate step between the two layers, physical and network, is similar to how mail is distributed by the U.S. Postal Service. The thousands of letters shipped from New York City to Los Angeles are aggregated into one mailbag before being shipped. Hundreds of letters flow into the post office in New York, are sorted according to mailing address, then shoved into the bag marked "To Los Angeles." The postal workers next determine if the bag needs to go by air and arrive the next day or go by train and arrive in three days—priority mapping. The key role the postal workers play is to sort the mail according to address, package the letters together in a bag or bags, and hand the bags to the transport vehicle for delivery.

The data link layer is responsible for the same functions. The Layer 3 protocol, such as IP, handles the data link layer packets assigned by IP address. The data link layer reads the address and shoves the packet or a portion of the packet into a frame—similar to the postal worker reading the mailing address and shoving the letter into a mailbag. Next, the data link layer has to determine what interface port or transport vehicle the frame should ride. It may be any of a number of interfaces, such as a T1 or a GigE port. Regardless, the data link layer bundles the packets together and hands them to the physical layer, the vehicle.

The reasoning behind requiring a step between Layer 3 and Layer 1 was to streamline the process at intermediate points along the network. If, for example, you shipped a letter from your home in Wilawanna, Texas, to Georgetown, Maryland, and postal workers did not bundle together letters heading in the same direction, each postal worker receiving your letter between Wilawanna and Georgetown would need to read the address on the envelope in order to know where the letter should be shipped. Obviously, this is not an efficient way to handle mail delivery. The same is true for data packets. Routers and switches, equipment used to route information, find it much easier to look at a frame containing partial packets or multiple packets than to decipher each packet individually. On the physical layer side, it is easier to handle a box holding multiple letters than shoving individual piece parts of a packet—or, in our analogy, letters—on a medium as shown in Figure 5–2.

5.1.3 The Network Layer

Layer 3 has multiple responsibilities. It assigns source and destination IP address, as, in our analogy, placing a mailing address and return address on the envelope. Layer 3 is also responsible for controlling the flow of the packets being shipped in order not to bombard and overload the network. Another task it has to deal with is to determine the best link to use at a given time. Layer 3 must also figure out from information it receives from Layer 2 what port to pull the packets from or, in the reverse, what port to send the packets on to. Finally, the network layer works with Layer 4 to ensure that the packets have arrived in the correct order and in a timely manner. In summary, the network layer is responsible for routing of packets, flow control, translating interface port information from Layer 2, and assisting packet delivery.

Using our U.S. Postal Service analogy, similar to the postal worker, Layer 3 is smart enough to know how to route the envelope according to the mailing address. In our example, the Layer 3 protocol is IP; therefore, mailing address compares to IP address. The postal worker reads the ZIP Code to determine if the envelope has to be shipped to a different post office. Using the same logic, the router looks at the network portion of the IP address to determine on what interface port the packet should be placed. IP addressing is discussed in detail in a later chapter. The key point to grasp is that the network layer handles internetwork routing through the use of addressing schemes.

Figure 5-2
Data flow is similar to how the post office handles mail.

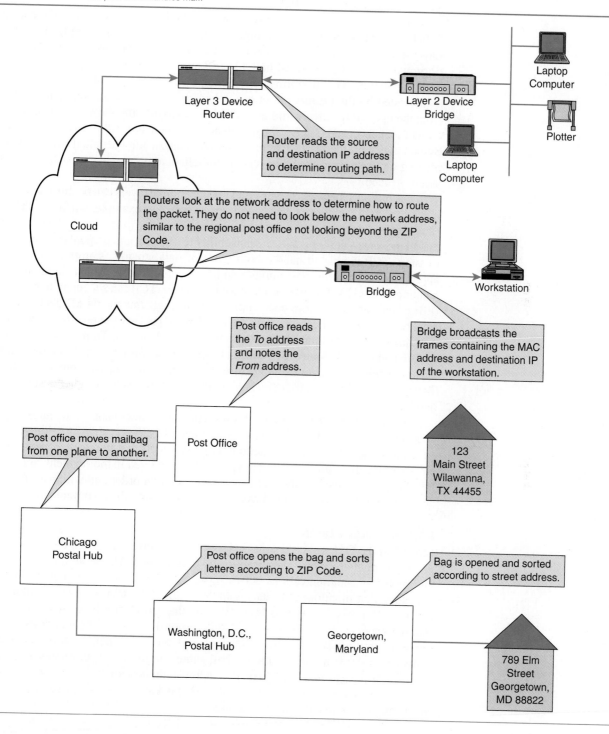

5.1.4 The Transport Layer

The transport layer is Layer 4 in the OSI stack. It sits between the network layer and the session layer. The term *transport* should not be confused with transport systems or transmission systems as these terms are used to describe SONET optical systems and other physical layer media. *Transport layer* in the OSI stack refers to the layer responsible for interfacing with the higher-order layers—session, presentation, and application—and the lower-order layers—network, data link, and physical. Common

transport
Refers to networks used to carry information.

transport layer
Fourth layer of the OSI stack responsible for establishing and maintaining information flow from device to device.

Layer 4 protocols include Transaction Control Protocol (TCP) and User Datagram Protocol (UDP) as defined by REF 768 STD 6.

The transport layer is responsible for setting up and maintaining the end-to-end flow between host devices. Data fragments from the upper layers are formatted by the Layer 4 protocol before being handed to Layer 3. The relationship between TCP and IP provides a good example of how Layer 3 and Layer 4 interface. The four most important functions TCP performs are defining the type of application being transmitted as indicated by the destination port number, bundling the data fragments it receives from the upper three layers into usable chunks, making sure the segments arrive in the correct sequence, and transmitting an acknowledgment or reading an acknowledgment. Some Layer 4 protocols, TCP in particular, also retransmit the data if necessary by employing a robust error-checking scheme to help determine when a packet retransmission is needed.

The interface between Layer 4 and the upper three layers can be illustrated by looking at how TCP handles multiple application types. For example, when two PCs on the same network transmit two different applications—for example, an FTP file and an e-mail message—it is Layer 4's responsibility to segment and prepare the data for transmission. TCP is able to handle both types of data streams, stamping each with the appropriate destination port number and segmenting each into usable chunks before handing it off to the Layer 3 protocol for packaging. TCP can accept up to 64 kilobytes (kB) of data at a time from each application. It fragments the 64 kB data chunks into smaller segments that comply with the standard IP packet. The end device looks at the port number to determine the application type, such as File Transfer Protocol (FTP), e-mail, and so forth. Many ISPs control traffic coming into their network by opening or closing specific ports. For example, throttling peer-to-peer traffic protocols such as Bit-Torrent is accomplished by rate shaping any incoming data with that destination port number.

A detailed explanation of TCP is presented in a later chapter. The most important concept to understand at this point is that Layer 4 is responsible for taking data from an application such as FTP, formatting it into segments that fit into standard-size packets, stamping the segment with the type of data contained in the segment, and ensuring the packet is received by the end device in the correct order and within a timely manner. Figure 5–3 illustrates how Layer 4 protocols fit into the communication flow.

5.1.5 The Session Layer

Layer 5 of the OSI stack, the session layer, is the interface between the transport layer and the upper two layers—application and presentation. The session layer establishes and monitors communications between two end devices. It is the behind-the-scenes watchdog that monitors the data stream to ensure both devices are still talking at the application level. The session layer converts the application layer address into a format the transport layer can read. It is also responsible for setting end device communications settings such as the number of data bits, port, parity, stop bits, and flow control. Similar to a referee at a baseball game, the session layer knows the rules and forces both sides to adhere to them. For example, the session layer determines what actions happen when the user presses the CTL break key. Two common session layer protocols include NetBIOS and IBM's SNA.

5.1.6 The Presentation Layer

The presentation layer—Layer 6—is responsible for translating and presenting the data to the user in a readable format. The presentation layer has three primary tasks. First, if necessary, it converts the data syntax into a format readable by the end device. For example, if the device is configured to read EBCDIC coded text, it is at the presentation layer that the file is translated from ASCII into EBCDIC.

A second function performed at the presentation layer is formatting data into files that comply with industry standards such as MPEG, Motion Pictures Expert's Group, used to format video streams; JPEG, Joint Photographic Experts Group, used to format

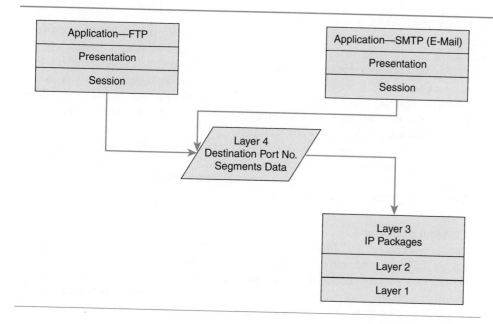

Figure 5–3
Applications are formatted by the Layer 4 protocol before being handed to Layer 3. Note how Layer 4 is able to multitask, accepting multiple application types.

graphical files; and MIDI, Music Instrument Digital Interface, used to format music files.

The third job the presentation layer tackles is encrypting and compressing the data. Encryption is performed to ward off intruders, and compression techniques are deployed to help reduce the amount of bandwidth needed on the network.

5.1.7 The Application Layer

The application layer is the seventh and final layer of the OSI stack. The end user indirectly interacts with the application layer when creating a document. Application layer protocols take the information created by the user and package it in a format acceptable to the lower-level protocols. In the reverse, the lower-level protocols hand the data segment to the application layer where it is unwrapped and presented to the end user through the computing program. The job the application layer performs is essential to sending and receiving information across a network. Common application layer protocols include SMTP, Simple Mail Transfer Protocol; Virtual Terminal; and HTTP, Hypertext Transfer Protocol, just to name a few.

The application layer of the network is often confused with the computer software program such as Microsoft Word. Our description of the application layer relates to networking and as such involves the process of interfacing with the computing program. Protocols such as FTP provide good examples of the roles and responsibilities of this layer. It formats the data into standard segments that are readable by the lower layers. It also is smart enough to read the file and work with the computing program to display the data accordingly.

■ 5.2 LINKING THE LAYERS THROUGH REAL-LIFE APPLICATIONS

Often, reading the individual definitions of each layer causes confusion as to what is responsible for what. For example, most OSI descriptions tell the reader that the session layer is responsible for end-to-end flow, then note in the next paragraph that the transport layer is responsible for end-to-end flow. Though both are correct statements, the differences in what "end to end" refers to need to be clarified. The best way to differentiate between what is doing what is to walk through a networking example that encompasses all seven layers.

5.2.1 Layer 7 to Layer 5

Henry is ready to send an e-mail to his friend Hose, as illustrated in Figure 5–4. He clicks on the *To* field in his e-mail program and types in Hose's e-mail address: Hose@Spidercom.com. He then clicks on the paperclip to attach a picture of his 1968 Ford Mustang that he took with his digital camera and downloaded onto his PC's hard drive earlier that day. He formatted the picture in JPEG to ensure that Hose would be able to display it on his machine. Before clicking on *Send*, Henry checks to make sure the file is attached and his message is complete.

The e-mail is packaged as data using the application layer SMTP and the JPEG file format, a presentation layer protocol. The data is received by the Layer 5 session layer protocol that acts like the program-to-program referee and organizes the segments before handing them to the transport layer.

5.2.2 Layer 5 to Layer 4

At the transport layer, the Transaction Control Protocol—TCP—takes the segment defined by Layer 5 and adds a header containing the port number, error-checking schemes, flow control checks, and sequence numbers. The TCP is similar to the postal carrier that sorts the mail according to house address. If the address is not correct, the TCP sends a message back to the source telling it to retransmit the message. The TCP is explained in detail in a later chapter.

5.2.3 Layer 4 to Layer 3

Segments of data carrying the TCP header are handed to the Layer 3 protocol for network routing. Internet Protocol—IP—divides the segments into packets and attaches an IP header that holds the source and destination address of the *from* and *to* devices; the device sending the information is referred to as the *source*, and the device receiving the information is referred to as the *destination*. The destination address is obtained by converting the e-mail address into an IP address through DNS resolution (a topic

source
Sending device.
destination
Receiving device.

Figure 5–4
Following the path of an e-mail sent from Henry to Hose.

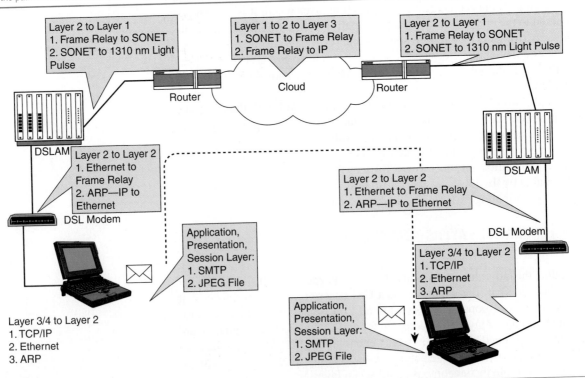

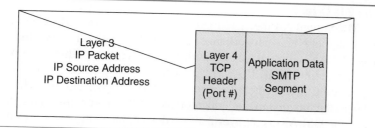

discussed in a later chapter). The destination IP address resolves to the device's Layer 2 MAC address through ARP (a topic also explained in a later chapter). The final result is a *datagram,* a term that defines a chunk of data that contains the TCP header, the IP header, and the data, as shown in Figure 5–5.

datagram
Chunk of data that carries the TCP header, the IP header, and the data.

5.2.4 Layer 3 to Layer 2

We now have our datagram, which contains an IP source and destination address, and a TCP port number, which defines the type of traffic the packet contains and a way to ensure that the packet is received at the far end. We also have upper-layer controls that establish the session between the end devices allowing the users to communicate. The next step in the process is to fragment the packets into frames. In our example, the Layer 2 protocol being used to hold the Layer 3 packet is 802.3, also known as Ethernet. The datagram—similar to multiple envelopes—is shoved into a frame—similar to individual envelopes being shoved into the mailbag—and prepared for its trip across the network. The frame header is attached to the packet, creating a frame containing the SMTP segment, the TCP header, the IP header, data, and the frame header, as shown in Figure 5–6.

5.2.5 Layer 2 to Layer 1

Henry interfaces to the Internet through a DSL broadband connection. Henry's PC Ethernet port is tied to the DSL modem LAN port via an Ethernet CAT5 cable. The frame is injected onto the Ethernet cable as $+3/-3$ V pulses sent in a bipolar format, that is, a positive pulse followed by a negative pulse. At the modem, the frame is bused to the WAN—wide area network—interface that ties the DSL modem to a copper loop tied to the serving central office. ADSL—asymmetric digital subscriber line—is the physical layer modulation scheme used to manipulate the bits into discreet frequencies in the same way your voice signal manipulates the current on a telephone line.

The one-zero patterns travel down the copper wire to the central office where they interface with the DSLAM, a device used to terminate DSL lines. The DSLAM demodulates the signal, looks at the frame header, and routes the frame to the outgoing network port according to the header information. The DSLAM performs the conversion between the physical layer protocol—ADSL—and the data link layer—Ethernet.

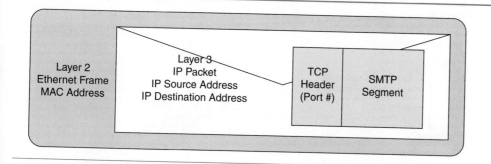

Figure 5–6
Simple view of how information is packaged across the OSI layers.

5.2.6 Layer 2 to Layer 2

The DSLAM demodulates the signal into one and zero frames identical to those sent by Henry's Ethernet NIC—network interface card. In the world of broadband, a tag referred to as a VLAN—Virtual LAN—is used to differentiate traffic flows between devices. In our example, the DSLAM reads the VLAN tag to determine the egress or outgoing port the frame needs to be shipped on. The frames carrying Henry's e-mail are placed onto the link tied to the Spidercom network router sitting at the POP—point of presence. The DSLAM is a Layer 2 aware device, meaning that it uses the Layer 2 protocols to figure out how to handle the traffic flows. A DSLAM, as with all networking devices, has ingress and egress ports, a topic discussed in detail in a later chapter. One interface port faces the DSL modem, as shown in Figure 5–7, and the other faces the network. In most scenarios, the Ethernet frame arriving from the DSL modem is converted into either a frame relay frame or an ATM cell and shipped out on the interface port tied to the network. Frame relay and ATM, Layer 2 protocols, handle large streams of data more efficiently than Ethernet and, as a consequence, are used as the aggregation protocol between the central office and the POP.

5.2.7 Layer 2 to Layer 1

The DSLAM is also responsible for converting the Layer 2 frame relay frame or ATM cell into a Layer 1 aware frame, such as a SONET or DS-1 bit stream. In either case, the DSLAM's network port takes the Layer 2 frame and packages it into a physical layer format. In our example, the DSLAM's physical layer port on the network side is an OC-3 SONET link. The port first converts the electrical signal into an optical signal and outpulses the bits onto the fiber optic line using a 1310 nm light signal. If a DS-1 circuit ties the DSLAM to the POP's network router, the DSLAM formats the frame into DS-1 time slots and outpulses the bits as voltage states, not light pulses. In either case, the DSLAM is responsible for manipulating the traffic into a format acceptable by the physical medium and readable by the end device. Figure 5–8 shows how the four layers relate to the protocols used in networking today.

5.2.8 Layer 2 to Layer 3

The router at the POP also has an OC-3 interface that receives the frame relay frame carrying Henry's e-mail. The interface first converts the optical signal into an electrical signal and strips off the physical layer frame information, reads the frame relay header information, and determines how to handle the data. It then strips off the Layer 2 frame, cracks open the packet, and reads the source and destination IP

Figure 5–7
Traffic flows from the subscriber through the DSL modem as a Layer 2 frame. The DSLAM repackages the Ethernet frames as Layer 2 frame relay or ATM cells. Note the value of being able to remove one type of protocol such as Ethernet and repackage the Layer 3 packets into a different Layer 2 protocol such as ATM.

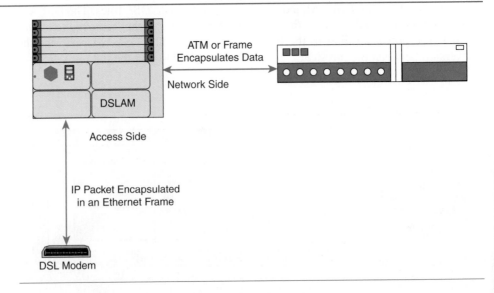

Figure 5–8
First four layers are responsible for formatting the data for transport.

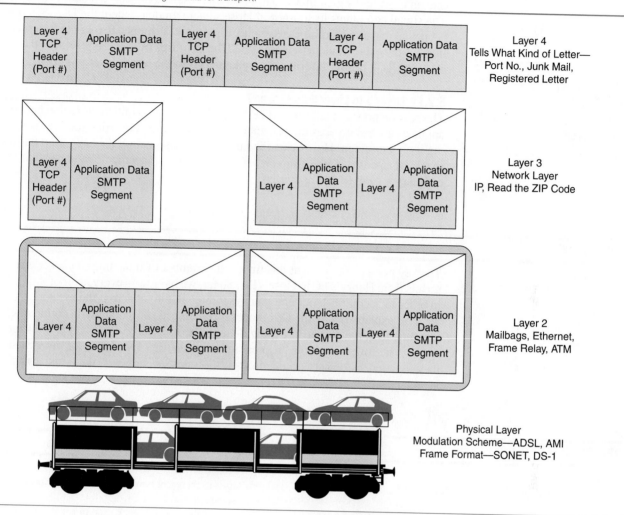

address. After looking up the IP address in its routing tables, the router routes the packet to the correct interface as determined by the source and destination IP address. In our example, the router looks up the address in its routing tables to determine what interface port is tied to Spidercom's core network. It needs to be noted that every routing device the signal passes through must crack the packet as explained here in order to route the packet.

5.2.9 Layer 1 to Layer 2 to Layer 3
The core network port on the router encapsulates the packet into an ATM cell, shoves the cell into a SONET STS frame, and outpulses the one-zero patterns onto the fiber optic link using a 1550 nm wavelength. The far end router receives the light signal, converts the one and zero bit pattern into the SONET frame structure, unpackages the ATM cell, opens the Layer 3 packet, and determines what interface port the packet should be routed toward.

5.2.10 Layer 2 to Layer 1
Hose's broadband connection is provided by the local cable television provider. The router sits at the CATV provider's head end where it interfaces with the outside plant network tied to the serving customers. The router encapsulates the packet into the DOCSIS frame and ships the one and zero patterns onto the optical port tied to a

distribution point in the outside plant. The distribution point converts the optical signal into an electrical signal that is outpulsed onto the coaxial cable feeding Hose's street. The signal completes its journey at Hose's cable modem where the DOCSIS protocol is opened revealing the Ethernet frame. The frame is shipped across the Ethernet cable that ties the cable modem to Hose's PC. The PC strips off the frame, cracks the packet, reads the TCP header information, and hands it to the upper-layer protocols.

5.2.11 Layer 5 to Layer 6 and Layer 7

The session protocol talks to SMTP Layer 7 protocols to initiate a session between the applications and invokes the presentation protocol into converting the file and displaying it on the screen. Hose hears the familiar jingle signaling that an e-mail has just arrived. He opens the mail from Henry, clicks on the attachment, and opens the picture of Henry's 1968 Mustang convertible.

SUMMARY

The key point to focus on at this time is the number of transitions a packet makes as it travels from Henry's PC to Hose's PC. So far, we have used every layer of the OSI stack in order to accomplish our end goal—sending an e-mail across the network. Now that we have explained why the OSI model is referenced so frequently in communications, it is time to delve deeper into individual protocols deployed in today's converged networks. The following three chapters focus on the first four layers of the OSI stack as they function as the networking protocols—transport, network, data link, and physical layers.

REVIEW QUESTIONS

1. What purpose does the OSI stack perform?

2. What is encapsulation?

3. What are the advantages of organizing the functions of the OSI module into layers?

4. What functions reside under the application layer?

5. What functions reside under the presentation layer?

6. What functions reside under the session layer?

7. What functions reside under the transport layer?

8. What functions reside under the network layer?

9. What functions reside under the data link layer?

10. What functions reside under the physical layer?

11. If you were asked to set up a routing protocol between five network routers, what layer or layers would you be working under?

12. Manipulating the voltage on the line falls under the data link layer.

    a. true   b. false

13. TCP and UDP are:

    a. layer 1 protocols used to identify the state of the ones pulse

    b. used to represent how the information is formatted on the screen

    c. part of the Layer 3 stack used to route packets between devices

    d. part of the transport layer responsible for end-to-end transmission of information

    e. none of the above

14. List two Layer 4 protocols.

15. List two Layer 3 protocols.

16. List two Layer 2 protocols.

17. List two Layer 1 protocols.

18. Explain the difference between the data link layer and the network layer.

19. A Layer 1 protocol:

    a. defines the physical characteristics of the medium

    b. defines the functional characteristics of the signal

    c. is a technique used to modulate the signal

    d. defines the frame format of the signal

    e. a, b, and d only

    f. a and b only

    g. a, b, and c only

    h. none of the above

20. The session layer establishes a session between devices.

    a. true   b. false

21. The Windows programs falls under the application layer of the OSI stack.

    a. true   b. false

22. A Layer 2 and Layer 3–savvy networking person will be knowledgeable in _____ protocol.

    a. SONNET   e. all of the above
    b. IP   f. a, b, and c
    c. IPX   g. b, c, and d
    d. Ethernet

23. What is the difference between end-to-end node device communications and end-to-end node device process communications?

KEY TERMS

transport (129)   destination (132)
transport layer (129)   datagram (133)
source (132)

6

Layer 3 and Layer 4 Networking and Transport Protocols: TCP/IP, UDP/IP, RTP/UDP/IP, IPX/SPX, DDP/ATP

Objectives

After reading this chapter, you should be able to

■ Define the relationship between TCP and IP.

■ Define TCP as it relates to networking.

■ Define UDP as it relates to networking.

■ Define RTP as it relates to VoIP.

■ Define IP as it relates to networking.

■ Define "other" Layer 3 and 4 protocols.

Outline

Introduction

6.1 TCP: Transmission Control Protocol

6.2 UDP: User Datagram Protocol

6.3 RTP: Real-Time Protocol

6.4 Clarification of Terms

6.5 IP: A Layer 3 Network Protocol

6.6 Other Layer 3 Protocols

■ INTRODUCTION

TCP
Transmission Control
Protocol: a Layer 4 protocol.

Whenever we hear the name Ginger Rogers, we instantly think of Fred Astaire. The same is true when we hear the term *TCP*; we instantly think IP. Like Ginger Rogers and Fred Astaire, TCP and IP work together as a protocol suite to package, prepare, and transport information across a network from device to device. One without the other is unnatural. Of course, Fred Astaire also danced with other partners throughout his career, for example, Debbie Reynolds. IP also partners with other protocols when needed—the most common being UDP, User Datagram Protocol, which is used when real-time applications such as digitized voice or peer-to-peer traffic are carried across the Net. TCP, like Ginger Rogers, may also have different dance partners, though IP is the most common.

The main point to grasp from this analogy is that in order to carry traffic across a network, whether the network is a small 5 PC LAN or the World Wide Web, Layer 4 and Layer 3 protocols have to work in unison and the two dominant partners are TCP/IP and UDP/IP. Chapter 6 defines the structure of the three protocols—TCP, UDP, IP—and explains how each relates to networking functions, specifically, how they work together to carry information end to end.

■ 6.1 TCP: TRANSMISSION CONTROL PROTOCOL

Thanks to the Internet, TCP/IP has become the most common protocol suite used to carry information across networks. TCP/IP is used to carry e-mail, file downloads (FTP), Web site exchanges, and on and on. TCP's claim to fame is as a connection-oriented protocol capable of handling multiple sessions simultaneously. It also has a flow-control mechanism, employs an error-correction scheme, and is smart enough to retransmit corrupted or lost packets. The following few paragraphs define TCP as it relates to networking.

6.1.1 TCP: A "Reliable" Protocol

Before dissecting the piece parts of the TCP, it is important to understand the role TCP plays in data transmission. First, TCP accepts large amounts of data from the application software and organizes it into bytes of data called *segments*. TCP is a connection-oriented protocol, meaning that the protocol establishes a connection or session between the source and destination devices prior to sending the data segment. As shown in Figure 6–1, the method used to establish the connection between the devices starts with the source host negotiating a connection with the destination host, synchronizing the two devices, and acknowledging that the connection has been established before sending the data segment. Therefore, a second function TCP performs is setting up a session between devices.

A third function TCP performs is to determine when the next segment can be sent and when a previously sent segment needs to be retransmitted. Throughout the session, the two devices continue to monitor the connection by exchanging acknowledgments as defined by the protocol. The beauty of TCP is that it can send multiple segments

segments
Organized bytes of data.

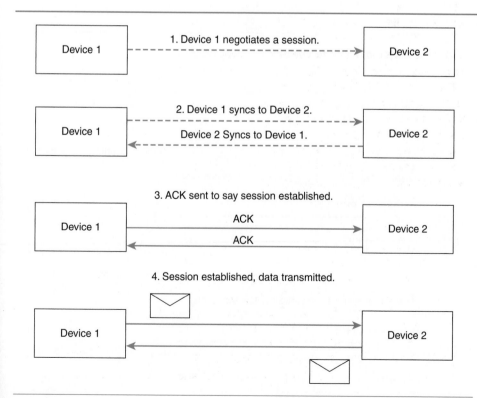

Figure 6–1
TCP is a connection-oriented protocol. Acknowledgments are used to verify the connection is up.

windowing
Process TCP uses to manage data sessions from multiple applications.

flow control
Control mechanism used to manage data flow from device to device.

packet loss
Packet is dropped or lost during transmission.

prior to receiving an acknowledgment. The term *windowing* is used to explain this process. A window is established between the two end devices allowing communications to flow. TCP looks at the type of data being exchanged and determines how large the window needs to be; that is, it determines the number of segments that can be sent before a response (ACK) is received from the far end. Some applications require an acknowledgment to be sent every time a segment is shipped, while others may allow the device to ship multiple segments before requiring an acknowledgment. The value of this feature is obvious: data transmission speeds are enhanced without compromising data reliability. The term *windowing* is used frequently by data engineers when discussing transmission thresholds and overhead utilizations. Consequently, TCP is responsible for controlling multiple application sessions across one link.

The fourth function that contributes to TCP's reputation as a "reliable" protocol is its ability to manage the flow of data across the network. *Flow control* is used to ensure that device buffers are not overwhelmed with data segments at any one point in time resulting in dropped packets. Buffers are similar to buckets placed at the interface ports to hold overflow data when the network is overloaded. You can compare a buffer to a holding pen used when loading cattle into box cars. You cannot load 500 cows into a box car all at one time when the loading ramp only holds twenty cows. A holding pen is used to contain the cows until the ramp has free space. A device buffer is essentially the same thing. The difference is that it holds bytes in queue, not cows in pens. TCP can be viewed as the cattle master responsible for controlling the number of cows allowed into the holding pens and onto the ramps. If too many cows enter the pen, some will be forced to jump over the fence while others will be stepped on and injured. Overflowing a buffer creates a similar scenario. The data may be lost, which is referred to as *packet loss*, or a data segment may be corrupted; both require TCP to retransmit the segment. Figure 6–2 illustrates how TCP handles overflow conditions.

A fifth function of TCP is to help ensure the integrity of data transmission through the use of error-checking schemes. Very elaborate data-checking algorithms are employed by TCP to guarantee that data arriving at the destination match data sent by the source. TCP is smart enough to be able to tell the sending device that the data were accepted intact or that the data were received in error or lost. Depending on what the

Figure 6–2
Flow control is one advantage a connection-oriented protocol such as TCP provides. The flow-control field in the header handles overflow conditions.

1. Packets overflow the receive buffers at Device 2.

Packets overflow and are lost.

2. Device 2 tells Device 1 there is an overflow condition.

Packets overflow and are lost.

3. Device 1 receives *Go* and retransmits lost packets.

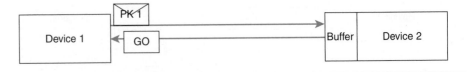

error-checking scheme determines, TCP sends ACKs or NACKs to the far end. The process used to accomplish this task is referred to as sending an acknowledgment (ACK) or asking for a retransmission (NACK). If the acknowledgment is received by the source device, all is good and the source continues to send new data. If, however, the receiving device determines the data are corrupted or missing, it will request its buddy on the source end to retransmit the data segments immediately, as shown in Figure 6–3.

The final job left to TCP is referred to as *data sequencing*. Data sequencing refers to making sure the data packets or segments arrive in order. It is TCP's responsibility to hand the upper layers the data segments in the same order they were shipped. This may seem trivial because the source device sends the data segments one after the other in sequence. But, as already explained, TCP has the ability to carry on multiple sessions at a time on the same network. In addition, the Layer 3 protocol sends packets across multiple links if necessary (a topic discussed in the next section), resulting in packets arriving with data segments out of order. The second data segment may arrive prior to the first. Remember, TCP is able to send multiple devices' data segments across the same link.

data sequencing
Responsible for keeping the packets or segments in order.

6.1.2 Defining Data Types

Now that we have established TCP as a "reliable" protocol, we discuss how it is used to carry multiple types of application data across the same network to multiple devices. TCP accepts data from the upper three layers of the OSI stack and packages the data into segments. Each segment can be defined by a source and destination port number that relates to the type of application or process residing in the segment. For example, when a user sends an FTP file, the TCP header destination port number is set and used to reference the type of file being carried. For example, the destination port number for HTTP is port 80. Devices on a network are configured to allow or disallow specific types of applications to pass through. ISPs must be vigilant at managing and maintaining port access. Customers often call complaining that they are not receiving a particular file or are unable to participate in the newest game. The ISP has to determine if the port is disabled and, if it is, whether it needs to open the port on the router in order to allow the traffic to flow.

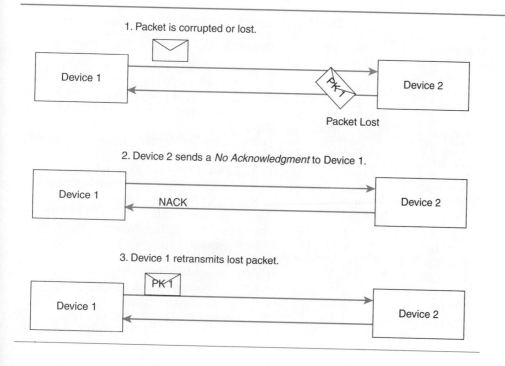

1. Packet is corrupted or lost.

Device 1   Device 2

Packet Lost

2. Device 2 sends a *No Acknowledgment* to Device 1.

Device 1   NACK   Device 2

3. Device 1 retransmits lost packet.

PKT

Device 1   Device 2

Figure 6–3
TCP retransmits lost or corrupted packets.

Figure 6–4
TCP header is used to control the flow of data from device to device.

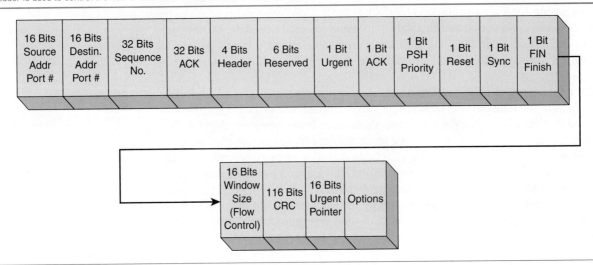

The TCP header, as shown in Figure 6–4, consists of multiple fields that relate to the functions that TCP controls. Note that the *source* and *destination* ports define the type of application being carried in the segment. Typically, as mentioned earlier, the source port is set at zero as this field is not mandatory. The destination port defines the type of data being carried in the segment to the end device. Each Layer 4 aware device in the network reads the destination port address before allowing the segment through. The term *well-known port* refers to port numbers ranging from 0 to 1023. Each port number is associated with a common application type such as FTP, SMTP, and so forth. Table 6–1 lists common port numbers and the applications they represent. The sixteen-bit address provides a total of 65,536 usable port numbers.

The IANA—Internet Assigned Numbers Authority—organization oversees port number assignments. The IANA divides the 65,535 port numbers available into three

well-known port
ports which have numbers preassigned by the IANA.

Table 6–1
Port numbers located in the Layer 4 header are used by network managers to restrict or allow different types of traffic onto the network. The list provides a few of the more popular port numbers and the associated application type.

	Short List of Layer 4 Port Numbers	
Protocol	**Port No.**	**Definition**
TCP, UDP	17	QOTD—Quote of the Day
TCP, UDP	18	MSP—Message Send Protocol
TCP, UDP	20	FTP Data—File Transfer Protocol Data
TCP, UDP	21	FTP File Transfer Protocol
TCP, UDP	23	Telnet
TCP, UDP	25	SMTP—Simple Mail Transfer Protocol
TCP, UDP	43	WHOIS—Who Is
TCP, UDP	50	RE-Mail-CK—Remote Mail Checking Protocol
TCP, UDP	53	Domain—DNS server—Domain Name System server
TCP, UDP	69	TFTP—Trivial File Transfer Protocol
TCP, UDP	80	WWW—World Wide Web
TCP, UDP	101	Hostname—NIC Host Name Server
TCP, UDP	107	RtelNet—Remote TelNet Service
TCP, UDP	113	Auth—Authentication Service
TCP, UDP	119	NNTP—Network News Transfer Protocol
TCP, UDP	161	SNMP—Simple Network Management Protocol
TCP, UDP	179	BGP—Border Gateway Protocol
TCP, UDP	194	IRC—Internet Relay Chat protocol

categories: well-known port—0 to 1024; registered port—1025 to 49,151; dynamic or private port—49,152–65,535.

Along with tagging the port number, TCP is responsible for allowing multiple applications to travel on the same network link destined for one or multiple devices. The software program attaches bits to the data before handing them to the application protocol. The application protocol formats the data with the identifying program tag and hands them to TCP, which uses it to set the destination port number and to initiate a session between the two devices. Data flowing from the application are always tied to the session through the destination port number. Additional applications can be started at the same time, thanks to TCP's ability to manage multiple sessions—multiple windows, as explained earlier.

Again, we can view TCP as the cattle master at the rail station. Each cow branded with the *flying* J is stamped as prime beef. Cows branded with the *upside-down* Y are stamped as milk cows. Both types of cows are shoved into the same box car and ride together to the destination city. But once they arrive, they are separated according to type—milk cow or beef cow. TCP functions in a similar fashion. It is responsible to tag the data segment according to the application. The far end device, similar to the cattle master, reads the brand—the destination port number—and separates the data segments accordingly.

Hopefully, at this point, the role TCP plays in data networking is clear. TCP is a connection-oriented Layer 4 protocol that is able to manage the flow of multiple types of data across a network to one or many devices. TCP works with both the upper and lower layers to accomplish this task and is referred to as a "reliable" protocol.

■ 6.2 UDP: USER DATAGRAM PROTOCOL

UDP is the second most popular Layer 4 protocol found on the Internet. UDP, similar to TCP, is a transport layer protocol that provides an interface between the application layer and the network layer. The main difference between UDP and TCP is that UDP is not a "reliable protocol." UDP does not have header fields for flow control, the ability to send acknowledgments, nor the ability to retransmit a datagram if necessary. UDP is a connectionless protocol, that is, it does not establish a session before transporting the data. The application layer hands UDP the data, UDP formats the data in segments and throws them out onto the network, crosses its fingers, and turns its attention to the next segment needing to be transmitted. The throw-and-hope method is a good way to describe how UDP handles data transmission.

You may wonder why such a protocol was developed. The answer is very simple. Real-time applications such as voice and video are extremely sensitive to any type of *delay*—a term used to define the time it takes a packet to traverse the network. Ensuring reliability adds delay. Acknowledgments cause delay. This, of course, is obvious. When you are talking to your friends, stop every second and ask them if they heard you. Does your sentence flow naturally? More than likely, your friends are very puzzled by your behavior.

delay
The time it takes a packet to traverse the network.

Voice transmission has a completely different set of requirements than data. Unlike data, voice segments must be received within a certain time frame in order for the end user to hear exactly what the sender is saying—real-time traffic flow. Nixing acknowledgments—beyond eliminating a good portion of the TCP header—helps reduce overhead and consequently reduces delay. As shown in Figure 6–5, the UDP header consists of four fields: source port, destination port, length, and checksum.

The source and destination port addresses are used to distinguish the type of application being carried in the segment and are similar to TCP's source and destination port fields. The remaining fields in the header are used to define the size of the packet and additional application information.

UDP has proven itself to be a very good fit for real-time data applications—voice and video. As the world migrates to digitizing voice across the Net, playing interactive video games, and downloading movies and radio and television programs,

Figure 6–5
The UDP header is simpler primarily because it is a connectionless protocol.

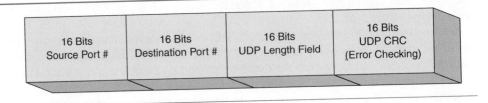

| 16 Bits
Source Port # | 16 Bits
Destination Port # | 16 Bits
UDP Length Field | 16 Bits
UDP CRC
(Error Checking) |

UDP becomes a key player in communications networks. The Debbie Reynolds and Fred Astaire dance team is in vogue.

■ 6.3 RTP: REAL-TIME PROTOCOL

Two years ago, UDP would have been the final topic discussed in this section. Today, with the emergence of VoIP, RTP—Real-Time Protocol—has to be included as one of the commonly used Layer 4 transport protocols. RTP can be viewed as a pseudo–Layer 4 protocol. It straddles the application layer and the transport layer. Though not able to carry out Layer 4 functionality on its own, RTP does make a great partner with UDP to prepare and package real-time applications—specifically, VoIP.

The major function RTP performs is to organize packets in a way that improves their chance of crossing the network in a timely fashion. RTP can be viewed as the specialized cattle herder employed to make sure certain cows are handled properly as they are transported from location to location.

RTP performs several functions in conjunction with the upper-layer application protocol. First, the RTP header, as shown in Figure 6–6, consists of ten fields. The protocol inserts a time stamp on each segment to help organize the packets according to when they were transmitted. RTP also monitors and identifies the data type—similar to UDP and TCP.

■ 6.4 CLARIFICATION OF TERMS

Before the discussion on Layer 3 protocols, it is important to clarify a few terms that may cause confusion. Figure 6–7 illustrates how the following terms relate to one another.

segment
Unit of information after it has been formatted by the upper three layers.

- Segment—Application data handed to the Layer 4 protocol. The term *segment* is used to describe the unit of information formatted by the application, presentation, and session layer protocols. Layer 4 receives segments from the upper layer or, in the reverse, strips off Layer 4 headers, handing the segment to the upper layers. Segments vary in size depending on the application used to format the segment.
- Datagram—Layer 3 header + Layer 4 header + data segment—A datagram includes information received from the upper-layer protocols, the Layer 4 header and the Layer 3 header. An example of a datagram is an FTP segment with a TCP header and an IP header attached. The size of the datagram varies

Figure 6–6
RTP header contains field used to assist UDP in transmitting real-time traffic such as VoIP.

| 2 Bits
Version
Identifier | 1 Bit
Padding
Field | 1 Bit
Extension
Flag
Field | 4 Bits
CSRC
Contributing
Source
Indicator
Field | 1 bit
Marker
Flag | 7 Bits
Payload Type
Field
(Encoding
Method) | 16 Bits
Sequence
No. Field | 32 Bits
Time Stamp
(Jitter
Indicator) | 32 Bits
Sequence
Source ID | 32 Bits
CSRC
Field |

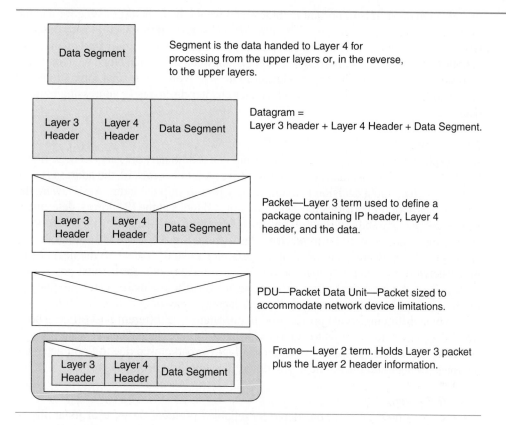

Figure 6–7
Common terms used in networking
are described in the diagram.

depending on the data link layer's frame format. An Ethernet frame can carry a 1500 byte datagram.

- Packet—A packet, similar to the datagram, is an envelope containing data plus a Layer 3 header used to define how the information will be routed from device to device. A packet varies in size depending on the amount and type of data being shipped and the Layer 3 protocol.
- PDU— Packet Data Unit—A PDU is a packet sized to accommodate network device limitations. An IP PDU is a variable-length chunk of data with a maximum size of 65,000 octets.
- Frame, Layer 2—A frame, Layer 2, is a unit of data containing a packet, a Layer 2 header and trailer, Layer 3 header, Layer 4 header, and the data. Typically, Layer 2 frames are defined by their MTU—Maximum Transfer Unit. An MTU is the maximum size frame a network can handle as defined by the protocol. For example, if you are shipping twenty-five books to a friend, you need to determine how many books you can package in one box. Say the box the carrier gives you can only hold ten books. You have twenty-five books to send. In order to comply to the carrier's rules, you need three boxes to hold the twenty-five books. In this example, you have just fragmented your book collection to accommodate the size of the box. The same is true of packets traversing the network. An IP network formats its PDU—packets— into 65,000 octet (or fewer) chunks plus header and trailer information. The Layer 2 protocol, such as Ethernet, says it will take 1500 of those octets at a time, fragmenting the 65,000 octet packet, packaging them into individual frames and shipping them across the network. Terms such as *fragment* and *reassembly* refer to the process of putting packets into frames—books into boxes and unpacking the books.

■ 6.5 IP: A LAYER 3 NETWORK PROTOCOL

IP—Internet Protocol—is the most popular network layer protocol deployed today. Similar to Julia Roberts, IP is universally accepted. In order to explain IP in a logical, organized manner, it is important to understand what jobs IP is responsible for and the different tools it uses to accomplish those jobs. Our goal in this section is to describe how IP is used in the network to carry all types of data. The section starts with a description of how packets are routed within and between networks. Next, we look at the IP header and focus on the fields relevant to how IP works. The final topic presented relates to IP addressing at a high level. A later chapter discusses IP addressing in more detail.

6.5.1 IP Functions

IP is responsible for routing data packets from network to network, host to host. IP can be compared to the post master at the post office. A letter comes in, the address is read and interpreted, and a decision is made on how best to ship the letter. An IP aware device such as a router performs the same functions. It looks at the IP address, determines where it is heading and if it requires any priority routing, and decides what port it needs to be shipped out on in order to reach its destination. Similar to the post office, IP is a device-to-device aware protocol; that is, the router ships the packet to the next device in the network in same way that the post office sends a letter to the next postal stop such as from the local post office serving the customer to the main postal hub that ties to other hubs. The network serving the end location may be hundreds of miles away served by a different service provider or, in our analogy, a different post office. Therefore, IP's main role in life is to determine where to route the packet, at what priority, and what to do if an interface port is not available—road closed due to flooding. What alternate road should I send the traffic down. The question becomes, how?

6.5.2 IP Routing

We discussed how TCP is considered a "reliable" protocol because of its ability to know when data have not been received or when they have been received but corrupted. In contrast, IP is considered an "unreliable" protocol in that it does not care if a packet has been received in good order with no errors. IP is concerned with how to route the packet according to the rules defined within the device, router, and configuration.

IP is the navigator. What do navigators need? First, of course, they need maps that show the topology of the network, the routes, and route distances. Second, they need to know the state of the route. Is the route closed? Is it under construction? Are there delays? Is it congested? Third, a navigator needs to know what vehicles are allowed on the road—size, speed, and so forth. The IP routing protocol, as the navigator, requires this information—topological map, state of the link, type of packet being transported—in order to decide what interface port to ship the packet onto. A later chapter provides detailed explanation on IP routing.

6.5.3 Routing Protocols

How does a device know who its neighbors are? Two routing methodologies are available to network administrators—static and dynamic routing. Static routes between two devices are built manually by the network administrators through the router's command line interface. The administrator logs into the device and enters specific commands to turn up a circuit nailing up the path between the two ports. Packet routing is simple. The device automatically sends packets across the static route to the second device. Map topology, converging the networks, and so forth are not required. Static routing works well for small networks with few routes or stub networks. Once a network interfaces to the World Wide Web requiring multiple routes and multiple devices, static routing becomes unmanageable.

Dynamic routing, a subject discussed in Chapter 12, routes packets according to rule sets defined by a routing protocol. The path between device A and device B is not nailed up as with a static route. Instead, the routing protocol, such as RIP or OSPF, selects the "best route" to take at that given time. The routing program uses algorithms and rules to determine which port the packet needs to be routed to in order to reach the destination address. The rules used to determine which path to choose vary per routing protocol. Some, such as RIP, use hop count, while OSPF looks at multiple variables—distance, least cost, and so forth. Regardless of the method used to select a route, all routing protocols exchange information with other devices and, in most cases, between networks. The ability to exchange routing table information between devices and networks allows routers to dynamically route traffic accordingly. Chapter 12 provides a detailed description of how dynamic routing works and the types of routing protocols used in networks today.

In summary, routing selection is handled through a nailed—"tacked"—up path defined by manual configuration or through implementing routing protocols that exchange route information between devices and networks. In either case, static or dynamic routing, the goal is to make sure the packet arrives at its destination intact and on time.

6.5.4 IP Packet

We have learned that IP and the routing protocol are responsible for telling the router how to route a packet from device to device through a network. It is now time to dig deeper and look at how the IP header is structured to accomplish routing functions. The IPv4 header may contain up to twenty-four octets of information depending on the type of data being carried. A detailed description of the key fields follows.

Source and destination address fields are the most important fields in the IP packet. They hold the information that allows the router to route the packet to the correct destination. The source address field holds the IP address of the source device, the device sending the packet. The destination address field holds the IP address of the destination device, the device the packet is being sent to. IP addressing, though complicated, allows multiple devices on multiple networks to exchange information in a connectionless environment using many types of transport media. The beauty of IP is its ability to work in many environments regardless of the type of equipment, the type of circuit, or the type of application data being sent.

The IP source and destination fields of an IPv4 header are thirty-two bits long divided into four octets or three decimal digits per segment: XXX.XXX.XXX.XXX. The address, as explained in the next section, is divided into two segments, network and host. The network portion of the address defines what network the packet belongs to, and the host defines what device the packet is being sent from or sent to. The IPv6 address field contains 128 bits for a source and a destination address, resulting in zillions of potential addresses for host devices.

Table 6–2 lists and defines all the fields found in the IPv4 header. The fields shown in the table include the protocol field, flag fields, identification and fragment offset fields, and time-to-live fields. All fields are used to assist in moving data across a network from device to device. The IP packet, as shown in Figure 6–8, contains the IP header plus the data.

6.5.5 Real-Time Traffic

Before moving on to the next topic, it is important that we include additional detail on how IP handles real-time traffic such as VoIP. VoIP is becoming a viable method for carrying digital voice throughout the world. The IP header is used to set priority or QoS values to improve the quality of a VoIP call. In the beginning, IP was viewed as the Layer 3 protocol used to carry data communications across the Internet. Due to IP's variable-length packets, most believed IP would not be able to carry time-sensitive traffic that had to maintain a relatively short end-to-end delay—that is, the time it takes a packet to travel from point A to point B. But, as processors became more robust, and networks cleaner (fewer problems), engineers started to manipulate the IP header to accommodate time-sensitive traffic. Specifically, the CoS (class of service) field is used to facilitate the transmission of different service types, voice included.

The CoS field—also called type of service (ToS) or service type field—is used to tell the device how to handle the packet. For example, if the packet contains RTP flows (voice), the CoS precedence bits will be set accordingly to ensure that the packet is given priority through the network. This is referred to as *tagging* the packet with a

tagging
Flagging the packet as high priority.

Table 6–2
The IP header contains multiple fields used to assist in the transport of packets across the network.

IP Header Field Definitions

No. of Bits	Field Name	Function
4	Version	Indicates the IP protocol version—IPv4 or IPv6.
4	Header Length	Header Length field is used to indicate the length of the IP header.
8	CoS	Class of Service used to set priority levels such as that required for real-time traffic—VoIP. Diff Serv, also known as the DSCP, bits are set to indicate traffic that should be expedited across the network.
16	Length	Tells the device how long the entire packet is, including header and upper-layer information. Total length of an IP packet is 65,536 octets.
16	Fragment Identification	Works with the three-bit flag and thirteen-bit fragment offset field in telling the device to look for a fragmented segment.
3	Flag	Flag marked to indicate fragmentation as described above.
13	Fragmentation Offset	Works with the Fragment ID and Flag to indicate data fragments.
16	TTL	Time-to-Live field is used to define the maximum number of hops a packet can travel around the network before it is discarded. This field helps reduce the chance of rogue packets endlessly hopping between devices.
8	Protocol-Type Header	Tells the device what protocol is embedded inside the packet. For example, it may be TCP or UDP or UDP with RTP.
16	CRC	Cyclic Redundancy Check—error-checking algorithm that helps guarantee the health of the network.
32	Source Address	The thirty-two-bit IP address assigned to the device sending the packet.
32	Destination Address	The thirty-two-bit IP address assigned to the receiving device—device the packet is destined for.
Variable	Options	Future

Figure 6–8
IP header consists of multiple fields, each responsible for different functions.

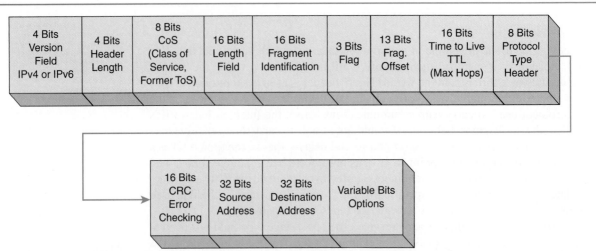

quality of service or QoS flag. Referring to our postal analogy, it relates to stamping the letter as air mail. The post office ushers the letter through the system, giving it priority over standard mail. The user is given the ability to set the CoS precedence bit in the same way the post office clerk stamps the letter as air mail or ground. The common terms used to describe priority settings are BE, best effort, similar to regular mail, and EF, expedited forwarding, similar to same-day delivery. The router reads the header field to determine if the packet is stamped as BE or EF before deciding what to do with it.

Voice over IP providers have the option of configuring the end devices—MTAs, media telephone adapters—to tag the packet as EF. If the MTA does not set the precedence bit to EF, the network has no clue it is carrying voice and, therefore, treats it just like all the other packets—BE or best effort—and may drop the packets if congestion is experienced resulting in choppy audio quality. One last term that has to be explained is DSCP or Diff Serv. When a packet has a DSCP tag set in the header, it is given priority or precedence on the network. Do not let the multiple terms used to describe QoS settings confuse you. They all relate to setting the CoS precedence bit to the correct priority: DSCP = EF. Remember, they all are telling you the packet has been stamped with a priority flag—DSCP, Diff Serv, CoS, EF, and the older ToS. Figure 6–9 illustrates the structure of the CoS field. In summary, the IP header has multiple fields that are used by Layer 3 aware devices responsible for routing packets end to end across and between networks.

6.5.6 IP Addressing: High-Level View

Presently, two versions of IP addressing are used in networks: IPv4 and IPv6. IPv4 continues to be the dominant addressing method deployed across networks. A slow migration to the newer standard, IPv6, is in progress.

IPv4 segments a thirty-two-bit bit address into five classes, each able to define a specific number of network and host addresses, as shown in Figure 6–10. Similar to the postal address (ZIP Code, city, street, and house number), an IPv4 address is divided into different segments used to differentiate between the regions. The network portion of the IP address relates to the city and ZIP Code, and the host portion of the address relates to the street address. *Network* refers to the network circuit routing from device to device such as router to router. The *host* refers to an IP address assigned to a host device such as a PC or a VoIP MTA. IP addresses are managed and controlled by InterNIC, a nonprofit government-run organization. AARON, the branch of InterNIC that doles out IP addresses, scrutinizes service providers, corporations, and any organization requesting IP addresses to assess whether the addresses are needed and how they are used.

network
Portion of the IP address associated with the network.

host
Portion of the IP address associated with the host.

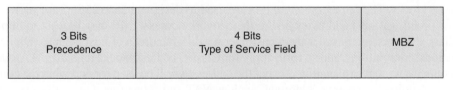

Older ToS Field Definition

Newer Diff Serv Field Format

Figure 6–9
Diff Serv is used to stamp the IP packet with a priority flag. Today, priority is set within the six-bit CoS, Class of Service, field formerly called the ToS, Type of Service, field.

Figure 6–10
Classful IPv4 addressing is defined
according to the number of networks
and hosts it supports.

Class	Division Network and Host	Range of First Octet	Maximum No. of Networks and Hosts
Class A	NNN.hhh.hhh.hhh.	0–127	128 Networks and 16,777,214 Hosts
Class B	NNN.NNN.hhh.hhh.	128–91	16,384 Networks and 65,534 Hosts
Class C	NNN.NNN.NNN.hhh.	192–223	2,097,152 Networks and 254 Hosts
Class D	NNN.NNN.NNN.hhh.	224–239	268 Million Multicast Addresses
Class E	NNN.NNN.NNN.hhh.	240–255	Reserved

N—Refers to Network Portion of IP Address
h—Refers to Parallel Host Portion of the IP Address

classful addressing
IP addressing method that
categorizes the address as
A, B, C, D, or E.

Companies must justify IP address requests by guaranteeing that their networks will be designed to optimize IP usage. Due to the limited number of public IP addresses, AARON is very frugal when it comes to handing out IP space—as it should be.

The IPv4 addressing standard categorizes addresses into five types—Class A, B, C, D, and E—according to the way the network or host portions are segmented. The term *classful addressing* is used to describe segmenting the IP address by class. Class A IPv4 addresses consist of eight bits representing networks and twenty-four bits representing hosts. Class A addresses can support 128 networks, meaning that only 128 organizations can be given a Class A address. If an organization is lucky enough to obtain a Class A address block, it can support up to 16.78 million hosts. Obviously, only very large organizations and governments warrant a Class A address block. If the address starts with an O, it is a Class A address.

Class B IPv4 address block uses the first two bits to distinguish it as a Class B address: 10. The next fourteen bits are used to provide 16,384 network address blocks. The remaining sixteen bits are used to designate the host addresses tied to the network address. Each network address has the ability to assign 65,536 host addresses. Again, it is easy to see why a network manager who has been given a Class B address block wants to call all his or her geeky friends and brag. Class B addresses are given only to those with large networks such as large ISPs and very large organizations.

The Class C IPv4 address block uses the first three bits to distinguish it as a Class C address—110. The next twenty-one bits are used to provide two million network address blocks. The remaining eight bits are used to designate 254 (256−2 for network and broadcast) host addresses tied to the network address. ISPs may request multiple Class C addressees in order meet customer needs—hundreds of host devices. Organizations, corporations, and so forth apply for Class C address blocks regularly. AARON requires a written explanation of what the addresses will be used for and what the company is doing to conserve addresses, such as NAT and subnetting a topic discussed in a later chapter. Most service providers initiate a request for addresses many months prior to the time they need to load them into the network, as the process is fairly long and cumbersome.

The Class D IPv4 address block uses the first four bits, 1110, to identify it as a Class D address. The remaining twenty-eight bits of the thirty-two-bit address are used for multicast sessions. A total of 268 million addresses are available for multicast sessions. Multicast addressing is used when a provider wants to send the same information to multiple hosts. Video conferencing may use multicast addressing.

The Class E or experimental IPv4 address block has a four-bit header followed by twenty-eight bits used to define 268.4 million hosts. Class E is used for experimental purposes.

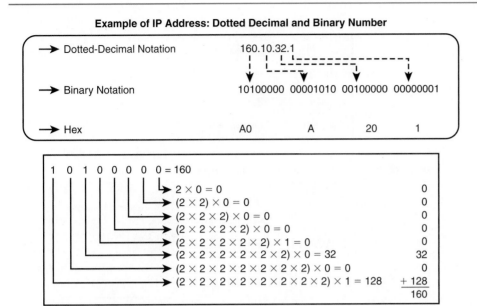

Figure 6–11
Dotted-decimal notation is used to
simplify the IP address for humans.

Figure 6–11 illustrates how the IP address is divided into four one-byte segments called octets. The first row shows the decimal equivalent to the binary number, referred to as a dotted-decimal notation address. In our example, addresses between 128 and 192 refer to Class B addresses. Note that the addresses have a binary equivalent showing the one and zero patterns. The last row illustrates the hex number equivalent. All IPv4 addresses are referenced in the dotted-decimal format to simplify management. The IP class can be determined quickly by looking at the first decimal, as shown in Figure 6–10. The first decimal of the IP address tells you if the organization is using a Class A, B, or C.

CIDR, NAT, DHCP, VLSM

Two problems have arisen over the past ten years. First, the Internet has grown at a phenomenal rate. The consequence of this growth has been an exponential increase in the number of routes traversing the Internet. Through the 1990s, the number of routes added to the Internet increased from the thousands, to more than 100,000. Obviously, the presence of too many routes causes network routers to become overwhelmed with route maps and routing choices. The routers' horsepower—referred to as processor power or CPU requirements—increased as the number of routes increased. A solution to this problem was introduced by the IEFT in the mid-1990s. CIDR, often referred to as classless routing, was introduced to help reduce the providers interfacing directly into the Internet. The IP address structure using CIDR allows providers to manage their networks according to subnetwork classifications. Today, only large ISPs interface directly to the Internet. In turn, they provide address space to smaller ISPs they service. Finally, smaller ISPs provide IP address space to organizations, companies, and so forth. Today, organizations, small ISPs, and so forth are not given their own address space. Classless addressing is presented in greater detail in Chapter 12.

Other methods introduced to help reduce the number of addresses needed to service the millions of devices on the Internet are NAT, Network Address Translation; DHCP, Dynamic Host Configuration Protocol; and VLSM, Variable Length Subnet Mask. Chapter 12 introduces each of these methods in detail. At this time, the point to grasp is that industry has invented ways to resolve the IPv4 address dilemma.

Private versus Public IP Addressing

The term *public IP address* defines addresses used throughout the Internet. You cannot have two devices on the Internet with the same address, such as 128.24.110.3. Three

public IP address
IP address space used for
public networks.

Figure 6–12
Private IP addressing is a common way network managers preserve IP address space.

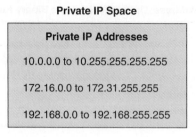

Private IP Space

Private IP Addresses
10.0.0.0 to 10.255.255.255
172.16.0.0 to 172.31.255.255
192.168.0.0 to 192.168.255.255

Example of How Private IPs Are Used in the Network

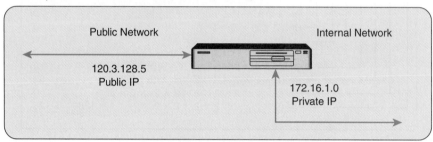

Public Network   Internal Network

120.3.128.5
Public IP

172.16.1.0
Private IP

private IP address
IP address space used for internal networks.

address blocks have been reserved for internal or private networks. *Private IP addresses* are used by many organizations as a way to conserve public IP space. A private address sits on the inside of an organization's network, such as a network LAN, Local Area Network, or an ISP's customer base—any device hanging off the core router. For example a port—also called an interface—on a router faces the outside world or the public Internet and is given a public IP address. All traffic destined for that network is pointed toward that router interface. On the opposite side of the router, an internal network interface is assigned a private IP address. This address routes to all the devices hanging off the internal network. Each device on that network is assigned a private IP address that ties it to the internal network interface port, as shown in Figure 6–12. Network managers assign private IP addresses to internal interface ports, that is, ports tied to internal devices such as Ethernet switches, printers, PC, and so on. If the router ties to the public network, in most cases a connection to the Internet, the network manager assigns a public IP address to the network interface port—interface port tied to the big wide world. Private IP addresses are assigned to ports facing the internal network—LAN, and so forth, and public IP addresses are assigned to ports facing the public network—Internet, VLAN, and the like. The use of private network addresses *throughout the world* help reduce the number of public IP addresses needed to support the millions of hosts hanging off internal networks.

■ 6.6 OTHER LAYER 3 PROTOCOLS

Though beyond the scope of this text, it is important to understand that there are many Layer 3 protocols deployed in networking, particularly LAN networking. IP is the predominant Layer 3 protocol deployed across LANs and the Internet, but other protocols are used to provide the same functionality. Following is a short explanation of the two most common protocols deployed today—not counting TCP/IP of course.

IPX, Internetwork Packet Exchange, is a network layer protocol used to route traffic across LANs, specifically Novell LANs. IPX, similar to IP, uses an addressing scheme divided into two segments—network and host. A Novell IPX address consists of a thirty-two-bit network address and a forty-eight-bit host or node address that in most cases carries the device MAC address. IPX is a connectionless, unreliable protocol that depends on higher-layer protocols to ensure information delivery. IPX routes traffic through the use of a routing protocol. RIP, a distance vector protocol, is deployed on IPX networks to determine the best route for the traffic to follow.

The IPX Layer 4 partner is SPX, Sequenced Packet Exchange. SPX is a Layer 4 protocol similar to TCP. SPX provides a connection-oriented, reliable connection through the use of acknowledgments and segmenting. IPX depends on SPX to guarantee delivery of packets in the same way that IP depends on TCP.

AppleTalk, the suite of protocols used to network Apple computers together into a LAN, depends on the Layer 3 protocol DDP, Datagram Delivery Protocol, to route traffic. DDP, similar to IP, is responsible for packaging the segments into packets that contain an address used for routing. DDP's Layer 4 partner within the AppleTalk protocol suite is ATP, AppleTalk Transaction Protocol. It provides the Layer 4 functionality similar to TCP.

SUMMARY

The most important concepts to grasp from this chapter are that networking requires protocols and that the protocols defined by the industry must define specific functions in order for information to flow end to end. A Layer 3 protocol must partner with a Layer 4 protocol in order to enhance its capability to ensure proper data delivery. A Layer 4 protocol must abide by the restrictions defined by the network deployed and work with its partner in networking. Finally, though the focus of this chapter is on TCP and IP, there are other protocol suites deployed that are developed around the Layer 3 and Layer 4 stack.

REVIEW QUESTIONS

1. TCP and IP work together to carry information across the Internet.

   a. true   b. false

2. What is the purpose of TCP and UDP?

3. What is the purpose of the IP?

4. Describe how TCP uses a three-way handshake to set up a connection.

5. IP partners only with TCP.

   a. true   b. false

6. What is the most significant difference between the Layer 4 TCP and the Layer 4 UDP?

7. Denying entry into a network by traffic type can be accomplished by:

   a. turning down the physical port connected to the Internet
   b. turning off the UDP port at the router
   c. rejecting packets according to source address
   d. both a and b
   e. all of the above
   f. none of the above

8. What is the purpose of the TCP and UDP port numbers?

9. Name two applications that utilize UDP.

10. What applications depend on the RTP?

11. RTP is a _____ protocol.

    a. Layer 1
    b. Layer 2
    c. Layer 3
    d. Layer 3 and 4
    e. Layer 4 and 5
    f. Layer 4

12. Explain the difference between a PDU and a datagram.

13. IP encapsulated data are referred to as a segment.

    a. true   b. false

14. Explain the difference between a frame and a packet.

15. Explain the different between a segment and a datagram.

16. The primary functions performed by IP are:

    a. packet routing, setting the priority level, determining alternate routes
    b. packet routing, setting the priority level, determining alternate routes, guaranteeing packet delivery
    c. packet routing, setting the priority level, guaranteeing packet delivery
    d. none of the above

17. List three conditions IP looks at before shipping a packet onto the network.

18. IP routing protocols are referred to as:

    a. static, dynamic, intermediate, combination
    b. static or dynamic
    c. static, combination, distance vector, time sensitive
    d. dynamic, split horizon
    e. none of the above

19. A dynamic routing protocol determines how to route a packet according to the Layer 2 frame address.

    a. true   b. false

20. The most important fields within the IP header used when determining how to route packets are:

    a. priority tag
    b. start flag
    c. source IP address
    d. destination address
    e. CRC byte
    f. both c and d
    g. a, c, and d
    h. c, d, and e
    i. all of the above

21. The IPv4 address field is divided into a(n) _____ and _____ segments.

22. How do you tell what class an IP address is?

23. List the overhead fields contained in the IP packet.

24. Why did the IETF introduce CIDR and what does *CIDR* stand for?

25. What protocol does Voice over IP or VoIP use to ensure the voice quality?

KEY TERMS

TCP (138)

segments (139)

windowing (140)

flow control (140)

packet loss (140)

data sequencing (141)

well-known port (142)

delay (143)

segment (144)

tagging (147)

network (149)

host (149)

classful addressing (150)

public IP address (151)

private IP address (152)

7

Layer 2 Protocols: Ethernet, Frame Relay, ATM

Objectives

After reading this chapter, you should be able to
- Define the Ethernet protocol.
- Define the frame relay protocol.
- Define the ATM protocol.

Outline

Introduction

7.1 Networking Basics

7.2 Ethernet

7.3 Frame Relay

7.4 ATM

■ INTRODUCTION

Layer 2 protocols are found throughout the communications networks. The job a Layer 2 protocol performs is simple. It is responsible for packaging Layer 3 traffic flows into easier-to-handle Layer 2 frames or cells. The Layer 2 protocols also interface with the Layer 1 protocol to compartmentalize and prepare the ones and zeros for transport. Chapter 7 focuses on the three most popular Layer 2 protocols: Ethernet, frame relay, and ATM.

■ 7.1 NETWORKING BASICS

7.1.1 Connection-Oriented Protocols versus Connectionless Protocols

In dealing with packet, frame, or cell technologies, the first parameter that must be defined is whether the protocol is *connection oriented* or *connectionless*. A connection-oriented protocol relies on predefined virtual paths configured at the time the circuit is turned up. The traffic travels along the virtual path the same way it travels along a physical path. The significant difference is that many virtual paths may reside on one physical link. In essence, a connection-oriented protocol provides a predefined path between the originating device and the terminating device. It requires that the path be preconfigured at the time of turnup and depends on the path to remain in place until the circuit is disconnected. Connectionless protocol defines the source and destination address of the originating and terminating devices in the overhead of the frame and then

connection-oriented protocol
Predefined connection is provisioned at the time of turnup or during call setup.

connectionless protocol
End-to-end transmissions are handled by each device reading the source and destination address. A permanent circuit is not defined.

relies on each intermediate device to route the packet to the correct destination. A pre-defined logical path is not established before the transmission. Instead, the information is thrown out into the network and depends on the source and destination address to route the information to the correct destination. IP is a connectionless technology, and ATM and frame relay are connection-oriented technologies. When you are introduced to a new packet technology, you should immediately ask whether it is connection oriented or connectionless.

7.1.2 PVCs versus SVCs

permanent virtual circuit
Logical circuit provisioned at the time of turnup.

There are two types of logical circuits used in connection-oriented protocols. A *permanent virtual circuit* (PVC) is nailed up when the circuit is first turned up. For example, if there is an ATM switch in Milwaukee and one in St. Louis, for traffic to flow between the two cities a logical circuit must be defined between the two nodes. If PVCs are used to carry the traffic, a PVC must be configured between the two nodes before traffic can be placed on the connection. It would then remain up indefinitely or until it was manually disconnected from service.

switched virtual circuit
Logical circuit established for the length of the transmission.

Switched virtual circuits (SVCs) are virtual circuits that are dynamically configured by request. SVCs are built when a session is initiated and torn down when the session is terminated. Special signaling protocols are used to establish the connection at the request of the customer, similar to a voice call being established through the PSTN. The signaling protocol notifies the originating equipment that a path is needed between specific locations. The packet or cell switch then builds the path for the duration of the session or call. PVCs and SVCs define the logical path the traffic will follow as it travels through the network.

7.1.3 Packet, Frame, and Cell Multiplexing

The difference between a virtual circuit (VC) and a physical TDM circuit is that the virtual circuit may be oversubscribed. Oversubscribing refers to assigning more customers to a circuit than there is bandwidth available for. Multiple logical VCs are defined within each circuit, similar to multiple lanes on a highway. The highway has multiple lanes depending on the projected traffic load. Many vehicles travel together on the shared highway, but during rush hour, too much traffic merges onto the highway, causing bottlenecks and traffic jams. Like traffic on a highway, information is multiplexed onto the VC only when a customer has information to send. When every customer sends information at the same time, bottlenecks and traffic jams occur, causing an increase in the amount of time it takes to transmit the data across the network. In communications, as in highway engineering, increasing the size of the highway reduces the number of slowdowns and traffic jams. The issue then is how much money should be spent to ensure that traffic jams do not occur. In data transmission, as in highways, building a thoroughfare that is large enough to eliminate bottlenecks is not practical. The cost would be enormous and wasteful considering the majority of hours that the circuit or highway would be underutilized. A compromise is found by determining the oversubscription ratio that ensures a certain quality of service for information being transmitted while containing the cost of the infrastructure to a reasonable level. Typically, a full study is performed to determine circuit size. The philosophy of most network engineers is that a pipe can be oversubscribed because, under normal circumstances, not all users will transmit information at the same time. Essentially, the reason fast-packet technology is encroaching upon TDM transport technologies is its ability to allow oversubscription on the links and thus gain efficiencies on the network.

TDM, on the other hand, is more like a railroad network than a highway. It allocates specific boxcars for a specific customer's traffic. If the boxcar is partially filled, it remains partially filled because it is not allowed to carry other types of goods. This results in wasted bandwidth.

Fast-packet technologies, like frame relay and ATM, can also be explained by using a highway analogy. Multiple lanes are used by multiple users, and each of the lanes is permanently constructed between locations. Traffic is multiplexed onto the highway and demultiplexed off the highway at different entrance and exit ramps. The established path, configured either permanently or during a switch setup, is responsible for making sure that the information reaches its destination.

Information being shipped using the IP protocol relies on the originating and destination address to make sure the information reaches its destination. The network is similar to an air transport network. Each plane is routed along the best path according to the air traffic controller at a particular airport. For example, if Kelly wants to travel to Boston from Los Angeles, she boards a plane in Los Angeles with a ticket that says "Los Angeles to Boston." The ticket also shows the hops she will make along the way. She rides on one plane from Los Angeles to Chicago, deplanes, and jumps onto a second plane that takes her to Detroit. From there, a third plane takes her to Boston. Kelly's ticket shows each of the hops between Los Angeles and Boston but defines the originating address, Los Angeles, and the destination address, Boston. If, in Detroit, the flight to Boston was canceled, Kelly could take a different plane to Pittsburgh. In Pittsburgh, she would board a plane heading to Boston. So, the route is defined from origination to destination but the path taken between the two ends may change depending on traffic congestion, circuit outages, and so forth. Each node along the route is responsible for making sure that the traffic is routed properly to the next hop in the network.

Therefore, the three general information transmission networks can be compared to the three transportation networks. TDM compares to a railway system, ATM and frame relay to a highway system, and IP to an air traffic system.

7.1.4 Ingress, Egress, Access, Edge, Backbone

The five terms used to define the new converging networks are *ingress, egress, access, edge,* and *backbone.* To understand how the network is partitioned, these terms must be defined. *Ingress* refers to signals entering the network. *Egress* refers to signals leaving the network. *Access* is local access network. *Edge* is the edge between the backbone and the access. *Backbone* refers to the long-haul portion of the network.

■ 7.2 ETHERNET

The Ethernet has been around for many, many years. It was first introduced as a LAN—local area network—protocol used to carry traffic between devices located in one local area such as within a business or across a campus. Over the past three or four years, the Ethernet has extended its reach into the access and metro network as a vehicle for carrying traffic across the WAN—wide area network. Our explanation of the Ethernet includes the following topics: Ethernet standard as defined by the IEEE; Ethernet network topology; Ethernet frame structure; Ethernet address; ARP; VLAN tagging; CSMA/CD protocol; Ethernet line speeds. The Ethernet section ends with a connect-the-dot explanation of how Ethernet interfaces with IP and with the physical layer protocol.

7.2.1 Ethernet Standard

In the 1970s the Xerox Corporation developed Ethernet. In the 1980s, Digital Equipment Corporation, Intel, and Xerox created the first *Ethernet* standard referenced in the "Blue Book" published by the three corporations. By the mid-1980s, *Ethernet II* was released. The IEEE, using Ethernet II, developed the *802 Ethernet* standard referenced as *802.3.* Today's networks use both the Ethernet II and the 802.3 IEEE Ethernet. A slight difference in the frame structure is what differentiates the two. Both define Ethernet as a CSMA/CD—Carrier Sense Multiple Access with Collision Detection—protocol, and both types of Ethernet ride on the same medium.

From the beginning, Ethernet—similar to Cher—has reinvented itself multiple times. Though still a *CSMA/CD baseband* technology, Ethernet has moved from the

ingress
Information entering the access network from the backbone network.

egress
Information leaving the access network into the backbone network.

access
Portion of the network that services the end customer—CO to the customer.

edge
Portion of the network that lies between the access and the backbone.

backbone
Portion of the network that connects main data or switch centers.

Ethernet
Layer 2 transmission protocol introduced in the 1970s and deployed today across both local and wide area networks.

Ethernet II
Ethernet LAN standard developed by Xerox, Digital Equipment Corporation, and Intel.

802 Ethernet
Family of Ethernet standards formalized by the IETF.

802.3
Ethernet LAN standard formalized by the IETF.

CSMA/CD
Protocol that uses a broadcast technique for placing information onto the wire (circuit). Device listens before it ships data onto the shared bus or pipe. Collision detection is used to handle all data collisions on the shared pipe.

baseband
Frequencies are not segmented; thus, traffic is shoved out onto one big pipe. Information coming from multiple devices is not multiplexed together onto the circuit. Devices share the big pipe.

LAN
Local area network refers to multiple devices tied together within a local region.

WAN
Wide area network refers to devices residing at remote locations, across cities, states, or continents tied together through circuit connections.

LAN side of the network into the *WAN*. Network equipment vendors have built line interface cards that support the Ethernet 802.3 protocol's frame structure and contention algorithms for high-speed access and metro networks. SONET MUXES have been revamped to carry Fast Ethernet and GigE signals, directly mapping them into STS concatenated frames across fiber optic cable. Fiber to the business is much easier to cost justify, thanks to small inexpensive multiplexers designed to carry Ethernet. In addition, network managers are less hesitant to deploy Ethernet WAN solutions than traditional TDM solutions, as they are comfortable with Ethernet—it is similar to employing a friend. The Ethernet standard is one that proves its flexibility, simplicity, and, most important, its longevity over and over as the communications world evolves and changes. Table 7–1 lists the Ethernet standard naming schema as they relate to the different interface speeds.

7.2.2 Ethernet at the Physical Layer

Ethernet in this text can be viewed as a combination Layer 1 and Layer 2 protocol. The Ethernet standard defines how the bits are placed on the wire, the voltage level of the bits, the line code, the pinouts of the Ethernet RJ-45 connector, and so forth. The key concepts related to the Layer 1 side of Ethernet are the types of physical media Ethernet rides on, the distances the frame is able to travel on the medium, the method used to terminate the medium, the interface speeds, and the way Ethernet puts information onto the wire—a baseband CSMA/CD protocol.

Physical Media Used to Carry Ethernet

Ethernet rides on four types of media: coaxial cable, twisted copper pair, fiber optic cable, and radio frequencies. The key points to remember are the cable type, the distance the signal can travel on the cable, and the network topology associated with the cable type. The following describes the six most common types found in the network. Before we define the different types of cable media, it is first important to define how they were named. The IEEE defines the media types using the term 10Base followed by the

Table 7–1

The Ethernet standard is defined by the medium used in the network. Each medium is specced according to the interface speed, the network topology, the medium, and the typical distance.

Ethernet Standard

Standard Name	Common Name	Interface Speed	Physical Network Topology	Medium	Distance
10Base2	Thinnet	10 Mbps	Bus	50 ohm coax	185 m (607 ft.)
10Base5	Thicknet	10 Mbps	Bus	50 ohm coax	500 m (1640 ft.)
10BaseT	Twisted Pair	10 Mbps	Star	UTP, STP—Category 5	100–150 m (328–500 ft.)
10BaseF	Fiber	10 Mbps	Point to Point	Multimode Fiber	400 to 2000 m
100BaseTX	Fast Ethernet	100 Mbps	Star	UTP—CAT5 STP—CAT1	100 m (328 ft.)
100BaseFX	Fast Ethernet Fiber	100 Mbps	Point to Point	Fiber Optic—Multimode or Single Mode	2 km
1000BaseTX	GigE	1 Gbps	Point to Point	Fiber Optic—Multimode or Single Mode	2 km

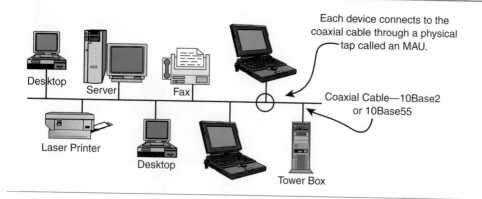

Each device connects to the coaxial cable through a physical tap called an MAU.

Coaxial Cable—10Base2 or 10Base55

Desktop

Server

Fax

Laser Printer

Desktop

Tower Box

Figure 7–1
Coaxial cable is used to connect devices on an Ethernet LAN. Devices tap into the coaxial cable. A physical and logical bus topology defines an Ethernet LAN using coaxial cable.

individual cable types indicator. For example, a 10Base2 network refers to a network running at 10 MHz baseband network. The *2* indicates it is Thinnet coaxial cable.

Coaxial cable falls into two categories: Thinnet and Thicknet. Thinnet cable, referred to as 10Base2, carries 10 Mbps of bandwidth up to 600 feet before requiring regeneration. Thicknet, called 10Base5, is able to carry the Ethernet signal 1640 feet. Both follow a linear bus topology, as shown in Figure 7–1, and both require special connection devices referred to as *MAUs*—media attachment units—to tie devices to the pipe.

A popular LAN cabling method used today is referred to as 10BaseT. 10BaseT defines how the signal rides on a twisted pair cable. The cable is further defined as *UTP*—unshielded twisted pair—or *STP*—shielded twisted pair. The 10BaseT standard carries 10 Mbps on the transmit-and-receive pair, as shown in Figure 7–2, in a star topology. This means the devices are tied back to a central hub or location through individual twisted pair connections. Once at the hub, the traffic is broadcast onto all segments forming the logical bus. The physical network topology looks like a star and

MAU
Media attachment unit is used to connect 10Base coaxial cable to the LAN computing device.

UTP
Unshielded twisted pair— unshielded twisted copper wire used in LANs.

STP
Shielded twisted pair— shielded twisted copper wire used in LANs. Deployed in noisy environments such as computer rooms.

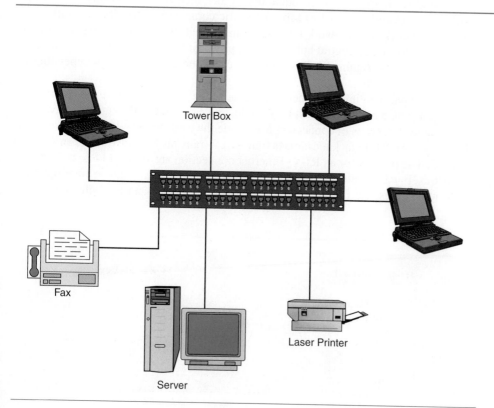

Tower Box

Fax

Server

Laser Printer

Figure 7–2
Category 5 twisted pair cable is the most common method deployed in Ethernet LANs. Each device is tied to one central location in a star topology while the logical network is configured as a bus. The devices all share the same baseband pipe.

Figure 7–3
Connecting devices to coaxial cable requires a media access device or tap. The device connects to the cable through BNC connectors similar to cable television connections.

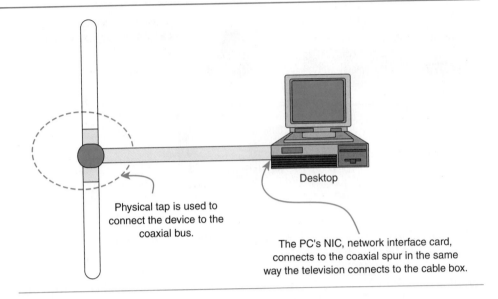

Desktop

Physical tap is used to connect the device to the coaxial bus.

The PC's NIC, network interface card, connects to the coaxial spur in the same way the television connects to the cable box.

the logical network topology is a bus as defined by the Ethernet standard. The 10BaseT network carries the signal around 100 to 150 m depending on the resistive factors of the cable.

Connecting devices to coaxial cable is accomplished through the use of AUIs and MAUs. An AUI is an attachment unit interface and refers to the connection between the transceiver and the controller on an Ethernet network. An MAU is a media attachment unit and is the transceiver on the network. A device or station connected via coaxial cable is tied into the network by tapping into the cable, as shown in Figure 7–3.

100BaseTX Ethernet, referred to as Fast Ethernet or FastE, also rides on twisted pair cable. Category 5 (CAT5) UTP or CAT1 STP is required for 100BaseTX networks, as the bandwidth speeds are substantially higher than those of 10BaseT networks. 100BaseTX networks carry 100 Mbps of traffic compared to 10 Mbps of traffic on a 10BaseT network. The network topology mimics that of 10BaseT systems. Each device is routed back to a central hub forming a physical star even though the logical network forms a bus. Again, UTP is typically used due to cost savings depending on distance requirements.

Twisted pair cabling is defined according to the number of twists per foot. Categories 1 through 5 are available in UTP cable, and categories 1 through 4 are available in STP cable. The types of connectors used to tie the cable to the NIC are referred to as RJ-45 jacks. An RJ-45 jack terminates four wires: transmit 1, receive 1, transmit 2, and receive 2. The pinouts for an RJ-45 Ethernet connection are shown in Figure 7–4.

10BaseF was developed to carry 10 Mbps of Ethernet traffic over fiber optic cable. In most cases, the fiber optic cable is multimode cable and the light sources are

Figure 7–4
RJ-45 modular connectors are used to terminate 10Base5 twisted pair in the connections. The transmit and receive pin designations are shown in the figure. Note that the 10Base5 standard requires a two-pair–four-wire connection.

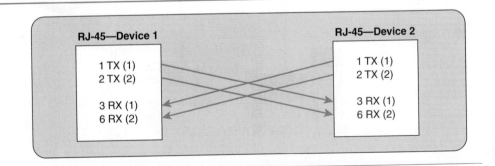

RJ-45—Device 1

1 TX (1)
2 TX (2)

3 RX (1)
6 RX (2)

RJ-45—Device 2

1 TX (1)
2 TX (2)

3 RX (1)
6 RX (2)

low-cost, low-power LEDs. Due to the changes in technology and the reduction in price of both fiber and light sources, most networks today deploy a standard that supports a higher bit rate such as 100BaseFX or even Gigabit Ethernet. 100BaseFX is also used to carry Ethernet over fiber optic cable. The main difference is that it carries 100 Mbps, making it more cost-effective. The last fiber optic standard defined for Ethernet is *GigE*—Gigabit Ethernet. GigE carries 1000 Mbps or 1 Gbps of information on fiber optic strands. All three fiber standards follow a point-to-point network topology. The value of these standards is that they all use the same Ethernet frame structure and the CSMA/CD network protocol. The only difference among the three is how fast the frames are outpulsed onto the wire.

The last type of medium used to carry Ethernet signal is RF—radio frequency—or wireless. The wireless Ethernet standard was developed by the IEEE and is called the 802.11 standard or *Wi-Fi*. The wireless standard is very similar to the wire-line standards. The physical topology follows the star configuration, while the logical topology stays true to a bus network. Similar to fiber optic cable, the distance and bandwidth of the wireless LAN are dependent on the type of transmitter and receiver deployed. A wireless LAN is typically rated at 100 Mbps. But, as with all transmission systems, the strength of the signal leaving the transmitter, the amount of loss experienced across the path from transmitter to receiver, and the sensitivity of the receiver determine the bandwidth and distance the signal can travel.

CSMA/CD

Carrier Sense Multiple Access with Collision Detection is the method used to place the frame onto the wire. CSMA/CD can be viewed as the Ethernet police force. The CSMA/CD protocol is responsible for maintaining traffic flows in the network and handling traffic jams as they occur. Collisions refers to instances when multiple frames grab the wire at the same time, similar to rush hour traffic in Los Angeles—too many vehicles driving on the same road. When a collision occurs, it is CSMA/CD's job to force both parties, stations, to back off before retransmitting the frame. Each party or station is given a different wait time by the CSMA/CD protocol to guarantee they will not collide a second time. Ethernet is a baseband protocol, meaning that it shoves the frames out onto one pipe that is viewed as one big channel or bus. Everyone tied to the network uses the same road. It is up to CSMA/CD to manage the traffic traveling on the road. When too many collisions occur, the network engineer has to figure out how to adjust or build out the network to reduce congestion.

CSMA/CD works because each device listens to the network before shipping its traffic onto the pipe. Again, similar to rush hour traffic, Frame A waits to pull out of its driveway until it sees an opening. When traffic jams do occur, CSMA/CD is able to quickly stop other frames from grabbing the wire by notifying them with a jam frame, similar to a traffic cop blowing his whistle. A jam frame is shipped out on the network telling each device to hold off for a minute in order to avoid a larger pileup. Each device waits a random amount of time before pushing frames back onto the wire.

Ethernet Speeds

Ethernet comes in three primary flavors: 10BaseT Ethernet, *Fast Ethernet*, and Gigabit Ethernet. All three abide by the Ethernet frame structure, the CSMA/CD protocol, and the ARP—address resolution protocol—method for resolving network layer and data link layer addresses and are designed to share the same channel. The differences between the three flavors are the location they are deployed in the network and the amount of bits they are able to carry every second—speed or bandwidth available.

Traditional LANs support 10Base Ethernet systems that are capable of handling 10 Mbps. Devices on the LAN share the 10 Mbps bus as just described. Fast Ethernet, a newer standard, increases the carrier speed to 100 Mbps and is commonly deployed across LANs and WANs. Fast Ethernet rides on CAT5 twisted pair cable, coaxial cable, and fiber optic cable. The third Ethernet speed found in the network is GigE and is most often deployed in the access and metro networks. GigE rides on

GigE
Gigabit Ethernet is an Ethernet standard that transmits 1 Gbps of information onto the wire. GigE uses the same frame structure and physical modulation techniques as the standard 802.3. The difference is the amount of information that is outpulsed onto the wire every second.

Wi-Fi
Wireless Ethernet, also defined by the IETF, is Ethernet packaged for RF transmission in a localized area.

Fast Ethernet
Ethernet standard defines the line rate of 100 Mbps. Rides on either twisted pair copper, coaxial cable, or fiber optic cable and is used across LANs and WANs.

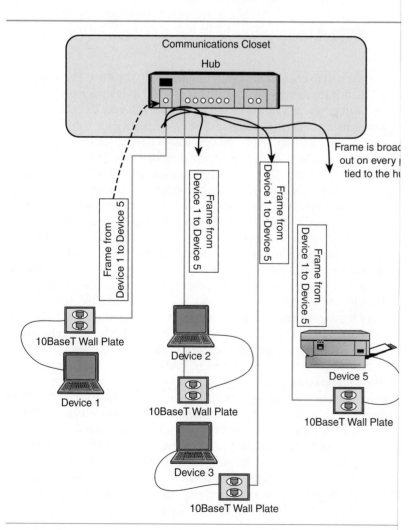

Communications Closet

Hub

Laptop Computer

Laptop Computer

10BaseT Wall Plate

10BaseT Wall Plate

Laptop Computer

10BaseT Wall Plate

Laser

10BaseT

Communications Closet

Hub

Frame is broad
out on every
tied to the hu

Frame from Device 1 to Device 5

Frame from Device 1 to Device 5

Frame from Device 1 to Device 5

Frame from Device 1 to Device 5

10BaseT Wall Plate

Device 2

Device 1

10BaseT Wall Plate

Device 3

10BaseT Wall Plate

Device 5

10BaseT Wall Plate

Figure 7–5
Ethernet is based on a bus network topology. All devices on the network sit on the same logical pipe.

Laptop

Serve

fiber optic cable as its line rate is
GigE are explained in more detail i

Ethernet Topology
Ethernet is based on a logical bus to
network on the same logical channe
vice on the network regardless of
10BaseT LAN connecting multiple
the signal, you would shake your he
ernet topology is no more a bus top
be correct in that the physical topol
vices are tied back to a central loca

If you work in an office with
blue cable connecting the PC to a d
ernet jack. You can look in any one
nected to a jack. If you happen to
the jack and notice the wires termin
wire connection—two pair, tip and
twisted pair CAT5 wires used to ti
nications closet, as shown in Figur

The Ethernet, whether it is a L
bus networks. The network is open
at one location and two nodes are
the end of the segment are discarde
the network, contending for space

7.2.3 Ethernet Equipment
Ethernet LANs depend on several
the Ethernet frames across the net
Ethernet are the hub, the Ethernet

Hub
Hubs are used for two purposes. T
distance the signal can travel on th
all the twisted pair CAT5 connect
network a logical bus topology by
lowing frames to travel to all dev
forward and repeat frames to eve

RJ-45
Eight-pin modular connector used typically in LAN applications that use twisted copper wire.

CAT5
Category 5 cable has a specific number of twists per foot and is constructed of a specific diameter copper wire. CAT5 is typically used in LAN applications.

hub
Devices placed in Ethernet LAN to extend the distance the signal can travel and to provide a connection point for multiple LAN connections.

Figure 7–8
Switches and routers are used to separate network segments through the use of routing and switching protocols. A router routes the packets onto the port on which the destination device resides. It does not broadcast the packets onto every port as is done with a hub.

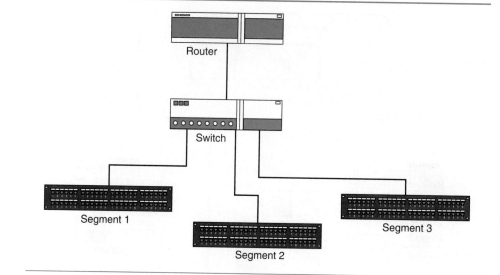

Router

Switch

Segment 1   Segment 3

Segment 2

A hub is not smart enough to read the Ethernet address to determine what port to ship the frame out on; it only knows it needs to forward the frame to all the devices on the segment. A segment is defined as the components bounded by a switch, bridge, or router. One segment includes the hub, the devices tied to the hub, and the medium tying the devices to the hub. A switch, bridge, or router is used to connect multiple segments together, as shown in Figure 7–8, using the hub as a connection point.

Ethernet Switch
An *Ethernet switch* is used to connect multiple LAN segments together. The switch is a Layer 2 aware device that learns what stations are hanging off its interface ports by reading and storing the source address of the incoming frames. Ethernet switches have the ability to learn who is living in its neighborhood. The switch, unlike the hub, sends an incoming frame out the interface port tied to the destination device. Interface tables, also referred to as bridging tables, hold the device information for a specified amount of time. If, for example, a device has been idle beyond the specified amount of time, its entry in the bridge table is flushed or removed. When a frame comes in destined for that device, the switch has to broadcast the frame out every port and wait for a response. The idle device grabs the frame and responds to the switch, telling it "yes, I am here and this is my address." The switch takes the device address and places it back into the table. The next frame destined for that device is sent out on the device's interface port as a unicast message, destined for that device only, meaning that only that device will grab and read the frame. The switch no longer needs to broadcast across all interfaces. The switch purges stale addresses from its bridging tables every so often in order to reduce the memory load.

The advantage the switch provides to the network manager is obvious. Network traffic is controlled and reduced between devices and between LAN segments, thanks to the switch's ability to recognize and learn where a device sits in the network. Ethernet switches act as interconnection points between LAN segments. The term *segment* is used to describe a group of devices sitting on the network tied to one interface port. For example, devices tied to a hub sitting in the communications closet on the first floor are tied to an Ethernet switch sitting in the fifth floor communications room. Other segments from floor 4 and floor 7 also tie to the fifth floor switch. A frame sent from a PC sitting on the seventh floor to a PC on the first floor travels into the fifth floor switch where it is analyzed and shipped out on the port tied to the hub located in the first-floor communications closet. The hub broadcasts the frame to all devices tied to it on the first floor. The PC on the first floor is the only one that grabs the frame, recognizing the MAC address as its own. Figure 7–9 illustrates this scenario.

Ethernet switch
Layers 2 aware device that connects multiple LAN segments together.

segment
Portion of the network tying specific devices together. Similar to a neighborhood, all houses belong to the town but only a subset of the homes belong to a neighborhood within the town. Devices belong to a segment, and segments belong to a network.

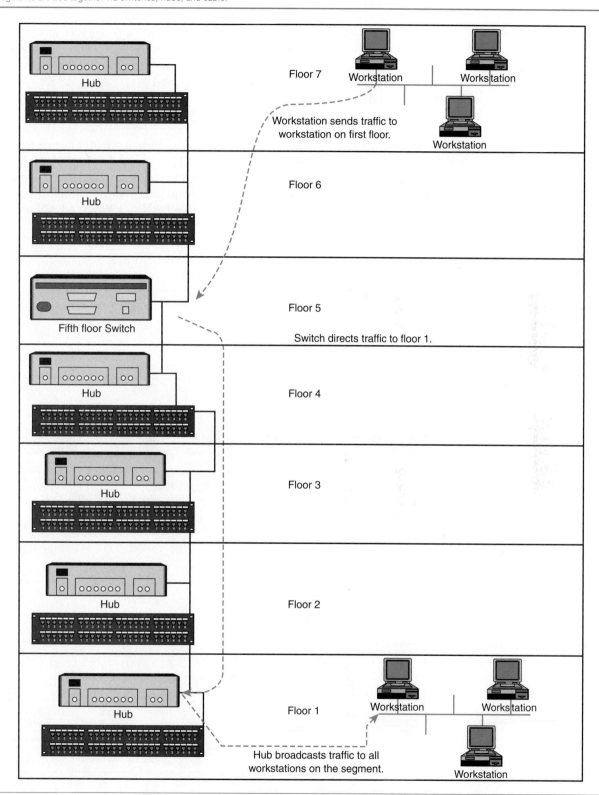

Floor 7   Workstation   Workstation

Workstation sends traffic to
workstation on first floor.

Workstation

Hub

Floor 6

Hub

Fifth floor Switch   Floor 5

Switch directs traffic to floor 1.

Hub   Floor 4

Hub   Floor 3

Hub   Floor 2

Hub   Floor 1   Workstation   Workstation

Hub broadcasts traffic to all
workstations on the segment.

Workstation

Bridge

bridge
Layer 2 aware device that
learns the network, similar to
an Ethernet switch.
Difference between a bridge
and a switch is primarily size
and cost.

An Ethernet *bridge* is similar to a switch in that it also learns a network and forwards frames according to the device's Layer 2 address, or MAC address. The primary differences between a bridge and a switch are size and cost. A bridge typically has two interface ports; an Ethernet switch has two or more interface ports. However, this definition may be disputed as manufacturers define devices differently. In general, when you hear the term *bridge* or *switch*, think of a Layer 2 aware device whose role in life is to connect multiple LAN segments together.

Router

router
Layer 2 and Layer 3 device
that reads both the Ethernet
frame overhead and the IP
packet overhead before
determining what to do with
the information.

The *router* is a Layer 2 and Layer 3 aware device, meaning that it is able to read the Layer 2 device address, translate it into a Layer 3 address, and maintain bridging tables accordingly. Routers are discussed in detail in a later chapter. For now, it is important to understand that LANs and WANs use routers to interconnect multiple networks together: LANs to LANs, a LAN to the Internet, a LAN to a WAN, a LAN to a server farm, and so forth.

In Summary

A bridge, switch, or router is used to tie multiple segments together to form one LAN or to extend the LAN across a WAN, as shown in Figure 7–10. A hub is used to connect devices together, allowing the network topology to be configured as a physical star but a logical bus.

Figure 7–10
LAN 1 in Detroit is connected across a DS-1 circuit to LAN 2 in St. John's. The devices on each LAN talk to each other in the same way the LAN segments talk to each other between floors. They do not know they are located hundreds of miles apart.

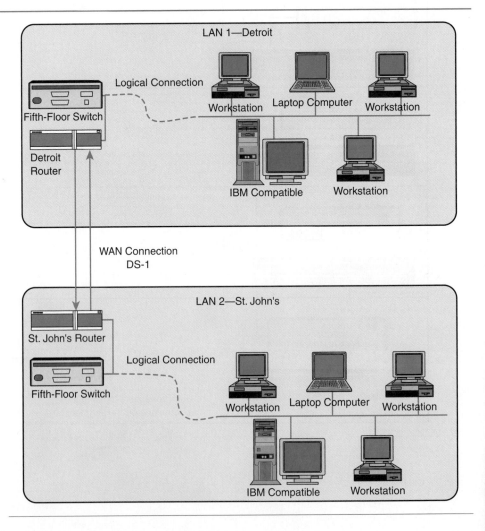

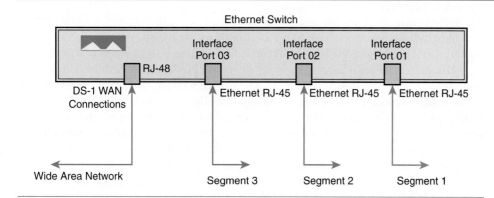

Figure 7–11
Switches are commonly used to
interface the WAN to the LAN using
different interface speeds. The
diagram shows a DS-1 WAN port.
DS-3, Fast Ethernet, and so forth are
also common WAN interfaces that are
deployed.

7.2.4 Ethernet WANs

The primary differences between Ethernet LAN equipment and Ethernet WAN equipment are size and the way each handles the interconnection to the physical layer. WAN switches, such as Cisco's Catalyst switch, have multiple ports tied to individual networks on the drop or line side and one or two high-speed ports tied to the WAN, as shown in Figure 7–11. The low-speed or line-side ports are typically FastE optical connections that aggregate into one large GigE pipe connecting to the POP or other aggregation site. Ethernet switches may also connect using CAT5 twisted pair cable and aggregate into a FastE connection upstream. In either case, the switch's job is to move frames between *interface ports*.

Routers are commonly used in the WAN to tie multiple switches together, as shown in Figure 7–12. A router is more sophisticated in that it is able to route Layer 3 packets between Layer 2 interface ports using routing protocols, as explained earlier. Routers today interface directly using FastE and GigE optical connections, allowing traffic to move freely between interface ports. Routers, as explained later, are Layer 3 aware devices, meaning that they are able to decipher the network layer protocols in order to route traffic to its destination.

interface port
Port or physical connection
on a piece of equipment
used to connect the circuit
May be a copper wire, fiber
optic cable, or wireless
transmitter/receiver.

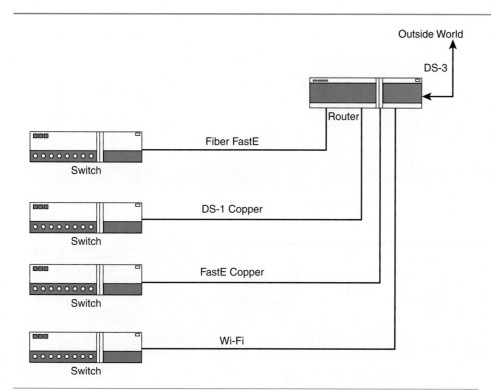

Figure 7–12
Aggregating traffic from multiple
switches is often handled by placing a
router in the network. The router has
the ability to crack open the packet
and route according to the network
layer protocol.

7.2.5 Ethernet Frame

The Ethernet frame is similar to a cardboard box that is used to carry multiple letters to a location. An address is placed on the top of the box indicating where the box is coming from and where the box is destined. The address also contains other information telling the receiver what is in the box, the number of letters in the box, and how the box should be handled during its travels. The Ethernet frame, similar to the address on the box, has a field for the source and destination address; a field indicating what type of information it contains and/or a field telling the receiver how much data the frame contains; a field used to determine if the frame arrived intact; and a variable-length field—from 46 to 1600 bytes—for the actual data—or, in our analogy, letters. Figure 7–13 depicts the structure of the Ethernet II and 802.3 frames.

The Ethernet frame mimics the two sublayers of the data link layer: MAC and the LLC. The MAC—medium access control—layer of the protocol is responsible for controlling who talks to whom and when. The MAC layer sits next to the physical layer, ensuring that the box or frame is properly loaded and unloaded from the transport vehicle in addition to figuring out where the box should go using the source and destination address fields.

The LLC—logical link control—sublayer interfaces with the network layer in establishing flows between the two. The LLC in certain cases, depending on which Ethernet frame format is deployed, sets a marker that indicates the type of information contained in the frame. The Layer 3 protocol uses this marker to determine how to hand the package to the upper-layer protocols. Ethernet II, used in TCP/IP networks, uses a *type* field to define the data being carried. 802.3 uses a *length* field to indicate the size of the header information in the data portion of the frame.

Both frame formats are segmented according to what portion of the protocol they service. For example, the address portion of the frame is used to interface to the MAC layer of the protocol and the type field of the frame is used by the LLC to improve the handoff between Layer 2 and Layer 3. Even though it is nice to know how an Ethernet frame is segmented, the key point to grasp is that Ethernet's purpose in life is to carry lots of packets the same way a shipping box carries lots of letters—a more efficient way to transport information.

type
Field within the Ethernet frame that indicates the type of data being carried and whether the data should be given priority over other frames.

length
Field within the Ethernet frame that indicates the size of the header information in the data portion of the frame.

Figure 7–13
Ethernet framing comes in two flavors—Ethernet II as defined by Xerox and 802.3 as defined by the IEEE. The main difference between the two standards is the fifth field in type field versus length field.

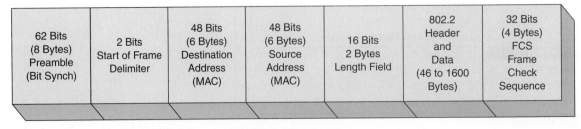

7.2.6 Ethernet Addressing

The Ethernet frame has a six-byte source address field—whom the frame is coming from—and a six-byte destination address field—whom the frame is going to. The first twenty-four bits of the address indicate the manufacturer's OUI—Organizationally Unique Identifier. The IEEE assigns each manufacturer an OUI that is burned into the NIC at the time it is built. NETGEAR, Cisco, Nortel, Lucent, and others have OUIs that differentiate them within the industry. The remaining twenty-four bits in the address are assigned by the manufacturer and used to identify the physical card or port. Each NIC has its very own twenty-four-bit address. Three terms commonly used to describe the address are *physical address*, *hardware address*, or *MAC*—medium access control—*address*. If you want to read your PC's MAC address burned into the NIC, go to the DOS prompt and type in "IPconfig/all." Note that the MAC address is listed along with the IP address and the IP gateway. You may over time be given different IP addresses depending on the ISP you use, but the MAC address stays the same unless you physically replace the NIC in the PC.

A MAC address is burned into the PC's ROM—read-only memory—and is referred to as the BIA or burned-in address. Once the NIC turns up for the first time, the address moves from ROM to RAM—random access memory—so future requests coming from the network can be processed quickly. Some devices are capable of spoofing MAC addresses, that is, copying a MAC address of a device onto a different device. Spoofing allows the device holding the copied MAC address to become part of the network.

physical address
Twenty-four-bit Layer 2 address used to identify the network interface port of the piece of hardware.

hardware address
Same as the physical address.

MAC address
Media access control address—essentially the same as the physical and hardware address.

Unicast, Multicast, and Broadcast

Ethernet addresses come in three flavors: *unicast*, *multicast*, and *broadcast*. Unicast addresses are used to send frames to one specific PC. The source address in the frame is always a unicast address. It holds the address of the device sending the frame. The unicast destination address is the MAC address of the destination device, such as the NIC of a PC. Each device on the network reads the destination address to determine if the frame is meant for it—like reading the address on the shipping container. If it matches the MAC address burned into its NIC, the device grabs the frame, reads the frame fields, and hands the frame to the network layer for further processing. The network layer reads the LLC portion of the frame in order to determine which upper layer needs to receive the information inside the frame. At this point, the Layer 2 Ethernet frame has been stripped off and discarded, leaving the Layer 3, Layer 4, and higher information.

Only destination addresses can hold multicast and broadcast addresses. A multicast address is used when a frame is destined for multiple devices on the network, such as when certain devices participate in Web conferencing. The third type of address is the broadcast address and is used when all devices on the network need to receive and accept the frame. All ones are used in the destination address of the frame to indicate it is a broadcast frame. Broadcast addresses are very useful, as is discussed in the next section on ARP, but they may also be very detrimental to the network if not properly managed. A network can be flooded by broadcast frames that force each device to stop, read the LLC, and open the frame. Remember that devices on an Ethernet share the same pipe, meaning that they all receive every piece of traffic flowing on that pipe. It would be similar to every person on the street receiving hundreds of pieces of junk mail at one time. The amount would be overwhelming to the home owner, not to say what it would do to the postal carrier. Every frame pushed out on pipe eats up a portion of the bandwidth. Unnecessary broadcasts can flood the network causing contention, collisions, and rebroadcasts, which only increase the effects of flooding. An example of a MAC address is shown in Table 7–2.

unicast
Ethernet address used to send information to one specific device.

multicast
Multicast address refers to an address used to send to multiple devices that have been specified.

broadcast
Ethernet address used to send the information to all devices on the network.

Address Resolution Protocol

ARP—address resolution protocol—is a protocol used to translate or resolve the MAC layer address to the LLC layer address. As we know, each device has its very own

ARP
Address resolution protocol is used to translate Layer 2 addresses to Layer 3 addresses.

Table 7–2
A source MAC address is always a unicast address, meaning that it is the unique hardware address burned into the device. Destination addresses may be unicast; multicast, destined for a group of devices; or broadcast, headed for all devices.

	Ethernet Address Types		
MAC Address Type	**Source Address**	**Destination Address**	**Definition**
Unicast	Always Unicast	May Be Unicast	A unicast address is the unique hardware address of the device. Forty-eight-bit address burned into the device.
Broadcast	NA	May Be Broadcast	A broadcast address is an all ones address that directs the device to broadcast the data to every device regardless of the unique hardware address.
Multicast	NA	May Be Multicast	Multicast addresses are used to target a group of devices. The packets are sent out to multiple devices that have been assigned to the multicast group.

RARP
Reverse ARP defines how devices respond to an ARP request.

unique MAC address. In the world of TCP/IP, each device also has its very own IP address. In order for a network device such as a switch or router to be able to send a frame out on the correct interface port, it needs to figure out what IP goes with what MAC. ARP messages are broadcast to every device on the network asking for each device's logical address. For example, a LAN supporting ten PCs receives traffic from the outside world and has to figure out which interface port it needs to be shipped out on. The router sends out an ARP request to all ten PCs on the network asking them to look at the destination message of the frame; if it matches, the PC needs to respond with its MAC and IP address. The PC complies by placing its MAC address inside the ARP response, or *RARP*—reverse address resolution protocol. The router looks at the MAC and logical address and places them in a bridging table, which is a table that associates the PC's IP and MAC addresses. The bridging table sits in the device's memory for future routing requests. When a frame arrives, the switch or router is able to determine which interface port to shove it out on by referring to its bridging table.

ARPs are periodically broadcast into the network in order to update the router's or switch's bridging tables. An ARP's out time is defined by the manufacturer and typically is a configurable value. Other means for translating MAC addresses into network layer addresses are available and used as required. TCP/IP, the predominant Layer 3 protocol used today, depends on ARP to associate the logical address to the hardware or MAC address.

7.2.7 Ethernet, the Interface between Layer 1 and IP

A simple way to visualize how the Ethernet moves traffic between Layer 3 and Layer 1 is to compare it to a candy factory that makes five different kinds of candy for five different kings. Five candy stations are located around a large warehouse. A shared conveyor belt used to carry the boxes of chocolates to the shipping department ties each of the candy stations together. Each candy station is identified by each king's coat of arms. As the chocolate candy is manufactured, it is placed inside a candy box. The candy box is stamped with the king's coat of arms and then placed on the conveyor belt. The conveyor belt carries all the boxes of candy destined for the shipping station. The candy boxes that reach the shipping department are packaged in large shipping containers and readied for shipment. The shipping personnel are able to determine to which carton the candy box belongs by reading the coat of arms. They do not have to open each box to see the type of candy inside in order to figure out which shipping box should be used. Though this is a simplistic view showing how traffic flows in one direction only, the basic concept remains the same. Ethernet, as with all Layer 2 protocols, simplifies moving information around the network. If each piece of chocolate was placed on the conveyor belt, the shipping department would quickly become overwhelmed by the volume and soon would start to throw out the delicious treats.

■ 7.3 FRAME RELAY

7.3.1 Frame Relay: A Definition

Frame relay is a protocol designed by information technology experts as an alternative method to transport data across digital circuits. X.25 was the standard packet transport technology throughout the 1980s, but the limitations associated with it, such as slow speeds, cumbersome routing algorithms, and less-than-desirable reliability statistics, warranted the introduction of a new protocol. The maximum data transmission speed available on X.25 networks ranged from 9.6 to 56 kbps. Another weakness of X.25 networks was a high latency rate—the time it takes a packet to traverse the network—because of the extensive error checking and retransmission schemes used to ensure the information was not corrupted. Additionally, a high error rate of 10^{-5} or 10^{-6} required that a large percentage of the packets be retransmitted.

The frame relay protocol was meant to solve many of the problems associated with the X.25 protocol and was built to take advantage of the new digital high-speed circuits allowing for higher-speed interfaces. Frame relay links run as high as 1.544 Mbps in North America and 2.03 Mbps in Europe. To reduce the latency through the network, the frame relay protocol leaves error checking and retransmission of packets to the higher-layer protocols such as TCP/IP. The link between the beginning and ending device is handled at the Layer 2 (data link) level, thus simplifying the connection. Frame relay is a very simple, unsophisticated protocol. Information encapsulated in a frame relay frame does not have to be opened by every device to know which circuit to route it on, as is the case with IP. This reduces the time it takes the information to travel the network, essentially providing a low-latency pipe.

Frame relay protocol requires that a VC—virtual circuit—be built between devices before the circuit is turned up for service. Information is encapsulated in a frame relay frame that identifies which VC the information will travel on. Many analysts predicted that ATM and IP would replace frame relay networks by the late 1990s, but so far they have been wrong. Customers like the simplicity of frame relay because of the low cell tax or overhead and the low cost of the equipment.

7.3.2 Frame Relay Networks

The frame relay protocol is a connection-oriented protocol. Unlike IP, which is a connectionless protocol, frame relay eliminates the need to read individual device addresses and routes across the best link. Logical circuits—PVCs (permanent virtual circuits) or SVCs (switched virtual circuits)—are used to build logical connections between multiple devices. As shown in Figure 7–14, multiple PVCs may be configured between multiple locations from one device. PVC 10 connects the Houston device to the Philadelphia device, while PVC 8 defines a logical circuit between Houston and Dallas. The alternative is an SVC that establishes the connection at the time the request is generated. Signaling messages are used to set up a connection when needed and to disconnect the connection when finished. SVCs are used more extensively in private networks at this point than in large public frame relay networks.

A frame relay network is composed of three general parts. The access piece interfaces with the customer premises equipment, also called the enterprise network; the switching equipment establishes connections between locations; and the backbone

Figure 7–14
Layer 2 connection-oriented protocols
such as frame relay use PVCs to carry
different signals across the same
physical interface. Four PVCs have
been assigned to the DS-3,
connecting the two frame switches.

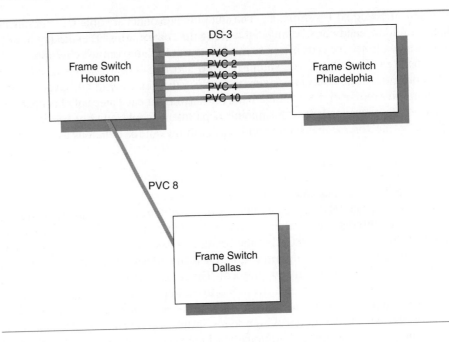

network carries the frame relay frames through the network. As shown in Figure 7–15, the customer premises router often contains a frame relay access device (FRAD) that is used as an interface to the access circuit of the service provider. The FRAD performs two tasks. The first is to act as an interface to a digital high-speed circuit of 56 kbps or more in North America. The second task is to encapsulate or de-capsulate information leaving and coming into that device, much like multiple envelopes are placed into one larger envelope. For example, when an IP packet is sent to a router, the router sends the packet to the T1 interface card that has a FRAD attached. The FRAD encapsulates the higher-layer protocols and readies them for transport at the Layer 2 level.

From the enterprise location, the access circuit travels from the customer's location into the local telephone network plant where it is terminated either to a long distance provider as a T1 physical circuit or into the local service provider's own frame relay switch. The frame relay switch determines how the information arriving from the

Figure 7–15
A FRAD—frame relay access
device—is used to interface into the
customer's terminating equipment.

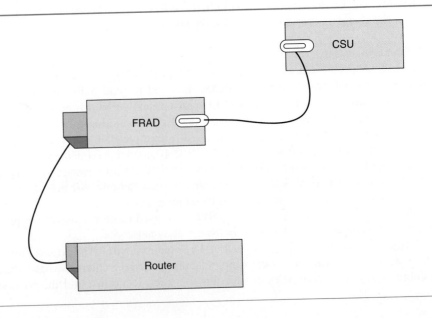

customer's location will be routed. When the switch decides on a route by looking at the header, which contains a link identifier in the frame, the information is shipped out onto the backbone network, often depicted as a cloud. The information enters the cloud and travels along a path that may or may not go through additional frame relay switches. Within the frame relay cloud, as shown in Figure 7–16, frame relay switches and a backbone network connect multiple devices. The size of the frame relay backbone depends on the total bandwidth required but typically ranges from multiple T1s to a DS-3 circuit. Working hand in hand, the three areas of a frame relay network ensure that the information is packaged properly, routed properly, and handed off.

Beyond the physical connections, a frame relay circuit is defined by three parameters—the speed of the port, the *data link connection identifier* (DLCI), and the *committed information rate* (CIR). The speed of the port is usually 56 kbps or 1.544 Mbps. The DLCI is the number assigned to that virtual circuit. A DLCI is needed because there are multiple VCs within a physical pipe. Each must be identified so that the frames can be routed onto the correct logical connections. A typical DLCI is ten bits long and resides in the frame relay header. It tells the equipment which device the circuit terminates on.

A CIR is a guaranteed bit-per-second rate defined for each DLCI. Each end user commits to the CIR. For example, if the customer has a port speed of 1.544 Mbps and a CIR of 50%, the frame relay provider guarantees that 756 kbps of bandwidth will be available on its network for the customer's traffic. The customer is able to go above the CIR rate of 756 kbps to 1.544 Mbps when the circuit is not being used by, or is not congested by, other customers' traffic. If the circuit is congested, the carrier has the right to discard any bits above the CIR of 756 kbps; but, if it is not, the end user can send a full 1.544 Mbps even though the CIR is only 756 kbps. Carriers and end users are very cognizant of the relationship between the CIR and the actual amount of bits being shipped. The network carriers must be careful not to overengineer their network to the point that they are providing free bandwidth to those who buy a low CIR but send large amounts of data. On the other

data link connection identifier
The number assigned to the frame relay circuit.
committed information rate
Guaranteed bit-per-second rate defined for each DLCI.

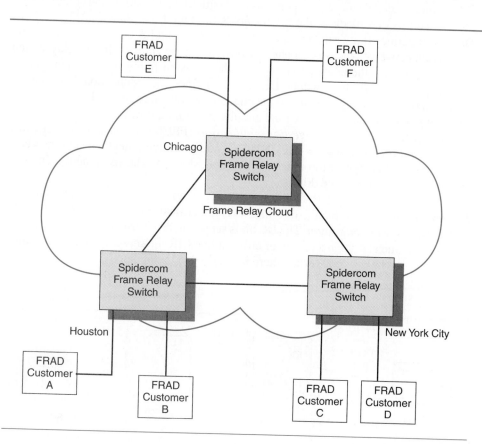

Figure 7–16
The frame relay network shows a common network architecture. The frame relay switches establish virtual circuits between cities. Spidercom's customers share the bandwidth between the cities, as shown inside the frame relay cloud.

Figure 7–17
A review of the overhead bits found in the frame relay frame.

FLAG 8 Bits	DLCI 16 Bits	A 8 Bits	B 8 Bits	Data $n \times 8$	FCS 16 Bits	FLAG 8 Bits

service level agreement
Agreement made between the customer and the service provider defining the level of service promised and the parameters such as uptime, bandwidth, and lost fees.

hand, the carriers must comply with *service level agreements* (SLAs) established between the customer and the carrier that place an obligation on the network provider to guarantee a specific amount of bandwidth for that customer's traffic. Customers often assume that carriers have overengineered their networks, decide to purchase a lower CIR, and burst above it thinking that they will not incur any adverse consequences. Thanks to sophisticated traffic engineering processes, frame relay networks are being watched more closely than they were a few years ago. Customers who purchase low CIRs expecting little congestion are finding that they need to rethink their strategy and adjust their agreements to increase their CIRs.

7.3.3 Frame Relay Frames

The structure of the frame relay frame is shown in Figure 7–17. It is fairly simple compared with other protocols such as IP and IPX. We begin by focusing on the most important blocks in the frame relay frame. The frame begins and ends with flags that define the boundaries of the frame. The portion of the frame that holds the overhead information is called the *header*. It contains the beginning and ending flags referred to as blocks. The second block in the header holds the DLCI. As shown in Figure 7–18, DLCI 100 defines the virtual circuit between points A and B or, in our example, the connection between the customer site in Houston and the Houston frame relay switch. DLCI 101 defines the connection between the Houston frame relay switch and the Philadelphia frame relay switch. DLCI 201 defines the connection between the Philadelphia frame relay switch and the end device at the customer's location. DLCIs are locally defined and identify virtual circuits throughout the network.

forward explicit congestion notification
Block that monitors congestion on the line.

backward explicit congestion notification
Block that monitors congestion on the line.

discard eligibility bit
Bit used to mark a frame as being discard eligible.

The *forward explicit congestion notification* (FECN) block and the *backward explicit congestion notification* (BECN) block notify the equipment when there is congestion on the line due to too much traffic being shipped. The equipment then automatically ratchets the speed down.

The *discard eligibility* (DE) *bit* is used to mark a frame as being discard eligible. *Discard eligible* means that the information is not critical and can be thrown out if the circuit experiences congestion. The DE bit is set when the information in the frame has low priority, such as when a customer orders a low CIR. Frames that do not have their DE set will not be discarded when there is congestion on the line.

Figure 7–18
Three DLCIs are needed to connect one FRAD to a second, shown here for one direction.

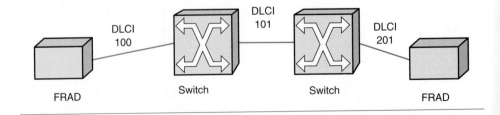

The portion of the frame that carries actual information, called the *data block*, consists of a variable-length frame similar to the IP frame that has a maximum length of 4096 octets or 32,768 bits. (One octet equals eight bits.) The overhead tax associated with a frame relay frame is lower than that of ATM and other protocols because of the large frame size. Cell tax is the amount of bandwidth used for header and trailer information. Frame relay's long, variable-length frame designates a small portion to overhead bits, thus improving the efficiency of the protocol. Therefore, it provides a very low cell tax or frame tax from overhead.

The final block in the diagram is the cyclic redundancy check (CRC) block. It is used to check the physical link for errors. A high CRC error rate causes the frames to be discarded. Retransmission is handled by the upper-layer protocol. The CRC tells the end device that there is a problem on the line, thus causing a critical, major, or minor alarm to appear at the equipment. The frame relay protocol is not intelligent enough to know that it has to retransmit information because the strength of the protocol comes from not having to think about that. The less time the frame has to think, the faster and more efficiently the information can be carried across the network. Conversely, TCP/IP is always looking, thinking and deciding whether or not traffic has been corrupted, which increases the time it takes for information to travel the network, hence the higher latency between start and end.

7.3.4 Frame Relay Switches

Technically the frame relay switch is, a switch that moves information between multiple locations. The difference between a frame relay switch and a voice switch is that the frame relay switch uses the DLCI in the frame header to determine where to send the frame, while the circuit switch looks at the telephone number punched in by the end user or caller. Persons using frames do not even know what a DLCI is. They hit the *send* key on their e-mail and let the frame relay equipment figure out where the information needs to go. A frame relay switch has interfaces coming in from subscribers, ingress circuits, and circuits interfacing into the network, or egress circuits. The subscriber circuits coming in may range from 56 kbps to 45 Mbps depending on the type of device placed at the end customer's premises and the speed of the circuit connecting the two locations.

The frame relay switch receives the frames from the end device, strips open the header, and looks at the DLCI that tells the equipment which device the frame should be sent to. Multiple DLCIs are used to define the path between two locations. The frame relay switch takes frames from one port and directs them to a second port, directing incoming traffic from access lines to the correct trunk or network port that carries the traffic into the cloud across the backbone network.

A frame relay switch has four components: the access interfaces, the network interfaces, the switching fabric, and the processors. The access interfaces are normally lower-speed ports where information is passed between the user network and the public network. The term used to define this interface is *user-to-network interface* (UNI). The connection just described would be a UNI. The term used to define the network interface between two carriers' networks is *network-to-network interface* (NNI). As shown in Figure 7–19, a frame relay switch has both types of interfaces. The access interface can be a 56 kbps circuit, a T1 circuit, or a 45 Mbps circuit; the network interface may be a T1 or T3 circuit. Determining the speed of the backbone network is one of the jobs of the traffic engineer. Building a backbone network for frame relay services depends on many variables that can affect service. For example, the type of switch fabric depends on the manufacturer. In addition, most frame relay switches are categorized by switching capacity. Frame relay switches are not as sophisticated as ATM switches or routers and therefore do not require as much switching fabric and processor power. If SVCs become more prevalent, frame relay switches will have to acquire additional processing power and more complex configurations. The switch is now built around the backbone network, the number of access ports, and the number of frames per packet required per second.

Figure 7–19
The frame switch is able to accept many types of signals. T1 ports for input and output are available, as are DS-3 and optical interfaces.

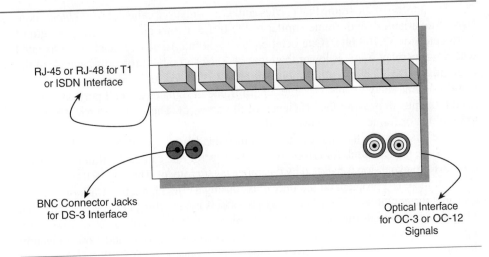

RJ-45 or RJ-48 for T1
or ISDN Interface

BNC Connector Jacks
for DS-3 Interface

Optical Interface
for OC-3 or OC-12
Signals

7.3.5 Frame Relay Interface Devices

The frame relay interface device is found at the end user's site or at the terminating access port. It encapsulates the data into the frame relay frame before placing it onto the digital circuit. With this service, a frame relay device must be placed at the end of the circuit. The frame relay access device, the FRAD, is often placed directly into the router or CSU—channel service unit—depending upon the end user's preference. The FRAD encapsulates the data in a frame relay frame and places a DLCI and other header information around the data. The FRAD can be seen as a place where information is placed in a box, an address is placed on the box, and the box is shipped out. Figure 7–20 illustrates where a FRAD is found in the network.

Figure 7–20
The FRAD is often built directly into the router, the CSU, or other data equipment. Routers may have internal FRADs as shown in (a). CSUs may also have internal FRADs as shown in (b). The FRAD may be a stand-alone unit as shown in (c) or reside in a shelf as shown in (d).

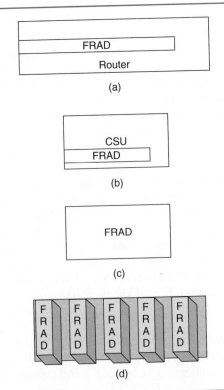

FRAD

Router

(a)

CSU

FRAD

(b)

FRAD

(c)

F R A D   F R A D   F R A D   F R A D   F R A D

(d)

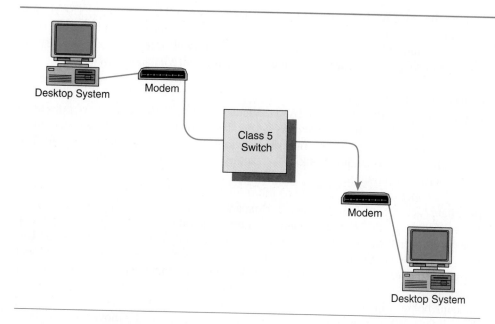

Figure 7–21
Common dial-up circuit-switched circuit. The benefit of using circuit-switched lines is the flexibility offered by dialing up any end. The connection is released after the session is complete. The downsides of dial-up connections are the low speeds available—maximum of 56 kbps— and lack of security.

7.3.6 Connecting the Dots
Defining the Network

The best way to understand frame relay as a networking protocol is to compare it with a traditional dedicated circuit transport method. We begin by describing a dedicated data circuit used today and then convert this network to frame relay, illustrating the advantages associated with using the frame relay protocol. The network consists of two links: one between Dallas and Houston and one from Dallas to Philadelphia.

Conventional Dedicated Circuit Network

Embassy Enterprises has multiple circuit-switched lines that carry data among its remote offices, as shown in Figure 7–21. Several dedicated 9.6 kbps circuits connect the four locations, where each 9.6 kbps circuit requires a CSU that interfaces into the DTE. The 9.6 kbps circuits are multiplexed into a router that is attached to T1 circuits that link each office to any of the other four offices. Three T1 circuits are needed to connect the three remote locations together. Denver does not have enough bandwidth to warrant a full T1 circuit and therefore has three 56 kbps digital circuits connecting it to Houston, Dallas, and Philadelphia.

The three components that make this scenario expensive are the multiple interfaces required in the router, the multiple access circuits, and the channelized backbone. Each circuit requires a DSU and an interface port in the router, as shown in Figure 7–22,

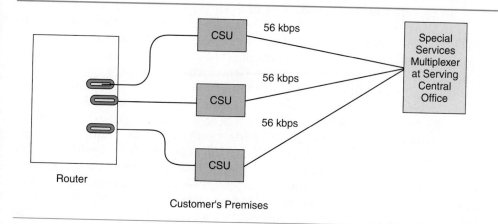

Figure 7–22
Typical point-to-point special service data circuit. The customer requires three CSUs and three individual ports on the router in order to complete the circuit.

making the price of the router and the deployment of DSUs fairly high. The circuits connecting the cities use channelized T1 circuits that do not allow information to be statistically multiplexed. This eliminates the possibility of oversubscribing the service. This is a fairly typical situation for small- to medium-size business customers.

Conversion to Frame Relay

Before explaining the conversion from dedicated circuits to frame relay, we present an analogy to help define how frame relay works. Henry owns a paper cup and paper plate factory and ships his product to multiple locations throughout the country via rail. The train pulls up to the loading dock at the factory, where product is loaded to be transported across the country.

The paper cups are loaded onto the train from loading dock 1, while the paper plates are loaded from loading dock 2. Boxcars carry either paper cups or paper plates but not both. The problem associated with this method of transport is that paper cup sales skyrocket during the summer, exceeding the capacity of the paper cup boxcar. Unfortunately, we cannot fill empty space in the paper plate boxcar with the excess paper cups, so we must lease another boxcar for the additional paper cups. Once the summer season is over, the demand for paper cups is reduced and one boxcar is able to hold all the cups being shipped. At the same time, orders for paper plates increase because of the holidays. We must now reconfigure our network to physically move the extra boxcar to the paper plate loading dock to carry the additional paper plates. Comparing our boxcars and products to dedicated circuits and information, we can see that this method of carrying data is very inefficient and costly due to wasted bandwidth, the need for additional ports, and time spent rearranging circuits.

The frame relay solution would change the scenario to a shared loading dock and shared boxcars for both paper plates and paper cups. This would eliminate the need for two loading docks, reduce the number of boxcars needed, allow different types of traffic to ride on one circuit without channel boundaries, and reduce the time it takes to rearrange circuits.

A more significant benefit of frame relay is the elimination of multiple circuits connecting the three remote sites. Returning to the train analogy, each remote site requires its own dedicated train. We needed a train from Houston to Philadelphia, one from Houston to Denver, and from Houston to Dallas. However, if we were to ship via UPS, the paper cups and paper plates could be packaged, addressed, and placed into a local UPS truck, which would carry the packages to the airport. They would then be placed onto a UPS plane with other companies' packages. The plane would carry the packages to the end destination by reading the address on the box. Like UPS, the frame relay switch sorts the incoming data frames and ships them out onto a frame relay plane that carries them to their end destinations. The use of switched frames and shared facilities eliminates the need for each location to be connected to every other in a full mesh topology. Embassy is then paying for space on the network, not for a dedicated circuit.

To convert Embassy's traffic over to a frame relay network, several decisions have to be made. The first is the rate of the access line—the circuit that connects the end user to the frame relay switch. Deciding between a 56 kbps and a T1 access line requires a traffic study by the customer. Embassy's traffic study shows that each site (except Denver) requires a full T1 circuit. Denver requires a 56 kbps circuit. The second decision is the CIR rate, which depends on how critical the data transmission is to the customer. Very often, end users buy whatever the carrier offers. Some carriers only offer 50% CIR while others offer levels of service that are tied to the cost of the service. In our case, Embassy orders the circuits with a 30% CIR. If Embassy was going to place its voice traffic on this circuit, the CIR would have to be much higher. The final decision is where the FRAD will be placed. Embassy buys a frame relay interface in its router that does not require an external box. Figure 7–23 shows Embassy's newly configured frame relay network. The frame relay network provides much more redundancy, better bandwidth utilization, and reduced cost of end user equipment. The

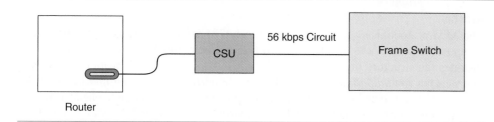

Figure 7–23
Embassy's frame network provides a cleaner, more efficient use of bandwidth by allowing users to share the 56 kbps pipe.

decision to move to a frame relay network from a TDM network is easy; the cost savings quickly justify the decision.

Voice Over Frame Relay

When frame relay was introduced, pundits claimed that voice would never ride on a frame relay network. They were wrong. Engineers have added special functions to the protocol that allow voice to travel reliably across a frame relay network.

However, because of the variable-length frame of the frame relay protocol, finding a way to carry voice traffic was not easy. Humans do not like gaps in voice conversations. Therefore, the structure of the frame had to be adjusted so that anytime voice was carried it took priority over data. Besides marking the voice frames with a priority bit, the frames had to be divided into defined-length segments because voice does not like long, variable-length frames. The result was equipment that could segment the long 4096 octet frame and add priority bits to signify voice traffic.

Many corporate frame relay networks carry data and voice between regional offices. The second biggest users of voice over frame relay circuits are international companies that connect directly into a foreign country's voice switch. One of the main reasons for this is to bypass the *public telephone and telegraph* (PTT) networks' expensive interconnection policies. Frame is viewed as data and therefore is exempt from the PTT's voice tariffs. In many cases, only one or two large carriers are given permission to interconnect into the PTTs' voice network. All other carriers must purchase bandwidth from one of them or depend on bypass carriers.

public telephone and telegraph
Telephone and telegraph service in most foreign countries. Usually run by government.

■ 7.4 ATM

7.4.1 Introduction to ATM

Asynchronous transfer mode (ATM) does not refer to the ATM you use to access your bank account. The use of ATM as a multiplexing/switching protocol for broadband services began in 1985 with standards organizations, manufacturers, and service providers. ATM is now overseen by the ATM Forum, the organization that helps define the standard, introduce new applications, and promote ATM as the broadband technology of the twenty-first century. From its inception, the founders of ATM wanted to create a protocol that could switch and multiplex broadband traffic across a network, regardless of the type of information being transported. Their goal was to incorporate the statistical efficiencies found in packet switching with the reliability found in circuit switching/TDM. ATM networks are growing in popularity. The quality-of-service options they provide allow multiple services to travel reliably on one circuit and offer a statistically multiplexed technology that helps improve the efficiency of the network.

Two factions have now emerged. One touts ATM as the choice for the new converged networks; the other claims IP as the dominant protocol for the new networks. The advantage of using ATM is the extensive quality-of-service features built into the structure of the protocol. The disadvantage is that ATM is not ubiquitous across the network, meaning most enterprise customers depend on IP to carry their LAN and Internet traffic into the access network. Though IP is ubiquitous, it is not able to provide a quality-of-service guarantee for time-sensitive traffic such as voice and video. Who will win? Probably both, because both provide benefits.

7.4.2 ATM Protocol

ATM Cell

ATM cell
Fifty-three-octet cell.

The ATM protocol is a cell-relay protocol. An *ATM cell* is defined as a fifty-three octet composed of a five-octet header and a forty-eight-octet block for data. Figure 7–24 illustrates the structure of a generic ATM cell. ATM is a connection-oriented protocol that uses the *virtual path identifier* (VPI) and the *virtual circuit identifier* (VCI) to define the connection end to end. Error checking is performed only on the header blocks. The ATM header holds the VPI, VCI, and error-checking blocks. The VPI and VCI define the logical circuit ID end to end similar to the DLCI in frame relay. The VCI defines the connection between two points, while the VPI defines the path of multiple VCIs from the beginning to the end. Error checking for the forty-eight-octet data is left to the higher-layer protocols such as TCP/IP. Similar to frame relay, ATM relies on the intelligence of upper layers to ensure that information arrives uncorrupted.

virtual path identifier
An eight-bit field in the ATM header, identifying the virtual path (virtual circuit) over which the transmitted data will flow.

virtual circuit identifier
A sixteen-bit field in the ATM cell header identifying the virtual circuit on which the data will travel from transmitting device to target device.

ATM Physical Layer

The ATM physical layer is responsible for preparing information for transport across the physical medium by defining the interface type, the interconnection rate, and the electrical characteristics of the signal. The interconnection rates that carry ATM cells depend on the manufacturer's interface. ATM switches and access concentrators commonly terminate signals ranging from 1.544 Mbps to OC-12, or 600 Mbps. Initially the ATM Forum defined two standard interface rates that carried a 150 Mbps and a 600 Mbps signal. Within a short time, the forum realized that the high-speed interfaces limited the deployment of ATM to the backbone network, thereby denying access and enterprise networks the use of ATM. ATM standards have since been built to include lower-rate signals such as T1, DS-3, and OC-3 interface rates. The new lower speed interfaces have helped push ATM further out into the access and enterprise networks, creating a protocol that serves all areas of the industry.

inverse multiplexing ATM
a protocol that facilitates combining multiple (up to 32) low-speed ATM circuits into a single logical pipe.

A special type of circuit interface was introduced in the late 1990s, allowing multiple T1s to be bonded together to produce one logical pipe from multiple physical links. *Inverse multiplexing ATM* (IMA) bonds T1 circuits together to produce one

Figure 7–24
The ATM frame. Each cell contains fifty-three octets comprised of overhead bits and information bits.

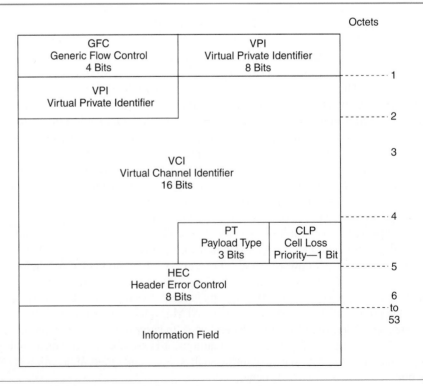

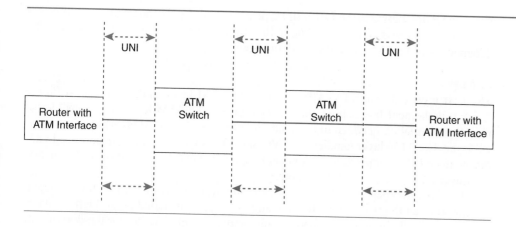

Figure 7–25
The user-to-network Interface between ATM devices.

composed of multiple T1 circuits. For example, if customers require 6 Mbps of bandwidth between two sites, they might install an IMA interface that will bond four T1s together to produce a pipe that allows traffic to burst up to rates as high as 6 Mbps. The advantage of combining individual pipes into one big pipe can be explained by comparing IMA to the water system. A large water pipe can handle a greater stream of water than many individual small pipes. The same is true when sending bits down a circuit. A larger circuit can handle a greater volume of bits at one time than multiple small circuits can.

The OC-3c interface provides a similar scenario. Instead of three OC-1s, an OC-3c provides one pipe of 155 Mbps. The advantage is that the overhead information associated with one OC-1 does not have to be duplicated three times. Only one header frame is needed for the entire OC-3c interface.

The physical layer also defines the type of handoff made between devices. There are two general device interfaces—*user-to-network interface* (UNI) and *network-to-network interface* (NNI). UNI, shown in Figure 7–25, is divided further into public UNI and private UNI. UNI defines all of the physical attributes of the connections between the two devices.

The NNI is shown in Figure 7–26, which illustrates the connection between two carriers' ATM networks. The NNI is slowly beginning to be deployed between the large carrier ATM networks. This interconnection of ATM network to ATM network is similar to phone company to phone company switch and transport interconnections. NNI helps define how each network will handle ATM cells, how signaling will be accomplished between networks, and how both networks will share network management. The interface standards ensure that if a user or a network is trying to connect,

user-to-network interface
Specifications for the procedures and protocols between user equipment and either an ATM or frame relay network.

network-to-network interface
A protocol defined by the Frame Relay Forum and the ATM Forum to govern how ATM switches establish connection and how ATM signaling requests are routed in an ATM network.

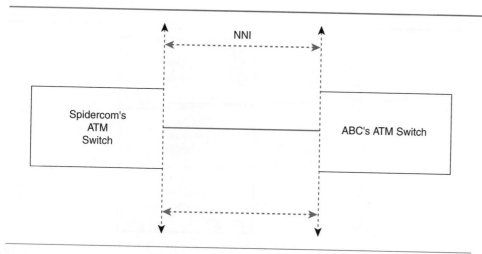

Figure 7–26
The network-to-network interface between two ATM networks. The NNI depends on signaling information to establish connections.

interconnection between the two will be possible and the attributes of the traffic will be maintained between the two. The ability to use SVCs across the network will be dependent upon the success of NNI deployment.

ATM Layer

The next layer defined in the ATM stack is the ATM layer, and, as shown in Figure 7–27, it is the next layer above the physical layer. The ATM layer is responsible for moving information between the adaptation layer (a higher ATM layer) and the physical layer. The ATM layer handles the PVC and SVC logical circuit connections. In addition, the ATM layer performs error checking on the header information to ensure that the Layer 2 overhead has not been corrupted.

PVC is established between the source and destination sites, providing a logical circuit between the two points. It is a static circuit that is nailed up and left in place. SVC is a logical circuit established through signaling information for the duration of the session and torn down by signaling information when the session is complete. Both types of VCs are being used today. While SVCs are just beginning to be introduced in large numbers across large public networks, private networks have been using them for some time. The decision about which to use depends on many factors that relate directly to the individual networks. Using SVCs reduces the provisioning time needed for building PVCs but increases the complexity of troubleshooting a problem. Therefore, a network using SVCs instead of PVCs is much more complex and requires skilled personnel to maintain.

Adaptation Layer

The adaptation layer is used to place information in the appropriate packages according to the type of traffic. ATM was built specifically as a switching and multiplexing method for broadband services. Broadband services include voice, video, and data. These types of data have to be packaged differently and given different priorities in the network for ATM to carry the information reliably across the network. For instance, because voice does not like any interruptions, voice and video must maintain continuous signals. They are time sensitive. Data, though, are not time sensitive. A five-second gap between pieces of information while a Web page paints across a PC screen will not corrupt the information.

ATM must take these characteristics into account when packaging information into a cell. The adaptation layer is where this differentiation is performed. ATM traffic is divided into five classes, and each class is used to define how the network will handle the cell—whether it will discard it if there is congestion, give it high priority, and so forth. The classes defined as Class A are also known as *AAL*1, AAL2, AAL3, AAL4, and AAL5. AAL1 adaptation is used to package time-sensitive traffic. It uses the *constant bit rate* (CBR), quality-of-service (QoS) parameter when formatting the cells.

AAL (ATM adaptation layer)
Accomplishes conversion from the higher-layer, native-data format and service specifications of the user data into the ATM layer.

constant bit rate
An ATM service that supports a constant or guaranteed rate to transport services that require vigorous timing control and performance parameters.

Figure 7–27
The ATM protocol is a Layer 2 protocol, meaning that the information is packaged into cells that conform to the data link layer standard. The type of service is defined by the AAL function, which resides above the data link layer.

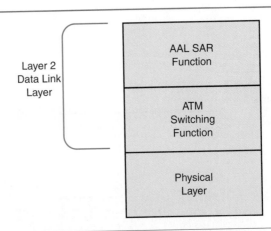

Layer 2 Data Link Layer

AAL SAR Function

ATM Switching Function

Physical Layer

Circuit emulation service (CES) nails up a portion of the bandwidth within the circuit to carry voice, much like nailing up channels in TDM. AAL1, therefore, relies on a quality-of-service measure that maintains a CBR through a CES pipe. The constant bit rate can only be maintained by allocating a specific portion of the bandwidth for AAL1 traffic. If that portion of the bandwidth is not used, it is not able to accept other types of traffic as is the case with AAL1 structure or AAL adaptations. Each PVC or SVC is told that the cell is CBR and must be given priority. The QoS is defined by CBR, the type of pipe is defined by the CES, and the AAL1 defines how the cell is packaged.

The second class of traffic, AAL2, uses the *variable bit rate* (VBR) QoS traffic measurement. VBR is further separated into real-time (*rt-VBR*) and non-real-time (*nrt-VBR*). AAL2 carries video and certain types of voice. Rt-VBR has a higher priority than nrt-VBR and demands that the link have enough bandwidth to ensure the delivery of the time-sensitive traffic encapsulated in the AAL2 format. The nrt-VBR is not as particular, because the information it carries is not as sensitive to delay. An example of nrt-VBR is a video clip being downloaded onto a computer.

The third and fourth classes of traffic, AAL3 and AAL4, carry data traffic but are dying adaptation layers due to increased overhead requirements and the evolution of the other AAL methods.

The fifth class of service, AAL5, is the adaptation layer that carries data traffic. It has two types of QoS available to attach to the cell—*available bit rate* (ABR) and *unavailable bit rate* (UBR). ABR provides a better *QoS* than UBR because it promises that the bandwidth will be delivered. If it is not delivered, you will know it has not been delivered. UBR is the most common QoS for data transmission in ATM networks. UBR does not provide any guarantees that there will be bandwidth available for the transmission and therefore does not guarantee that the data will be delivered. Because data are not sensitive to time, UBR is a good QoS choice for non-time-sensitive data. If the data must be guaranteed, ABR should be used. Table 7–3 lists the ATM classes of service and the common uses of each.

Every ATM cell goes through a construction process called segmentation and reassemble (SAR). It actually builds the cells, and it is during the SAR function that the AAL is defined within the cell header information. The ATM device segments and reassembles the cells as they arrive and depart and prioritizes traffic according the AAL attached to the cell. Figure 7–28 illustrates the entire process from packaging to shipping and unpacking.

variable bit rate
A voice service over ATM switch.
rt-VBR
Real-time VBR.
nrt-VBR
Non-real-time VBR.

available bit rate
ATM layer service category for which the limiting ATM layer transfer characteristics provided by the network may change subsequent to connection establishment.
unavailable bit rate
An ATM service category that does not specify traffic-related service guarantees.
QoS
A measure of the telephone service quality provided to a subscriber.

ATM Class	AAL Type	QoS Type	Delay Parameter	Services
Class A	AAL1	CBR—constant bit rate	Delay sensitive	Voice, video, time-sensitive traffic.
Class B	AAL2	Rt-VBR—real time variable bit rate	Delay sensitive	Voice, Video, data.
Class C	AAL3	Nrt-VBR—non-real-time variable bit rate	Delay tolerant	Non-time-sensitive data services. AAL3 and 4 are normally grouped together.
Class D	AAL4	Nrt-VBR	Delay tolerant	Non-time-sensitive data.
Class C	AAL5	VBR–UBR–unavailable bit rate; ABR–available bit rate	Delay tolerant	Data–non-time-sensitive traffic.

Table 7–3
ATM Adaptation Layer Summary.

Figure 7–28
The ATM signal can carry multiple types of traffic. This shows voice and data being combined onto one circuit.

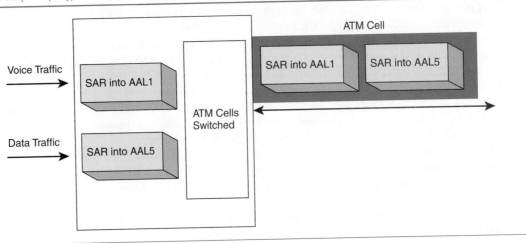

The adaptation layers may be viewed as similar to sending mail via priority overnight or by third class. A box is packaged, a label indicating priority overnight is placed on the outside, and the package is shipped. At each stop between the source and destination, someone reads the priority label and reacts accordingly. The overnight package is guaranteed to arrive by the next day. CBR traffic is similar to overnight priority mail. The third-class package is marked *third class* and shipped. At each stop between source and destination, the third-class package is given low priority compared with the overnight package. If there are no overnight packages, it might arrive within twenty-four hours if there is extra room on the mail truck. If there are numerous overnight packages or even many third-class packages, there is no guarantee when the package will arrive. The third-class package is similar to UBR traffic.

The purpose of the ATM protocol was to provide a way to converge all types of traffic onto one network. The ability to prioritize information and define the size of cells allows all types of traffic to travel on one network. As the protocol evolves, additional variations of each class are being introduced to improve the efficiencies and reliability of ATM.

7.4.3 ATM Equipment

The ATM switch is the core of the ATM network. It is responsible for directing cells to their destination, defining the type of QoS for each virtual circuit, switching traffic between interfaces, policing the network, and shaping traffic to help reduce congestion. An ATM switch is defined by the speed of its switching fabric, the number of terminations possible, and the sophistication of the traffic shaping and policing algorithms.

Most ATM switches provide interfaces ranging from DS-1 up to OC-48 speeds. The choice of an ATM switch often revolves around the type of interfaces provided because many networks are tied to the type of circuits that are in place. The ATM switch includes interface cards, processor cards, timing cards, and supplemental cards capable of performing IP routing or other auxiliary functions. The interface card normally performs the SAR function while the switch processor controls the cell switching. The switching matrix of an ATM switch varies depending upon the switch manufacturer and is responsible for directing cells from one interface to another. Figure 7–29 depicts a typical ATM switch.

The switching matrix is defined by the number of bits per second it is capable of carrying between interface ports. For instance, a 2.5 Gbps switch means there is a 2.5 Gbps bus available for cells to travel on between ports. Switch rates vary from just over

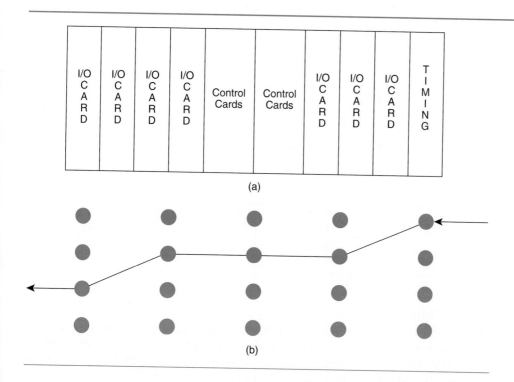

(a)

(b)

Figure 7–29
ATM switches vary in the number of interface ports, the number of controller cards, and the switch matrix used to carry the cells from ingress to egress. (a) Shows a common ATM switch shelf with I/O (input/output) cards and common cards. (b) Shows a distributed switch matrix—a common method used to carry traffic through the shelf.

1 Gbps to as much as 45 Gbps. Again, the choice is dependent upon the carrier's network requirements. Small networks find a 2.5 Gbps switch sufficient for their needs, while large carriers might require a 45 Gbps switch capable of handling large volumes of traffic. ATM engineers are constantly designing around volume of traffic compared with cost of the switch fabric and cost of processors.

Beyond the switch fabric, ATM switches are rated on their ability to police and shape traffic. Policing traffic means that the switch is able to watch the incoming traffic and determine whether congestion has occurred or will occur. The switch acts like a traffic cop that tells the cells to slow down or leave when the highway is too full. The main parameter most engineers look at is the size of the buffers—the storage areas for overflow cells. The buffers can take in the extra traffic and hold it until the switch has room available on its bus. The ATM switch also polices the egress network to see if it is congested, in which case it buffers the ingress traffic until the congestion is relieved. Traffic shaping, another function of the switch, balances the traffic across the network. It shapes the traffic before it is handed off to the network, reducing the chance for congestion and, thus, reducing the problems associated with too much traffic vying for the same pipe. Traffic shaping and policing work hand in hand using switch buffers to help improve the flow of cells across the traffic.

Most ATM switches are fairly costly because they are able to perform sophisticated traffic shaping algorithms in addition to being able to switch millions of cells every second. However, the cost of ATM switches is now dropping, thanks to less expensive processor chips and better use of software programs.

ATM Access Concentrator

The ATM concentrator is used at the edge of the network to aggregate all types of traffic onto an ATM pipe. The concentrator normally does not perform ATM switching but does provide traffic shaping and policing options. As shown in Figure 7–30, the ATM access concentrator is used to interface different streams of traffic such as voice, video, and data into one outgoing ATM-formatted signal.

The ATM concentrator consists of interface cards, switch matrixes, processor cards, timing cards, and in some cases auxiliary cards used for routing. Most ATM

Figure 7–30
The ATM concentrator does not switch cells. Instead, it acts as an aggregation point for incoming circuits. Multiple circuits interface into the concentrator, which combines them onto one high-speed circuit.

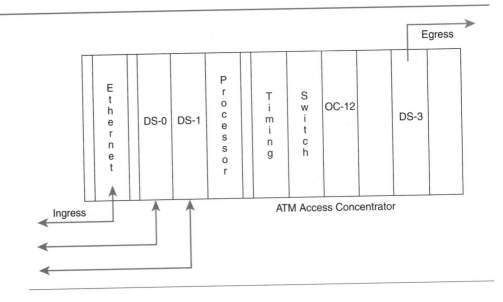

concentrators are less expensive than full-blown ATM switches and are able to interface with lower-speed circuits. For instance, in Figure 7–30, the ATM concentrator shows a T1 interface, an Ethernet interface, and a DS-0 interface being fed into the concentrator and one DS-3 leaving on the egress side of the box.

Similar to the ATM switch, the ATM concentrator is defined by the Gbps back plane. The concentrator normally has a lower gigabit-per-second rating because it is used farther out in the network and does not actually switch cells. A typical gigabit-per-second rating ranges from 1.2 to 12 Gbps. There are, of course, exceptions. Customers' needs normally drive manufacturer designs. The concentrator's main purpose is to convert other signal types such as TDM, frame relay, and IP into an ATM cell and to aggregate all of the cells together into one signal.

ATM End Device

Initially, the ATM Forum did not see ATM as an end-customer transport method because of the high-speed interfaces first defined by the standards organizations and the high overhead required by the ATM cell. ATM is now being deployed to the customer, and some analysts believe it will become the transport method from end to end, beating out IP. ATM to the customer site requires a device there that will convert the customer traffic—whether voice, data, or video—into an ATM cell. The device is similar to a modem or CSU in that it is an interface between the DTE and the circuit. Therefore, the device is considered a DCE.

The new *VoDSL* product, using ATM as its transport protocol, relies on the ATM protocol to package the customer's voice and data traffic at the customer's site. The product uses ATM's VBR classes of service to ensure that the voice traffic is given priority on the network, while the data traffic is encapsulated using UBR QoS. The main advantage of VoDSL is that one copper loop can be used to carry multiple voice conversations in addition to the customer's data traffic. Other such products are being offered that will help push ATM farther out into the access and enterprise network and, consequently, increase its acceptance in the industry.

VoDSL
Protocol defining how voice rides on DSL signal.

7.4.4 ATM Network
UNI to UNI

UNI to UNI is one of the most common ATM networks deployed today. As mentioned earlier, there are two types of UNI—public UNI and private UNI. Figure 7–31 depicts a public UNI, which refers to an end customer buying an ATM service from a service provider. A private UNI refers to a privately owned ATM network such as a

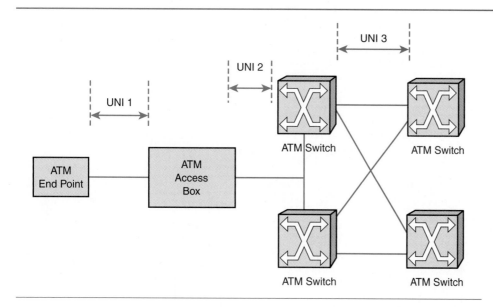

Figure 7–31
The ATM end point connection to the
ATM access box is under UNI 1. The
access box connecting to the ATM
switch creates UNI 2. The ATM switch
to ATM switch connections create
UNI 3.

large customer or a service provider's network. The standard is the same for both types
of networks.

As shown in Figure 7–31, each network has ATM switches that are connected using a mesh topology. The ATM switches feed access concentrators that interface with
customer signals. The handoff between the customer's end device and the access concentrator is the first UNI. A second connection is shown between the access concentrator and the ATM switch, and a third connection between the ATM switches. This
represents a typical UNI, whether public or private.

NNI

The purpose of the NNI is to provide a seamless ATM network between carriers the
same way the PSTN provides a seamless network between circuit switches. Several issues must be dealt with in order to connect different providers' networks. One is the
compatibility between the providers' ATM equipment. Even though a standard exists
that defines the parameters that manufacturers must comply to, allowing every switch
to talk to every other switch is not always easy. In addition, signaling between the
switches must be accounted for because it is responsible for making sure the networks
pass information. A third issue is the ease with which a circuit can be provisioned end
to end. Multiple processes must be followed between the two companies to ensure that
the two network circuits are provisioned properly. Provisioning causes the most problems when trying to implement an NNI.

7.4.5 Connecting the Dots

A typical ATM network combines PVCs, SVCs, UNIs, NNIs, and customer ATM devices. Figure 7–32 illustrates such a network. An example will help clarify the information in this chapter. Spidercom, our all-purpose carrier, has built a large ATM
network to carry all the traffic between its many locations. In addition to carrying its
own traffic, Spidercom decides to offer an ATM product to its customers. It has also established an interconnection agreement with several other ATM carriers allowing them
to interconnect to its network.

Figure 7–33 illustrates Spidercom's entire ATM network from coast to coast. First,
we look at Spidercom's own traffic to see how it is carried through the network. Next,
we look at a customer-to-customer connection using an NNI between Spidercom and
ABC Communications. Finally, we look at a customer-to-customer link using UNIs
and PVCs.

Figure 7–32
When SVCs are deployed in an ATM network, the connection between two points is established by Q.93B signalling. Once the session is done, the SVC is dropped.

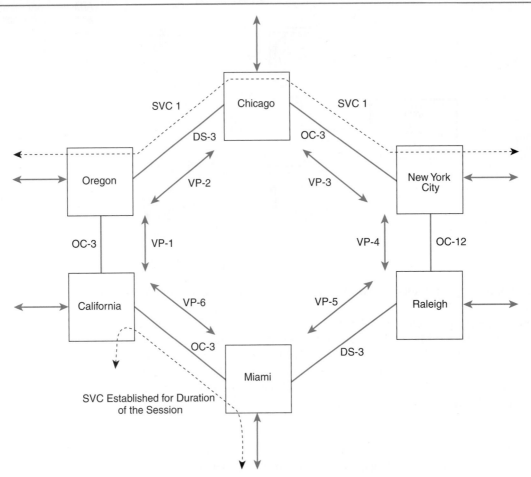

Spidercom's traffic includes TDM, IP, and frame relay. The traffic arrives at the Spidercom switch site as shown in Figure 7–34. In certain instances, access concentrators are placed at remote colocation cages where the incoming TDM, frame relay, and IP traffic is aggregated into an ATM signal. Other traffic arriving at the switch site comes directly in as TDM, IP, or frame relay. The ATM switch receives all of the traffic and converts non-ATM protocols into ATM by segmenting and reassembling the information. The traffic is then switched by the ATM switch to the correct interface and transported out into the network. SVCs are built as needed, thanks to the signaling protocol based on the Q.2931 signaling standard defined by the ITU and accepted by the ATM Forum. SVCs are set up whenever a call is initiated and torn down when the call is completed, similar to the circuit-switched world.

The second type of call is from the end customer residing on Spidercom's network to a customer on ABC Communications' network. ATM traffic is able to travel from

Figure 7–33
Spidercom has deployed a nationwide ATM network.

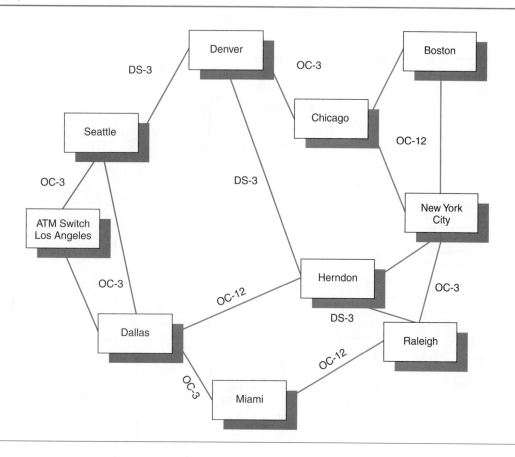

end customer to end customer using SVCs that are built upon request from the customer and torn down once the call is complete. SVCs between the carriers' networks are built using the ATM signaling protocol that establishes the end-to-end connections between the two networks. The NNI allows internetwork signaling to pass between carriers' networks.

The final type of connection found in Spidercom's network is ATM connections all the way out to the end customer. As shown in Figure 7–35, the ATM cell is formed at the customer site and is nailed up between SpiderCom's network and ABC Communications using a PVC instead of a dynamic SVC.

ATM provides a way to carry multiple types of information—voice, video, and data—on one network. The ability to statistically multiplex information onto the pipes helps in the deployment of a full ATM network. The introduction of multiple classes of service along with the ability to carry ATM cells on lower-rate circuits such as T1s has helped establish ATM in the industry. The future of ATM is promising, because much of the access network is starting to take advantage of ATM's ability to provide QoS assurances that enable the transport of time-sensitive information.

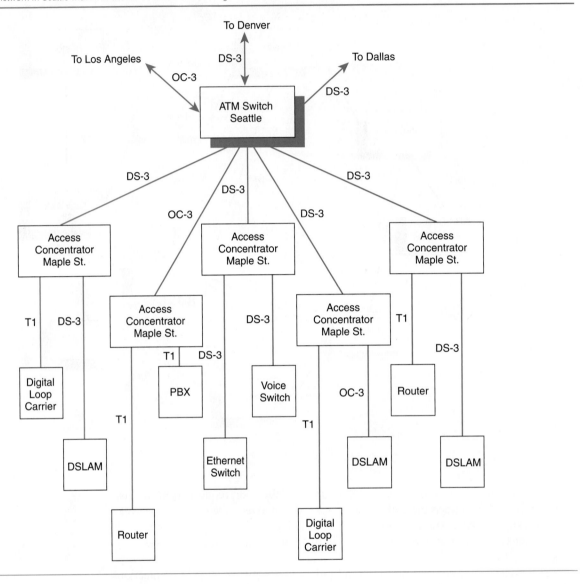

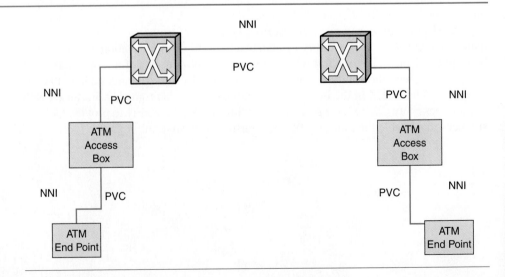

SUMMARY

Layer 2 protocols are used in communications to ensure that information is efficiently carried from location to location. The role of the Layer 2 protocol is similar to that of the parcel post, UPS, FedEX, or DHL carrier. Multiple pieces of information are packaged into one box and transported to the end destination where they are unpacked and distributed. It is not up to the UPS driver to determine which person in a building should receive the letter. It is the driver's job to make sure the box arrives at the building. Layer 2 protocols function in a similar manner. They take the multiple pieces or parts of information from Layer 3, package them into larger frames or cells, and carry the package to the end device, which pulls the information out of the box and hands it to a higher protocol to determine to which person or which end device it belongs. Though each of the protocols presented handles how it puts the information into boxes and carries the package across the network differently, each has the same end goal: deliver the information in the most efficient, least cost manner as possible. Chapter 7 focused on the three most common Layer 2 protocols: Ethernet, frame relay, and ATM.

Ethernet, though often referred to as a Layer 2/1 protocol, depends on simple baseband CSMA/CD functions to establish and carry information end to end. Ethernet is deployed in both LANs and WANs and offers a variety of interface types and speeds. The newest of the Ethernet interfaces are GigE, 1 Gig across fiber, and Wi-Fi—wireless 802.11 WAN interface standard.

The frame relay protocol, defined as a fast-packet protocol, defines the interface speed, the frame structure, addressing from interface to interface, and frame delivery assurance. Frame relay, though not as old as Ethernet, has been around for several years and continues to be deployed for WAN applications.

ATM (Asynchronous Transfer Mode), a standard defined as the fast-packet technology for ubiquitous networks, is one of the newest Layer 2 protocols found in the network, though it even has become known as an "old horse." Unlike frame, ATM, is a connection-oriented protocol and thus requires additional network planning to establish links between carriers. ATM's strong point in the Layer 2 world is its ability to carry all types of data without jeopardizing quality.

All three Layer 2 protocols are commonly deployed in the communications network by all types of providers and end users. Though each has gone through periods of obsolescence—or, more appropriately, pundits predicting obsolescence—each continues to maintain its place in the network.

CASE STUDY

You are now familiar with the types of media used in the network and also with the use of multiplexers to decrease the number of facilities needed in an area. Using the map and the information on population distribution provided in Appendix A, put together a preliminary design showing what areas require multiplexers. Place a star on the map showing where you will place the multiplexers. Determine the break point between the cost of adding cable pair and the cost of placing a multiplexer. Again, this is a preliminary design based on the assumption that you may place multiplexing equipment on the end of the circuit, such as in the hospital or in the factories.

The main purpose of this exercise is for you to become familiar with the number of residents in the area and the number of signals that will be carried on the cable. Break points will be supplied to determine whether or not a multiplexer using fiber optic cable is needed or if copper cable can be used instead. Information on the capacity per location and the cost of each type of multiplexer is given in Appendix A.

Rules:

- All residential customers will be served by copper cable.
- A distance of greater than ten miles may warrant an RF system.

Determine the type of connection that should be made between Green Grass and Grass Hopper. The distance is thirty-five miles.

REVIEW QUESTIONS

1. Why is statistical multiplexing a more efficient way to transport information across a network than traditional circuit-switched TDM networks?

2. What is the difference between the ingress and egress of a network? Draw a picture showing where both reside in a network.

3. Explain the term *connectionless* and list one connectionless protocol.

4. What does the term *backbone* refer to in telecommunications?

5. Which part of the network is considered to be the access portion?

6. Explain the term *connection oriented* and list one connection-oriented protocol.

7. Draw a picture of the frame relay frame and label each of the components.

8. Frame relay has a fixed-length frame of fifty-three octets. True or False? Explain.

9. Frame relay is able to carry time-sensitive traffic such as voice and video as easily as it carries data. True or False? Explain.

10. What does the acronym *DLCI* stand for and where is it found in the frame relay network?

11. Why can a circuit carry multiple PVCs?

12. What does *ATM* stand for?

13. ATM has a fixed-length cell that can hold fifty-three octets of information. True or False? Explain.

14. ATM has multiple levels of service. Name and define the four most common QoS standards used in today's ATM networks.

15. What is the CBR standard used for?

16. Explain the difference between a PVC and an SVC.

17. Why is ATM able to carry all types of traffic?

18. The Ethernet 802 standard was developed by the Defense Department in the early 1970s.

    a. true   b. false

19. CSMA/CD is used for:

    a. pushing data out onto the wire
    b. Ethernet frame structure
    c. physical interface pinouts of an RJ-45 connector
    d. handling traffic collisions on an Ethernet network
    e. baseband technology
    f. broadband technology
    g. a, b, and c
    h. a, d, and e
    i. a, b, c, d, and e
    j. all of the above

20. Ethernet frames can ride on what types of media?

21. GigE carries how many more bits per second than Fast Ethernet?

22. Ethernet protocol was designed as a logical star and physical bus topology.

    a. true   b. false

23. The Ethernet switch is used to:

    a. repeat the signal and extend the distance the signal can travel
    b. reduce traffic on a segment by maintaining bridge tables that contain device information
    c. connect multiple stations together thus improving the efficiency of the physical wire plant
    d. a and b
    e. b and c
    f. all of the above

24. The MAC portion of the Ethernet frame sits closest to Layer 3 and the LLC portion of the Ethernet frame sits next to Layer 1.

    a. true   b. false

25. List two terms that mean the same thing as MAC address in communications.

26. Explain the difference between unicast, multicast, and broadcast addresses.

TROUBLESHOOTING

You have received word that the ISP you normally send all Internet traffic to has had a major catastrophe and has shut down. Assuming you have access to a backup ISP, what should you do next?

KEY TERMS

connection-oriented protocol (155)

connectionless protocol (155)

permanent virtual circuit (156)

switched virtual circuit (156)

ingress (157)

egress (157)

access (157)

edge (157)

backbone (157)

Ethernet (157)

Ethernet II (157)

802 Ethernet (157)

802.3 (157)

CSMA/CD (157)

baseband (157)

LAN (158)

WAN (158)

MAU (159)

UTP (159)

STP (159)

GigE (161)

Wi-Fi (161)

Fast Ethernet (161)

RJ-45 (161)

CAT5 (161)

hub (161)

Ethernet switch (164)

segment (164)

bridge (166)

router (166)

interface port (167)

type (168)

length (168)

physical address (169)

hardware address (169)

MAC address (169)

unicast (169)

multicast (169)

broadcast (169)

ARP (169)

RARP (170)

data link connection identifier (173)

committed information rate (173)

service level agreement (174)

forward explicit congestion notification (174)

backward explicit congestion notification (174)

discard eligibility bit (174)

public telephone and telegraph (179)

ATM cell (180)

virtual path identifier (180)

virtual circuit identifier (180)

inverse multiplexing ATM (180)

user-to-network interface (181)

network-to-network interface (181)

AAL (ATM adaptation layer) (182)

constant bit rate (182)

variable bit rate (183)

rt-VBR (183)

nrt-VBR (183)

available bit rate (183)

unavailable bit rate (183)

Qos (183)

VoDSL (186)

8

Layer 1 Networking Protocols: TDM Protocols—DS-1, DS-3, SONET

Objectives

After reading this chapter, you should be able to

- Define Layer 1 TDM transmission protocols.
- Define network timing.
- Define the DS-1 signal.
- Define the DS-3 signal.
- Define the SONET transmission protocol.

Outline

Introduction

8.1 TDM: Time Division Multiplexing Protocols

8.2 Time Division Asynchronous Protocol: DS-1 Circuit

8.3 DS-3 Circuit

8.4 SONET Transport Protocol/SONET Standard

■ INTRODUCTION

Physical layer protocols carry all types of information—voice, video, data—across all types of telecommunication networks and are commonly referred to as transmission protocols. The job of a transmission protocol is to package the information and manage its flow across networks. A simple way to understand where transport protocols fall within the communications system is to compare them to a vehicle. A car is built to carry all types of passengers, short, tall, thin, heavy. The car has a predefined structure: the doors open the same way on the same model cars; the engine starts by turning a key; safety devices such as turn signals and flashers are used as warnings; the horn may be blown to notify cars nearby that they are too close. Also, the car is driven on many different types of roads from superhighways to dusty dirt roads. The transport protocol functions in the same way. It is used to encapsulate information according to a predefined framing format; it has common safety features such as alarm conditions and flow control; and it travels on all types of circuits from DS-1 links to OC-192 systems.

All types of service providers—cellular/PCS, CATV, ILECs, CLECs, ISPs—depend on TDM protocols to carry the millions of bits of traffic across their networks. TDM-based protocols are deployed throughout the industry.

Chapter 8 is divided into four primary sections. The first defines TDM and notes the North American Digital Hierarchy and the SONET line rates. The second section defines the DS-1 signal by looking at the frame structure. The third section defines the DS-3 signal, and the last explains SONET.

■ 8.1 TDM: TIME DIVISION MULTIPLEXING PROTOCOLS

The best analogy to use to describe *TDM*—time division multiplexing—protocols is a train pulling boxcars between two locations. Visualize same-size boxcars capable of carrying the same amount of stuff. The TDM protocol is similar in that it puts data into same-size frames that hold the same number of ones and zeros. Unlike fast-packet protocols such as frame relay and IP, the boxcar or frame may travel half full because it cannot vary in size according to the amount of stuff or traffic being shipped. You may ask, Why then are TDM protocols used at all? Traffic riding in same-size frames arrives at the end device within a predictable time frame. An end device is not trying to decipher variable-length frames or packets such as a frame relay frame or an IP packet. It knows the exact time that information will arrive and in what order. Therefore, TDM fits perfectly in the world of voice and video transmission—real-time applications. There is no need to stamp the frames with a priority flag. The devices do not have to take time to crack the packet and figure out what to do with the information. TDM protocols have low overhead requirements and are easy to manage. Time-based protocols, such as the DG-1 standard were the first digital transmission protocols developed in the late 1970s. The protocols were designed to carry voice across the Public Switched Telephone Network (PSTN) more efficiently than analog circuits.

TDM remains a viable transmission protocol because of its ability to efficiently carry higher-layer protocols across all types of media. TDM manages the physical layer functional needs of placing one and zero pulses onto the line. TDM transmission circuits have built-in error-checking schemes that are much less complicated than higher-layer protocols, thus reducing overhead burdens. The funny fact that many data-centric techies do not like to discuss is that a digital voice call placed onto a DS-1 TDM circuit takes up 64 kbps of bandwidth. A VoIP call requires 80 kbps of bandwidth to support the same quality voice service.

Over the years, the TDM protocol types have evolved dramatically. The first TDM circuits deployed in the PSTN were asynchronous DS-1 and DS-3 links. *Asynchronous* refers to a nontimed digital circuit, meaning that an external time source is not required. The DS-1 and the DS-3 are categorized as asynchronous protocol signals. The North American Digital Hierarchy is the defacto standard used in North America. The second TDM standard deployed throughout North America was built as *synchronous* protocol—*SONET*—meaning that it requires an external clock source to sync up each end; each end is listening to the same drummer or clock. Table 8–1 lists both the North American Digital Hierarchy line rates and the common SONET line rates.

8.1.1 Digital Circuit Timing

Before we begin our discussion on physical layer transmission protocols, it is important to explain the role that network timing plays in digital communications circuits. Timing or synchronization is responsible for keeping all of the bits of a digital signal in line, somewhat like a drill sergeant keeping his or her troops in step. With the advent of digital circuits, timing in telecommunications networks became a hot topic at all the user conferences. Initially, atomic clocks located in far off caves and maintained by AT&T and the government supplied the central timing references for communications networks. After deregulation and the influx of multiple carriers, it was no longer

TDM
Time division multiplexing is a physical layer transmission protocol that places bits into time slots similar to railroad cars.

asynchronous
Nontimed digital circuit.

synchronous
Digital circuit that utilizes an external clock source.

SONET
The Synchronous Optical Network is a Layer 1 protocol developed in the early 1990s to standardize optical transmission methods.

Table 8–1
Digital Signal Levels As Defined by
the North American Digital Hierarchy.

Digital Signal Level	Bit-Per-Second Rate	Number of DS-0s	Number of DS-1s	Number of DS-3s
DS-0	64 kbps	1	n/a	n/a
DS-1	1.544 Mbps	24	1	n/a
DS-2	6.312 Mbps	96	4	n/a
DS-3	45 Mbps	672	28	1
560 Systems	560 Mbps	8064	336	12

feasible for AT&T to supply timing across all carriers' networks. Today, the most common clock source is derived from the GPS—Global Positioning System—satellite maintained by the U.S. military. Clock sources coming from far-off caves are telecom legends.

8.1.2 Effects of Improperly Timed Networks

Before we explain how the clock signal is distributed through the network, we need to explain why timing is important and what happens when timing fails. Every piece of equipment that interfaces with a digital circuit has a buffer or holding room. The buffer is used to hold one or two frames at a time allowing the device to adjust to the incoming bit stream. Timing problems are caused when the frequency of the incoming signal and the equipment's receiver clock vary, resulting in the buffer overflowing or underflowing. Overflow conditions cause frames to be thrown out, and underflow conditions cause the frames to be repeated. When the lost or repeated frames exceed a predefined threshold, a *slip* occurs; *slip* is a term used in telecommunications to represent an error condition generated by a timing problem. A slip is a result of a difference in the frequency and phase of the device's receive clocks or a result of a problem in the network.

One slip on its own will not crash a system. A slip produces a sudden burst of errors on the line that results in clicks in a voice signal or missing lines in a fax or a frozen screen in a video transmission or a retransmission on a data circuit. In contrast, when slips occur constantly due to a major timing problem, the system will go up and down causing a troubleshooting nightmare. Figure 8–1 shows timing slips. For example, when a SONET carrying a local telephone company voice traffic experiences too many slips, the phone calls being made may arrive at the wrong destination, similar to spinning a roulette wheel—nobody knows where it will stop and whom it will connect to. Therefore, it is important to realize that timing is a critical component of digital networks and should never be taken lightly.

Two additional conditions monitored to ensure that the system is timed properly are jitter and wander. *Jitter,* happens when there is a short-term variation due to bit stuffing into the bit stream and faulty transmission clocks. *Wander* happens when there

slip
A timing slip is a loss or drop of bits because of timing issues.

jitter
Pulses experience slight shifts in phase as a result of timing problems in the network.

wander
Pulses move out of position over a long period of time resulting in timing errors.

Figure 8–1
Example of timing slips.

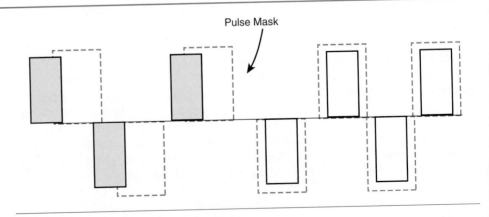

Pulse Mask

Timing Source	Precision	Number of Frame Slips
Global Positioning System	1×10^{-12}	0.2 per year
Loran—C	5×10^{-12}	1.25 per year
Stratum 1	1×10^{-11}	5 per year
Stratum 2	1.6×10^{-8}	11 per day
Stratum 3	4.6×10^{-6}	132 per hour
Stratum 4	3.2×10^{-6}	15.4 per minute

Table 8–2
Timing Standard Hierarchy.

is a long-term fluctuation in temperature or when some other environmental variations occur. Equipment vendors and design engineers provide specs documenting the maximum and minimum thresholds allowed for slips, jitter and wander in a given system.

8.1.3 Timing Hierarchy

Timing architectures seldom vary between service providers. At the top of the timing hierarchy sits a stratum 1 clock source referred to as the PRS—Primary Reference Source. The PRS can be one of the following: GPS, LORAN-C; or a stand-alone cesium oscillator. All three clock sources provide a timing reference that guarantees that no more than one slip will occur every 144 days, as shown in Table 8–2. The Stratum 1 source is the matriarch of the entire timing clan. All lower stratum levels look to the Stratum 1 clock for guidance and stability. If the Stratum 1 clock fails, the Stratum 2 clock—normally located at the switch or POP site—takes over as the timing reference for the lower clock levels. A Stratum 2 clock is able to maintain synchronization for a number of days without experiencing large numbers of timing slips—estimated at eleven slips per day. A Stratum 3 or 3E clock is the third level of the timing hierarchy and, on average, if running without a Stratum 2 reference, experiences 3179 slips per day. The final clock source available is the Stratum 4 that, if left on its own, experiences 22,118 slips per day. Typically, Stratum 3, 3E, and 4 clock sources are built into transmission devices specifically as backup timing supplies to maintain the equipment during global timing outages. Service providers do not use Stratum 3, 3E, or 4 as primary or secondary reference clocks. Therefore, if the Stratum 2 source is compromised, the Stratum 3 or 4 is able to keep the ones pulses in step for around seventy-two hours before they mutiny. In summary, the timing hierarchy flows from Stratum 1 to Stratum 4 with Stratum 1 providing the supreme clock reference or Primary Reference Source.

A good way to understand this concept is to imagine that all the clocks in your house maintain time by listening to one super-duper clock timed off the GPS satellite. The super-duper clock or grandmother clock feeds the parent clocks a timing reference that they grab and use to maintain synchronization. The parent clocks in turn feed their reference to the children clocks. If the grandmother clock fails for some reason, the parent clocks have an internal clock source and are able to continue keeping time, though not quite as accurately as before. As is obvious, if the parent clocks lose their internal clock source, the children clocks have to move to their own internal clock, which is much less accurate than its parent and grandparent. Figure 8–2 depicts the digital clock distribution topology.

8.1.4 Digital Clock Equipment and Signal Distribution at the Switch Site or POP

UTC—Coordinated Universal Time—is the timing source derived through the air from a GPS signal or a LORAN-C signal. The Stratum 1 clock source—grandmother of the clan—is pulled from the air by a receiving antenna typically mounted on top of the roof. A coaxial cable ties the antenna to a piece of equipment sitting in the switch or transmission area as shown in Figure 8–3. The coaxial cable is connected to a PRR— Primary Reference Receiver, the parent of the clan—that is used to distribute the clock

UTC
Coordinated Universal Time is the standard clock used to synchronize synchronous digital circuits across all networks.

Figure 8–2
Every communications network
requires a timing source. Timing is
fed through the network from the
Stratum 1 source to the network
equipment.

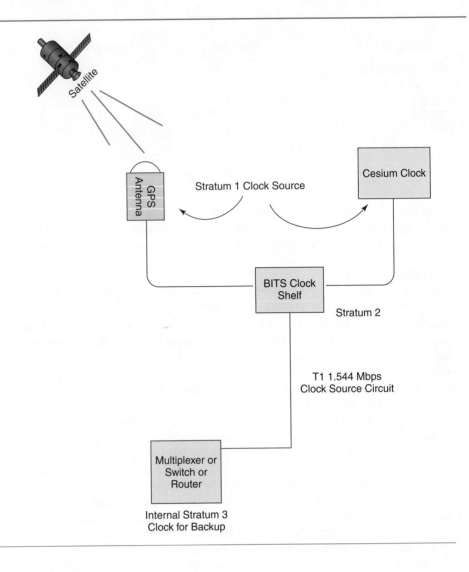

Satellite

GPS
Antenna

Stratum 1 Clock Source

Cesium Clock

BITS Clock
Shelf

Stratum 2

T1 1.544 Mbps
Clock Source Circuit

Multiplexer or
Switch or
Router

Internal Stratum 3
Clock for Backup

Figure 8–3
GPS rooftop antenna.

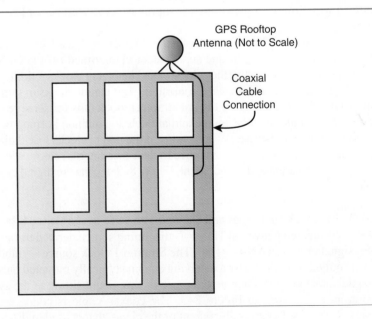

GPS Rooftop
Antenna (Not to Scale)

Coaxial
Cable
Connection

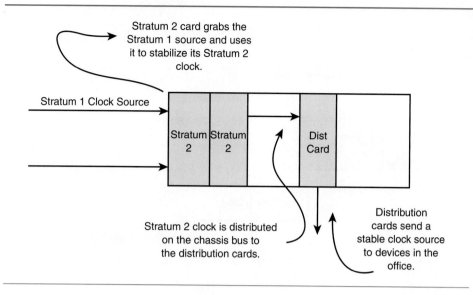

Figure 8–4
The incoming clock signal from a Stratum 1 source is fed into the Stratum 2 cards in the BITS shelf. The Stratum 2 cards use the Stratum 1 signal to stablize the clock signal. They then feed the Stratum 2 clock to the distribution cards that distribute the signal to devices in the site.

source to all the digital devices in the office. The PRR is often referred to as the *BITS*—Building Integrated Timing Supply—shelf and is typically no larger than six inches high and nineteen inches wide.

BITS
Building Integrated Timing Supply is the equipment used to distribute timing circuits to transmission equipment.

Within the BITS shelf, the Stratum 1 clock is fed into a Stratum 2 clock that grabs it and uses it to synchronize itself. The BITS shelf Stratum 2 cards feed the clock source to distribution cards, as shown in Figure 8–4. Physical hardwired connections are made on the back of the distribution cards tying them to each digital device in the office, as shown in Figure 8–5. The distribution cards outpulse a clock signal that references back to the grandmother clock, Stratum 1. All equipment being fed by this signal is guaranteed to stay in sync according to the Stratum 1 specifications—one slip every 144 days—even though they are not directly connected to the GPS receiver.

The digital devices sitting in the office receive one of two types of clock signals. The first and most common is referred to as a DS-1—digital signal 1—clock signal. All this means is that from the BITS shelf distribution circuit packs, a DS-1 is used to feed

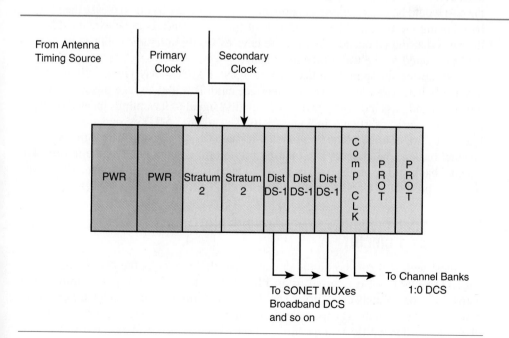

Figure 8–5
BITS shelf is used to distribute timing through a switch or communications site. Clock DS-1s and composite clock signals feed the different devices at the site.

Figure 8–6
Timing SONET MUX.

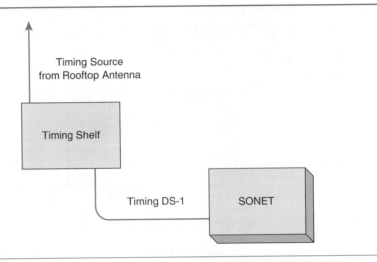

the digital device. Typically, the devices requiring DS-1 clock source signals are digital switches, digital cross connects, and SONET multiplexers. The second type of clock distributed by the BITS shelf is the composite clock. The composite clock is a 64 kbps or 8 kHz signal used to time DS-0 devices such as channel banks, CSUs, and MUXes in order to maintain the phase of the signal between devices across building cable. Most digital devices contain an internal clock—usually a Stratum 3 or 4—that is used if the DS-1 clock source fails. This gives the technician a window to fix the problem before timing goes awry and the network becomes confused and crashes.

Some providers purchase a cesium oscillator that—like the GPS signal coming from the satellite—offers a Stratum 1 timing source; grandmother lives in the building. The shelf holding the cesium oscillator is referred to as the PRS—Primary Reference Source—and is hardwired to the PRR. In either case, using the GPS or your very own Stratum 1 clock, timing is distributed in the same fashion.

We have discussed how timing is grabbed and distributed throughout an office to the devices physically sitting in the office. You may wonder if a timing shelf has to be placed in every location a digital device is found. The answer of course is no, because the cost would be overwhelming to poor network engineers trying to please their bosses by reducing the budget. The second part of timing that needs to be explained is how timing is distributed outside the office to devices sitting in remote cabinets or channel banks mounted on the wall of the local bank.

The device sitting in the office—say, a SONET MUX—has a timing circuit, DS-1, physically hardwired into its timing interface card. It is able to take this external timing signal and pass it on through the information signal to its siblings located throughout the network, as shown in Figure 8–6. The remote MUXes pull timing off the line—they are configured as line timed—to keep the bits in step. The same is true of channel banks. The channel bank sitting in the office is optioned as the master. The channel bank sitting on the wall of the local bank is optioned as the slave. The master or parent provides a stable time source to the child sitting in the bank.

■ **8.2 TIME DIVISION ASYNCHRONOUS PROTOCOL:**
   DS-1 CIRCUIT

DS-1
Digital signal 1 references a 1.544 Mbps circuit.

The most widely deployed digital circuit in North America is the *DS-1*—digital signal 1. The DS-1—also referred to as T1—is used to carry all types of information across all types of networks. If we could open up a fiber optic strand and peer inside beneath the higher-rate signals, we would observe hundreds of DS-1s carrying millions of channels filled with every type information—voice, video, and data. It is safe to state that DS-1 digital circuits form the backbone of the communications networks in North

America. Before we discuss the structure of a DS-1 circuit, we first define the structure of a DS-0, the lowest channel rate of the North American Digital Hierarchy.

8.2.1 The DS-0 (64 kbps) Channel

The *digital signal 0* (DS-0) channel is equivalent to one channel of information. When we speak on the telephone, the sound is placed in one DS-0 channel after another to travel across the country. Digital-to-analog conversion, as discussed in Chapter 4, separates words into eight-bit one and zero "words" that are placed in specific time slots on a T1 circuit. The 64 kbps DS-0 is one channel on a T1 circuit. Twenty-four 64 kbps channels can travel on one DS-1 circuit, and there are 672 DS-0s on one DS-3 45 Mbps circuit.

> **digital signal 0**
> 64 kbps.

One data or voice signal is transported inside each DS-0 channel, which has room for 64,000 ones and zeros traveling down a circuit every second. The following example should help explain how a DS-0 carries information from one point to the next.

Rod and Jake are talking to each other in an Internet chat room. Rod types "Did you see the game on Sunday?" and hits the Send button. Following digital conversion, we determine that there are thirty-one characters in the sentence. (Remember to count letters, spaces, and all punctuation!) Because each character is represented by eight ones and zeros, there are 248 ones and zeros that need to travel between the two computers. The network between Rod and Jake is one DS-0 channel that allows 64,000 bits to be carried every second. Our sentence composed of 248 ones and zeros easily fits into one 64,000 bps channel; so the sentence "Did you see the game on Sunday?" will reach Jake within one second. This is a simplified example of how information travels through the network. The key concept is that information travels in the form of ones and zeros; and, for all of those ones and zeros to make sense at the opposite end, a standard bit sequence is needed to identify where the information begins and ends.

Increasing the content traveling between the two computers allows us to expand on our example. Imagine that Jake wants to send Rod a data file with a short video clip of his fishing trip. The file contains one million bits of information.

To determine how long it will take to transport the file, we first need to divide the 1,000,000-bit file into 64,000-bit sections.

$$1,000,000 \div 64,000 = 15.6.$$

Therefore, it will take sixteen seconds to carry 1,000,000 bits between the two sites. The DS-0 channel will be filled and will traverse the network sixteen times in order to deliver the entire file.

8.2.2 TI (DS-1) Circuits

DS-1 signals travel on all types of media, copper twisted pair cable, coaxial cable, microwave systems, and fiber optic systems. A DS-1 may be the only circuit riding on the medium, as with twisted pair copper cable; or it may be riding with other signals, as those found on a fiber optic strand. In either case, the format of the signal and the method used to multiplex or demultiplex lower-rate signals into a DS-1 remain the same.

Twisted pair copper cable, as discussed in a later chapter, carries the digital signal across the copper circuit as voltage pulses. The distance the signal can travel depends on the gauge of the cable and the design of the circuit. Typically, a traditional *AMI*—alternate mark inversion—T1 signal can travel about one mile on standard twisted pair T-Screen cable before requiring signal regeneration. The HDSL *2B1Q*—formatted T1—is able to travel 12,000 ft. before requiring signal regeneration; 2B1Q, as explained in Chapter 4, halves the transmission frequency and, as such, increases the distance the signal can travel. T1s riding on coaxial cable are typically multiplexed into DS-3 or STS-1 signals and, as such, comply to distance limitations placed on those signals. The same holds true for fiber optic systems and long-haul microwave systems: distance is dependent on the transport signal. In summary, the T1 circuit is the workhorse of the network used to carry all types of applications across all types of media.

> **AMI**
> Alternate mark inversion defines a line-encoding scheme in which the pulses are alternate—positive followed by a negative.
>
> **2BIQ**
> Line encoding scheme that produces a 4-level bit pattern—two positive, two negative.

Figure 8–7
Signal voltage.

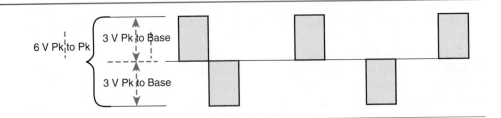

The following section focuses on DS-1 signal characteristics, analog-to-digital conversion, DS-1 framing format, and DS-1's place in communications network.

8.2.3 Signal Characteristics

All digital signals are constructed around a defined set of parameters used to maintain specific signal characteristics. The key signal characteristics are signal level, frequency, pulse shape, and polarity. Signal level refers to the signal strength of the DS-1's bipolar-return-to-zero bit stream. A DS-1's strength is measured by reading the peak-to-peak voltage level or the base-to-peak voltage level, as shown in Figure 8–7. The peak-to-peak voltage level of a standard DS-1 is specced at 6 V peak to peak. The base-to-peak voltage level is specced at 3 V base to peak. A second signal level measurement commonly taken is the decibel level of the signal at interface points in the network. A technician will often read the decibel level at a DSX panel to determine if the signal is too low (long) or too high (hot).

Monitoring the frequency range of a signal helps isolate issues related to timing problems. The DS-1 must fall within the frequency range of 1,543,923 to 1,544,077 Hz in order to maintain synchronization. Technicians use test equipment designed to read the incoming frequency values when checking T1 circuits' health.

Pulse shape is the third parameter that is used to determine if the DS-1 is behaving properly. A pulse must fall within assigned boundaries referred to as its *pulse mask* as shown in Figure 8–8. A pulse that has shifted in either direction—before the leading edge or after the trailing edge of the pulse mask—indicates there is a problem with the signal and that further action needs to be taken. The ones and zeros marching along the pipe must stay within their allotted space in order for the receiving equipment to be able to read and assemble them correctly.

Finally, digital signals are transported across the link in a bipolar-return-to-zero format. The polarity of the bit stream must follow the bipolar format, meaning that every other ones pulse is of the opposite polarity, and a zero pulse is represented by a zero voltage state called a *space*, as shown in Figure 8–9. A unipolar format refers to a signal composed of all positive or all negative pulses, as discussed in an earlier chapter. In the world of transmission, the bipolar-return-to-zero format is used for several

pulse mask
Pulse mask defines the boundaries within which the ones pulse must reside the bit stream.

Figure 8–8
Pulse shape.

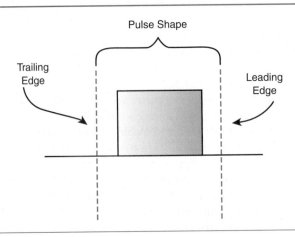

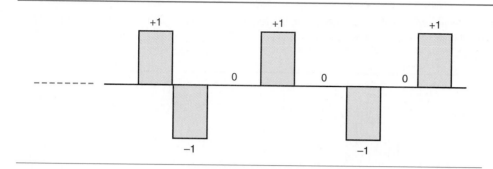

Figure 8–9
Bipolar-return-to-zero signal format.

reasons. Reversing the polarity of the ones pulse halves the frequency of the signal; and, as discussed in a previous chapter, the lower the frequency, the greater the distance the signal can travel before petering out. This holds true for bipolar DS-1s. An auxiliary benefit of the bipolar format is the ability to check the signal for errors through a polarity error-checking scheme. The end equipment reads the pulses noting any polarity violations, with two consecutive pulses of the same polarity indicating a bipolar error. Bipolar formatting also eliminates the DC characteristics of the bit stream; a unipolar signal mimics a DC signal. A continuous stream of pulses of one polarity tends to blend together, making it very difficult for the end equipment to decipher the beginning and end of a pulse; essentially, it looks like a DC signal. In addition, because a unipolar format mimics a DC signal, it would be impossible to use DC power to energize the repeaters in a span, as the power component would interfere with the information signal. Therefore, a DS-1 bit stream on a transmission line is always formatted in the bipolar-return-to-zero format.

8.2.4 Analog-to-Digital Conversion

A DS-1 is built by combining twenty-four DS-0 channels together into one 1.544 Mbps signal. A DS-0 channel is made up of 64,000 ones and zeros that represent information, such as a voice signal or an analog signal generated by a modem or a digital signal from a PC. The procedure used to convert an analog signal into a digital signal involves a three-step process involving sampling, quantizing, and encoding. A chip called a *Codec,* coder/decoder, is responsible for the analog-to-digital and digital-to-analog conversion and can be found in all types of communications devices such as ISDN phones, channel banks, and multiplexers.

In order to preserve the quality of the analog signal end, to end a sampling rate of 8000 times per second was adopted as the standard for carrier class voice transmission. The 8000 samples per second value is derived from the Nyquist theorem. The Nyquist theorem assumes that in order to recreate a comparable analog signal the sampling rate has to be twice the highest frequency of the originating signal. Since voice signals on telephone lines use frequencies between 300 and 4000 Hz, the sampling rate for traditional voice traffic is set at 8000 samples per second: 2 × 4000. Newer techniques have been developed that reduce the amount of bandwidth required for one voice call by reducing the sampling rate, but the older standard of sampling 8000 times per second continues to dominate the industry.

The second step in the process requires that the sample taken, which is nothing more than a voltage level, is measured against a scale similar to a yardstick with 255 levels, each representing an eight one-zero pattern called a PAM sample. The term used to describe this process is *quantizing.* Once the sample has been quantized, it moves on to the third step in the process called *encoding* where the PAM sample is translated into an eight-bit *PCM word.* The PCM words are converted into a bipolar signal and multiplexed into the DS-1 accordingly. In the reverse, the signal is decoded and quantized and the samples are strung together similar to a connect-the-dot puzzle forming an analog signal that is comparable to the original. The entire process of analog-to-digital and digital-to-analog conversion is presented in detail in Chapter 4. Figure 8–10 shows the

Codec
The coder decoder is used to convert an analog signal into a digital signal.

quantizing
Converting a voltage level into a one and zero bit stream. Quantizing is the second step in digital-to-analog conversion.

encoding
The third step in analog-to-digital conversion. Encoding formats the quantized sample into an eight-bit word.

PCM word
Pulse code modulation word is an eight-bit word produced during analog-to-digital conversion.

Figure 8–10
T1 signal format. (a) One eight-bit word from one information source. Information source may be a telephone, computer, or video server. (b) Twenty-four eight-bit words fill one 192-bit frame. Each 192-bit frame holds twenty-four eight-bit words from twenty-four different information sources. The information source may be a voice signal converted into a digital one/zero word, data information from a computer, or a picture from a video clip. (c) Twenty-four 192-bit frames each holding twenty-four eight-bit words. $8000 \times 192 = 1.536$ Mbps.

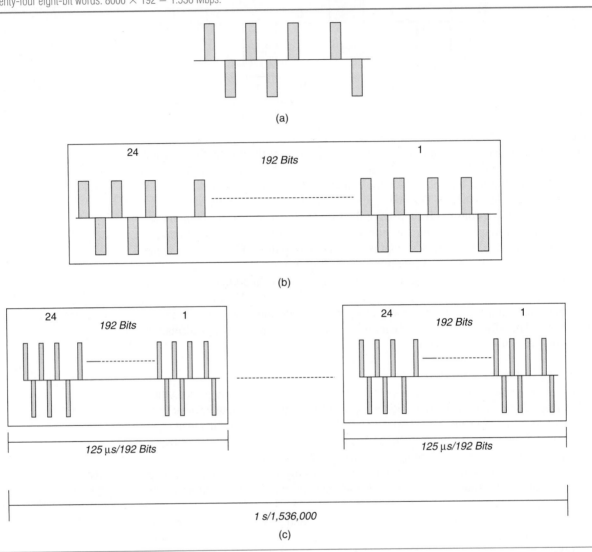

eight-bit PCM word, how the word is aggregated into one 192-bit frame and how twenty-four 192-bit frames are multiplexed into one 1,536,000 bps signal. Figure 8–11 shows how adding framing bits increases the line rate to 1,544,000 bps.

8.2.5 DS-1 Framing Format

As you can imagine, a signal made up of twenty-four channels each containing 64 kbps could be very difficult to decipher, especially since the channel contains only ones and zeros. You might compare it to a room filled with two types of clones, one wearing purple and one wearing red. How would you know whom to choose if you were asked to separate the 1,544,000 clones into groups containing a variety of both red and purple? Digital signals are very similar in that there are only two types of conditions, a one and a zero, and, as with the purple and red clones, all ones and zeros look the same. Therefore, how does a receiver determine where the eight-bit PCM word coming from your telephone station starts and ends? The answer to this dilemma is framing or simply an agreed-upon standard that designates the beginning of a group through the use of one-zero patterns. The two most common types of DS-1 framing are SF—superframe—and ESF—extended superframe.

Figure 8–11
T1 signal format.

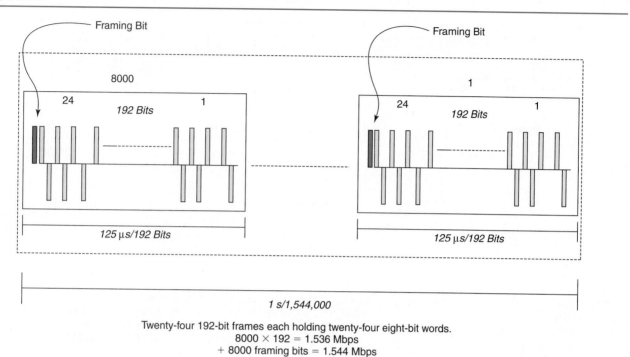

Twenty-four 192-bit frames each holding twenty-four eight-bit words.
8000 × 192 = 1.536 Mbps
+ 8000 framing bits = 1.544 Mbps

Superframe

Superframe is the older of the two and as such has slowly been replaced by *extended superframe* though should not be dismissed as it still continues to be used throughout the industry. SF is based on a twelve master frame sequence that uses framing bits to form a pattern that is read by the receiving device allowing it to sync up to the bit stream correctly. For example, when a receiver locks into the framing pattern, it is able to determine where the individual words start and end. Each frame consists of 192 bits plus one framing bit equaling 193 bits per frame as shown in the following formula:

$$24 \text{ channels} \times 8\text{-bit words} = 192 \text{ bits}$$
$$192 \text{ bits} + 1 \text{ framing bit} = 193 \text{ bits per frame}$$
$$193 \text{ bits/frame} \times 8000 \text{ frames/sec} = 1.544 \text{ Mbps}$$

The pattern used to designate a master frame in SF is 100011011100, eating up ten of the twelve available frame bits. The frame bits are referred to as *Ft bits*. The receiving device continuously scans the incoming bits ready to latch onto the 10-bit framing pattern. Once it syncs up to the pattern, it is able to determine count and map the PCM word positions.

To increase the confusion, additional patterns have been developed to represent CAS—channel associated signaling—states. The framing bits found in frames 6 and 12 form a pattern used to represent signaling states—on-hook, off-hook, and so forth. The term *A- and B-bit signaling* refers to the bit patterns associated to the signaling states. End equipment reads the patterns and reacts accordingly, as shown in Figure 8–12.

Figure 8–13 illustrates how a master frame contains twelve frames, each made up of 192 bits and one framing bit equaling 193 bits per frame. The diagram also shows where the framing bits and signaling bits are located in the frame.

Extended Superframe

The second type of DS-1 framing is extended superframe or ESF. ESF extends the number of frames per master frame from twelve to twenty-four, thus expanding the number of framing bits available to build patterns out. In addition, thanks to the advances in digital

superframe
DS-1 framing method that uses twelve frames per master frame.

extended superframe
A DS-1 framing method that uses twenty-four frames per master frame.

Ft bits
Framing bits of a DS-1 signal.

A- and B-bit signaling
Bits within the DS-1 used to designate signaling states.

Figure 8–12

In order for ringing, off-hook, or idle channels to be communicated across digital circuits, special patterns are used. The least significant bit of each eight-bit word in the sixth and twelfth frame are used to build the pattern.

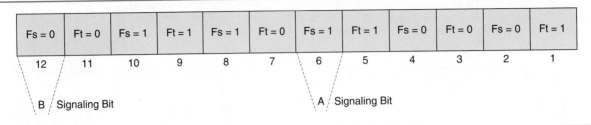

Fs = 0	Ft = 0	Fs = 1	Ft = 1	Fs = 1	Ft = 0	Fs = 1	Ft = 1	Fs = 0	Ft = 0	Fs = 0	Ft = 1
12	11	10	9	8	7	6	5	4	3	2	1

B / Signaling Bit   A / Signaling Bit

Figure 8–13

Superframe format for a T1 signal. (a) Superframe format has twelve 192-bit frames that equal one master frame. There are twelve framing bits that may be used to synchronize or line up the signal between devices. (b) Framing pattern sequence is used by the end devices to synchronize the bit stream. The end device buffers the twelve frames of data and picks out the framing pattern. Once the equipment reads the frame pattern sequence 101010, it is able to read the ones and zeros coming and translate them into information. (c) Signaling bits are designated in all even frames and are used to flag the sixth and twelfth frames. The end equipment uses the signaling bit pattern to find the sixth and twelfth frames.

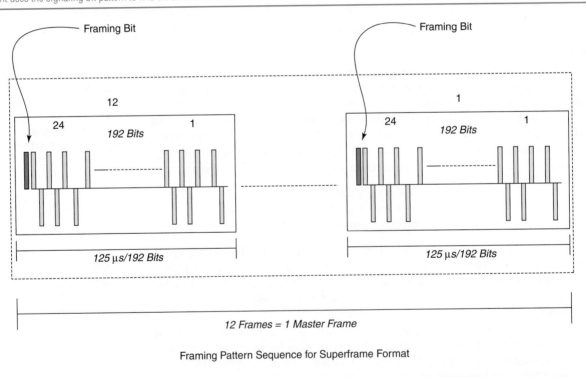

12 Frames = 1 Master Frame

Framing Pattern Sequence for Superframe Format

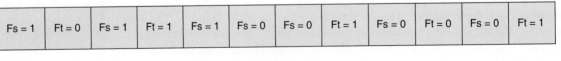

Fs = 1	Ft = 0	Fs = 1	Ft = 1	Fs = 1	Fs = 0	Fs = 0	Ft = 1	Fs = 0	Ft = 0	Fs = 0	Ft = 1

(b)

Fs = 000111

Figure 8–14

Extended superframe format for a T1 signal. (a) Extended superframe format has twenty-four 192-bit frames that equal one master frame. There are twenty-four framing bits that may be used to synchronize the signal between devices, carry maintenance bits, and carry an error-checking algorithm. (b) Framing pattern sequence is used by the end devices to synchronize the bit stream. The end device buffers the twenty-four frames of data and picks out the framing pattern. Once the equipment reads the frame pattern sequence 101010, it is able to read the ones and zeros coming in and translate them into information. (c) Signaling bits may now reside in the sixth, twelfth, eighteenth, and twenty-fourth frames. A, B signaling happens in the sixth and twelfth frames. A, B, C, and D signaling information is found in the sixth, twelfth, eighteenth, and twenty-fourth frames. (d) D bits are used to carry link information. 8000 bits are available for link status information such as whether there are errors being received at the far end. (e) C bits are used to carry an error-correction algorithm. The algorithm is referred to as CRC (cyclic redundancy check), which guarantees the health of the circuit to 99.95%.

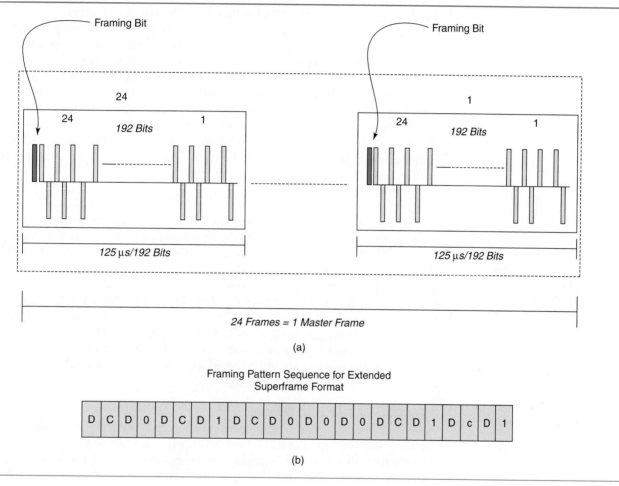

24 Frames = 1 Master Frame

(a)

Framing Pattern Sequence for Extended
Superframe Format

| D | C | D | 0 | D | C | D | 1 | D | C | D | 0 | D | 0 | D | 0 | D | C | D | 1 | D | c | D | 1 |

(b)

circuitry, devices no longer need a full twelve-bit frame sequence for synchronization. Thus, the increase in the number of framing bits, twelve additional bits, and the reduction in the number of Ft bits leave an ESF T1 with 8000 bits to build patterns from. As we all know, the digital world is all about one and zero patterns—the more the better.

As a result, signaling options have also been expanded in ESF. A, B, C, and D signaling bits located in the sixth, twelfth, eighteenth, and twenty-fourth frames are available to carry signaling states as defined by the application.

Beyond framing and signaling, ESF also offers a very robust error-checking algorithm, again thanks to the availability of the additional bits. The CRC—cyclic redundancy check—error-checking algorithm employed on an ESF circuit measures the line to a 98.4% accuracy rate. Translated, that means that when there are zero CRC errors, there is a 98.4% chance that the line is green—good.

The remaining twelve bits are reserved for data link bits used by the equipment vendor. Figure 8–14 illustrates the frame format of an ESF DS-1. Note the location of the Ft bits; the A, B, C, and D signaling bits; the CRC bits; and the data link bits.

8.2.6 DS-1 Line Code: B8ZS and Bit7/AMI

A DS-1 signal composed of one and zero states in a bipolar-return-to-zero format must comply to the ones density rule mandated by the FCC and standards bodies. The ones density rule states that a DS-1 signal must consist of 12.5% ones pulses with no more than fifteen consecutive zeros. In traditional voice transmission, thanks to the PAM samples, this rule was rarely violated. In contrast, data transmission varies so widely that there is a good chance that sixteen consecutive zeros or more will occur on the line. Two methods are available to reduce the chance of violating the ones density rule. The first and oldest is called Bit7 coding or AMI coding and refers to a scheme that jams the least significant bit of the eight-bit word making it a ones pulse whenever there are more than eight consecutive zeros. The second and most popular coding method called B8ZS (bipolar with eight-zero substitution) handles the consecutive zeros problem by replacing the zeros with a four-bit code containing a bipolar violation that is detected by the end equipment. The end equipment reads the code and replaces it with all zeros. Figure 8–15 shows both zero suppression methods, Bit7 and B8ZS.

The main purpose for devising a new coding scheme to eliminate the all-zeros condition on the line was to increase the bit rate of the channel from 56 to 64 kbps. The consequence of jamming one bit of the eight-bit word in Bit7/AMI coding is the reduction in the bandwidth of each channel from 64 to 56 kbps:

$$7 \text{ bits} \times 24 = 56 \text{ kbps}$$

B8ZS, on the other hand, substitutes a code that can easily be replaced with the original signal sequence thus leaving the least significant bit alone. Therefore, the term given to B8ZS encoding is *clear channel coding*, referring to the full channel being clear or available for use. Though Bit7/AMI encoding has little effect on voice frequency channels, including those produced by a modem, it does reduce the information carrying capacity of data-only channels. In order for providers to offer a full 64 kbps clear channel to their customers, B8ZS is configured end to end on all T1 circuits. Older systems in the network may still be optioned as Bit7/AMI, as changing them requires a circuit pack replacement—often an unnecessary expense if the circuit is carrying voice. Today's systems are optioned using software commands, making it much easier to change the encoding scheme.

8.2.7 Channel Associated Signaling

As explained earlier, signaling information is carried within the information signal of the DS-1 and is referred to as *channel associated signaling* or CAS. Some in the industry refer to this as *in-band signaling*, as it uses bits reserved for information to carry signaling patterns. The signaling patterns carried are used to indicate off-hook/on-hook, ringing, and busy conditions. Channel associated signaling designates the sixth and twelfth frames in SF and the sixth, twelfth, eighteenth, and twenty-fourth frames in ESF to carry signaling information. The seventh bit of the eight-bit word is robbed—thus, the term *robbed bit signaling*—and is used to indicate a signaling condition as

clear channel coding
Clear channel DS-1 provides a full 64 kbps worth of bandwidth due to not needing to use signaling bits for voice information.

channel associated signaling
Channel associated signaling robs the sixth and twelfth frames and uses them to designate signaling states: busy, ringing, and so on.

robbed bit signaling
Stealing the bits from the signaling frames to designate signaling states.

Figure 8–15
(a) The B8ZS code inserted into the all-zeros bit stream. (b) Jamming one bit is the method used in the Bit7 zero suppression scheme.

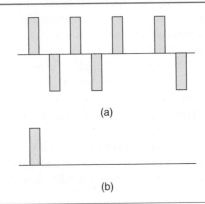

(a)

(b)

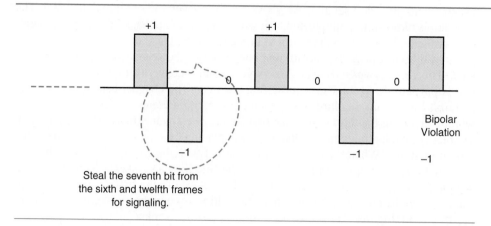

Figure 8–16
Robbed bit signaling.

shown in Figure 8–16. Because one bit of the eight-bit word is being used for signaling, it cannot be used to carry information; as a consequence, the bandwidth of the signal is reduced from 64 to 56 kbps. Voice signals are not adversely affected by losing one bit; therefore, a 56 kbps channel provides enough capacity for a clear, high-quality signal. On the other hand, losing one bit significantly reduces the bandwidth available for digital data, a concern for customers sending data information. In addition, a digital data channel does not require the traditional on-hook/off-hook signaling conditions that a voice channel requires. In order to correct the problem, most providers option a channel (the 64 kbps channel that carries the voice or data) accordingly: a voice channel is optioned for robbed bit signaling, and data is optioned for digital data channel. Channel associated signaling is discussed in more detail in Chapter 11.

8.2.8 End-to-End DS-1

Figure 8–17 shows an end-to-end DS-1 circuit used to tie two channel bands together.

The fundamental concepts associated with how a circuit is framed, the line code, signaling, and so forth can be used to explain the end-to-end multiplexing process. We have described bit stream, framing bits, line code, and how the bandwidth of a DS-0 is calculated. The next step is to pull all these concepts together and explain the end product—multiplexing channels onto a T1 digital circuit. To do so, we follow the multiplexing, bit-interleaving process from *channel bank* A to channel bank B.

Amy's phone is connected to port one of channel bank A; Carol's phone is connected to port one of channel bank B. Amy and Carol are talking to one another over the T1 circuit. Amy asks Carol, "Have you studied for your calculus test?" The

channel bank
Multiplexer that muxes DS-0 level signals such as voice or data into DS-1. Channel banks are most often found at customer premises and terminate T1 circuits.

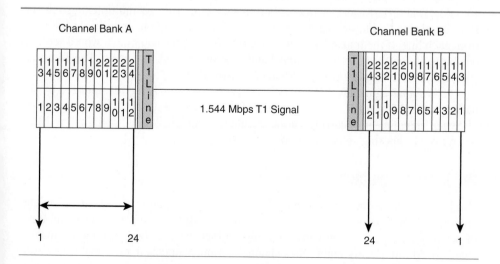

Figure 8–17
Two back-to-back channel banks connected by a T1 circuit. Twenty-four information sources feed into each channel bank.

channel bank line card samples Amy's voice and converts it into a digital signal. The question is turned into multiple eight-bit words. The channel bank takes the first eight-bit word and places it into the first seat in the first train car (i.e., frame). It then grabs an eight-bit word from the second line connected to the second port on the channel bank. This line belongs to Hank, whose voice signal is converted into an eight-bit word that is put into the second seat in the first train car. The channel bank then grabs an eight-bit word from the third line, which connects to a computer. The eight-bit data word is placed in the third seat of the first frame. The channel bank continues to grab samples from the remaining twenty-two lines, placing each in consecutive seats in the boxcar or frame. When the twenty-fourth seat is filled, the channel bank attaches the framing bit to the frame, jumps back up to the first line, and takes a second eight-bit sample from Amy's voice signal. The sample is placed in the first seat of the second frame, while the channel bank continues to multiplex the eight-bit word samples onto the T1 circuit. Figure 8–17 shows back-to-back channel banks.

Channel bank B receives the incoming bit stream from channel bank A. Bank B synchronizes according to the framing bit patterns and begins to demultiplex the incoming bit stream. It takes the first eight-bit word in the first seat of the frame and sends it to the line card that Carol's phone interfaces with. The line card converts the eight-bit word into an analog voltage signal and outputs it on the line connecting to Carol's phone. Channel bank B then takes the second eight-bit word in frame 1 and points it to the line card attached to Rod's phone. The line card converts the digital one and zero bits into an analog sample and outputs the signal on the line connecting to Rod's phone. The third seat in the frame is pointed toward the line card attached to Hillary's computer. The line card receives the digital eight-bit word and outputs it on the line attached to Hillary's computer. The process continues until all twenty-four eight-bit samples have been demultiplexed. When channel bank B receives the framing bit, it immediately grabs the eight-bit word in the first seat of the second frame and points it toward the line card attached to Amy's phone. The process then repeats itself.

Signaling is handled as follows. Hillary logs off her computer and decides to call Ralph. She picks up her telephone handset and waits for dial tone. Terminating equipment sends a dial tone to Hillary's receiver. Hillary dials Ralph's number and waits to hear ringing. The terminating equipment changes the eighth bit in the sixth frame to a one and the eighth bit in the twelfth frame to a zero to signify ringing. The end equipment has a special conductor in the sixth and twelfth frames that walks through the boxcar and watches each of the twenty-four eight-bit samples for signaling information. The frame conductor notes whether the eighth bit in each of the seats is a one or a zero. The head conductor takes the information from the frame conductors to determine if it needs to ring the phone, send dial tone, or do something else. In this case, the head conductor sees that in the third seat in frame six, the eighth bit is a one, and in the third seat in frame twelve, the eighth bit is a zero. This tells the head conductor to ring channel 1's telephone. The format just described is a *byte-interleaved* multiplexing scheme. Byte-interleaved multiplexing is the process of shipping out one byte (eight-bit word) from each line. If the transmission were *bit interleaved*, the equipment would send out one bit from line 1, then move to line 2 and send out one bit, and so on. The T1 circuit standard is built around a byte-interleaved format.

A, B signaling means that the sixth and twelfth bits are being manipulated by signaling states. Most test equipment has LEDs labeled A, B, C, and D that turn on and off when signaling is occurring on the channel. Technicians monitor signaling LEDs when troubleshooting signaling problems.

byte interleaved
Multiplexing scheme that interleaves an eight-bit word from one channel into the higher-level signal.

bit interleaved
Multiplexing scheme that interleaves one bit from one channel, then moves to a second channel and interleaves one bit.

■ 8.3 DS-3 CIRCUIT

The DS-3 circuit is the workhorse of the communications network. Service providers in North America depend on DS-3 circuits to carry all types of traffic across all types of networks. A DS-3 circuit carries 45 million bits every second. The circuit may be channelized into twenty-eight DS-1 channels or unchannelized and capable of carrying 45 Mbps. In either case, several definitions related to DS-3 circuits need to be discussed. The following topics are discussed in this section: physical media, signal characteristics, and framing format.

8.3.1 Physical Media

DS-3s are high-frequency, high-speed signals. As such, they require a medium with low loss characteristics. As discussed in an earlier chapter, the higher the speed of the signal, the shorter the distance it can travel before becoming indistinguishable to the receiving equipment. In order to compensate for this restriction, DS-3 signals do not ride on twisted pair copper cable. In fact, the only copper cable capable of carrying a DS-3 is coaxial cable and even then, the distance is limited—450 to 600 feet. DS-3s travel on coaxial cable inside a building and no farther. Typically, most systems are engineered around the 450-ft. standard, allowing for an additional 450-ft. extension to be added if a repeater is placed mid-span. DS-3s also travel through the air on microwave systems as explained in a later chapter. Traditional microwave systems are spaced on average thirty-five miles apart, meaning that the DS-3 is able to travel thirty-five miles before requiring regeneration. Fiber optic cable, the most common medium used to carry DS-3s, is able to carry the signal substantial distances before requiring amplification or regeneration. In extreme cases, a signal may travel as far as seventy-five miles before needed to be amplified.

If it were possible to observe a DS-3 as it rode down the medium, it would quickly become obvious that there is no difference between the signal traversing a fiber optic link and the one traversing a coaxical connection. The format, structure, and appearance are identical.

8.3.2 Signal Characteristics

All communications circuits are designed according to the signal's characteristics. Signal characteristics refer to the structure of the bit stream or simply the height and width of the pulse, the number of pulses generated per second, and the mode or polarity of the pulse. A DS-3 transmitter outputs pulses that comply with specific parameters that define the shape of the pulse, that is, the minimum/maximum height and width. If, for example, the pulse falls below the minimum height as defined by the standard pulse shape, it can be concluded the signal level is too low—the signal amplitude falls below the allowed threshold. If the pulse spreads over the *leading edge* or *trailing edge* of the pulse shape, terms used to describe the beginning and ending of the pulse, the signal is out of phase or out of sync. Additionally, if a signal is a bipolar formatted bit stream, as with a DS-3, when two consecutive pulses of the same polarity occur, an error is logged indicating that the signal is corrupted. It is important to realize that all digital signals comply with standard signal characteristics and that these characteristics are used to design and maintain circuit quality.

leading edge
Leading edge refers to the start of the one pulse.
trailing edge
Trailing edge refers to the end of the one pulse.

Similar to measuring a dog's heart rate, blood pressure, and temperature to determine if the dog is healthy, technicians measure a DS-3's frequency, signal level, and pulse shape to determine if a DS-3 circuit is healthy. A DS-3 runs at a frequency of 44.736 MHz $+/-$ 880 Hz, carrying 45 million bits every second. The frequency of the DS-3 must fall within the specified range in order to be considered healthy. Signal level measured in dBm relates to the strength of the signal, signifying whether it has too much power or too little, both being detrimental to the health of the circuit. Power, a third key parameter, measures the peak-to-peak voltage of the signal, similar to signal level. If the power level falls below or above specs, the circuit must be analyzed further in order to determine the origin of the trouble.

The pulse shape of a DS-3 must comply to the defined height and width. Test equipment uses a template called a *pulse mask* that defines the boundaries a pulse needs to adhere to, as shown in Figure 8–18. If, for example, the pulse were to shift to the left, the end equipment could interpret the incoming signal incorrectly because it would mistake a one for a zero or a zero for a one. Remember that all digital information consists of patterns that are translated into information. If you were watching a ball game on television and the middle of the picture suddenly showed a portion of a sewing bee, would you be able to figure out what pitcher was throwing the ball?

Figure 8–18
Pulse mask.

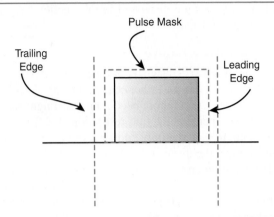

The same is true when the signal loses synchronization and shifts. The end equipment is seeing incorrect patterns that form a collage of information that is worthless to the end application. A term used to describe this phenomenon, the pulse shifting off of its center, is jitter. Jitter occurs when the pulse shifts either to the right or to the left away from its prescribed position. Jitter in a circuit, if severe, can cause timing problems, such as cross talk, and other debilitating conditions. As discussed in the earlier section on timing, jitter should be avoided, especially as the circuit speeds increase.

Similar to DS-1s, DS-3s are built using the alternate mark inversion, bipolar-return-to-zero formats, as shown in Figure 8–19. Every other pulse must be of the opposite polarity to ensure the integrity of the signal. Similar to the DS-1 ones density rule explained earlier in the chapter, a DS-3 must also adhere to a similar rule set by the standards bodies and the FCC. The standard states that the ones pulses must comprise 33% of the signal. In order to comply with this rule, a line-code scheme called B3ZS (bipolar with three-zero substitution) is used that produces a bipolar violation code anytime too many zeros are shipped down the line. The end equipment is able to read the code and replace it with the actual values—zeros. Only one zero suppression scheme is available on DS-3 circuits.

8.3.3 Framing Format
All digital signals depend on framing formats defined by the standards bodies: ANSI, the ITU, and Bellcore. A frame is used to house the bits in an orderly fashion, making it possible for equipment components to synchronize with each other and correctly read the incoming bit stream. Without a defined frame structure, the ones and zeros arriving from the network would look like a garbled bunch of nothing. It would be impossible for the receiving equipment to determine what bit belonged to what channel and so forth. Discussions on framing often focus on the position of the bits in the frame, forgetting to pull all the elements together in order to show how they relate to the signal flow. Our goal is to not only detail the frame format but also provide a view of how the

Figure 8–19
DS-3 AMI format.

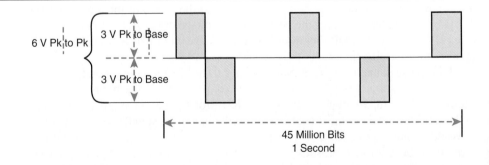

DS-3 frame actually travels across the network end to end. We begin with a description of the frame structure.

The three types of framing used to format a DS-3 are M13, C-bit parity, and unchannelized. The oldest and most prevalent in the networks today is the M13 framing scheme. Understanding the structure of the M13 framing scheme makes it very easy to learn the other two, because both of the others are built on the M13 frame, differing only in the number and assignment of overhead bits.

M13 Framing Format

The *M13 frame* is more complicated than the DS-1 frame format. A DS-3 frame consists of a DS-2 frame used to aggregate the DS-1s into manageable groups. Four DS-1s are bit interleaved into the DS-2 frame. *Bit interleaved* refers to feeding bits one by one from DS-1 number 1 into the frame, then from DS-1 number 2, from DS-1 number 3, and so forth. The DS-2 frame does not have one entire DS-1 intact within one channel. Just think of what a jigsaw puzzle would look like if it were made up of a bit-interleaved signal. Yuk.

The DS-2 frame is made up of four frames called M subframes, as shown in Table 8–3. Each M subframe is further divided into six blocks capable of carrying forty-nine bits. The overhead bits used in DS-2 framing are F bits, M bits and C bits. The total number of overhead bits per frame is twenty-four, and the number of payload bits is 1772 per frame. F bits (framing bits) are found in the third block, first position, and sixth block, first position. The pattern used to designate a frame is 01. This means that position 1 of block 3 should always be a zero and position 1 of block 6 should always be a one. A framing pattern of 01010101 is created when the framing bits of each subframe are singled out and placed next to each other. The equipment's receivers are very intelligent devices

M13 frame
DS-3 framing format.

Table 8–3
DS-2 Frame Format.

Subframe Total Bits	Block 1 49 Bits	Block 2 49 Bits	Block 3 49 Bits	Block 4 49 Bits	Block 5 49 Bits	Block 6 49 Bits
1st	M0	C11	F0	C12	C13	F1
2nd	M1	C21	F0	C22	C23	F1
3rd	M1	C31	F0	C32	C33	F1
4th	MX	C41	F0	C42	C43	F1

Description of DS-2 Frame Structure

Each DS-2 has six blocks containing forty-nine bits. Forty-eight of the bits are T1 signal bits. The remaining bits per block are overhead bits. The DS-2 decides whether or not to stuff the T1 signal in order to synchronize the bit streams.

M Bits Used to Align the Multiframes

M0 = 0
M1 = 1
MX = Either 1 or 0

F Bits Used to Align the Frame

F0 = 0
F1 = 1

C Bits Used to Indicate Stuffing or No Stuffing

C11, C12, C13 = indicate stuffing for DS-1 1
C21, C22, C23 = indicate stuffing for DS-1 2
C31, C32, C33 = indicate stuffing for DS-1 3
C41, C42, C43 = indicate stuffing for DS-1 4

that are able to quickly read this pattern and sync up to the incoming bit stream. Therefore, framing bits act as flags or markers that are used to delineate the bit stream into readable chunks. Table 8–4 shows the DS-3 M13 frame structure. Additional detail follows.

M bits (multiframe bits) are found in the first block, first position of the subframe. All four subframes require an M bit in the first position. The job of the first M bit is to designate the beginning of the subframe. The first M bit is used in conjunction with the other subframe M bits to form a pattern, 011X, where X is either a one or a zero. The X bit is common to all asynchronous devices and changes according to the condition of the line. If the upstream equipment goes into alarm, the X bit is changed to a zero and the equipment generates a yellow alarm—a topic discussed in a later chapter. If the upstream equipment is not in alarm, the X bit is a one.

Timing a network requires that all of the ones and zeros arrive and leave according to a very rigid time schedule. The C bits in the DS-2 frame were designed to help guarantee that all of the DS-1s would be multiplexed into the DS-2 frame at the same rate. Because DS-1s interface into the M13 multiplexer from various pieces of equipment in the network, it is common that they all do not arrive at precisely the same rate. In order to synchronize each DS-1 being multiplexed into the DS-2 frame correctly, a method called *bit stuffing* was devised to adjust all of the bit streams to the same rate, 1,545,796 bps. The job of a C bit is to tell the network that bit stuffing has or has not taken place. Within the DS-2 frame, C bits are found in each subframe in the first position of block 2, block 4, and block 5. There is a total of twelve C bits in the entire DS-2 frame. The pattern used to designate bit stuffing is 111, and no bit stuffing is 000.

The DS-2 frame is bit interleaved into the DS-3 frame. The DS-3 frame really does not care about where the DS-1s reside within the DS-2 as they are invisible to higher-ups in the DS-3 frame. The first step taken by the DS-3 multiplexer is to multiplex the seven DS-2s into its frame structure. Similar to the DS-2 frame, the DS-3 frame is composed of overhead bits and payload. The frame is configured with seven subframes, each holding eight blocks containing eighty-nine bits. One bit per block is used for overhead and eighty-eight bits for payload. The overhead bits include F bits, M bits, P bits, X bits, and C bits.

bit stuffing
Additional bits are stuffed into the DS-2 frame to match incoming DS-1 circuits to make sure they are synchronized properly.

Table 8–4 DS-3 M13 Frame Format.

Subframe Total Bits	Block 1 85 Bits	Block 2 85 Bits	Block 3 85 Bits	Block 4 85 Bits	Block 5 85 Bits	Block 6 85 Bits	Block 7 85 Bits	Block 8 85 Bits
1st	X	F1	C11	F0	C12	F0	C13	F1
2nd	X	F1	C21	F0	C22	F0	C23	F1
3rd	P	F1	C31	F0	C32	F0	C33	F1
4th	P	F1	C41	F0	C42	F0	C43	F1
5th	Mo	F1	C51	F0	C52	F0	C53	F1
6th	M1	F1	C61	F0	C62	F0	C63	F1
7th	Mo	F1	C71	F0	C72	F0	C73	F1

Description of DS-3 M13 Frame

The DS-3 M13 frame has 7 M subframes. Each of the subframes has 85 bits per block. Within each block, 1 bit is used for overhead and the remaining 84 bits are used for information. The overhead bits include X bits, P bits, M bits, F bits, and C bits.

Description of Overhead Bits

1. F bits are used to align the frames. The code used is 1001.
2. M bits are used for multiframe alignment signal.
3. P bits are used for parity checking.
4. X bits are used for messages.
5. C bits are used as stuffing bits.

F bits (framing bits) form a frame pattern 1001 that is used to line up the frame, similar to a sergeant lining up his or her platoon. The framing pattern is repeated over and over and over, similar to a small child repeating the question "Why can't I? Why can't I?" The mother is able to immediately recognize that it is her child in the same way the equipment is able to recognize the bit stream as a framed DS-3. The F bit is found in the first bit of the second, fourth, sixth, and eighth blocks. The total number of F bits per DS-3 frame equals forty-nine because there are seven subframes and four F bits per subframe.

Again, M bits are used to help designate the subframes. They are found in the fifth, sixth, and seventh subframes—first block, first bit. P bits (parity bits) are found in the third and fourth M subframes, the first bit in block 1. The job of the P bit is to provide an end-to-end error-checking scheme that is read by the end equipment and used to signal trouble on the line—bipolar violations. Unfortunately, the M13 P bits are fairly useless because the fiber optic termination (FOT) and microwave equipment strip out the P bits as soon as they receive the signal.

X bits are used to indicate a yellow alarm condition or a no alarm condition as explained earlier in the DS-2 frame description. The X bits are located in the first and second subframes, the first position in block 1. X bits are service provider configurable, though typically the default is used as it generates a yellow alarm if needed.

Again, similar to the DS-2 frame, the C bits in the DS-3 frame are used to indicate that bit stuffing has occurred or has not on the DS-2. The C bits are located in subframes 1 through 7 in block 5, position 1. When DS-3 framing was first designed, it was assumed that DS-2s would interface into the M13 multiplexer from various pieces of equipment, similar to the DS-1s. At the time, no one realized that the DS-2 transmission system would not be widely accepted. Instead, the DS-2 remained an intermediate MUXing step used primarily for the purpose of providing synchronization to the incoming twenty-eight DS-1s. Therefore, the C bits became useless and unused. Due to this phenomenon, their conditional state is always set to a no stuffing condition.

C-Bit Parity Framing Format

The second DS-3 framing method is called *C-bit parity framing*. C-bit parity framing closely resembles the M13 frame structure. The difference between the two is how the overhead bits are allocated and used. As shown in Table 8–5, the C-bit parity format includes X bits, F bits, M bits, P bits, and CP bits. The F bits and M bits are identical to those in the M13 frame and therefore require no description. The new way of using the X bits and the C bits is significant in that they improve the circuit's ability to identify problems and notify the network elements when a trouble condition occurs.

The job of the X bit has not changed significantly other than generating a new alarm called an *FEOOF*, Far End Out of Frame, which is the same as a *yellow alarm*. The X bit resides in the same location as in an M13 frame and changes to a zero when an AIS (Alarm Indication Signal) or Loss of Frame from the upstream equipment is received. The equipment responds to the condition by sending an FEOOF to the upstream equipment, letting it know that it is receiving an All Ones signal. The term *FEOOF* is harder to remember than *yellow alarm;* but, because it is used in the industry, it is good to remember what it stands for.

The C bits have been changed significantly in the C-bit parity framing scheme. As we discussed earlier, the bits proved to be idle and unused because they were not needed to indicate a bit stuffing condition—their main purpose in life in the M13 framing scheme. Consequently, the very intelligent design engineers decided to use the idle, unemployed bits and put them to work. Their first decision was to use them as parity bits, referred to as CP bits, because the microwave or fiber optic equipment does not manipulate the C bits in the frame. The CP bits carry the same bit sequence as the P bits, and the end equipment reads the C bits in the same way it reads the P bits. Determining the integrity of the end-to-end DS-3 path is accomplished from deciphering the CP bits to determine if parity has been violated. Therefore, the C-bit parity framing method

C-bit parity framing
Framing method used on DS-3 circuits. Provides additional overhead bits that can be used to monitor and maintain the circuit.

FEOOF
Far End Out of Frame is an error code generated when too many out-of-frame errors are received by the end device.

yellow alarm
An alarm indication sent by the receiving device when it receives too many FEOOFs.

Table 8–5 DS-3 C-Bit Parity Frame Format.

Subframe Total Bits	Block 1 85 Bits	Block 2 85 Bits	Block 3 85 Bits	Block 4 85 Bits	Block 5 85 Bits	Block 6 85 Bits	Block 7 85 Bits	Block 8 85 Bits
1st	X	F1	AIC	F0	Na	F0	FEAC	F1
2nd	X	F1	DL	F0	DL	F0	DL	F1
3rd	P	F1	CP	F0	CP	F0	CP	F1
4th	P	F1	FEBE	F0	FEBE	F0	FE BE	F1
5th	Mo	F1	DL	F0	DL	F0	DL	F1
6th	M1	F1	DL	F0	DL	F0	DL	F1
7th	Mo	F1	DL	F0	DL	F0	DL	F1

Description of DS-3 C-Bit Parity Frame

The C-bit parity frame structure differs from the M13 frame format. The main difference between the two frames is found in blocks 3, 5, and 7. The overhead bits for C-bit parity signals are X bits, P bits, M bits, F bits, ALC bits, DL bits, CP bits, FEBE bits, Na bits, and FEAC bits.

Description of Bits Unique to C-Bit Parity Signals

1. AIC bits are used for Application Identification Channels.
2. N bits are reserved for future uses.
3. FEAC is used for Far End Alarm and Control channels.
4. DL is used as a data link.
5. CP is the C-bit parity bit.
6. FEBE is used for Far End Block Error.

offers a much more robust and effective way to monitor and maintain signal quality. CP bits are found in the second, fifth, and seventh blocks, first bit of the third M subframe.

The CP bits eat up three of the twenty-one C bits available. An FEBE, Far End Block Error, is an alarm indication carried by the first bit in the fourth M subframe, second, fourth, and seventh blocks. The FEBE pattern is used to tell the downstream equipment that a frame error or a CP bit error has or has not occurred, depending on the situation. Loopback codes and alarm status are two functions carried by the FEAC (Far End Alarm and Control) bits located in the first bit of the first M subframe and the seventh block. The C bits in this block are used to carry alarms or signal status and loopback codes.

The final code used in place of C bits is the AIC (Application Identification Channel) used to designate the type of framing used, M13 or C bit. It is located in the first M subframe, first bit of the third block. The Network Application Bit is reserved for future use and resides in M subframe 1, bit 1 and block 5. The remaining C bits are used as data link bits capable of carrying specific information about the equipment, the application, and the carrier. The data link A bits located in M subframes 2 and 7 are most often used as a communications channel between video Codecs. The data link T bits located in M subframe 5 are used as a 28 kbps terminal-to-terminal data link. The data link 1 bits in the sixth M subframe are used to carry an interexchange carrier code that is used to identify the originator of the DS-3 as it passes through multiple carriers' networks.

Unchannelized Framing

unchannelized
DS-3 formatted as one pipe used to carry data or other unchannelized information.

The final type of DS-3 framing is called *unchannelized*. This refers to a DS-3 that is not divided into twenty-eight DS-1s or seven DS-2s. Instead, it carries a full 44.5 Mbps of information. Visualize a giant straw with seven smaller straws inside and four even smaller straws inside each of the seven. Now visualize a second identical giant straw that does not have the smaller straws inside. Both are shown in Figure 8–20. The straw containing the smaller straws is similar to M13 or C-bit parity configured DS-3, and the second big straw is similar to an unchannelized configured DS-3. What types of information do you think the unchannelized DS-3 would carry? *Video or data* is a good

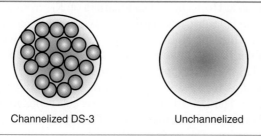

Figure 8–20
Channelized and unchannelized DS-3.

Channelized DS-3   Unchannelized

answer. Both could care less about being multiplexed up from smaller signals, DS-1 into DS-2 and then DS-3. Their information is already formatted in packages, for example, ATM cells. Though unchannelized framing eliminates the multiplexing/ demultiplexing step, the overhead bits remain the same as those described in the M13 framing scheme. A hard concept to understand for many is that a signal cannot be shoved onto a pipe without some sort of defined bit sequence. Remember that the framing bits are needed to let the end equipment know what part of the bit stream it is receiving and when. Because the end equipment is built to accept a DS-3 interface, the bit sequence must meet the DS-3 standard even when set for unchannelized; unchannelized does not mean unframed. Figure 8–21 shows an end-to-end view of a DS-3 circuit connecting two M13 multiplexers.

8.3.4 Connecting the Dots

The past few sections have focused on individual systems used to carry information between two points. We can combine that information and build an asynchronous fiber network from end to end.

The example we use starts at the local telephone company's central office. Our customer, HoBo Inc., has ordered a point-to-point circuit between its headquarters in Albany, New York, and its branch office, also in Albany. It requires one full DS-0 of bandwidth and would like to be able to use the entire 64 kbps channel. HoBo Inc. headquarters is served by the local phone company's Black Oak central office, and the branch office is served out of the Red Rover central office in Albany. HoBo ordered the circuit from the local phone company, which will need to provision a circuit from HoBo's Plymouth Avenue site to its Forest Street location.

Figure 8–21
A DS-3 M13 multiplexer. Within the multiplexer, the T1 signals are multiplexed into DS-2. Each DS-2 carries 6.176 Mbps. The DS-3 contains seven bit-interleaved DS-2s.

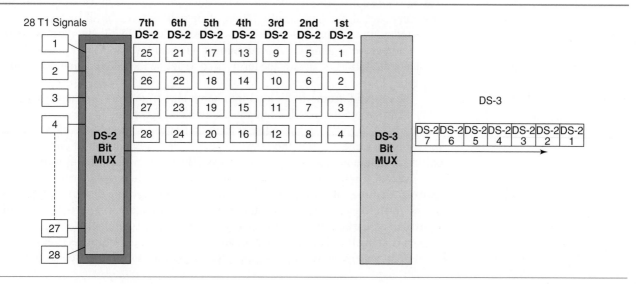

The circuit begins at HoBo Inc.'s headquarters on River Street in Albany where a 64 kbps four-wire line is terminated. HoBo's network manager connects the 64 kbps DS-0 four-wire circuit into the HoBo network router. The four-wire circuit goes from HoBo's location to the Black Oak central office, where it is fed into a channel bank where twenty-four DS-0 circuits are multiplexed into one T1 circuit. The T1 signal from the channel bank is fed into an M13 multiplexer that combines twenty-eight T1 signals into one DS-3. The DS-3 is then fed into a 560 Mbps multiplexer that multiplexes twelve DS-3s into one 560 Mbps signal. The multiplexer also performs an electrical-to-optical conversion. The optical 560 Mbps signal is shipped out on fiber optic cable that connects to the local telephone company's access tandem office. At the access tandem office, the DS-3 channel carrying HoBo's DS-0 is demultiplexed from the 560 Mbps signal into twelve DS-3s. The DS-3 carrying DS-0 is fed into a piece of equipment called a *digital cross-connect system* (DCS). The DCS electronically grooms the DS-3 carrying HoBo's DS-0 down into twenty-eight DS-1s. The DCS then grooms out the twenty-four DS-0s riding in the DS-1 and cross connects them to a DS-1 pointed toward the Red Rover central office. The DS-1 is then groomed into a DS-3 that feeds an external fiber multiplexer. The fiber multiplexer receives all traffic heading to the Red Rover central office. The twelve DS-3s are converted into an optical signal, then transported on a fiber optic cable to the Red Rover central office. The twelve DS-3s arrive at Red Rover central office where they are converted into an electrical signal and demultiplexed down into twelve individual DS-3s. The DS-3 carrying HoBo's DS-0 is fed into an M13 multiplexer, which demultiplexes the DS-3 into twenty-eight individual T1 signals. The T1 carrying HoBo's DS-0 is fed into a channel bank. The channel bank demultiplexes the twenty-four DS-0 channels out of the T1 signal. Each of the DS-0s is transported out on individual copper loops to the different businesses in the area. In our case, HoBo's DS-0 is carried on a four-wire link that connects the Red Rover central office to HoBo's branch office. There, the four-wire 64 kbps circuit is connected to a router.

digital cross-connect system
Electronic cross connect used to cross connect circuits without having to physically wire the circuits together at a DSX panel.

■ 8.4 SONET TRANSPORT PROTOCOL/SONET STANDARD

Once fiber optic systems proved to be a viable means to carry information, the standards bodies realized that they needed to create a nonproprietary transport protocol for optical transmission. Their goal was to develop a protocol that would allow fiber optic strands to carry large amounts of information, build in overhead bits for network maintenance, and maintain a standard frame format that would allow all types of vendors' equipment to talk to one another. The Synchronous Optical Network (SONET) standard emerged in the early 1990s as North America's optical transmission standard. SONET was developed by ANSI; and the ITU, Bellcore, equipment vendors, and service providers started the rapid proliferation of optical systems in all network segments. Even today, though often scorned as a dinosaur, the SONET standard continues to evolve and grow, making it the predominant physical layer transport protocol in communications. Table 8–6 details the terms used to define the SONET line rates and interface types.

The SONET frame is divided into same-size time slots or channels similar to other TDM protocols. SONET line rates are named according to the number of DS-3s they carry. A SONET signal is like a train that pulls only one type of boxcar. Every boxcar is the same size and the same shape, has the same number of cubby holes, and carries the same number of ones and zeros. The asynchronous DS-3 train pulls boxcars that vary in the number of ones and zeros carried. In simple terms, a DS-1 is not an exact multiple of a DS-3; a DS-3 is not an exact multiple of a 560 Mbps system; and so forth. The variation in the number of ones and zeros is a result of bit stuffing, as explained earlier. SONET's dependence on an external clock source to keep the signals in step eliminates the need to stuff bits to maintain timing. Relieving the protocol of bit stuffing allowed developers to design line rates that are exact multiples of

Table 8–6 SONET Digital Rate Chart.

Electrical Interface Name	Optical Interface Name	Bit-Per-Second Rate	Number of DS-0 Channels
VT-1.5		1.78 Mbps	24
VT-6		6.912 Mbps	96
EC-1 (STS-1)	OC-1	51.84 Mbps	672
	OC-3	155 Mbps	2016
	OC-3c	155 Mbps	
	OC-12	622.08 Mbps	8064
	OC-12c	622.08 Mbps	
	OC-48	2.5 Gbps	32,256
	OC-192	10 Gbps	129,024

each other. All equipment vendors build interface ports knowing that the boxcars they build can be carried on any railroads' network seamlessly. A standard is a great thing.

As you review the line rates shown in Table 8–6, you should note three terms used to represent the same value: OC, STS, and EC. *OC*—optical carrier—refers to the SONET optical signal; *EC*—embedded channel—refers to the electrical signal; and *STS*—Synchronous Transport Signal—refers to the SONET frame. The use of each depends on the location in the network. A signal in its optical form—laser light—is an OC signal. A signal converted to the electrical equivalent inside the multiplexer is referred to as an EC signal. The term used to describe the signals riding inside an OC or EC signal is STS. For example, an OC-48 carries forty-eight STSs, an OC-3 carries three STSs, and so forth.

Beyond offering uniform line rates, a second significant difference between the asynchronous format and the SONET format is the method used to time the network. The SONET requires that an external clock source be used to line the bits up in their proper position. Without a solid clock source, such as one derived from the Stratum 1 clock as explained earlier in the chapter, the SONET network would slowly lose synchronization and crash. The term *synchronous* in SONET gives us a good clue as to why the network requires external timing.

8.4.1 Network Structure Defined by the SONET Standard

The SONET standard also defines the structure of the network. Three segments make up the SONET network—section, line, and path—as shown in Figure 8–22. The section portion of the network includes the elements between two regenerators or a regenerator and Line Terminating Equipment. The term *LTE* is used to represent a

EC
Embedded channel is the term used to describe the electrical channel of a SONET signal. The EC is equivalent to an STS.

STS
Synchronous Transport Signal represents one channel within the SONET standard.

LTE
A term defined by the SONET standard, Line Terminating Equipment refers to a device that terminates a SONET circuit.

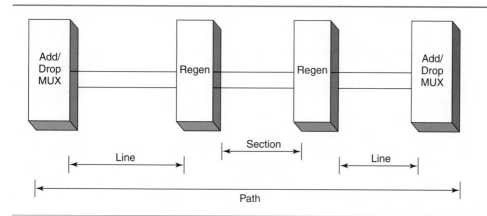

Figure 8–22
The three main segments of a SONET—section, line, and path.

Figure 8–23
STS frame.

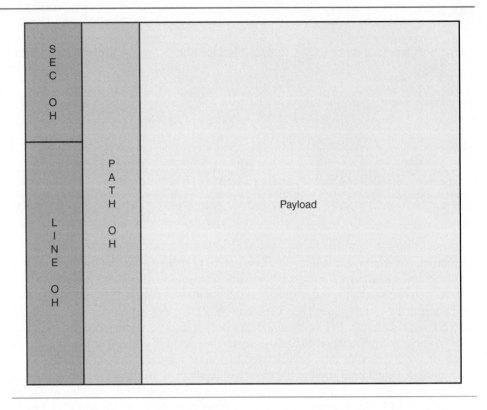

SONET multiplexer. The line portion of the network includes the section between two LTEs. The path portion of the network includes the end-to-end path of the circuit from the point the signal enters the SONET multiplexer to the point it leaves the SONET multiplexer at the far end. The path does not include the portion of the signal beyond the MUX such as the DS-3 circuit terminating at the M13. Building the network in this manner helps technicians during trouble conditions sectionalize and isolate problems more quickly. Each segment of the system generates alarms specific to that portion of the network, similar to the yellow alarm generated by an M13 MUX. A term used frequently in place of LTE is *ADM*, add-drop multiplexer. An ADM is an LTE that has the capability to pass through traffic add and drop traffic, and groom signals accordingly.

ADM
An add-drop multiplexer is used to terminate and pass through traffic in a SONET.

8.4.2 STS-1 Frame

The foundation of the SONET protocol is the STS-1 frame. The frame is built around two functional areas, overhead bytes and payload, as shown in Figure 8–23. Again referring to our train analogy, the frame is the boxcar, the overhead is the conductor, the steward, and the porter; and the passengers make up the payload. The overhead—like the conductor, the steward, and the porter—oversees the train making sure customers arrive at their destination in tact and on time. Customers—or payload—sit and enjoy the ride.

The overhead bytes correspond to the structure of the network, the section, line, and path. The conductor and porter on a train do not ride with the passengers in the boxcar. They sit outside the boxcar making sure the train stays on track. The section and line overhead bytes are the same. They reside outside the payload portion of the frame and are responsible for maintaining the section and line portion of the network. Remember that the SONET standard divided the network into usable segments in order to assist Turing outages—section and line. The final group of overhead bytes, the payload, like the steward sits with the passengers in the payload section of the train. The path overhead bytes make sure the passengers board and deboard or enter and exit in good order. Figure 8–24 illustrates the location of the overhead bytes and payload in an STS frame. Each overhead segment takes care of its network equivalent; for example, the

Figure 8–24

Structure of an STS-1 frame of the SONET hierarchy. The frame has two main sections. The transport section includes bytes responsible for the section overhead bytes and the line overhead bytes of the network. The path overhead resides in the synchronous payload envelope portion of the frame and contains the overhead bytes for the path portion of the network.

	Transport Overhead			1 Byte	87 Bytes		
	Framing A1	Framing A2	STS-1 ID C1	Trace J1			
	BIP-8 B1	Order Wire E1	User F1	BIP-8 B3			
	Data Comm D1	Data Comm D2	Data Comm D3	Signal Label C2			
	Pointer H	Pointer H2	Pointer Action H3	Path Status G1	SPE		
	BIP-8 B2	APS K1	APS K2	User Chan F2	Synchronous Payload Envelope		
	Data Comm D4	Data Comm D5	Data Comm D6	Multi-frame H4			
	Data Comm D7	Data Comm D8	Data Comm D9	Growth Z3			
	Data Comm D10	Data Comm D11	Data Comm D12	Growth Z4			
	Growth Z1	Growth Z2	Order-wire E2	Growth Z5			

90 Bytes

Section — Section Overhead

Line — Path Overhead

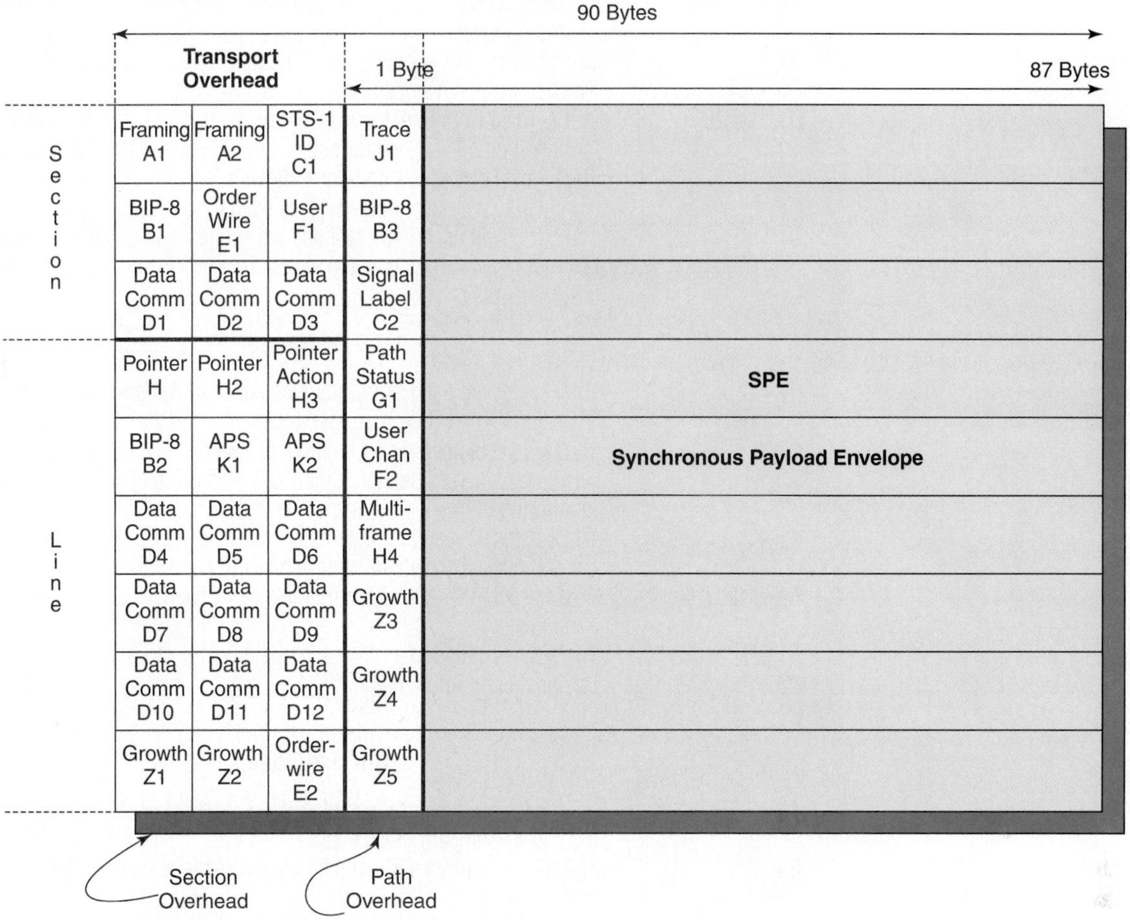

path overhead carries all alarms and information about the path portion of the network, the section carries section information, and so forth.

The STS frame consists of ninety columns and nine rows of eight-bit bytes equaling 810 bytes per STS frame. Since 8000 frames can be shipped in one second, the total line rate for an STS-1 is:

$$8000 \times 810 = 51,840,000 \text{ bps}$$

The STS-1 rate is often said to be 52 Mbps, though in reality it is 51.84 Mbps. We will, for simplicity's sake, use 52 Mbps. An important point to mention is that even though the line rate is 52 Mbps, the actual information carrying capacity is 45 Mbps due to the cost of overhead—the conductor, porter, and steward. The payload portion of the frame—in our analogy, the seats—consists of eighty-seven columns and nine rows, each able to hold eight bytes. One column of the eighty-seven is used for path overhead, reducing the total space for information to eighty-six columns and nine rows, resulting in a total bandwidth capacity of 45 Mbps—amazingly, the same rate as a DS-3. The payload portion of the frame is called the SPE—synchronous payload envelope.

The first two overhead segments, section and line, make up the transport overhead as already explained. The path overhead resides within the SPE. It is important

Table 8–7
SONET Frame Overhead Bytes.

Specific Byte Section Overhead	Description of Byte Overhead
A1, A2	Frame Synchronization—The beginning of the STS-1 frame is defined by A1 and A2 bytes.
B1,	Section parity is handled by the B1 bytes. Parity is used to determine whether errors have occurred.
D1, D2, D3	DCC—Data Communications Channel is used to carry management information plus to carry information on the health of the transmission.
E1	The E1 byte provides a path for orderwire.
F1	Assigned to users for whatever they wish to use it for.
C1	The C1 byte is used to carry a defined sequence of bits that represent a text message used to trace the signal from end to end, often referred to as a Section Trace.
Line Overhead	
H1–H2	The H bytes are used as pointer bytes used to designate where the path OH begins. The H bytes are responsible to keep the frame lined up.
H3	The H3 byte is used for bit stuffing if needed.
B2	BIP-8 is the name used to define the B2 byte. The byte is used for error monitoring of the line portion of the frame.
K1–K2	The K bytes are used for automatic protection switching. The bytes send out and receive orders that a switch has to happen.
D4–D12	Provides a DCC—Data Communications Channel—between two line terminals.
S1 (Z1)	The S byte is used to monitor the quality of the clock source.
Z2	Designated for future use.
E2	Line side orderwire.
Path Overhead	
J1	Path Trace is a text message placed in the J1 byte and is used to identify the end-to-end path of a signal.
B3	The B3 byte carries a BIP-8 path error scheme to monitor the end-to-end path.
C2	The C2 byte is called the Signal Label and is used to identify the type of signal in the payload. For example, 02 represents Floating VT mode. 04 represents Asynchronous DS-3.
G1	The G1 byte carries information on the status of the path.
F2	The F2 byte (path user channel) is reserved for user-specific information.
H4	The H4 byte (multiframe indicator) is used to indicate the VT payload.
Z3–Z5	Designated for future use.

to understand the functions performed by each of the overhead segments as defined in Table 8–7. The segments within the STS frame are defined in the following paragraphs.

The section overhead is comprised of nine bytes located in the first three rows of the transport overhead section of the frame. Similar to the porter, each byte is responsible for specific tasks:

- A1 and A2—Framing bytes used to tell the equipment that it is the beginning of the STS-1 frame.
- C1—C1 bytes, also called J1, are used as STS-1 identification bytes and function as a section trace: *trace* refers to tracking the signal through the use of a special identifier.

- B1—BIP-8, bit-interleaved parity code, is used to calculate errors on the section portion of the network. The calculation is performed on the previous STS-1 frame after scrambling.
- E1—Orderwire byte is used for the local orderwire channel. Orderwire is a communications channel used by technicians.
- F1—User channel byte is defined by the user for whatever purpose is desired. Because the F1 byte resides in the section overhead, it terminates at each regenerator element, making it less useful than a path channel.
- D1, D2, D3—The DCC (data communications channel) bytes are used to carry maintenance messages for OAM—operations, administration, and maintenance purposes. Alarms, conditions, control messages, and so forth are carried on the DCC bytes, between LTEs. The DCC is very handy especially in a large multinode network.

Line overhead bytes are responsible for the line portion of the network, between LTEs, and as such are one of the most important sets of bytes in the frame. An alarm reading a line error indicates trouble between two line terminating devices. The following are the bytes assigned to the line portion of the STS-1 frame; the jobs they are responsible for are also listed.

- H1, H2, H3—Pointer bytes are used to indicate the payload envelope in the frame. The envelope floats in the frame and must be located—thus, the reason for the H bytes. Asynchronous signals fit into the frame without cumbersome conversions. The ability to float in the frame allows the STS-1 frame to synchronize an asynchronous signal.
- B2—The BIP-8 byte is responsible for finding transmission errors on the line portion of the circuit. The calculation is the same as that performed by the BIP-8 byte in the overhead section.
- K1 and K2—APS, automatic protection switching—between LTEs is handled by the K1 and K2 bytes. Once errors are detected using the BIP-8 byte, a line switch is initiated by the K bytes referred to as a *soft failure*. When an AIS (alarm indication signal), an all-ones signal, or an loss of signal condition arises, a hard failure also also causing the K bytes to initiate a protection switch on the line affected. The final type of protection switch occurs when a manual switch, either through a hardware action or programming, causes the K bytes to initiate a switch.
- D4 to D12—DCC bytes reside in the line section of the overhead and are used to carry alarm, condition, and maintenance information about the line segment of the network. A full 576 kbps is available for DCC information.
- S1—The Synchronization Messaging Byte defines the health of the timing source and may not be available on older devices.
- MO—STS-1 Line FEBE—Errors detected by the BIP-8 byte ride in the MO byte telling the upstream line device the number of errors received.
- E2—Orderwire byte provides a 64 kbps channel for communications between the two LTE devices. Similar to the orderwire in the section overhead, special optical headsets are used by the technicians to communicate across the span.

soft failure
Failure initiated by a software switch.

The path overhead resides in the SPE, not the transport overhead, and is responsible for making sure the end-to-end path is functioning properly. The path overhead includes:

- J1 Trace—The trace byte is used to confirm that the information is being transported between the correct termination points. The byte carries a code specific to the circuit. The byte is user configurable; and, often, the CLLI—Common Language Location Identification—is used to represent the path name.
- B3—The BIP-8 byte is used to identify path errors. A parity calculation is performed by the PTE—path terminating equipment—and used to determine if errors have occurred on the path. If they have, an alarm is raised letting the network devices know of the trouble.

- C2—The signal label is used to indicate whether the SPE is equipped or unequipped. It also warns the downstream equipment of any problems with the payload.
- G1—Path performance monitoring information such as the FEBE and RDI are carried in the G1 byte. Far End Block Errors are used to notify the upstream equipment when errors are detected at the network element. If the path has failed completely, a PRDI, yellow alarm, condition rides in the G1 byte notifying the upstream equipment of the failed condition. Alarm conditions are described in detail later in the maintenance portion of the text.
- F2—Channel Path User Channel—The F2 byte is used to carry communications between the path network elements.
- H4—The Multiframe Indicator carries information about a concatenated signal or when the VTs (Virtual Tributaries) are in floating mode.
- Z3—Has been set aside for future growth.
- Z5—The Tandem Connection Maintenance/Path Data Channel byte is responsible for carrying the IEC, Incoming Error Count, for tandem connections and for carrying the path data channel.

The final overhead descriptions that need to be addressed are the VT Path Overhead bytes. Within the SPE the payload is segmented further into 1.7 Mbps time slots referred to as VT, *Virtual Tributary,* channels. Like a DS-1 riding inside a DS-3, the VT signals ride inside an STS-1. Additionally, the VT-1 signal has its own overhead bits to maintain its integrity or, in our world, make sure the passengers arrive on time, at the right location and intact. A VT-1 consists of the following overhead bytes.

Virtual Tributary

Virtual Tributary is a 1.7 Mbps signal that is equivalent to a DS-1.

- V1, V2—VT Pointer bytes are used to identify the VT.
- V3—The VT Pointer Action byte is used for bit stuffing.
- V4—Future.
- V5—The Path Overhead byte is used for error checking and performance monitoring. The BIP-2 error-checking scheme is deployed as is carrying the error condition within this byte. FEBEs and RDIs are transmitted within this portion of the VT overhead. Note that V5 is not used when the VT is not floating within the payload.
- J2—VT Path Trace—Future.
- Z6 and Z7—Future growth.

As is obvious from the previous descriptions, the SONET overhead bytes are responsible for a large number of tasks, ranging from keeping the network synchronized to monitoring it for troubles. Though in this text it is not expected that you memorize the position of each byte in each overhead segment of the STS-1 frame, it is important that you note the significance of the overhead bytes. For example, it is nice to know the Z bytes in the VT overhead are there for future growth but it is not imperative that you know this when troubleshooting a circuit. However, it is essential that you understand that the BIP-8 byte performs error checking and the G1 byte in the path overhead carries the condition of the incoming signal that is first detected by the B3 byte. Luckily, test equipment does not display that 111 code as read from the B3 byte in the path overhead. Instead, the equipment will display a "PRDI" condition. You need to know what PRDI refers to, not the bit pattern. In conclusion, time spent learning what PRDI stands for is more constructive than reciting the bit pattern of each overhead byte.

8.4.3 Network Architecture

Along with additional line rates, the SONET standard introduced new ways to design the network. The most common SONET architectures are point-to-point, linear, ring, and hub or, in today's terms, mesh. The point-to-point architecture is as it sounds—two LTEs, Light Terminating Equipment, connected together via fiber optic strands. The LTEs are optical multiplexers that use the SONET standard protocol to format and transport information, that is, the STS frame format.

The second type of architecture commonly deployed is the linear network consisting of multiple LTEs connected together in a row, as shown in Figure 8–25. The

Figure 8–25
The path portion of the SONET refers to the end-to-end circuit path. Path errors received at a test box signify that problems are occurring on the path portion of the network.

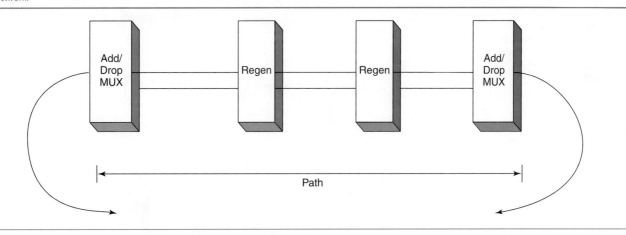

linear network LTEs must have add/drop capability in order to pass through traffic, add traffic, or drop traffic as needed.

The third and most popular type of SONET architecture is the SONET ring. The SONET ring architecture consists of LTEs (ADMs) connected together in a closed ring, as shown in Figure 8–26.

Two types of SONET rings have been developed: the UPSR—Unidirectional Path Switched Ring—and the BLSR—Bidirectional Line Switched Ring. Both offer fully redundant protection paths capable of switching from working to protect in less than 50 milliseconds (MS)—a requirement built into the SONET standard. The difference between the two stems from the way they handle protection.

The UPSR architecture assigns a working path, traveling in one direction either clockwise or counterclockwise around the ring, and a protection path, flowing in the opposite direction around the ring. During a network failure, the traffic riding on the working path immediately switches direction and moves over to the protection path as shown in Figure 8–27. The protection scheme is very simple and easy to implement. One drawback associated with a UPSR design is the amount of bandwidth sitting idle. The protection path is used only when there is a network outage. In addition,

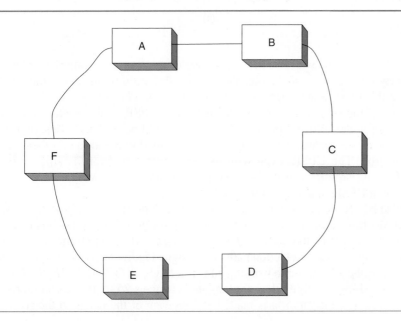

Figure 8–26
SONET ring architecture is one of the most common methods used to connect locations together. The ring architecture provides diverse paths when fiber troubles occur on the network.

Figure 8–27
(a) Traffic is flowing in a
counterclockwise direction. (b) The
fiber cut between C and D causes all
traffic traveling in a counterclockwise
direction around the ring to switch
and travel around the ring in a
clockwise direction.

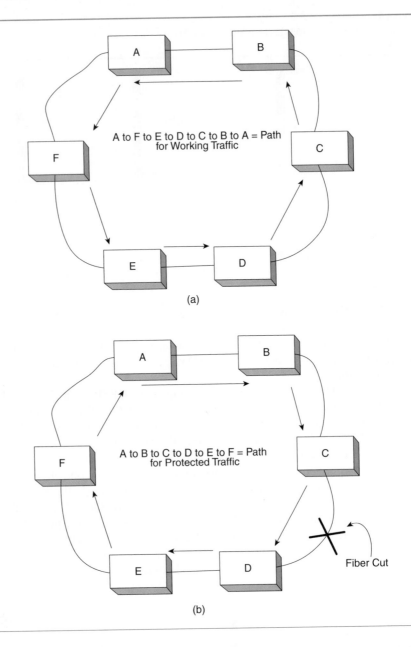

A to F to E to D to C to B to A = Path
for Working Traffic

(a)

A to B to C to D to E to F = Path
for Protected Traffic

Fiber Cut

(b)

the total bandwidth available is the line rate of the ring, for example, an OC-48 ring
has forty-eight STS-1s available. That means that even though multiple OC-48s are
used in the network and each is able to terminate or add forty-eight STS-1s, the total
capacity of the ring is forty-eight channels. The UPSR does not place a limit on the
number of nodes allowed per ring, though it would be foolish to design a ring with
too many nodes because each LTE would be limited to the amount of traffic it could
add or drop. The UPSR, even with its bandwidth limitation, is still a very useful
SONET ring architecture in the metropolitan space. Very, very seldom would you
ever find a UPSR in a long-haul network.

The BLSR architecture handles protection in a much more sophisticated way than
the UPSR. Instead of allocating one whole path for protection, it assigns protection on
the line level, that is, the portion between the LTEs—thus, the term *lineswitching*. As
shown in Figure 8–28 protection paths are assigned individually between each node by
segmenting the channels into working and protect. On an OC-48 BLSR, the first twenty-
four channels are configured as working and channels 25 to 48 are configured as protect. The same configuration is made between each of the nodes in the ring. Routing

lineswitching
Switching the path between
two LTE devices.

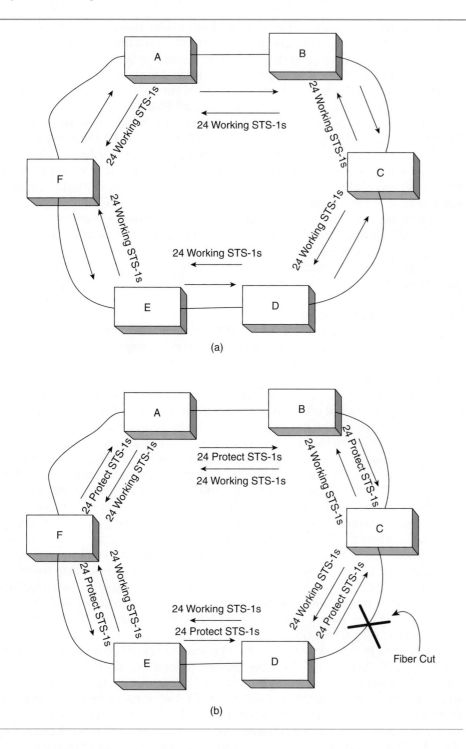

(a)

(b)

Figure 8–28
(a) The BLSR varies from the UPSR in that traffic is carried between each node in both the clockwise and counterclockwise directions. For example, A to B in the clockwise direction has twenty-four working STS-1s and twenty-four protect STS-1s. (b) When a fiber cut occurs, the twenty-four STS-1s traveling between C and D reverse their direction and travel on the protect STS-1s in the opposite direction.

tables are stored in the LTEs' processors and used during trouble conditions to reroute the traffic accordingly. As an example, when the six-node OC-48 ring experiences a network outage between nodes 3 and 4, traffic traveling between nodes 4, 5, and 6 jumps on the protect path channels traveling in the opposite direction. Moving the traffic traveling between nodes 1, 2, and 3 is unnecessary, as that traffic is not affected by the network outage. When configuring paths in a BLSR, a choice of either shortest or longest path is given. In almost all instances, the shortest path should be selected.

Traffic flows are assigned per line segment in a BLSR. When ten DS-3s travel between nodes 4 and 5 on channels 1 through 10, channels 1 through 10 are still available between the remaining nodes. The amount of bandwidth available in a BLSR is, of course, not one for one because most traffic travels between multiple nodes, thus eating up channels between multiple nodes. If all of the traffic coming from nodes 2 through 6 feeds back to node 1, the bandwidth available is the same as that in a UPSR. On the other hand, if traffic flows between nodes and is not homed back to just one, the bandwidth of the total ring is much greater than that of a UPSR.

Other renditions of the BLSR and UPSR have been introduced to accommodate data traffic, specifically, packetized protocols such as Ethernet or IP. The individual STS channels have been merged together allowing packetized traffic to burst if necessary, thus allowing the engineer to oversubscribe the network. RPR—Resilient Packet Ring—is the term given to a SONET ring with this capability. Making the payload bandwidth flexible in order to accommodate data traffic has helped improve SONET's life expectancy.

A final improvement to the ring architecture helps reduce the amount of bandwidth wasted due to protection paths. Many vendors have built systems that allow the use of protection paths for low-priority working traffic or non-time-sensitive data traffic. During a network outage, the low-priority traffic is dropped and the traffic on the working side is shoved onto the protection path.

8.4.4 Connecting the Dots

We have only touched the surface of the SONET architecture; entire texts have been devoted to the subject. To review how the total network connects, we will walk through a call traveling from a subscriber's home in Savannah, Georgia, to a subscriber living in Indianapolis, Indiana.

Carmen picks up his phone and dials Paula's number. The local class 5 switch in Carmen's serving central office sees that the call is a one-one long distance call and therefore has to be routed to Carmen's PIC, Spidercom Telecommunications. The switch routes the call to the local access tandem by way of a trunk leaving the class 5 switch. The trunk is multiplexed into a T1 signal, then into a DS-3, and then a 560 Mbps signal. The trunk rides on the point-to-point asynchronous optical network to the access tandem at Oak Street. There, the 560 Mbps multiplexer demultiplexes the 560 Mbps pipe into twelve DS-3s, which are fed into M13 multiplexers where the twenty-eight individual T1s are demultiplexed. In this case, the T1 is fed into the access tandem where the trunk carrying Carmen's request for call setup is received. The access tandem knows the call has to travel to Carmen's long distance carrier and knows that Spidercom is connected to it on T1 number 500. The access tandem routes the trunk onto T1 500, which is multiplexed at the Oak Street central office into a SONET OC-12 signal. Carmen's long distance carrier, Spidercom, connects to the Oak Street access central office using an OC-12 pipe. T1 500 rides on the OC-12 pipe into Spidercom's point of presence (POP).

Once in the POP, Spidercom maps the T1 circuit onto its OC-48 network that carries it to Raleigh, North Carolina, where Spidercom's regional long distance switch resides. The OC-48 signal breaks out the STS-1 carrying Carmen's voice signal and feeds it directly into Spidercom's long distance toll switch. The toll switch interprets the telephone number Carmen punched in and determines the call has to travel to Indiana. The route the switch takes is on trunk group 100 that rides on Spidercom's OC-48 network connecting Raleigh to Indianapolis where Spidercom's regional long distance switch resides.

Carmen's call leaves the Raleigh regional switch and heads out on the OC-48 ring network that connects Raleigh to Indianapolis. Because it is ring architecture, the signal passes through other cities along the way; there are four nodes between Raleigh and Indianapolis. The call arrives in Indianapolis and is demultiplexed down to T1 signals. The regional long distance switch accepts T1 100 and looks at the information in channel 10 where Carmen's call resides. The switch sees that the call has to travel to the local telephone company's network in order to reach Paula's home. The call is routed out onto T1 300 that connects Spidercom's long distance switch to the local telephone company's access tandem. (An access tandem is a switch that is used to interface long distance switches into the local telephone network.) The call travels into the local telephone network's access tandem in Indianapolis where it is routed to the Wayne Street central office serving Paula's home. The T1 leaving the access tandem is multiplexed into an OC-48 signal that rides on a local OC-48 SONET. The T1 is demultiplexed out of the OC-48 multiplexer at the Wayne Street central office and fed into the class 5 end-office switch. The end-office switch looks at the trunk carrying Carmen's call and knows that the call has to be routed to line card 10 where Paula's telephone line terminates. The switch looks to see if Paula's line is idle; when it sees that it is, it places ringing voltage on the line to ring Paula's phone. When Paula picks up the handset of her phone, a circuit connection is completely cut through and Carmen can talk to Paula.

Information travels many different routes as it traverses the world's telecommunications network. Many different types of networks are tied together using a tremendous amount of equipment, signal rates, and switches to carry information in the most efficient and reliable way possible.

SUMMARY

Physical layer protocols were presented in Chapter 8. The physical layer protocols are responsible for a broad range of tasks. Not only do they provide electrical, functional, and physical guidelines such as voltage levels, physical pinouts, and so forth; they also define how the information should be packaged according to the type of medium the signal travels. Similar to a vehicle, the physical layer defines the size of the vehicle, the speed the vehicle can travel, and the type of road the vehicle travels on. The physical layer protocols presented in Chapter 8 were the DS-0, DS-1, DS-3, and SONET. In conjunction, the terms *asynchronous* and *synchronous* were discussed as they relate to timing the network. The synchronous timing network was explained as it relates to handling timing functions.

REVIEW QUESTIONS

1. Explain why the North American Digital Hierarchy was developed.

2. List the four bit-per-second rates defined by the North American Digital Hierarchy.

3. Define each of the following terms and explain what each is used for.

   DS-0
   DS-1
   DS-3

4. What are the framing bits used for in the T1 signal?

5. Which frames in the T1 frame are used to carry signaling information?

6. Explain the purpose of overhead bits in digital signals.

7. How many DS-0s are multiplexed into a T1 signal?

8. Are DS-0s byte interleaved into the T1 signal?

9. How many DS-2s reside in a DS-3?

10. Explain how a T1 signal is multiplexed into a DS-3. Define the role of the DS-2 within the DS-3.

11. Explain the term *time slot* and list other terms used to refer to time slot.

12. Why was the SONET standard developed?

13. List the SONET bit-per-second rates that compare to a T1 and a T3.

14. Explain the difference between an OC-1 and an STS-1.

15. Draw the SONET STS frame and label each section.

16. Define the terms *line*, *section*, and *path* as they relate to a SONET.

17. How many DS-0s ride inside an OC-192 signal?

18. How many DS-3s ride inside an OC-48, an OC-12, and an OC-3 signal?

19. What is meant by the term *OC-3c*?

20. Explain the difference between an asynchronous fiber optic network and a SONET.

21. Draw a typical SONET ring and explain why most carriers prefer the ring architecture.

22. Explain the difference between a UPSR and a BLSR.

TROUBLESHOOTING

Problem 1: A T1 circuit connecting the Emory Street CO and Corner Street CO has gone into alarm, showing a loss of frame. You review the design layout record to determine the types of equipment the circuit is traveling through. You note that the T1 travels from a channel bank in Emory Street into an M13 DS-3 multiplexer in Emory Street to a second M13 in Corner Street and then to a second channel bank in Corner Street. You drive your truck to Emory Street first and look at the channel bank to see how the line card is optioned. Following are the options set for each device. Using the information presented, correct the problem and turn up the T1.

- Channel bank 1 = ESF framing, B8ZS line code
- M13 1 = extended superframing, B8ZS line code
- M13 2 = superframing, AMI line code
- Channel bank 2 = superframing, AMI line code

Note: The customer requires access to all 64 kbps of each DS-0 channel.

Problem 2: The DS-3 carrying the T1 in problem 1 is showing a loss of signal at the Emory Street M13 multiplexer. Using Diagram 8–1 in Appendix B, show how you would initiate a loopback at DSX-3 panel in Corner Street toward the Emory Street M13 and explain what the loopback will prove. The technician sitting in Corner Street sees a yellow alarm. Explain the significance of receiving a yellow alarm and list reasons why it is generated.

Note: The term *yellow alarm* is defined in the chapter marginal notes.

KEY TERMS

TDM (195)

asynchronous (195)

synchronous (195)

SONET (195)

slip (196)

jitter (196)

wander (196)

UTC (197)

BITS (199)

DS-1 (200)

digital signal 0 (201)

AMI (201)

2B1Q (201)

pulse mask (202)

Codec (203)

quantizing (203)

encoding (203)

PCM word (203)

superframe (205)

extended superframe (205)

Ft bits (205)

A- and B-bit signaling (205)

clear channel coding (208)

channel associated
    signaling (208)

robbed bit signaling (208)

channel bank (209)

byte interleaved (210)

bit interleaved (210)

leading edge (211)

trailing edge (211)

M13 frame (213)

bit stuffing (214)

C-bit parity framing (215)

FEOOF (215)

yellow alarm (215)

unchannelized (216)

digital cross-connect
    system (218)

EC (219)

STS (219)

LTE (219)

ADM (220)

soft failure (223)

Virtual Tributary (224)

lineswitching (226)

Switching and Routing

Global communications through switching and routing.
Photo courtesy of the Library of Congress

9

The Digital Circuit Switch

Objectives

After reading this chapter, you should be able to

- Describe digital switches.
- Route traffic through networks.
- Explain traffic engineering.
- Describe switch features.

Outline

Introduction

9.1 Defining the Digital Switch

9.2 Routing Calls through the Public Switched Telephone Network

9.3 Engineering Trunks, Lines, and Switch Ports Using Traffic Engineering Methods

9.4 Switch Features

■ INTRODUCTION

The digital stored program control switch introduced in the early 1980s is one of the reasons a call between Albany, New York, USA, and Katmandu, Nepal, is established within seconds. In addition to routing traffic more efficiently, the digital switch provides a multitude of other features. Call waiting, call forwarding, and ring back are just a few of the features made possible by the digital switch. *Circuit switching* is explained in detail in this chapter, and building a connection from one point to another point for the duration of a call is explored. The local class 5 switch, the tandem switch, and the toll switch are described in detail; and routing traffic between the different types of switches is covered. Traffic engineering concepts are presented concerning the method used to size a switch and the associated transport network. Other customer features are also discussed.

circuit switch
Telephone switch that switches circuits or calls. The circuit switch establishes a route using circuit information such as the telephone number between two end points.

■ 9.1 DEFINING THE DIGITAL SWITCH

9.1.1 Digital Switch Structure

Nortel Networks introduced the first digital switch in 1979. Soon, AT&T (now Lucent Technologies) introduced a digital stored program control switch to compete with Nortel. Many other companies such as Ericson, Alcatel, and Siemans then joined the race to provide digital stored program control switches. Digital switches provided a more efficient way to handle the many voice calls traveling through the network. It took advantage of microprocessor technology to perform functions such as routing traffic, converting incoming analog signals into digital signals, and providing features on a per line basis.

The switch could perform all these tasks because of the millions of lines of computer code residing in every switch. In addition to providing features and faster call processing, the digital switch improved the reliability of the communications network. The digital switch guaranteed 99.999% uptime, often referred to as the *5–9s* by vendors and engineers. Redundant processors, line modules, and signaling links helped ensure that 99.999% uptime was met.

The digital switch structure varies somewhat among manufacturers, but the basic function is identical—they switch calls. A high-level view of digital switching reveals five main functional areas: the stored program control system (SPCS), line module, trunk module, ancillary control, and maintenance controls that encompass both the hardware and software components. We will discuss the quasi-distributed control-based architecture, to which the majority of today's digital switches conform.

The quasi-distributed control-based architecture uses the following functional groups: the *central processing system* (CPS), *network control processor* (NCP), *interface controller* (IC), and the *line* and *trunk interface modules.*

The CPS can be thought of as the "big brain." It has access to all the functional groups from the NCP to the line and trunk modules. The main functions of the CPS are global call processing, network control, signaling control, maintenance, and administration. It is responsible for system recovery, software upgrades, storage of global routing information, and distribution of information to all subsystems. The CPS is also responsible for billing. Billing information is passed to a special interface port connected to external billing systems that process the call detail records (CDRs) logged every time a call is placed. As detective movies show, CDRs are handy when trying to catch a criminal. The numbers the person is calling from and to, the date, and the times the call started and was completed are recorded. The CPS can hold a great deal of information and is also used to interface external video displays and terminals used for maintenance, provisioning, and switch control. It is similar to a corporate CEO. The CEO's function is to delegate tasks to specific groups with the expectation that they will function on their own. The groups have access to the CEO if a decision affecting the entire organization needs to be made. In turn, the CEO has control over every employee in the company. The CPS also handles all global issues, has control over all subsystems, and is contacted when a system decision must be made.

The NCP is connected to the CPS and the line and trunk modules. The NCP receives questions from the line and trunk units regarding call routing, feature requests, and so forth. Again, the structure is similar to that of a corporation. The NCP is the vice president who has the power to make decisions without consulting the CEO; however, the CEO can be consulted if necessary and the orders passed down the line to the workers. The NCP holds all subscriber line information and refers to it whenever the line module requests a path through the switching fabric. It first determines from the subscriber line information whether the call is allowed. Once it determines the call is valid, the NCP determines which path it should take through the switching fabric. It does this by either looking up the destination route in its own routing tables or requesting a digit translation from the CPS. When the CPS provides the translation, the NCP selects a path through the switch fabric and sends this information to the IC. The IC completes the call path from the line module through the switch fabric to a second line module or trunk module, depending on the type of call. The NCP continues to track the call path until the call is disconnected.

The IC sits between the switching fabric and the line or trunk modules and is used to connect the two. It also talks directly to the NCP. The IC can be considered the highway that is used to pass messages between the different modules. It does not interface directly with the subscriber line, nor does it hold translation information; rather, its

5–9s
Refers to 99.999% uptime. Reliability measurement equipment vendors and service providers strive to provide.

central processing system
Brains of the switch.

network control processor
Mini brains of the switch.

interface controller
Interface between the NCP and the line modules.

line interface module
The portion of the switch where the individual POTS lines terminate.

trunk interface module
The portion of the switch where the trunk circuits terminate.

main purpose is to provide a communications path between the processors, the line/trunk modules, and the switch fabric. Many IC modules use optical links to interface the processors and line and trunk units in order to pass huge numbers of messages.

The line module is, as it sounds, the module that interfaces with the subscriber's line. *Line side* is the analog interface into the line module. Line modules convert incoming analog signals into digital signals or digital signals ready to leave the switch into analog signals. The line modules also continuously scan the idle lines for off-hook/on-hook conditions.

Trunk modules terminate trunks into the switch. Trunks arrive from other switches—long distance, tandem, and so forth—and terminate at the trunk module. Whenever a call arrives at a switch and needs to be processed and shipped to a different switch, the call travels through the trunk module or, as it is commonly called, the trunk side. The *trunk side* of a switch has multiple digital circuits that interface directly into the trunk unit circuit packs. Multiple trunks ride on the digital circuits. The signal coming into and leaving a trunk module is a DS-0. Because there are twenty-four DS-0s within one T1, most trunk modules terminate a T1 digital line. Today, switches also terminate higher-rate links directly into their trunk modules. For example, the Lucent 5ESS terminates a full STS-1 link in its DNU-S (Digital Networking Unit–SONET) trunking module. However, even when a higher-rate link is used, the trunks still ride on T1 circuits that reside within the higher-rate circuit. Figure 9–1 details the components of the switch.

9.1.2 Class 5 (Local) Switches

The *class 5 switch*, manufactured by many vendors, is one of the most complex digital switches in the network because it is responsible for all line-side terminations. Depending on the size of the CPS, the class 5 switch is capable of terminating thousands

line side
Term used to designate the line interface side of the switch.

trunk side
Term used to designate the trunk interface side of the switch.

class 5 switch
End-office level switch as defined by the hierarchical switch architecture.

Figure 9–1
The logical structure of the digital stored program control switch.

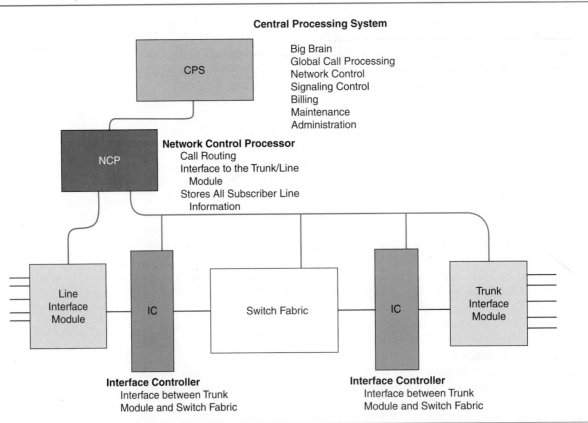

of subscriber lines. The size of the CPS determines the size of the class 5 switch. Because processing has become faster and microprocessors smaller and less expensive, a typical class 5 switch can now handle as many as 50,000 line-side subscribers. Each line can be individually provisioned for different features in addition to having a long distance carrier assigned to it. A *primary interconnect carrier* (PIC) is a long distance carrier chosen by the subscriber. The switch not only has to manage thousands of lines, it also has to route calls to destination switches. Every piece of information associated with the subscriber resides in the class 5 switch's memory. It also has to route many different types of calls—local within its own network, local into another phone company's network, long distance, and 911 calls to the public safety answering position (PSAP). The multitude of features in the class 5 switch increases its complexity and, of course, its cost.

A class 5 switch has line modules, trunk modules, interface communications modules, a switch fabric, network switch processors, and central switch processors. It is also referred to as an end-office switch because, as it sounds, it sits in the central office. The line modules terminate subscriber lines while the trunk modules connect the switch to other class 5 switches, tandem switches, or long distance switches. Today, most class 5 switches can serve as both a local switch and a toll long distance switch. Microprocessors can handle the millions of transactions required to route both local and long distance traffic.

The class 5 switch provides features specific to individual lines. The features are added to the line during the provisioning of the subscriber circuit, and the switch bills the customer for every feature. It is also capable of providing per call features such as ring back, call block, or turning off call waiting. The subscriber dials *70 and the switch recognizes that as a request to turn off call waiting on that line for the duration of the call. This is a must when dialing in using a modem because call waiting tones will disconnect the user.

The class 5 switch can connect directly to a long distance switch or to another class 5 switch. Many long distance carriers and CLECs now prefer direct end-office trunking that allows them to bypass the local tandem switch. The class 5 end office can accept trunks directly from a long distance switch or from a CLEC's class 5 switch.

The class 5 local switch may be found in several places throughout the network—in the serving central office of the ILEC, in a CLEC's regional switch center, in a university telecommunications building, in large corporations, and in military compounds. When you pick up your telephone and hear the dial tone, you are interfacing with a class 5 switch.

9.1.3 Class 4 (Tandem) Switches

The tandem switch is also called the *class 4 switch*. Figure 9–2 shows where the class 4 switch resides in the switching hierarchy. Its main purpose is to switch trunks, routing them between class 5 switches, between the toll switch and the local telephone network, between CLEC regional center class 5 switches, between the ILEC's end-office class 5 switches, and between operator services toll switches called OSPSs (Operator Services Position Stations). Many class 5 switches now have tandem switch capabilities blurring the line between class 4 and class 5 switches. Trunking functionality is added to a class 5 switch by loading tandem switch software to the class 5 switch operating system. ILECs still use tandem switches, mainly because of their need to route multiple carriers' traffic through the local network.

A typical class 4 switch has a CPS, an NCP, interface communications modules, and trunk modules. The flow through a class 4 switch is very similar to that through a class 5 switch. The main difference is the lack of line modules in the class 4 tandem switch. The class 4 switch accepts trunks from many different switches and routes them according to a predefined routing scheme. When you pick up your telephone and dial a friend who lives across the country, your call will be routed through one or more tandem switches. Tandem switches are the interface between networks and between switches.

primary interconnect carrier
Each telephone subscriber is asked to choose a long distance carrier. The switch holds the PIC information in the customer record.

class 4 switch
Tandem switch as defined by the hierarchical switch architecture.

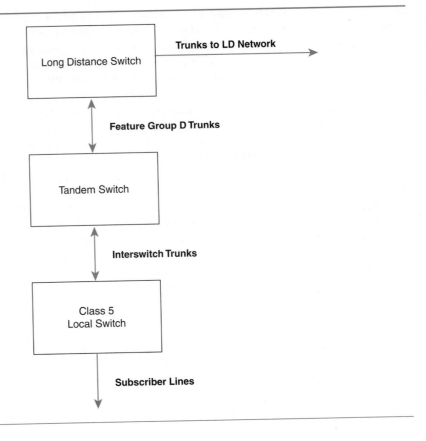

Figure 9–2
The position in the network of the tandem class 4 switch.

9.1.4 Class 3 (Toll) Switches

class 3 switch
Toll switch as defined by the hierarchical switch architecture.

The *class 3 switch* switches trunks between toll switches, between class 4 or 5 switches, and between international gateway switches. The toll switch uses a routing matrix to determine where to switch trunks. For example, when a call enters a toll switch in St. Louis that requires a path to New York, the toll switch must determine which trunk is idle, which trunk offers the shortest or least costly route, and which trunk provides the correct format for the call—all within a split second.

The toll switch is composed of a CPS, NCP, interface control module, and trunk modules. The physical structure of the class 3 toll switch varies little from that of the tandem class 4 switch. The main difference between the two can be found in the software. For example, the toll switch has to hold all numbering plans throughout the toll network. A toll switch in New York has to know how to route a call destined for the California area code 917. The routing matrix in the toll switch differentiates it from the class 5 and class 4 switches. The toll switch can also interface international switches. It provides billing software that counts the minutes of usage per call and interfaces with calling card databases. When you pick up your phone and dial a 1+ number (toll number), the call is routed to the long distance carrier's toll switch where it is routed to the destination toll switch.

Differences are now very few among class 5, 4, and 3 switches. A class 5 local switch and a class 4 tandem switch can both contain a toll switch software load and technically become a toll switch. New switches are making it possible to combine all functions in one physical structure, but the functional areas of the switch remain divided. The class 5 switch continues to switch local calls, the tandem switch switches local trunks, and the toll switch switches long distance calls. The difference is that they all reside in one physical entity.

class 2 switch
Sectional switch as defined by the hierarchical switch architecture.

class 1 switch
Regional switch as defined by the hierarchical switch architecture.

9.1.5 Class 2 (Sectional Toll Center) and Class 1 (Regional Toll Center) Switches

The remaining switch classes are the *class 2* (sectional toll center) and the *class 1* (regional toll center) *switches* as shown in Figure 9–3. The switching hierarchy no longer

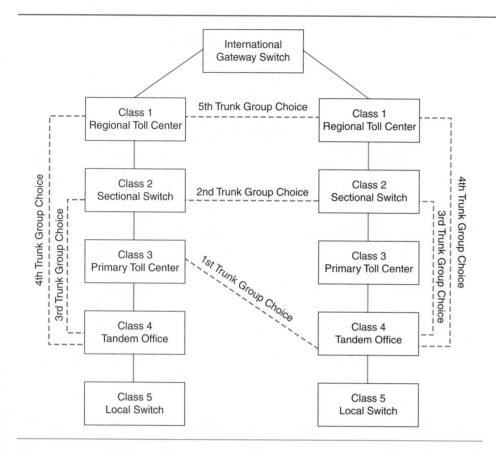

Figure 9–3
The switch classes interconnect to provide diversity through the switching network. The switch hierarchy was defined by AT&T before divestiture.

includes class 2 and class 1 switches, and the architecture follows a distributed topology instead of the hierarchical topology initiated by AT&T before divestiture. The purpose of the class 2 and the class 1 switches was to provide simplified and redundant routing throughout the national network. If a class 4 switch in Albuquerque needed to connect to a class 4 switch in Denver and all of the trunks were busy, the class 4 switch could connect into the class 2 switch that would then connect to the class 4 switch in Denver. Within AT&T's network, there were many class 4 switches, fewer sectional class 2 switches, and only a handful of class 1 regional switches.

Currently, the number of class 5 switches is increasing while the number of class 2 and class 1 switches is decreasing. The advent of the super tandem toll switch makes routing trunks that use a distributed architecture much more viable. In addition, deregulation eliminated the need to switch only through the AT&T long distance network and therefore allowed for newer distributed networks to evolve as shown in Figure 9–4.

9.1.6 International Gateway Switches

The international gateway switch performs three main functions. The first is to switch trunks between the domestic toll network and a foreign country's telephone network. The second function, and the one that truly differentiates it from a typical toll switch, is signal conversion. For calls to travel between countries, they need to understand one another's call setup, tear down, supervision, and routing protocols. The gateway switch performs this task. For example, a call coming in from Nebraska destined for France travels through a gateway switch that interprets the international dialing plan and translates the SS-7 messages into *C7* European signaling messages. The third function of the gateway is to convert the T1 physical interface to the European E1 physical interface.

9.1.7 Cellular Mobile Telephone Switching Offices

The *mobile telephone switching office* (MTSO) is a digital switch at the cellular provider's switch center. Three types of connections interface into the MTSO. The first

C7
Out-of-band signaling standard for Europe.
mobile telephone switching office
Cellular switch center where all cellular calls are switched between cell sites or between a cell site and a wireline switch site.

Figure 9–4
New distributed switch hierarchy
shows how each type of switch
interconnects into either itself or a
higher-class switch.

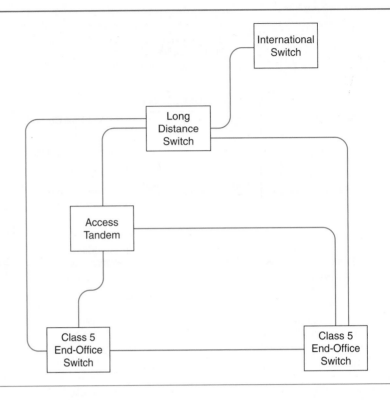

is from cellular towers. The second is from the trunk circuits that connect the MTSO to a tandem or class 5 switch within an ILEC or CLEC's network. Third are the long distance trunks connecting to other cellular MTSOs and long distance carriers. The MTSO also interfaces with the 911 public safety answering points (PSAPs) and can interface with an operator service tandem.

The cellular switch is a class 5 switch with line-side and trunk-side interfaces. The dial tone you hear before placing a call on your cell phone is coming from the class 5 switch residing at the MTSO site. The call is also routed through the MTSO to the called party. Cellular switches contain the same functional groups as local class 5 switches.

9.1.8 PBX Switches

The *private branch exchange* (PBX) resides at the customer's premises. Each employee's telephone line in each building terminates at the PBX instead of traveling all the way to the central office class 5 switch. The PBX, like the class 5 switch, routes the incoming and outgoing calls, provides features on the line, and keeps call records on each individual line. The PBX reduces the number of lines needed between the business and the telephone company because many of the calls placed are intrabusiness calls. An intrabusiness call travels between two telephones that terminate on the same PBX. For example, worker Jane needs to talk to worker Phil. She dials his four-digit extension and waits as the PBX switches the call to Phil's line. Therefore, the PBX does not need to tie up external trunks to the telephone company. The voice signal travels on Jane's line into the PBX line interface and the PBX routes the call to Phil's line. Essentially, it is a small class 5 switch. Those deployed today are digital stored program control switches. Analog PBXs are rare because the feature set is limited and the switching structure is less tolerant to faults.

private branch exchange
Small switch placed at a
customer's premises to
provide call switching and
features. Similar in structure
to the telephone switch, but
smaller.

■ 9.2 ROUTING CALLS THROUGH THE PUBLIC SWITCHED TELEPHONE NETWORK

9.2.1 The North American Numbering Plan

North American numbering plan
National numbering plan for
direct distance dialing.

Direct distance dialing created a need for a national *North American numbering plan* (NANP). The switch required the use of a systematic method when determining how

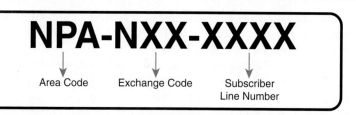

Figure 9–5
The telephone numbering plan is based on an area code for the region, an exchange code for the wire center, and a line number unique for the NXX of the switch center.

Area Code	1st Digit	2nd Digit	3rd Digit	Total Nos. Available
Initial Plan	1–9	0 or 1	0–9	1 Billion
New Plan	1–9	0–9	0–9	6 Billion

to route a call. A ten-digit number was proposed containing a three-digit area code—*Numbering Plan Area (NPA)*, a three-digit *number exchange (NXX)* or central office code, and a four-digit subscriber code *(XXXX)*. The area code identifies a region within the country, the exchange code defines the central office within that area code region, and the subscriber code identifies the customer.

When the numbering plan was first devised, the area code consisted of a three-digit number in which the first number was any digit from 1 to 9, the second number could be a 1 or a 0, and the third number could be any digit from 0 to 9. Using this scheme, the total number of area codes possible was 152 with 640 exchange codes per area code and one billion customer numbers. Due to the increased use of second lines, cellular telephones, and fax lines, the number pools were quickly being exhausted. In 1994, the numbering plan was revised to increase the number of area codes, which would in turn increase the number of telephone numbers available. The second digit in the area codes was revised to include any digit from 0 to 9. This increased the number of area codes from 640 to 792, which automatically increased the number of telephone numbers to six billion. Figure 9–5 illustrates the NANP as it is defined today. Unfortunately, many areas are running out of telephone numbers, causing the local commissions to mandate number pooling or forcing an NPA split. Number pooling divides the 10,000 available NXX exchanges into 1000 number blocks, allowing each carrier a certain number of blocks of 1000 in place of the entire 10,000 number block. Adding a second NPA in an area is the other method used to increase the number of telephone numbers. This is referred to as an NPA split.

9.2.2 The International Numbering Plan

The benefit of the digital switch is the ability to direct dial a person living three continents away. *International direct distance dialing* (IDDD) is made possible by digital switches and international standards organizations. Similar to the NANP, the International Telecommunications Union (ITU) devised an international numbering scheme. The standard includes a three-digit country code and a fourteen-digit national significant number (NSN). The country code for the United States and Canada is 1. The United Kingdom's country code is 44; Singapore's is 65. To call someone in Singapore, you would dial 011 (three-digit notification code—similar to dialing 1), the 65 country code, and then the NSN fourteen-digit number.

9.2.3 Local Access and Transport Areas (LATAs)

In 1982, the divestiture of AT&T changed the structure of the telephone network in the United States. The country had to be divided into regions served by newly formed Regional Bell Operating Companies (RBOCs) and those served by the existing independent telephone company. Using the statistical metropolitan studies, 161 local access and transport areas (LATAs) were created to define regional boundaries. The seven RBOCs and the independent local telephone companies were assigned specific LATAs for which they were exclusively responsible for providing local telephone service. Initially,

Numbering Plan Area (NPA)
Area code for North America.
number exchange (NXX)
Identifies the central office or servicing wire center exchange code.
XXXX
Subscriber number.

IDDD
International direct distance dialing.

intraLATA call
Call traveling between two
locations within the same
LATA.

the long distance carriers were restricted from transporting *intraLATA* long distance *calls*, but they are now allowed to in most states. State public utilities commissions (PUCs) determine the rules within the LATA. A three-digit number is assigned to each one identifying its geographic region.

Today, CLECs are changing the way calls are transported through the LATA. The CLECs and the independent telephone companies (except GTE) can transport long distance traffic across LATA boundaries; however, the ILECs such as Ameritech, Bell South, and Bell Atlantic are not allowed to until they prove they have opened up their local networks to competition. At this time, none of the ILECs except Verizon have been granted permission to carry long distance traffic across LATA boundaries. Section 271 of the 1996 Telecom Act defines the criteria an ILEC must meet to enter the long distance market. ILECs are eager to offer long distance because they will then be able to provide end-to-end telephone service. Once they are granted the right to carry traffic anywhere they wish, LATA boundaries will be less significant.

9.2.4 IntraLATA Routing

Within each LATA, rate centers are established to determine which calls are intraLATA toll and which are free. A rate center normally encompasses several central offices. Local telephone providers sometimes send ballots to customers asking what areas they would like to be able to call without being charged. Sometimes, rate centers are drawn that seem illogical to subscribers, such as when people who live across the street from one another are required to pay local toll charges whenever they call each other. Here, too, the local telephone provider has drawn up the rate center boundaries with the approval of the state's PUC or public service commission (PSC).

A CLEC that moves into an ILEC's territory normally accepts those rate center boundaries in order to provide the same service as the ILEC. Technologically, this is much easier for the CLEC.

The term *wire center* is often used interchangeably with *rate center*. While each represents a particular area defined by the central office, they do have different meanings. The wire center is the physical central office where the lines terminate; a rate center may include several wire centers.

Calls that travel within a rate center are free of charge. Calls that travel between rate centers are intraLATA toll calls. Telephone subscribers may choose the local telephone provider, a long distance provider, or a CLEC as their intraLATA long distance carrier. Customers do not have to use the same carrier for intraLATA toll as they do for long distance toll (interLATA toll).

To explain how a digital class 5 switch and a class 4 tandem switch handle intraLATA switching, we must first define the types of trunks, called intraLATA trunks, that are used to carry intraLATA interswitch calls. They are classified according to how the traffic flows—one-way incoming, one-way outgoing, or two-way incoming/outgoing. One-way incoming carries traffic into the switch from another switch. One-way outgoing carries traffic out of the switch and into the destination switch, so traffic flows from one switch to another switch. Two-way incoming/outgoing allows traffic to flow into and out of the switch on the same trunk. One-way trunks were used extensively before SS-7 signaling evolved, and inband signaling methods were used before the SS-7 out-of-band signaling network evolved. Because inband signaling cannot control *glare*, which happens when both switches seize a trunk at the same time, trunks were optioned as one way. With SS-7 signaling, one-way trunks are not as necessary but are still used to help control trunk use. The trunk types used to route calls within the LATA are called interswitch trunks. They connect the ILEC's tandem to either a CLEC's class 5 switch or an intraLATA long distance carrier.

glare
When two calls collide on the
same trunk during call
initiation.

9.2.5 Local Number Portability

Local number portability (LNP) is one reason the 1996 Telecom Act was instituted. It allows telephone subscribers to keep their telephone numbers when they switch to a different local telephone provider. If you decided to change your local telephone provider to a new start-up CLEC, your telephone number would be ported to your new service provider where it would be added to the CLEC's routing tables. Thanks to LNP, telephone numbers are no longer tied to one switch. Eventually, geographical number portability will allow you to keep your telephone number even if you move across the country.

9.2.6 InterLATA Calling

InterLATA calling refers to calls that travel between LATAs. Long distance telephone companies, called interexchange carriers (IECs or IXCs), route long distance calls to the appropriate LATA and then connect to the local service provider's network.

interLATA call
Call traveling between two LATAs.

9.2.7 Different Types of Trunks

Each trunk is defined by the type of signaling used—MF, CAS, or SS-7; whether it is one way or two way; and whether it is incoming or outgoing. The originating and terminating switches must be configured with the same type of signaling protocol for the two of them to exchange information. A *one-way trunk* carries traffic in one direction. If it is incoming, it means that the far end switch B is sending traffic into switch A. Outgoing means switch A is sending traffic to far end switch B. A *two-way trunk* carries traffic in both directions. Switch A can send traffic to switch B, and switch B can send traffic to switch A on that trunk, but not at the same time. They share the trunk. Figure 9–6 illustrates the different types of trunks.

A *trunk group* is a group of trunks that share the same circuit. For instance, trunk group 10 carries as many as twenty-four trunks between switches. The switch assigns a trunk group between itself and the far end switch, and traffic can then ride on any of the trunks within that group. Trunk groups are shown in Figure 9–7.

To understand how traffic flows between the different types of switches, we now need to define the different types of trunks available.

one-way trunk
Trunk set up to transmit information in one direction.
two-way trunk
Trunk set up to transmit information in both directions simultaneously.
trunk group
Trunks grouped together and defined as one group.

Figure 9–6 The various trunk connection between the different switch types.

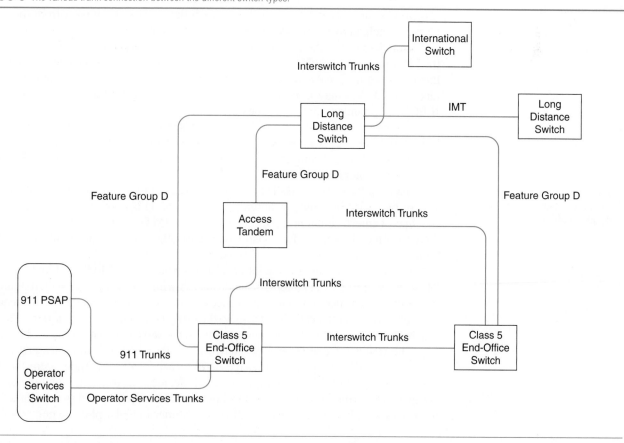

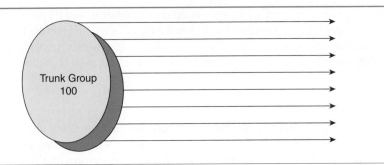

Figure 9–7
A logical view of how trunks are grouped together into a trunk group. The trunk group on one end of the switch matches the opposite end switch. These trunks belong to one-way trunk group 100.

Trunk Group
100

interswitch trunks
Trunks established between two switches.

tandem switch trunks
Trunks used to connect class 5 switches.

1996 Telecom Act
Act signed in 1996 allowing competition in the local telephone market.

access tandem switch
Local ILEC switch used to interconnect other carriers to the ILEC's local network.

intermachine trunks
Trunks established between two long distance switches.

Feature Group trunks
Trunks used to interconnect IXCs to the local telephone network in order to ensure equal access.

Interswitch trunks are used to connect two class 5 switches. They are either incoming or outgoing, one way or two way. The signaling is normally SS-7 and CAS and the type of interface circuit is a 1.544 Mbps T1. Interswitch trunks are able to carry multiple trunk groups between the two switches.

Tandem switch trunks are also used to connect class 5 switches. They provide an intermediate switch point for traffic traveling through the local network. Calls coming into the Maple Street central office switch destined for the Elmwood central office switch are routed onto trunks that connect the Maple Street office to the tandem switch. The tandem switch receives the call from the Maple Street office and routes it onto trunks connecting into the Elmwood central office. The tandem switch eliminates the need to connect every switch in the local network to all the others. Many local telephone areas have hundreds of central office switches. Trying to connect every central office switch to every other central office switch would be costly and would create a maintenance nightmare.

Since the *1996 Telecom Act*, tandem switch trunks are also used to connect CLEC class 5 switches to ILEC *access tandem switches*. The CLEC and the ILEC establish trunks between the ILEC access tandem or tandems and the CLEC regional class 5 switch. When a call comes into the CLEC switch that requires termination in the ILEC Elmwood central office, the call flow is as follows: The call travels from the CLEC class 5 switch, across the tandem trunk connections, into the ILEC access tandem. The access tandem routes the call onto the tandem trunks that connect the access tandem to the Elmwood central office's class 5 switch. The number of access tandems depends on the density of the local calling area. Some areas have as many as seven or eight access tandems. CLECs may terminate at each of the tandems or establish trunks into the ILEC's supertandem that distributes the calls to the correct tandem. The ILEC determines where the CLEC will terminate. CLECs may also opt to establish direct end-office trunking. The main difference is that the CLEC establishes trunks between the ILEC's class 5 switch and its own regional center's class 5 switch. The trunk types in this case are interswitch trunks.

Intermachine trunks (IMTs) are used to connect two long distance switches (one way or two way) incoming or outgoing. The signaling is normally SS-7 and CAS and the type of interface circuit is a 1.544 Mbps T1. IMTs connect a long distance company's switches together. They also connect switches between different long distance companies.

Feature Group trunks came into being as a result of the Modified Final Judgment. They are used to connect a long distance switch with a local tandem switch or an end-office switch in order to provide equal access. After divestiture, the RBOCs were told to develop a way to connect their network to the many new long distance carriers. The FCC required that they connect to the new carriers in the same way they connected to AT&T.

The RBOCs faced several problems. First, they had to figure out how the two switches would pass signaling messages to ensure that call setup times were consistent and equal to that of AT&T. Next, they had to decide how to charge the long distance company for terminating into their local network. Finally, they had to determine a way to provide equal access that would allow a customer to dial 1 plus the number instead

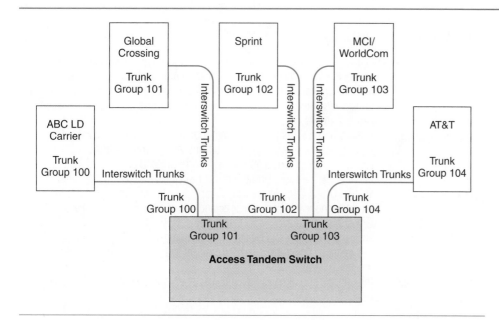

Figure 9–8
A local network access tandem interconnects with multiple long distance carriers.

of a long access code and PIN. The regulatory bodies and the newly formed IXCs realized that customers would hesitate to leave AT&T if they were forced to dial additional digits or enter a PIN every time they called a long distance number.

Feature Group A trunks were the first trunk types developed to connect the two networks. They solved the problem of passing signaling information between the two networks so that traffic could travel somewhat transparently. However, Feature Group A trunks did not solve the problem of 1+ dialing for all IXCs. The customer still had to dial an access code and PIN.

Feature Group A trunks quickly evolved into what we use today, Feature Group D or equal access trunks. They solved the problem of 1+ dialing because they did not require special access codes. Equal access trunks connect the local telephone company's access tandem or local end-office switch to various long distance carriers. Figure 9–8 depicts a typical access tandem connecting to multiple long distance companies.

Feature Group D trunks perform several functions. They charge the long distance company an access charge defined by tariffs and approved by the PUCs or PSCs. They direct the incoming or outgoing traffic to the correct trunk leaving the access tandem. Finally, and most important, the Feature Group D trunk allows customers to choose their long distance carrier. The carrier of choice is called the customer's primary interconnect carrier (PIC). Customers' PICs are entered into their line attributes record in the switch, so when a customer picks up the telephone and dials a 1, the switch automatically directs the outgoing call to the Feature Group D trunk connected to the customer's PIC. Now almost all of the switches in the United States are capable of Feature Group D termination. The Modified Final Judgment decreed that RBOCs and GTE would provide equal access by 1985, but the independent telephone companies were exempt from the ruling. The incentive for local telephone companies, even the smallest, was the billions of dollars of equal access charges paid to them by long distance companies for calls they terminated. Equal access is the main reason that 99% of the switches in the United States are digital. Table 9–1 describes the differences between Feature Group trunks.

911 trunks connect the class 5 switch or tandem switch to the 911 public safety answering point (*PSAP*). The regional PSC defines how and where 911 trunks will interface. ILECs and CLECs must provide redundant trunk groups into the primary PSAPs as defined by the 911 commission.

In most areas, the 911 trunks still use multifrequency (MF) signaling. The trunks normally ride on digital T1 1.544 Mbps circuits and the PSAP usually requires two trunks, as shown in Figure 9–9.

911 trunks
Trunks established between the 911 center and the local switch.

PSAP
911 centers. Each region has PSAPs developed by the municipal committees.

Table 9–1
Switch-to-Switch Termination
Definitions.

Trunk Type	Definition
Feature Group A	Feature Group A terminations were first deployed after divestiture and advent of equal access. The Feature Group A is technically a line-side termination that allows for the billing of originating and terminating traffic for carriers other than AT&T. Feature Group D trunks have replaced Feature Group A trunks.
Feature Group B	Feature Group B trunks were also introduced after the advent of equal access. Feature Group B provides a trunk termination dissimilar to the Feature Group A line-side termination. Feature Group B trunks are billed on a flat rate basis. Feature Group B is used when flat rate billing is desired.
Feature Group D	Feature Group D trunks are connections used by IXCs to interface to the local telephone company's access tandems or class 5 switches. Feature Group D trunks are the most commonly deployed trunk types between IXCs and local carriers—CLECs and wireless companies.
IMTs	The intermachine trunk is used to connect two long distance switches together.
Interoffice Trunks	Interoffice trunks are used to connect class 5 switches together in a local network.

operator services trunks
Trunks established between an operator services switch and the local switch.

Operator services trunks connect the class 5 switch to the operator services switch. Operator services trunks use MF signaling to handle call supervision and to create a billing record when a subscriber dials 411 for directory assistance. Normally, two to four trunks are needed to carry the traffic load between a class 5 switch and an operator services switch. The class 5 or tandem switch may have the operator services switch built into its architecture. When a subscriber dials 0 or 411, the class 5 switch must have trunks connecting to the operator services platform.

When you pick up the phone and dial 0, your call travels through your local central office switch, into the tandem, and on to the operator services platform, which may be part of the tandem. Your telephone number and the local telephone carrier are displayed on the operator's video screen. ILECs and CLECs establish trunks between their networks and the operator services tandem switch. They also build agreements to pass the automatic number identification (ANI) to the operator's workstation, where they

Figure 9–9
The designs used to ensure that 911 service is available 100% of the time. The 911 networks in metropolitan areas vary across the country.

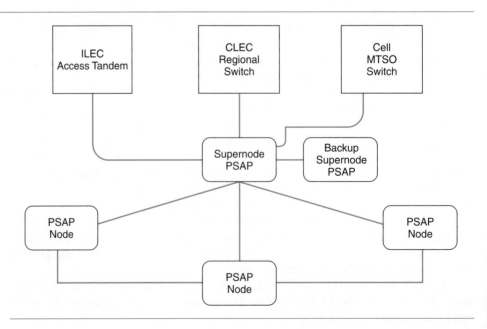

use the number to verify what company the subscriber belongs to and that the subscriber is allowed to access their network.

The second type of operator services trunks are those that connect to the local Toll Operator Position Station (TOPS) switch. The TOPS handles all NPA-555-1212 requests for information by routing the callers to the correct directory assistance region. The TOPS is functioning as a toll switch.

Analog trunks are used to connect PBXs to the local class switch or to a second PBX. The PBX may be viewed as a small switch that sits at a customer's location to help reduce the number of lines needed between the customer and the serving central office switch. Analog trunks are a common method used to connect the two switches.

Analog trunks are similar to typical telephone lines in that they carry one analog signal. The difference is in the signaling. Most analog trunks that connect a PBX to a class 5 switch use special signaling techniques that both switches understand. Instead of the switch scanning the line for off-hook conditions, the PBX notifies the switch when a telephone or a station goes off-hook. The signaling method used to establish a trunk connection between the PBX and the central office switch is referred to as inter-switch signaling. Various signaling types perform this function—E&M, DTMF, and DP. The advantage of using trunks instead of lines is their ability to hub large numbers of telephone stations onto one trunk connecting the PBX and the class 5 switch. The trunk can be designated as a *direct inward dial* (DID), which has the ability to bypass a receptionist station and ring the station directly. Though not as common as inter-switch trunks, analog trunks are one of the required trunk types.

ISDN PRI circuits are included in the mix of trunk types because of the Internet. Internet service providers (ISPs) establish multiple circuits directly into a telephone company's class 5 or tandem switch. Dialing into the Internet via a dial-up modem is a very simple task. You click dial on your Internet icon and wait as the modem outpulses the digits. Both modems then talk and establish a connection. In no time, you are surfing the Net.

Take a closer look at how the call travels through the network. The class 5 switch examines the incoming telephone number to determine where to send the call and establishes a circuit between the class 5 and the intended ISP. In this case, the outgoing trunk group is a digital ISDN line that connects the class 5 switch to the ISP's modem pool. A modem pool is a rack of modems located in the ISP's POP that are used to connect to your modem. ISDN PRI is the favorite type of circuit for connecting ISPs to the local telephone network because the signaling is performed by the D channel within the circuit. The main point is that routing through a switch to an ISP is most commonly handled via ISDN PRI circuits that, in essence, represent shared trunks between the local telephone company and the ISP.

9.2.8 Switch, Trunk, and Line Interfaces
Class 5 Switch
Class 5 switches are able to terminate interswitch trunks; analog trunks; 911 trunks; operator services trunks; Feature Group A, B, and D equal-access trunks; ISDN PRI circuits; and cellular MTSO trunk connections. In addition, a class 5 switch terminates analog lines, digital lines, ISDN BRI lines, and ISDN PRI lines. Figure 9–10 depicts different switch interfaces.

Access Tandem
An access tandem may be viewed more as a function of a switch than as a stand-alone switch. Today, for example, many class 5 switches have the ability to act as an access tandem. The type of routing the switch performs and the type of interfaces allowed to terminate into the switch define the access tandem. Its main purpose is to connect two switches together. It terminates interswitch trunks; Feature Group A, B, C, and D trunks; operator services trunks; 911 trunks; ISP ISDN PRI circuits; and MTSO cellular trunks. It does not route line-to-line or line-to-trunk connections because it does not have line-side connections, only trunk connections.

analog trunks
Trunks established between a telephone switch and a PBX. Special signaling is used to establish calls on analog trunks. Often the trunks are defined as DID or DOD depending on the customer's needs.

direct inward dial
Trunk type that provides a path for calls coming into the premises.
ISDN PRI
ISDN Primary Rate Interface circuits are used often as trunks between the local switch and a PBX or ISP.

Figure 9–10
Different line and trunk interfaces as
they relate to an end-office switch.
Most switch types have separate line
and trunk modules.

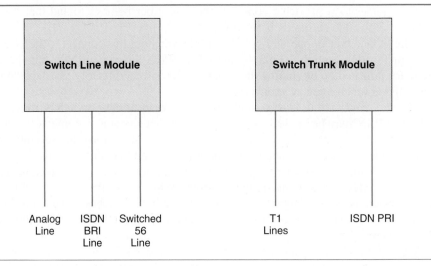

Long Distance Toll Switch

The long distance (LD) switch routes traffic through the long distance network. The types of interfaces that terminate on a long distance switch are intermachine trunks (IMTs); Feature Group A, B, C, and D trunks; 800-number routing; and TOPs trunks. Long distance switches also connect to databases that are used to validate calling card calls. As with the tandem, it does not have line modules so it only terminates trunk connections. The LD switch routes long distance traffic between LD switches and routes traffic to the local telephone network—trunk-to-trunk connections.

PBX

PBX switches sit at a customer's premises. They connect to the local telephone end-office switch through shared trunks, thus reducing the number of telephone lines needed to supply dial tone to all of the employees in the company. A PBX is able to terminate lines on analog, interswitch, DID/DOD, and ISDN PRI trunks. The main difference between a PBX and a class 5 switch is the number of lines supported and the processing power, which is much less for a PBX than for a full class 5 switch. Processing power is defined as busy hour call attempts (BHCAs) and determines the number of calls a switch can handle at one time. The PBX does not require a substantial amount of processing power because it serves fewer lines and does not connect to multiple networks.

MTSO Cellular Switch

The mobile telephone switching office is where the cellular switch resides. This switch is similar to the class 5 end-office switch but it connects to its subscribers differently. T1 digital circuits connect directly into the cellular switch feed radio antennas that are located in a honeycomb fashion throughout a local region. Trunks are established between the antenna and the switch, and the antenna receives and transmits calls through the air to the cellular telephones. The calls ride on the T1 circuits that connect the switch to the antenna. The cellular switch also has interswitch trunks that connect to the local telephone company's access tandem or end offices. LD Feature Group trunks may also terminate into the cellular switch.

9.2.9 Connecting the Dots

We will now walk through several call types to illustrate how the PSTN routes calls through the network.

Intraswitch Calls

Intraswitch calls are those that travel between two subscribers served out of the same central office switch. The calling party picks up the telephone handset and waits for dial tone, which the line module in the switch immediately sends once it detects the off-hook condition. The calling party enters the telephone number, and the line module sends the digits to the interface modules. They ask the NCP for a path to route the call across, and the NCP looks up the digits in the routing table to determine the call path. Because the telephone number that was dialed is served out of the same switch, the NCP provides the route information to the intermediate communications module (ICM). The ICM receives the information from the NCP that tells it which line module terminates the called party's telephone. The NCP then checks to see if the called party's line is busy or idle and, because it is idle, the NCP seizes the called party line and completes the connection through the switch fabric. Essentially, it establishes a route between the calling party's line card and the called party's line card. The switch sends ringing voltage to the subscriber's telephone causing the telephone set to ring and notify the subscriber of an incoming call. The switch also sends ring back to the calling subscriber notifying her or him of the state of the call. Once the called party picks up the telephone, the two lines are connected and the talk path is established. The switch marks the lines as busy and provides a busy notification if another call tries to seize the line. Once the two parties go on-hook, the switch immediately releases the connection through the switch matrix and the lines become idle.

IntraLATA Calls

An intraLATA call is routed from a line module on the originating switch to a trunk group that connects the originating switch to the terminating switch. At the terminating switch, the call is routed from the trunk side of the switch to a line port. As described in the previous section, the calling party goes off-hook, receives dial tone from the originating switch, dials the called party's telephone number, then waits for ring back. The line side of the switch receives the digits and passes them on to the ICM. The ICM passes the digits to the NCP which determines the correct response to the request. The NCP tells the ICM that the telephone number does not belong to a subscriber served out of that switch and therefore must travel to the local tandem to be routed. The routing from that point on is handled via the out-of-band signaling method referred to as SS-7. SS-7 looks through the network before allowing the switch to establish a route between the line side of the originating switch and the line side of the terminating switch. The SS-7 message tells the originating switch to seize trunk 100, which connects to the local tandem. The SS-7 message also tells the tandem switch to establish a connection between itself and the terminating switch. The tandem seizes trunk 200, which connects to the terminating end-office switch. The terminating switch receives the call request, then scans the subscriber's line to determine if it is idle or busy. The line is idle, so the switch seizes the circuit and sends ringing voltage to the called party's phone. The originating switch interprets the message from the terminating switch that the line is idle and immediately sends ring back to the calling subscriber.

The signaling messages that pass among the three switches set up the call. The advantage of using a separate signaling network is the speed at which calls are established. The switches use small messages that travel on separate signaling circuits to establish routes through the network. You may want to visualize them as scouts or messengers that ride ahead and scope out the trail.

The tandem switch has to route the call, mark the trunks as busy, and listen to the SS-7 messages in order to know when to release the trunks and make them idle. In this situation, the switch does not have to establish a billing record because the call travels within the same rate center. If the call had traveled within the LATA but to a switch resident in a different rate center, the tandem would have had to mark the call as intraLATA toll, determine the customer's intraLATA toll provider, route the call on the correct trunk group, and create a billing record for that call. Routing calls through the PSTN is not easy.

IntraLATA calls that travel from one service provider to another have become very common since the 1996 Telecom Act. CLECs are entering the local telephone company's network in record numbers. A facilities-based CLEC is a company that provides line-side termination to its subscribers. In order for its customers to talk to the ILEC's customers, the CLEC must establish trunks between the ILEC and itself. The routing of calls is further complicated by the adoption of local number portability (LNP). The two methods used to connect the switches between ILECs and CLECs are tandem trunking and direct end-office trunking.

The best way to understand the routing of calls from one service provider to another within the same LATA is by tracing a call through the two networks. CLEC Spidercom has moved into the Minneapolis, Minnesota, market and is competing with the local ILEC—Qwest. For Spidercom to have access to the local POTS (plain old telephone service) loop, it must colocate special equipment in Qwest's serving central offices. This establishes a direct connection between Spidercom's network and the local loop that connects to all the subscribers served by that serving central office. Qwest's equipment in the colocate space connects to Spidercom's regional class 5 switch. Spidercom's equipment in the colocate cages is connected to the class 5 switch in its regional switch center via T1 circuits. Spidercom uses Qwest local loops to complete the connection from its colocate space to the customer. When one of Spidercom's customers picks up the telephone, the call travels on the local loop into the central office and down to Spidercom's colocate cage where it interfaces into the special access equipment. It then travels on shared dedicated circuits that connect the colocate equipment to Spidercom's regional class 5 switch. Figure 9–11 illustrates this connection. The equipment in the cage serves as the line-side termination for the Spidercom regional switch.

Now that we know how Spidercom accesses local customers, we can begin our call trace. The example starts with a Spidercom customer calling a Qwest customer.

Spidercom's customer picks up the receiver. Spidercom's regional switch sees the off-hook condition and immediately sends out dial tone. The customer hears the dial tone and dials the telephone number of the called party. Spidercom's regional class 5 switch receives the digits, looks them up in its routing table to determine if it owns the number, and, because it does not, performs an LNP query to determine who does own it.

An LNP query sends a message via the SS-7 signaling network to the LNP database asking for information on that telephone number. The LNP database returns the location routing number (LRN)—a switch address in the form of a telephone number. The switch knows that LRN 716-226-0000 is the only address that defines the Qwest switch in the Elmwood central office and that the number just dialed lives at that switch. The call must be routed to that office, so the SS-7 message takes this information and builds a call path between Spidercom's switch and the terminating switch via Qwest access tandem.

The access tandem receives the switch address within the SS-7 message and knows that it needs to establish a trunk connection to its end-office switch in the Elmwood

Figure 9–11
The CLEC switch is able to provide dial tone for the local subscriber by connecting to a remote digital loop carrier placed at the subscriber's serving wire center. Colocation is a common way for facilities-based CLECs to provide service in the local telephone network.

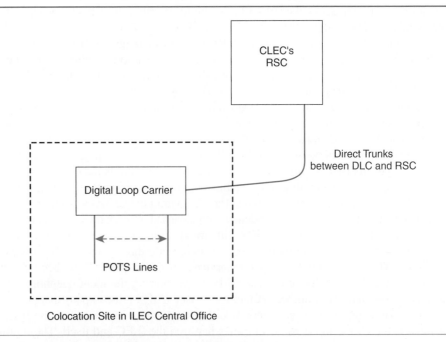

Colocation Site in ILEC Central Office

central office. Once the Elmwood-to-Qwest tandem trunk has been established, the Elmwood central office switch reads the SS-7 message to decipher the incoming called number. The switch's NCP looks up the number in its translation tables to determine which line module the customer is connected to. The NCP passes this information on to the ICM that establishes a path through the switch matrix between the trunk and the customer's line.

The switch sends ringing out onto the called party's line and sends a message back to Spidercom's switch informing it that the line is idle and the telephone is being rung. Spidercom's switch sends ring back to the calling party. Once the called party picks up the phone, the call path is established between the two networks. Figure 9–12 demonstrates the entire call path of a CLEC-to-ILEC call.

InterLATA Calls

InterLATA calling is much less complicated since the inception of out-of-band signaling. SS-7 can scout a path through the maze of networks, eliminating the need for multifrequency tones to be sent from switch to switch. The next call type is a long distance call between New York and California. Our example will follow a call from a New York City resident to a friend in Los Angeles. The caller picks up the telephone, the local end-office switch sees the off-hook condition and sends dial tone to the calling party's telephone, and the customer dials 1-418-555-5789 and waits for a response. The originating switch receives the number and knows that a 1+ call must be routed to a long distance carrier. The end-office switch then determines the type of trunk the call must be placed on. If the LD carrier has established direct end-office trunking, the call will be routed directly to the LD's toll switch from the local end office. If the carrier has established trunks to the ILEC's tandem office, then a trunk must be seized between the local switch and the local access tandem. The latter trunk type requires that a connection be established between the local access tandem and the LD's toll switch. We will assume that the LD carrier has established end-office trunking with the local service provider.

The local switch determines which LD carrier to send the call to by looking up the PIC in the customer's record. In this case, the customer has chosen Spidercom as the

Figure 9–12
A call being placed from subscriber A travels to the CLEC switch. From the CLEC switch, it travels on shared trunks to the ILEC access tandem. From the ILEC access tandem, the call flows to the end-office switch of the serving wire center. From the end-office switch, it travels out on the copper line to subscriber B. The ILEC owns the copper pair serving subscriber A and the pair serving copper pair B. The ILEC cross connects the copper POTs line at its MDF to the CLEC's tie cable that feeds the CLEC DLC in the colocation site.

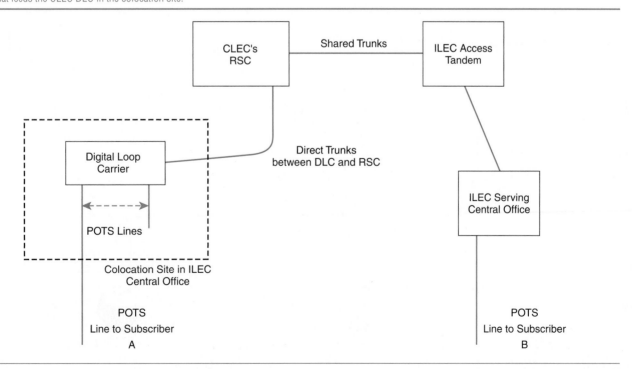

long distance provider. The switch establishes a connection through the switch fabric between the line and the trunk groups linking the local ILEC switch to Spidercom's toll switch. Spidercom's toll switch asks the SS-7 network to find a route between New York City and Los Angeles to carry the call. Once the path is built using intermachine trunks (IMTs), the call is routed from Spidercom's New York City toll switch directly into Spidercom's Los Angeles toll switch. IMTs are used to connect two toll switches together. In this case, there are trunks available between the two switches, but that is not always the case. The call may have to travel through intermediate switches before connecting to the final terminating toll switch. IMTs are also used to connect to the intermediate switches. Figure 9–13 depicts a typical LD network comprised of multiple IMT connections.

The Los Angeles toll switch receives the incoming digits from the New York City toll switch and looks at the ten-digit number to determine how to route the call. Again, thanks to LNP, there are several possibilities. The LD switch looks in its translation tables to determine if the NXX-555 has ported numbers, meaning that someone within the 555 exchange has changed local telephone carriers causing his or her telephone number to be ported to a new switch. There are 10,000 numbers available within one exchange group; when one of those 10,000 numbers is ported, the LD switch has to perform an LNP lookup. The 555 exchange has at least one ported number, so the toll switch launches an LNP query. The LD switch sends out an LNP query on the SS-7 links to the LNP database and receives a response that contains the LRN—the switch address of the terminating switch. Spidercom's LD switch in Los Angeles does not have direct trunking between itself and the class 5 switch end office serving the customer. Therefore, the LD switch asks to set up a path between itself and the ILEC's access tandem. The call is received by the local access tandem where it is routed onto a trunk group that connects to the terminating end-office class 5 switch. The class 5 switch connects the trunk to the called party's line and sends ringing out onto the local loop to ring the called party's phone. When ringing is established, the class 5 switch sends a message back to the originating class 5 switch telling it that the called party's phone is ringing. The originating switch sends ring back to the calling party's phone.

The SS-7 network is used to relay all of these supervision messages to the many switches. Once the called party picks up the phone, the call path is cut through and the two parties can talk. It is notable that the time it takes to establish a call end to end is

Figure 9–13
The end-to-end call flow between two long distance switches and two ILEC networks.

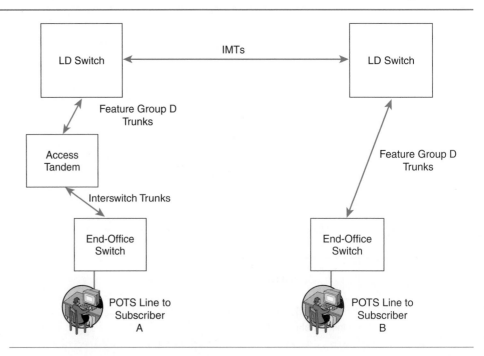

shorter than the time it takes to say hello. The trunk and line connections are established within seconds. In fact, the standard call setup time defined by Bellcore is five to seven seconds.

911 Trunking

911 trunking is one of the most important trunk connections made in the local telephone network. People in the United States depend on 911 for emergency services such as ambulance, fire, and police. If a 911 call fails, the consequences are more than a lost conversation—they may directly affect the outcome of the emergency. There is no simple, clean routing matrix that applies to all 911 calls. Each region has its own 911 topology, determined by the Public Safety Administration. Some regions require service providers to terminate all calls at one public safety answering point (PSAP), which is then responsible for routing the calls to the correct region. Others have multiple PSAPs and expect the local service providers to connect and route to each accordingly. We will describe a typical call handoff to the 911 PSAP, assuming that it will route to the correct PSAP region on its own.

A subscriber picks up the telephone. The switch sees the off-hook condition, sends dial tone, and waits for the digits. The subscriber then dials 911. The switch's NCP looks up the 911 digits and tells the ICM to build a route between the line and the 911 trunk. The ICM builds the trunk by sending MF signaling tones, which are similar to the DTMF used by regular telephones to provide the called number. MF tones are used to establish connection between the two switches. In this case, one end is the class 5 switch; the other is the PSAP's terminating equipment. Once the trunk is up, the line-to-trunk connection is made through the switch matrix and ringing and call setup information is passed to the PSAP. End-office switches occasionally connect into the 911 center through the local access tandem. The only difference between the two methods is the need to build a trunk between the tandem and the end office or the tandem and the 911 center.

Operator Services Trunks

Operator services trunks are similar to 911 trunks. They are used only when customers dial a specific code—411 for directory assistance—or 0. The trunks built between a class 5 switch and the operator services tandem also use MF signaling and are always built as one-way outgoing trunks because the operators will not use the trunks to call the switch. A 0 or 411 call travels through the class 5 switch out onto a trunk that connects to the operator services switch. From the operator services switch, the call comes into an automatic call distributor (ACD) that distributes calls to various operator stations. The customer's name, number, and service provider appear on an operator's screen. Again, the call flow for operator services trunks matches that of 911 trunks.

ISDN PRI Circuits

The integrated services digital network (ISDN) PRI circuit, which is used to connect ISPs to the PSTN, is a protocol established in the early 1980s for the purpose of providing digital links to customer premises. It is a digital circuit that carries twenty-three information channels and one signaling channel called the D channel. ISDN PRI is a popular interface between the ISP and the local telephone company because it is able to carry signaling out of band within the D channel.

ISDN circuits allow the two entities to talk to one another the same way a switch talks to another switch. ISDN PRI acts like a shared trunk between the two ends. Multiple users can only access the ISP's modem ports through the ISDN circuit for the time the call is up. As soon as the user disconnects, the channel on the ISDN line is available for use by other subscribers.

Routing a dial-up call to an ISP goes something like this. The subscriber clicks on the connect icon on the PC. The modem outpulses the digits to the class 5 switch, where the switch looks up the digits in its translations table and determines that the call should be routed to the ISP ISDN PRI trunk. The signaling channel within the ISDN circuit establishes the connection between the class 5 switch and the ISP's terminating equipment. Ringing is sent digitally on the D channel to the modem pool at the ISP. The D channel tells the switch that ringing is being transmitted, and the class 5 provides ring back to the calling party's modem. Once the call has been picked up by the modem, the switch connects the line to the channel on the ISDN circuit. The data call rides on the ISDN channel until the customer disconnects. Once the connection is released, the trunk being used to carry the call is marked as idle and may be used by a different subscriber. ISDN PRI allows shared trunking arrangements to be built between itself and the local class 5 switch.

MTSO Cellular Switch Center

To route a call from a class 5 switch to the MTSO cellular switch center, trunks are established between the two networks, similar to the trunks between an ILEC and a CLEC network. The trunk connections are used to carry the calls from the cellular network to the wire-line network and vice versa. Imagine making a call from your cellular phone to your friend's wire-line phone. You go off-hook on your cellular phone and wait for dial tone. The cellular phone switch at the MTSO detects the off-hook condition and sends dial tone. You punch in your friend's telephone number and wait while the digits pass through the network.

The MTSO receives the digits and looks them up in its translations table to determine where the call needs to be routed. The switch sees that the call has to terminate in the local telephone network and must therefore be routed to the service provider's access tandem. The SS-7 network sets up, tears down, and monitors calls between the local service provider's network and the cellular network. The trunks are seized between the MTSO and the access tandem. It receives the digits and looks up the LNP. The LNP query returns the terminating switch LRN (address) and a trunk is grabbed on the trunk group. The digits are analyzed by the end-office switch, which checks to see whether the called party's line is busy. Seeing that the line is idle, it sends ringing voltage down the line to ring the phone. The end-office land line switch notifies the originating switch, via the SS-7 network, that the line is being rung. The MTSO cellular switch provides ring back to the cellular phone. When the phone is answered, the call path is established.

The cellular network and the land line network are very similar. They connect through trunks, and the SS-7 signaling network is responsible for establishing the path between the two networks. The interconnection of multiple cellular networks with local service providers is creating a very complex, interconnected web of trunks.

■ 9.3 ENGINEERING TRUNKS, LINES, AND SWITCH PORTS USING TRAFFIC ENGINEERING METHODS

9.3.1 Traffic Engineering

Traffic engineering tools have been developed to help determine the number of trunks, the number of switch ports, and the capacity of the switch needed to handle a traffic load. There is no need to build a switch with enough trunks to handle every subscriber line terminating there. We do not use the phone twenty-four hours a day. We use it randomly for varying time periods. Therefore, if telephone companies built switch capacity to match the population of telephone subscribers, the cost would be outrageous and wasteful. Instead, switches are built to handle the percentage of random calls of variable lengths.

9.3.2 Sizing the Switch and Network

Traffic analysis for telecommunications systems is similar to determining capacity for a railroad. Both provide a transportation path between two points. When a train has more passengers than seats, passengers must be turned away or blocked from boarding the train. When the train has too many cars with empty seats, the railroad loses money. The balance between providing good service and keeping costs down is the job of traffic engineers. Telecommunications traffic engineers are no different. Their job is to manage network capacity by maximizing the use of the switch ports and trunks while maintaining the quality of service for the customer.

9.3.3 Traffic Engineering Methods
Grade of Service and Blockage

What quality of service will suffice for the network? When determining switch port numbers, the quality of the service is defined as the probability that the call will be blocked. *Grade of service* (GOS) is the term used to define the quality of service on a network. Traffic tables are indexed by GOS value, providing a tool to the traffic engineer. The engineer first decides what GOS to use and, in doing so, determines the probability that the switch may block the call. For instance, if the GOS is *P.01*, the probability of a call being blocked is less than 1%. Conversely, a P.01 grade of service means that 99% of the calls will be switched. A GOS of P.05 says that for every 100 calls that are placed, five calls may be blocked. Using a P.05 GOS means that the engineer has built enough switch ports to handle 95% of the customers at one time.

What does this mean as the call travels through the switch hierarchy—local class 5 to tandem to toll back to tandem to terminating class 5? If a switch is built for a P.01 probability of blockage and a call stays within its network, it is safe to say that it will provide a P.01 GOS. However, if the call leaves the network, blocking may occur at any point along the call path—on the trunks and also through the switch matrix. GOS is cumulative through the network, so multiple P.01 grades of service increase the chance for blockage at each new switch.

Several factors contribute to calls being blocked in the network, but four areas are responsible for most of the call failures. They include the following:

- DTMF receivers
- Too few trunks or circuits
- Inadequate switch processing
- Too few SS-7 links

Traffic engineers are responsible for ensuring that these areas are sized correctly and are consistently monitored to ensure the guaranteed GOS.

DTMF receivers provide dial tone when a phone goes off-hook. Slow dial tone is a result of too few DTMF receivers. This type of blockage is referred to as a delayed blockage system and causes the customer to wait for service. A more severe type of blockage is referred to as loss systems, which denies service to the customer altogether. The customer receives a fast busy tone forcing him or her to hang up and try again. The types of blockage we will discuss are based on a loss system.

A second type of blockage occurs when the switch lacks trunking capacity or switch ports. Callers experience a fast busy when all trunks or switch ports are being used. A fast busy indicates problems in the network. If the customer repeatedly receives fast busies, additional trunks need to be added.

Too few switch ports are the result of an overly aggressive concentration ratio being used by the telephone company. Most concentration ratios range from 4:1 up to 6:1. The ratio stands for four or six customers sharing one switch port. One variable that has skewed the commonly used 6:1 ratio is the increased popularity of surfing the Web. Users stay online for hours instead of minutes and consequently have corrupted the standard traffic formulas.

A third form of blockage happens when the switch's processing power or CPU is too small to handle the call volumes. The term used to define the number of calls a switch can handle is the *busy hour call attempt* (BHCA). Switch spec sheets specify the BHCA value; it is one of the first things an engineer needs to know before selecting a switch. Resolving a blockage problem caused by too few trunks is fairly simple. The solution is to order more trunks. Resolving blockage issues that are due to switch overload is much more complicated and costly. Therefore, it is very important that the switch's BHCA capacity is examined thoroughly before deciding how to solve the problem.

Common channel signaling links, or SS-7 links, are the fourth problem area that can cause calls to fail. For calls to traverse through the network, signaling links must be in place to carry the call setup and teardown messages. Remember the scouts that went out to look for an idle call path? If the link used to carry the scouts is congested,

grade of service
Service level establishing the percentage of blockage provided by the telephone company.

P.01
Term used to define a 99% nonblocking rate for the switch.

DTMF
Dual-tone multifrequency—signaling method used primarily between a touchtone telephone and the switch.

busy hour call attempts
A traffic engineering measurement used to size the number of trunks required or switch ports needed.

calls will fail due to missing routing information. Specially trained engineers calculate SS-7 link use. A typical percentage for SS-7 links is 40% use, meaning that when the link reaches 40%, it is time to order an additional SS-7 link.

Traffic Modeling Tools—Poisson and Erlang Traffic Tables

The central premise used by mathematicians when developing traffic formulas is that callers are unpredictable. Normally, all subscribers do not go off-hook and dial at the same time. Calls arrive in a random fashion. Callers dial at different rates and react differently when the call is blocked. Traffic models have been built around the phenomenon that calls arrive randomly and that at no time other than during a local or national disaster will every subscriber go off-hook at the same time. Beyond assuming random call arrivals, the formulas also assume an average hold time, or time a caller remains on the telephone talking. Using these facts, software programs, traffic tables, and numerous other tools have been developed to help engineer the network. Two common traffic models have been used for years in telecommunications.

French mathematician S. D. Poisson formalized the first statistical relationship between telephone usage and sizing the network. His model is called the *Poisson* distribution of call arrivals. The second model was devised by A. K. *Erlang*, a Danish mathematician. From the two models, traffic tables have been developed to calculate the number of trunks needed for a specific traffic load or number of arriving calls. Conversely, the tables can help determine whether there are too many trunks for a given traffic load. The Poisson and the Erlang traffic tables call distribution formulas that assume calls will arrive at random intervals during the busiest hour of a twenty-four-hour day.

The difference between the two models is how they handle call-block retries, or how they expect callers to behave when they hear a fast busy. When a caller immediately redials, it is referred to as blocked calls held (BCH). If a caller hangs up and retries at a later time, it is called blocked calls cleared (BCC).

The Erlang B formula assumes all calls will be cleared when blocking occurs; therefore, it does not count all call attempts because it is not including the blocked calls in its formula. For example, if a P.01 GOS is used and the traffic load equals 100 calls, the Erlang B table sizes the trunk groups using 99 call attempts instead of the actual 100 call attempts.

The Poisson formula treats blocked calls differently. It expects that callers will immediately redial once they receive a fast busy tone. The Poisson table includes the retries in the total number of call attempts. Using our previous example of 100 call attempts at a P.01 GOS, the Poisson table sizes the trunks for 100 call attempts because it includes the one blocked call. The Erlang C table handles blocking similar to the way the Poisson table does.

Traffic tables have now been incorporated in easy-to-use software programs. The hours spent calculating traffic loads using a calculator, slide rule, pencil, and paper are gone. Traffic engineering software packages are available that can determine the number of trunks required for a specific connection in seconds. In fact, many of the programs use the raw data from the switch when determining the number of trunks required. All these programs are based, however, on either Poisson or Erlang's formula.

Determining the Busy Hour

When the GOS has been determined and the model to calculate the traffic load chosen, the traffic engineer must estimate the traffic load. Traffic load is normally determined by looking at statistics from the busiest hour during the busiest week. Mother's Day is one of the busiest traffic days of the year and is often used for the busy day busy hour (BDBH) calculation. Busy hours for residential users and business users vary dramatically. A good network design will take this into account. Accumulating traffic statistics can still be a tedious, time-consuming job.

Average Hold Time

Average hold time must also be determined when calculating trunking requirements. It refers to the average amount of time a subscriber remains on the telephone and it is one

Poisson table
Traffic table used to determine the number of trunks required to maintain a desired GOS.

Erlang
A measurement standard used by traffic engineers to size the number of trunks needed.

average hold time
Average time a subscriber remains on the telephone.

of the hardest values to predict. The Internet revolution has increased the complexity of determining the average hold time. Voice calls average about three minutes in length, while Internet calls last an average of one and a half hours. The issue has become what to use for holding times because the calls vary between connections to the Internet and voice conversations. Additional switch ports have been added to alleviate port blockage caused by Internet subscribers.

Quantifying Call Volumes—CCS, Erlangs

Two common units of measurement are used in telecommunications. The unit most commonly used in North America to express call volume is *centum call seconds* (CCS). The formula used to derive CCS is

centum call seconds
A measurement standard used by traffic engineers to size the number of trunks needed.

Number of calls per hour × average call duration in seconds / 100 = CCS

The unit primarily used in the rest of the world is Erlangs, or TUs (traffic units). The formula used to determine the number of TUs or Erlangs is

Number of calls × average call duration in seconds / 3600 = Erlangs

Some examples will help clarify these formulas. If one circuit receives thirty calls, and each call lasts sixty seconds, how much of that circuit is being used?

30 calls × 60 s average call duration / 100 = 18 CCS

or

30 calls × 60 s average call duration / 3600 = 0.5 TUs or Erlangs

How do we determine the total capacity used? First, we need to determine how many CCSs and Erlangs fill up one trunk. CCS is saying that it represents a 100-second call. Therefore, full capacity in CCS is

60 calls × 60 s / 100 = 36 CCS

Therefore, 36 CCS represents a fully loaded trunk. Looking at the previous example showing thirty calls at sixty seconds call duration, we see that the call volume is at 50%. Full capacity using Erlangs can be calculated as

60 calls × 60 s / 3600 = 1 Erlang

One Erlang represents 100% capacity. From our example we see that call volume is at 50% capacity. Comparing the two methods shows us that 36 CCS represents 100% circuit use and that 1 Erlang represents 100% use. If we need to convert CCS to Erlangs, we would need to divide CCS by 36.

18 / 36 = 0.5 Erlang

To convert Erlangs to CCS, we would need to multiply Erlangs by 36.

0.5 × 36 = 18 CCS

While you will probably not need to calculate CCS or Erlangs, you should understand what is meant by a circuit being at 50% use or 18 CCS or 0.05 Erlang. You should also be able to read a simple Erlang B or Poisson traffic table as shown in Table 9–2.

Connecting the Dots

Connecting the dots in traffic engineering means that you are able to complete a simple traffic example. The following one should help pull together the concepts presented in the previous sections. You have just been hired by Spidercom Telecommunications as a traffic engineer. Unfortunately, Spidercom did not receive the amount of venture capital it had hoped for and was unable to buy a traffic analysis program. Consequently, you will need to determine the traffic load using Erlang B tables and a calculator. Information that you will need to begin your analysis follows:

- Spidercom is offering a P.01 GOS.
- Spidercom is clearing all blocked calls.
- Busy hour and average hold times have been calculated.

Table 9–2
Examples of Traffic Tables.

Erlang B Table				
Trunk	**Erlangs**	**P.01/CCS**	**Erlangs**	**P.05/CCS**
1	0.01	0.04	0.05	1.8
2	0.15	5.4	0.38	13.7
3	0.46	16.6	0.90	32.4
4	0.87	31.3	1.52	54.7
5	1.36	49	2.22	79.9
6	1.91	68.9	2.97	107
7	2.50	90	3.75	135
8	3.14	113	4.53	163
9	3.78	136	5.36	193

Poisson Table				
Trunk	**Erlangs**	**P.01/CCS**	**Erlangs**	**P.05/CCS**
1	0.01	0.4	0.05	1.9
2	0.15	5.4	0.36	12.9
3	0.44	15.7	0.82	29.4
4	0.82	29.6	1.36	49.1
5	1.28	46.1	1.97	70.9
6	1.79	64.4	2.61	94.1
7	2.33	83.9	3.28	118
8	2.92	105	3.97	143
9	3.50	126	4.69	169

Your first task is to determine the number of trunks needed between Spidercom's St. Louis and Austin switches. The call volume estimated between the two switches is 180 CCS. How many trunks will be needed? Your second task is to determine whether or not the number of trunks between Spidercom's Denver switch and the local access tandem will support the call volume. The switch is showing 224 CCS riding on four trunks. Will customers experience blocking?

By using your calculator and the traffic table provided in Table 9–2, you should have determined that Spidercom has to build five trunks between St. Louis and Austin in order to provide a P.01 GOS. You also should have determined that you would need to add two trunks between Denver and the local access tandem in order to alleviate blockage.

Determining the number of trunks needed between two switches, the number of switch ports, and the size of the switch processor (BHCA) is the job of telecommunications traffic engineers. The ability to predict correct trunk numbers, switch port needs, DTMF receivers, switch capacity, and signaling links is essential for every communications carrier. Network congestion causes loss of customers and ultimately loss of revenue. Overbuilt networks waste precious capital and cause loss of revenue. The attempt to build a congestion-free network and a fully utilized network will continue indefinitely.

■ 9.4 SWITCH FEATURES

9.4.1 Where Features Reside

Features in the digital circuit switch may reside in the NCP and CPS. Features are simply software code-written to instruct a switch to perform a particular task. For example, when you want to forward calls to your cell phone, the switch allows you to do this by attaching a call-forwarding script to your line. You, in turn, enter in the number you wish to send your calls to. The script is not written for you exclusively. It is a feature that resides within the switch's computer code and is available to all subscribers. Service providers may purchase the software version containing the feature if they wish. Features can be purchased at the same time the switch is purchased or later, after the switch is up and running. The number of features a switch is capable of supporting depends on the switch defined by the software load. Upgrading the software load in a switch is very costly.

Some features do not reside in the switch. Intelligent peripherals controlled by the signaling network are often attached to the switch to provide a reservoir in which features can reside. These peripherals are often used for calling card systems, voice mail, and other advanced services.

Features may also be found within a shared database located in a separate network. An Advanced Intelligent Network (AIN) can be used to access feature databases outside the service provider's network. The LNP uses the AIN to access the LNP database. Similar to the intelligent peripheral, the AIN allows remote servers to be accessed and used as feature reservoirs.

9.4.2 Categories of Features

Manuals that describe each feature can be five inches thick. There are thousands of features available, but they can be categorized into two groups—call-processing features and customized local area signaling service (CLASS) features.

Call-processing features reside in a switch's CPS. Features such as three-way calling, speed calling, and call waiting are examples of call-processing features. The revenue generated from the sale of features has allowed even the smallest telephone companies to justify upgrading their networks to digital switches.

The second category of features are CLASS features. CLASS features also reside in the switch's CPS, but they work with the SS-7 network to allow features to travel between carrier networks. Automatic recall, calling number identification, and distinctive ringing are common CLASS features. CLASS features depend on digital switching technology and the SS-7 signaling network.

SUMMARY

Circuit switching can be defined as a switching method that establishes a connection through the network using a telephone number as the routing address. The connection remains up until one of the parties terminates the call. During the time the call is established, no other individual can use that circuit in the switch, in the network, or in the wire used in the local loop.

The chapter also defined the different types of digital switches and where each resides in the PSTN. The method used to size a switch using traffic engineering statistics, along with determining traffic load on the network, was explained. The final topic was a definition of switch features.

CASE STUDY

Each of the remaining chapters concludes with an activities section that will require you to design, engineer, and troubleshoot telecommunications networks. As is the case in the real world of telecommunications, there is not just one answer to any problem. Several solutions may be correct. The purpose of each exercise is for you to gain experience in designing, troubleshooting, and planning communications networks.

Each case study is designed around a hypothetical town called Green Grass, Wyoming. As explained in Appendix A, the council members of the town have just decided they would like to join the twenty-first century. For them to accomplish this, they require a state-of-the-art telecommunications network. The goal by the end of the text is to build Green Grass a functioning communications network that will allow its citizens to access long distance networks; regional networks; local networks; and, most important, the Internet.

Appendix A provides information you will need to complete each part of the case study. You should, though, search the Web for additional information and insight.

Using the information in Appendix A, determine where local end-office class 5 switches should be placed. You will need to size the switch according to the number of subscribers and services. You will also need to list the call features you will purchase. Once you decide where to place the switch, you will need to determine the type of trunks needed to connect the two switches together and to connect the class 5 switch to the many long distance carriers that will need interconnection into the local network.

A numbering plan and a connection to NPAC (Number Portability Administration Center) for LNP queries should be established. NANC (North American Numbering Council) has given you two exchange numbers—560 and 232—that provide a total of 20,000 numbers that can be given to Green Grass telephone subscribers. The following list outlines each of your assignments.

1. Determine the number of switch ports needed to serve all of the residents and businesses in Green Grass. The switch ports should be divided into line ports and trunk ports. You will need to define the standard concentration ratio for a customer-to-switch port. Also list all of the call features that will be offered.

2. Draw a schematic showing the different parts of the switch and label each functional part.

3. Determine the types of trunks you will need to interconnect to long distance carriers, 911, and operator services. Develop a trunk diagram showing the from and to locations for the trunk, the type of trunk, and the rate of the circuit required.

4. Assume the 911 bureau has been established by the town board and is located in Town Hall. The Grass Hopper PSAP will serve as the backup 911 site. Make a diagram of the 911 network in Green Grass.

5. Operator services will be handled by the Grass Hopper telephone company in Grass Hopper, Wyoming. Draw a diagram showing how the operator services trunks will interface into the Grass Hopper operator services switch.

6. Five long distance companies have requested termination in the network. Illustrate how the LD companies will interconnect into the local network.

7. Establish a numbering plan for the Green Grass area. You have submitted a request to NANC for an NXX and have been given 232 and 560. You have decided to establish two rate centers—one for all residents on the east side of Green Grass and one for all residents on the west side.

REVIEW QUESTIONS

1. What is the significance of a digital switch and when was it first introduced?

2. Define the following terms:

   CPS   trunk module
   NCP   line module
   ICS

3. Explain the difference between a tandem switch and a class 5 switch.

4. Explain the difference between a tandem switch and a long distance switch.

5. Explain the difference between a long distance switch and an international switch.

6. Explain how a class 5 switch routes a call traveling between two lines that are served from the same switch.

7. Explain how a call is routed between an end-office switch and the local tandem switch.

8. Explain how a call is routed between an end-office switch and a long distance switch.

9. Explain how a call is routed between an end-office switch and an international switch.

10. Why do business customers purchase private branch exchanges?

11. What is different about the switch used to route cellular calls?

12. Explain how a call is routed between an access tandem and a cellular class 5 switch.

13. What is meant by the term *PIC?*

14. Define each part of the following number as defined by the North American numbering plan: 818-555-9877.

15. Explain the purpose of a Feature Group D trunk.

16. Explain the purpose of an intermachine trunk.

17. Explain the purpose of an interswitch trunk.

18. Explain the difference between using the Erlang traffic table and the Poisson traffic table.

19. What does P.01 grade of service mean?

20. Define the following terms:

    busy hour
    average hold time

21. Name five features offered by the class 5 switch.

22. List three of the CLASS features offered by the class 5 switch.

23. Explain what is meant by LNP and why it is an important feature in telecommunications today.

TROUBLESHOOTING

Several customers have called and complained that they are receiving fast busy tones when placing a call. The customers are calling into the Angel Street CO, which is a class 5 end-office switch. You are given the assignment of traffic engineer and, as such, must determine where in the network the blockage is occurring.

The first step is to pull down traffic reports from the traffic reporting program. The traffic reports provide you with the following

information. Once you have read through the information, explain where in the network you will need to add trunks.

- Angel Street has trunk connections to the access tandem, Oak Street end-office switch, and Red Street CO.
- Angel Street to the access tandem shows a P.1 GOS.
- Angel Street to Oak Street shows a P.01 GOS.

KEY TERMS

circuit switch (232)

5–9s (233)

central processing system (233)

network control processor (233)

interface controllers (233)

line interface module (233)

trunk interface module (233)

line side (234)

trunk side (234)

class 5 switch (234)

primary interconnect carrier (235)

class 4 switch (235)

class 3 switch (236)

class 2 switch (236)

class 1 switch (236)

C7 (237)

mobile telephone switching office (237)

private branch exchange (238)

North American numbering plan (238)

Numbering Plan Area (NPA) (239)

number exchange (NXX) (239)

XXXX (239)

IDDD (239)

intraLATA call (240)

glare (240)

interLATA call (241)

one-way trunk (241)

two-way trunk (241)

trunk group (241)

interswitch trunks (242)

tandem switch trunks (242)

1996 Telecom Act (242)

access tandem switch (242)

intermachine trunks (242)

Feature Group trunks (242)

911 trunks (243)

PSAP (243)

operator services trunks (244)

analog trunks (245)

direct inward dial (245)

ISDN PRI (245)

grade of service (253)

P.01 (253)

DTMF (253)

busy hour call attempts (253)

Poisson table (254)

Erlang (254)

average hold time (254)

centum call seconds (255)

10

Signaling

Objectives

After reading this chapter, you should be able to

- Define signaling.
- Define in-band signaling.
- Define digital signaling.
- Define AIN.

Outline

■ INTRODUCTION

How does the telephone know when to ring, when to stop ringing, and when to let the far end switch know that someone has answered or that the line is busy? How do the switches between the originating and the terminating subscribers know that they need to establish a connection? Signaling is responsible for establishing connections, gathering information, and controlling how calls and information are routed through the network.

It is one of the most complicated processes in telecommunications. It not only establishes calls, it also controls peripheral devices that perform special tasks and handles voice mail response systems and prepaid calling cards. To truly understand call processing, it is critical to understand how signaling controls the network. Chapter 10 describes the different signaling methods, explains how signaling actually works, and provides a brief description of what to expect in the future.

■ 10.1 IN-BAND SIGNALING

10.1.1 Signaling between End Stations and Switches
Ringing
A knock on your front door indicates that someone is at your front door. Early pioneers yelled "Hello!" before entering a neighbor's cabin. If a person stood at your door and

did not knock or yell, you would not know anyone was there. A knock or yell is a signaling method. You respond by opening the door and greeting your visitor. A ringing telephone is equivalent to a knock on the door. When you pick up the handset, you are "opening the door" and greeting your visitor.

The first signaling method used by telephone companies to notify subscribers that a call was waiting to be answered was a yell, tap, or scream from the operator. However, screaming for the subscriber proved to be awkward and difficult. Resourceful telephone engineers soon discovered a way to ring a bell on a subscriber's telephone set to notify the subscriber that someone was calling. The operator generated a voltage on the line by turning a crank attached to a magneto on the switchboard. The voltage caused the bell attached to the subscriber's telephone to ring; thus, the term *ringing voltage* was chosen to define the voltage used to activate the ringer. Ringing voltage is still used in modern telephone sets, but electronic ringers have replaced bells.

The terminating end-office switch knows that it needs to send ringing voltage to a called party's telephone when it receives a message from the signaling links specifying the telephone number of the incoming call. The subscriber line is found by cross-referencing the routing tables in the switch. The switch first scans the line to check for a busy condition. If the line is idle, the switch outpulses a 70 V, 20 Hz AC ringing voltage, produced by a ring generator at the switch, onto the ring lead of the two-wire pair connecting the subscriber to the switch. The telephone set receives the 70 V, 20 Hz signal, which causes the ringers to ring.

Dial Tone

Every time a telephone is picked up, the switch generates dial tone. Two types of signaling methods tell the switch that the telephone has gone off-hook: *loop start* signaling and *ground start* signaling. Loop start signaling is the most commonly used method of signaling between a telephone switch and a subscriber's telephone. The telephone circuit in idle state has 48 V on the ring lead and no voltage on the tip lead. It is an open circuit, but once the handset is picked up, the circuit is completed between the tip wire and the ring wire. The 48 V on the ring lead is dropped across a 430-ohm resistor (the telephone) and current flows through the circuit. The loop is closed. The switch detects the current, knows the circuit is complete, and sends dial tone. The switch is programmed to respond anytime it sees a loop closure. Dial tone is only on the line when the telephone is off-hook. Figure 10–1 depicts a loop start signaling scenario.

The second method used to notify the switch that dial tone is needed is called ground start signaling. Business customers that have a customer premises switch called a private branch exchange (PBX) use ground start trunks. Occasionally, we pick up the telephone expecting to hear dial tone but instead hear a voice from an incoming caller. Think about the circuit between your house and the central office switch. One two-wire circuit carries one call at a time—either an incoming call or an outgoing call. For an *incoming call,* the switch induces ringing voltage on the idle circuit, which makes the telephone ring. With an *outgoing call,* the subscriber picks up the handset and dials the telephone number of the person he or she wants to call. When the incoming caller just happens to seize the circuit at the precise time you pick up the handset, a phenomenon called glare occurs. For an individual telephone user, glare does not affect service. In fact, glare happens very infrequently when a telephone has only one connection to the switch, as is the case with most residential telephone subscribers. However, business systems such as a PBX or key system are adversely affected by incoming, outgoing call collisions and therefore require a special signaling method—ground start signaling—to eliminate the chance of glare.

The PBX is the middleman between the telephone subscriber and the central office switch. Telephones are connected to the PBX rather than the central office switch. Trunk circuits connect the PBX and the central office switch in the same way they connect two central office switches. Trunks are shared circuits. Two-way trunks—those that carry both incoming and outgoing calls—are very susceptible to glare. The PBX

ringing voltage
Voltage sent down the telephone line to ring the telephone or terminating device.

loop start
Type of signaling found on telephone lines that initiates dial tone when the handset is picked up and the circuit is closed.

ground start
Type of signaling found on telephone lines that initiates dial tone when one lead is touched to the ground.

incoming call
Call coming into a switch.

outgoing call
Call leaving a switch.

Figure 10–1
(a) A loop start line with two-wire conductor that produces a complete circuit that allows current to flow. (b) DTMF signaling. (c) Ringing voltage signaling the telephone subscriber that a call is coming in.

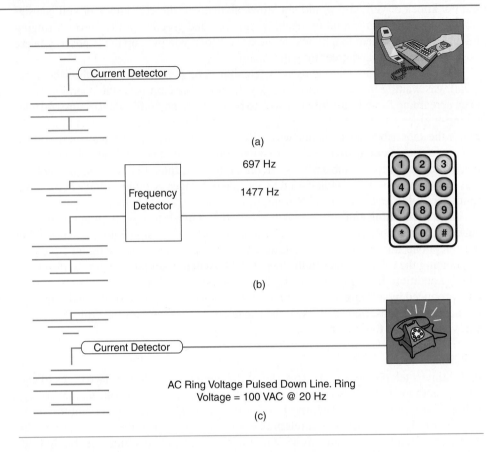

receives an off-hook condition from one of the stations and seizes an idle trunk connecting to the central office. At the same time, the central office switch seizes the same trunk toward the PBX. A collision occurs causing both calls to fail and the PBX to become confused. The solution is either to option separate incoming and outgoing trunks—one-way trunks—or to option the trunks for ground start signaling.

Optioning the trunks for ground start signaling requires that the switch send dial tone only when it sees a ground condition. A ground occurs when one lead of the two-wire pair is connected to ground. Think of birds on a wire, as illustrated in Figure 10–2. Birds can sit on a power line without being electrocuted because they are not grounded and, therefore, are not creating a complete circuit. For current to flow, the circuit must be completed. Therefore, ground acts as the return path, creating a complete circuit and consequently allowing current to flow. When one of the two wires is briefly connected

Figure 10–2
Birds on a power line. Birds are not affected by the electricity flowing through the wires because they are not grounded.

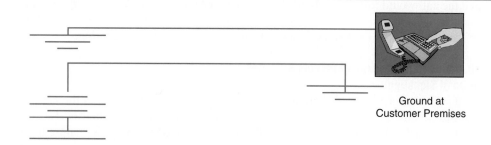

Ground at
Customer Premises

Figure 10–3
Ground start signaling is used to
eliminate the effect of glare and is
often found in circuits terminated into
a PBX. One side of the pair is
grounded to signal the switch end that
the telephone has gone off-hook. The
switch end detects the ground and
sends out dial tone.

to a ground reference, the switch recognizes the ground condition, current momentarily flows (called a wink condition), and the trunk is immediately seized.

Tracing two calls—one outgoing and one incoming—will help provide a more detailed description of ground start signaling. First, we will look at an outgoing call. The condition of an idle line is that the tip-side wire is open to the central office and the ring-side wire is open to the PBX. Picking up the handset causes the PBX to place a momentary ground on the ring lead. The central office sees this ground—a wink—and places a ground on the tip lead. The PBX sees the ground on the tip lead and connects the tip and ring leads to the 600-ohm termination, allowing loop current to flow. The switch also cuts through the tip lead causing the circuit to complete. With an incoming call, the switch receives a request from the network to connect a call through to the PBX. The switch grounds the tip lead and induces the ringing voltage on the ring lead to the PBX. Answering the telephone completes the circuit. Figure 10–3 illustrates a typical ground start circuit.

Dialing

The two methods used to carry a telephone number to the switch are touch tone, *dual-tone multifrequency* (DTMF) and rotary dial (pulse code dialing).

DTMF is the most common way to send digits to the switch. It uses a combination of tones to represent digits. For instance, if we press the number 3 on a touchtone pad, we are actually sending tones composed of 697 Hz and 1477 Hz down the line, as shown in Table 10–1. The switch has a tone detector that deciphers the frequencies and changes them into binary digits that represent the number 3. Whenever you use a touch-tone phone, you provide the serving central office switch with the called-party number.

The other way a subscriber sends a called-party telephone number to the switch is by dialing a rotary dial telephone and creating pulse codes. Pulse code dialing, used since the late 1800s, uses voltage pulses to identify the different numerals. For instance, the number 1 is very close to the resting position on the dial. When it is dialed, a specific pulse sequence is generated. Each digit on a rotary dial has a defined number of pulses that click when dialed and are interpreted by the switch.

When the handset is picked up, current flows continuously. The pulses created by a rotary dial are breaks or interruptions in the current flow. The switch has a current detector that interprets the breaks as defined by the dialed digits. If the number 3, for example, on the rotary dial was broken, the switch would not be able to interpret the digit

dual-tone multifrequency
Signaling method using tones
to establish a connection.

	1209	1336	1477	1633
697	1	2	3	A
770	4	5	6	B
852	7	8	9	C
941	*	0	#	D

Table 10–1
DTMF.

and the call would fail. Even though touchtone signaling has replaced rotary dial telephones, the central office switch can still recognize rotary dial pulses. More than 90% of the telephones in the United States now use touchtone DTMF signaling.

10.1.2 In-Band Signaling between Switches
In-Band Signaling

As discussed in Chapter 9, there are many different types of switches in the network and most calls travel between at least two switches before reaching their destination. For the switches to talk to and set up call paths between one another, signaling information must be used to define the route. Switch-to-switch signaling is accomplished in one of two ways—*in-band signaling* and *common channel signaling.*

"In band" defines the path that signaling information will follow when traveling between switches. The talk path is used to carry signaling information, which is passed on the same circuit that will be used to carry the call.

The problem associated with using the call path as the signaling highway is that a reduced number of trunks is available, due to the extra time the trunk needs to carry signaling information. Remember that not every call will complete. Some will be busy, some will be the wrong number, and some will drop out. Consequently, in-band signaling eats up precious bandwidth on shared trunks. For years, though, in-band signaling has provided a viable way to send signaling information across the network. We will discuss three types of in-band signaling deployed in North America: *single frequency* (SF), *multifrequency* (MF), and *ear-and-mouth* (E&M) *signaling.*

Single-Frequency Signaling

Single frequency (SF) is an in-band signaling format that uses one frequency to notify the next switch that the telephone has gone off-hook or on-hook. SF works like this. A call from a toll switch in Albany, New York, to a toll switch in Syracuse, New York, requires that a trunk be built between the two switches. Switch A in Albany generates a frequency, normally 2600 Hz, that it sends on a trunk circuit toward switch B in Syracuse. The Syracuse switch receives the single-frequency tone and recognizes that switch A is requesting that a trunk circuit be established between the two switches. Switch B seizes the trunk, switch A recognizes the seizure, and a call path between the two switches is established.

One disadvantage of using SF signaling is that fraudulent calls can be placed using a black box tone generator. With a black box, signaling tones are placed on the line causing the switch equipment to establish a path across the network. Once the path is established, free long distance calls can be placed to anyone on the network. Figure 10–4 shows a black box. Another disadvantage associated with SF signaling is the amount of bandwidth used to carry the signaling information. Today SF signaling is less common, thanks to newer, more sophisticated signaling methods.

Multifrequency Signaling

Multifrequency (MF) signaling is a common in-band signaling method used to carry digits across the network between switches. Multiple tones, similar to DTMF, are used to represent the dialed digits. The originating switch sends MF tones through the network to the terminating switch where the digits are processed. MF signaling is still used to convey messages between switches that are not SS-7 capable. For example, many operator services switches and 911 switches are not SS-7 capable. Therefore, they need MF signaling to be optioned on the trunks that connect them to the adjoining end-office or tandem switch.

MF signaling is less susceptible to fraud because multiple tones are used. However, MF does use a lot of bandwidth as it journeys through the network. For example, switch A sends MF tones to switch B. Switch B realizes the call must be routed to switch C and sends the tones on a trunk to switch C. Switch C processes the MF tones and routes them to the subscriber's phone. Because the subscriber is already talking on the telephone, the circuit is marked as busy. Switch C recognizes that the circuit is busy

in-band signaling
Signaling rides on the same circuits as the information signal.

common channel signaling
General term that refers to out-of-band signaling methods. SS-7 is the out-of-band signaling method deployed in North America.

single frequency
Signaling tone used to establish connection between switches.

multifrequency
Signaling tones used to establish connection between switches.

ear-and-mouth signaling
Method used to establish connection between switches.

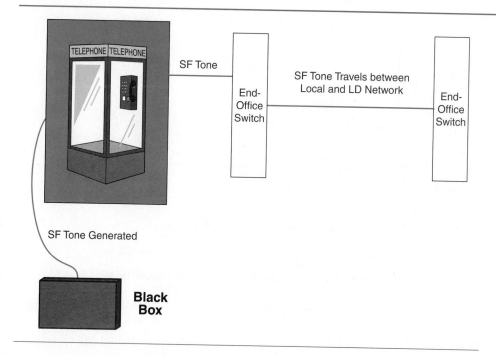

Figure 10–4
Black box tone generator used to simulate signaling tones in order to establish illegal calls through the "before SS-7" network.

and sends the busy condition to switch B, which in turn sends switch A a notice that the telephone is in a busy state. The trunks between switches A and B were seized during the call setup process as were the trunks between switches B and C. Consequently, the trunks were not available to carry other customer traffic during the call setup process even if the terminating telephone is busy, as shown in Figure 10–5. Though the normal call setup times are very short, several combined tie up a substantial portion of the available bandwidth.

Ear-and-Mouth Signaling

POTS signaling—off-hook loop closure—is a DC signaling method in its simplest form. The phone goes off-hook, the loop is closed, and DC flows across the loop. A second type of DC signaling that is used primarily between switches is ear-and-mouth (E&M) signaling. E&M signaling is often used to carry signaling states between PBXs and is also used in conjunction with tie lines that physically connect PBXs together.

E&M signaling uses a separate wire pair to carry signaling information, thus requiring a four-wire circuit between switches—one pair for normal voice conversation and the other for signaling. The wire pair that carries the signaling information is designated as the E&M lead. E&M stands for ear and mouth. Ear represents the receiver of information and mouth the sender. The E lead of one switch is connected to the M lead of the opposite switch. When −48 VDC or ground is applied to the M lead of one

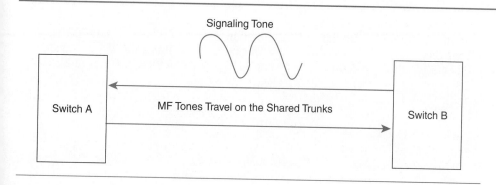

Figure 10–5
MF signaling uses two different frequencies similar to DTMF tones to establish connectivity between the two switches for the call. The tones ride on the talk circuit, thus the term *in-band signaling*.

Figure 10–6
E&M signaling is often found between a PBX and a central office switch. E&M circuits contain four wires.

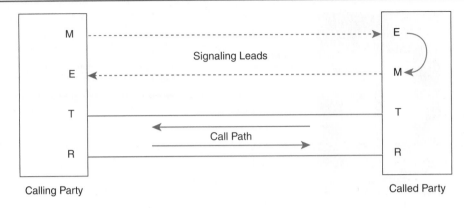

1. M lead at calling party sends signal to E lead at called party end.
2. M lead at called party responds with acknowledgment to E lead at calling party end.
3. Once the E&M leads establish the connection between the two ends, the call path is available for communications.

of the switches, the E lead of the opposite switch receives information and closes a relay contact. Once the contact is closed, current begins to flow out on the M lead of the second switch to the E lead of the first switch, causing that switch to close a relay contact and complete the circuit, as shown in Figure 10–6. The E&M pair carries all signaling and supervisory information. Six different types of E&M signaling use voltage, ground, and current to establish call paths between switches.

■ 10.2 THE DIGITAL SIGNALING NETWORK

The ability of our current telecommunications network to provide a multitude of features, short connect times, accurate telephone bills, and the opportunity to select a long distance and local carrier is a result of advanced digital signaling methods. Digital signaling and common channel signaling are two types used in our network to carry signaling information from one point to another.

channel associated signaling
Signaling method used on digital T1 circuits.

In *channel associated signaling* (CAS) or digital signaling, bit sequences from the digital signal represent signaling states. On-hook, off-hook, and ring are carried within the bit stream of the digital channel. CAS uses robbed bit signaling to steal the least significant bit of the eight-bit word in every sixth frame to represent the signaling state. For instance, an off-hook condition is represented by a two-bit pattern taken from the sixth and twelfth frames. Table 10–2 lists the binary states used to represent signaling.

The following example shows how CAS carries information. TCY Corporation has a PBX serving 200 employee telephones. A T1 circuit connects the PBX switch to

Table 10–2
One Example of a Loop Start Circuit AB Signaling Bit State.

Event Loop Start FX Line	Station Transmits A Bit	Station Transmits B Bit	CO Transmits A Bit	CO Transmits B Bit
On-Hook	0	1	0	1
Off-Hook	1	1	0	1
Dial Tone	1	1	0	1
Dialing	1	1	0	1
Conversation	1	1	0	1

the serving central office switch on Maple Avenue. Jesse Glass works in the public relations department at TCY Corporation and uses his telephone continuously. He picks up the handset to place a call to Tom, his favorite golf buddy. The PBX senses the off-hook condition, in the same way a central office 5ESS switch does, and instantly sends dial tone to Jesse's phone. Jesse dials the seven-digit number and waits to hear Tom's phone ring. The PBX receives the number, looks it up in its routing table, and determines that the number is not in the table. Therefore, it must route the call over one of the shared trunks on the T1 circuit that connects the PBX to the serving central office switch. The PBX seizes a trunk on the T1 circuit and waits for a reply from the serving central office switch. The connection between the two switches is established. The serving central office switch takes the digits, looks them up in its routing table, routes the call to the correct outgoing line, and waits for a ring-back message from the terminating switch. Once the serving central office switch receives the ring-back message, it automatically changes the signaling bits of the eight-bit word in the sixth and twelfth frames of the trunk that has been selected to carry Jesse's voice traffic for the duration of his call. Because all trunks are shared, that trunk will still be available to carry other employees' information when Jesse goes on-hook. The PBX receives the two-bit pattern that represents ringing and sends a ring-back tone to Jesse's telephone. When Tom picks up his handset to answer the call, the serving central office switch changes the signaling sixth and twelfth bits in the channel or trunk to represent no ring. The PBX reads the bit pattern and knows it has to stop ringing and allow the call path to complete. Figure 10–7 shows T1 bit stream signaling bits and different bit states.

CAS has been in use since the first digital circuit was introduced in the network. The sixth, twelfth, eighteenth, and twenty-fourth frames, designated as the signaling frames, are referred to as the A, B, C, and D bits. Many technicians troubleshoot T1 carrier circuits by watching the AB or the ABCD bits of the different channels. Fortunately for technicians, most types of test equipment provide an analysis of the bits as being on-hook, off-hook, ring, or no ring.

The one problem associated with CAS is its inability to carry signaling messages, with special feature information, between switches or between the switch and the telephone. An example is a message waiting light or stutter dial tone placed on the line when a voice mail message has been left in the system. If the PBX provides the voice mail system, it is not a problem. But if the voice mail system resides at the phone company, the customer will not receive the message waiting light or the stutter dial tone. Equipment has now been modified, however, to allow this message to be sent through to the end telephone when the customer is fed via a T1 circuit.

Another problem with CAS is the need to turn off the robbed bit signaling function when transmitting high-speed data. When it is enabled, the number of bits that can be transported in one channel is reduced from eight to seven. Point-to-point data circuits

Figure 10–7

Channel associated signaling (CAS) is the method used to carry signaling information on T1 digital circuits. (a) Framing pattern sequence used by the end devices to synchronize the bit stream. The end device buffers the twelve frames of data and picks out the framing pattern. Once the equipment reads the frame pattern sequence 101010, it is able to read the ones and zeros coming and translate them into information. (b) Signaling bits are designated in all even frames and are used to flag the sixth and twelfth frames. The end equipment uses the signaling bit pattern to find the sixth and twelfth frames.

Framing Pattern Sequence for Superframe Format

| Fs = 1 | Ft = 0 | Fs = 1 | Ft = 1 | Fs = 1 | Ft = 0 | Fs = 0 | Ft = 1 | Fs = 0 | Ft = 0 | Fs = 0 | Ft = 1 |

Ft = 101010
(a)

Fs = 000111
(b)

do not connect using on-hook, off-hook, and ringing states. In addition, CAS uses the trunk to carry the signaling information and consequently eats up precious bandwidth, much like SF and MF signaling techniques.

◼ 10.3 COMMON CHANNEL SIGNALING

10.3.1 Definition: Common Channel Signaling

The advent of the digital network launched the introduction of common channel signaling (CCS), a way to carry signaling information on a separate network that parallels the switch network. The two types of CCS used in the United States are *Signaling System Seven* (SS-7) and *integrated services digital network* (ISDN). Both use a special digital network that interfaces with every SS-7 capable switch. The signaling links form a robust network of paths that connect switches and carrier networks together. Earlier, we compared SS-7 messages to scouts looking for an open trail through the mountains. The messages travel between the switches and networks providing information that can be used to route calls, select a path, or receive information.

> **Signaling System 7**
> Out-of-band signaling protocol used in North America.
>
> **integrated services digital network**
> Out-of-band signaling protocol used in North America.

CCS provides several advantages over in-band signaling methods. The first is the speed at which the call is set up or torn down. Because the signaling network is digital and the messages travel at very high speeds, call setup time has been reduced to milliseconds. Anyone old enough to have made long distance calls in the 1960s may recall hearing the clicking of relays as calls traversed through each switch on the network. Thanks to CCS, calls now connect almost as soon as the last digit is dialed. The second advantage CCS provides is the near elimination of fraudulent calls. Because the signaling message rides on a separate path, it is impossible for anyone to place a tone generator that mimics the in-band signaling frequency that sets up a long distance connection. The third and most important advantage of CCS is its ability to carry information from external databases. Not only does the SS-7 network carry database information, it also retrieves the information when asked. The databases may contain subscriber information such as features, PINs for calling cards, and which carrier owns an 800 number, as well as the LRN from the LNP database. CCS is as significant a development as that of the digital stored program control switch, the invention of fiber optic cable, and the Internet.

10.3.2 Signaling System 7
Origins of SS-7

The introduction of digital technology changed the composition of the telecommunications network. In the late 1960s, radical views of what the signaling network could look like surfaced in research labs at AT&T and the international standards bodies. The International Telecommunications Union (ITU) developed the first CCS standard called signaling system 6 (SS-6), which was built around a packet switch network that carried packetized messages on 4.8 kbps digital links. The network formed a mesh of links connecting PSTN switches to the SS-6 packet network. The messages were used to set up and tear down calls between switches. In the mid-1980s, Signaling System 7 (SS-7) replaced the SS-6 standard in the United States. Similar to the SS-6 network, SS-7 was built to ride on packet network architecture. SS-7 is currently the primary signaling method used in North America.

SS-7 Network Architecture

Numerous telecommunications networks carry information from one point to another. The SS-7 is the auxiliary network that supports those networks by helping establish connections between devices within the network and between networks. See Figure 10–8. It is comprised of packet switches that move packets of information to and from network devices. Individual carriers have their own SS-7 network that connects to other carrier SS-7 networks. The national SS-7 network connects at a global level, allowing for uninterrupted communications around the world.

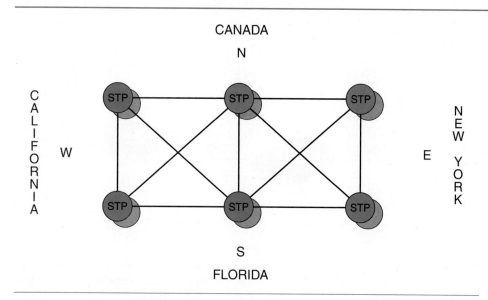

Figure 10–8
Abstract depiction of multiple regional gateway STPs connected together to form a national SS-7 between multiple companies.

The architecture of a typical SS-7 network is comprised of *signal transfer points* (STPs), *service control points* (SCPs), and *service switching points* (SSPs). The SSP is an end-office, tandem, or toll switch. An STP is a packet switch. It receives messages from SSPs, SCPs, or other STPs and routes the packets accordingly. The SCP also performs packet routing functions similar to the STP, but it is also able to route packets from the SS-7 network to external databases.

SSPs, STPs, and SCPs are connected to one another by digital circuits. Normally T1 or DS-0 links are used to carry signaling messages between the SS-7 network devices. The information is chopped up into variable-length packets and sent out onto the digital link to the adjacent packet switch—STP—where they are routed to other STPs or to the terminating switch. Figure 10–9 shows a typical signaling message traveling between two carriers' networks. The packet switches determine where to route the message by reading the destination address field of the packet. When a connection needs to be established, the originating switch sends out a request for a connection. The SS-7 message arrives at the terminating switch on the digital links that connect the switch to the SS-7 network. The terminating switch determines if a connection is possible and returns its answer on the signaling links. The originating switch receives the response and reacts accordingly.

If the call requires a database lookup, the STP routes the query message to the SCP. As shown in Figure 10–10, the SCP routes the message to the correct database. The information from the database is returned to the switch via the SS-7 links. The digital signaling links that connect the switch to the STP, or the SCP to the STP, are access links, or A links. The links that connect the STP to a second STP are bridge links, or B links. Cross links, also called C links, are used to connect to mated-pair STPs and are

signal transfer points
Packet switch used in the SS-7 network to route signaling messages to the correct destination point.

service control points
Signaling equipment used to direct queries to the correct database.

service switching points
Signaling point located on the switch that handles all signaling messages to and from the SS-7 network.

Figure 10–9
Information being transported between two signaling points. The messages contained inside information packets are of varying lengths. Each packet holds information on the calling party, the called party, the circuit ID, the local routing number, point codes, and pointers toward the correct databases if needed. The messages travel through the signaling network in the same way information travels around the Internet. Addresses are used to route the messages to the correct end points.

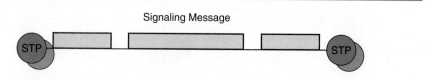

Signaling Message

Figure 10–10
The STP accesses data by sending its request to the SCP. The SCP looks at the request and determines which database the message is destined for. The database returns the information to the SCP, which then routes it back through the SS-7 network's STPs.

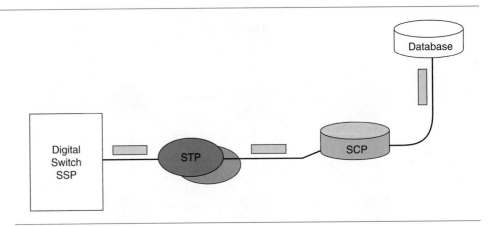

only used during periods of congestion. Diagonal links, or D links, also connect STPs together in a redundant fashion. Extended links, E links, connect remote STPs together; and fully associated links, F links, connect an SSP to another SSP. A links and B links are the most commonly used connections. Figure 10–11 shows the different types of links used in the SS-7 network. The combination and types of links are engineered to ensure 100% uptime, thus guaranteeing that phone service will never be disrupted due to a signaling problem.

SS-7 Message Format

The SS-7 message format is composed of several fields used to carry information about the call being placed. The message contains the originating switch address, the terminating switch address, the originating telephone number, the terminating telephone number, the next node the message will travel to, and the higher-level addresses used to direct the message to the correct database.

The SS-7 protocol is composed of four sections—MTP, ISUP, SCCP, and TCAP. The message transfer protocol (MTP) establishes the connection through the network. It uses the physical node address, called the point code, to route the message to the appropriate destination. This is a ten-digit number that acts as the address for a piece of equipment—the switch, the STP, and the SCP all have point codes. A field within the SS-7 message carries the point code of the originating switch. Each telephone company

Figure 10–11
Different types of links established between signaling points. The B links connect remote STPs; A links connect STPs with SSPs or SCPs; D links are diagonal links between gateway STPs; C links are cross links between backup STPs in a common location. The links are digital T1 or DDS circuits.

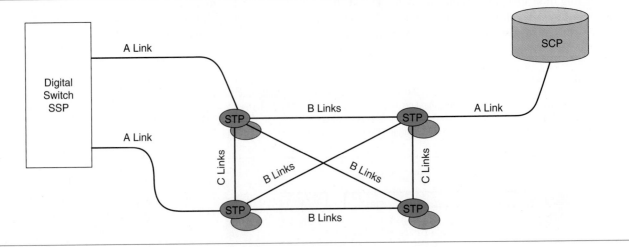

is given a specific number range to use in assigning these codes to each piece of network equipment. Table 10–3 shows point code assignments. The service providers are also responsible for updating information in the administrative routing tables that hold point codes for each adjacent node. For instance, as shown in Figure 10–12, when a new SCP is added to the network, the adjacent STP's routing database must be updated. Stated simply, the addressing information is placed in the MTP portion of the message and the MTP uses point codes to determine the route.

The second segment within the SS-7 protocol, the integrated services user part (ISUP), uses the called number and calling number to set up and disconnect the call end to end.

The third and fourth parts of the SS-7 protocol normally go hand in hand. The signaling connection control part (SCCP) and the transaction capabilities applications part (TCAP) work together to route and retrieve information from an external database. The SCCP carries the address of the database that needs to be accessed, and the TCAP is responsible for accessing the databases and retrieving the required information. In other words, the SCCP routes the TCAP message to the correct database and the TCAP carries the information back to the originating switch. The TCAP and the SCCP are only used when a database needs to be accessed, but the MTP and the ISUP are used for normal call connection. Therefore, the TCAP and the SCCP depend on the ISUP and the MTP portion of the protocol to establish the connection, but the ISUP and the MTP can function without the TCAP or the SCCP. Figure 10–13 depicts the SS-7 protocol message.

The path that a message travels depends on the type of call. If it is an 800 call, the message first travels from the originating switch to the adjacent STP. The STP looks at

Signaling Network	Network ID
AT&T	254
Sprint	253
BellSouth	252
Pacific Bell	251
Ameritech	250
Southwestern Bell	249
Bell Tri-Co Services	248
Nynex	247
Bell Atlantic	246
Stentor (Canadian Co.)	245
MCI	244
SNET	243
Allnet	242
Defense Infor Systems	241
GTE	240
United Telephone SYS	239
Independent Telecom Network	238
Cable & Wireless	237
CNCP Telecom	236
Contel	235
ALLTEL	234
Rochester Tel	233
Century Tel	232
AT&T (VSN)	231
Centel	230
Test Code AT&T	229
Spare	6
Reserved for Point Code Block	5
Reserved for Small Networks	4

Table 10–3
Point Code IDs for the United States and Canada.

Figure 10–12

The addition of an SCP into the network. A links connect the STP to the new SCP. The point code 245 1 3 is added to the routing tables in the network. Each message traversing the network has an origination point code and a destination point code. Point codes are node-to-node specific.

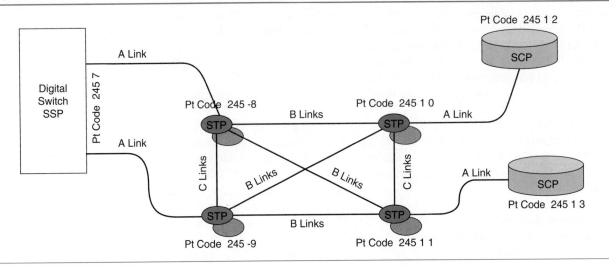

the message and sees the packet has to be routed to an SCP in order to access the database. The SCP receives the message, cracks the packet, and looks for the database address (the subsystem address). Similar to the point code used to identify the different devices on the network, the subsystem address identifies the database that the message needs to access. The SCP contains a routing table that holds subsystem numbers of special databases such as the LNP and 800 number databases.

Therefore, the message leaving the switch (the SSP) carries an originating point code in one field, a terminating point code in another field, the called party number, the calling party number, and, if needed, a subsystem number, and other control characters. The MTP portion of the protocol looks at the originating and terminating point codes to establish the physical connection through the network. The ISUP portion looks at the called and calling telephone numbers to establish the call end to end, and the TCAP and

Figure 10–13

SS-7 protocol architecture. The ISUP and the MTP portions of the protocol are used for normal call setup through the network. The SCCP and the TC are responsible for all AIN features such as LNP. The TU portion is used in the European out-of-band signaling network.

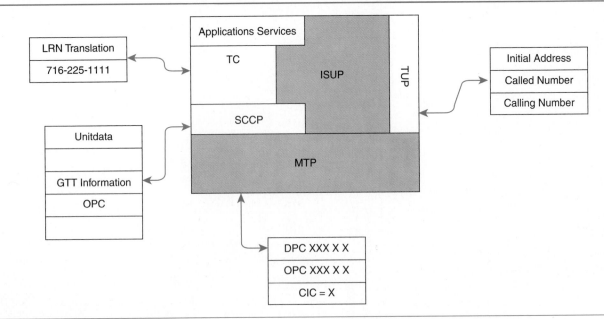

the SCCP use the subsystem number to route to, and to retrieve the correct information from, the appropriate database.

This is a simplistic view of how SS-7 works. It is one of the most complicated, but fascinating, areas of telecommunications. The most important thing to remember is how the network establishes calls and passes information between switches.

Connecting the Dots

To further clarify the process, we will follow a simple call through the network, as illustrated in Figure 10–14. Virginia picks up her telephone handset and dials Ned's telephone number—833-2783. Virginia's serving central office switch receives the dialed digits as DTMF tones, which it translates into binary digits. The switch compares the digits to the numbers stored in the routing tables to determine whether the called number resides in the switch or needs to be switched onto an outgoing trunk to another switch. In other words, the switch has to decide whether Ned's telephone line is one of its own.

Once the switch determines that Ned's telephone is not served from its line units, an SS-7 message is sent out onto the SS-7 network. Its purpose is threefold. First, the message asks Ned's switch whether or not Ned's line is idle. Second, if the line is idle, the message makes sure a connection between Ned's switch and Virginia's switch is

Figure 10–14

(a) Two subscribers' lines interface two switches within the same network. (b) Virginia goes off-hook causing her serving end-office switch to send out an SS-7 message through the SS-7 network to establish a trunk between the two switches. (c) Trunk established between Virginia's switch and Ned's switch.

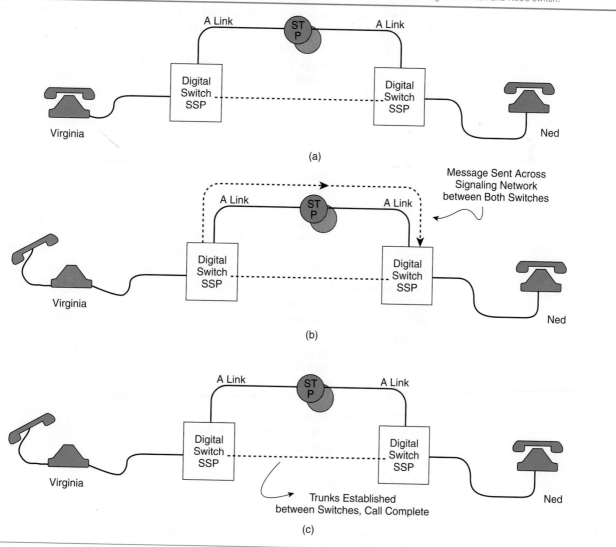

available. Third, the message tells Ned's switch to ring Ned's telephone. The call setup message is sent out on the A link of Virginia's switch that is connected to the adjacent STP. The STP looks at the point code, then correctly routes the message through the SS-7 network to the terminating switch. The terminating switch does not necessarily belong to the same telephone company as Virginia's switch, but for the call to travel between the two companies' networks, both must be able to pass messages in order to establish an end-to-end connection. In our example, Ned's switch does belong to the same phone company as Virginia's switch. His switch returns the message by way of the SS-7 network, telling her switch that it has a talk path available. His switch sends ringing voltage to his phone, along with an SS-7 message back to her switch, stating that Ned's phone is available and ringing has begun. Once his phone begins to ring, the SS-7 network establishes the talk path between the two switches. As soon as Ned answers the phone, an SS-7 message travels from his switch to Virginia's switch notifying it that Ned has answered the phone. Her switch terminates the ring-back message and the talk path is established between the two switches. The SS-7 network marks the trunk as busy in order not to double book a call onto that trunk. Virginia and Ned begin their conversation completely unaware of the work that was performed by the switch and the SS-7 packet network in order to establish the connection.

If Virginia hangs up first, her switch sends a disconnect message out onto the A link. The message traverses the SS-7 network to Ned's switch. The trunk established between the two switches is released, and when Ned goes on-hook, his line is marked as idle. If Ned hangs up his phone first, a message is sent from his switch to Virginia's telling it to disconnect the talk path and mark the trunk as idle.

What happens if Virginia calls a friend served by a different telephone company? This time Virginia dials 1-716-555-2456. The switch translates the received DTMF tones into binary digits, reads the 1+ digit, and realizes that the call is destined for the long distance network. Before the switch decides where to send the call, it references the customer's record located in the switch to see which long distance carrier (PIC) Virginia has chosen. Each interexchange carrier is identified by a unique PIC code. The first portion of the connection is established between Virginia's switch and the local access tandem. The access tandem uses the PIC code to direct the call to the correct long distance company. The connection between the end-office switch and the long distance company's switch is established through SS-7 messages traveling between the two networks via gateway STPs. In this example, Virginia's long distance carrier of choice is Spidercom Telecommunications. Direct connections have been established between the local Detroit tandem switch and Spidercom's toll switch. The information placed in the SS-7 message includes the called number, the calling number, and a carrier identification code (CIC) that references the trunk group identification number. The SS-7 message moves ahead through the network, building a call route between the two networks. When a call path has been identified and Adam's phone is idle and ready to accept a call, the call setup message asks Adam's serving central office switch to ring Adam's phone. Virginia's switch is then notified that it needs to send ring back to Virginia's phone. When Adam picks up the handset, a message rushes across the network to tell Virginia's switch to stop sending ring back to Virginia's phone. The talk path is established allowing Virginia, in Detroit, and Adam, in Nashville, to carry on a conversation.

Setting up a long distance call requires SS-7 messages to traverse across carriers' SS-7 networks. Special codes are used to identify which network the call needs to travel through. Carriers have recently been subverting the carrier of choice by offering lower rates to customers who dial a special access code. For example, 10-10-220 skirts the normal equal access routing by dialing the full PIC code of the bypass carrier. The 220 is the PIC of the promotional carrier.

10.3.3 ISDN Signaling
Definition: ISDN Signaling
ISDN is used to extend the digital network to a customer's location. Initially, ISDN was built as a signaling protocol to expand the features of the PSTN, but eventually the

ISDN created a way to extend sophisticated signaling protocols all the way to the customer site. These protocols carry supervisory information along with messages, similar to the SS-7 protocol, without tying up the talk path or adding a separate network. ISDN has received a lukewarm welcome in the United States due to the cost of the end equipment and the expense incurred by carriers when upgrading their network; however, it is a very popular transmission method in other parts of the world.

ISDN is simply a digital loop that reserves one channel for signaling information. In-band signaling uses the talk path to send signaling information in the form of tones, and CAS manipulates the bits within the information signal to send signaling information. CAS and in-band signaling methods cannot carry information messages and therefore offer a limited number of features.

Signaling information such as call setup, teardown, and supervisory information, as well as feature information such as caller ID, ride on the designated information channel. The channel is very similar to the A links that connect the switch and packet SS-7 network. The channel forms its own signaling link and, in essence, its own signaling network parallel to the information channels.

ISDN BRI Signaling

ISDN comes in several forms, of which the two most common are basic rate *ISDN BRI* and primary rate *ISDN PRI*. The advantage of ISDN BRI is that it provides two working channels and one 16 kbps data channel on one pair of copper wires. The two channels are completely separate, thus allowing two separate telephone conversations to take place at the same time on the same wire pair. The 16 kbps D channel is used to carry signaling information between the terminating device, such as an ISDN telephone, and the serving switch—whether that is a simple PBX or a full class 5 switch. In either case, the ISDN BRI D channel, similar to SS-7, provides out-of-band signaling functionality all the way to the customer's premises. For instance, if Virginia owns an ISDN phone and picks up the handset, the switch does not see a closed loop state but instead receives a request from a message riding in the signaling channel that Virginia needs dial tone. The switch understands the message and returns dial tone to her phone. The ability to send messages allows the switch to talk to Virginia's telephone and to send special features, such as caller name and address, through to the LCD screen on the phone. The ISDN signaling channel also carries all supervisory information including ringing, busy tone, fast busy, and dial tone. In addition to signaling information, the D channel may also carry packet data.

ISDN BRI
Out-of-band signaling channel used on ISDN links.
ISDN PRI
Out-of-band signaling channel used to connect a digital switch to a PBX.

ISDN PRI Signaling

ISDN PRI is similar to a T1 circuit in that there are twenty-four 64 kbps channels equaling a 1.544 Mbps signal. The main difference between the T1 circuit and an ISDN circuit is the way signals are carried. T1 circuits use CAS to carry on-hook, off-hook information between switches. ISDN provides a separate signaling channel, the D channel, to carry messages between the terminating equipment and the digital switch. PRI circuits are commonly used to connect a digital switch to a PBX. The PRI circuit carries additional information in the signaling channel that allows the efficient use of the circuit. ISPs also use ISDN PRI circuits to connect to the PSTN, because it allows them to establish shared trunks between their sites.

■ 10.4 ADVANCED INTELLIGENT NETWORK

10.4.1 Definition: Advanced Intelligent Network

Advanced Intelligent Network (AIN) has been discussed by standards organizations for years. In Europe, the term *intelligent network* (IN) is used to define the new advanced signaling network. The North American AIN standard has been evolving for the past ten years, its goal being to define a network that is able to manipulate a call at any point in the call process. Essentially, AIN will allow database queries to occur anywhere in the network, with information being gathered and distributed as needed. The two most common uses of AIN today are inbound 800 number routing and local number portability (LNP).

local number portability
LNP refers to transferring the customer's existing telephone number to the new service provider's network allowing the customer to retain his or her number.

10.4.2 Local Number Portability

Local number portability (LNP) has allowed CLECs to compete with ILECs by allowing telephone subscribers to keep their telephone numbers when they change service providers. Think of the ramifications of changing service providers if you had to change telephone numbers. A small business would have to redo all its advertisements, loyal customers would have to be told about the new number, and phone book listings would need to be reprinted every week. The success of the 1996 Telecom Act, to create competition in the local telephone market, is directly dependent on the success of LNP. Fortunately, LNP is working well throughout the country.

LNP is one of the first widespread services to take advantage of the AIN standard, a cooperative effort between the FCC, long distance companies, and local service providers that was launched after President Clinton signed the 1996 Telecom Act. From the efforts of the consortium, an LNP standard was developed. The LNP places switch addresses in regional databases maintained by Lockheed Martin. The regional databases hold all ported telephone exchange numbers and distribute those numbers to telephone companies' databases, often housed in their SCPs.

The SS-7 network now includes an SSP, which is the switch; an STP, which is the packet switch used to route the SS-7 messages; and an SCP with an LNP database module attached. In addition to the normal links that connect the devices together and connect to other service providers' SS-7 networks, other links are needed to connect to Lockheed Martin's regional databases. Figure 10–15 depicts a typical SS-7 LNP capable network.

10.4.3 Connecting the Dots

We will now examine three types of calls. The first call originates at the ILEC's end-office switch; the second call travels from the ILEC's end-office switch to a CLEC regional switch; and the third is a long distance call from the ILEC to an IXC to a CLEC.

Ab decides to call his friend Louis for some advice on how to fix his 1936 truck. Ab dials Louis's telephone number, 756-2232, and waits to hear his phone ring. Ab's end-office switch, located on East Street, receives the 756-2232 digits and looks up the digits in the routing table to see whether that number is one of its own. Finding that it is not, a query is sent to the LNP database. There are now two possibilities. Either the switch launches a query to the database asking for the LRN of the terminating switch or the switch passes the call to the tandem switch and lets the tandem launch a query to the LNP database. In this case, Ab's switch will launch the query. The switch sends the SS-7 message out to the STP. The STP sees that it needs information from the SCP and sends a message to the SCP, which in turn launches a query to the LNP database. The database looks at the number and determines that it is not a ported number and therefore does not require any additional information. The response is handed back to the SCP, which passes it back to the STP, which sends it back to the originating switch. The switch reads the response and sets up a trunk connection.

The second type of call does require an LNP translation before the call can be routed. Ab picks up his handset and dials his friend Erma to discuss her upcoming Halloween party. Ab dials 457-8976. His end-office switch receives the digits, looks in its routing tables, and determines that it is not one of its own numbers. The switch launches a query to find out whether or not the number has been ported. The STP performs a special translation called a global title translation that provides the point code of the correct SCP destination. The message is routed there and the SCP queries the database and determines that the number has been ported to Spidercom's local service provider. The LRN of Spidercom's switch is placed in the called party field of the TCAP message and returned to the originating switch. The SCP prepares the return message by placing the LRN in the called number field of the SS-7 message. The telephone number is moved to another field in the message, where the message is packaged and returned to the querying switch. The originating switch routes the call using the LRN instead of Erma's telephone number to set up the connection to the terminating switch.

The LRN is a ten-digit number unique to the switch, such as 716-232-9999, that is chosen by the switch engineer and programmed into the switch. Each switch in the

Figure 10–15

(a) Two subscribers' lines interface two switches, one served by the ILEC and one served by a CLEC. (b) Virginia goes off-hook, causing her serving end office to send out an SS-7 message through the SS-7 message in order to establish a trunk between the two switches. The message is marked as a ported number and causes the STP to query the LNP SCP for the correct switch address. The NPAC database returns the CLEC switch's LRN. (c) The trunk established between Virginia's switch and Ned's switch by using the information returned by the SS-7 network. The called party number is pulled out of the SS-7 message at the CLEC switch where it is routed to Ned's switch port.

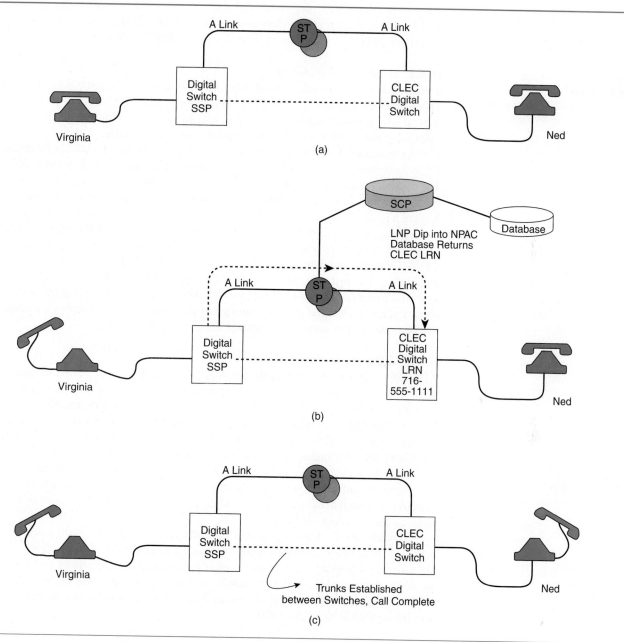

network has its own LRN chosen by the company's switch engineer. The LRN looks just like a telephone number and, in fact, is a telephone number that has been designated as the switch device number. Therefore, any number ported to that switch will be routed using the LRN instead of the dialed number.

The third type of call is from an ILEC to an IXC to a CLEC. Ab decides to call his friend Hal who lives in Pittsburgh, Pennsylvania. He dials Hal's number—1-401-555-2345—and waits for the ring to seize. The originating switch recognizes that it is a long distance call by the 1+ digit and routes it to Hal's carrier of choice, Spidercom. Spidercom carries the call to its long distance switch in Pittsburgh where it launches a query to the LNP database connected to its SCP. The number has been ported to Spidercom's local service. The database returns the LRN of the terminating switch in the field that normally carries the called party number. The LD switch sets up the connection to Hal's end-office switch using the LRN to determine the route. Hal's end-office switch then replaces the LRN with Hal's true telephone number and routes the call to Hal's phone. The entire process takes milliseconds to complete. LNP is one of the first widely deployed AIN applications in the network today. A future step for LNP is a geographically independent solution that will allow users to own their telephone numbers for life and carry them wherever they decide to live.

10.4.4 800 Number Portability

800 number
Toll-free number that requires special routing through AIN signaling.

Similar to LNP, *800 number* portability uses AIN messages to route 800 number calls. This allows subscribers to change carriers but keep their 800 numbers even when they change long distance carriers. Lockheed Martin maintains a public database that provides a location for each 800 number and the associated carrier.

An example of an 800 number call follows: Abby decides to call her favorite cooking company to order a special frying pan. She picks up her phone and dials 1-800-555-8576. The number travels through the local network and is routed to Abby's long distance carrier, Spidercom. Spidercom's long distance switch looks at the 800 number and sees that it is not a normal long distance call. The LD switch sends out a query on the SS-7 links. The message travels through the SS-7 network to an SCP. The SCP points the request to the correct database where the 800 number is converted into a telephone number called a POTS routable number. Every 800 number has a ten-digit telephone number associated with it. For instance, the cooking company is located in Syracuse, New York, and its POTS routable number is 315-434-4455. Abby does not realize that the 800 number she is calling is changed midway through to a normal ten-digit telephone number that can be routed in the same way a normal telephone call is routed.

10.4.5 Additional Uses of AIN

AIN is the signaling network of the future; LNP and 800 number portability are just the beginning. AIN provides access to peripheral devices that will control actions within the network. For example, instead of every switch in the country needing to list every 800 number and associated POTS routable number, AIN allows access to a regional database that centralizes the information. The advantage of depending on a centralized repository for information is that the individual switches will not require constant provisioning. In addition, the processing power needed to support one switch is reduced because of the information being stored in the centralized database. The task of updating a few regional databases is minor when compared with the alternative of having to upgrade thousands of switches.

Special features traveling across different carrier networks, such as caller name, rely on AIN, as does unified messaging, which will be discussed later. Voice-mail systems are intelligent peripherals that hang off the switch but are controlled by signaling messages. Geographic number portability will be possible because of AIN. Due to the AIN standard, the network will change as significantly as it did when the digital switch was introduced in the early 1980s.

SUMMARY

Chapter 10 provided an overview of signaling in the PSTN. Signaling is responsible for ringing our telephone, interpreting the digits, and connecting the call end to end. The SS-7 signaling network is the standard deployed in North America. Signaling between the end-office switches was described, as were the two ways to dial a phone. In-band signaling methods were defined, and MF, SF, and DC signaling were explained as the most common methods of in-band signaling. Out-of-band signaling methods such as

SS-7 and ISDN were explored along with their advantages: a reduction in fraudulent calls, the reduced bandwidth requirements, and the special messages allowed. Finally, LNP and 800 number routing were explained and features offered by the AIN standard were discussed.

CASE STUDY

Now that the switch location has been determined, you must design the signaling network. We will give you several hints because signaling is one of the most complicated networks to design. First, signaling links must be placed between the switches and the signaling network in order to switch traffic between the local Green Grass Telephone Company and the outside world. Now you need to determine the type of links needed and the number of links required to ensure diversity.

1. Draw the SS-7 network connections that will provide signaling for the Green Grass network. Label each of the signaling elements.
2. Determine the point code structure for the Green Grass telephone network. The number 5 has been given as the point code ID. List all of the codes for each of the signaling elements.
3. Determine how Green Grass will interface into NPAC for LNP provisioning.

REVIEW QUESTIONS

1. Explain the importance of signaling in the telephone network.

2. What method is used to tell the subscriber that someone is calling?

3. Explain how touchtone dialing sends digits to the switch. What is the name given to define touchtone dialing?

4. Explain how rotary dialing sends digits to the switch.

5. Explain how the switch knows when the telephone receiver has been picked up.

6. Why do business customers order ground start signaling lines?

7. Define ground start signaling as it pertains to notifying the switch that someone has gone off-hook.

8. Define in-band signaling and list the three in-band signaling methods used to initiate trunk connections between switches.

9. List one of the problems associated with in-band signaling methods.

10. What is CAS and where is it used?

11. List the A, B bit patterns used to signify ringing, no ringing, busy, on-hook, and off-hook.

12. Define out-of-band signaling.

13. Explain why out-of-band signaling is able to connect calls faster than in-band signaling.

14. Define the following terms:

    SSP
    STP
    SCP

15. What is meant by the term *common channel signaling?*

16. Draw the SS-7 message format by showing the MTP, the ISUP, the TCAP, and the SCCP, and explain the function of each.

17. Explain what a POTS routable number means as it refers to 800 number routing.

TROUBLESHOOTING

Problem 1: The network operations center has just received numerous alarms from Oakland showing that all calls being placed at the access tandem are failing. The list of alarms printing out from the switch shows that both A links have failed. First explain why calls will not switch when the A links fail; then list the procedures you will work through to determine why the A links failed.

Problem 2: A customer is complaining that his or her telephones do not ring when a call comes in. The customer has a T1 circuit feeding a channel bank. The analog lines extend from the channel bank to the telephones in the building. Using the test equipment shown in Appendix B, explain how you would set up a test to prove the A, B bits were working properly. (Appendix B explains how to use a typical transmission test set.)

KEY TERMS

ringing voltage (261)

loop start (261)

ground start (261)

incoming call (261)

outgoing call (261)

dual-tone multifrequency (263)

in-band signaling (264)

common channel
    signaling (264)

single frequency (264)

multifrequency (264)

ear-and-mouth signaling (264)

channel associated
    signaling (266)

signal system 7 (268)

integrated services digital
    network (268)

signal transfer points (269)

service control points (269)

service switching points (269)

ISDN BRI (275)

ISDN PRI (275)

local number portability (276)

800 number (278)

11

Distributed Switching Architecture—
Voice Over IP

Objectives

After reading this chapter, you should be able to

- Define distributed switching architecture and equipment.
- Define RTP and RTCP.
- Define the quality measurements for VoIP.
- Define switch features.
- Define an end-to-end VoIP flow.

Outline

Introduction

■ INTRODUCTION

Think about what happens when you pick up the telephone handset. What do you hear? What happens when you make a call? How is the call routed? What do you think happens when you pick up a telephone serviced by a VoIP provider? Where does the dial tone come from? What happens when you place a call? How is the call routed? All these questions need to be answered in order to understand how VoIP telephony works. The following sections focus on the VoIP telephony architecture; the types of equipment used to carry VoIP calls; the signaling protocols; the feature sets; auxiliary services such as voice mail; and, most importantly, how the two networks—the traditional circuit switched network and the VoIP telephony network—interconnect.

Voice over IP is changing the face of the telecommunications industry. Similar to the Western Union attempt to pull market share from the Bell Company in the early 1900s, today cable companies are attempting to pull market share from traditional telephone service providers by offering VoIP telephony through their broadband connections. Chapter 11 is divided into four general sections—Equipment and Architecture;

Signaling; Packet Flows; and Applications and Connecting the Dots—how VoIP telephony interfaces with the PSTN.

■ 11.1 EQUIPMENT AND ARCHITECTURE

Several devices are used in VoIP communications. The functionality of each is often combined into one hardware device, thus blurring the line between what does what. In our examples, we will explain each as a separate entity, making sure to note which devices may reside together in one chassis. Figure 11–1 provides a high-level view of the VoIP telephony network. Note that the residential customer ties to the network through a broadband Internet connection such as cable modem or DSL. From there, the call flows through the big Internet cloud until it reaches a device called a Media Gateway where it is routed to the PSTN, Public Switched Telephone Network, or to another Media Gateway somewhere in the world. The following sections zoom in on this topic, showing the role of each device in the network.

The standards groups involved in establishing standards for VoIP telephony include the IETF—Internet Engineering Task Force; the IPCC—International Packet Communications Consortium; and the ITU—International Telecommunications Union. The standards groups define protocols, establish technical parameters such as signaling method, and advocate for the adoption of the standard to ensure compatibility across networks.

11.1.1 Media Gateways

The *Media Gateway* (MG), as defined by the IETF, is responsible for media conversions; resource allocation and management; and interfacing with the soft switch, also called the *Media Gateway Controller* (MGC). In lay terms, the MG is what interfaces with the end user and the PSTN. An MG is nothing more than a sophisticated proxy server that is able to interpret incoming IP flows carrying voice packets and determine what to do with them.

The majority of large carriers separate the MG's function into residential side or access side and network or trunking side. Residential or access side relates to the end

Media Gateway
The Media Gateway interfaces between the Media Gateway Controller (soft switch) and the end device. The MG handles routing the VoIP flows across the IP network.

Media Gateway Controller
The Media Gateway Controller is responsible for call setup and teardown, managing traffic flows between networks, and controlling multiple MGs.

Figure 11–1
High-level view of the distributed switch architecture. The connections between the MGC and the MGs are logical connections. The MGC may reside hundreds of miles away from the MGs. The connections between the MGs and the ATAs are broadband connections such as DSL or cable modem.

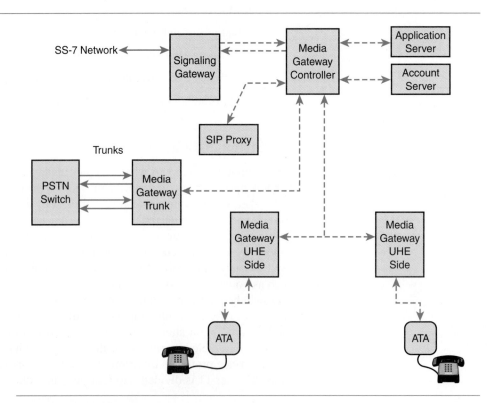

user interface or line side, the term used in traditional circuit switched telephony. The network or trunk side relates to the network interface to the network, which in this instance refers to the PSTN or other IP telephony networks. This definition compares directly to how a class 5 switch is defined. The portion facing the end user is viewed as the line side or access side of the switch. The portion facing the PSTN is the class 4 or trunk side, meaning trunks or circuits tied to the outside world.

In the world of distributed switching, the functions required to interconnect to the access device, the customer's telephone, and the functions required to connect to the outside network may reside in one box. The network architecture is decided by the network engineer during the planning phase of the design. The MG may therefore be used to connect to multiple carriers' networks, similar to a tandem switch with trunk connections to a list of carriers. An MG may also be used to interface with the line-side or access-side connections tying the end user's telephone to the network. Finally, the MG may be capable of handling both, particularly in a situation in which it is acting as an IP PBX for a corporation using VoIP for its internal network. Figure 11–2 shows a typical network design using an MG for line-side terminations and network interconnection.

The MG, though very important in the VoIP telephony world, is not able to establish packet flows from one interface port to the next. Simply stated in voice terms, the MG does not perform digit recognition, switch translations, or maintain routing algorithms. The MG is not able to determine if the trunk is available, which trunk provides the least cost route, and so forth. Additionally, the MG is not responsible for setting up and tearing down a call, accessing applications such as voice mail, or handling events such as network outages. All of these functions are handled by the MGC (Media Gateway Controller). The MG is supervised by an MGC that performs the standard class 5 functions. The MG, though, is responsible for processing the packets according to the configuration defined in the class 5 switch. For example, the MG performs voice compression such as *CS-ACELP*, referred to as G.729. It also processes the packets according to the IETF standard. In addition, the MG may have echo cancellation devices built in to reduce the effects of echo, a common problem experienced in VoIP telephony circuits.

CS-ACELP
Conversion, compression algorithm used to convert analog information into a digital signal. Referred to as the G.729 standard.

Media Gateways are distributed across a service provider's network. For example, in order to service customers located in Los Angeles, an MG is located within the Los Angeles region. When a customer picks up his or her phone and places a call, the flow is directed to the nearest MG, which in this instance is the MG in Los Angeles. If a person living in Hoopa, California, picks up his phone and dials a friend in Red Cloud, Minnesota, his call is directed to the nearest gateway, the Los Angeles MG. Network planners must determine where to place MGs in the network to reduce the time it takes for the call flow to reach it. Too many hops can cause too much delay because of the time it takes for a packet to flow from mouth to ear, a topic discussed later in the

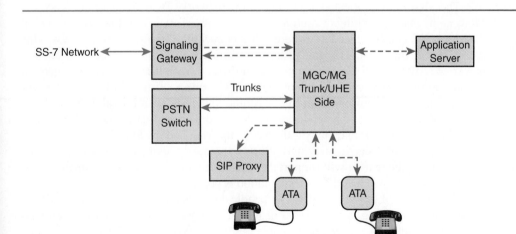

Figure 11–2
A call agent, also called a soft switch or an MGC, may be combined with the residential MG and the trunk MG to provide VoIP service on a small scale, such as for a corporate internal VoIP network.

chapter. Luckily for the planner, most gateways fit easily into one seven-foot relay rack or a portion of the relay rack. This gives the planner the option of leasing space in a carrier hotel, eliminating the need to build out a large switch site; or, if the carrier has its own POP, installing the gateway will not require additional space buildouts.

The volume of traffic an MG can handle varies depending on the vendor. Typically, MGs are designed to handle from a few hundred to twenty or thirty thousand concurrent calls. An MG is modular in design, allowing the engineer to start out small and add on to the device in the future. This allows smaller carriers to enter the marketplace without having to shell out hundreds of thousands of dollars up front. In fact, some very smart networkers have developed their own MGs and MGCs in their garages, offering scaled-down VoIP services. Though rare, there is open source code available for this function.

11.1.2 Media Gateway Controllers

The Media Gateway Controller (MGC) is also called a call agent or a soft switch. Its main purpose in life is to direct the MG, telling it what to do and when to do it, through the use of commands carried on control links. Signaling protocols are used to carry information between the MG and the MGC. The MGC rules over the MG in the same way the administration complex of a traditional class 5 switch rules over the switching modules. The relationship is really no different, other than the fact that MGs may reside hundreds of miles away from the MGC—distributed throughout the network. Additionally, since the MG is nothing more than a sophisticated proxy server, the two devices may be manufactured by different vendors. The software loaded into the servers is what defines the functionality; the hardware is standard.

The MGC responsibilities, as defined by the IPCC (International Packet Communications Consortium), relate directly to class 5 switch functions. They include digit interpretation; routing decisions through translation tables; line treatments such as message responses; and, most important, call setup and call teardown using signaling protocols such as SIP—Session Initiation Protocol—or Megaco (H.248). In summary, the MGC is responsible for call setup and teardown, managing traffic flows between networks, and controlling the multiple MGs in the network. You may want to view the MGC as the queen of the network who is responsible for all her subjects. She tells them what they need to do and when. If she becomes ill, the subjects sit around doing nothing.

application gateways
An application gateway's primary function is to provide voice mail and other application services.

The MGC also interfaces with *application gateways* that interface to application servers such as voice-mail systems: IP Centrex, Hosted PBX, Unified Messaging, and so forth. Beyond the application servers and the MGs, the MGC ties to the Signaling Gateway (SG) in the same way a class 5 switch has A links to the SS-7 network. Every queen needs her couriers, which—in the case of the soft switch architecture—the signaling protocol provides. Overall, the MGC is the center of the VoIP telephony network.

The MGC can be located anywhere in the network. Typically, carriers place the MGC at the main switching or routing center for convenience; but it needs to be noted that the MGC can be placed anywhere it has connectivity to the MGs and SGs. This means that if the engineer's garage has interconnection to the Internet and enough power, the MGC could technically be placed there, though it is unlikely management would approve. Most carriers deploy two MGCs, one a primary MGC and one for backup. The space required for an MGC, similar to the MG, is minimal; it fits within a seven-foot relay rack. Some vendors have designed devices that contain all functions needed to provide VoIP telephony—the MGC, the MG, and the SG. The value of these devices is obvious. A carrier may start offering voice service in an area on a small scale to determine if the market will support the investment; or, if the market is small, an all-in-one small soft switch and MG combination is perfect.

BHCAs
Busy hour call attempts define the number of calls a switch can handle in one hour.

The volume of traffic an MGC can handle depends on the manufacturers. MGCs are specified according to the number of *BHCAs* (busy hour call attempts) it can handle and the number of subscribers it is able to service. Typically, today's MGCs handle from a few thousand BHCAs up to a million. The MGCs are specced to service as few as 30,000 subscribers up to and above 300,000 subscribers. Most of the MGC vendors are quoting values within this range, and all are promising higher BHCAs and subscriber numbers in the near future. Some are saying they will be able to handle above 2 million BHCAs and as many as 400,000 subscribers. The value the distributed switching architecture gives the network engineer is its modular design, its low space requirements, and the lower power requirements than a full-blown carrier switch. An MGC can support a customer base of a few hundred to over a million.

11.1.3 MG and MGC Placement

Most large VoIP carriers place MGs strategically throughout the country. The MG talks to the MGC using a control protocol such as MGCP—Media Gateway Control Protocol. It is important to understand that a call carried on the VoIP network is not routed physically to the MGC. The MGC has control links to the MGs. The control links are similar to the SS-7 network in that they carry control information, not call traffic. The types of links used to connect the MGC and the MGs vary depending on the provider. Most establish redundant dedicated circuits between the devices. In certain cases, the Internet is used to connect the MGC and MGs, though this method is less desirable.

The MGC manages the traffic flows by telling the MG how to route the packet. For example, a call made in Los Angeles destined for New York City flows from the MG in Los Angeles and is directed to the MG in New York City, where it is handed off to the PSTN through a trunk connection, as shown in Figure 11–3. The MGC told the MG to ship the packets on interface port 1, which links Los Angeles to New York. It also told the MG in New York to ship the packet on interface port, which connects to the local ILEC's tandem switch. Figure 11–3 illustrates how the two devices interface with each other.

11.1.4 IPCC Definitions

SC-F—Signaling Conversion Function—is the term the IPCC has given signaling tasks performed by a Signaling Gateway. The SG routes signaling messages between devices such as STPs—signaling transfer points—and MGCs. The SG is also responsible for

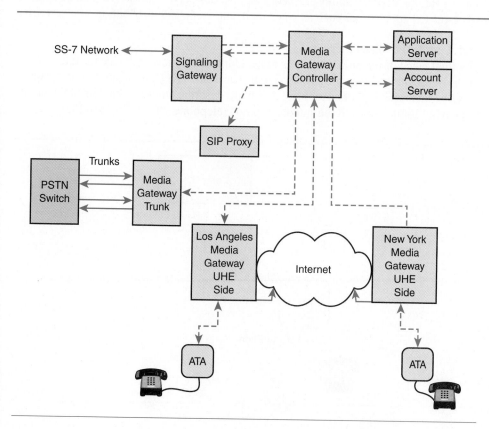

Figure 11–3
A call flowing from Los Angeles to New York City is set up by the MGC. The call path flows from the MG in Los Angeles to the MG in New York.

converting signaling protocols such as MGCP to SS-7, and SS-7 to MGCP, or SIP to SS-7 and so forth. The SG is typically placed at the same location as the MGC and trunking gateway, though the SG may reside anywhere in the network. Similar to the SS-7 network, redundancy is a must in order to guarantee network reliability.

The IPCC has also defined three additional functions that are used in a distributed IP network. The first, called AS-F or Application Server Function, works with a Signaling Conversion Function, called an SC-F, to handle applications such as voice mail, Automatic Call Distributors, Media Services, and so forth. An Accounting Function, called an A-F, is used to handle billing transactions and accounting tasks. All three functions—AS-F, SC-F, and A-F—are used to provide ancillary services. Obviously, service providers would not be able to charge customers.

Note that the IPCC definition does not tie the function to a physical device. The SG may reside within the MGC, the AS may reside in the MGC, the trunk gateway may reside in the MG, and so forth. The functions are required in order for the call to be routed across the network. The location where that function resides depends on how the vendor designed the device.

11.1.5 Application Servers

The application server (AS) is responsible for services such as voice mail, automatic call directors, universal messaging, and on and on and on. An AS, similar to the MGC, sits in one location and talks to the MGC which then talks to the MG. In some cases, the AS talks directly to the MGs. The AS can be seen as a peripheral or auxillary box, similar to a voice-mail system found in traditional circuit switched voice networks. The beauty of the AS is that it does not have to sit in a serving central office. Each site does not need its own AS. Instead, one AS is sufficient in most networks.

In summary, a VoIP telephony network consists of multiple MGs dispersed throughout a network tied to a central MGC. The MG's trunk and line side may reside in separate boxes—even separate locations—or may reside together in one device. The network must also be able to perform a signaling conversion between the PSTN SS-7 protocol and the various VoIP telephony protocols. Application servers are used to handle specific voice-related applications such as voice mail, automatic call distribution, and so forth.

11.1.6 MTA/ATA—End Device

The final piece of the puzzle that has to be addressed is the end device used to convert the VoIP packet into an analog voice signal. The term *MTA* (Media Telephone Adapter) or *ATA* (Analog Telephone Adapter) is used to describe a device that sits between the broadband connection and the standard analog telephone set, as shown in Figure 11–4.

MTA or ATA
The Media Telephone Adapter or the Analog Telephone Adapter is the device that sits at the customer's site and is used to convert the digital VoIP signal into an analog signal.

Figure 11–4
A broadband device such as a cable modem or DSL modem interconnect to the ATA via an Ethernet connection. The ATA connects to the standard analog telephone set via an RJ-11 analog telephone port.

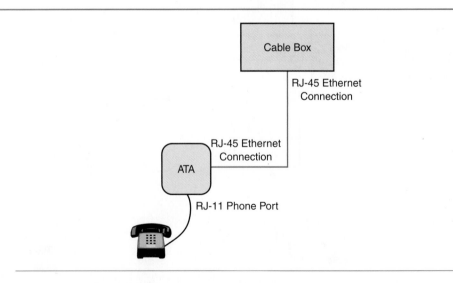

In a general sense, the MTA is an extension of the switch in that it talks directly to the MGC through the MG. The MTA generates dial tone when the MGC tells it to generate dial tone. The biggest difference between the standard analog telephone set and the MTA is the distance the signal has to travel before it is converted into a digital signal. The MTA is a host sitting on the broadband network in the same way a PC sits on the broadband network as an end device. When the telephone handset is picked up, a message formatted in an IP packet is generated and shipped across the same pipe used when surfing the Web. The packet is directed to the MG, a server sitting on the Net; but other than that, it is treated the same as any other IP packet crossing the network. (As discussed in Chapter 6, voice traffic may have tags set giving it priority over best-effort data.)

The second type of device found at the end user's premises is a soft phone. The soft phone connects to the local PC using the audio card in the PC to convert the digital signal to analog. Today, some vendors define soft phones that are ATAs essentially built into the phone, meaning that the phone connects directly to the broadband Ethernet port on the cable modem or DSL modem. In any of these cases, the IP packets arriving at the customer's premises—or, in the reverse, leaving the customer's premises—end up as or start out as analog signals. Until scientists figure out how to train our brains to perform an analog-to-digital conversion in our heads, every device interfacing to a VoIP telephony network will have to have an A to D Codec—coder/decoder.

■ 11.2 SIGNALING METHODS IN A DISTRIBUTED SWITCH ARCHITECTURE

We have discussed how traffic flows are controlled and processed by MGCs and MGs. We have also discussed how the MGC and MG talk to each other through control links. The question we need to answer in this section is how call supervision, routing information, call setup, and so forth are handled in the VoIP telephony network. Explaining how signaling is handled in this new distributed switch architecture is not an easy task, as there are multiple ways to accomplish the same goal. This section has been divided into three general areas: definition of the signaling protocols, where they are implemented, and an end-to-end example.

During the late 1990s, several methods were developed to address signaling within the IP network. At a high level, the protocols fall into one of two categories: peer to peer or Master/Slave. The term *peer to peer* refers to a protocol that gives the end point enough smarts to handle call processing—end point to end point. Session Initiation Protocol and H.323 fall into the peer-to-peer signaling category. The second group, Master/Slave, refers to a protocol that depends on a central controller for signaling functions. Master/Slave protocols include Megaco/H.248. Though being able to categorize the protocols into neat little buckets is nice, it is much too simplistic for the IP world. In reality, a VoIP call may encounter multiple types of signaling protocols on its path from end to end. Unfortunately for the student, understanding all the possible variations associated with signaling in the VoIP telephony world is daunting. Therefore, our goal in this section is to introduce each protocol and explain how it is typically deployed in the network. It is important to emphasize that our explanation is simplistic and may not include all the possible scenarios existing today.

11.2.1 Session Initiation Protocol
Session Initiation Protocol—SIP—has become one of the most popular protocols in the VoIP world. In our example, we will look at how SIP handles signaling functions to establish voice calls across the network. SIP was developed by the IETF as a multimedia application layer protocol capable of establishing sessions end to end across the Internet. SIP, unlike SS-7, is easy to decipher as it uses text messages to describe the action. As a client/server protocol, it provides the ability to handle large volumes of

Session Initiation Protocol
SIP is a signaling protocol used to set up, tear down, and handle call supervision on VoIP networks.

traffic by simplifying the functions of the application. An SIP proxy is used to establish connections across the network by establishing multimedia flows using IP routing through domain names. An SIP proxy consists of a registration and location server that holds the domain names and IP addresses to all the MGs and SGs in the network. The registration and location servers perform DNS lookups to convert the domain name in the SIP message into an IP address. The SIP protocol is responsible for setting up the session, monitoring the session during the time it is up, and tearing down the session when one end terminates.

An SIP aware ATA (Analog Telephone Adapter) establishes a call by sending out an invite message when a call is made. The SIP packet travels from the ATA to the SIP proxy, which reads the message that includes the telephone number. The MGC directs the message to the SG, which converts the message into an SS-7 ISUP. Once the route is determined by the signaling network, it is fed back through the MGC, which tells the MG how to route the traffic flow. The call path is built from the ATA through the Internet to the trunking gateway to trunks tying to the PSTN. SIP messages continue to flow in order to maintain the session. When the call ends, a Bye message is sent from the ATA through the network to tear down the call session.

One of the best ways to understand how the signaling protocol is used to establish a connection is to walk through a call flow by following the SIP message exchange, as shown in Figure 11–5. The diagram shows a ladderlike structure with messages flowing between the customer end device, referred to as the *user agent*, and the MGC; in this example, the MGC is handling the PSTN connection. A call is set up as follows:

user agent
End VoIP device located at the customer's premises.

1. User agent goes off-hook. SIP message sent is called an INVITE + telephone number and domain address. Ex. Invite 12123452929@gwspidercom.
2. The MGC recognizes the INVITE, sends message to SG that translates the request into an SS-7 message.
3. MGC sends ringing to the user agent (183 session message).
4. User agent responds with a PRACK (provisional response ACK).
5. When the end device goes off-hook, a message is sent to the MGC. MGC sends a message to user agent (200 message).

Figure 11–5
Simple SIP ladder showing how a call is established and disconnected.

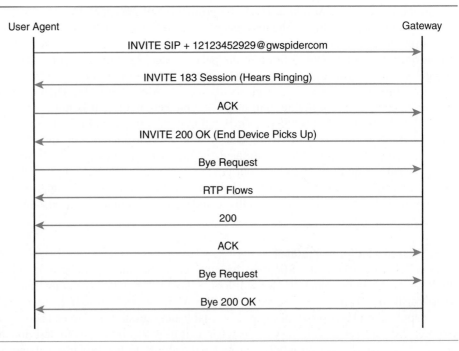

User Agent   Gateway

INVITE SIP + 12123452929@gwspidercom

INVITE 183 Session (Hears Ringing)

ACK

INVITE 200 OK (End Device Picks Up)

Bye Request

RTP Flows

200

ACK

Bye Request

Bye 200 OK

6. RTP flows are established on the IP network between the user agent and the MG.
7. User agent acknowledges the call flow.
8. End user hangs up the phone. MGC sends a Bye to user agent.
9. User agent responds with a Bye.
10. Call is disconnected.

11.2.2 Megaco/H.248 and MGCP

Megaco/H.248 is the Master/Slave protocol defined by the ITU and IETF in the late 1990s under RFC 3015. *MGCP* is a procotol proposed by the IPPC and SGCP in I-RFC 2705. The protocols Megaco/H.248 and MGCP are deployed today as control protocols used to create connections between the MG and MGC and exchange signaling information related to setting up trunk connections in the outside network—primarily, the PSTN. Though Megaco/H.248 and MGCP may be used between the end device and the controller and SIP used to establish trunk connections, the most common deployment is an SIP signaling methodology in the access network and a Megaco/H.248 and MGCP methodology in the network.

Converting Megaco/H.248 and MGCP to SS-7 is performed by the *Signaling Gateway* (SG), a device placed in the network to handle signaling functions. The SG may reside within the MGC or even the MG, depending on the design and manufacturer. In either case, similar to the MGC, the SG's function is to tell the MG what to do with the call. Traffic does not flow through the SG in the same way that traffic does not flow through a signaling transfer point in the traditional SS-7 network. Instead, it directs the conversions on the links in order to make sure that when one party hears ringing on the VoIP telephony phone, the phone is truly ringing on the phone connected to the PSTN.

Similar to how a call flows using the SIP signaling methodology, the following steps happen when an MGC talks to the PSTN using Megaco/H.248:

1. MG tells the MGC the phone has gone off-hook.
2. The MGC tells the MG to tell the end device to play dial tone.
3. The MGC tells the MG to collect the DTMF digits.
4. The MG sends the MGC the DTMF digits.
5. The MGC tells the MG to connect to such and such an IP address.
6. Connection is made.

■ 11.3 DEFINING VOICE OVER IP PACKET FLOWS: RTP, RTCP

11.3.1 RTP: Real-Time Protocol

As discussed in Chapter 6, RTP—*Real-Time Protocol*—is a transport protocol used to carry real-time traffic across the IP network. If we opened up a VoIP packet, we would see an IP header, a UDP header, an RTP header, and the 64 kbps PCM word created during the analog-to-digital conversion. As explained in Chapter 6, RTP is a pseudo Layer 4 protocol used to carry real-time applications such as voice over an IP network. The IETF developed the RTP in the mid-1990s under RFC 1889 and 1890. When a person talks into a telephone connected to an ATA, the voice signal is first converted into ones and zeros—analog-to-digital conversion, as explained in Chapter 4. The ones and zeros are encapsulated into the packet. RTP and UDP work together to establish live flows across the Internet, providing a way to carry time-sensitive traffic such as voice or video. The major function RTP performs is to organize packets in a way that improves their chance of being delivered in the right order and at the right time. As shown in the RTP header, the packet carrying real-time traffic such as VoIP has a time stamp, a version type, a way to monitor if the packet is delivered, and a way to monitor if the packet arrives in the correct sequence. The key fields important in VoIP are the version

Megaco/H.248 and MGCP
Protocols developed to handle signaling between MGs and MGCs.

Signaling Gateway
A device in the network that translates signaling protocols such as SS-7 to Megaco or SS-7 to SIP.

Real-Time Protocol
RTP is a Layer 4 protocol used to carry real-time traffic across an IP network.

Figure 11–6
RTP header contains field used to assist UDP in transmitting real-time traffic such as VoIP.

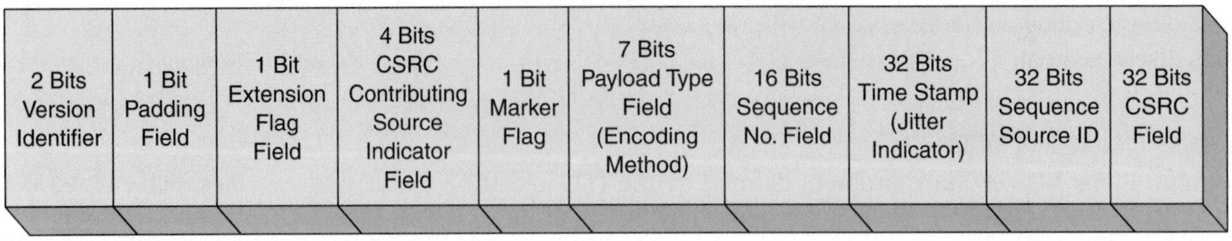

2 Bits Version Identifier	1 Bit Padding Field	1 Bit Extension Flag Field	4 Bits CSRC Contributing Source Indicator Field	1 Bit Marker Flag	7 Bits Payload Type Field (Encoding Method)	16 Bits Sequence No. Field	32 Bits Time Stamp (Jitter Indicator)	32 Bits Sequence Source ID	32 Bits CSRC Field

G.711
The encoding scheme that converts the analog signal into 64 kbps.

identifier, which notes the RTP version deployed; the sequence field, which notes the position of the packet in the flow; the time stamp, which is used to determine if the packet arrived in a timely manner; and the payload type, which indicates the type of coding such as *G.711*. Figure 11–6 shows the RTP header. Table 11–1 lists the most common encoding algorithms deployed in VoIP networks. *Note:* The difference between the encoding types is primarily the compression rate deployed. A G.711-encoded signal refers to a 64 kbps PCM word; a G.729-encoded signal carries a 16 kbps word; and so forth. The remaining fields in the header are used primarily for other real-time applications, such as video conferencing, and are beyond the scope of this chapter. The important points to grasp are that RTP is a protocol developed for the sole purpose of carrying real-time applications across an IP network and that RTP has become the protocol deployed in VoIP networks. When you hear a networking-savvy person say "RTP," you should immediately realize that he or she is probably talking about VoIP flow.

An RTP traffic flow refers to the voice information being carried from source to destination inside an RTP packet. In the world of VoIP, there are always two RTP flows—one upstream and one downstream—from the calling party to the called party and from the called party to the calling party, as shown in Figure 11–7. Each flow needs 80 kbps of bandwidth. This includes the 64 kbps voice PCM word, the RTP header, the UDP header, and the IP header, as previously explained. In total, an uncompressed VoIP call eats up 160 kbps of bandwidth from ATA to ATA.

11.3.2 RTCP: Real-Time Control Protocol

Real-Time Control Protocol
RTCP is used as a network monitoring troubleshooting tool. Overhead packets are read and interpreted by devices.

RTCP—*Real-Time Control Protocol*—is used in conjunction with RTP to maintain network reliability. RTP on its own is not able to guarantee network metrics such as latency, packet loss, and so forth. RTCP is deployed to carry network information related to the RTP flow. For example, a network technician can capture RTCP information from an ATA and determine if the RTP flow experienced delay. Though RTCP is a newer protocol, it is beginning to be deployed in carrier networks to assist during trouble resolution.

Table 11–1
Encoding Schemes.

Encoding Scheme	kbps	Also Called
G.711	64 kbps	PCM u-Law Encoding
G.726	32 kbps	ADPCM
G.728	16 kbps	LD-CELP
G.729	8 kbps	CS-ACELP
G.723.1	5.3, 6.3 kbps	ACELP

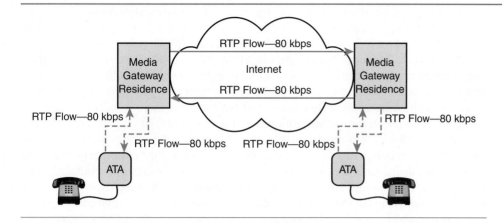

Figure 11–7
RTP flow requires 80 kbps of bandwidth in one direction: 80 kbps path from sender to receiver and 80 kbps path from receiver to sender. One call between two people eats up 160 kbps of bandwidth across the network.

■ 11.4 APPLICATIONS AND CONNECTING THE DOTS

11.4.1 Measuring Quality

Quality of voice service is the primary reason skeptics have dismissed the possibility that VoIP would overtake traditional land line telephony. Due to this contentious issue, engineers have defined quality metrics that can be used to measure voice service. The primary measurements taken on a voice line are delay/latency, jitter, packet loss, sequence errors, and echo. Signaling metrics have also been defined for VoIP calls. They include post dial delay and initial response time.

MOS: Mean Opinion Score

The industry has adopted the standard *MOS* or R score to provide an overall measurement of a call's quality. The MOS takes into account jitter, delay, packet loss, and so forth to calculate a value that can be used to rate the quality of the call. The ITU-T P.800 recommendation specifies a five-point quality scale on which 5 is excellent, 4 is good, 3 is fair, 2 is poor, and 1 is bad. However, it is difficult to assign a definite MOS value range, as measurement tools use different algorithms to calculate the end score. In most situations, a MOS above 3.5 is considered a "good" quality call. A MOS such as 2.5 is considered poor. Before interpreting a MOS given to you by a carrier, ask the carrier to give you the algorithms' minimum and maximum MOS values.

Programs such as Ethereal capture RTP traffic flows and use the header information to formulate quality-of-service measurements. For example, Ethereal notes jitter, delay, latency, and packet loss, giving the technician a way to determine where in the network the trouble is happening. As mentioned earlier, RTCP also provides enormous amounts of call quality data that can be used when trying to troubleshoot a voice quality issue. Other programs similar to Ethereal have been developed by vendors and are often purchased by carriers to assist in maintaining a reliable network.

Latency, Jitter, Packet Loss, and Echo

The four most common problems associated with VoIP service are choppiness, dead air, echo, and talking over. Each can be associated to the quality metrics already discussed. A call traversing the network should not experience more than 150 ms of delay from mouth to ear. If it does, the user will experience what he or she perceives as delay. Jitter, as well as delay, may also result in talking over the person on the other end. Typically, network engineers do not want to see delay over 150 ms or jitter higher than 10 ms. Packet loss may also cause voice choppiness or dead air. Most networks require less than 1%

MOS
The mean opinion score provides a quality measurement for voice signals.

packet loss over the length of the call. In some cases, a call may show less than 1% packet loss even though the customer experiences dead air due to the loss happening all at one time. A burst of loss is much more harmful to voice quality than loss spread across the length of the call. Sequence errors—packets arriving out of sequence—may result in the end user experiencing dead air. The cause of the sequence errors should be investigated, as it may point to a device or routing problem within the network. Echo is one of those phantom troubles that cause technicians to pull their hair out. Echo is defined as near-end echo, which is echo heard by the customer off the VoIP phone, or far-end echo, which is echo heard by the person on the far end. In either case, echo is very difficult to troubleshoot. Echo may be caused by a speakerphone or headset at the customer premises but is more often the result of a two- to four-wire conversion in the network resulting in a reflection back to the originating source. Echo cancellation devices are deployed in MGs to assist in nullifying out the echo effect caused by two- to four-wire hybrids. Additionally, it is important to keep packet loss and delay in check as both can cause echo.

11.4.2 Quality Assurance

QoS/CoS

Quality of service or *class of service* refers to placing priority levels on the packets.

QoS, quality of service, or *CoS*, class of service, are the terms used to define mechanisms that combat the previously discussed problems of delay, jitter, and packet loss. The most popular way to add QoS to voice-call packets is to use priority tags attached through the Layer 3 IP precedence bits. The bits flagged as precedence are located in precedence fields of the IP header. Several terms are used to define priority tagging in VoIP: Diff-Serv, CoS, QoS, and DSCP. Though each term refers to slightly different functions, in general, they all refer to Layer 3 priority tagging for real-time traffic. It is common in the industry to hear the terms used interchangeably, even though they technically have different meanings. For our purposes, the main point to remember is that the packet has been tagged as priority traffic.

When a packet has a QoS tag, it is given priority only if the equipment receiving the packet recognizes the tag. That means if a voice call is made between New York and Los Angeles and a portion of the call rides on the Internet, it is unlikely the priority tag will be adhered to throughout the network. However, if a call travels on an internal network, it is easy to ensure that the devices on the network recognize the priority tag.

In this instance, the IP packet is pushed out onto the circuit in an expedited manner, thus the term *expedited forwarding*. Traffic such as data traffic is viewed as best effort and is treated as such. For example, if you have been given VIP passes to a concert, you will not have to wait in queue to enter the building. The doorkeeper will allow you to move to the front of the line and enter the auditorium first. The people holding standard concert tickets will be held back until you are through the entrance and seated. Setting the priority bit in the IP packet is the same as handing the packet a VIP ticket.

You may question why engineers devised a way to tag a packet carrying voice with a QoS tag. IP, unlike traditional TDM protocols, is a fast-packet technology that cannot guarantee exact delivery times as voice packets compete for bandwidth with other packets traversing the network. Unlike data, voice requires a real-time connection as is obvious because voice needs to flow in sequence and within a set time interval.

11.4.3 Switch Features

Similar to a class 5 switch, the soft switch or MGC provides features such as call return, call waiting, and so forth. Most MGCs today are capable of providing all the standard features, as defined in Chapter 9. The CLASS features are available, as are voice mail and unified messaging.

Where Features Reside

Most features reside on the MGC or reside on a peripherial server controlled by the MGC. Features are developed by software engineers that try to mimic traditional class

5 feature sets. Where they reside is less important in a distributed network. In addition, services such as Web services or unified messaging reside on application servers that interconnect to the MGC. The MGC is the gatekeeper for all functions on the network, including auxiliary services such as voice mail.

Feature Categories

Features offered on the VoIP network mimic those deployed on the PSTN. CLASS features such as call waiting reside on the MGC and are built to function in the same way as traditional class 5 feature sets. Enhanced features such as unified messaging and Web services are simple to initiate, as they reside on application servers that tie directly to the MGC.

11.4.4 End-to-End Call Flow

The following two sections provide simple examples of an end-to-end call flow. The first illustrates what happens when a call is made from an ATA on an IP network to a telephone hanging off the PSTN. The second example walks through a call made on an IP network terminating on an IP phone on the same IP network.

PSTN to IP End Point

Joe, a Spidercom VoIP user, picks up his handset and initiates a phone call to Carol, as shown in Figure 11–8. The SIP message is sent to the SIP proxy, which looks up the domain name in the registration and location server. The server provides the IP address of the MG—trunk gateway tied to the PSTN. The MGC reads the request, talks to the SG to determine which trunk the call has to be routed on, and begins the call setup process. The SG converts the SIP into an SS-7 protocol such as ISUP (ISDN User Part).

Figure 11–8
A call traveling between Joe and Carol traverses the IP network as RTP flows and the PSTN as traditional DS-0 64 kbps voice traffic. Shaded boxes exist in the PSTN. Nonshaded boxes make up the VoIP network.

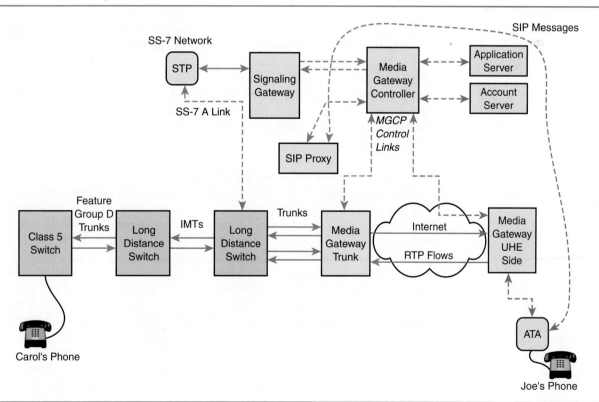

Figure 11–9
A call made between customers on the same network travels between MGs to the end devices.

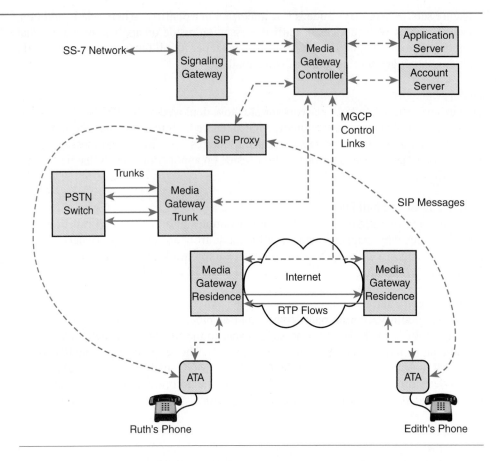

The SS-7 message travels across the SS-7 network as explained in Chapter 10 and arrives at the end-office class 5 switch feeding Carol's phone. The class 5 switch tells the phone to ring and returns a ringing message to the SG via the SS-7 network. The MGC relays ring tone to Joe's phone. Carol picks up the handset. Once the handset is picked up, the class 5 switch returns an SS-7 message through the SS-7 network to the SG telling it to establish a connection. The SG nails up the port tied to the PSTN and the port tied to the IP network, allowing the RTP flows to pass. It also lets the MGC know the connection has been established. The end device receives the SIP message, and the RTP flows begin. In simple terms, the conversation between Carol and Joe begins, as shown in Figure 11–8.

IP End Point to IP End Point on the Same IP Network

A caller, Ruth, using Spidercom's VoIP service, calls her friend Edith, also a Spidercom VoIP user. Ruth picks up her handset, hears dial tone, and punches in the DTMF digits. The SIP proxy reads the digits and domain name and determines that the call needs to be directed to the MG serving Edith's phone. The MGC directs Ruth's MG to set up a path between itself and Edith's MG, as shown in Figure 11–9. The SIP messages initiate ring tone on Ruth's phone. Edith picks up the handset, which causes the SIP message to run through the network telling the MGC to open the path for a call session. The MGC directs the MG to route the RTP flows between the two MGs and out to the end devices—Ruth's ATA and Edith's ATA. Though simplistic, the call path between devices on the same network is fairly simple. In a sense, it is similar to traffic traveling across a LAN. The ATA is nothing more than an IP aware device similar to a PC. The SIP messaging is telling the ATA what to do, that is, initiate ring voltage that rings Edith's phone. The MGC controls the flow of the call by watching the SIP messages, and the SIP proxy uses its registration and location function to build a path across the network.

SUMMARY

The change from the circuit switched architecture to the distributed switching architecture associated with IP voice is one of the most dramatic changes in telephony since the first digital switch was introduced the late 1980s. No longer is it necessary to place a large, expensive class 5 switch a few miles from the end users' homes. The customers may connect their telephones to a device attached to a CATV broadband connection and receive dial tone. The Internet now has become more than a World Wide Web of data servers. It is now a network used to carry real-time voice traffic around the world.

CASE STUDY

Hector, an independent filmmaker, has set up shop in Green Grass, hoping to make it the next Independent Film Capital of the world. He immediately sees an opportunity to make some money to fund his film endeavor. He wants to set up a Voice over IP network in Green Grass that will piggyback on the new broadband network being deployed. In order to do this, he will need to design a distributed switching network that will provide class 5 functionality to the end customer and connect to a larger VoIP carrier. You have been asked to put together a design packet that Angel, Hector's younger sister, will use to pitch the idea to investors.

Use the information in Appendix A to complete the design. Remember, you are only responsible for putting together the class 5 side of the network and assessing how you would connect to a larger VoIP carrier that handles the class 4 trunks.

REVIEW QUESTIONS

1. The Media Gateway is responsible for setting up and terminating calls.

    a. true   b. false

2. The MG has to be placed in the same location as the MGC.

    a. true   b. false

3. List two names used in place of MGC.

4. Explain why a company purchases two MGCs even when call volume does not warrant the purchase.

5. List two advantages the distributed switch architecture has over the traditional voice network.

6. An MTA provides all functions except:

    a. terminating the broadband connection
    b. converting the digital signal into an analog signal
    c. connecting to the analog telephone set
    d. connecting to the local central office switch
    e. none of the above

7. The SIP handles what functions?

    a. establishes call connection through the network
    b. tears down a call through the network
    c. rings the phone
    d. applies features to the line
    e. converts the IP signal into a G.711 flow
    f. a, b, and c
    g. a and b
    h. a, b, c, and e
    i. all of the above

8. Megaco/H.248 and MGCP are primarily used to set up trunks between Media Gateways.

    a. true   b. false

9. What OSI layer does RTP fall under?

10. What Layer 4 protocol does RTP team up with when carrying VoIP traffic across the Net?

11. RTCP is used to watch and evaluate an RTP flow.

    a. true   b. false

12. The key parameters measured to determine if the call flow meets standards are:

    a. jitter
    b. latency
    c. packet loss
    d. MOS

e. a, b, and d

f. a, c, and d

g. all of the above

13. MGCs typically do not support class features.

   a. true   b. false

14. Voice mail is commonly handled by the _____ server in distributed switch architecture.

15. A call made between Philadelphia and New York City may travel on which of the following networks?

a. IP to PSTN

b. IP to IP

c. PSTN to IP to PSTN

d. a and c

e. all of the above

f. none of the above

KEY TERMS

Media Gateway (282)

Media Gateway
  Controller (282)

CS-ACELP (283)

application gateways (284)

BHCAs (284)

MTA or ATA (286)

Session Initiation
  Protocol (287)

user agent (288)

Megaco/H.248 and
  MGCP (289)

Signaling Gateway (289)

Real-Time Protocol (289)

G.711 (290)

Real-Time Control
  Protocol (290)

MOS (291)

QoS/CoS (292)

12

IP Routing Fundamentals

Objectives

After reading this chapter, you should be able to

- Define IP aware devices.
- Define static routing.
- Define dynamic routing.
- Define IP addressing.
- Define routing across a LAN.
- Define routing across a WAN.
- Define routing across the Internet.

Outline

Introduction

■ INTRODUCTION

IP—Internet Protocol—is the most widely deployed Layer 3 protocol around the world. It is what routes information across the World Wide Web, making it possible to send information to any corner of the globe. Today, the section on switching has to include a chapter on IP routing as the distributed switching architecture is built on an IP network. No longer is IP just a data transport protocol used to carry Internet and LAN traffic. It has been adopted by networking professionals around the world as a way to carry communications flows to many types of devices ubiquitously. *Ubiquitous* is a favorite term of telecom marketing professionals; it refers to something that exists everywhere—in our terms, a protocol that can be used everywhere for all types of applications and services.

Chapter 12 presents IP as a protocol used to route packets between devices, not just a protocol used to carry traffic across the Internet. It begins by describing IP routing devices or interfaces. Next, it looks at how packets are shipped between the devices through static or dynamic routes. Following the discussion of routing methods, we look at IP addressing at a high level as there are many texts devoted to the subject of IP addressing. The chapter concludes by walking through three common routing scenarios: routing within a LAN, routing between a LAN and a WAN, and routing across the Internet.

■ 12.1 ROUTING DEVICES OR INTERFACES

IP aware—also called Layer 3 aware—devices are used to look inside an IP packet and determine through the use of rule sets what port to send the information out on. The most common IP device found in the network is the router. The important concept to grasp from this section is that there are multiple router configurations, interface types, and CPU or horsepower and management control interfaces. Some routers are used to connect small LANS through 10BaseT Ethernet ports. Other routers are used by Tier 1 ISPs to connect to NAPs—network access points—on the Internet. In either case, the router's main job in life is to open the packet and figure out how it should be handled and where it needs to go. Before describing the two basic types of routers—core and noncore—we begin with an overview of the in and out interface ports found on all routers.

12.1.1 Interface Ports

The term *interface port* is used to describe the physical port that the outside medium connects to. A router may have a 10/100 Ethernet port, a DS-3 interface port, a fiber OC-3 port, and so forth. It is like a door allowing traffic to enter and exit. Each door or port on a router can be configured according to physical layer transport requirements. For example, if a router has a DS-1 interface port tied to the outside network, it can be configured as an ESF—extended superframe—DS-1 or as an SF—superframe DS-1. The port is also configured according to the Layer 2 protocol used in the network. Often, DS-1 interfaces use frame relay as the Layer 2 protocol, setting up VLANs between one router and a second. A DS-3 interface may use ATM as the Layer 2 protocol, while a 10/100 Ethernet connection will use, of course, Ethernet.

Ports are also viewed as *ingress* or *egress,* depending on the location of the router in the network. A router sitting on a LAN has multiple ingress ports that tie to the devices—stations—sitting in the hundreds of cubes in the office. The egress port ties to the outside world, such as a DS-1 tied to the ISP. In contrast, a network router used to connect multiple DSLAM uses DS-1 interface ports on the ingress side of the router and an OC-3 interface port on the egress side connecting to the Tier 1 ISP. As with all facts related to IP, there is no one concrete design rule. A port's position in the network is determined by what it is tied to, not by the type of circuit connected to the interface.

A good way to understand this concept is to compare the router to a post office. A post office has a front door where customers walk in and deposit their mail. The post office also has a loading dock in the back that receives and sends mail to/from the "big post office network." The router is the same. It sends and receives packets from devices

---

ubiquitous
A term often used in telecommunications to describe an application that is everywhere, available throughout the network.

interface port
Term used to define the physical port on a router that ties to the circuit.

ingress
Refers to the port where the traffic flows into the internal network.

egress
Refers to the port that the traffic flows out of into the outside network.

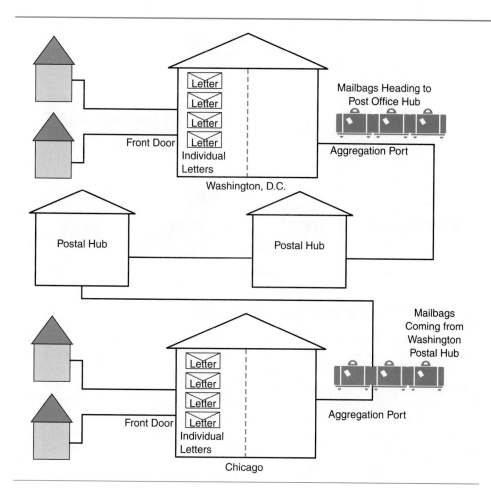

Figure 12–1
IP packets arrive in one door and are packaged together and shipped to their destination through a back door.

on one side, aggregates them into larger mailbags, and ships them out into vehicles destined for the post office hub, which is linked to other post office hubs, as shown in Figure 12–1. A post office hub selects the best path for the mail to travel to the destination post office—the post office serving the end recipient. Similarly, a network router, like the post office hub, selects the best route for the packets to travel on to the destination router—the router connected to the end device or end recipient.

12.1.2 Two Types of Routers

In general terms, routers can be categorized as core and noncore, network and host, gateway and host, or border and nonborder. A network or core router is one that connects to other core routers such as, in our analogy, one post office hub connecting to a second post office hub. The network router is aware of the outside world's network. The host router is only aware of the devices it services. The one loading dock or interface port connecting to the big post office network is the only port it knows about; and, therefore, the host router is not responsible for knowing how to route to other networks. A noncore or host router is one that connects to the hosts—PCs and other stations— and to a core router, as shown in Figure 12–2.

The core or network router, on the other hand, may have multiple egress or loading docks of all different interface speeds—a DS-3, a Gig-E, an OC-48, a DS-1. The core router has to be smart enough to know which port to send the packet onto under specific conditions. The core router has to be aware of the network beyond its borders in order to route traffic properly. In fact, most routing protocols define core routers as border and noncore routers as nonborder. A core router is one that connects to other networks, as shown in Figure 12–3.

Figure 12–2
Routers interfacing to host devices are
referred to as host routers. Routers
connected to the network are called
network routers if they interconnect to
an internal network and *core* if they
interconnect to the outside world's
core network.

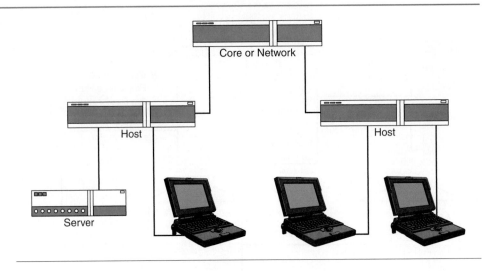

■ 12.2 ROUTING VERSUS ROUTED PROTOCOLS

It is important to be able to understand the difference between a routed and a routing
protocol. A protocol such as IP or IPX—Internetwork Packet Exchange—is used to
carry data between networks using addressing schemes, packet header information,
and general Layer 3 rules as defined by the OSI model. A routing protocol is a tool set
of sorts that enables routers to calculate the paths the packets will follow. Routing protocols are configured in a router to direct the packets according to algorithms and rule
sets. IP may use any of the several protocols to direct traffic flows accordingly. Common routing protocols include RIP, OSPF, IGRP, BGP, and so forth. Routed protocols
include IP, IPX, AppleTalk, and others.

Figure 12–3
Routing between networks requires
host routers to interface with end
devices. The host router also
interfaces with a core or network
router to connect to other networks.

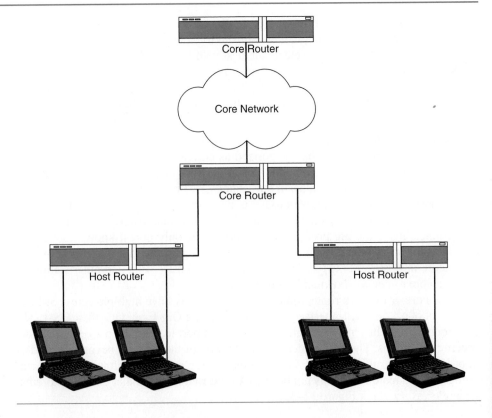

```
RouterC<enable
RouterC#config t
RouterC<config>#ip route 110.12.10.0 255.255.255.0 10.0.0.1
RouterC<config> 110.12.10.0 255.255.255.0 10.0.0.1
RouterC<config>#ctlz
```

Figure 12–4
Configuring a static route on a Cisco router is fairly simple. The commands needed to set up the route are shown here.

■ 12.3 STATIC ROUTING METHODS

Static routing is the simplest way to establish a path between two routers. A static route is, as it sounds, a route nailed up—tacked up—manually by a human being smart enough to log into the router. Static routes are configured using command line interface strings that are entered prior to turning up the interface port. The interface port directs traffic according the static route, similar to a railroad track carrying traffic between two locations. The router does not need to figure out what sits in the world outside. It does not care. Someone was nice enough to build a path to the far end device and take the burden of deciding how to best route the packet. Static routes, though appealing in their simplicity, are not scalable and often become more cumbersome to manage as the network grows. Figure 12–4 shows the command line used to build a static route in a router using Cisco IOS command line interface.

■ 12.4 DYNAMIC ROUTING METHODS

Dynamic routing is, as it sounds, a way to route packets according to specific rule sets and algorithms defined by the routing protocol. Routing protocols are smart enough to know what port to ship the packet out on by analyzing the IP address inside the packet and comparing it to information stored in routing tables. At a high level, routing protocols are categorized as internal or external. Internal gateway protocols such as OSPF and RIP exchange routing information on the internal network such as from device to device—router to router. External gateway protocols such as BGP and EGP are complex routing protocols capable of exchanging network maps with external networks. Tier 1 ISPs exchange routing tables by employing EGP routing protocols.

Dynamic routing protocols fall into three categories: distance vector, link state, and external gateway protocols. Each follows algorithms developed to improve the IP's ability to carry traffic efficiently across multiple networks within a specified time frame. The algorithms differ in the rules they use to accomplish this task, though the end goal is the same: ship the packet on the most appropriate path at that moment in time—dynamic routing. The four key concepts associated with dynamic routing protocols are (1) the type of decision tree used to select the "best" route, (2) the amount of routing information exchanged between devices, (3) how often the information is exchanged, and (4) the number of hops or devices allowed between source and destination.

12.4.1 Internal Gateway Protocols
Distance Vector Routing Protocols
Distance vector routing protocols use distance as the primary metric or rule set when determining which route to select. A router set up to use the distance vector protocol called RIP—routing information protocol—sends out a request asking for the number of hops, clicks, or ticks between it and the end router as shown in Figure 12–5. Router 2 sends information to router 1 about router 3. Router 1 knows that router 3 is one hop away and that router 4 is one hop from router 2 yet three hops from itself and so forth. Router 1 may then look at its second interface port to discover the distance it is from the final destination. In order for a distance vector protocol to work, a router must be able to communicate with its neighborhood. Distance vector protocols are not typically used in large, multiprovider networks.

Distance vector protocols eat up *cycles,* a term used to define the amount of CPU or horsepower required by the router to execute the function. The two most common

cycles
Computing cycles within a device.

Figure 12–5
Routing protocols deploying distance
vector routing algorithms count the
number of hops referring to devices
between the initiating router and the
terminating router.

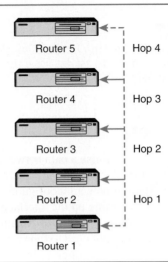

convergence
Process in which routing
tables converge or update
each other with new network
routing information.

distance vector protocols are RIP and Cisco's IGRP—Interior Gateway Routing Protocol. RIP updates its routing tables every thirty seconds, placing a load on the router every time a change occurs. In contrast, IGRP updates its routing tables every ninety seconds. However, both protocols exchange their entire route map or table, not just the changes, thus eating up bandwidth between devices and again placing stress on the router's processor. The term used to define the amount of time required to exchange routing table information is *convergence* time. Convergence is defined as the time it takes for a network to adjust to changes such as adding a new device or route. The longer it takes for a network to converge, the greater chance traffic is compromised and, consequently, transactions lost.

All routing protocols define a maximum hop count allowed between source and destination. A significant difference between RIP and IGRP is hop count. RIP can handle fifteen hops, meaning fifteen devices between the source and destination. IGRP allows 255 hops. Network designers need to make sure the network fits the routing protocol or vice versa.

Link State Routing Protocols

Link state routing protocols make up the second category of internal gateway protocols used to route IP packets. A link state routing protocol, unlike distance vector protocols, gathers a full view of the network, beyond its nearest neighbors. It holds a topological map that shows what is sitting where. From this map, it is able to determine what port to ship information out on according to the link state metric—distance, hops, least cost, and so forth. It does not count hops, ticks, or any other state before deciding what to do with the packet. Instead, it reads the map—again, similar to a navigator plotting his or her path.

The most common link state protocol is OSPF—open shortest path first. OSPF uses the SPF—shortest path first—algorithm to establish routes. It accomplishes this by first exchanging an LSP—link stat packet—that travels neighbor to neighbor until the area is flooded. Using the information gathered from the LSP, topological maps are built giving the router a view of the network, neighbors, and beyond. Once the maps are drawn, a network tree is built showing the SPF paths branching out from the source showing every router on the network. Every router builds a tree with itself as the source. The best path through the network is stored in the routing tables and used to route packets accordingly.

Link state protocols send routing information only when changes have occurred using the LSP, as explained. Convergence times are reduced because only the changes are exchanged, not the entire route map. In addition, bandwidth is spared as the LSP is transmitted to each router as a note of sorts explaining the changes. Routers update the routing tables and adjust the best route lists whenever a change occurs. In most networks, link state protocols define the best routes according to the shortest path and least cost. When one of the variables changes, a routing update is required. As you can imagine, a dynamic network will experience frequent routing table updates as the network and internetwork change. The reduction in the amount of information exchanged during updates makes link state routing protocols such as OSPF a better choice than distance vector protocols when dealing with large networks.

The number of devices or the hop count between the source and destination is not as easy to define when looking at a link state protocol like OSPF. OSPF is able to route according to areas or network segments. Though beyond the scope of this text, it is important to understand that link state routing protocols use sophisticated routing algorithms that allow them to route within an internal network referred to as an Intranet and an external network referred to as an Internet.

Hybrid Routing Protocols

Hybrid routing is the third category of internal gateway protocols. Hybrid routing protocols use the best of both distance vector and link state algorithms. The hybrid routing protocol uses distance and a topological network tree to form the algorithms rule sets. The advantage of combining the two protocols is an improvement in the way routes are selected, offering efficiencies in convergence times and CPU cycles. Common hybrid routing protocols include Cisco's proprietary EIGRP—Enhanced Interior Gateway Routing Protocol—and IS-IS—Intermediate System to Intermediate System.

12.4.2 External Gateway Routing Protocols

Border Gateway Protocol—BGP—connects gateway routers to other service providers' gateway routers through intelligent, "dynamic" route maps. BGP was developed as a way to simplify routing between networks often referred to as internetworking. ISPs exchange route maps that allow each border router to know the network beyond its borders.

This is accomplished by linking routers together according to domain, also called *autonomous system numbers,* a subject discussed later in the chapter.

autonomous system number
Number given to a network.

■ 12.5 CONNECTING THE DOTS

In summary, the function of the routing protocol is to query adjacent nodes for routing information, compile network maps from the responses, and put the maps into the router's memory. Routing protocols differ in the way they determine what the best route is between source and destination, typically determined by choosing the path with the least number of hops. The term *hops* is used to define the number of devices the packet has to travel between the source device and the destination device. For example, RIP—routing information protocol—selects the route that has the least number of hops between the source and destination, but OSPF—open shortest path first—uses a much richer algorithm that could select the route by number of hops, least cost, link speed, and link congestion. Deciding which protocol to use depends on the type of network being deployed, the design engineer, and the skill set of the workforce.

Simply, the routing protocol is similar to the address book in a post office that tells where every person lives, the best path to take to reach the person's house, and who should deliver the mail to that location. The address book is updated when new customers move into town or when a new ZIP Code or new city or town is added to or deleted from the post office network. Routing tables are also updated when new devices are added to or deleted from the network.

■ 12.6 COMMON TERMS USED TO DEFINE ROUTING ISSUES

Implementing routing protocols in networks is one of the most complicated tasks that technicians and engineers perform. Several common problems relating to routing protocols need to be defined, as they are discussed frequently in networking circles. The first is referred to as *route flapping.* Route flapping occurs when a routing change happens too frequently due to a router being misconfigured. A second common term,

route flapping
Trouble condition caused by too frequent route changes.

routing loop
Trouble condition that occurs when route table changes do not happen in a timely manner.

routing loop, is typically associated with distance vector routing protocols. Routing loops occur when a device does not receive a route table change in a timely manner. A distance router has not received the update, causing it to send its packets onto the wrong route. It is somewhat like a driver turning right instead of left causing him or her to become lost, driving around back roads indefinitely. The packet experiences the same phenomenon. It reads an old map, turns according to what the old map indicates, and ends up wandering around the network. Networks do not like aimlessly wandering packets, as they (1) eat up bandwidth and (2) cause confusion. Networks like things in order—very fussy fellows.

In summary, routing protocols are used in IP networks to efficiently route information between routers and between networks. Routing protocols can be defined according to the method they use to select a route, the amount of information and time it takes to exchange routing information, convergence time, and the number of devices or networks the protocols can accommodate.

■ 12.7 IP ADDRESSING

IP addressing is a subject most people try to avoid. One reason that people run when you mention the subject is the changes made to accommodate an exponentially growing Internet. If you have ever asked a carpenter if he or she would rather build a brand new house or put an addition on an old house, you would more than likely hear "Build a brand new house, please." The same is true of IP addressing. In the early 1980s, when IP addressing was first standardized, it was believed that the addressing scheme would suffice past the next Ice Age. Unfortunately, or fortunately, the boom of the Internet blew out everyone's assumptions. The classful addressing scheme that was developed before the boom and the classless addressing scheme that was developed later are discussed in this section. Similar to an old house that has been remodeled, the IP address has been restructured to compensate for the ever-growing networks in the world. Therefore, before we discuss the IP address structure of today, we will explain the original addressing scheme and why it was remodeled in order to avoid a network meltdown.

12.7.1 Classful Addressing

classful addressing
IP address space divided into five primary classes: A, B, C, D, and E.

Classful addressing is the term used to describe the first IP address structure used in networking. The term *classful* refers to defining the address according to classes that are divided by network and host. Simply, classful addressing refers to the class A, class B, and class C addressing blocks as discussed in an earlier chapter. Each address class is divided according to the number of network and host addresses, as shown in Table 12–1.

As shown in Figure 12–6, a dotted-decimal format is used to distinguish the four address octets. Each octet consists of possibly three decimal digits or, in binary, eight bits; each octet is separated by a period. The decimals are used to help humans decipher the address without having to read in binary, as shown in Figure 12–6. Note the difference between the three major address classes. Class A specifies the first octet as the network address and the remaining three octets as the host address. This equates to 126 class A network addresses with each class A address servicing 16,777,214 hosts. Class B uses the first two octets as the network address and the last two as the host address. The number of class B network address blocks available is 16,384 with each able

Table 12–1
IP Address Classes.

Address Class	Address Range	Number of Networks	Number of Hosts
Class A	1–126	126	16,777,214
Class B	128–191	16,384	68,532
Class C	192–223	2,097,152	254
Class D	224–239	Multicasting	
Class E	240–255	Reserved	

Decimal 164.64.1.1

Decimal   Binary Equivalent
164   = 10100100
64   = 00010000
1   = 00000001
1   = 00000001

Figure 12–6
The dotted-decimal system is used to
define IP addresses. Each decimal
value has an equivalent binary value
that is used by the device. The
decimal value is used by humans as it
is easier to understand.

to support 65,534 hosts. The class C block uses the first three octets as the network address and the last octet as the host address. This means that there are 2,097,152 network blocks available and each network block can support 254 hosts.

The network number often is referred to as the *network prefix,* indicating that portion of the address defines the network the devices reside on. All classful addresses can be segmented according to their network prefix and the host number. Multiple hosts on the same network can be identified quickly by noting that they all have the same network address. Every host on the network, though, will have its own unique host address. Using both the network prefix and the host address, routers are able to figure out how to direct the packet to the destination device—Layer 3's job in life.

network prefix
Portion of the IP address that
defines the network address.

The network prefix is identified after the IP address; often, the number of positions the address occupies is used as a quick identifier for the address type—A, B, or C. For example, a class A address has an eight-bit network prefix and is called a */8*—slash 8. A class B address has a sixteen-bit prefix and is called a */16*—slash 16. A class C address has a twenty-four-bit address and is called a */24*—slash 24. Though it is rare today to route using a classful /8 address, the concept of defining an address by the number of bits in the network prefix is common. Subnetting, a topic discussed shortly, also uses this method for defining the network portion of the address.

Early in the 1980s, addresses were granted freely according to the size of the organization. Large corporations, governments, and communications providers were given class A addresses or multiple class A addresses. The problem with this strategy was that no one expected the Internet to flourish as it did, causing a need for thousands of IP addresses for new networks and devices hanging off those networks. There are over 4 billion addresses available under the classful IP blocks. More than half of the 4 billion are attached to class A address blocks. Of those attached to the class A block, less than 100 million are actually assigned. The estimated address space actually being used today is a small percentage of the 4,294,967,296 address total.

A second reason networking professionals decided to revise classful addressing was the huge increase in the number of network routes on the Internet. Over the past ten years, the number of routes established on the Internet grew from a few thousand to over 100 thousand. Core routers were experiencing overload as they struggled to adjust their routing tables as new routes were added to the network. As discussed in the section on routing protocols, the time required to exchange routing information—called convergence time—is a key consideration when designing a network. Network outages commonly occur when convergence times exceed maximum thresholds or when router CPUs spin out of control due to excessive route maps. ISPs struggled with the balance between router size or horsepower and maintaining a reasonable cost ratio. As a result of too few IP addresses available and too many routes, standards groups introduced methods for shoring up the network until other methods, such as IPv6, are adopted and deployed.

IP addresses are very valuable commodities, almost as valuable as a barrel of oil. In order to compensate for the shortfall, network designers have devised ingenious ways to extend IPv4 addressing schemes by creating variations or additions to the old IP house structure. The first method developed to tackle the IP address dilemma was classless addressing. Though beyond the scope of this text, classless addressing increases the number of addresses available to an organization and reduces the number of routes a network router has to deal with on the internetwork.

12.7.2 Unicast, Multicast, and Broadcast Address Formats

Before explaining classless addressing, we need to explain the difference between unicast, multicast, and broadcast addresses. It is important not to confuse Ethernet unicast, multicast, or broadcast addresses with IPs unicast, multicast, and broadcast addresses. An IP unicast address refers to a specific device address, an address destined for one device talking to one other device. A multicast address refers to an address destined for multiple addresses, such as when one device talks to many devices. A broadcast address is an all-ones address that is used to reach one device talking to all other devices on the network. A full broadcast address in IP is an all-ones address defined in dotted-decimal notation as 255.255.255.255.

12.7.3 Classless Addressing
Subnetting

Classless addressing redefines the boundaries of the dotted-decimal segments of the classful address. The host portion of the address is segmented into a subnet and host fields. The network prefix is extended into the host portion of the address, allowing the administrator to build more secure, easier to manage networks from one address block. As noted in the section on classful addressing, the notation used to indicate the length of the network prefix is shown after the IP address as a /24 or /16. The same is true in classless addressing. If, as in our example, we have taken three bits of the host address to define the subnetwork address, we end up with a /27 instead of a /24—assuming we started with a class C address. The internal IP network, tying all the hosts to an interface port, routes according the subnetted address structure knowing that a /27 requires that it read the first twenty-seven bits to determine to what network the packet belongs. The value of subnetting is obvious. It allows the administrator to build multiple networks from one *address block,* which is a block of IP addresses that fall within a specified range.

A second key concept related to subnetting is the *subnet mask.* A subnet mask is used to define the number of bits used for the network and host portion of the address. If you pull up your IP configuration on your PC, you will notice a subnet mask entry between the IP address and the gateway. Devices use the subnet mask as a template, saying "Hey, the subnet mask of this class C address is 255.255.255.0," meaning that the network address consists of the first three octets and the host is defined by the last octet, as is the case with classful addressing without subnetting. The term *default mask* is used to describe subnet masks that match a full classful address. For example, a class A address has a default mask of 255.0.0.0. A class B's default mask is 255.255.0.0. Therefore, when you see a subnet mask of 255.0.0.0, or 255.255.0.0, or 255.2555.255.0, you know immediately that the address has not been subnetted. Binary ones in the subnet mask signify the network number and zeros the host number.

Again, the purpose of a subnet mask is to identify the number of bits used to define the host and network portions of the address. All a subnet mask does is tell you—and the router—what bits to read as the network prefix.

Classless Inter-Domain Routing

In the early 1990s, the IETF—Internet Engineering Task Force—introduced a standard called *CIDR*—Classless Inter-Domain Routing. (Networkers pronounce *CIDR* as *cider.*) CIDR changes the way a router reads an IP address. Routers configured for CIDR read the network prefix as defined by the subnet, not as defined by the classful address—class A and so forth. In fact, it ignores classful addressing. The term *supernetting* is often used to describe CIDR addressing as it combines multiple networks into one big supernetwork. A router looks at the network prefix as defined by the CIDR rules and routes accordingly. The advantage this provides is huge. The network or core router is now able to define a route according to a domain or supernetwork, not multiple individual routes for every classful address.

Large ISPs are given multiple class C addresses that they aggregate into one domain or supernet address. A Tier 1 ISP doles out addresses from the large supernet to

---

address block
A block of IP addresses assigned to a network or portion of a network.

subnet mask
Method used to extend the number of devices that can be served by an address block. Extends the network portion of the IP address by defining a mask.

default mask
Subnet mask that matches a full classful address.

CIDR
Classless Inter-Domain Routing is an IP address method that assigns a supernet address to identify many networks within one network. Purpose is to extend the IPv4 address method.

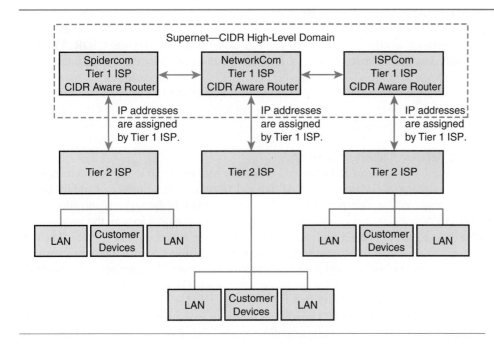

Figure 12–7
Classless IP addressing such as CIDR improves network efficiency by reducing the number of routes needed across the Internet through a hierarchical domain structure. Route tables are built between Tier 1 ISPs. Lower-level networks communicate only with the higher-level Tier 1 ISP.

the smaller ISPs it services. The smaller ISPs segment the address further onto their networks. Eventually, the end user receives addresses tied back to the supernet at the Tier 1 ISP. The Tier 1 ISP builds one route into the Internet for the supernet or domain. As is obvious, CIDR provides a much cleaner, hierarchical routing topology that reduces the number of routes needed to support the ever-growing number of networks and devices on the Internet, as shown in Figure 12–7.

To clarify this method, we again look at the U.S. Postal Service. The USPS learned ages ago that routing to individual streets or addresses was inefficient and unreliable. Every city has multiple ZIP Codes, street names, and house addresses. Every state has cities with the same name as cities in other states—for example, Rochester, New York, and Rochester, Minnesota. Cities have common street and house names; Cincinnati, Ohio, has a Main Street, an Elm Street, and a Maple Street. Louisville, Kentucky, also has a Main Street, an Elm Street, and a Maple Street. When we ship a letter from New York City to 222 Main Street, Cincinnati, Ohio, the post office uses the ZIP Code to route the letter to the post office serving 222 Main Street. Similarly, a CIDR aware device routes according to a high-level address identifier similar to a ZIP Code. It depends on the router to take that incoming packet and further identify through reading the address where it should go next. The mail service early on realized it would be impossible to set up individual routes from one post office to a second post office using full address strings. The same is true of network routers today using CIDR.

The ability to reduce the number of entries in the egress routing tables has extended the life of IPv4 addressing and indirectly averted a major network catastrophe on the Internet. Additionally, subnetting allows the administrator to build additional networks and to service many more hosts than allowed when using classful addressing schemes alone. One catch associated with CIDR is that a router has to be CIDR aware or capable in order to route according to the IETF standard. As most routes directly connected to the Internet are tied to Tier 1 and Tier 2 ISPs, this is not a problem as their networks comply with the most recent standards. It is, though, important for smaller networks to determine if they need to be CIDR capable or if it is okay to leave it up to their higher-order ISP. Classless addressing—specifically, subnetting—is beyond the scope of this text. There are, of course, multiple resources that detail subnetting and classless addressing schemes. The Internet also provides a great resource to learn more about this topic.

12.7.4 Variable-Length Subnet Masking

Though beyond the scope of this text, it is important to understand that there are other methods available to help conserve IP address space. Beyond CIDR, *VLSM*—Variable-Length Subnet Masking—is used to conserve addresses according to subnet schemes developed to extend the life of IPv4. The variable-length mask helps allocate the appropriate number of addresses according to the need. VLSM builds the mask according to the need, allowing network managers to adjust their addressing schemes accordingly.

12.7.5 Network Address Translation

NAT—Network Address Translation—is another tool available to the network engineer when designing an IP network. NAT is used to reduce the number of public IP addresses required by a corporation, an ISP, or any other IP-hungry organization. NAT works by translating a private address assigned to the host devices on the internal network into a gateway or network address defined in the router or proxy server. The outside world routes to the gateway or network address, not knowing there are multiple devices sitting behind the network interface using private IP addressing; the hosts on the internal network are invisible to the outside world. The router is configured to translate according to the schema defined by the engineer. The value of using NAT is obvious. A company with 100 PCs needing to connect to the Internet requires a public IP address from its ISP. If the company did not use NAT, it would require 100 public IP addresses, one for each device. The ISP would balk—as it should—and more than likely would refuse the request.

As with all terms used in telecom, there is some confusion in the industry when discussing NAT. Though beyond the scope of this text, NAT comes in three flavors—static NAT, dynamic NAT, and *PAT*—Port Address Translation; the terms are often used interchangeably. Technically, most networks deployed today use PAT, which is, as we have explained, a method used to translate multiple private IP addresses into a few public IP addresses. Commonly, the term *NAT* is used to describe this process even though technically it is *PAT*.

The advantage of using NAT is obvious. The ISP is able to build routes to the outside world using a few public IP addresses. The inside network is designed using the private IP space available to anyone. Again, it relates perfectly to how the post office handles addressing. A customer living on Second Street in Dallas will not receive mail for a customer located on Second Street in Anchorage, Alaska. In addition, Roy Rogers living in Houston will not receive mail destined for Roy Rogers living in Atlanta. Multiple devices can have the same private IP address as long as they live on different networks. If both of our Roys lived on Second Street, house 222, in Houston, the post office would be confused; and, of course, both Roys would be confused.

12.7.6 Static and Dynamic IP Addressing

IP routing requires devices to have an IP address that is used to receive packets routed to them as we have discussed. Similar to house addresses, anytime a new device goes online, the network manager must assign an IP address to the device. When a device moves to a new location on the network, the network manager wants to make sure the PC will come online without having to be reconfigured with a new address. Two methods are available for IP address assignment: static addressing and dynamic addressing.

Static addressing is done by manually typing entries in through a command interface, or GUI. Though it is not difficult to enter an IP address, it is difficult to manage hundreds of addresses—especially when the devices move or change within a network. Static addressing is required on devices such as Web servers, a router interface, an FTP server, and so forth because their addresses need to be set and not change. On the other hand, devices such as PCs, printers, and VoIP phones are given addresses that may or may not change from time to time through a dynamic address method such as DHCP—Dynamic Host Configuration Protocol.

VLSM
Variable-Length Subnet Masking is a method used to extend the address space associated with IPv4 addressing.

NAT
Network Address Translation is a method that translates a private IP address into a public IP address. Helps conserve IP address space.

PAT
Port Address Translation is a method used to translate multiple private IP addresses into a few public IP addresses. Often used interchangeably with NAT.

static addressing
Addresses are assigned permanently to a device. In order to change the IP address, someone must manually update the device by entering commands through the command line interface.

The problem of assigning IP addresses may seem trivial unless you are a network manager handling multiple networks and multiple devices. Especially impacted by this issue are ISPs that service thousands of devices. The IETF realized that IP address management required special consideration in order to address the high-maintenance demands placed on network providers. In the mid-1990s, the IETF introduced DHCP. Other protocols such as BOOTP—Bootstrap Protocol—preceded DHCP as a means for managing IP addressing through manually entering pools of addresses into tables. DHCP was developed to dynamically hand out IP addresses to devices, reducing the need for manual intervention. Though there are multiple DHCP methods available, the one that is used most frequently by ISPs, CATV providers, and the like is *dynamic allocation*. Dynamic allocation hands out an IP address for a set period of time referred to as a lease. When the PC powers up for the first time on the network, the TCP/IP grant-and-request process goes out and asks the server for an IP address lease. The DHCP server—or possibly the router running DHCP—returns an address that is automatically configured in the PC allowing communication on the network. This way, only PCs that are on and connected will use addresses, thus using fewer addresses than if all PCs had their own permanent address.

dynamic allocation
IP addresses are temporary according to a defined lease time. DHCP is a dynamically allocated address method.

Using DHCP leases allows ISPs to manage address space to its fullest. Some providers hand out a new IP address through DHCP each time the PC powers up. Others issue a semi-permanent address that sits on the NIC in the PC even when the device is powered down. If the server notices a device has been offline for a certain period of time, it will reclaim the address and give it to someone else. When the PC is turned on, it will be given a new IP address from the DHCP pool.

Other methods available for IP address allocation are BOOTP, the predecessor to DHCP, and PPPoE (Point-to-Point Protocol over Ethernet), a protocol used by DSL providers. DHCP is not deployed with IPX and AppleTalk, as they have built-in address allocation methods.

12.7.7 Domain Name Server

When you type "*http://www.google.com,*" your request is first processed by a *domain name server* (DNS). Domain name servers are used to translate domain names, such as the one you typed, into an IP address. As already explained, the IP address is used to route packets across the Internet. If DNSes were not deployed in the network, you would need to type in the entire four-octet IP address, such as 192.28.15.2, in order to reach the Google Web page instead of the URL. The DNS makes life for humans surfing the Net much more enjoyable. The key concepts to learn related to DNS include how the domain name is structured and how the DNSes are distributed throughout the Internet.

domain name servers
Devices that hold host and device IP address information.

Domain name servers are databases that hold host and device IP address information and their respective domain names. Domain name systems function in a hierarchical structure. A domain name can be segmented into three major sections: top-level domains, second-level domains, and third-level domains. Notice that the top-level domains include .com, .edu, .gov, and so forth. The term *zone* is used to define a subarea controlled by one or more domain name servers. The zone may encompass many hosts using the same name server to resolve IP addresses. The second-level domains may be country specific or Web page specific. Top-level domains are used to define the entity that owns the Web site. For example the ".gov" in *FCC.gov* tells you the Web site is part of the government's domain. The ".edu" in *RIT.edu* tells you the Web site is categorized as a higher educational Web site. The ".com" in *Google.com* tells you the Web site is a commercial site. The total length allowed between each octet is sixty-three characters. *Labels* is the term used to define the text between the dots. The maximum length allowed for the address is 255 bytes or 2040 bits. The address may consist of contiguous text fields separated by dots.

zone
Defines a subarea controlled by one or more domain name servers.

The Internet Assigned Number Authority (IANA) is responsible for assigning domain names such as *RIT.edu*. The host names within the domain name are handled by the network administrator.

Domain name servers are placed throughout the Internet in a distributed fashion. Typically, large corporations, large ISPs, educational institutions, and so forth manage their own DNSes. They interconnect into the big wide DNS world. Smaller corporations, small ISPs, and so forth use larger ISPs for domain name resolution services in order to avoid maintaining their own servers. The value of distributing the name servers throughout the network is twofold. First, it provides redundancy in the network. If one server fails, the entire network is seldom affected. Second, managing the number of names in one registry is unfathomable. Pushing the responsibility to the host's administrator or the zone administrator is much more doable and realistic. In summary, the DNS works with other IP addressing functions such as DHCP to help simplify the administration of the Internet.

■ 12.8 SECURITY: ACCESS CONTROL LISTS AND FIREWALLS

access control lists
Used to filter traffic by an IP address coming into or leaving a router.

Security on an IP-based network requires a vigilant network manager and some handy tools designed to ward off intruders. *Access control lists* (ACLs) are one of the most common methods deployed on router interfaces to filter out unwanted traffic in a network. The network manager sets up the access lists through a command line interface message. The ACL is used to determine if traffic coming from a particular IP address or network is allowed or not allowed to enter or leave a router interface. The ACL watches the allowed, not allowed list, similar to a bouncer at a very popular club. If the person's name is not on the list or, in the case of ACLs, on the list, he or she is not allowed to enter.

The second method used to secure a network is a firewall. Firewalls are devices that screen the traffic coming into the network to determine if it is corrupted or is not allowed. Firewalls provide additional security to a network by performing a stateful watch of the traffic. This means that packets are inspected as they flow through the gate for any gremlins that may be harmful to devices residing on the network. If the type of data contained in the packet is banned from the network, the firewall stops it before it enters.

■ 12.9 ROUTING ACROSS A LOCAL AREA NETWORK

A local area network (LAN) is a network built to carry traffic within a localized area such as within a business or around a group of buildings called a campus. The network is considered an internal network in that it ties local internal devices together allowing them to share information such as documents, e-mail, voice communications, and so forth. It is unlikely that a business in this day and age exists without a LAN. The following example illustrates how a LAN uses IP routing to carry traffic between the internal devices.

Andrea manages Butterfly Inc.'s information services. The company employs 100 people, residing on three floors of an office building in Detroit, Michigan. Andrea's job is to ensure that everyone in the office is able to access systems sitting on servers, share information, manage e-mail accounts, and interface to the outside world through the Internet. An Ethernet 100BaseTX connects the users' workstations using CAT5 twisted pair cable.

Each station on the LAN has its very own private IP address, as shown in Figure 12–8. Assume that the LAN has four printers, three servers, and 400 PCs. We start by following a transaction between two employees. Jesse sends an e-mail to Arthur even though Arthur sits two cubes away. The message is placed into an IP packet that is then encapsulated into a Layer 2 Ethernet frame prior to being pushed out onto the wire. The frame travels across the LAN to a router located in the third-floor communications closet. The router cracks open the packet, looks at the source and destination IP address, and again encapsulates it into an Ethernet frame. The message is then sent to the recipient's e-mail server where the message is accessed by the recipient—owner of the e-mail address—whenever he or she wishes.

Though the process sounds simple, it is important to understand that there are many piece parts to the end-to-end flow. For example, the router required a technician

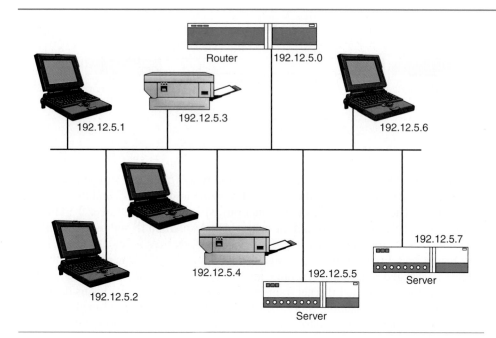

Figure 12–8
IP routing within an internal LAN
depends on IP addresses attached to
the devices hanging off the network.

to set up DHCP addressing to assign IP addresses to each of the hosts. The technician also had to configure static IP addresses for the Ethernet router ports. Routing between devices on a LAN is a microcosm of the larger external network. There are routers, switches, and multiple devices sending information using all seven layers of the OSI stack. The router has to determine routes according to IP address and routing protocols. The host device has to acquire an address and make sure it responds correctly when a packet belongs to it. Therefore, managing a LAN is not an easy task.

■ 12.10 ROUTING ACROSS A WIDE AREA NETWORK

Continuing with our scenario, Butterfly Inc. has a remote office in St. John's New Brunswick, Canada. A T1 circuit ties the two networks—the Detroit office and the St. John's office—together, allowing both offices to exchange traffic as if they were in the same building. The router sitting in Butterfly Inc.'s communications room has a DS-1 interface port tied to a local carrier. Butterfly Inc.'s router sitting in St. John's also has a DS-1 interface tied to the local carrier. The interface ports on both routers are configured with a Layer 2 frame relay DLCI and a VC (or virtual circuit) used to tie the two ends together.

In our scenario, Ruth, sitting in her office in St. John,'s needs to access the company's Web services server sitting in Detroit. In her browser she types in the address associated to the Web services server: "*https://butterflyStore/aspx.*" Once the GUI appears on her screen, Ruth is able to access the inventory database residing on the server. Ruth places an order by entering the required information and pressing the submit button at the bottom of the page. Once Ruth submits the order, she receives an e-mail verifying that she has deleted one butterfly net from inventory.

The transaction just described flows like this. A packet is created and packaged using TCP/IP. The server responds by opening up the GUI on Ruth's PC. The transaction, moving back and forth between the server and the PC, is directed across the DS-1 circuit because it is an internal secure correspondence. Andrea has set up the routing tables to point any internal traffic between St. John's and Detroit to travel across the DS-1 via a Layer 2 frame relay frame, as shown in Figure 12–9. The router sitting the in St. John's communications closet reads the request coming from Ruth, resolves the address, and determines the packet needs to be pointed toward the DS-1 interface tied to

Figure 12–9
Packets traveling between Detroit and
St. John's travel on a DS-1 within a
frame relay frame within an IP packet.
Each router interface has an IP
address assigned to it along with a
frame relay DLCI.

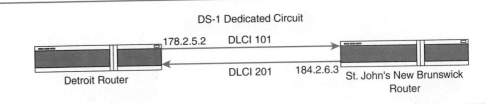

Detroit. The packet is encapsulated into a frame relay frame and shipped across the link. Once it arrives in Detroit, the frame relay frame is stripped off and the IP packet interrogated. The router handles both functions, stripping off the Layer 2 frame and cracking the packet in order to find the source and destination IP addresses. Similar to the previous scenario, the router looks at the address, realizes it belongs to the inventory server, and directs it to the port tied to the Ethernet LAN. The packet within the router is packaged in an Ethernet frame and shoved out onto the 10BaseT wire. Once it arrives at the server, the Ethernet frame is stripped off and the upper layers determine how to handle the transaction. Since the type of information being shipped is TCP/IP, an acknowledgment is returned for every packet indicating that the communication was successful.

12.11 ROUTING ACROSS THE INTERNET

Andrea understands the importance of tying both the St. John's office and the Detroit office to the Internet. Butterfly uses the Internet to receive and place orders, exchange information with vendors and colleagues working remotely, and, of course, track the migratory path of butterflies. All of these actions require the staff at both sites to have access to the Internet.

Andrea contracted Spidercom, a national ISP and more, to provide Internet services at both sites. After careful calculation and intense price negotiation Andrea decided to interface with Spidercom via a DS-1 circuit. Spidercom installed the DS-1 into Butterfly Inc.'s communications room, where Andrea asked her tech to connect it to the network router—in the case of Butterfly, the only router. The DS-1 port is the egress port to the outside world. It connects to Spidercom's POP—point of presence—in Detroit. At the POP, traffic is routed to the core or network router that ties to the circuits interfacing to the Internet, as shown in Figure 12–10.

When Bruce, a marketing type, decides to search for information related to butterflies on the Internet, he opens an Internet browser and connects to a search engine such as Google or Yahoo! and types in "butterflies." Bruce's request is packaged into a TCP/IP packet; inserted into an Ethernet frame; and sent through the switch to the ingress router port where the packet is cracked, the IP address read, and the packet routed to the Spidercom DS-1 interface facing the Internet—the egress port to the Internet.

In the reverse, Google responds with a listing of sites related to the key word— *butterflies*. The information displayed on Bruce's screen was packaged by Google into a TCP/IP packet, encapsulated into a Layer 2 frame, or cell, and shipped across circuits to a Spidercom network. From here, the packets are carried on Spidercom circuits to the Spidercom POP. At the Spidercom POP, the network router reads the destination IP address and figures out it needs to go to Butterfly Inc.'s network. Spidercom's router points the packets to the DS-1 interface nailed up to Butterfly's router sitting in the Detroit communications room. Once the traffic reaches the router, the IP address is interpreted, that is, translated using NAT into the private IP address given to Bruce's PC. The packet is sent out on the LAN as an Ethernet frame that makes its way through the Butterfly Inc. switch sitting in the closet of the first floor and onto the wire pair, into the NIC or Ethernet port located in Bruce's PC's input/output slot.

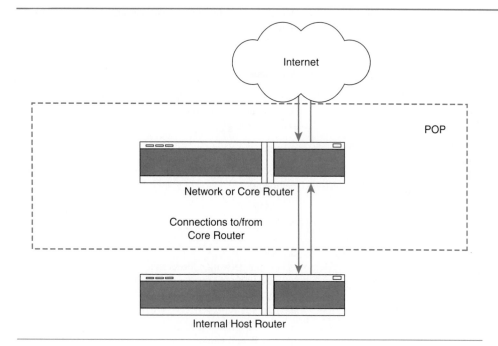

POP

Network or Core Router

Connections to/from
Core Router

Internal Host Router

Figure 12–10
The internal router ties to the ISP's POP. Traffic travels through the interface port on the internal router that is tied to the core router located at the ISP's POP. The core router ties to the supernet. Traffic is routed between the routers using standard routing protocols such as RIP. Traffic is routed across the Internet between Tier 1 ISPs with supernet routing using sophisticated protocols such as BGP. In both instances, there are interface ports or doors for traffic to come in and go out.

SUMMARY

Though the examples presented are simple, the significance of IP routing is clear. Internal networks and external networks depend on the Internet Protocol to navigate and direct traffic flows from device to device, from network to network. IP routing is one of the most important subjects in telecommunications today as IP has become the dominant protocol in the network. IP routing is responsible for directing traffic across the Internet, as well as routing traffic between almost every type of computing device. IP is the star network protocol of the twenty-first century.

CASE STUDY

You have been asked by the city council to design a LAN network for the internal county offices. Appendix A provides information on the number of users, the number of physical locations, and the application requirements.

REVIEW QUESTIONS

1. IP, Internet Protocol, is only used to route information across the Internet.

   a. true   b. false

2. A routing device is one that:

   a. routes packets between devices
   b. cracks the packet and reads the header information
   c. provides access to networks through interface port connections
   d. all of the above

3. List three types of interface ports common to network routers.

4. The following best represents the terms *ingress* and *egress:*

   a. Ingress port on a LAN router takes traffic from the computing devices hanging off the LAN and moves it to an egress port tied to an outside network.
   b. Egress refers to the port facing the outside network; ingress refers to the port facing the internal network.
   c. Egress refers to the port facing the internal network; ingress refers to the port facing the outside world.
   d. a and b
   e. a and c
   f. none of the above

5. A static routing protocol dynamically establishes routes between devices and networks.

   a. true   b. false

6. IP, IPX, and AppleTalk are referred to as routing protocols.

   a. true   b. false

7. Dynamic routing protocols are used to:

   a. establish paths' real time between routers
   b. define overhead bits within the packet
   c. encapsulate the IP packet into a Layer 2 frame
   d. reroute traffic during circuit failures and network congestion
   e. a and d
   f. a, c, and d
   g. all of the above

8. List two dynamic routing protocols.

9. *Route flapping* and *routing loops* are terms used to describe problem situations that occur on dynamically routed networks.

   a. true   b. false

10. Classful addressing describes the IP address scheme that:

   a. divides the address according to network and host
   b. defines addresses as A, B, C, D, and E
   c. is structured in a dotted-decimal format
   d. is referred to as an IPv4 addressing method
   e. all of the above
   f. none of the above

11. Classless addressing was introduced to extend the life of the IPv4 address scheme.

   a. true   b. false

12. Subnetting an address refers to redefining the boundaries of the network/host portion of the address.

   a. true   b. false

13. Default mask refers to a broadcast address assigned by the network device to initiate a route.

   a. true   b. false

14. What does *CIDR* stand for?

15. Large ISPs aggregate multiple class C addresses into one domain or supernet address and dole out the addresses to their users accordingly.

   a. true   b. false

16. Methods used to extend the life of the IPv4 addressing method are:

   a. CIDR
   b. VLSM
   c. static routes
   d. NAT
   e. private network address space
   f. subnetting
   g. VCI/VCI addressing scheme
   h. DHCP
   i. a, b, d, e, f, and g
   j. a, b, d, e, f, and h
   k. all of the above

17. DNS refers to DHCP name servers.

   a. true   b. false

18. What are .gov, .edu, and .com examples of _____.

19. List two methods that are used to stop unwanted traffic flows from entering or exiting a network.

20. IP routing is the star network protocol of the twenty-first century.

   a. true   b. false

KEY TERMS

ubiquitous (298)	autonomous system number (303)	address block (306)	PAT (308)
interface port (298)	route flapping (303)	subnet mask (306)	static addressing (308)
ingress (298)	routing loop (304)	default mask (306)	dynamic allocation (309)
egress (298)	classful addressing (304)	CIDR (306)	domain name servers (309)
cycles (301)	network prefix (305)	VLSM (308)	zone (309)
convergence (302)		NAT (308)	access control lists (310)

Telecommunications Networks

Courtesy of AT&T Archives and History Center,
Warren, N.J.

13

Transmission Media: Copper, Fiber, Wireless

Objectives

After reading this chapter, you should be able to

- Define how copper cable is used as a transmission medium.
- Define how fiber optic cable is used as a transmission medium.
- Define how wireless systems are used as transmission media.

Outline

Introduction

13.1 Basic Concepts of a Transmission Medium

13.2 Copper Cable As a Transmission Medium

13.3 Fiber Optic Cable As a Transmission Medium

13.4 Wireless Technologies

■ INTRODUCTION

Every type of information rides on a medium. When I chat with my next-door neighbor, my voice travels across the airwaves from my mouth to his ear. The sounds traveling through the air are the same sounds that travel across the telephone wire when I call my friend using an analog telephone set. The physical medium, similar to a highway, provides a path for the signal to ride on whether that is a voice conversation or a movie being downloaded from the Internet. The objective of Chapter 13 is to present an overview of the different types of transmission media—copper, fiber optics, wireless systems—used to carry information across communications networks. It is important to realize that the three dominant communications networks in place today—telephone, cable television, and cellular—use the same types of transmission media to carry traffic end to end. Even though the type of information being transmitted differs in the way it is formatted and packaged, the logical attributes associated with each of the media types remain the same. The chapter is divided into four general sections: basic concepts of a transmission medium, copper cable as a transmission medium, fiber optic cable as a transmission medium, and wireless as a medium.

■ 13.1 BASIC CONCEPTS OF A TRANSMISSION MEDIUM

All communication is dependent on each of two parties being able to understand what the other is saying. This applies whether the parties are two human beings or two electronic devices such as computers exchanging information. Communication rides on a physical medium similar to a vehicle driving down a highway. As with all roads, the medium must be designed according to the types of signals that will travel across it. Road construction must comply with the basic concepts associated with traffic engineering rules: the number of lanes, the slope of the hill, and the angle of the curve. Telecommunications engineers must deal with similar design rules: how much information needs to ride on the circuit, the distance between point A and point B, how much energy is lost between point A and point B, and what the topology of the terrain is that the circuit has to traverse. Before discussing the types of media used in communications networks, we first need to learn what variables are used to determine the type of medium to deploy. The next few paragraphs detail the basic communications parameters used by network engineers during circuit design and help to explain the differences between the three types of communications media.

Engineers design communications circuits around two general transmission parameters: *signal loss* and *bandwidth*. Signal loss, also called *attenuation*, is dependent on circuit bandwidth; and, in turn, bandwidth is dependent on the amount of signal loss associated with the physical medium. The five common factors used when designing a circuit are (1) the distance between transmitter and receiver, (2) the strength or power of the transmitted signal, (3) receiver sensitivity, (4) the type of frequencies being transmitted, and (5) interferences or disturbers on the line such as noise from power stations or crossover from an adjoining circuit.

The greater the distance between the transmitter and receiver, the greater the attenuation or signal loss. In order to compensate for greater distances, the transmit power of the signal must be increased or the distance between the transmitter and receiver must be reduced. For example, if I am a low talker and I try to converse with my friend who is standing three feet away, my words may be unintelligible due to the low volume or low signal strength coming out of my mouth. If, on the other hand, I am a loud talker, my friend will easily hear what I am saying. However, if I am a loud talker standing 1000 feet from my friend, it is very unlikely that he will be able to decipher my comments.

A second variable that plays into circuit design is how much the receiver can hear. Receiver sensitivity can be compared to how well a person can hear. As people grow older, they often lose their ability to hear lower-volume signals. A child typically has much better hearing than an adult does. Children are able to pick up a wider range of sounds—a wider band of frequencies—and a lower-volume signal, meaning that their ears are more sensitive to sound. A receiver in a communications circuit is built to accept a range of frequencies at a specified level—defined in decibels (dB). One of the first pieces of information a network engineer will ask an equipment vendor for is receiver sensitivity specifications. The purpose of the engineer's question is to determine if the equipment is capable of handling the incoming signal. In our example, if I talk softly to a child, it is likely the child will hear what I am saying. If I talk softly to a 100-year-old person, it is probable that the person will ask me to repeat myself.

Though distance is the key parameter used when determining total loss across a circuit, loss can also be caused by noises or disturbers that bleed into the signal. Trying to talk to my friend at a party, I have to shout to be heard or move closer to his ear. Conversations, music, and other noises create interference, drowning out my voice signal and making it impossible for him to hear what I am saying. In order to be heard, I have three choices: I can move closer to my friend and talk right into his ear, I can talk to a friend standing between us and ask her to repeat what I say, or I can use a microphone to amplify or increase the volume of my voice allowing me to be heard over the other noises. Unfortunately, I may not be able to move closer to my friend; the friend between us may change what I am saying, repeating the wrong information; or the

signal loss
The amount of energy that is lost as the signal travels down the medium measured in decibels.

bandwidth
The amount of information a circuit can carry.

attenuation
Term used to describe signal loss measured in decibels.

microphone may overpower the system and cause a very annoying squeal. The same is true in transmission systems. Changing one variable may cause other issues to surface that affect the quality of the signal. Though the examples just presented are simplistic, they hold true for all three communications media.

Circuit design is also dependent on how the transmission medium handles different frequencies. All signals are comprised of multiple frequencies. A good way to visualize this concept is to compare the sound coming from an orchestra playing in a well-designed amphitheater to the sound coming from an orchestra playing in a high-school cafeteria. The amphitheater has been designed in a way that allows all of the frequencies produced by the musical instruments to travel effortlessly around the hall. The amphitheater has been tuned to carry a wide range of frequencies and, as such, the audience hears rich, vibrant sounds. Our ears, being very sensitive receivers, hear the high and low frequencies projected by the musicians, making the listening experience much more enjoyable.

The balance a design engineer faces is deploying a circuit capable of carrying the signal to the far end in an efficient, reliable manner. It would be foolish to build an amphitheater in a high school even though the band teacher may think otherwise. The same is true when determining what medium to use to carry information across a communications network. Fiber optics can carry the widest range of frequencies (highest bandwidth) across great distances in the shortest amount of time. Twisted pair copper cable is capable of carrying high frequencies very short distances, or lower frequencies greater distances. The question the engineer has to answer is, Is the cost to deploy the fatter pipe warranted? When designing a voice circuit from my house to the local central office, should I choose fiber optic cable as the transmission medium? The human voice is comprised of frequencies that range from 300 to 3400 Hz, equating to a 64,000 bps signal. A fiber optic strand has the capability of handling billions of bits per second. Obviously, if all I needed was a phone line into my home, I would never deploy fiber; a copper circuit would be more than sufficient.

Again, the best analogy to use is that of our highway system. No one would build a four-lane highway to service a farmhouse located in a rural countryside. On the flip side, it would be foolish to build a dirt road between Los Angeles and San Diego. As shown in Table 13–1, the three types of media used to carry information can be categorized according to the frequencies they can carry, the average loss per mile, and the available bandwidth.

The quality of the signal and the distance it can travel between the transmitter and receiver are dependent on the type of signal—frequency range—being transmitted, the strength of the signal, the amount of interference from adjacent disturbers, and the loss characteristics of the medium. Communications media can be differentiated by how far they can carry a signal and the range of frequencies they can handle. Therefore, designing a circuit to carry a signal requires that the engineer take into account the type of medium being used, the power of the transmitter, the receiver sensitivity, adjacent

Table 13–1
Media Carrying Capacity.

Medium	Frequency	Bandwidth	Average Distance Signal Can Travel
Copper—twisted—22 to 26 gauge	0 to 1.4 MHz	7 Mbps	10,000 feet
CATV	46 GHz	11 Mbps	2000 feet
Fiber optic—1310 nm	3–30 THz	2.5 GHz per wavelength (trillions of bits per fiber possible)	1 mile up to over 100 miles (dependent on laser, frequency, bandwidth)
Microwave—point to point	660 MHz to 23GHz	45 Mbps/ channel	35 miles

disturbers, and the frequencies being transmitted. Each medium—copper, fiber, or wireless—adheres to these rules of transmission.

■ 13.2 COPPER CABLE AS A TRANSMISSION MEDIUM

13.2.1 Signal Transmission on Copper Cable

Copper is the oldest medium used in the communications network. The physical structure of copper makes it an excellent electrical conductor, in addition to being fairly lightweight when compared to other conductors such as steel or iron. A signal travels down a copper wire in the same way it travels through the air: a transmitter sends a signal, the medium carries the signal, and a receiver receives the signal. The main difference between copper and air is that a copper wire is energized with electrical power—voltage—and the signal depends on that power to carry it along—current. Simply, once the copper conductor is energized with a voltage, the electrons are disturbed in the copper creating a vehicle of sorts that carries the signal along the pipe similar to a river carrying a boat downstream. The engineer asks two key questions: (1) How far can the signal travel? (2) How much information can the circuit carry?

Determining how far the signal can travel on a copper wire depends on several factors, one of which is the resistive value of the wire. The gauge or diameter of the copper wire determines the resistive value of the wire. A larger diameter—or, in cable terms, a smaller gauge—produces a lower resistance value than a smaller-diameter, high-gauge wire, as shown in Figure 13–1. The rule of thumb is, the smaller the diameter, the more resistance, consequently, the higher the signal loss; and, in turn, the higher the signal attenuation, the shorter the distance the signal can travel on the wire.

A second factor that affects how far the signal can travel on a copper conductor is the frequencies being transmitted. For example, a high-frequency T1 signal peters out after about a mile, requiring it to be regenerated while a standard POTS telephone signal made up of lower frequencies can travel many miles before fading out. Therefore, the engineer must design the circuit around the frequency characteristics of the signal—similar to designing an amphitheater for an orchestra. Copper circuits are able to handle frequencies that range from 0 Hz up to 1 MHz, though the distance the signal can travel directly relates to the frequency range and the gauge of the cable. Higher-frequency circuits such as a T1 circuit—746 kHz—attenuate much faster than lower-frequency circuits such as POTS lines—3 kHz—and, as such, are typically deployed on a lower gauge-cable—24 AWG, depending on company guidelines. If the engineer is not able to adjust the gauge of the cable, he or she must instead adjust the distance between the transmitter and the receiver or reduce the speed of the circuit. The higher the speed, the higher the frequency; the higher the frequency, the higher the attenuation; subsequently, the higher the attenuation, the shorter the distance the signal can travel.

Figure 13–1
Comparison of the different cable gauges most commonly used in telecommunications networks. The higher the number, the smaller the diameter of the wire and, consequently, the higher the resistance per 1000 feet.

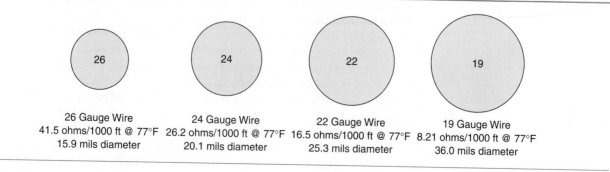

26 Gauge Wire
41.5 ohms/1000 ft @ 77°F
15.9 mils diameter

24 Gauge Wire
26.2 ohms/1000 ft @ 77°F
20.1 mils diameter

22 Gauge Wire
16.5 ohms/1000 ft @ 77°F
25.3 mils diameter

19 Gauge Wire
8.21 ohms/1000 ft @ 77°F
36.0 mils diameter

13.2.2 Electrical Characteristics of Copper Wire

When designing an amphitheater, multiple variables are taken into consideration, such as the height of the ceiling, the type of walls, the slope of the floor, and so forth. The same is true when designing a circuit that uses copper wire. The electrical characteristics used when calculating the distance a signal can travel and the amount of information a wire can carry include the voltage level, the current level, and the total impedance—resistance, capacitance, and inductance.

The amount of water coming out of a faucet is controlled by how far the faucet has been opened. If we turn the handle only slightly, we allow just a trickle of water through. If we turn the faucet on all the way, the water gushes out. The same is true of communications circuits. We place a voltage on the line that energizes the flow of electrons and produces a current. If we place a very low voltage on the line, the current is weak and it is unlikely it could carry a signal very far. A higher voltage will increase the flow of current, thus making it possible to carry the signal a greater distance.

A third electrical characteristic that must be considered when designing a copper circuit is the total resistance value of the circuit. Resistance, as mentioned earlier, is inherent in all electrical conductors. Resistance pushes on the electrons, slowing their flow—similar to what happens if you place your hand in front of a water spout, damming the flow of water. Engineers use Ohm's law to determine loop resistance, as shown in the following equation. Note that when resistance increases, current will decrease. If voltage increases, current will increase.

$$\text{Ohm's law:} \quad I = V/R$$

How much resistance is the question engineers must determine when designing a circuit. If, for example, we push -48 V down the line, we start to push those little electrons along creating a current. The standard copper loop used to carry analog voice signals requires 23 mA of current. The engineer has to make sure the water spout outputs enough voltage to drive a current across the wire in addition to maintaining an acceptable resistance value that will not impede the current flow and ultimately degrade the signal. In summary, a voice signal is carried down a copper line requiring a minimum of 23 mA of current driven by -48 V. If the resistance on a copper pair is too high, the current will drop to a point that is below the minimum and the signal will not be discernible at the receiver. Therefore, when designing a copper loop, the gauge of the copper wire is a critical piece of information used by the engineer. The standard loop resistance for a copper analog circuit is 1300 ohms.

At this point, we have discussed how circuit engineers use Ohm's law to calculate loop resistance. Two additional variables that must be considered when designing a loop are capacitance and inductance. The term *line impedance* refers to how resistance, capacitance, and inductance interact on a copper wire. Capacitance—inherent in copper wire, like resistance—affects signal flow. A capacitor is composed of two conductors placed in parallel with an insulating material—referred to as a dielectric material—in between. A telephone circuit composed of two wire strands separated by air acts like a capacitor. The two wires make up the conductors, and the air separating the two wires is the dielectric insulator. Capacitance builds up on the line as the loop distance increases. The farther the circuit travels, the higher the capacitive value. Capacitance affects higher frequencies first and more intensely than lower frequencies. A voice signal made up of 300 to 3400 Hz is fairly immune to the capacitive effect on a copper wire for the first 12,000 ft. or so depending on the wire gauge. After a certain point, though, a voice signal will begin to fade or lose strength.

Inductance is the third impedance musketeer. Inductance is capacitance's counterpart—the yin of the yang. Inductance has the ability to zero out or reduce capacitance on a copper loop. The term *balance a circuit* refers to adding inductance into the line to help eliminate the negative effects of capacitance after 12,000 ft. When a circuit length exceeds 12,000 ft (again depending on gauge), the loop must be loaded. For peak performance, loading a circuit means 88 mH inductors called *load coils* are placed every 6000 ft. apart. (A millihenry [mH] is the unit of measure for inductance.) The

line impedance
Capacitance, impedance, and resistance values associated with a copper circuit.

balance a circuit
Balance the copper circuit capacitance using inductive load coils.

load coils
Inductors placed in a copper circuit to cancel out the capacitive effect on the line.

load coils cancel out the capacitive effect of the two copper wires, thus allowing the signal to travel farther before experiencing signal degradation. Consequently, the distance the voice signal can travel down a copper line is increased. The term *loaded circuit* refers to a circuit loaded with 88 mH load coils.

Load coils work great for signals composed of frequencies below 4000 Hz. As shown in Figure 13–2 inductance flattens the frequency response curve, helping offset the negative effects of capacitance—the yin and yang of electrical properties. Unfortunately, load coils block frequencies above 4000 Hz; and because digital ISDN, T1, DDS, and DSL signals employ frequencies well above 4000 Hz, loads must be removed on all digital circuits. The deployment of DSL and ISDN forced telephone companies to de-load large portions of the outside plant copper cable. Luckily, newer switching technologies help increase the distance a voice signal can travel before requiring loading.

Designing a circuit requires the engineer to take into consideration all the variables that affect how a signal flows down a copper wire—resistance, capacitance, and inductance. Beyond calculating line impedance, the engineer must take into account the frequency range of the signal. Frequencies behave differently according to the impedance characteristics of the line. Higher frequencies are affected more profoundly by impedance than lower frequencies and, as such, are not able to travel as far down a line. A standard rule of thumb in communications, even when dealing with other physical media such as fiber optic and microwave systems, is the higher the frequency, the higher the loss per foot. A copper wire is able to carry very high frequencies but only so far. Very-High-Data-Rate DSL (VDSL) is a technology touted as the telephone company's answer to offering video services on the existing local loops. Unfortunately, the distance a VDSL signal can travel is still limited due to the high frequencies involved, thus making it an impractical solution for large deployments. However, if the business sits close to the central office, VDSL may provide a great way for the customer to receive video services from the phone company. A good example of how frequency and cable gauge relate to the distance a signal can travel is shown in Table 13–2, which lists the various DSL flavors and the related parameters—frequency, distance, cable gauge.

Decibels

Attenuation on a copper wire is measured in decibels (dB), a logarithmic value deployed to compensate for the large variations between the low and high values, as discussed earlier in the text. The most common decibel measurements are shown in Table 13–3. The decibel is used by the technician when running level tests on a medium. A good decibel reading tells the technician the line meets standard requirements

loaded circuit
A copper line that has load coils placed every 6000 ft.

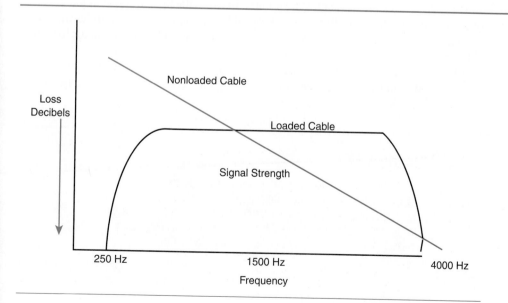

Figure 13–2
Graphical view of loaded and nonloaded copper cable. Loading a cable creates a flattened frequency response in the 250 to 4000 Hz frequency range. As shown in the graphic, loading a cable reduces the loss in frequencies used in voice communications.

Table 13–2
Copper Cable Line Rate–Estimated Parameters.

Line Type	Cable Gauge	Frequency	Average Distance—Bandwidth
Traditional POTS two-wire	24/26	0–3400 Hz	18kft. unloaded—56 kbps >18kft. loaded—56 kbps
Basic rate 2B1Q ISDN two wire	24/26	0–80 kHz	18kft. unloaded—160 kbps
2B1Q IDSL two wire	24/26	0–80 kHz	26kft.—160 kbps
T1 AMI four wire	22/24	0–1544 kHz	6kft.—1.544 Mbps
T1 2B1Q HDSL four wire	24/26	0–392 kHz	10kft.—1.544 Mbps
T1 CAP HDSL four wire	24/26	10–175 kHz	12kft.—1.544 Mbps
T1 CAP SDSL two wire	24/26	10–175 kHz	11kft.—1.544 Mbps
CAP RADSL two wire	24/26	0 to 3.4 kbps = POTS 26kHz to 1.4 MHz = data	18/15kft.—1.544 Mbps DN —64 kbps UP 12/9kft.—7.1 Mbps DN —640 kbps UP
DMT ADSL two wire	24/26	0 to 3.4 kbps = POTS 26kHz to 1.1 MHz = data	18/15kft.—1.544 Mbps DN —64 kbps UP 12/9kft.—7.1 Mbps DN —640 kbps UP
SDSL two wire	24/26	10–175 kHz	29kft.—128 kbps 21kft.—768 kbps 12kft.—3 Mbps

and, as such, should carry the signal end to end without a problem. Though the decibel is not a direct measurement of resistance, capacitance, and inductance, it indirectly implies that all three of the impedance values are behaving properly.

Twisted Pair Copper Cable

Factors other than resistance contribute to how far a signal can travel down a copper wire. Manufacturers follow rigid guidelines when constructing copper cable. The number of twists per foot, the type of insulation used to encase the copper wire, and the method used to splice the ends together contribute to how well the signal will travel down the line. The information signal traveling down the copper wire must be isolated from outside interferences such as noise or signals from adjacent circuits. The outside jacket and shield of the cable plus the number of times the tip and ring conductors cross over or twist, as shown in Figure 13–3, should be reviewed before purchasing cable. Both the type of insulation and the twists per foot helps shield the conductor from outside interferences referred to as *disturbances*.

Both inside and outside copper cable require twisting the tip and ring wires over each other in order to reduce induction—signal bleeding—from adjacent circuits

disturbances
Electrical interference coming from outside or adjacent circuits and electrical components.

Table 13–3
Decibel Unit Description.

Unit	Description
dB	Decibel is the unit measure used to measure link degradation or loss. Measured against 1 mW, 600 ohm impedance, and .775 V.
dBm	Standard measurement used in communications systems to represent the transmit and receive levels. Measurement tools are designed to measure the signal coming from the transmitter and entering the receiver at dBm.

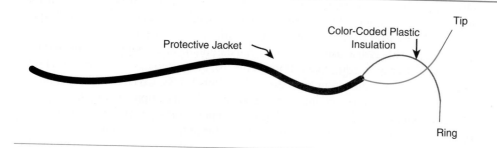

Figure 13–3
Example of twisted pair copper cable. The tip and ring wires typically cross over each other twenty times every foot.

referred to as *cross talk*. The number of twists per foot is determined by the type of signal the wire pair will carry. The most common inside cable types deployed are referred to as CAT3 and CAT5. Category 5 cable is used to carry high-frequency signals such as LAN Ethernet traffic. Category 3 cable is often used for phone systems and other lower-frequency signals. The main difference between the two cable types is the number of twists per foot. Category 5 cable has more twists per foot and a better shield surrounding the conductors. Category 3 cable is less expensive than CAT5 cable. Higher-frequency signals are especially susceptible to outside interference from adjacent high-speed signals and from electrical disturbances such as fluorescent lights or AC conductors. Consequently, most LANs deploy CAT5 wiring to help ward off outside interferences.

Outside plant copper wire is also twisted. The main differences between inside and outside cable are size and outside jacket. An outside cable requires that the outside jacket be able to withstand harsh environmental factors such as rain, snow, and sun. The second major difference is the size of the cable. Outside plant twisted pair copper cable comes in a variety of sizes from small, two-pair drops up to 3500 or more pairs in one outside plant cable. The composition of the cable is fairly standard, as each wire has a colored plastic coating that relates to its position in the cable, as explained in a later chapter. The outside of the cable is made up of a tough outer jacket that is used to ward off environmental factors such as weather and rodents.

Coaxial Copper Cable

Copper cable may be divided into two general categories, twisted pair and coaxial cable. Though the preceding description refers to standard telephone cable, the physical attributes associated with copper hold true for coaxial cable as well. Coaxial cable, unlike twisted pair cable, has two metal copper conductors as shown in Figure 13–4. Coaxial cable is composed of a solid inner copper conductor centered in an insulating material. A braided metal jacket surrounds the insulation, and a plastic polyurethane cover encases the jacket. Originally, cable television networks deployed coaxial cable from the head end to the customer premises. Today most CATV networks are a combination of fiber optics and coaxial cable. The coaxial cable extends from a central distribution point located near the residential neighborhood, and fiber optics connect the distribution point to the CATV head end, discussed later in the text. As with all

cross talk
Signal from an adjacent or nearby circuit bleeds onto the signal causing interference.

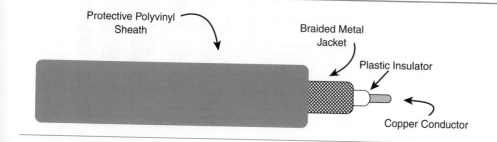

Figure 13–4
Coaxial cable consists of an outer sheath surrounding a braided metal jacket that surrounds an insulating material such as plastic that encompasses the copper conductor.

copper transmission systems, the design of a hybrid fiber distribution system must take into account the cable's resistance value, impedance value, and signal frequencies in order to meet the engineering specifications for CATV signals.

Coaxial cable is also used to carry digital communications signals inside buildings and between transport equipment. Prior to the use of twisted pair CAT5 cable, Thinnet coaxial cable was deployed in business and campus LANs. Coaxial cable also plays a very important role in connecting high-speed networking equipment together within an office. Similar to twisted pair copper, high-frequency (high-speed) signals can travel on copper conductors. The limiting factor is the distance the signal can travel.

Copper cable, whether it is twisted pair or coaxial cable, continues to be used in telecommunications networking. Copper's versatility, longevity, and flexibility ensure its role in networking for many years to come.

13.2.3 Physical Connectors Used to Terminate Copper Wire

Wire wrap pins, metal strips such as those used on punch-down blocks, and modular connectors with copper pins or metal screws sunk into a jack are used to terminate and bridge copper conductors. As is obvious, in order for the signal to travel from the copper wire to the connector, a metal conductor must be present to act as a bridge between two components. For example, a wire pair coming from an outside plant cable is bridged to an inside plant copper pair by terminating both on a metal punch-down block, as shown in Figure 13–5. The metal punch-down pin is used as the bridge

Figure 13–5
66 blocks are used to connect the inside cable to the outside cable. In the diagram, the wires are punched down on the same transmit and receive pins.

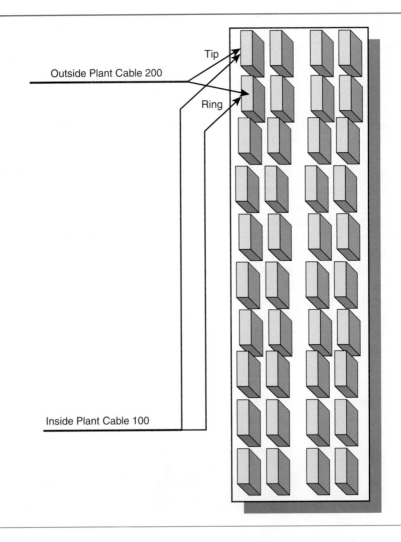

between the wires, allowing the signal to travel from the outside plant wire pair to the inside plant wire pair.

Large varieties of termination devices are available in the market today. The most common are blocks with either wire wrap pins or punch-down connections, jacks with modular or screw terminals, and termination panels with punch-down connections or screw terminals. Termination blocks, commonly called *66 blocks*, are used to connect copper wire pairs together. For example, a cable coming in from the outside plant is demarced at a 66 block. Each of the twenty-five pairs shown can be cross connected to a particular location in the building by placing an inside jumper between the outside plant cable and the inside house cable.

The 66-block connection is typically a punch-down connection that requires the technician to hook the end of the wire between two metal barbs, then using a punch-down tool and pushing the wire between the metal hooks. Punching the wire between the hooks removes a portion of the plastic insulation exposing the copper conductor. The exposed copper conductor rests against the metal pins or hooks, creating a metal-to-metal connection or path. Both the tip and ring wire are punched down in this fashion.

Typically, three methods are used to terminate wires on a 66 block. The first method, as shown in Figure 13–6, bridges the two cables through a jumper wire used to connect two separate pin locations. A second method used to connect the cables is to punch down each cable on adjacent pins and place a metal bridging clip over all four pins, as shown in Figure 13–7. A third method commonly used to connect the cables is to punch down each cable pair on separate adjacent pins that have been internally cross connected, making them essentially one conductor, as shown in Figure 13–8. Some 66 blocks are made with wire wrap pins instead of punch-down pins. Whether the block uses hooked pins or straight wire wrap pins, the function remains the same—a passive connection point. The most significant difference between the two styles is that a wire wrap design requires the wire to be stripped before spinning it onto the pin, as shown in Figure 13–9. The punch-down method strips the wire when it is pushed between the barbs.

66 block
Physical block used to terminate multiple copper wires.

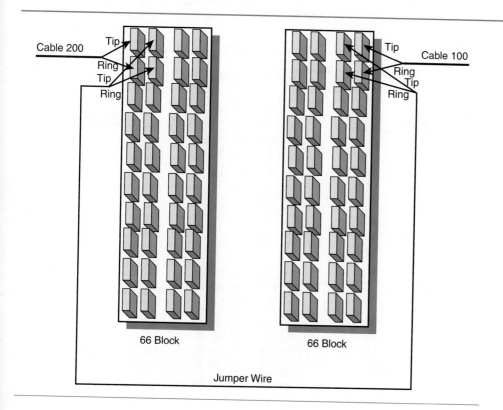

Figure 13–6
Connecting two-wire pairs together using 66 blocks as termination points and jumper wire to connect the two pairs.

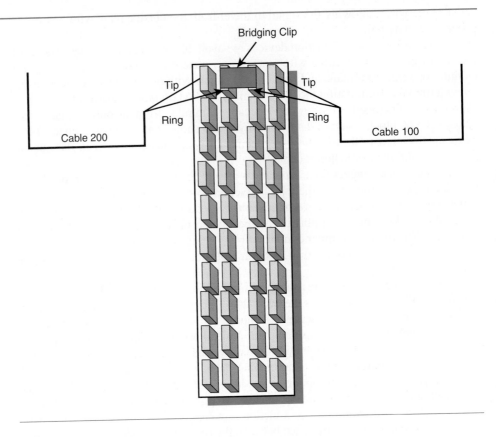

Figure 13–7
View of how two cables are connected using a bridging clip on a 66 block.

Bridging Clip

Tip

Ring

Cable 200

Tip

Ring

Cable 100

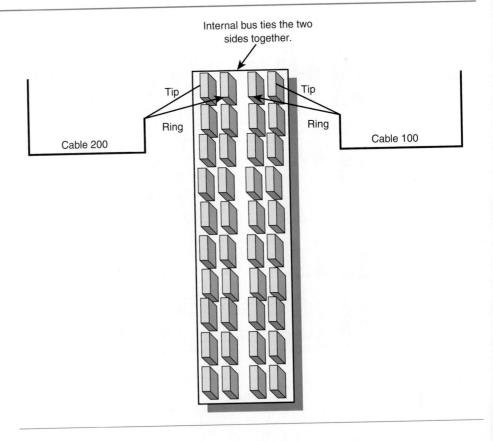

Figure 13–8
View of how two cables are connected using a 66 block with internal connections between the two sets of pins.

Internal bus ties the two sides together.

Tip

Ring

Cable 200

Tip

Ring

Cable 100

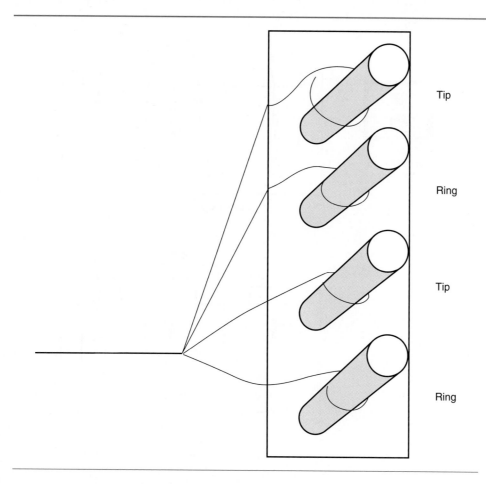

Figure 13–9
Wire wrap pins used to terminate twisted copper wire.

Jacks are also used to terminate copper wire. The most common is the RJ-11 jack used to terminate the standard telephone wire pair. An RJ-11 jack comes in many styles, sizes, and shapes. Figure 13–10 shows a typical RJ-11 jack termination, illustrating how the tip and ring are terminated on metal screws. The screws are loosened, the insulation on the wire conductor stripped back, and the exposed wire wrapped around the termination screw. The RJ-45 jack, also used to terminate copper wire, is most

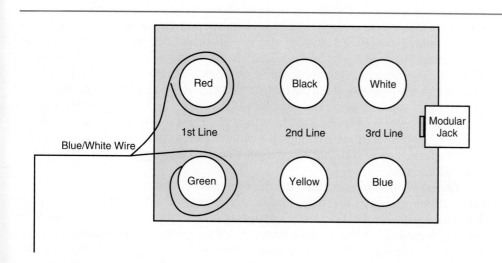

Figure 13–10
Connecting inside telephone wire to an RJ-11 jack using termination screws. The tip and ring conductors are stripped, then wrapped under the screw as shown. The jack shown is capable of terminating three two-wire telephone lines.

Figure 13–11
RJ-45 termination jacks are eight-
conductor jacks used to terminate
high-speed digital circuits. Typically,
RJ-45 jacks are used to terminate
Ethernet or ISDN services.

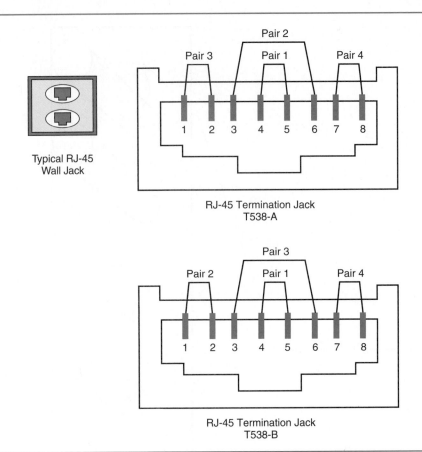

Typical RJ-45
Wall Jack

RJ-45 Termination Jack
T538-A

RJ-45 Termination Jack
T538-B

commonly associated with four-wire Ethernet or ISDN T-interface connections, as shown in Figure 13–11. RJ-48 terminal jacks are used to connect two four-wire T1 circuits, as shown in Figure 13–12.

The RJs—registered jacks—described terminate the copper wire on screws, punch down barbs, or wire wrap pins, similar to the 66 block. For example, the wire pair is stripped of its outside insulation and wrapped around the termination screws, thus forming a connection between the wire pair and the conductors in the modular connector, as shown in Figure 13–13. Once the modular end to the cord is inserted into the modular jack, a connection between the pins in the connector and the jack is made. Looking closely at a modular plug reveals strips of copper embedded into the plastic case. The copper pins bridge directly to the copper pins embedded in the modular jack. The copper pins on the termination block are tied to the screws. For example, a telephone is connected to the outside world by plugging an RJ-11 modular jack into an RJ-11 modular plug. The modular plug is tied to the screw terminals that terminate the copper wire leading to the outside plant and, eventually, to the telephone switch at the central office. Making up a modular plug is fairly simple as long as the pinout information is available. The standard copper color code is used when attaching modular ends.

Other copper termination interfaces that are commonly found in communications networks are those that interconnect the DTE—Data Termination Equipment—to the DCE—Data Communications Equipment. RS-232 and V.35 interfaces are used to carry traffic between, for example, a CSU—channel service unit—and a file server. Pinout information for both RS-232 and V.35 may vary depending on the vendor's specifications but most often follows the specifications

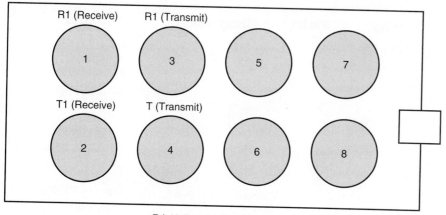

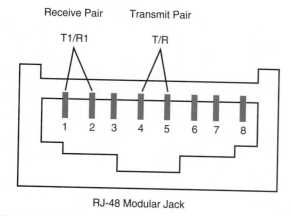

Figure 13–12
The RJ-48 jack terminates a four-wire T1 circuit. Pins 1, 2, 3, and 4 are tied to the modular jack. The TR or side 1 pair is connected to pins 1 and 2 and the T1/R1 pair is connected to pins 3 and 4. The lower diagram shows a view of the modular jack pinouts.

shown in Table 13–4. RS-232 and V.35 interfaces are used for a variety of reasons in communications networks.

Connecting large outside plant copper cable will be discussed in a later chapter. Splice cases, cross-connect terminals, and cable vaults are used to terminate large cables in the outside plant, at customer locations, within switching centers, or within POPs—points of presence.

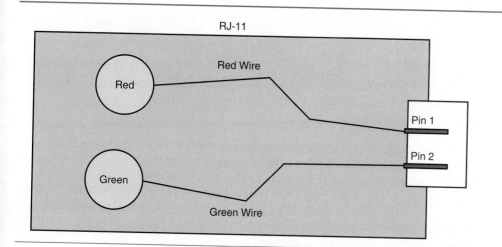

Figure 13–13
View of how the termination screws are tied to the modular jack. Note the connection between the copper pins and the screw termination. All modular conductors are built using the same technique. The difference is in the number of connections and which screws are tied to which pins.

Table 13–4
RS-232 and V.35 Wiring Designation.

Connector Type	Pin No.	Designation	Description
RS-232 DB25	1	GND	Ground
	2	TX	Transmit data
	3	RCV	Receive data
	4	RTS	Ready to send
	5	CTS	Clear to send
	6	DSR	Data set ready
	7	SG	Signal ground
	8	DCD	Data carrier detect
	20	DTR	Data terminal ready
	22	RI	Ring indicator
RS-232 DB9	1	DCD	Data carrier detect
	2	RCV	Receive data
	3	TX	Transmit data
	4	DTR	Data terminal ready
	5	SG	Signal ground
	6	DSR	Data set ready
	7	RTS	Ready to send
	8	CTS	Clear to send
	9	RI	Ring indicator
V.35	A	GND	Chassis ground
	B	SG	Signal ground
	C	RTS	Request to send
	D	CTS	Clear to send
	E	DSR	Data set ready
	F	DCD	Data carrier detect
	G	Not used	
	H	DTR	Data terminal ready
	I	Not used	
	J	Loop	Local loop back
	K	LT	Local test
	L	Not used	
	M	Not used	
	N	Not used	
	P	TX-A	Transmit data A
	R	RCV-A	Receive data A
	S	TX-B	Transmit data B
	T	RCV-B	Receive data B
	U	T timing-A	Terminal timing-A
	V	R timing-A	Receive timing-A
	W	T timing-B	Terminal timing-B
	X	R timing-B	Receive timing-B
	Y	S timing-A	Send timing-A
	Z	Not used	
	AA	S timing-B	Send timing-B
	BB–NN	Not used	

Terminating multiple high-speed circuits such as DS-1s or DS-3s requires passive termination devices called DSXes—digital signal cross connects. A DS-1 DSX panel is able to terminate multiple T1 circuits. The T1 circuit is terminated as shown in Figure 13–14 on the back of the DSX panel using a wire wrap gun to spin down the two-wire pair. The front of the DSX panel is used as a cross-connect point to tie the T1 circuits together. DS-3 DSX panels serve the same purpose as DS-1 DSX panels. They terminate and cross connect high-speed DS-3 signals riding on coaxial cable, as shown in Figure 13–15. A BNC connector is attached to the end of the coaxial cable

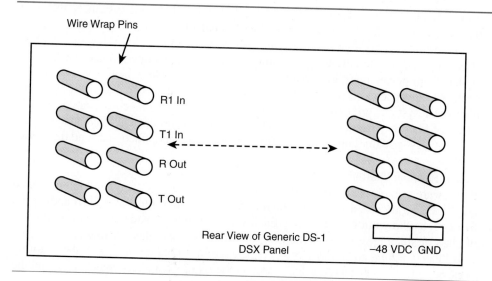

Wire Wrap Pins

R1 In

T1 In

R Out

T Out

Rear View of Generic DS-1
DSX Panel

−48 VDC   GND

Figure 13–14
Rear view of generic DS-1 DSX panel.
T1 shielded twisted pair is wire
wrapped onto the pins as shown in
the diagram.

and terminated on the back of the DS-3 DSX pane. Short cross-connect cables are used
to cross connect one DS-3 to a second.

◼ 13.3 FIBER OPTIC CABLE AS A TRANSMISSION MEDIUM

Many analysts believe that the most significant invention since the telephone is fiber
optic cable. The first fiber optic system deployed in telecommunications in the late
1970s consisted of two multiplexers connected together by two strands of fiber—a

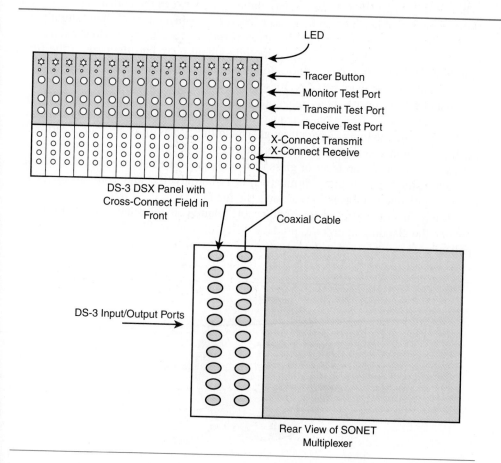

LED

Tracer Button

Monitor Test Port

Transmit Test Port

Receive Test Port

X-Connect Transmit
X-Connect Receive

DS-3 DSX Panel with
Cross-Connect Field in
Front

Coaxial Cable

DS-3 Input/Output Ports

Rear View of SONET
Multiplexer

Figure 13–15
Connecting a DS-3 DSX panel to a
DS-3 port on a SONET multiplexer
using coaxial cable.

transmit and a receive—forming the first point-to-point fiber optic system. At that time, the system was able to carry 672 simultaneous calls compared to a copper system's ability to carry twenty-four simultaneous calls. Fiber also provided a slew of other benefits, the most significant being its immunity to extraneous interferences such as AC signals from power lines; environmental factors such as water, heat, and fog; and so forth. The advancements made over the past forty years have been phenomenal. Today, fiber optic systems are capable of carrying billions of simultaneous bits on a single strand of fiber the size of one human hair. Fiber optic advancements have made it possible to carry vast amounts of information—voice, data, and video—around the world.

13.3.1 Physical Characteristics

If you were to look at a fiber optic strand under a microscope, you would notice three layers: a central core, a layer surrounding the core, and a third layer surrounding the first two, as shown in Figure 13–16. Each strand of fiber is made up of a central layer of glass or silica referred to as the *core*, a second layer of glass surrounding the core called the *cladding*, and a color-coded plastic casing called the *coating*. Fiber's ability to carry light great distances is dependent on the relationship between the core and the cladding and on the purity of the glass.

The light pulse sent down the core bounces against the cladding and reflects back into the core, forcing the light to continue its travels down the fiber strand. In order for the cladding to act as a wall, the composition of the glass is altered using boron or germanium to create a material with a different refractive index than that of the core. It is the difference in the refractive index that creates light's bouncing effect. If you look at light traveling from the air into a pool of water, you will notice it bends as it traverses from air to water. Water has a different refractive index than air, thus causing the light photons to behave differently when they hit the barrier between air and water. In scientific terms, refractive index refers to how fast light travels in the medium compared to how fast it travels in a vacuum. The relationship between the core and the cladding is the same as that between air and water. Therefore, when materials have slightly different refractive indexes, the light bends or refracts when it hits the lower refractive index medium, the cladding, forcing the light signal back into the core of the fiber, thus contributing to the fiber's ability to keep the light signal inside the fiber strand.

A display at the Corning Glass museum in Corning, New York, illustrates how the scientists first discovered how light followed a stream of water as it flowed down a fountain. From this revelation, scientists started to imagine ways they could harness light by forcing it to follow a path through a medium such as glass. Further experiments frustrated scientists as light radiated out the edges of the glass pipe. Eventually, it was discovered that a second layer of glass surrounding the central core kept or contained the light inside the glass pipe, eliminating the radiating effect and, consequently, forcing the light signal to travel farther down the medium.

Today, manufacturers of fiber optic cable manipulate the relationship between the core and the cladding in order to produce different types of fiber for different types of applications.

core
Glass center of a fiber optic strand.

cladding
Outer layer of glass that surrounds the center glass core. Cladding has a different refractive index than the core.

coating
Plastic coating that surrounds the fiber cladding on the fiber strand.

Figure 13–16
View of a fiber strand showing the core, cladding, and coating.

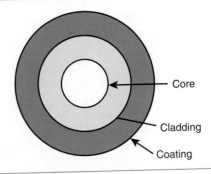

Core

Cladding

Coating

13.3.2 Single-Mode versus Multimode Fiber

Light contained in a very small diameter core can travel many miles before needing to be boosted or regenerated. The rule of thumb is, the smaller the core, the farther the signal can travel. A good way to visualize this concept is to shoot a light through two tubes, one with a smaller diameter and one with a larger diameter. Turn out the lights and point the tubes at a blank wall. The light coming out of the smaller-diameter tube will be brighter than the light coming out of the larger-diameter tube. Why is this significant? Simple. We now know that a fiber strand works as a transmission medium or pipe for light. We also have learned that the fiber strand contains light in a small space directing it in the same way a water pipe contains and directs the flow of water. In addition, we can conclude that the size of the pipe, the core diameter, is directly related to how strong the light will be at the output. A fiber with a small core size will always carry light farther than a fiber with a larger core size.

The size of the core is what differentiates multimode and single mode fiber. Multimode fiber is used inside buildings, across a campus, or in networks of spanning less than a few kilometers. Single-mode fiber is deployed in networks that travel beyond a few kilometers such as in the local telephone network, the long distance network, CATV, cellular, and the international networks. A multimode fiber has a much larger core diameter when compared to single-mode fiber. Multimode core ranges from 50 μm to 62 μm while single-mode core diameter ranges between 7μm and 10 μm. As the names imply, single-mode fiber has one mode or path for light to travel while multimode has multiple modes or paths for light to travel, as shown in Figure 13–17. Light traveling down a multimode fiber meanders back and forth in the core, expending energy as it works its way down the pipe. Consequently, signals riding on single-mode fiber experience lower loss per kilometer than those traveling across multimode fiber.

Multimode fiber can be further categorized as graded index or step index. Graded index refers to a fiber with a nonuniform refractive index core, which causes the light to propagate in a wavelike fashion. Step index refers to a fiber with uniform but different core cladding refractive indexes. As the core is large, many paths are available for the different light waves to travel. Both the step and graded index fibers suffer from multimode dispersion which is the primary reason that the distance the signal can travel is limited. Dispersion is discussed in the next section.

13.3.3 Signal Characteristics

Similar to attenuation caused by impedance in copper, fiber optic cable also has inherent physical phenomena that contribute to how far a signal can travel down the medium.

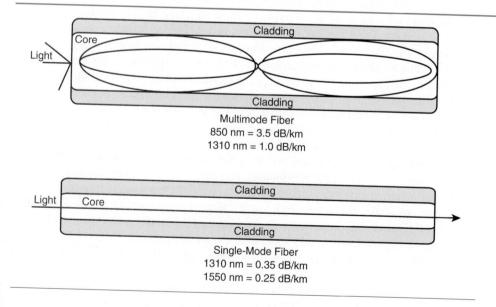

Multimode Fiber
850 nm = 3.5 dB/km
1310 nm = 1.0 dB/km

Single-Mode Fiber
1310 nm = 0.35 dB/km
1550 nm = 0.25 dB/km

Figure 13–17
The two diagrams illustrate the difference in the way light travels down multimode versus single-mode fiber. Note how, because of the larger core diameter, the light bounces off the cladding as it travels down multimode fiber and stays focused in a single beam in single-mode fiber.

Now that we are comfortable with the idea that an invisible light can travel down a piece of glass, it is time to delve a bit deeper to determine what properties in the light's path affect the strength of the signal. The three general causes of signal loss in fiber are scattering, absorption, and radiation. Clear as mud!

Scattering happens when the light disperses from its path because of imperfections inherent in the fiber. The light disperses—similar to the light leaving a flashlight. This scattering causes the signal to lose power or strength and, thus, is one reason that light cannot travel indefinitely without needing to be regenerated or amplified. The engineer cannot eliminate the effects of scattering because the manufacturing process causes it. Luckily, thanks to the modern manufacturing techniques, loss due to scattering is now at an acceptable level.

Absorption is a second cause of signal attenuation. The light traveling down the fiber strand may encounter impurities in the glass, that absorb and dissipate the light as heat. Because certain wavelengths are more prone to absorption, design engineers have avoided building lasers which are used to transmit light into the fiber using these wavelengths.

Radiation is a loss characteristic resulting from the geometric structure of the fiber strand. There are two types of radiation loss: microbending and macrobending. Microbending is the result of imperfections at the core/cladding interface, similar to bumps in the road. These bumps cause the light to reflect in ways it should not and, consequently, affect the strength of the signal. As with scattering and absorption, there is little an engineer can do to correct this problem as it is part of the makeup of the fiber. A macrobend occurs when bends are large in comparison to the diameter of the fiber. Macrobends can be reduced through proper design and installation. *Rayleigh backscatter* and *Fresnel reflection* are terms used to describe phenomena occurring within fiber that affect how light travels through the strand.

13.3.4 Signal Loss Caused by Man-Made Events

We have discussed the general cause of loss due to properties inherent to the fiber. This information is very important because it tells us that even if there were no human-made problems, light still dissipates as it travels along a fiber strand. As is obvious, the events we should be most concerned with when dealing with signal loss are those that we can prevent or at least reduce. The most common human-made causes of signal loss in fiber are improperly mounted connectors, misaligned splices, dirt, and—as explained earlier—macrobends.

As light travels down a fiber, it prefers to travel in a straight line. If the strand bends or twists too dramatically, the light signal, similar to a car going around a curve, can exit the road. Bending the fiber strand beyond the *bend radius*—a term used to define the maximum allowed bend in a fiber—can result in signal level problems.

A second condition that causes light loss is dirt. Dirt on a splice, on a connector, or on any other component reduces the level of the signal—often significantly. Therefore, you must be careful whenever you handle jumpers, splices, or any other component. Commonly, systems fail to turn up due to one small fingerprint smudged on the end of a fiber jumper. It is amazing to think that one invisible speck of dirt or a smudge from a fingerprint could interfere with a light signal, but it does.

Though well intentioned, a technician may cause the signal to attenuate dramatically by cleaning the end of the fiber with the wrong compound. Cleaning the fiber with the incorrect type of alcohol leaves a film on the end of the fiber similar to the film left on your windows when you accidentally use the wrong window cleaner. Light is blocked or obstructed due to the film left on the window; the same is true of the fiber.

When calculating the end-to-end loss of a system, the connector loss and the splice loss must be included in order to obtain the total loss across the span. There is always a certain amount of loss associated with a splice even when the splice is good. Calculating the total splice loss is important when designing the network. If the splice does not fall within the allowed spec, it should be broken and respliced. Connectors also contribute to span loss. Normally, a company defines how much loss it is willing to tolerate per connector. If the loss across the connector exceeds this value, the connector must be redone.

---

scattering
Phenomenon in which light disperses from its path as it travels down the fiber strand. Inherent property of fiber.

absorption
Impurities in the glass absorb the signal causing signal loss.

radiation
Loss happens across fiber optic strand due to bends.

Rayleigh backscatter
Phenomenon caused by imperfections in the fiber caused during the manufacturing process. Light scatters as it hits impurities.

Fresnel reflection
Reflection that occurs when the incident light travels between two media of differing refractive indexes. Simply, it is how the light behaves when it crosses between two media such as glass to air or air to water.

bend radius
Allowable bend of a fiber strand. Too great a bend exceeds the suggested bend radius.

In conclusion, similar to attenuation on a copper line, fiber optic systems must provide a path with minimal loss characteristics in order to carry the signal end to end. Fiber systems experience loss from splices, connectors, dirty fibers, and extreme bends in the fiber strand.

Dispersion, a second cousin to loss, is the spreading or distortion of the light pulse. The three types of dispersion that affect the light signal are modal, chromatic, and polarization mode dispersion. Modal dispersion affects light traveling on multimode fiber, while chromatic and polarization mode dispersion affect signals on single-mode fiber.

Higher-bandwidth systems such as OC-192 and OC-768 are much more susceptible to dispersion than lower-speed signals. The higher the bit rate, the closer the pulses are to each other and, consequently, the more vulnerable they are to blurring when the pulse spreads. A second factor that contributes to dispersion is distance. The farther the signal travels, similar to loss, the greater the dispersion and, thus, the greater the chance the signal will deteriorate.

Signal attenuation in fiber is caused by several factors, each relating directly to the wavelength being used to carry the signal and also the type of fiber carrying the wavelength. As with all media, attenuation is measured in decibels and the loss across a fiber strand is noted as dB/km or dB/mile. Bad splices or connectors are caused by improper installation and can be fixed fairly easily. Fiber impurities occur when the fiber is being produced and, as such, are impossible to correct. Physical anomalies such as water molecules cause certain wavelengths to attenuate dramatically, making them useless for transmission. Since the manufacturing process creates water molecules in the fiber, newer fiber types have succeeded in reducing the water molecules in the fiber to a point that reduces the attenuation in that particular wavelength window.

Dispersion similar to attenuation affects how far the signal can travel before needing to be amplified or regenerated. Dispersion causes the light pulse to spread or disperse, making it difficult for the end equipment to determine if it is seeing a one or a zero. Engineers have designed fiber optic systems and fiber cable around dispersion and attenuation values, as each wavelength can tolerate different amounts of attenuation and dispersion. The first fiber optic systems used 1310 nm and 1550 nm lasers because they produced the best combination of low-loss, low dispersion on SMF—single-mode fiber.

Three types of dispersion cause concern: modal, chromatic, and polarization mode. Modal dispersion only affects multimode fiber and, as very little multimode fiber is deployed in the network other than inside buildings, it rarely causes any concern. The two remaining types, chromatic and polarization mode, do cause concern as they affect single-mode fibers. Chromatic dispersion happens when the pulse traveling through the fiber strand begins to spread—similar to a light that is projected from a flashlight. Different wavelengths travel at different rates of speed causing a catch-up syndrome to happen at the end of the fiber. This catch-up effect causes the wavelengths to overlap each other making the pulse spread beyond the allowed pulse mask, thus becoming blurred or indistinguishable to the receiver resulting in a corrupted signal. Remember that a signal is made up of only two variables, a one and a zero.

Chromatic dispersion (CD) is measured in picoseconds/kilometer/nano seconds. Each fiber type has a CD value according to the wavelength deployed and should be reviewed before purchasing the fiber. In addition, a chromatic dispersion test is often taken during turnup to ensure that the system will function properly and to provide a reference measurement. If the system exceeds the allowed chromatic dispersion level, chromatic dispersion compensation devices may be added to nullify the condition. Typically, only long distance networks require CD measurements and compensation as the shorter-distance metro and access networks rarely experience high levels of chromatic dispersion; chromatic dispersion increases with distance and bit-per-second rate. The exception to this rule occurs when very high rate systems are deployed in the metro space such as OC-192 or OC-768 because CD can be severe at these rates.

The third type of dispersion—polarization mode dispersion (PMD)—is as difficult to explain as it is to isolate. Polarization mode dispersion is a phenomenon caused during the manufacturing process. The cable's geometric structure is altered slightly,

Figure 13–18
PMD causes the x and y axes to shift,
resulting in pulse spreading as shown
in the figure. As the signal travels
down the line, the signal disperses,
resulting in a corrupted signal.

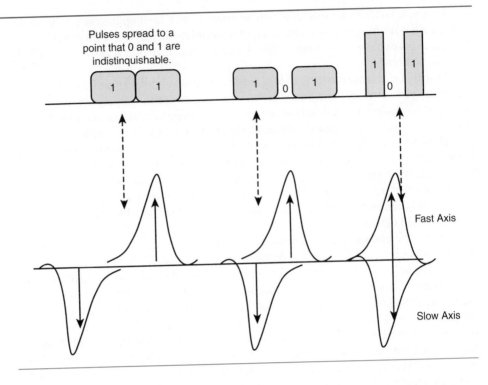

causing light to gravitate off its polar axis, as shown in Figure 13–18. Becaues of this movement off its natural course—orthonogolical structure—the pulse spreads, becoming unrecognizable to the end equipment. Because PMD is introduced during the manufacturing process, a PMD value is always taken after the fiber is placed on the reel and noted in the fiber specification sheet for future reference. Hanging fiber between poles or pulling it through the underground invariably affects the PMD value as it puts pressure on the fiber and, thus, may change its geometric composition. Therefore, a PMD value should always be taken after the fiber is in place to make sure it still falls within the acceptable range according to the manufacturer's PMD specifications. The values taken should always be recorded and saved, as PMD can change over time. PMD is measured in picoseconds/square root kilometers. In the real world, measuring for PMD is a fairly simple task, thanks to very sophisticated test sets that provide a pass/fail value.

Today, manufacturers have worked very hard to produce fiber types with lower dispersion characteristics. Lucent's True Wave and Corning LEAF are two examples of low-dispersion fiber referred to as NZDF—non-zero dispersion fiber. NZDF is primarily deployed in long haul networks or in very high bit rate metro networks that deploy DWDM—dense wave division multiplexing.

One of the main objectives fiber manufacturers strive for is to develop fiber with low loss and low dispersion characteristics. The lower the loss and the less dispersion on a fiber strand, the farther the signal can travel before requiring amplification or regeneration.

13.3.5 Wavelengths Used in Optical Systems
Three wavelengths dominate network deployments: 850 nm, 1310 nm, and 1550 nm. LANs and intrabuilding networks use 850 nm LEDs as light sources because the distance the signal has to travel is minimal. The 850 nm LED is most often deployed on multimode fiber—again, only across short distances.

The 1310 and 1550 nm wavelengths have been used in the industry since the first optical system was deployed in the early 1980s. Both 1310 and 1550 nm wavelengths experience low loss and low dispersion across single-mode fiber. The 1310 nm wavelength averages .35 dB/km and 1550 nm averages .25 dB/km. Today wavelength division multiplexing requires manufacturers to produce lasers capable of transmitting many wavelengths. The main concept to grasp from this section is that wavelengths behave very differently from each other as they travel down a glass fiber. It is up to the engineer to determine which wavelength has the least amount of loss and dispersion in addition to keeping the cost factor in control.

13.3.6 Types of Fiber

Because dispersion and attenuation affect how far the signal can travel on a fiber strand and different wavelengths attenuate and disperse differently, scientists designed fibers according to their loss and dispersion characteristics. The first fiber developed, SSMF—standard single-mode fiber, has zero dispersion at the 1310 nm wavelength. Therefore, the first fiber optic systems were built with 1310 nm lasers. Because 1550 nm lasers realized the least loss on SSMF, scientists decided to build 1550 nm lasers to increase the distance the signal could travel. One problem encountered when using the 1550 nm laser is the need to place repeaters closer together because of the effect of chromatic dispersion. As we all are aware, the more repeaters needed, the more money needed; and this, of course, is not acceptable. Manufacturers resolved this issue by introducing a new fiber called DSF—dispersion shifted fiber—that shifted the zero dispersion point from 1310 nm to 1550 nm. DSF was deployed in long-haul networks in the early 1990s. The fiber worked fine when just the 1550 nm wavelength was being transmitted; but, with the introduction of WDM technology in the long-haul network in the mid- and late 1990s, it soon became apparent that DSF experienced problems. The wavelengths traveling down the DSF blend together creating an entirely new wavelength—a phenomenon called *four-wave mixing*. The cause of four-wave mixing turned out to be a lack of chromatic dispersion at the transmission wavelengths, thus causing the entire system to experience cross pollution of the channels, as shown in Figure 13–19. Scientists went back to the drawing board and produced a new fiber called NZDSF—non-zero dispersion shifted fiber. NZDSF added a slight amount of dispersion at the 1500 nm window, resulting in the elimination of the four-wave mixing. Today, almost all long-haul core networks used NZDSF and continue to benefit from the longer distances the signals can travel before requiring amplification or regeneration. The most common NZDSFs are Lucent's True Wave, Corning's LEAF, and Alcatel.

> **four-wave mixing**
> Adjacent wavelengths cross over each other creating a new wavelength.

The last type of fiber to be discussed is the newest called ESMF—enhanced single-mode fiber. ESMF was developed for the metropolitan network to specifically address the water peak issue associated with fiber. As mentioned earlier, fiber has inherent impurities that cause certain wavelengths to attenuate more rapidly than others do. Water molecules found in the fiber produce significant loss at certain frequencies, which is referred to as the *water peak*, as shown in Figure 13–20. The water peak affects the wavelengths between 1370 and 1423 nm causing them to attenuate dramatically, thus rendering them useless for communications systems. As a result, manufacturers do not build electronic systems using these frequencies. The new enhanced or extended

> **water peak**
> Phenomenon in fiber in which water molecules cause excessive loss—around the 1400 nm wavelength.

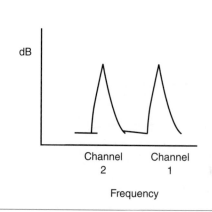

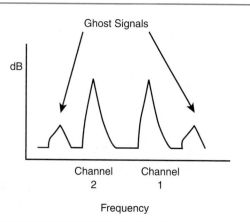

> **Figure 13–19**
> Four-wave mixing refers to the phenomenon that occurs when ghost channels appear within the spectrum causing the signal to become disrupted.

Figure 13–20
Graphic view of single-mode fiber showing the water peak at around 1380 nm.

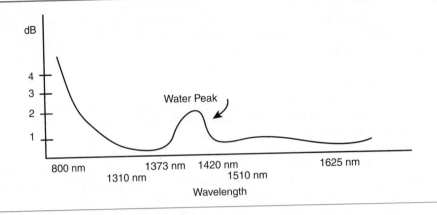

single-mode fiber is manufactured in such a way that the water molecules are removed, eliminating the water peak and in turn increasing the number of usable wavelengths, as shown in Figure 13–21.

Today's ESMF includes Lucent's AllWave, Alcatel's E-SMF, and Corning's SMF-28e. The significance of using enhanced fiber in a metropolitan network is that it allows the user to place many more wavelengths on the fiber, as there are now many more wavelengths available for use. A second advantage of using ESMF is that it allows the vendors to build WDM systems with greater channel spacing, thus reducing the cost for the system. The terms *CWDM*—coarse wave division multiplexer—and *fat WDM* refer to systems built with wide channel spacing, though they are deployed on traditional SMF as well as ESMF. A note should be made regarding ESMF. The issue many vendors are facing today is that the majority of fiber deployed is SSMF, raising the concern of whether there would be a market for their troubles if they were to build a WDM system to include the new water-peak wavelengths. As one vendor put it, it is the old chicken and egg question: Who is willing to go first—will the service providers replace the SMF with ESMF first, or will the vendors first build the equipment to service the new wavelengths? Figure 13–22 shows the standard frequency spectrum as defined by the ITU and the terms used to describe each of the bands. As is obvious from the diagram, no longer are we dealing with two frequencies and one type of fiber.

CWDM
Coarse wave division multiplexing refers to muliltiplexing multiple wavelengths onto one fiber. *Coarse* refers to fewer wavelengths than DWDM, dense.

fat WDM
Wavelength multiplexing fewer channels with greater spacing between the wavelengths.

13.3.7 Fiber Connectors/Jumpers
The physical connectors spliced onto the ends of a fiber strand come in many different shapes and sizes. All serve the same purpose, from the older D4 screw-type connector to the new small form factor LC connectors, connecting a fiber strand to a termination point. The most common types of connectors are D4, FC, SC, ST, and SFF. Precise

Figure 13–21
Graphic view of ESMF—enhanced single-mode fiber—showing usable frequencies between 1373 and 1420 nm.

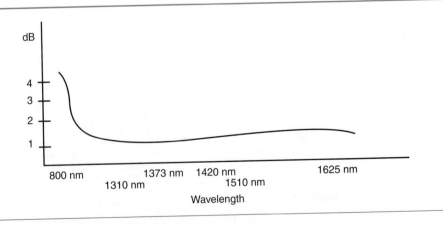

Original Band O-Band 1260–1360 nm	Extended Band E-Band 1360–1460 nm	Short Wavelength S-Band 1460–1530 nm	Conventional C-Band 1530–1565 nm	Long Wavelength L-Band 1565–1625 nm	Ultralong Wavelength U-Band 1625–1675 nm

Figure 13–22
The ITU has divided the frequency spectrum for optical transmission into six primary bands. Note that the E-band contains the frequencies affected by the water molecules—water peak.

specifications are used to manufacture the connector in order to guarantee that the ends are flush, keeping the light signal from escaping into the air.

Connectors are classified as UPC—ultra-polished—or APC—angled polished. The difference between the types is how the fiber is mounted in the ferrule of the connector The UPC meshes evenly to the fiber in the bulkhead, and the APC meshes at an angle, as shown in Figure 13–23. The purpose of an angled connector is to reduce the potential for the light to reflect as it hits the flush end of the fiber. Currently, UPCs are the most commonly deployed in the network. Whether it is a UPC or an APC, the basic structure is the same. Three components make up a connector: ferrule, body or cap, and strain boot. The ferrule is where the fiber is placed and is made of ceramic, steel, or plastic. The cap or body is the portion sitting just behind the ferrule and is used to secure the connector to the bulkhead either by a threaded barrel, a snapping mechanism, or a twist-lock spring. The boot of a connector is used to relieve the strain between the connector and the jumper. The ferrule, the body or cap, and the boot are shown in Figure 13–24.

Fiber jumpers are used to connect fiber optic equipment ports to FDPs—fiber distribution panels—or to connect a port on an FDP to another port on an FDP. A flexible outside jacket made of vinyl protects the fiber strand inside. Most jumpers are connectorized at the factory according to the engineer's specifications for distance and the type of connector. In some cases, especially when the distance between the devices is not known or when the fiber has to be pulled through a narrow opening not large enough for a connector, only one end of the fiber is connectorized. In these cases, the technician must splice in the connector once the jumper has been pulled through the conduit. A yellow colored jacket is used to represent single-mode fiber, and an orange jacket is used to represent multimode fiber.

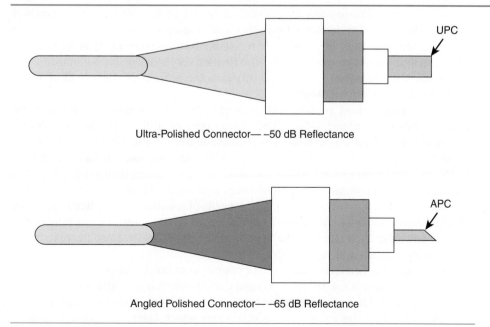

UPC

Ultra-Polished Connector— –50 dB Reflectance

APC

Angled Polished Connector— –65 dB Reflectance

Figure 13–23
The two most commonly deployed connectors are designed with either a flush ultra-polished end as shown in the top picture or an angled end as shown in the lower diagram. The APC has a better reflectance value, −65 dB, compared to the UPC, −50 dB. The UPC, on the other hand, has a greater surface area and is less likely to experience chipping than the APC. Both are used in optical networks. The APC is green, helping distinguish it from the UPC, since a barrel designed for a UPC will not mate to an APC and vice versa.

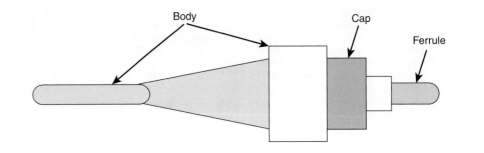

Figure 13–24
A typical fiber optic connector consists of a ferrule, a cap, and the body. The diagram is similar to an ST connector, and has a connector loss of around 0.5 dB. There are several other types of connector available, though each is constructed with the same three components: ferrule, cap, and body.

Fiber jumpers are placed in protective fiber gutters, helping ensure that they will not be accidentally pinched or cut. Fiber pinching, particularly from tie wraps, is a common cause of signal degradation and system failures.

13.3.8 Fiber Optic Transmitters

Though fiber attenuation and dispersion characteristics play a critical role in how far a light signal can travel down a fiber strand, the light source also must be taken into consideration, as it is responsible for the end-to-end distance a signal travels. Today, there are several types of light sources on the market. In general, light sources are categorized as lasers or LEDs. An LED is used for very short distance spans, emitting a fairly low powered, dispersed light that is limited to around 2 km. LEDs are less expensive than lasers, making them useful for LANs and short access loops.

Lasers have evolved over the years into very application-specific components that are built to meet the needs of a particular network. Long distance networks such as core long-haul networks depend on high-powered, well-engineered, expensive lasers. Short, last-mile access networks typically use low-powered, less precise, less costly lasers. Before selecting a laser, the engineer must look at the loss budget of the span, the type of fiber being used, and the type of traffic being carried.

The most common lasers on the market currently are the Fabry-Perot or FP, the VCSEL or LW-VCSEL—Long Wavelength Vertical Cavity Surface Emitting, and the DFB—distributed feedback—cooled or uncooled. The FP laser operates at the 1310 nm wavelength range and can be used on single-mode and multimode fiber across fairly short distances, such as around a campus or within high-rise buildings. The FP laser has been around for some time and has proven itself as a very cost-effective, reliable light source. VCSELs operate at the 850 nm range and are designed to function on both single-mode and multimode fiber. Often this type of laser is used to connect equipment located on the same floor or within a short distance such as connecting an ATM switch to a DCS, digital cross-connect switch. VCSELs are very inexpensive but transmit very short distances. The newer LW-VCSEL 1310 nm lasers are used to carry GigE traffic across access and metro networks.

The final and most prevalent laser in communications networks today is the DFB—distributed feedback—laser. DFB lasers operate in the 1310 nm to 1550 nm range and function on single-mode and multimode fiber in the 1310 nm window and single-mode fiber in the 1550 nm window. DFB lasers are used in local, long-haul, CATV and other networks that require the signal to travel farther than a few miles. DFB lasers are the most commonly deployed lasers in the network and, as such, have been tweaked and refined to match the particular needs of the different applications. When DFB lasers are thermally cooled and isolators are added, the lasers are much more precise and powerful, thus allowing the light to travel much farther down the pike before requiring amplification or regeneration.

The trade-offs between lasers are, of course, cost and function. If you were designing a metro network with short internodal distances, it is doubtful you would purchase a long-reach DFB cooled laser. Instead, you would look at the loss across the link and determine from the equipment specifications which laser fits your needs. Most vendors classify lasers as short reach, medium reach, and long reach, referring to the

Laser Type	Application
Short reach	CATV, BPL, access networks, short metro
Medium reach	Metro, regional
Long reach	Long haul, international

Table 13–5
Typical Laser Categories.

distance the light wave can travel across standard single-mode fiber. Table 13–5 provides a summary of the different lasers.

13.3.9 Transmit Power and Receiver Sensitivity

An optical output leaving a transmitter, laser or LED, is able to travel a certain distance before it becomes unrecognizable to the receiving equipment. In order to determine how far the light can travel, it is important to understand the roles the light source and the light receiver play in the end-to-end design of the system.

If you point a flashlight into an empty field, objects within a certain range of the light are illuminated. Objects beyond this range are not visible. However, if you turn on a floodlight and point it into the field, the distance the light travels is substantially greater and the objects out of range of the flashlight are now visible. The same is true in optical transmission. The optical power source is specced at a certain power level—similar to the flashlight compared to a floodlight. If, for example, the signal has to travel a long distance, the optical source must be strong enough to reach the far end receiver.

Optical light sources are very lucky in love. They always have a partner at the opposite end of the span—a receiver. The receiver's job is to receive the incoming signal and decipher it. It is similar to how our eyes are able to perceive the objects in the field when the flashlight illuminates them. Our eyes only see objects illuminated with light. There are, though, nocturnal animals that enjoy roaming about after dark (I'm not talking about teenagers) whose eyes have adapted to less light and are able to make out those objects we cannot see.

The same method is used to build optical receivers. The receiver's job is to look at the light pulse coming in, determine if it falls within the specified level—not too strong, not too weak—and interpret the one and zero pulses. A very sensitive receiver is able to "see" or interpret a lower level signal than a less sensitive receiver. The term *receiver sensitivity* is used to define the range of light a receiver is willing to accept. A receiver is also sensitive to the strength of the signal. Imagine how a cute little deer feels when a hunter shines a spotlight at it in the middle of the night. Do you think the deer can see the sinister smile on the hunter's face? Of course not; the light is much too bright. Even though this is a simple analogy, it makes the point perfectly. The optical receiver, similar to the deer, is blinded when the signal exceeds its maximum threshold and, consequently, the receiver is not able to decipher the pulses of light. If, though, the deer puts on specially designed spotlight glasses that reduce the intensity of the light, it is able to see the hunter's sinister smile and run. The same is true of the receiver. An attenuator is placed before the receiver in order to pad down—reduce—the signal to an acceptable level. Once the signal is padded down, the receiver is able to properly read the incoming pulses.

receiver sensitivity
Minimum and maximum level the light needs to fall into in order for the receiver to be able to decipher the incoming light signal.

13.3.10 Measuring Optical Power

We are now ready to take what we have learned so far and apply it to a loss calculation. Optical transmitters project a light at a specific signal strength into the fiber, and the receiver accepts or rejects the signal accordingly. Before we can determine how far the light will travel, we have to define the third piece of the equation—the loss associated with the system.

An end-to-end optical system consists of a fiber strand connecting the optical gear at each end. In order to feel comfortable that the light signal will travel from point A and arrive at the proper level at point B, a loss budget calculation must be made. The calculation may be made using an estimate or from actual loss measurement readings. As explained in an earlier section, loss in fiber is caused by inherent properties in the fiber in addition to human-made contributors such as splices and connectors. Loss or attenuation is measured in decibels, a unit of measurement used to quantify the signal level in the same way a foot is a unit of measurement used to quantify distance. In order to evaluate an optical system it is imperative that you are able to interpret decibel readings properly.

In telecommunications, the decibel is the unit of measure used to express the difference between power levels. For example, the difference between the loudness of your voice when you are whispering compared to the loudness of your voice when you are yelling can be expressed in decibels. The variations between the highest and lowest volume represent trillions of values. It would be impossible to state the difference between any one of these levels because there is such a wide range between the lowest and the highest. Therefore, the decibel is used to interpret these variations and provide a value that is understandable to mere humans. Though this is not a concept you need to remember, it is important to understand that the reason we measure power levels using the decibel is that it is able to adjust to and simplify the huge range of possible values.

dBm
Decibel referenced to 1 mW.

Fiber system loss is measured using the decibel and dBm scale. The term *dBm* is used to express a measurement value referenced to 1 mW. In simple terms, this means 1 mW of power is equal to 0 dB. OK, so it is not that simple. As the power level in milliwatts increases, so does the decibel reading. In order to take a measurement, a reference point or level had to be adopted in order to designate a starting point. If, for example, they had wanted to use 5 mW, the 0 dB reference would be 5, and so forth. It is a human-made reference that helps simplify the measurement process.

Understand that test equipment such as a power meter has a photodetector that receives and reads the light pulse before converting it into an electrical signal. The electrical signal is then referenced to the decibel scale producing a dBm value. The meter compares the milliwatt reading to the decibel. Therefore, when you take a power reading measurement, you are reading the light level at that point in the circuit in dBm. The light source pulses out a light signal into the fiber strand. It travels through the core of the fiber as we discussed in the beginning of this section, losing some of its strength along the way as it crosses splices; traverses each connector; and encounters bubbles, density changes, and so forth. At the far end, the dBm reading shows up on the test equipment screen and you feel very proud that you have found the total loss of the system. There is one catch. Are you correct in assuming that you have found the end-to-end loss of the system?

If, for example, the transmitter is sending a 5 dBm signal and the receiver at the far end is receiving a 15 dBm level, the total loss across the span is:

$$\text{Span loss} = +5 - (-15) = 20 \text{ dB}$$

As you can see from the calculation, a real loss of 20 dB has occurred across the span. The dBm reading is the actual level compared to 1 mW while 20 dB signifies the amount of loss between A and B. The loop budget in our example shows a loss of 20 dB from transmitter to receiver.

13.3.11 Reflectance and Optical Return Loss

Other factors that can affect the quality of the light signal are caused by reflections bouncing off components in the span. Two measurements, optical return loss (ORL) and reflectance, are used to calculate the amount of reflection in the system. ORL is a measure of the total system reflectance while reflectance looks at the reflectance of a specific component in the span such as a connector.

A simple way to visualize this concept is to think about what would happen if you were to shine a bright light into a mirror. The mirror would reflect the light back toward you. You, being the transmitter, may experience confusion due to this reflected light hitting the original light and thus changing its properties. However, this is a somewhat flawed analogy. If a physics teacher were in the room, he or she would shake a finger at us and begin explaining in excruciating detail the properties of light reflection. Nevertheless, for our purposes, it works fine.

ORL measures incident power over reflected power, while reflectance is a measure of the reflected power of a specific component over the incident power. Though this sounds fairly complex, it is simply a measurement of the light signal over the reflected signal or, as in reflectance, the reflected light signal divided by the light signal.

It is apparent that the ORL result will always be positive, and, because we want the reflective signal to be low the higher the value the better. In fact, most companies require that ORL values exceed 30 dB. Reflectance, on the other hand, is always a negative number. Similar to ORL, it also should be fairly large because the greater the value, the lower the reflectance.

Should we really care what the ORL and the reflectance value are in a span? Similar to tuning a guitar before playing it, it is important to make sure the system is in tune or in spec before putting an information signal across it. This is particularly true for DWDM systems or very high bit rate systems, as they are much more sensitive to reflectance.

◼ 13.4 WIRELESS TECHNOLOGIES

13.4.1 Radio Frequencies

Though not as tangible as a piece of copper or fiber, air is also used as a medium to transport information. From Marconi's first transatlantic radio transmission on top of Signal Hill outside St. John's, Newfoundland, wireless technology has steadily evolved into a compilation of wireless networks worldwide. Today the wireless industry includes many different types of technologies, each built to serve a particular part of the market. The airwaves are used by many types of wireless technologies such as microwave transmission systems, satellite communications, cellular, PCS, new short-reach protocols such as Bluetooth and Wi-Fi in addition to optical wireless systems. Figure 13–25 associates the frequency spectrum to communications technologies showing where each falls within the bands.

13.4.2 Microwave Systems

Though we cannot see it, our world is surrounded by an electromagnetic force that is able to carry information such as sound, radio signals, television signals, and microwave frequencies. Microwave has much higher frequencies and shorter wavelengths

Figure 13–25
Frequency spectrum is divided into sections assigned by the ITU internationally and the FCC nationally.

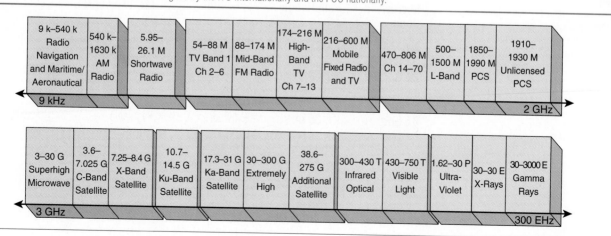

than radio and television and, as such, is able to travel a in very straight line. This ability, traveling in a straight line, is what allows microwave signals to travel significant distances before fading out. The first use for microwave transmission was of course in the military, an offshoot of radar. Soon communications companies such as AT&T in the United States realized they could use this new technology to carry telephone traffic over long distances, thus eliminating the need to run endless miles of copper wire.

Analog microwave systems were the first wireless long-haul systems deployed to carry information coast to coast. As with all other technologies in telecommunications, the analog systems were replaced by digital microwave systems that substantially increased the amount of bandwidth available and also increased the distance the signal traveled. With the advent of fiber optic transmission systems, microwave links were gradually replaced by fiber optic cable in the long-haul and local networks, slowing the growth of microwave technologies.

Today, microwave systems are again in vogue as the new, smaller, lower-capacity systems serve the end user market providing last-mile services at very cost-effective rates. Microwave systems are still being used by carriers for overflow and backup links. Microwave frequencies are still a valuable commodity in the market.

Frequency Spectrum Used

Microwave transmission falls within the C-band of the electromagnetic spectrum ranging from 1 to 40 GHz. Within this band, several subdivisions have been made to accommodate different types of applications, as shown in Table 13–6. The FCC controls the point-to-point licensed frequency spectrum and allocates frequencies according to availability. Unlicensed spectrum is available to anyone who is willing to risk encountering interference from other such systems. Today, both licensed and unlicensed—also called license-exempt—systems are deployed throughout the local market connecting small businesses to each other, cellular towers to the hub site, railroad switch points to the base, and on and on and on. Microwave's new popularity has brought about competition for all frequencies being sold by the FCC causing fierce bidding sessions between users. The FCC Web site, http://www.fcc.gov, lists the frequency auctions currently in progress and provides an interesting view of the actual process.

Not all microwave frequencies are auctioned off; many are granted through an application process that requires the user to fill out several forms and submit them to the FCC wireless group. Several companies have built a business on designing, obtaining the frequencies, and turning up the microwave services for the end user—turnkey operations. Currently, frequencies used for point-to-multipoint systems are handled at the regional and local levels and, as such, more easily—most cases—gain approval.

Table 13–6

Microwave Frequency Allocation.
From BandLocator Table:
http://www.comsearch.com/bandrang
e/service8-p2.jsp

Microwave Name	Path Range	Application	Licensed or Unlicensed
960 MHz	2 to 31 km	12 FM/FDM voice	FCC regulated
2.1 GHz	1 to 50 km	4 DS-1s	Reserved for emerging technologies
2.4 GHz	1 to 50 km	1.0 FM video	Reserved
4 GHz	17 to 38 km	2 DS-3s	Unlicensed-Part 15
6.1 GHz	1 km	1 OC-3	Regulated
6.5 GHz	9 to 36 km	1 FM video	Regulated
6.7 GHz	1 to 34 km	1 DS-3	Regulated
10 GHz	2 to 14 km	16 DS-1s	Regulated
11 GHz	1 to 5 km	1 OC-3	Regulated
13 GHz	1 to 22 km	1 FM video	Regulated
18 GHz	1 to 6 km	2 DS-3s, 1 DS-4	Regulated
23 GHz	1 to 5 km	1 OC-3	Regulated

Multiplexing/Encoding Schemes

The first microwave analog systems deployed frequency division multiplexing (FDM) schemes, combining several channels into one frequency. Traditional digital microwave systems use a TDM—time division multiplexing—method to combine many channels into one and often depend on QAM or FSK to encode data before transport. The newest method used to combine channels and encode information in digital microwave systems uses spread spectrum technology similar to PCS telephone networks. Spread spectrum technology is discussed in detail later in the text.

System Components

Microwave systems consist of antennas with transmitters and receivers, a radio frequency (RF) oscillator with a digital modem, and waveguide or coaxial cable as an interconnection medium. The RF oscillator takes an intermediate frequency generated by the internal digital modem and blends it with a microwave signal produced by the device. The output of the RF unit is fed directly to the antenna, which projects it into the atmosphere for transport. At the receiving end, the antenna accepts the incoming RF microwave signal and feeds it to the RF unit that strips off the microwave signal leaving the intermediate frequency. The intermediate frequency is demodulated by the digital modem back into an electrical signal that can be terminated on a DS-1 or DS-3 DSX panel, where it will be cross connected to a land line, as shown in Figure 13–26.

If the RF generator is located a distance from the antenna, a waveguide, as shown in Figure 13–27 is used to carry the signals between the two devices. A waveguide is a rectangular copper duct—similar to heating ducts, though smaller—designed to carry high-frequency microwave signals between the electronic equipment and the antenna with very little loss. Occasionally, when the antenna and the RF generator are close to each other, coaxial cable may be used to interconnect the two. Designing a system revolves around determining the total loss of the system and, if possible, where loss can be reduced.

Two final components often found in microwave systems—especially the larger transport units—are branching filters and directional couplers. Directional couplers similar to bidirectional fiber couplers filter both the transmit and receive signals into one waveguide, thus reducing the cost by eliminating the construction of one waveguide in addition to freeing up physical real estate space. Branching filters are used to

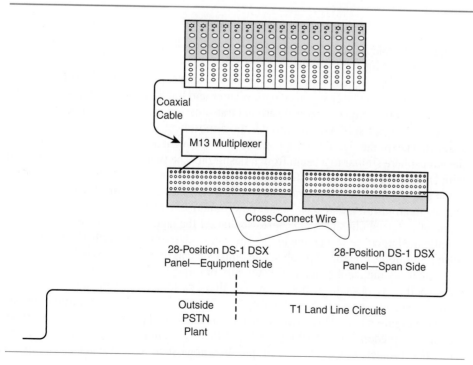

Coaxial Cable

M13 Multiplexer

Cross-Connect Wire

28-Position DS-1 DSX Panel—Equipment Side

28-Position DS-1 DSX Panel—Span Side

Outside PSTN Plant

T1 Land Line Circuits

Figure 13–26
Microwave signal connects to the outside plant PSTN through DSXs.

Figure 13–27
View of a microwave transmission site. The land line connections are tied to a DS-1 DSX where they cross connect into an M13 multiplexer. From the M13 multiplexer, the signal travels across coaxial cable into a DS-3 DSX panel where it is cross connected to the RF multiplexer. From the RF multiplexer, the signal is converted into a radio frequency that is sent across the waveguide up to the microwave antenna.

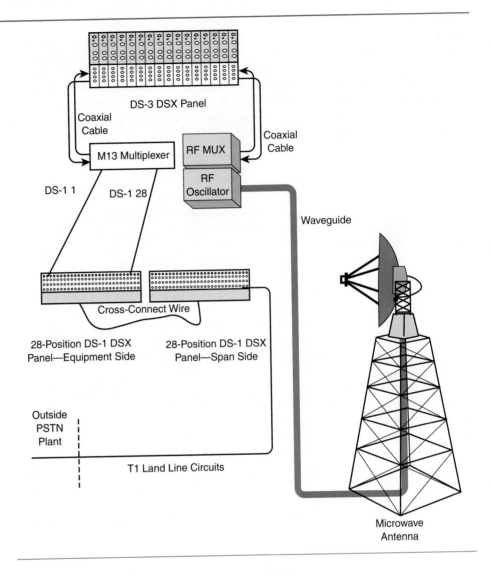

antenna gain
Measure of how much the antenna increases the transmit and receive signal power.

directional
Antenna projects the signal in one direction toward one receiving antenna.

omnidirectional
Antenna projects the signal to multiple receiving antennas.

combine multiple transmitters and receivers into one waveguide. Again, this helps reduce the cost of the total system.

Microwave antennas vary in size, geometric shape, and transmitter radiation. Because the microwave signal wavelength is short and the transmit power fairly low, the antenna must be designed precisely around transmission parameters in order to be able to send the signal successfully through the atmosphere. An antenna has a parabolic shape that keeps the signal from spreading, thus forcing it to project a narrow beam into the atmosphere similar to a beam from a flashlight. The two key parameters associated with an antenna that directly affect how far the signal can travel are the gain of the antenna and the directionality. *Antenna gain* is a measure of how much the antenna increases the transmit and receive signal power and is the result of the physical size and shape of the parabolic antenna sphere. The larger the antenna, the higher the gain; or, simply; the larger the saucer, the higher the gain and, in turn, the farther the signal can travel. This holds true for all parabolic antennas such as those seen in satellite networks.

Antennas radiate the signal directionally or omnidirectionally—hitting one point on the bull's-eye or multiple points on the bull's-eye, as shown in Figure 13–28. A directional antenna projects the signal in one direction, thus, the term *directional*, as shown in Figure 13–29. Directional antennas are used in all point-to-point applications and are the predominant topology deployed today. *Omnidirectional* describes an antenna that projects the signal in multiple directions such as in a point-to-multipoint

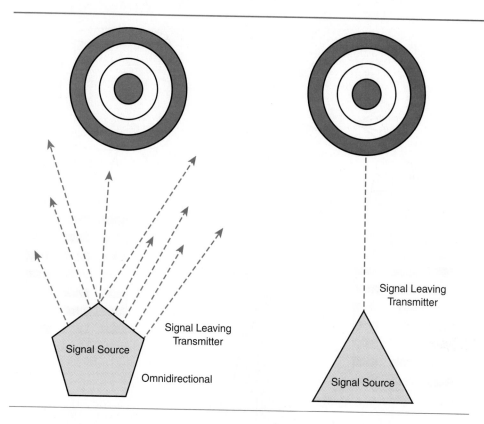

Figure 13–28
In order for light to travel in a single beam, the transmitter must be designed to focus the output signal. When light projects out of a wide lens, the rays scatter and dissipate much faster than when they are focused in a single beam.

network, as shown in Figure 13–30. Photo 13–5 shows a typical directional antenna mounted on a freestanding tower. Microwave antennas are found on top of high-rise buildings, on freestanding or guy wire towers, billboards, farmers' silos, and any other high up places.

A microwave signal is able to travel, depending on its transmitting power, a maximum of thirty-five miles before requiring the signal to be regenerated. Two types of regenerators are used to extend the distance a signal can travel. The first type of regenerator, referred to as a baseband regenerator, demodulates the signal, recreates the signal, and transmits the signal on to the next microwave site. The second type of regenerator, called a heterodyne regenerator, uses an intermediary frequency that eliminates the need to demodulate the signal. The purpose for either method is to increase the distance between sender and receiver.

Interference on a Microwave Link

Interference on a microwave link contributes to signal degradation and possibly circuit failure. Therefore, during the design phase, it is very important for the engineer to be aware of any disturbers that can affect signal quality. The most common disturbers are

Figure 13–29
Long-haul microwave systems are commonly used to carry traffic between two distant cities.

Figure 13–30
Omnidirectional antenna sending the signal to multiple buildings on a campus.

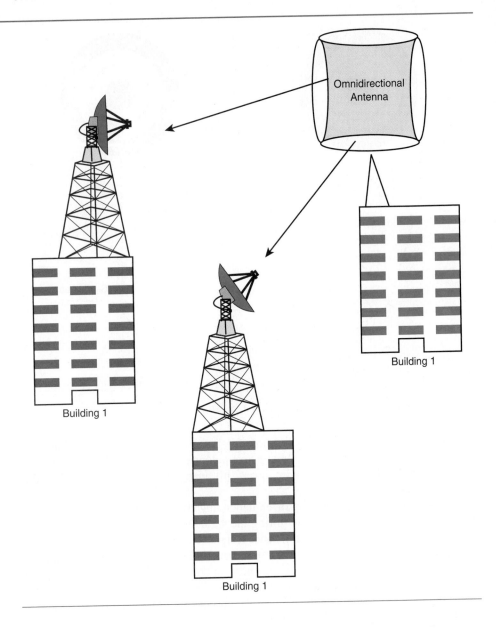

rain fade, atmospheric fade, cross talk, and reflections. Rain is one of the most common culprits that cause the microwave signal to dissipate, as the size of a rain drop can match the wavelength of a microwave, some being as short as 0.5 cm. When a very heavy rain falls, the microwave signal is washed out and the communications signal lost until the rain lets up. Design engineers should research the rainfall statistics of a region before deciding to deploy a microwave system. If the area experiences hard rain, not necessarily long rains, the engineer may opt to lessen the distance between the two antennas in order to prevent the possibility of the system failing during heavy rainfalls.

Fade caused by temperature inversions, fog, and so forth also should be taken into account when designing a microwave link. A fade margin should always be used when calculating the total path loss, as atmospheric conditions can increase the loss experienced on the line. A system should never be engineered for clear air only, as very seldom does an area experience clear, dry conditions. The term *fade margin* refers to a buffer given to compensate for any problems encountered on the path at any given time. It is similar to hedging your bet by making sure you always have a stash of cash for emergencies.

fade margin
Loss value referred to as a *buffer value* added to the loss budget calculation to compensate for unknown loss components.

Reflections caused by the signal bouncing off an object located within the signal's path is called Fresnel effect, similar to the Fresnel effect described in the section on optics. A Fresnel zone is constructed around the signal path mapping out the area that must be kept clear in order for the signal to travel straight and not reflect back. Objects such as smoke stacks, mountains, and so forth may cause the microwave signal to bounce or echo, resulting in a distorted signal and loss of communications.

Interference similar to cross talk can be caused by adjacent signals overlapping into each other's frequencies making it difficult for the receiver to decipher who is what and what is who. Multipath fading happens when signals arrive out of phase, thus causing the signal to attenuate at a higher rate than anticipated. A final type of interference called *overreach interference* happens when a signal shoots past the antenna and reaches an antenna farther out, thus causing the poor little antenna to try and figure out what it is receiving. The term *frequency hopping* is used to describe the function of equipment that is capable of correcting this problem by forcing the signal to jump to a new frequency before being repeated to the next tower.

Microwave Link Loss Budget

Microwave links are designed according to signal level measured in dBW, referring to decibels referenced to 1 W. Determining the total loss across a microwave link is accomplished through a loss budget calculation. Microwave link loss is calculated by adding up the total loss across the link—connector loss, and waveguide loss, loss per kilometer. In addition, engineers always include a fade margin or loss values used to compensate for environmental factors such as rain fade.

Before the link budget is calculated, the transmit power, antenna gain, and receiver sensitivity are determined by reviewing equipment specifications. Transmit power and antenna gain relate to the strength of the signal as it leaves the antenna. Transmit power is regulated by the RF license granted by the FCC. The term *RSL*—received signal level—is used to define the signal threshold at the receive side of the link. If the signal is too weak, below the RSL, the receiver will not be able to decipher the incoming ones and zeros. If the signal arrives above the RSL, it will overpower the receiver and cause reflections and echo. Once these values have been determined, a very simple calculation is used to estimate the link loss.

Transmitter power/antenna gain − (component loss + waveguide loss + link loss + fade margin) = dBW level at the receiver.

As is obvious, the design of a microwave path requires that the antennas are sized properly, the transmit power meets specifications, the connectors are constructed correctly, the waveguide is engineered with the lowest loss, and the environmental factors for the area such as heavy rain or temperature inversions are taken into consideration.

A typical long-haul microwave link is able to travel around thirty-five miles before the signal fades to an unrecognizable blur. Today, smaller systems such as those carrying one or two T1 circuits are designed to travel shorter distances. The shorter the distance, the less powerful the transmitter, the less powerful the transmitter, the less costly the system. Whether the link has to travel many miles carrying large amounts of data or only a few blocks, the design remains the same.

Microwave Network Architecture

Two types of architectures are deployed in microwave systems—point to point and multipoint. The most common systems are point to point and are used to connect two sites together by pointing two antennas toward each other, as shown in Figure 13–31. Multipoint architectures use omnidirectional antennas that broadcast the signal to many other antennas, as shown in Figure 13–32. A multipoint design has become much more common, specifically in the access network, as data-only wireless services have become popular. A good example of using multipoint design is when a campus has to connect several buildings to a LAN. An omnidirectional antenna is

overreach interference
Signal shoots past the antenna and reaches an antenna past that being targeted.

frequency hopping
Receiver forces the signal to hop to a new frequency before being transmitted to the next antenna.

RSL
Received signal level refers to the signal threshold at the receive side of the link.

Figure 13–31
Point-to-point systems carry traffic
between two microwave antennas.

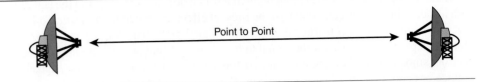

Point to Point

placed strategically in the center of the campus, and directional antennas are mounted on each of the buildings. Typically, the omnidirectional antenna also receives its upstream link from the data center from a point-to-point microwave link, as shown in Figure 13–33.

Large, point-to-point, long-haul networks deploy line-of-site connections used to carry traffic many miles across the country, as shown in Figure 13–34. Repeaters are placed every so many miles along the path helping to extend the distance the signal can travel end to end, as shown in Figure 13–35. Local telephone companies also use point-to-point microwave architectures to interconnect central offices. Rural areas often depend on microwave systems to cost-effectively carry traffic between remote offices.

End users have also started to deploy microwave links—both point to point and multipoint. The newest microwave technologies LMDS—local multipoint distribution systems—and MMDS—microwave multipoint distribution system—define how the new lower-rate microwave technologies function. In addition, the introduction of unlicensed frequencies has spurred on the adoption of small, easy-to-turn-up systems, eliminating the headache of submitting a request to the FCC for radio spectrum.

The types of circuits deployed on microwave systems range from T1 digital circuits up to 155 Mbps STS-3s. The most common signal deployed is the DS-3, which is used to carry multiple DS-1 circuits. The types of traffic riding on microwave signals are the same traffic types that ride on copper and fiber systems. Remember, all three are media used to carry information. They do not care what they carry.

Figure 13–32
Microwave networks are designed as
either point to point or multipoint.

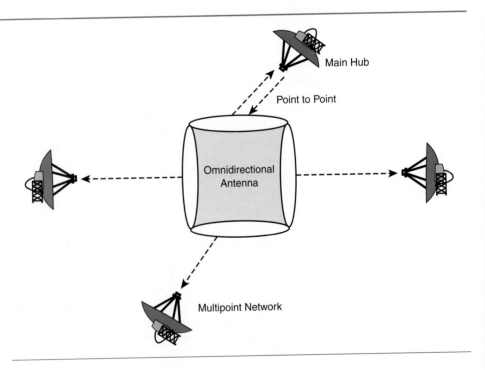

Main Hub

Point to Point

Omnidirectional
Antenna

Multipoint Network

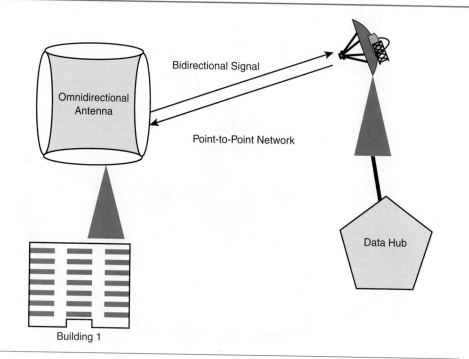

Figure 13–33
Within a multipoint network, a point-to-point leg is built to carry traffic from the main hub site to the omnidirectional antenna.

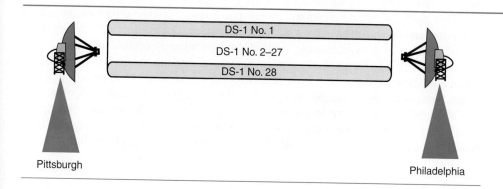

Figure 13–34
Long-haul microwave systems carry DS-3 signals comprised of twenty-eight DS-1 signals.

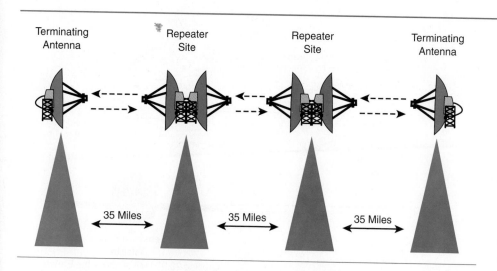

Figure 13–35
Microwave repeaters are used to regenerate the microwave signal about every thirty-five miles. All microwave systems must be designed as line of sight.

Figure 13–36
A satellite hovering 22,300 feet above the earth is able to cover 41% of the globe excluding the North and South poles.

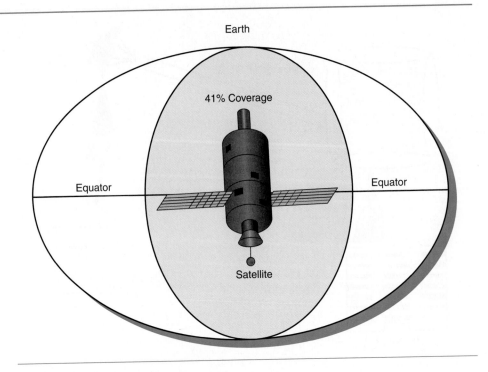

13.4.3 Satellite As a Wireless Technology

Satellites have been hanging over our heads since the early 1960s. Today hundreds of satellites used for communications, global positioning, defense purposes, broadcast television, and on and on have been launched into space. The majority of satellites in orbit today hover in geosynchronous orbit, meaning stationary orbit, above the equator 22,300 miles up. The footprint of one geosynchronous satellite covers 41% of the earth's surface, as shown in Figure 13–36. In order to cover the entire world, minus the polar regions, three satellites are placed strategically around the globe. This allows for live television broadcasts anywhere in the world. A person living in Katmandu, Nepal, can watch the World Series at the same time as people sitting in their living room in Cleveland, Ohio. Figure 13–37 simplistically illustrates satellite coverage of the United States.

Figure 13–37
Geosynchronous satellites are used to transmit communications signals around the globe. Each satellite covers 41% of the earth. North America falls inside the footprint of one satellite.

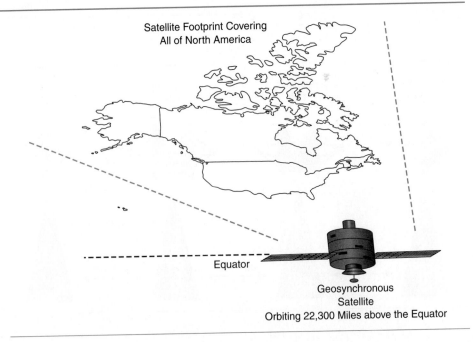

Common Band Name	Frequency Range MHz
C-band	3650–6875
Ku-band	10,750–14,500
Ka-band	18,300–30,000
V-band	37,500–38,500

Table 13–7
Satellite Frequency Allocation. *From BandLocator Table: http://www.comsearch.com/bandrange/service8-p2.jsp*

AT&T launched the Telstar satellite in the early 1960s to carry communications traffic across the ocean. The evolution of the satellite helped spur the growth of telecommunications services internationally. The advent of fiber optic transoceanic cable significantly reduced telecom carriers' dependence on satellite systems. Today, satellite transmission systems are used mainly for data services, typically connecting global corporations' remote locations. Also, new low-orbiting satellites have been launched around the globe and are being used for satellite telephone service, touting an anywhere-in-the-world service.

Frequencies Assigned to Satellites

Three bands have been assigned to satellite services by the standards organizations. The C-, Ku-, and Ka-bands cover frequencies ranging from 4 to 36 GHz, as shown in Table 13–7. The C-band encompasses a lower frequency spectrum than the Ku- and Ka-bands making it more desirable because it experiences less loss per kilometer. (Lower frequencies have less loss than higher frequencies.)

Due to the radio spectrum's finite number of frequencies, satellite services are subject to the same bureaucratic rules placed on microwave and cellular services. The C-band, though offering lower loss characteristics, does border terrestrial microwave frequencies, making it vulnerable to interference. Even though the C-band may experience interference, it still is the most desirable frequency spectrum to use for satellite services. The Ku-band is the second most desirable band, even though it experiences a higher loss per mile than the C-band. The Ku-band is also much more susceptible to interference such as moisture, rainfall, snow, and temperature fluctuations. Recently, systems have been designed for the Ka-band, thanks to solid-state electronics and their ability to compensate for loss and interference. The Ka-band is the least desirable, as the frequencies are higher and more sensitive to interference, in addition to experiencing higher loss per mile than the other two bands.

Satellite System Structure

A satellite system used for communications consists of several elements: two earth stations, an uplink and a downlink, and the satellite. A satellite has a transmitter, a transmitting antenna, a receiver, a receiving antenna, transponders, an uplink/downlink interface, solar panels, solar batteries, and a stabilizer. The satellite receiver accepts signals coming in from the earth station across the uplink. The signal travels from the receiver across an interface tying the receiver to the transmitter. The transmitter accepts the signal, processes it, and sends it to the transmitting antenna. The transmitting antenna shoots the signal out into space (downlink) at a power level ranging from a few watts up to 200 W. The signal may be either broadcast or directional depending on the network requirements. A broadcast signal is sent to every receiver in the footprint; a television signal is a good example of a broadcast link. A directional signal aims or points at one receiver; a VSAT—very small aperture terminal—satellite link is a good example of a direct signal.

The *transponder* in a satellite is used to generate or receive a frequency, completing the circuit between the uplink and the downlink. The higher frequency used on the uplink as the output power of the transmitter is much easier to supply on the ground than the solar-powered satellite transmitter. Most satellites have several transponders,

transponder
The term used to describe the receiver or transmitter of a satellite system.

some as many as forty-eight, each able to handle an individual communications link. Companies that do not own their own satellite—the majority, as launching a satellite is not cheap—may purchase time on a transponder and build their own point-to-point satellite connection. In addition, a satellite may hold both C- and Ku-band transponders to provide flexibility to service providers.

Solar panels and batteries are used to power the satellite. The life of a satellite is often tied to the life of the power system, not the structural device. The sun, of course, supplies the power source to the solar panels that are used to generate enough power to transmit signals. When the sun is not available, the battery system kicks in providing full power backup—similar to land line systems.

Telemetry is needed to control and monitor the satellite. Technicians sitting in nice comfortable network operations centers—NOCs—are able to log into the satellite and check the links for bit errors check battery power, and so forth. Stabilizing the bird using either spin stabilizers or three-axis stabilizers helps to keep the orientation of the satellite to the earth station secure. Geosynchronous satellites are less affected by orientation problems, as they remain stable at 22,300 miles up. In contrast, the low-orbiting satellites must be kept in line as the changes in terrain can alter the satellites' view of the earth.

Depending on the application, several types of earth stations are deployed on the ground. Large television stations depend on very large satellite antennas with very high output power and high gain; the larger the antenna, the higher the gain. Smaller antennas are also common today and very useful as they are lower in cost and require smaller footprints or space. The satellite antenna is similar to the microwave antenna in that the design is critical to the application. Satellite antennas also follow the same link budget rules explained earlier. Gain and directionality are variables that can be adjusted depending on the functional requirements of the network. Network topology can be point to point or multipoint. When a network has many earth stations, one station is designated as the master and is used to poll or control the flow of information to the others using an access polling scheme such as TDMA—time division multiple access.

Each earth station has an antenna, an RF generator or oscillator similar to microwave systems, a waveguide or coaxial cable interface between the generator and antenna, and multiplexing equipment used to aggregate the traffic into a single signal. The RF generator accepts the intermediate frequency and blends it with a microwave signal. The modulated signal is shipped out onto the interface, waveguide, or coaxial cable to the satellite antenna. The transmitter projects the signal onto the uplink, that is, shoots it into the atmosphere.

In the reverse, the antenna receives the signal from the downlink, sends it to the RF generator that demodulates it, and sends it on to the multiplexer. The multiplexer pushes the signal onto the land line. Essentially, a microwave system is the same as a satellite system minus the satellite floating around over our heads in space. Figure 13–38 shows each of the components associated with a satellite link.

Interference Affecting Satellite Transmission

Rain and moisture cause the greatest number of problems on satellite links. As mentioned earlier, the Ku- and Ka-bands are much more sensitive to rain fade and temperature changes than the frequencies associated with the C-band. Due to this phenomenon, a C-band transmitter may output as little as 20 W of power while a Ku-band transmitter outputs 120 W of power. As with microwave systems, a full path analysis should be made during the design phase of the network. Transmit power, receiver sensitivity, what frequencies to use, and the weather patterns of the areas should all be studied. A link may experience as much as 200 dB of loss between the earth station and the satellite.

A second type of interference that can affect a signal is similar to cross talk or frequency crowding and is associated most often with frequencies in the C-band as they sit closest to terrestrial microwave. Spectral analysis is often done before a frequency is selected, helping reduce the chance of frequency crossover. The final type of

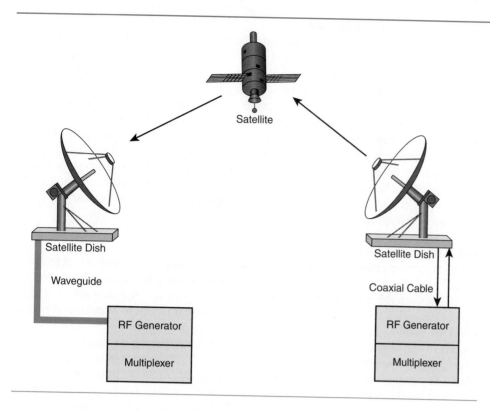

Figure 13–38
The typical structure of a satellite system includes the satellite antenna, the satellite RF receiver that accepts the incoming signal from either the waveguide or coaxial cable, and a multiplexer.

interference affecting satellite systems is radiation from the sun, occurring most often in the spring and fall. Though this issue cannot be corrected—because the sun cannot be moved at will—design engineers have learned to adjust the satellite's position during these seasons.

Satellite System Types

Today, three types of satellite systems are commonly deployed around the world: broadcast, directional, and VSAT. Broadcast satellites are used to broadcast information to multiple locations: the best example is broadcast television. The second type of satellite service is the directional service that sends signals between defined points using traditional satellite technologies. The most common examples of this service are large communications networks such as AT&T that sends communications information around the world between designated earth stations. The third and newest is the VSAT—very small aperture terminal—that has become popular in the past ten years. A VSAT is mounted on a rooftop or other fixed point and used to send signals up to the satellite, which relays them to one or more terminals. The large box retail chains often deploy VSAT systems to tie their stores to their distribution point. They place an antenna on top of each store's roof and send information such as inventory, balance sheets, and so on to the main headquarters hundreds of miles away through a bird in the sky.

Satellite Control

Unlike microwave systems, which are regulated by a country's communications organization such as the FCC in the United States, satellite control is divided into domestic and international. International control is handled by Intelsat. Intelsat is comprised of a monopoly of companies created by a treaty of agreement. The group is responsible for allocating frequencies, setting standards, and overseeing issues associated with satellite transmission globally. Domestic control is handled by several companies such as COMSAT, AT&T, GTE, and so forth that help set and regulate the industry in North America.

13.4.4 Optical Wireless Systems

Free Space Optics (FSOs) emerged onto the market in early 2000. FSOs have been deployed as last-mile, point-to-point systems used to carry all types of traffic, voice, video, and data in the local network. Using the air as a medium, FSOs shoot a narrow beam from a laser toward a receiving device that has a very sensitive lens. The laser transmits at the same frequencies as those in fiber optic systems, primarily at the 1310 nm and 1550 nm wavelengths. The distance between the two FSOs is dependent on the speed of the signal and the local climate conditions. The advantage of deploying an FSO compared to pulling fiber into a building is that physical construction issues such as digging up the road to place the entrance cable or vying for municipal right of way or building access are eliminated. Often, an FSO link is used in place of a diverse route and is used only when the primary fiber optic route is disrupted.

FSO System Components

An FSO system consists of two devices similar to antennas, though not referred to as antennas. Each device has an optical transmitter and a highly sensitive receiver. The light from the laser is projected through a telescope into the atmosphere and received through a telescope at the opposite site where it is sent to a receiver that connects to inside plant fiber optic cable, as shown in Figure 13–39. Once at the receiver, the signal is fed to a fiber optic multiplexer similar to that of a physical fiber system. The FSO telescope is mounted on the roof or some other fixed location above any impediments such as a building, billboard, or such. Between the device mounted on the roof and the equipment placed in a relay rack in the communications room, an inside Mic cable is run using SMF and fiber optic connectors such as those used in typical fiber systems. In addition to the fiber cable, a power feed must be run to the antenna for power. Most vendors offer both AC and DC electrical interfaces. FSOs are fairly simple devices and should be considered as a replacement for fiber optic transmission cable in the last mile of the network—especially for diverse entrance.

Figure 13–39
A Free Space Optics system has two antennas or telescopes located on top of the roof or antenna, fiber optic cable used to connect the telescope, and the transmit/receiver. Power is also run between the power source and the telescope.

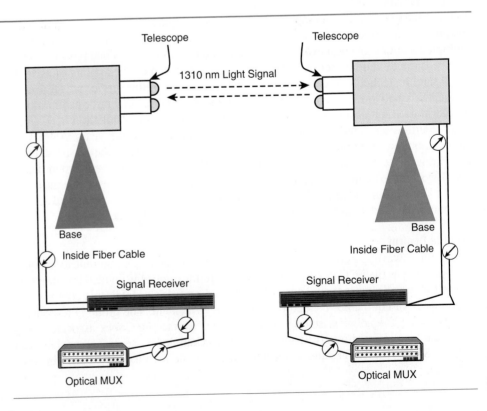

Interference Affecting FSOs

FSOs, similar to microwave and satellite systems, experience interference that can affect the quality of the signal and possibly cause it to degrade to a point that it is indistinguishable from noise at the far end. The most problematic type of interference is fog. FSOs' signal is very susceptible to fog as it absorbs the light causing it to scatter—attenuate—more rapidly. If an FSO is placed in a city such as Seattle, the engineer must take into account the number of foggy days and the intensity of the fog. In these situations, such as in Seattle, the solution is to move the antennas closer to each other allowing for additional signal degradation similar to increasing the fade margin. Luckily, rain or snow does not affect FSOs' transmissions. However, they are affected by moisture in the atmosphere that may cause the light wavelength to be absorbed and subsequently degrade the signal. Scattering is another type of interference that is caused when the signal hits particles and scatters, dispersing the light throughout.

In addition to atmospheric interference, objects flying through the beam can momentarily interrupt the signal forcing it to resynchronize with the transmitter. Birds often are the cause of this type of interference, though many vendors have added additional beams—referred to as spatial diversity—to eliminate this problem. Heat rising from the rooftop also may interfere with the light beam. Again, adding spatial diversity helps lessen this problem. The final issue that should be taken into consideration before designing a system is building sway. Building sway may cause the two antennas to misalign, thus allowing the beam to miss the appropriate target. Compensating for building sway is often accomplished through the use of a tracking device.

FSO Advantages and Disadvantages

The two most significant advantages provided by an FSO are the higher amount of bandwidth available when compared to traditional RF transmission systems and the fact that the frequencies used are not regulated, meaning that the user does not have to file for, bid on, or purchase frequencies. Additionally, FSOs are fairly economical as the system is simple and the components used are common in the industry. Most customers use FSOs to service those hard-to-reach places such as office towers that have restricted access or for backup links. Most FSO vendors offer systems that are able to carry bit rates as high as 2.5 Gbps and expect to handle rates up to 10 Gbps in the near future. The system is usually specified to carry the standard optical rates starting with DS-3, OC-3, OC-12, and OC-48. An engineer should always remember: the higher the bandwidth, the shorter the distance the signal can travel.

Some of the disadvantages of FSOs are the cost to rent roof space or inside riser space and interference issues such as those mentioned earlier. Many building owners have learned that their roof space is a valuable commodity and that charging as much as $700 to $1000 a month is acceptable. Additionally, many building owners have opted to charge for riser space for every cable pulled. Interference from atmospheric disturbances may force an engineer to avoid using FSOs such as in locations that experience high amounts of fog as in coastal regions. Even when the climate experiences heavy fog, FSOs can still be deployed as long as the engineer lessens the distance between sites, though the practicality of doing so may not exist, such as if there is no alternative location available. Overall, though, the introduction of FSOs has helped improve last-mile connectivity and, as predicted by many analysts, provides a viable way to optically link to a building.

13.4.5 Other Wireless Technologies

Several types of wireless technologies have emerged onto the scene over the past couple years. The two most commonly discussed are Bluetooth and Wi-Fi. Bluetooth is used to connect devices together over fairly short distances, such as PCs sitting on the same floor of an office building. Wi-Fi, on the other hand, is used to connect devices within the access network and is used to carry data transmission using wireless protocols. Both use radio frequencies as their medium.

SUMMARY

Communications traffic must have a medium to travel across just as vehicles need a road to drive on. The three types of media used in the communications networks are copper cables, fiber optic cables, and wireless technologies. Chapter 13 focused on the physical attributes of all three and the most common components used by each. The first section of the chapter looked at copper cables, both twisted pair and coaxial cable. The causes of signal attenuation on copper wire were discussed, mentioning the presence of resistance, capacitance, and inductance on all copper lines. The need to twist the wire was explained and the category nomenclature noted. The uses and physical structure of coaxial cable were presented. Methods used to terminate both twisted pair and coaxial cable were talked about, listing the different methods used to terminate copper wire such as punch-down blocks and wire wrap pins. The chapter also noted the pinout specifications for several connectors.

Fiber optic transmission cable was presented next in the chapter, including a look at the physical structure of the fiber strand: core, cladding, and coating. Attenuation and dispersion on fiber cable were explained, noting that both vary according to wavelength and fiber type. The different types of fiber were also discussed, making sure to relate each to dispersion. The different types of lasers used in networks today were presented, noting the differences between the Fabry-Perot, the VSCEL, and the DFB. Finally, fiber optic termination connectors were described and illustrated.

Wireless technologies discussed in the chapter included microwave systems, satellite systems, and FSOs—Free Space Optics. Microwave path loss was explained, as were antenna gain and directionality as they relate to transmit power. Topics such as microwave frequencies, licensed versus unlicensed, and the architecture of microwave networks were described.

CASE STUDY

Using the maps in Appendix A, determine what type of media should be deployed in Green Grass to support the communications services requested. Make a copy of the map and draw in the type of media you would deploy in that area. Like a puzzle, you should try to place each of the media types in the network before selecting the final transmission media for that portion of the network. Your selections are fiber cable, copper cable, and wireless transmission systems.

REVIEW QUESTIONS

1. What are the two primary transmission parameters engineers use when designing a circuit?

2. Circuit bandwidth is dependent on _____ of the physical medium.

3. Define the term *attenuation* as it relates to a communications medium.

4. List the five common factors used when designing a circuit.

5. The shorter the distance between the transmitter and receiver, the greater the loss due to higher resistance.

   a. true   b. false

6. A twenty-six-gauge wire has a lower resistive value than a twenty-four-gauge wire.

   a. true   b. false

7. Explain how the telephone line acts like a capacitor.

8. Engineers place load coils in the circuit to:

   a. amplify the voice signal in order to extend its reach down the line
   b. reduce the amount of loss in the circuit
   c. filter out cross talk coming from adjacent circuits
   d. counteract the capacitive effect on the circuit allowing the signal to travel farther down the line
   e. none of the above
   f. all of the above

9. Load coils are placed on all circuits greater than 18,000 ft. including DSL and ISDN circuits.

   a. true   b. false

10. The decibel is a unit of measure used in communications to measure the strength of a signal.

    a. true   b. false

11. List three methods used to terminate copper wire.

12. *DTE* refers to

    a. Data Terminal Equipment
    b. a computer
    c. a server
    d. a router
    e. c and d
    f. a, c, and d
    g. a and b
    h. all of the above

13. A fiber strand consists of a central _____, surrounded by a second layer of glass referred to as the _____, encased inside a plastic _____.

14. Explain why light travels farther down a single-mode fiber strand than a multimode fiber strand.

15. Inherent properties found in fiber that contribute to signal loss—light losing power—are:

    a. EMF
    b. absorption
    c. scattering
    d. radiation
    e. all of the above
    f. b, c, and d
    g. a and d

16. Properties produced by human-made events that contribute to signal loss in fiber are:

    a. dirt on the end of the fiber
    b. violating the bend radius
    c. EMF
    d. misaligned splice
    e. a, c, and d
    f. a, b, and d
    g. all of the above

17. Explain the term *dispersion* as it relates to fiber optic transmission systems.

18. Explain the difference between SSMF and NZDSF fiber.

19. The 1310 nm wavelength has a _____ dispersion point and on average experiences _____ dB/km of loss.

    a. 10 and .15
    b. −0.5 and 0.15
    c. 0 and 0.25
    d. 0 and 0.35
    e. none of the above

20. *UPC* refers to an unpolished connector and *APC* refers to an angled polished connector.

    a. true   b. false

21. List the two types of light sources used in communications networks.

22. Receiver sensitivity refers to the light level the receiving equipment needs to see in order to decipher the signal.

    a. true   b. false

23. Loss across a fiber span is measured in _____, and the power level from the optical transmitter is measured using _____.

    a. dBm and dB
    b. dB and dBm
    c. dB and dB
    d. dBm and dBm
    e. mW and dBm
    f. dB and mW
    g. none of the above

24. Microwaves are much longer than radio waves and, as such, are able to travel much greater distances.

    a. true   b. false

25. A microwave system consists of the following:

    a. transmitter
    b. receiver
    c. waveguide
    d. radio frequency oscillator
    e. digital modem
    f. antenna
    g. a, b, c, and f
    h. all of the above

26. Antenna gain is a measure of how much the antenna increases the signal power and relates directly to the size of the antenna.

    a. true   b. false

27. Circle the most common disturber that cause signal loss on a microwave link.

    a. atmospheric fade
    b. optical dispersion
    c. cross talk
    d. reflectance
    e. fog
    f. rain fade
    g. splice loss

28. What application might use a point-to-multipoint microwave system?

29. The _____-band, _____-band, and _____-band are used for satellite systems.

30. What does *VSAT* stand for, and where might it be deployed?

KEY TERMS

signal loss (317)

bandwidth (317)

attenuation (317)

line impedance (320)

balance a circuit (320)

load coils (320)

loaded circuit (321)

disturbances (322)

cross talk (323)

66 block (325)

core (332)

cladding (332)

coating (332)

scattering (334)

absorption (334)

radiation (334)

Rayleigh backscatter (334)

Fresnel reflection (334)

bend radius (334)

four-wave mixing (337)

water peak (337)

CWDM (338)

fat WDM (338)

receiver sensitivity (341)

dBm (342)

antenna gain (346)

directional (346)

omnidirectional (346)

fade margin (348)

overreach interference (349)

frequency hopping (349)

RSL (349)

transponder (353)

14

Telecommunications Networks' Physical Infrastructure

Objectives

After reading this chapter, you should be able to
- Define the copper cable infrastructure.
- Define the fiber optic cable infrastructure.
- Define inside building structured wiring.

Outline

Introduction

■ INTRODUCTION

As you drive down the road, notice the large black cable strung between the telephone poles. Try to follow the cable as you drive along, noting junctions where it creeps down the pole or shoots off to a metal cabinet mounted on a cement pad or into a green metal case stuck in the ground. Aerial telephone cable, the lowest cable hanging on the pole below power and CATV, traverses through terminal boxes, splice cases, and other devices. The goal of Chapter 14 is to present the components that make up the physical infrastructure of the communications network referred to as the outside plant. The topics covered include the copper cable plant, the fiber optic cable plant, and inside structured wiring. Each topic is divided further into specific areas such as cable composition, splice cases, terminals, and so forth.

Figure 14–1
Poles are secured using guy wires
and anchors. Poles located at the end
of a span are referred to as "dead-
end" poles, and poles found in the
span are called "straightaway" poles.

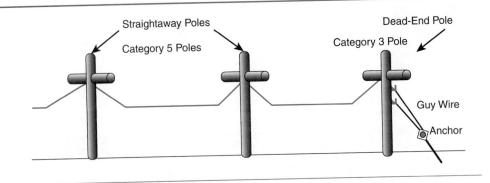

Telephone poles are categorized
according to their height and diameter. Poles may hold power,
CATV, and telephone cable.

dead-end pole
Corner telephone pole
requiring a higher grade
referred to as a level 3.

■ 14.1 COPPER CABLE DISTRIBUTION METHODS

Copper cable distributed throughout the plant is hung between poles, pulled through ducts running underground, or buried directly below the earth. Aerial cable refers to cable that is hung between telephone poles; buried refers to cable plowed under the earth; and underground refers to cable pulled through a labyrinth of duct work. Each of the three methods is used throughout the telephone plant.

14.1.1 Aerial Placement

Aerial cable is hung between telephone poles that are made of solid wood, typically formed from one tree with the bark stripped and the wood sealed with creosol or some other wood-protecting solution. The height and diameter of a telephone pole vary depending on the size and weight of the cable or cables to be mounted on the pole and whether other services such as CATV and electrical lines will share the pole. The height of most telephone poles is in the range of 30 to 45 ft.; telephone poles are purchased as 30, 35, 40, or 45 ft., depending on the design. As an example, a 5½ ft. hole is needed to hold a 30 ft. pole and a 6½ ft. hole holds a 45 ft. pole, depending on the soil and so forth. Power lines are typically strung on 50 ft. poles, as the electrical forces can cause real problems to both communications lines and human beings.

Telephone poles are categorized according to size and structural attributes. The majority of the telephone plant is made up of category 3 and 5 poles where the 3 is the stronger and is used on corners referred to as *dead-end poles*, while category 5 poles are used on straightaways, as shown in Figure 14–1. Spacing between poles depends on several factors, including the size and weight of the cable, the size of the pole, the number of houses or drops within the section, and future growth. Telephone poles in rural areas strung with small, 50-pair or less cables on a straightaway may be spaced as far apart as 300 ft. while poles found in residential areas carrying large 1800-pair cable may be spaced as close as 50 ft. Engineers must use all the criteria available when determining cable placement.

Guy wires and anchors are used to stabilize the poles—similar to a tie line used to hold up a pole supporting a badminton net, as shown in Figure 14–2. If the tie line is cut, the pole immediately leans in toward the net and consequently causes the pole on the opposite end to also lean in toward the net. Telephone poles are freestanding,

Figure 14–2
Tethering wire between poles requires
the poles to be anchored at the ends.
If the guy wires and anchors were
removed, the poles would collapse in
on each other.

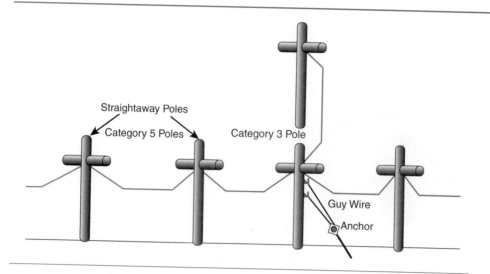

Straightaway Poles

Category 5 Poles

Category 3 Pole

Guy Wire

Anchor

Figure 14–3
Guy wires and anchors are also used to secure a pole used at a cable change.

meaning the pole depends on the adjacent pole to keep it from crashing down; the cable plays no role in keeping the pole plumb. Instead, poles located at a corner—or the poles involved when a cable change happens—are tethered using guy wires, anchors, and push braces that act similar to the badminton net tie line, keeping the entire row of poles straight and solid. Corner poles have guy wires and anchors used to keep the pole straight and taut. Additionally, whenever there is a direction change in cable, such as that shown in Figure 14–3, guy wires and anchors are needed to stabilize and secure the pole in order to support the cable's direction change. Push braces are also used to keep the pole straight and are referred to as *push brace poles*. Many outside plant technicians have encountered situations in which a vehicle crashes into the guy or pole brace, dislodging it and, consequently, causing the adjacent poles to immediately lean in or even snap. In extreme cases, several poles right in a row may snap off—similar to a domino effect—causing the cable to crash to the ground, disrupting service. Proper pole anchoring is critical to ensure that the cable remains taut and stable. Guys and anchors must be sunk deep enough in the ground to keep the pole from leaning.

Numbers and letters are used to designate what pole is what in the network. Manufacturers stencil or burn into the wood the height and category of the pole, such as

push brace pole
Pole with a brace used to support the pole.

Corner poles use guy wires to keep them straight.

Pole numbers are used as identifiers. Maintenance crews use pole IDs to determine where the poles are within a span.

30-5, meaning that the pole is 30 ft. high and is a category 5 pole. The service provider also marks the pole using a route number, the company logo, and a pole number. A common naming method used in the industry to mark poles is to position the route number on every fifth pole and the company logo and a pole number on every pole. A *P* following the company logo refers to a private pole, while a pole with nothing but the company logo refers to a highway pole; these are designations used for right-of-way and tax purposes. Below the company logo, a pole number is tacked on using metal numbers similar to those used as house address labels; similar to a house address, the number should be visible to those walking or driving by. As-built drawings showing the outside plant infrastructure include the pole numbers for the purpose of helping technicians quickly identify areas in the cable span before venturing out to find the location.

The placement of telephone cable is determined according to several factors, one of which is whether there are power lines present. If power lines are mounted on the pole, a minimum distance of 40 in. between the power and telephone cable is imperative for both safety and signal quality reasons. Cable television cable is hung below the power wire and above the telephone cable, as shown in Figure 14–4.

Hanging cable between poles requires the use of heavy winches located on large construction trucks or on trailers. Two types of aerial cable are most often deployed in the local network. The difference between the two is the type of *messenger*, a galvanized steel strand used to hold the cable in place. One type has a strength messenger embedded inside the cable, and the second lashes the cable to the messenger. The messenger, also called a *strand*, is a heavy galvanized steel cable with metal clamps attached every so many feet. The first task performed when hanging new cable is to run the strand, making sure to place the correct amount of tension in order to secure the cable the standard distance above the earth. Most

messenger
Galvanized steel strand used to hang telephone cable to.

strand
A term used to describe the messenger. Steel cable that is used to hold the telephone cable.

A strength messenger may be embedded inside the wire; or a steel messenger, as shown, may be used to secure the cable.

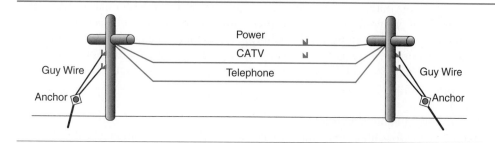

Figure 14–4
The three main utilities—power, CATV, and telephone—share the telephone poles, helping lessen the number of poles needed. Power is always on top, CATV below power, and telephone below CATV. A 12 ft. distance must be left between power and the other two cables.

construction technicians let the strand sit overnight in order to let it settle before lashing the telephone cable to it. Lashing the telephone cable to the strand is done using a special device called a *spinner*. The spinner consists of two wheels and a drum that ride along the strand, similar to a tightrope walker. The cable is pulled along the strand, and the spinner lashes the cable to the strand using metal lashing wire. Strand sizes vary depending on the weight of the cable and whether the engineer feels there will be additional cables needed in the future. Strand comes in 6, 10, 16, and 25 in. increments. When aerial cable is strung between poles, the technicians purposely place several twists in the cable between the two poles in order to help reduce the amount of dancing caused by wind and storms. The general rule is to place twists in the cable of every other section, as shown in Figure 14–5, as this proves to be the best method for stabilizing the span. Cable clamps anchored with nails or lags are used to secure the cable to the pole.

spinner
Device used to lash the telephone cable to the messenger.

Cable reels are ordered by outside plant engineers to accommodate the build; for example, if the section is very long such as in a rural area, reels holding as much as 50,000 ft. may be purchased, thus reducing the number of splices needed. On the other hand, if the section is in a busy metropolitan area and many drops are needed, a reel may hold a few hundred feet, thus allowing the technicians to pull short sections placing terminals and splice cases according to the drops. Each job is tailor-made for the situation; and the outside plant engineer is the person responsible for

Cable lags and anchors are used to secure the cable onto the pole.

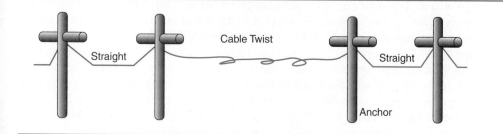

Figure 14–5
Twisting the cable as it is spun onto the messenger every other pole section helps reduce the amount of cable swing caused by wind.

determining the correct cable size, reel length, and so forth. Water causes the greatest number of problems in the outside plant, and technicians must be very careful not to nick or cause an opening when anchoring the cable to the pole as water will seep in and corrode the strands. Cable rising up a pole is covered with a protector tube called a *U-Guard*™, and the point where the cable leaves the duct is covered with a *bell cap*.

14.1.2 Underground Placement

Within the metropolitan area, underground cable is the most common type deployed as it is less susceptible to damage caused by weather, does not eat up valuable real estate, and is much more aesthetically pleasing when hidden from view. Manholes are square or rectangular rooms placed underground and are used to hold cables and other communications devices. If you have ever peered down inside a manhole, you would have noted that the walls are cement or cement blocks and that thick black cables run through end to end coming out of and going into round duct holes embedded into the walls. The air smells dank, and the floor looks wet—a very inviting place indeed.

In order to enter a manhole, the heavy, round, saucer-shaped cover must be popped off, revealing a tubelike entrance called a *manhole chimney*. The chimney serves as an interface between the ground and the manhole. For example, if the road has to be resurfaced, the outside plant engineer will extend the chimney to accommodate the extra feet of macadam added to the road surface. Climbing down into the manhole requires either a manhole ladder or permanent ladder rungs mounted in the wall.

One of the most dangerous tasks performed by telephone technicians is climbing into manholes, as gases such as methane may seep into the manhole extinguishing the oxygen. By the time the technician realizes that he or she is lacking oxygen, it is too late. Before entering a manhole, a gas test using a calibrated certified gas tester should be performed to guarantee that there are no toxic fumes. Additionally, most service providers mandate that the manhole be aired out before entering and that a fresh air supply be pumped in to help offset any gases. This is accomplished using a large fan with a flexible tube that is hung into the manhole, directing outside air into the hole. OSHA has also issued safety guidelines such as requiring that one person stay above while work is performed in a manhole. As a note, manhole safety is something that should not be taken lightly as more people die each year from manhole gases than from falling off a telephone pole.

Beyond aerating a hole before entering, many manholes require that the water be pumped out. Special pumping equipment that is run off gas motors similar to a generator is used to suck out the water letting it run onto the ground nearby. Some manholes fill up so quickly with water that constant pumping is required even during the time the technicians are working in the hole. Water in manholes is one of the reasons the cable must be covered with special airtight jackets with air pressure pumped into the cable. In most cases, a manhole is wet or damp, smells of stale dirty air, and may house rodents—a pleasing thought as you climb down the chimney. As you will learn, the stories comparing the size of the rodent seen in the manhole to that of a basset hound are similar to the stories that fishermen tell when describing the size of the fish they caught in the lake.

The size of the manhole varies depending on the number of cables and other devices residing in it. A manhole serving a congested intersection with many businesses and used for large feeder cables can range between 35 and 40 ft. long. Manholes placed in sparsely populated areas or carrying small distribution cables are fairly small. The distance between manholes, similar to telephone poles, is dependent on the situation. For example, it is not uncommon to place a manhole at every intersection along a busy route as cable often splits off in many directions. Multiple manholes may reside close to each other depending on the needs and the layout of the network. If a manhole has to be extended, the walls are pushed out and rebuilt almost like adding on to a basement. Stacking manholes on top of each other is also done when additional space is required such as in large metropolitan areas that have a limited amount of real estate to work with. Size and shape are dependent on the location, the need, and, of course, the engineer's preference.

U-Guard
Metal or plastic cover placed over riser cable mounted on a pole.

bell cap
Cap that protects the duct entrance.

manhole chimney
Manhole entrance tube that is used to access the manhole.

A duct is a tubular pipe placed underground between manholes, similar to underground sewer pipes, though much smaller in diameter—typically 4 in. As it is very costly to construct underground ducts, placing multiple ducts at the same time is common even though they may lie idle for years. In a Tier 2 city such as Buffalo, New York, the cost to construct underground cable ranges from $300 per foot to $900 per foot as the streets must be dug up and other services interrupted, not to mention disrupting traffic flow. Putting in duct in Tier 1 cities is even costlier, as the amount of underground space is limited due to heavy congestion. Some construction companies have made millions of dollars selling duct systems to communications companies, as it is much easier to buy duct than to construct duct. The high cost associated with the build comes not only from the physical construction but also from the cost to the municipality such as right-of-way costs, special taxes, and, often, a percentage of the profits earned on the communications circuits. If you were to own ducts between two or three blocks in downtown Manhattan, you would be considered a millionaire, as the value of underground ducts in the largest metropolitan areas is priceless.

Pulling the cable through the duct can be very challenging depending on the condition of the duct and the distance of the pull. One of the most common methods is to shoot a ball tied to a very tough flat nylon rope through the duct using air pressure, similar to an air gun. The rope is used to fish a rod attached to the cable through the duct. Once the rod is in place, the cable sitting on a reel above the manhole is fed across a device called a *shoe;* it looks like a shoehorn, similar to a slide hanging over the side of the manhole. The cable is pulled from the opposite end using a winch located on a trailer or on the back of a large construction truck. Often, before the cable is pulled, a unit called a *pig* is pulled through the duct for the purpose of cleaning out any debris such as mud, rocks, or anything else that may obstruct the cable from sliding through. Many outside plant technicians have faced the challenge of removing a cable that has become stuck between two manholes—not a pretty sight. Today, pulling cable through ducts is much easier than years ago thanks to specialty devices such as a rotating spinner that helps adjust for the inherent spin that occurs when cable is pulled. Other devices are also used but only if the company has the funds to purchase them. In the end, pulling cable requires lots of strength, persistence, and patience.

Once the cables are in place, racks anchored to the manhole wall are used to suspend the cable on special hooks, as shown in Figure 14–6. If you peer down inside a manhole, you should be able to see the racks and hooks that secure the cables. Carrier housings similar to those mounted on poles are also found mounted on the walls in a manhole. Special airtight carrier housing covers are used to ensure that water does not seep in and short out the electronic repeaters. A mistake commonly made by new outside plant engineers is to order the wrong carrier housing, not realizing that watertight

shoe
Device used to feed underground cable through the manhole and duct system.

pig
Device used to clean the duct prior to pulling the underground cable through.

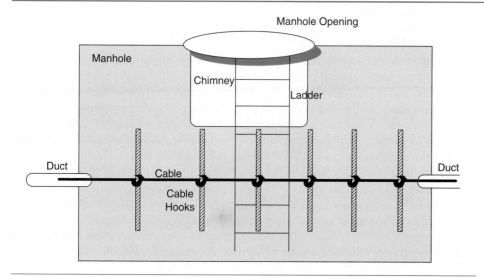

Figure 14–6
Cable hooks are anchored onto the walls of a manhole and used to hold the cable.

Figure 14–7
Lead tags inscribed with the cable and pair numbers are tied to the cable.

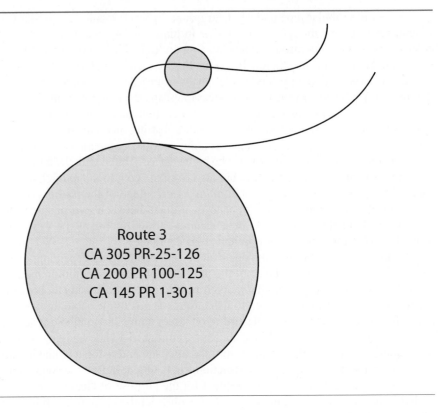

Route 3
CA 305 PR-25-126
CA 200 PR 100-125
CA 145 PR 1-301

housings are available and should always be used in manholes. When a heavy rain occurs and the manhole fills up, the engineer quickly realizes there is a difference between carrier housings and rarely makes the same mistake twice thanks to the endless ribbing from the outside plant crew. During troubleshooting, the last location a technician will look is in the carrier housing in a manhole as the job of opening up the manhole, pumping out the water, and aerating it to dispel any harmful gases is very time consuming.

Cable traveling through the manhole is marked using lead tags tied to the cable or splice case and inscribed with the cable and pair number, as shown in Figure 14–7. The numbers are tapped into the lead disks using lettered stamps. The cable passing through the manhole is tied together in a splice case. Underground cable relies on splice cases to interconnect cable segments between manholes in addition to supplying a junction point for branching off smaller cable segments. Splice cases built to withstand the water that often fills a manhole—similar to a bathtub—are secured to the racks and hooks extending from the wall and are used to connect the two ends of the cable together, as shown in Figure 14–8.

14.1.3 Buried Placement

Buried cable is laid in a trench dug using a backhoe or other digging device or is plowed under using a special plow. Digging trenches to hold communications cable is common in new developments; often, individual plastic ducts hold the cable, helping reduce the chance of environmental catastrophe. Burying cable is common in the access network and rural areas as underground duct systems are hard to cost justify, as the return on investment is very long. Today, most housing developments have trenches dug for power, telephone, and CATV cables, helping eliminate the need to dig every time a new building goes up. The television and telephone cables should never be placed in the same ducts within a trench as the power lines, because induction from the electrical fields may cause cross talk or noise on the communications signal.

Buried cable is most commonly filled with a gellike substance and is referred to as *gel-filled cable*. Water causes the wires to corrode and the insulation to deteriorate making the transmission of electrical signals difficult. The viscous transparent gel

gel-filled cable
Cable filled with gooey gel used to protect the copper cable from water and other environmental hazards.

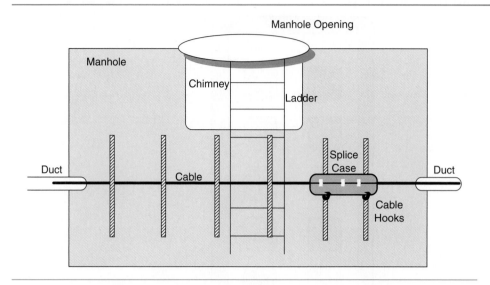

Figure 14–8
Splice cases are found in manholes, where they are used to connect two sections of cable together.

surrounding the wires helps protect them from moisture. Technicians, on the other hand, have few good words to say about the gooey liquid as it numbs their hands in the winter and covers everything that comes in contact with it, making their job much messier. Underground cable may be gel filled but more often has air pressure pumped in to help block water from infiltrating the jacket. Buried cable, like aerial and underground, comes in a variety of sizes and lengths. The main difference between it and underground is the way it is installed. Splice cases are not buried under the ground; therefore, the cable must rise above the earth and terminate in a physical device such as a pedestal or a cross box, as shown in Figure 14–9, in order to be distributed or spliced to another cable segment. Buried cable in the access network is terminated into pedestals, and the cable is cut in, as shown in Figure 14–10, to fixed-count terminals. Naming a buried cable follows the same naming convention as aerial or underground; it, like the others, is marked with a tag that has the cable and pair inscribed.

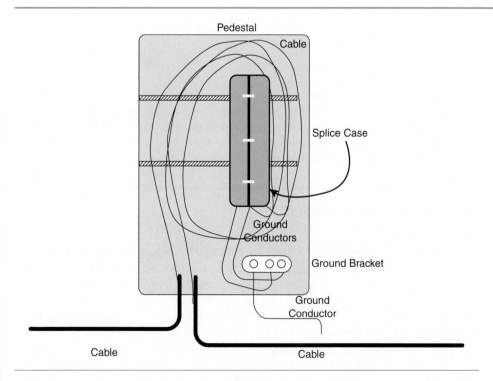

Figure 14–9
Splice cases may be found in outside pedestals secured to the ground.

Figure 14–10
Pedestals are also used to hold fixed-count terminals. A fixed-count terminal is used as a connection point between the distribution cable and the drop wire.

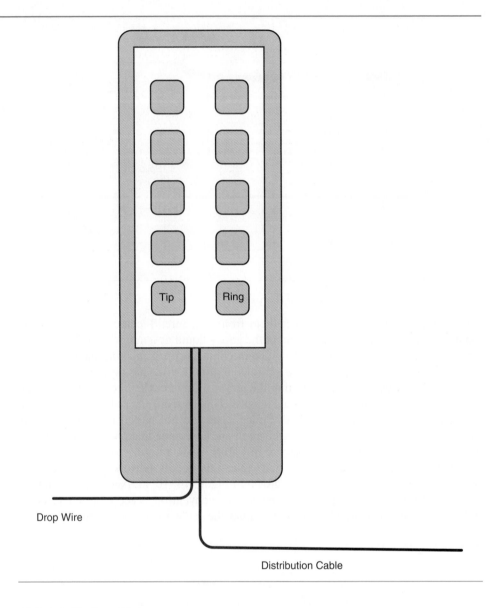

Drop Wire

Distribution Cable

14.1.4 Cable Transitions

Cable traveling through underground duct or buried transitions to aerial by emerging from the earth and feeding up the telephone pole is referred to as *riser cable*. A covering called a U-Guard is placed around the cable on the pole for the purpose of protecting it from car bumpers, lawn mowers, and so forth. The transition from buried to underground is fairly simple in that the cable enters the underground through a duct submerged in the earth.

■ 14.2 COPPER ENCLOSURES

14.2.1 Splice Cases

As copper cable does not consist of one long continuous strand of cable, methods have been developed to efficiently connect the sections together. Splice cases have been designed for this purpose; and, as with all telecommunications equipment a variety of styles and sizes are available. Splice cases are used to connect two sections of aerial cable together. Mounting a splice case in a cable section is either *in-line,* meaning a section of the cable must be cut out to accommodate the case, or *bridged,* meaning two cables are connected at a corner point. The technician must be very exact when marking the section of cable to be cut out in order to accommodate the splice case. A stiff brush called a *curding brush* is used to rough up the cable jacket before it is cleaned

riser cable
Cable that rises from underground up the pole to an aerial span.

in-line
In-line splice refers to two cables spliced along the cable span. A section of cable is removed in the span and a splice case is installed and used to connect the two sections.

bridged
Splice case is placed at the corner point of the span where it is used to bridge two cables together.

curding brush
Stiff brush used to rough up the cable jacket prior to sealing the opening.

A splice case is used to connect two cable segments.

and sealed with rubber cement and two layers of tape. The end pieces of the splice case must also be marked according to the size of the cable before being drilled out. Again, being exact is critical. Once the end holes are drilled, the cable is fed through the opening and secured in place. Tightening all of the nuts and bolts and making sure the banding strips are pulled snug help seal the splice case against the elements. Note that preparing, mounting, connecting, and closing splice cases vary according to each manufacturer's specifications. Most service providers require their technician to be certified by the manufacturer before attempting an installation. When both ends of the cable have been secured, the individual wires are connected according to the engineering run sheet and the standard color code.

Several types of connectors are used to splice the cable pair together. Some of the most common types of splicing connectors are Scotchloks, B-connectors, PICABONDs, modular connectors, and spin connectors. The purpose of each is the same: to connect two wires together making sure the copper conductors mate, allowing the signal to travel across with little loss and zero interference. One of the most common connector types is made by 3M and is called a Scotchlok. The wires from both cable sections are pushed into the Scotchlok and a special tool that looks like a pair of pliers—though with flat tips like a duck's bill—is used to press the round part of the connector down. When the connector is squeezed together, small metal teeth puncture the insulation on each wire forming an electrical bridge between the two conductors. The manufacturer also places the same type of gel that is injected into gel-filled cable inside the connector. The clear gel is used to keep moisture away from the exposed copper conductors. The color code is used to line up the wire pair accordingly, and the binder groups—a subject explained later in the chapter—are kept intact helping ensure proper alignment of the cables in the case.

Most splice cases are made of heavy plastic or metal with carefully designed ends and seals to help reduce the chance of water entering the case. Older splice cases made

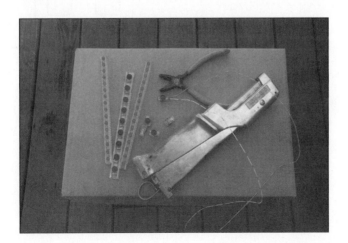

Connectors called Scotchloks are used to connect two copper wires together. Special connector tools are available.

of lead and usually found in manholes require the technician to be specially trained to open and reseal the case making sure to avoid exposure to the fumes, as lead is very toxic. Today, lead cases are not installed, though many networks still have hundreds throughout their network. Even plastic cases require the technician to go through special training classes typically put on by the vendor in order to know how to properly mount and place cables and reseal the cases. Some companies scrimp on training but usually pay the price when a splice case fills with water because of the technician not having closed it or installed it properly. Many cables have been lost from water seeping into the splice case and eventually being sucked throughout the cable, similar to a straw pulling liquid up.

14.2.2 Terminals

Terminals are also found in the outside plant and are used to tie the distribution cable to the drop wire. As with splice cases, there are many types and varieties of terminals. The *ready-access terminal* usually hangs from the cable and is accessed by unsnapping the closures and popping the plastic cover off. Once open, the technician has to find the correct wires by associating the color code to the cable pair number. The technician cuts the assigned pair, rendering the outgoing conductors as dead, or *dd*, as shown in Figure 14–11. Next, the technician splices the pair coming into the drop wire feeding the customer site, as shown in Figure 14–12.

The *fixed-count terminal* may be mounted on a pole or in a pedestal stuck in the ground where the buried cable protrudes. Similar to the ready-access terminal, the technician finds the proper pair by counting through the color code and connects it to the position on the terminal designated to that customer. The drop wire also connects to the same position in the terminal, consequently tying the distribution cable pair to the drop

ready-access terminal
Telephone terminal case that provides access to all the wires and allows the technician to attach the drop wire without having to splice into the main cable.

dd
Symbol used to designate dead cable pair referring to unconnected cable pair.

fixed-count terminal
Telephone terminal case that ties cable pair to a fixed termination point. Technician attaches the drop line to the specified termination point.

Figure 14–11
Splice cases are used to distribute smaller feeder cables throughout the network. Terminals are located close to a splice case and connect to the distribution cable. The wire pairs that are split off are marked as *dd* or *dead*, indicating that they no longer connect all the way through to the CO.

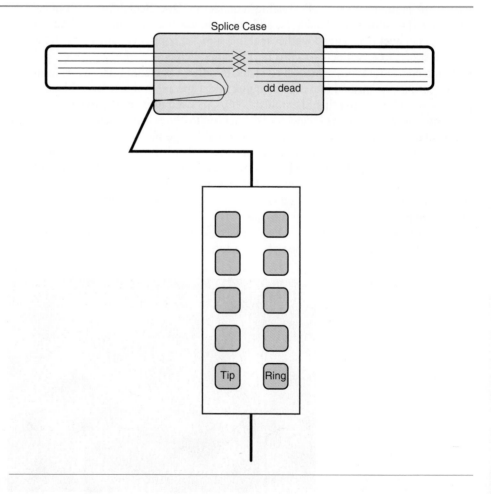

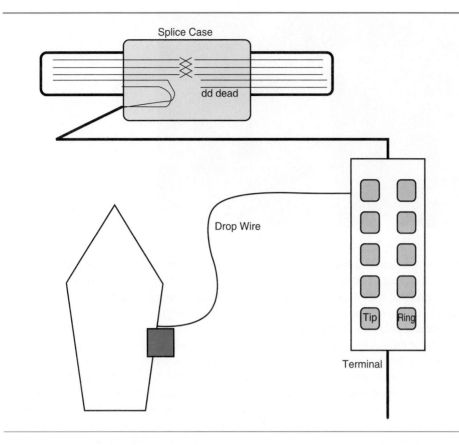

Splice Case

dd dead

Drop Wire

Tip   Ring

Terminal

Figure 14–12
From the terminal, a drop wire is used to complete the circuit connection between the distribution cable and the protector on the subscriber's residence.

line. Installing ready-access and fixed-count terminals is similar to installing splice cases. Using the engineering diagrams, the technician calculates the length of cable that should be cut out to give room for the ready-access unit. Fixed-count terminals have only the pairs assigned to that location cut in, making it easier to locate and connect the pair to the drop.

14.2.3 Demarcation Devices

At the customer location, the service provider is responsible for terminating the cable at the *demarcation point*—the handoff between the service provider's network and the customer's inside plant. Demarcation termination devices vary depending on the type of cable, the number of pairs, and the termination location. Protectors used to terminate standard two-wire POTS lines are used to tie the drop wire to the inside house wire. A

demarcation point
Handoff between the service provider's network and the customer's inside plant wire.

The ready-access terminal is used to terminate service drop lines.

Fixed-count terminals are used to terminate service drop lines.

typical residential subscriber's protection block consists of tip, ring, and ground terminals, as shown in Figure 14–13. The ground terminal should always be connected to a solid ground, meaning one that is tied to the same ground as the power feed. A standard prescribed by most service providers recommends that a copper rod be driven into the ground at least 6 ft. Some service providers use two copper rods placed a minimum of 6 ft. apart to connect grounds. In certain cases, when access to ground rods is not available, the ground wire is attached to metallic water pipes. This method is acceptable only if the water pipe is 100% metal; even one segment of PVC pipe is not allowed.

The method used to demarcate cables at multitenant buildings or businesses varies depending on the size of the cable; where it is demarcated; and, in some cases, the customer's preference. Today 66 blocks commonly are used to terminate copper outside plant cable, as shown in Figure 14–14. Pedestals, PIC closures, and ready-access terminals may all be used as demarcation devices especially when the engineer wants to place gas protectors to ensure that the equipment on both sides of the demarcation—central office and customer premises—is safe from electrical spikes. The purpose of each is to act as both a termination and a cross-connect point. Large campus or corporate locations may terminate telephone cable on main distribution frames similar to

Figure 14–13
Protectors placed on the side of a building or inside a basement or other termination room consist of tip, ring, and ground lugs.

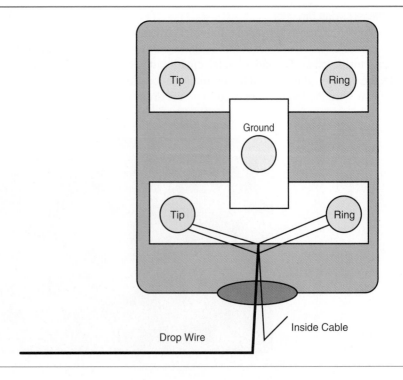

Most drop lines are terminated on a protector that is mounted on the side of the house. The protector is used to connect the inside wire to the outside plant drop line.

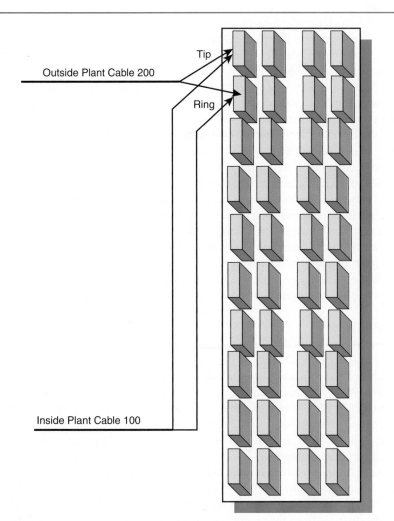

Outside Plant Cable 200

Tip

Ring

Inside Plant Cable 100

Figure 14–14
66 blocks are used to connect the inside cable to the outside cable. In the diagram, the wires are punched down on the same transmit and receive pins.

lateral
Term used to describe the portion of the network between the feeder or distribution plant and the customer's premises. Typically used to describe cable feeding into a business or business complex.

those found in a central office. Though the demarcation point is inside the customer's premises, most service providers consider it part of the outside plant infrastructure.

Pulling cable into a customer's premises can be very time consuming and costly. The term *lateral* is often used to describe the portion of the network between the distribution or feeder cable and the customer's premises. Often, in metropolitan areas, ducts must be purchased or built between the main network and the building. If the duct has to be built, the cost can run in the hundreds of thousands of dollars because the street and sidewalk have to be dug up and ultimately replaced. Often, fiber optic companies are asked how many buildings they have wired, referring to the number of laterals they have spurring off their main cable. Once the cable is pulled into the building, the owner of the building must specify where to place the demarcation terminal. In many cases, there is a room in the basement, within a parking garage, behind a boiler, or in some other dark and hard-to-get-to space. The cable is opened and cut into the demarcation terminal according to the color code. Each terminal is tagged with the cable and pair number using cloth tags that are tied around the cable. In addition, the terminal is labeled and any cable sheets attached to the terminal are filled out. Properly labeling the demarcation terminal is one of the most important tasks a technician performs. The time spent wandering around a building trying to find the correct cable (many buildings have more than one room where cable terminates) is more than the time spent watching the World Series each year. Jobs can easily run into overtime if the cable has not been properly labeled.

■ 14.3 COPPER CABLE PACKAGING

14.3.1 Size

The size of a copper cable depends on the application, the number of subscribers, and the expected growth rate for the area. For example, if a cable is used to service homes in a rural area with few residents and little chance of growth, a small cable—possibly 12- or 24-pair—is suitable. In contrast, if a new cable is run to service a large number of subscribers in a downtown central business district of a large metropolitan city, the cable may be as large as 3500 pair or greater. Engineers determine cable size according to the number and the type of customers in an area. A high-tech business park would require a larger cable than a senior citizen's home, as the users in the business park are more likely to require more services and thus more connections than the residents of the home. Drop wires used to connect the subscriber to the network range from small, one-pair cables up to larger 100-pair cables used in high-density sites. Today, the typical drop wire feeding a residential subscriber has two pair, as many users have two telephone lines. Outside plant engineers straddle a line between having ample pairs of wires when needed and keeping down the cost for infrastructure. Many outside plant engineers attend local municipal meetings trying to stay informed of any new developments that may require a large infrastructure built. In addition, the engineer should also keep up with changes in the municipal code and establish a good working relationship with the board members, as pushing right-of-way issues through quickly often requires tact and good communications.

14.3.2 Wire Gauge

Wire gauge expressed in AWG, American Wire Gauge, is another factor the engineer must take into consideration when deploying copper cable. Generally, smaller-gauge cable such as 26 AWG is deployed close to the central office and higher-gauge cable such as 24 AWG is deployed farther out due to the difference in their resistive values. Traditionally 19, 22, 24, and 26 gauge cable are the most commonly deployed in the local telephone plant. As shown in Figure 14–15, the resistance per foot directly relates to the gauge of the cable and, as such, is the main reason 24 gauge cable is used when the span runs farther from the office. Today, some applications may warrant the use of higher-gauge cable even when located close to the central office. A good example of this is VDSL, a technology that uses high frequencies and consequently requires a low resistance loop. When designing the CSA, carrier serving area, the engineer must take into account the distance from the serving equipment and the type of service being offered.

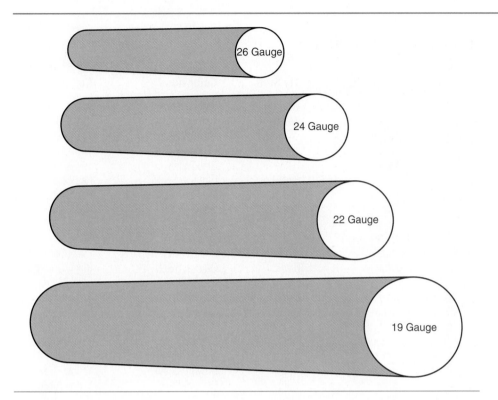

Figure 14–15
Cable gauge values relate to the diameter of the copper wire. The higher the gauge number, the smaller the diameter. In turn, the smaller the diameter, the higher the resistance per foot.

14.3.3 Load Coils

Devices called *load coils* are used to extend the distance a voice signal can travel on a copper pair. Cable pair extending farther than 18,000 ft. from the central office is loaded using 88 mH load coils. A load coil consists of wire wrapped around a core conductor and introduces inductance on the line, a subject explained in a later chapter. The first load is placed 3000 ft. out of the office and every 6000 ft. thereafter. Load coils may be individual firecrackerlike devices used to load a single pair, or they may reside in a load case. Cable and pair counts should be marked as loaded in order to help the technician during troubleshooting or installation. Load coils must be removed when digital circuits such as T1, ISDN, or DSL are deployed.

load coils
Inductive devices placed strategically along a copper wire pair to increase the distance a signal can travel on the cable.

14.3.4 Color Code

The color code associated with copper cable is one of the pieces of information that should be learned by anyone involved in the technical side of telecommunications. With the introduction of plastic insulation, a color code was introduced specifically to help technicians determine what wires should be connected together, eliminating the need to send tone as was the case with paper insulated wires. The color code used in outside plant cable follows the standard color code shown in Table 14–1. Large cables

Cases holding load coils are found next to copper splice cases.

Table 14–1
Standard Copper Cable Color Code.

Cable Count	Solid Color		Banded Color	
	Ring Side	Tip Side	Ring Side Color/Band	Tip Side Color/Band
1	Blue	White	Blue/White	White/Blue
2	Orange	White	Orange/White	White/Orange
3	Green	White	Green/White	White/Green
4	Brown	White	Brown/White	White/Brown
5	Slate	White	Slate/White	White/Slate
6	Blue	Red	Blue/Red	Red/Blue
7	Orange	Red	Orange/Red	Red/Orange
8	Green	Red	Green/Red	Red/Green
9	Brown	Red	Brown/Red	Red/Brown
10	Slate	Red	Slate/Red	Red/Slate
11	Blue	Black	Blue/Black	Black/Blue
12	Orange	Black	Orange/Black	Black/Orange
13	Green	Black	Green/Black	Black/Green
14	Brown	Black	Brown/Black	Black/Brown
15	Slate	Black	Slate/Black	Black/Slate
16	Blue	Yellow	Blue/Yellow	Yellow/Blue
17	Orange	Yellow	Orange/Yellow	Yellow/Orange
18	Green	Yellow	Green/Yellow	Yellow/Green
19	Brown	Yellow	Brown/Yellow	Yellow/Brown
20	Slate	Yellow	Slate/Yellow	Yellow/Slate
21	Blue	Purple	Blue/Violet	Violet/Blue
22	Orange	Purple	Orange/Violet	Violet/Orange
23	Green	Purple	Green/Violet	Violet/Green
24	Brown	Purple	Brown/Violet	Violet/Brown
25	Slate	Purple	Slate/Violet	Violet/Slate

binder groups
Large cables are divided into sections referred to as binder groups. Binder group colors are used to differentiate each cable group.

transposition
Crossing one wire pair to a different wire pair within the cable span.

frog
Transposing a cable pair or wire within a cable is crossed midspan.

are divided into color-coded groups called *binder groups*. A color-coded thread that is wrapped around the bundle of wires is referred to as a binder group. The thread follows the tip-side color code so that the first binder group consisting of five 25-pair cables is wrapped with a white thread; the second binder group consisting of the next five 25-pair cables is wrapped with a red thread; and so on, as shown in Table 14–2. Five binder groups are tied together—again, with a color-coded thread, and referred to as a *superbinder group* consisting of 600 pairs of wires—$(5 \times 25) \times 5$. Unlike inside plant cable, which works with smaller increments of wire pairs, outside plant cable must follow this strict standard in order to allow technicians to find the correct pair that is hidden deep inside the cable. In reality, there are only ten colors used to distinguish the cable pair, though it seems much greater when you open an 1800-pair cable.

Splicing an 1800-pair cable while sitting in a manhole in the dead of winter requires the technician to precisely coordinate the binder groups according to the cable running list, as shown in Table 14–3. As is obvious, proper connection of wire pair is essential in order to produce an end-to-end circuit. Though the color code makes it much easier to track a circuit through the span, often changes are made during outages that violate the standard color code. Subsequently, many service providers are faced with the dilemma of mismatched wire pairs thanks to *transpositions* and *frogs* performed midspan during troubleshooting. Transpositions happen when a cable repair person needs a good pair between a certain section of cable and opts to splice, let's say, an orange/white wire to a green/white wire. Because the green/white wire is not being used, this poses no problems as long as the cable technician marks the change on the engineering records. Unfortunately, too often the record change is not made and the next time a splicer goes to throw the cable or cut the pair into a terminal, the customer riding on the orange/white pair loses service. Frogs are similar to transpositions in that one

Table 14–2
Copper Cable Binder Group Color Code.

Super Binder Group	Bindings Color Code	Binder Group	Binder Color	Cable Count
White	White/Blue	1	Blue/White	1–25
	White/Orange	2	Orange/White	26–50
	White/Green	3	Green/White	51–75
	White/Brown	4	Brown/White	76–100
	White/Slate	5	Slate/White	101–125
	Red/Blue	6	Blue/White	126–150
	Red/Orange	7	Orange/White	151–175
	Red/Green	8	Green/White	176–200
	Red/Brown	9	Brown/White	201–225
	Red/Slate	10	Slate/White	226–250
	Black/Blue	11	Blue/White	251–275
	Black/Orange	12	Orange/White	276–300
	Black/Green	13	Green/White	301–325
	Black/Brown	14	Brown/White	326–350
	Black/Slate	15	Slate/White	351–375
	Yellow/Blue	16	Blue/White	376–400
	Yellow/Orange	17	Orange/White	401–425
	Yellow/Green	18	Green/White	426–450
	Yellow/Brown	19	Brown/White	451–475
	Yellow/Slate	20	Slate/White	476–500
	Violet/Blue	21	Blue/White	501–525
	Violet/Orange	22	Orange/White	526–550
	Violet/Green	23	Green/White	551–575
	Violet/Brown	24	Brown/White	576–600
Red	Repeats	Repeats	Repeats	601–1200
Black	Repeats	Repeats	Repeats	1201–1800
Yellow	Repeats	Repeats	Repeats	1801–2400
Violet	Repeats	Repeats	Repeats	2401–3000
Blue	Repeats	Repeats	Repeats	3001–3600
Orange	Repeats	Repeats	Repeats	3601–4200
Green	Repeats	Repeats	Repeats	4201–4800
Brown	Repeats	Repeats	Repeats	4801–5400
Slate	Repeats	Repeats	Repeats	5401–6000

color wire is spliced to a different color. The difference is that the frog usually jumps back to the original later in the span, thus using only one segment of the wrong color wire. A *tone generator* is used in these situations to determine what is connected to what.

Twisting the tip and ring conductors over each other is done to help eliminate induction and to keep other noise from disrupting the signal. Most cable is twisted according to very specific engineering rules that take into account the frequency, resistance value, and the impedance associated with the cable type. The number of twists per foot is a value stated on the cable specification sheet and should always be noted before purchase. Beyond twisting the individual wires, larger cables twist the binder

tone generator
A tone generator is connected to the end of the wire and a tone is pushed down the wire and used to identify the wire pair at the opposite end.

Table 14–3
Cable Running List Example.

Cable No.	Pair No.	Color	Connect To	Cable No.	Pair No.	Color
201	5	White/Blue		110	10	Red/Slate
201	6	Red/Blue		110	15	Black/Slate
201	8	Red/Green		110	25	Violet/Slate
101	1	White/Blue		110	30	White/Blue

groups over each other again to help reduce the chance that signals will ride parallel with each other and cancel themselves out.

■ 14.4 LOCAL TELEPHONE CABLE PLANT ARCHITECTURE

Local telephone networks are engineered to supply communications services to the residential and business subscribers. The physical infrastructure of the plant is made up of copper cables, fiber optic cables, all of the connecting components such as splice cases, terminals, and so forth. The architecture is similar to a city's road system. Large cables connect key locations such as central offices, and smaller cables spur off the large cables and feed less-used roads and alleys. Copper cable in the local network is still the predominant means to carry information to the customer's premises, known as the last-mile or the access network. Copper laid in the early twentieth century may still be used today, as the life of a copper cable is long and its usefulness still valued.

An outside plant engineer is responsible for designing the layout of the local network and, as such, strives to build the most cost-effective, reliable plant possible. His or her first step is to determine the CSA, carrier serving area, the area served by the CO—central office—as shown in Figure 14–16. A CO is the building where all of the cables feeding the subscribers within that area terminate and where the circuit switch and data switching equipment reside. A CO is the hub of the subscribing area and is often called the wire center, service exchange center, or exchange office. A city requires multiple COs to service the geographical area, while a small town may have only one CO depending on its size. Today, it is unusual for an engineer to design a CSA, as most central offices have been around for many years. However, the engineer may need to redefine the CSA as newer technologies emerge making it possible for the CO to serve areas farther and farther out. In either case, every outside plant engineer uses the transmission specifications available when determining the cable plant in the CSA, a topic to be discussed later in the chapter.

Figure 14–16
Central offices located strategically throughout a local telephone network are designed to service an area referred to as a CSA. All locations residing within the CSA are served by the central office.

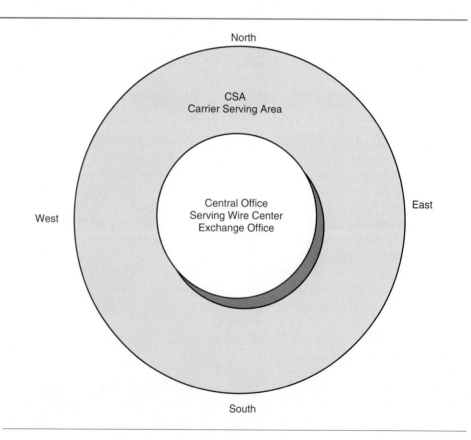

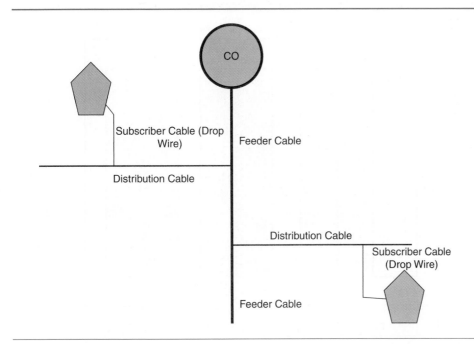

Figure 14–17
Cable feeding the local telephone plant consists of a large feeder cable, distribution cable, and subscriber or drop wire.

The structure of the local telephone cable plant is divided into three sections: feeder cable, distribution cable, and the subscriber cable or drop wire. The feeder cable is the large cable interfacing into the CO and extending into the network feeding hundreds of copper pairs strategically throughout the CSA. Smaller distribution cables spur off the feeder cable, similar to exit ramps on an expressway that link the highway to the secondary, thus extending the plant to the customer. Subscriber cable—also known as the drop wire—spurs off distribution cable terminating at the customer's premises. Figure 14–17 illustrates how the cables—feeder, distribution, and subscriber—interface with each other to form the connection from the CO to the subscriber. Though today fiber optic cable is the primary way central offices are connected together, copper cable continues to be used for this purpose, as shown in Figure 14–18.

Devices are placed strategically in the network to help distribute the cable more efficiently. Cross boxes are placed on a cement pad in a logical location in the network where the main F1 (feeder) cable is split into smaller units called *branch feeder cables or F2 cables* that disperse in different directions, as shown in Figure 14–19. A cross box is a passive unit that has termination blocks where the feeder cables terminate, and cross-connect wire is used to join the wire pair accordingly, as shown in Figure 14–20. The advantage of installing a cross box in the network is apparent, as it is able to efficiently distribute the cable throughout the plant.

A second device found in the outside plant network used to aggregate multiple subscriber circuits is called a *digital loop carrier* and is discussed in detail later in the text. For now, our focus is on how the copper cable plant is distributed so at this time we discuss how the DLC connects to the copper plant. A DLC has two sides, a line

branch feeder cables or F2 cables
Terms used to describe a cable that feeds off the primary feeder cable.

digital loop carrier
A multiplexing device that is used to aggregate multiple subscriber lines in the outside plant. A DLC may be placed on a cement pad close to the subscriber.

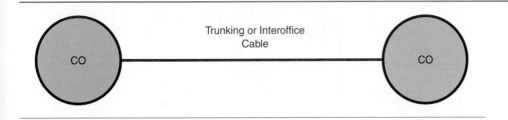

Figure 14–18
Trunking cable is used to connect central offices together. Traffic traveling between offices rides on the interoffice cable.

Figure 14–19
A cross box is used to tie together and distribute cable pairs from different cable segments. Cable feeding from the central office is referred to as F1 cable. Cable extending off the F1 cable is called F2 cable.

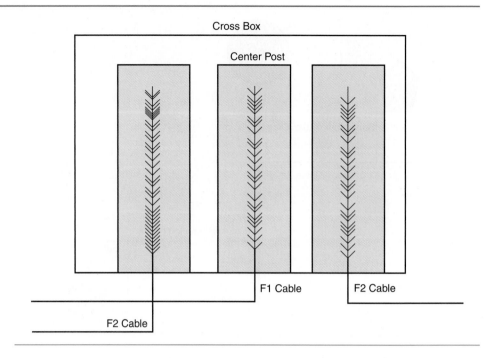

side and a trunk side, similar to a switch. The line side terminates copper subscriber pair coming from the customer, and the trunk side terminates digital carrier circuits that feed back into the switch. The copper pair from the subscriber rides on the same distribution and feeder plant as regular POTS lines terminating directly into the CO. In most cases, the subscriber pair first terminates at a cross box located next to the DLC and is tied to the DLC via a cable with a special pre-connectorized end that is plugged into a block on the DLC. When the trunk connection is copper, the cable terminates at the DLC and is tied to the feeder plant through the cross box, as shown in Figure 14–21.

Figure 14–20
Jumper wire is used to connect a wire pair from the F1 cable to a wire pair in the F2 cable. The jumper wire tip and ring are wrapped around the termination lugs creating a connection between the two cables.

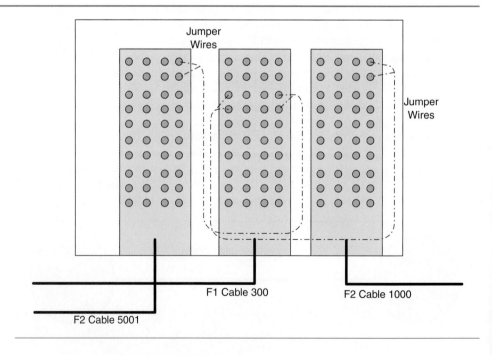

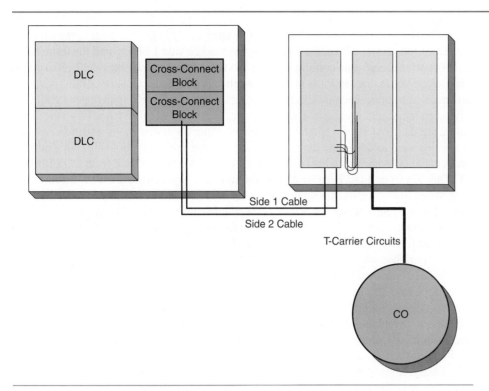

DLC

Cross-Connect Block

Cross-Connect Block

DLC

Side 1 Cable

Side 2 Cable

T-Carrier Circuits

CO

Figure 14–21
Digital loop carriers are tied to the central office by T1 Carrier circuits. Typically the T-Carrier lines first terminate into a cross box that is tied to the DLC cabinet where they terminate onto a cross-connect panel.

■ 14.5 SPECIAL SERVICE CIRCUITS RIDING ON COPPER CABLE

Copper carrier circuits ride on the same feeder and distribution cable infrastructure as standard POTS lines though the way they are laid out differs significantly. A traditional *T-Carrier span*, the term used to define the end-to-end system, requires the wires carrying the high-frequency 1.544 Mbps signal to ride together in one section of the cable. Every 6000 ft., the wires designated and designed specifically for T-Carrier circuits spur off and feed into repeaters located in a T-Carrier housing. Because a T1 circuit is a four-wire circuit composed of a transmit and a receive pair, special care has to be taken by the splicing technician when cutting in a housing in a span. Often the transmit of one side is crossed with the receive of the other, or vice versa, causing the signal to loop around and head back in the direction it just came from. In addition, the high frequencies of a T-Carrier circuit require that the transmit signal is separated by at least 25-wire pair specifically to avoid crossover and signal cancellation. Therefore, in traditional T-Carrier loops, the transmit pair and receive pair can always be found in different binder groups.

T-Carrier span
Copper T1 lines used to carry multiple signals between end points. *Span* refers to the portion of the network between the two end points.

T-Carrier housings hold repeaters that regenerate the DS-1 every 6000 ft.

side 1
Side 1 refers to the side of the circuit that carries traffic from the designated side 1 office to the opposite end or side 2 designated office. Used to verify the transmit and receive directions of a four-wire circuit.

side 2
Side 2 refers to the side of the circuit that carries traffic from the designated side 2 office to the opposite end or side 1 office. Used to verify the transmit and receive directions of a four-wire circuit.

Special terms have been assigned to different parts of the span in order to help simplify the ins and outs of the housing. The term *side 1* is used to refer to the transmit side from the central office, the CO designated as the side 1 office; and the term *side 2* refers to the receive pair coming in from the CO designated as the side 2 office. This may sound confusing, and it is at first, though it is one of the most important pieces of information to grasp because troubleshooting a copper T-Carrier span depends on identifying transmit and receive at each test point. Figure 14–22 illustrates how side 1 and side 2 are designated on a span. In addition to using the terms *side 1* and *side 2*, A and B are also used to designate the transmit and receive cables at a housing. For example, the A cable, also shown in Figure 14–23, carries the transmit signal from CO Elm St to CO Oak Street and the B cable carries the transmit signal from CO Oak Street to Co Elm Street. If the technician were to connect the A cable pair coming into the housing to the B cable pair leaving the housing, she would cause the signal coming in from Elm to loop around back to Elm and the signal coming from Oak to drop off at the can. It is important for the technician to realize that there are four splices that have to be made at the carrier housing, as shown in Figure 14–24. Side 1 transmit *In* is spliced into the A-side tail connected to the repeater, and side 1 *Out* is tied to the B-side tail connected to the repeater. The next two splices are side 2 tied to the A-side tail and side 2 tied to the B-side tail. Technicians dealing with T-Carrier circuits should always note which side is designated as side 1 and which is side 2. In the case of a span feeding a business subscriber, 99% of the time side 1 is from the CO to the business and side 2 is from the business to the CO, as shown in Figure 14–25.

Today, a good portion of copper T1 Carrier circuits ride on HDSL-coded lines that eliminate the need for the pairs to be separated from the POTS lines. No longer are specially designed T-Carrier spans built between locations every time a T1 is deployed. HDSL T1s ride in the same cable as POTS lines and, as such, are less expensive to provision. HDSL circuits do not require a carrier housing every 6000 ft. They do, though, similar to ISDN, require a repeater housing every 12,000 ft. depending on the cable gauge. The repeater housing must be cut in similar to the T1 Carrier circuit as it, too, has four wires, two pair. The method used to cut in the

Figure 14–22
T-Carrier circuits are designated as side 1 and side 2. Each pair has a transmit and a receive coming into and going out of each repeater.

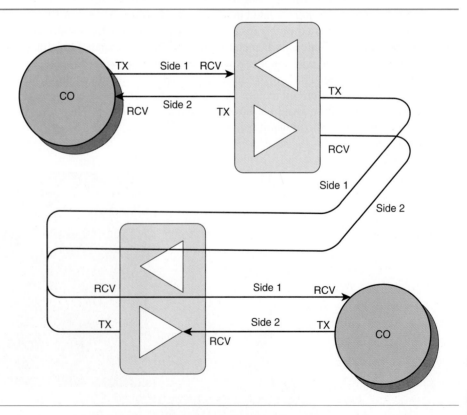

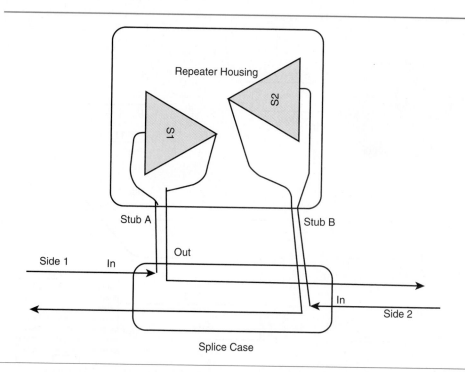

Figure 14–23
Cables coming into and going out of a repeater housing are designated as A and B stubs. Side 1 and side 2 normally correspond to the A and B stubs. The diagram shows a bi-directional, one-cable-fed repeater housing.

T1 housing holds true for the HDSL repeater housing, though typically there are fewer circuits terminated—from six to twelve, compared with twenty-five to fifty in traditional carrier spans. An ISDN, on the other hand, rides on one pair and, as such, is less complicated to connect.

Other special service circuits such as digital data 56 kbps circuits, FX, OPX, and so forth are indistinguishable from regular POTS lines in the outside plant. They ride, as do HDSL T1 lines, on the same copper wires as regular voice frequency signals. DSL circuits also ride on traditional POTS lines and, as such, are terminated in the same manner as a voice circuit—at an RJ-11 jack at the customer premises. The only visible difference between special service circuits and POTS lines is the color of frame wire used at the central office and at the cross box. The most common defacto standard followed is red/white for special service circuits and orange/white for an ISDN. Red caps are often placed over the termination connectors in a cross box and on a frame signifying a special circuit.

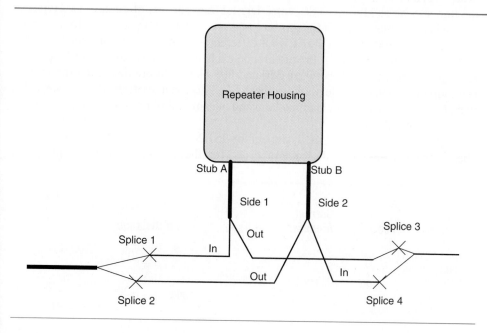

Figure 14–24
Cutting in carrier cable is confusing because four splices are needed to tie the Ins and Outs together. Often a flop is made by accident causing the signal to loop back out the same way it came in.

Figure 14–25
Side 1 is almost always designated as
the transmit pair leaving the central
office, and side 2 is designated as the
transmit pair leaving the subscriber.

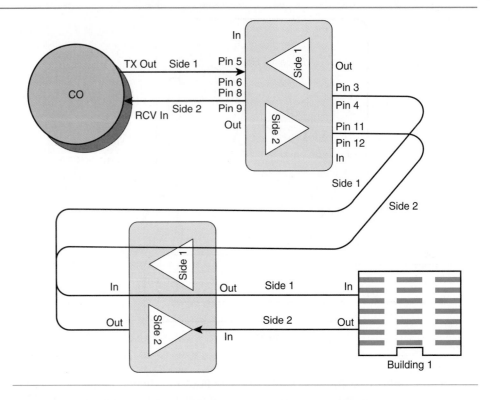

■ 14.6 COPPER CABLE LOOP RESISTANCE AND LOOP POWER

loop resistance
The total resistance value of a
copper loop that connects
the central office to the
customer's premises.

Loop resistance is a term used to define the total resistance value attached to a copper loop between the line termination at either the CO switch or the DLC channel card and the subscriber's equipment. Loop resistance is determined by adding up the resistance associated with the copper loop in both directions, cross connections at the CO and other points, drop wire, subscriber equipment, and the resistance associated with the battery feed, as shown in Table 14–4. Today's switching equipment allows for a loop resistance ranging between 900 and 1300 ohms, meaning that the copper loop must not exceed the maximum loop resistance as specified by the equipment manufacturer. When the other resistive components such as the telephone station, the drop wire, and so forth are added into the equation, the total amount of resistance on the line may be as high as 2086 ohms. Digital loop carriers often have two types of channel cards, one designed for shorter loops around 900 ohms and one for extended loops around 1300 ohms. Cost is what differentiates the two, the shorter loop card being cheaper than the long. Private branch exchange switches also have loop resistance parameters that fall between 350 and 800 ohms. Before designing the loop, it is important that the outside plant engineer take into consideration all of the variables that affect the total resistance of the loop and the amount of resistance allowed by the switch or other device.

Special circuits such as T-Carrier or ISDN lines all have loop resistance characteristics that must be met in order for the circuit to function properly.

Table 14–4
Copper Loop Design.

Loop Resistance	Current Amperes	Voltage	Cable Distance
1300-ohm loop design	23 mA	−48 V	19, 22, 24 gauge = 12,000–18,000 ft.
1300-ohm loop design	23 mA	−48 V	26 gauge = 12,000 ft.
1300-ohm loop design	23 mA	−48 V	Load coils added to extend loop design past 18,000 ft.

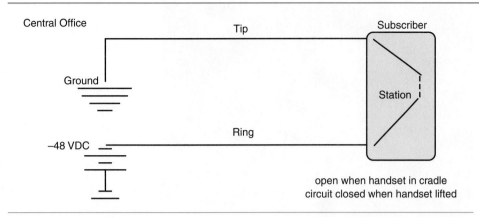

Engineers must use the defined parameters whenever designing a copper loop for special circuits.

Powering copper spans is the responsibility of the central office and has been so since almost the first telephone set was installed. The switch puts out −48 V on the ring lead, as shown in Figure 14–26, on standard voice frequency circuits. When the telephone handset is picked up and the switch hook closed, a 23 mA current flows if the loop resistance falls within the defined specifications as we have described. Powering copper cable with negative voltage helps to reduce the corrosion that occurs when wire is exposed to the elements. T1 Carrier circuits are also line powered, meaning that voltage rides on the circuit itself and is referred to as a simplex current loop. The amount of voltage needed to power the repeaters depends on the distance of the span; the number of repeaters; and, of course, the wire gauge. Repeaters located in the CO generate the line power for the span. The most common powering schemes for T-Carrier circuits are −48 V and ground; −120 V and ground; −120 V and +120 V; and, for very long spans, −240 V and ground. Whether the voltage applied to the circuit is −48 or −240, the current flowing through the span and energizing the repeaters remains 60 mA. Older repeaters used 100 mA to energize the span. Today, 60 mA is the standard. As shown in Figure 14–27, for very long spans a midpoint is designated and the span is split into two segments creating two spans, each powered by CO repeaters sitting at each end.

Figure 14–27
T-Carrier spans are line powered by a central office repeater. The voltage is applied onto the transmission line powering the line repeaters. A simplex loop is shown at the far end of the circuit at the CO repeater.

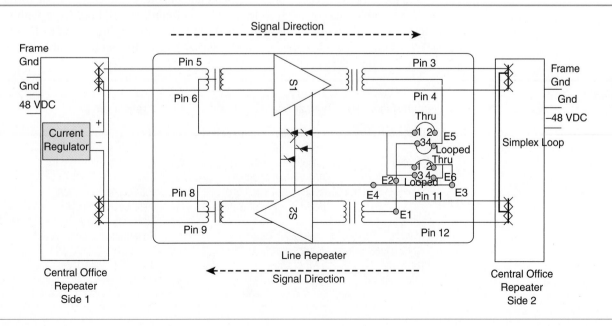

**Telco, cable, and other carriers'
cables hang on the same pole.**

■ 14.7 FIBER OPTIC CABLE IN THE LOCAL AND LONG-HAUL TELEPHONE NETWORK

Fiber optic cable is found throughout the communications plant, strung between poles, pulled through underground ducts, and buried directly in the earth. Discerning fiber optic cable from copper cable can be tricky because both are encased in black polyethylene jackets and both are lashed to the same messenger. Fiber optic cable is the predominant type of cable deployed in the local and long distance networks today. The methods used to distribute fiber optic cable compare to those used for copper cable deployment and consequently help promote the spread of fiber optic technologies because the cable is strung using the same installation equipment, thus reducing its cost. The following few sections detail the cable types, how fiber is distributed in the network, how strands are spliced, and the methods used to terminate the strands.

14.7.1 Fiber Optic Cable Types

If you were to cut open a fiber optic cable and a copper cable, you would immediately see that the similarities between the two stop beneath the outside jacket. Fiber optic cable contains strands of transparent silica glass, each used to carry light pulses from high-speed optical transmitters. Over the years manufacturers have produced a variety of fiber optic cable types designed with specific attributes that help classify the cable according to how it should be used, as shown in Table 14–5. As noted in the table, beyond the general category of inside plant and outside plant, cable can be further divided into cable makeup such as loose tube, buffered, and ribbon. Even within each of the subgroup classifications, additional categories specifying the type of protective jacket, strength member, bend radius, and so forth differentiate cable types. All

Table 14–5
Fiber Optic Cable Types.

Cable Type	Typical Use	Application
Loose tube	Outside plant—direct buried aerial or underground	TELCO
Armored	Outside plant—buried	CATV
Distribution	Underground or aerial or riser	CATV
Aerial	Outside plant aerial or underground—large networks	TELCO and CAPs, long distance networks
Distribution	Riser cable	Riser, short conduit runs
Breakout	Inside cable, riser	No termination panel needed
Tight buffered	Jumpers, patch cords	
Plenum	OFNP or OFCP	Run in air ducts, meets fire code
Nonplenum	OFNR or OFCR	Used between floors, not in air ducts

manufacturers describe their cables in detail on their Web sites or within product brochures. Reviewing diagrams showing how the fiber is packaged within the cable and reading through the specification sheets are very important because different scenarios warrant different packaging methods. A good rule of thumb is to take advantage of the manufacturer's expertise before purchasing a fiber type for a specific job.

The two most common outside plant cable types are loose-tube cable and ribbon cable. Both have been designed specifically to adjust to the bends, movement, and stresses of cable deployed in the outside plant. *Loose-tube cable*, the older of the two, as shown in Figure 14–28, is built using color-coded buffer tubes that surround the bare fiber strands. The tubes, 2 to 3 mm in diameter, are hollow, allowing for the fiber strands to give and take as the cable is pulled and maneuvered around bends. The individual strands are also color coded with a plastic coating according to the fiber optic color code as shown in Table 14–6. A steel member called a *strength member* is built in the center of the cable surrounded by the buffered fiber strands and is used specifically to reinforce and stabilize the strands. A gel substance, the same as that in gel-filled copper cable, is often added to help retard moisture from building up inside the cable.

The strand count of loose-tube cable ranges from just a few to over 250 strands and, as is obvious, physical size of the cable directly relates to the fiber count: the greater the number of strands, the larger the diameter. As with copper cable, a strong outer jacket encases the buffer tubes and acts as a shield against outside disrupters such as water, rodents, infrared rays, and so forth. The material used to jacket the strands varies depending on where the cable will be deployed. For example, a cable buried underground requires a much hardier outside jacket than one pulled through an innerduct and a cable laid across the ocean floor requires an additional steel mesh jacket that should deter sharks and other disruptive elements. Cable jackets vary according to where the cable is deployed, as shown in Table 14–7.

Ribbon cable, the newest on the market, was developed for two reasons—first, to increase the strand count of the cable and, second, to reduce the time it takes to splice two cables together. Ribbon cable is, as it sounds, a cable that looks like a multicolor ribbon made up of several strands, typically 12 or 24, riding parallel to each other forming a flat ribbon similar to ribbon candy. The ribbons of fiber are either placed in loose-tube buffers referred to as *standard loose-tube ribbon cable* or huddled together

loose-tube cable
Fiber optic strands are placed inside a larger tube primarily to secure and protect the strands from outside elements.

strength member
Steel cable built into the center of the fiber optic cable to provide support.

standard loose-tube ribbon cable
Ribbon fiber is divided into individual buffer groups and each buffer group is placed inside a larger diameter tube. The multiple tubes are encased inside the cable jacket with a strength member in the center.

Figure 14–28
Loose-tube cable consists of tubes that hold fiber strands, allowing them to move freely within the tube.

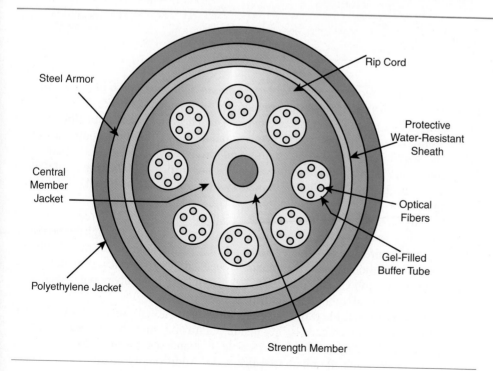

Steel Armor

Central Member Jacket

Polyethylene Jacket

Rip Cord

Protective Water-Resistant Sheath

Optical Fibers

Gel-Filled Buffer Tube

Strength Member

Table 14–6
Fiber Optic Cable Color Code.

Binder Group	Buffer Tube	Color	Strand Count
Blue	Blue buffer tube	Blue	1
		Orange	2
		Green	3
		Brown	4
		Slate	5
		White	6
		Red	7
		Black	8
		Yellow	9
		Violet	10
		Rose	11
		Aqua	12
	Orange buffer tube	Blue	13
		Orange	14
		Green	15
		Brown	16
		Slate	17
		White	18
		Red	19
		Black	20
		Yellow	21
		Violet	22
		Rose	23
		Aqua	24
	Green buffer tube	Repeat 12 colors	25 to 36
	Brown buffer tube	Repeat 12 colors	37 to 48
	Slate buffer tube	Repeat 12 colors	49 to 60
	White buffer tube	Repeat 12 colors	61 to 72
	Red buffer tube	Repeat 12 colors	73 to 84
	Black buffer tube	Repeat 12 colors	85 to 96
	Yellow buffer tube	Repeat 12 colors	97 to 108
	Purple buffer tube	Repeat 12 colors	109 to 120
	Rose buffer tube	Repeat 12 colors	121 to 132
	Aqua buffer tube	Repeat 12 colors	133 to 144
Orange	Repeat buffer tube colors	Repeat 12 colors	145 to 288

central buffer ribbon cable
The strands of ribbon are grouped together in the center of the cable. Buffer tubes are not used to isolate the groups of fiber strands.

plenum cable
Cable designed to be run through the plenum ducts in a building. Plenum cable meets the fire code associated with plenum builds.

NEC
National Electrical Code. Code defined by the National Fire Protection Association.

in one section referred to as a *central buffer ribbon cable*, as shown in Figure 14–29. Ribbon cable allows the manufacturer to increase the number of strands per cable without having to dramatically increase the cable size. The strand density of ribbon cable varies from 200 to 300 strands up to 864 strands. One of the most common ribbon cables deployed consists of 432 strands and, as is obvious, is used where high cable counts are warranted. Several companies are also manufacturing cables with 864 strands specifically targeting large metropolitan areas such as New York City and Los Angeles.

Mass fusion-splicing machines, discussed later in the chapter, help reduce the time it takes to splice up to 864 strands thus making it practical to deploy a high strand count. Ribbon cable requires special splice cases and closures in addition to special splicing equipment. Even though new equipment is needed, large service providers have adopted the use of ribbon cable in their networks typically in highly populated areas or regions with high bandwidth needs and space restrictions.

Inside plant fiber cable differs from outside plant cable in several ways. One of the most important differences is that inside plant cable must be encased in a fire-retardant jacket referred to as *plenum cable* as mandated by the *NEC—National Electrical Code*®—and the municipal fire code. Additionally, inside plant fiber is

Cable Type	Description	Deployed
Distribution cables	Multiple buffer tubes encased inside a single jacket. Sturdy strength member used to stabilize cable. Low number of strands to keep cable size small. Require method for breaking cable out for termination.	Short innerduct runs, riser cable
Breakout cable	Fiber strands are separated into individual simplex-type cables and bundled together eliminating the need to add a breakout box to terminate the fiber. This cable works well when small fiber count required and termination panels not available. May be individually connectorized.	Conduit and plenum installations
Loose tube	Fibers bundled together into small groups, with a strength member and outer jacket. May be filled with gel to suppress water absorption.	Outside plant cable, both aerial and buried builds
Ribbon	Largest numbers of fibers per cable reducing the size of the cable—more fibers per cable diameter. The individual fibers placed next to each other, typically twelve strands, are formed into a ribbon. The standard color code pertains to the twelve strands	Outside plant cable, used both aerial and underground
Armored	A metal jacket is inserted between two plastic jackets. Prevents rodents from chewing through the protective casing.	Outside plant—typically used for direct buried applications
Aerial	Cable that has a strength member that can be used to support the cable hanging between poles or is designed to be lashed to existing cable or an existing strength member.	Outside aerial plant cable
Submarine	Cable with special watertight and mesh jacket used to keep the inside components dry and ward against sharks.	Cable placed underwater such as translantic runs and so forth

Table 14–7
Fiber Optic Cable Types.

typically tight buffer tube cable, meaning that a buffer surrounds the individual fiber strands, as shown in Figure 14–30. Similar to loose-tube cable, a strength member is placed in the center of the cable and the buffer is placed around it with a stiff, hard protective fire-retardant jacket covering the buffers. Tight buffer cable is spliced into either a fiber-distribution panel, as shown in Figure 14–31, or inside a splice case or tray. When the cable terminates directly into a piece of equipment or the front of a distribution panel, it may be connectorized and is referred to as MIC cable or fan-out cable.

14.7.2 Deploying Fiber Optic Cable in the Outside Plant
Aerial Placement

Outside plant aerial fiber optic cable is lashed to the steel messenger using the same devices—spinner and lashing wire—that are used to construct aerial copper cable. Lashing the cable to the messenger is done in the same way as explained earlier in the chapter and therefore is not repeated in this section. In addition, fiber optic cable—like copper cable—may come with the messenger embedded next to the cable as with

Figure 14–29
Two types of ribbon cable. Loose tube and central buffered.

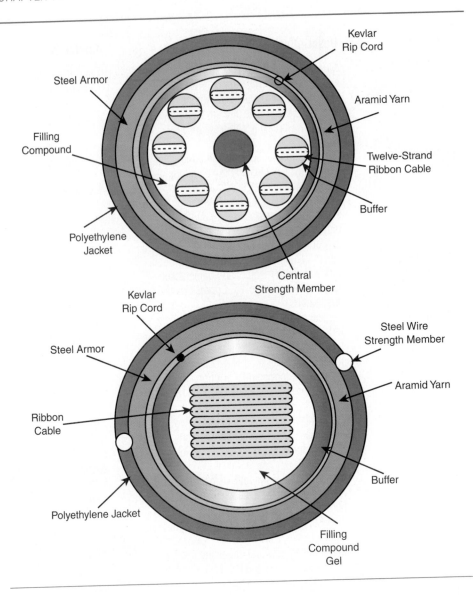

Figure 14–29
Two types of ribbon cable. Loose tube and central buffered.

snowshoe
Device used to hold slack fiber optic cable on an aerial span.

route miles
Physical distance of the fiber span from point A to point B. Route miles do not take slack loop lengths into consideration. Route miles are always shorter than fiber miles.

fiber miles
Length of the fiber deployed in the span from point A to point B. This includes all slack loops. Fiber miles are always longer than route miles.

figure-8 copper cable, as shown in Figure 14–32, and therefore is anchored at each end eliminating the need to lash to a separate messenger. Though deploying aerial fiber cable is almost identical to deploying copper cable, there is one significant difference. Fiber optic cable always contains a certain amount of slack cable built in at strategic points along the route. The slack is looped around a device called a *snowshoe* that is sized according to the cable diameter. If the snowshoe size is too big for the cable, problems may arise in the future as the cable shifts causing cable stress; or, if the snowshoe is too small, the cable eventually may exceed the minimum bend radius as it strains and shifts. Manufacturer specifications should always be followed when selecting snowshoes, anchors, and so forth. The extra cable in slack is used for maintenance or future growth as it is much easier for the technician to pull from the slack loop than to cut in an entirely new section.

One issue that arises from the placement of slack loops in fiber optic cable is the difference between the route miles and the fiber miles, a seemingly insignificant problem that always becomes a big issue. *Route miles* refers to the distance between the start and end points of the fiber span, and *fiber miles* refers to the number of feet of fiber between the start and end of the span. For example, the OTDR—optical time domain reflectometer—measures the fiber miles or kilometers including the slack loop, thus the discrepancy between route and fiber miles. Maintaining accurate records is crucial

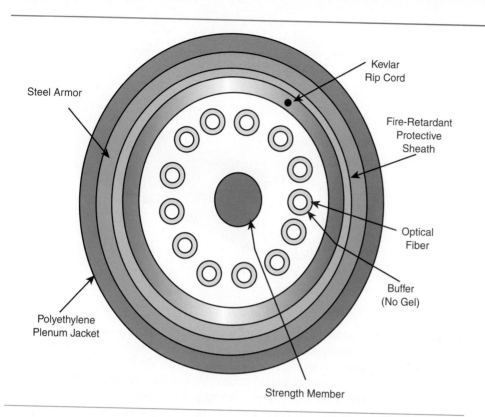

Figure 14–30
Tight-buffer-tube cable is often used inside buildings.

Kevlar
Rip Cord

Steel Armor

Fire-Retardant
Protective
Sheath

Optical
Fiber

Buffer
(No Gel)

Polyethylene
Plenum Jacket

Strength Member

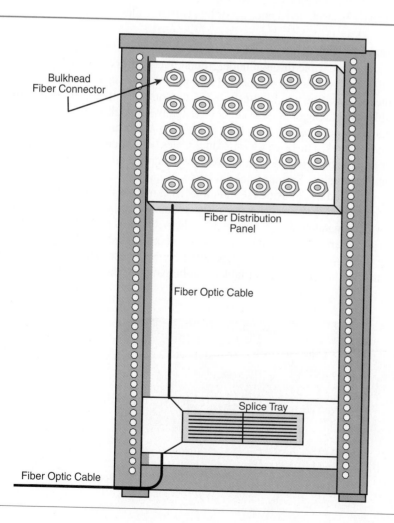

Figure 14–31
Fiber distribution panels are the main termination points for fiber optic cable in a building. A splice tray is typically placed at the bottom of a relay rack and used to splice the inside building fiber cable to the pigtail cable extending from the FDP.

Bulkhead
Fiber Connector

Fiber Distribution
Panel

Fiber Optic Cable

Splice Tray

Fiber Optic Cable

Figure 14–32
Similar to copper cable, fiber optic cable may also have the strength member joined to the cable. The cable is referred to as figure-8 cable as it looks like a figure 8 when looking at it from the front.

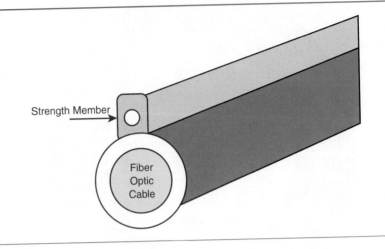

especially when the network is carrying trillions of bits of information. When a fiber cut occurs, the technician compares the OTDR reading to the as-built records in order to find the location of the break. If the records fail to account for slack loops, the distance displayed on the OTDR will be interpreted incorrectly and the technician will misinterpret the location of the break and consequently spend precious time hunting for the trouble. Additionally, when purchasing fiber from a fiber provider, it is very important to specifically ask whether the mileage is for route miles or fiber miles. In some cases, service providers have been quoted a price according to route miles, then charged for fiber miles.

Underground Placement

Pulling fiber optic cable through underground ducts differs from pulling copper cable only in that each conduit is segmented into innerducts—smaller tubular pipes placed inside the duct, as shown in Figure 14–33. Each innerduct may carry one or more fiber optic cables depending on its size and the company's guidelines. As with ducts, service providers often purchase an innerduct in order to avoid having to build an entire underground network. The price for innerduct varies per region and often costs more than the fiber optic cable itself, depending on how complicated the build is and how much innerduct is available. Similar to aerial cable, slack loops are added to the cable span and are coiled and hung inside the manhole, as shown in Figure 14–34. Underground cable provides a very secure, weather-resistant means for cable deployment. Many service providers mandate that at least half of the network is underground. The trade-off between aerial and underground cable deployment is cost, as shown in Table 14–8. Building underground duct varies depending on the landscape, the available duct, the geological makeup, and other variables that help dictate the cost to build.

Fiber slack loops are fed around devices called snowshoes.

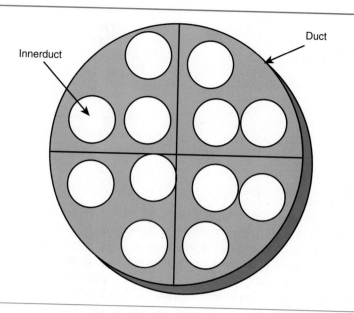

Figure 14–33
Underground ducts have plastic tubes called innerducts. Fiber optic cable is pulled through an innerduct, helping increase the amount of cable run in a duct.

Running fiber optic cable through sewer pipes is a new and innovative way to deploy underground cable. Robots equipped with small cameras scoot through the municipal sewer pipes mapping out the route for the fiber optic cable. The camera relays information back to the technician sitting in the manhole on the structure of the pipe such as whether there are breaks, collapses, or other disruptions that would hinder the placement of the fiber. Once the technician feels confident that the pipe is clear, the robot is equipped with a special device that attaches hooks on the inside of the pipe, similar to clamps on an aerial cable. Next the fiber optic cable rides on the robot and is fed through the pipe and attached to each of the hooks while the technician maneuvers the robot through the ducts—similar to maneuvering a remote-control car. The use of the sewer pipes eliminates the need to build underground ducts for the backbone network, thus helping reduce the cost associated with disrupting the thoroughfares and business traffic, digging, boring through bedrock, and all the other heavy construction tasks necessary to build underground duct systems.

Even though the sewer pipes eliminate the need to build new underground ducts for the backbone network, the final piece of the network—the lateral—must still be built as the sewer pipes into buildings are too small to accommodate a fiber optic cable. Even

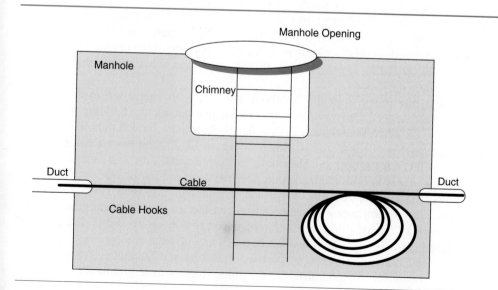

Figure 14–34
Slack loops are secured in manholes and are used when additional cable is needed in the span, such as when a cable break occurs or a new leg is added.

Table 14–8
Fiber Construction Estimated Costs.

Estimated aerial cost per mile	$20 per foot
Estimated underground cost per mile	$100 per foot
Estimated buried cost per mile	$35 per foot

when this ingenious method for deploying underground cable is used, the cost to build the lateral can still be substantial—the most significant obstacle to wiring every business with fiber.

Buried Placement

Buried fiber optic cable can be found in suburbs, in rural areas, and especially in long-haul networks. Burying fiber optic cable in the local market depends on the same method used to bury copper cable—plowing the cable in using special trenching equipment. Fiber optic cable buried by long-haul providers may vary depending on where the fiber is being buried. For example, many long-haul fiber optic cables ride next to railroad tracks as the location provides a clear, safe route. Special trenching equipment designed to sit on a rail car is used to dig in the fiber making the process less costly and more efficient. Burying cable next to a highway or through a farmer's field depends on the same methods as those used to bury copper cable and fiber optic cable in the local network.

When fiber optic cable is buried, a separate copper wire is laid above the fiber cable and is used as a tracer wire, that is, a marker signifying where the cable is buried. Special tone-sensing devices are used by stakeout crews when trying to find where the cable is buried. The device is run across the ground similar to a metal detector until it picks up the tracer wire. When digging has to take place, for example, along a thoroughfare, a stakeout person is dispatched to mark the location where there is cable. He or she uses fluorescent orange flags and spray paint to mark the area where fiber is located. If a contractor digs without contacting the local utility and hits a fiber optic cable, the cost to the contractor may be in the thousands of dollars. Digging before checking is not a smart thing to do.

14.7.3 Transition from the Backbone to the Building

Pulling cable into a building, referred to as a lateral, is—as mentioned earlier—the wild card when costing out a project. In some cases, building a lateral exceeds the cost to build the entire network. For example, a build that requires the construction company to bore through solid rock, dig up a busy thoroughfare, and resurface the top can run in the hundreds of thousands of dollars. In addition, building owners and municipal governments have become very savvy as to the value of lateral builds—often requiring a percentage of the profits gained from the communications circuits. When the word *fiber optics* is bantered around at a municipal meeting, board members see dollar signs. At this point, extending the fiber optic network to the building is the toughest piece of the network to cost justify, especially when there are only one or two tenants that will lease bandwidth on the strands.

Cable transitions from the backbone network to a lateral are made in small cement-lined holes called *handholes* that reside just outside the building or near the junction of the network and the lateral. A handhole is large enough to hold a fiber optic splice case and fiber slack, as shown in Figure 14–35, and has a small manhole-cover-like lid over the opening. Handholes perform several functions, one of which is to provide access to both the backbone fiber network and the branch or spur off the backbone. Slack loops are coiled and placed in handholes, also shown in Figure 14–35.

Though some metropolitan networks are constructed of 100% underground cable, most networks consist of a compilation of underground, aerial, and buried cable, as shown in Figure 14–36. The transition between any of them requires the engineer to design in transition locations such as a riser on a pole, or a handhole in a sidewalk, or so forth. The most common transitions are going from an underground duct system to

fiber optics
Term used to describe a communications medium used to carry light pulses that contain information signals. Fiber optics is a transmission medium made of glass.

handhole
Small, cement-lined vault used to hold a fiber optic splice case and fiber slack loop. The handhole is placed at the junction point between the main cable and a spur or lateral cable that needs to feed a particular location.

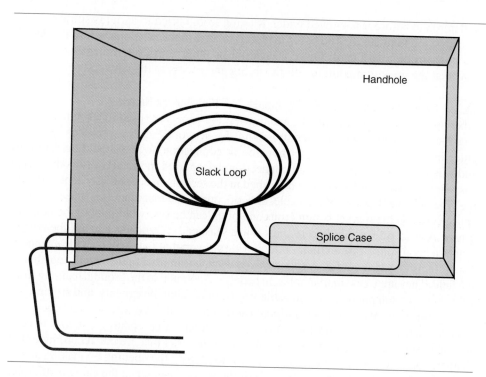

Figure 14–35
Handholes are small boxlike
structures buried beneath the ground
for the purpose of holding excess
cable and splice cases. Typically, a
handhole is found close to the
entrance of a building or near a
manhole that has reached capacity. In
some cases, a handhole is placed by a
service provider that needs to
interconnect to another service
provider's cable located in an adjacent
manhole.

an aerial span, underground duct to a buried section, or a buried section to an aerial section, or vice versa. A transition from underground to buried requires the cable to emerge from the earth into a weatherproof enclosure. The cable creeps up the pole inside a U-Guard and is anchored to the pole, as discussed earlier in the chapter. Moving from underground to buried cable is common, especially in suburban areas that interconnect to metropolitan centers. The underground cable again emerges from the

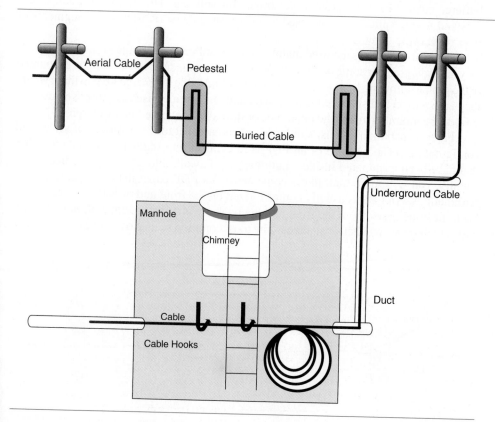

Figure 14–36
Aerial, buried, and underground
cables are all found in the telephone
plant.

underground before it is plowed under. Moving from buried to aerial is similar to that of underground to aerial. The cable is placed in an airtight envelope before it rides up the pole inside a U-Guard. Often a pedestal or other outside termination device is placed at the transition point to add flexibility and access to the network.

14.7.4 Fiber Optic Splice Cases, Terminals, and Distribution Panels

Fiber optic splice cases, terminals, and distribution panels are used to terminate, splice together, and cross connect fiber strands in the network. Service providers draw up guidelines stating what type of termination or connection device should be deployed and where. The most predominant connection device used in the network is the splice case. Splice cases similar to those described in the section on copper vary depending on the type of cable, the size of the cable, and, of course, the manufacturer. The variety of splice cases on the market is numerous and often can be overwhelming to the new engineer. A good rule of thumb is to track down seasoned technicians and ask them what makes a good splice case in their opinion. I assure you that they will provide a laundry list of items to watch out for. A splice case is used to join fiber optic strands together in the field. They are found in manholes, in pedestals mounted in the ground, hanging from cable between telephone poles, in cable vaults, in building basements, and so forth.

A splice case used to splice buffered cable consists of three sections: end openings, outside enclosure, and the inside splice tray or support. End openings, similar to copper cable splice cases, are made of plastic or metal disks that are cut according to the diameter of the cable. Technicians are trained to prepare the opening by making an exact cut for the cable opening in the ends of the splice case. Once the cable is in place, the hole is sealed as specified by the manufacturer to prevent water or debris from entering the case. A sturdy cover or door, usually on some type of hinges or snap-off latches, opens exposing the inside of the splice case, as shown in Figure 14–37. Again, the opening is weathertight and should be sealed according to the manufacturer's specs. Inside the splice case, the cable is secured to the inside walls and splice trays are used to hold the individual fiber strands, as shown in Figure 14–38. The technician must cut open the outer jacket of the cable and expose the inside buffer tubes—in the case of buffered cable—before readying the strands for splicing. The important point to understand is why splice cases are used in a fiber optic network: to join two segments of fiber optic cable together.

Pedestal enclosures are used mainly to house splice cases on the ground. Typically, they are light green or some other neutral color and are made of metal. They are mounted in the ground allowing for the buried or underground cable to emerge up into them where they are secured and sheltered against the elements. Splice cases as those we have described are mounted inside the pedestals, or the cable is encased in a waterproof jacket.

Fiber distribution panels, similar to splice cases, come in a variety of styles and configurations. The purpose of an FDP is to connect two segments of cable together that have been terminated inside a building. If you walk into any central office in the country, you will more than likely come across an FDP panel that has multiple fiber terminations. An FDP is usually connectorized on the front and directly spliced on the back. In some cases, both the front and the back are connectorized; and, in some rare cases, both front and back are spliced. The most common scenario involves a splice tray

Figure 14–37
On the outside, fiber optic splice cases look very similar to copper splice cases.

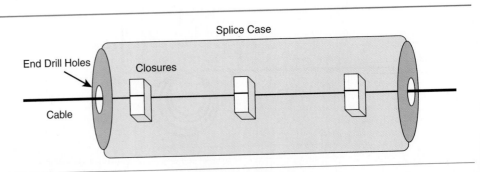

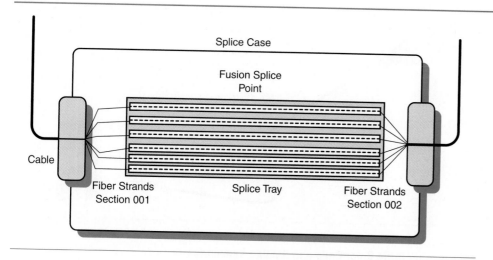

located at the bottom of the rack and specially made fiber cables feeding from the splice tray up to the back of the termination panel, as shown in Figure 14–39. The front of the termination panel is connected together using fiber optic jumpers with connectors on both ends, as shown in Figure 14–40. For example, two fiber spans coming from two different offices terminate in a third office; a connection between the two is made at an FDP in the third office using fiber optic pigtails or jumpers, as shown in Figure 14–41. Another example is when a span coming in connects to a piece of fiber optic equipment such as a multiplexer. The fiber optic jumpers connect from the FDP to a termination point on the FDP that is connected to the fiber optic multiplexer, as shown in Figure 14–42. Fiber distribution panels are very useful devices. Not only do they act as cross-connect devices, but they also are used as test points during turnup and maintenance.

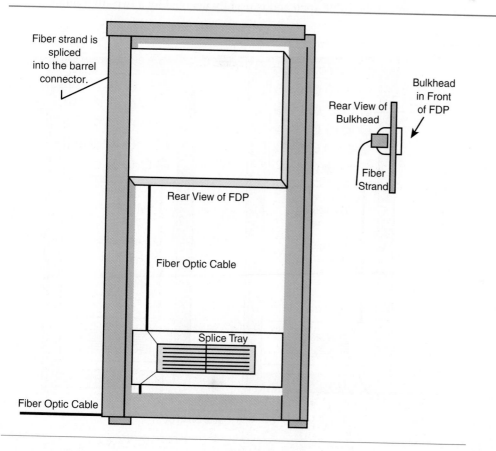

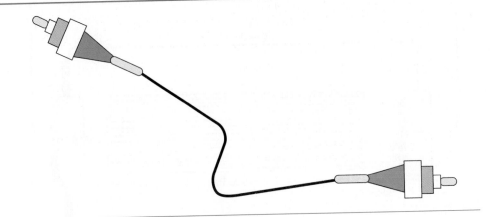

Figure 14–40
Fiber optic jumpers are used to connect fiber optic equipment to fiber optic cable or cable spans to cable spans. Fiber optic connectors are spliced onto each end of the fiber strand. A flexible plastic covering encases the fiber.

14.7.5 Fiber Optic Color Code and Naming Convention

As with copper cable, a color code has been devised to assist in determining what fiber strand is what. The color code, as shown in Table 14–6, consists of twenty-four colors used to identify both the fiber strand and the buffer tube. Without the color code, which is infused onto the fiber strand—similar to dipping an ice cream cone in chocolate topping, must be stripped back before splicing though only far enough to accommodate the splice. Ribbon fiber adheres to the standard fiber color code even though there is less chance to incorrectly splice the wrong strands together because they are lined up and fused together.

Knowing where the fiber cable is heading and understanding what systems are riding on the cable are critical as engineering and operations depend on being able to quickly determine locations and so forth. Naming conventions are devised to help organize and specify cable and strands in a network. For example, the terms *segment*, *span*, and *system* are commonly used to identify the cable. The cable *segment* refers to a subsection of the cable span and is usually denoted by a two-digit number such

segment
The section of fiber between splice points.

Figure 14–41
Connecting fiber spans together at a POP or switch site is typically done at an FDP using fiber optic jumpers.

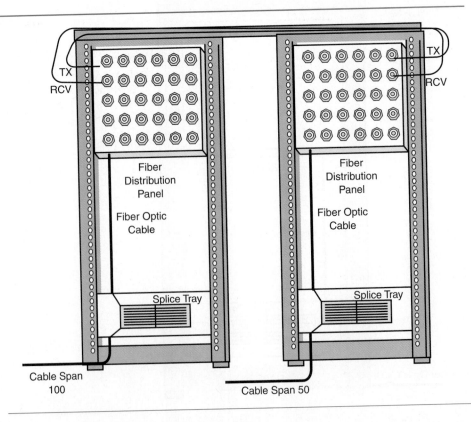

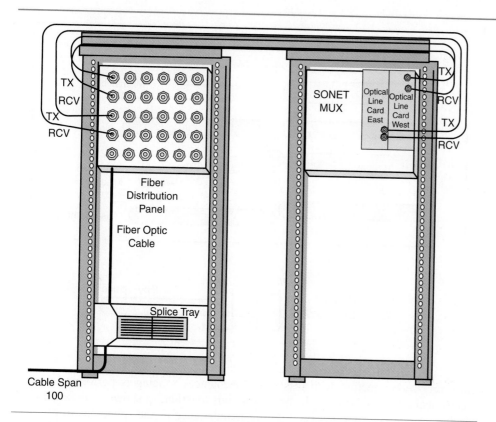

Figure 14–42
Fiber distribution panels are used as intermediate meeting points between a fiber span and an optical MUX. Fiber optic jumpers with connectors on each end are run between the FDP and the optical line card of the MUX.

as 01, 02, and so forth. A *span* is made up of multiple segments and stands for that portion of the cable running between offices such as span 100 running between offices A and C. *System* refers to the end-to-end optical link and is used to designate the path of the optical system such as optical system 205 containing spans 100, 40, and 330. Figure 14–43 illustrates how segments, spans, and systems are denoted on an engineering drawing. Beyond naming the cable segments, spans, and systems, each cable and strand is given a cable and strand number similar to copper cable numbering. A 288-strand cable may be denoted as cable 100, strands 1–288. Technicians must label FDPs using the segment, span, system, and cable and strand number. Engineering records also use the naming sequence to identify the different pieces of the network.

span
The fiber cable between two end points such as end point A to end point B. Multiple segments are contained in one span; and, in turn, multiple spans are contained in one system.

system
End-to-end fiber optic system. A system includes multiple spans.

14.7.6 Splicing Fiber Optic Strands
Two methods are used to splice fiber optic strands together: mechanical splicing and fusion splicing. Today, fusion splicing is the preferred method as it provides a cleaner,

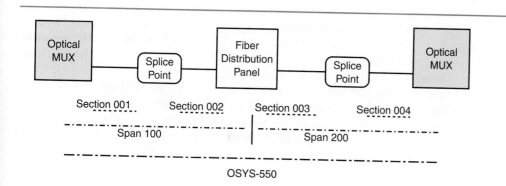

Figure 14–43
Most service providers logically layer divide their network making it easier to pinpoint a location. Sections, spans, and optical systems are used to identify specific areas in a network.

Figure 14–44
Splicing fiber using a mechanical splice requires the two ends of the fiber to be joined together and anchored into position using specialized splicing hardware. In order for the splice to be good—less than 0.1 dB of loss—the fiber ends must line up perfectly. The diagram in the box shows several common problems encountered during splicing.

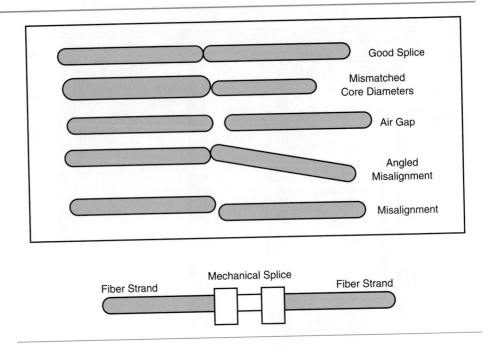

more precise connection with less loss. Mechanical splicing is performed using a mechanical connector that holds the two strands together, as shown in Figure 14–44. Mechanical splicing is often used during maintenance as the splice has to be made quickly and waiting for a fusion splicing machine may not be an option. Mechanical splices have been shown to cause a greater amount of reflection because of the chance that the cores may not be aligned exactly; as such, mechanical splices are often not used in networks that carry very high data rates such as OC-48 and above. In addition, a mechanical splice typically has a higher decibel loss than a fusion splice.

Splicing fiber optic strands using a mechanical splice first requires that the outside jacket be opened and the inside buffer tubes exposed. Next, the individual strands are exposed and the coating removed using a fiber stripper. Once the bare fiber is visible, it is cleaned using 99.99% isopropyl alcohol and a soft, low-lint wipe. Standard drugstore rubbing alcohol leaves a film on the fiber end resulting in a high-loss splice. Therefore, it is very important to only use alcohol from fiber optic supply houses that guarantee its purity. After cleaning, the bare fiber ends are aligned—making sure the core of each fiber meshes exactly—before securing them in place with the mechanical connector. A loss reading should be taken to ensure that the splice is good. Typically, a mechanical splice ranges from a 0.5 dB up to a 1 dB loss. If the cleave is not clean or the mechanical connector not seated properly, the loss may exceed 1 dB requiring it to be respliced.

Fusion splicing is similar to mechanical splicing in that the fiber cable must be open, stripped, and cleaned. Once the two fiber ends are exposed, they are placed in a fusion-splicing machine and lined up using a microscope attached to the box. The technician is able to adjust the position of each fiber end until the cores are aligned precisely. Once aligned, the technician pushes the fuse button and the machine melts or fuses the two ends together. A special safety jacket is placed around the splice before placing it in the splice tray in order to keep it secure. Fusion splicing typically has a very low loss and low reflectance value ranging from 0.05 up to 0.08. If the splice loss is too high as shown by an OTDR trace, the splice is broken and redone. It is important to read through the technical specification portion of any fiber contract to see what loss values have been specified for splice loss. In many cases, an average splice loss, meaning the average of all splices in the span, is stated in the

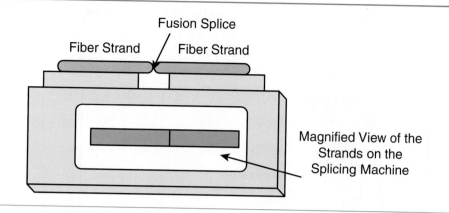

Figure 14–45
Fusion splicing is accomplished by lining up the two ends of the fiber strands in the tray and checking the alignment on the monitor.

document and is used as the benchmark for acceptable loss values. For example, if a span has five splices equaling 0.7 dB in average splice loss and the contract states that the loss must not exceed 0.8 dB average splice loss, the span would be acceptable to the customer. There may be instances in which one splice has a very high loss such as 3 dB or more. In these cases, the contract usually specifies that the splice has to be broken and respliced at least three times and then noted. A splice with such a high splice loss may indicate a problem in the fiber itself and, therefore, should be brought out to the customer.

Mass fusion splicing uses fusion-splicing equipment that is built to splice multiple strands at one time, specifically, ribbon cable. As mentioned earlier, ribbon cable is normally structured in strands of twelve or twenty-four; therefore, mass fusion splicers are built to hold twelve or twenty-four strands at one time, as shown in Figure 14–45. The ribbon cable is prepared similar to single-strand cable by opening, cleaving, stripping, and cleaning the strands before placing them in the fusion tray. The value of mass fusion splicing is obvious: it reduces splicing time; and, because one of the biggest costs when building a network is labor, equipment that reduces time spent is very popular.

14.7.7 Bonding and Grounding Cable

Bonding and grounding of cable are done to reduce the chance that stray electrical current will affect the cable and the people exposed to the cable. The term *bonding* refers to tying the network components together to form a common bond or potential that helps protect them from electrical interference. Bonding is performed on copper cable, fiber optic cable, pedestals, splice cases, and so forth. The following explains where and how bonds should be made, though it is important to emphasize that each provider has specific recommendations that should be followed. Cable shields, fiber optic strength members, and buried service wire are bonded using braided or stranded tinned six-gauge copper wire at each splice location and pedestal, as shown in Figure 14–46. The bonding connectors should also be tinned copper in order to avoid corrosion.

Grounding all electrical components in a network to an acceptable ground is imperative not only for safety reasons but also for service quality. A poorly grounded cable may experience interference from power lines, radio stations, and so forth, in addition to posing a threat to those using their telephones and those working on the lines. A ground rod must be solid copper or another highly conductive metal and must be driven—not placed in a dug hole—into the ground at least 5 to 8 ft. depending on the frost level in the area; the deeper the frost level, the deeper the rod must be driven in. Attaching the network components to ground requires a solid copper six-gauge conductor. Again, it is important to emphasize the need to follow the service provider's and the equipment manufacturer's grounding and bonding policies before beginning a job.

bonding
Tying the network components—all components that may be affected by stray electrical forces—to one common neutral or ground.

Figure 14–46
Grounding cable is one of the most important tasks performed during installation. As shown in the figure, a pedestal is bonded and grounded by securing a bond connector to the cable shield, then attaching it to a ground bracket that is tied to a proper ground such as a ground rod or power company ground.

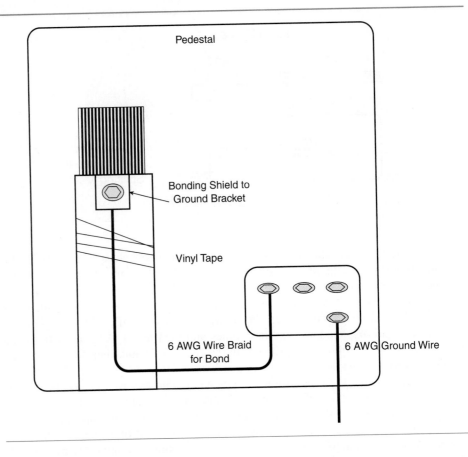

Pedestal

Bonding Shield to Ground Bracket

Vinyl Tape

6 AWG Wire Braid for Bond

6 AWG Ground Wire

■ 14.8 CABLE ENTRANCE

Before moving on to structured wiring, we offer an explanation of how cable enters a building. Cable entrance, whether into a central office or into a multitenant building, requires that an entrance hole be made in the side of the structure, as shown in Figure 14–47. The cable, typically feeding from a manhole located just outside the building, is fed through a duct used to connect the manhole to the inside of the building. In the case of a central office, the manhole is designated as manhole 0 and the duct extending from the manhole enters the building into a room called a *cable vault*. The cable vault is where all of the cable coming into and leaving the building resides. Inside the vault, splice cases are anchored to the wall and used to connect the inside cable to the outside cable.

Buildings other than central offices connect to the outside plant cable in the basement. Again, splice cases or terminal boxes are used to distribute the cable and connect to the inside wiring. Unlike central offices, whose cable vaults are fairly standard buildings, wiring schemes vary as much as the buildings themselves do. Figure 14–48 illustrates how cable enters a multitenant building, a single-tenant building, a central office, and a residential telephone subscriber.

cable vault
The term used to describe the room in a central office where all outside cable enters and leaves the building. Typically, a cable vault is located in the basement of the building.

■ 14.9 INSIDE PLANT WIRING

structured wiring
The method used to wire a building for communications.

Structured wiring is the term used to describe how inside plant wiring is distributed. Today, companies specializing in structured wiring provide services to building owners and contractors offering a variety of services such as pulling cable, engineering cable runs, and installing termination blocks. Inside plant cabling may be as simple as a telephone line fished through the walls of the home of a residential phone customer or as complicated as wiring an entire thirty-story building for voice, data, and video. The following sections describe the various aspects of inside plant cabling.

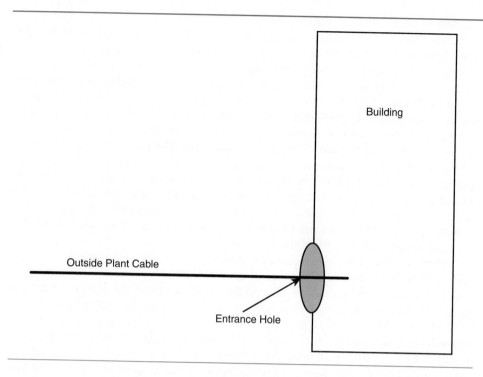

Figure 14–47
Cable enters a building through an entrance hole constructed during the initial build.

14.9.1 Physical Components of Inside Plant Wiring

Wiring a building for communications services involves the deployment of inside plant materials such as termination blocks; outlets; conduit; and, of course, the physical wiring—both copper and fiber. A variety of styles are available for all inside plant components.

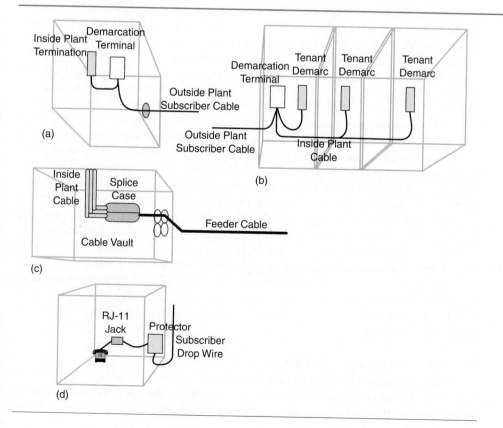

Figure 14–48
(a) This figure shows how cable enters a building. (b) This drawing shows how cable may be distributed in a multitenant building. (c) The third diagram shows how cable enters a central office. (d) The fourth diagram shows how cable enters a residential subscriber location.

Cabling—Copper and Fiber Optic Cable

Whether cabling a new building or a building over 100 years old, the inside wire plant must conform to certain industry standards and meet defined distance and topological structures. For example, wiring a building for data requires that category 5 or 6 cable be used to ward against outside disturbers such as fluorescent lights, other communications circuits, or electrical buses. The *NEC—National Electrical Code®*—and the manufacturer's specifications should be consulted before selecting cable, laying out cable topology, and so forth. Following industry standards when constructing inside plant wiring is as critical as building outside plant networks. The most common types of inside plant copper and fiber optic cable used to carry voice and data services are categorized according to their loss in decibels, weight, diameter, type of shield, the maximum hertz or bits per second they can carry and their fire code rating. Deciding what type of cable to use depends directly on the type of services it will carry. In most instances, CAT3 cable is not used when the circuits carry data while category 5 or 6 is not used to carry voice or other low-frequency signals.

Fiber optic cable, though, is chosen to match that of the incoming outside plant fiber and the fiber optic equipment interface. If the MUX asks for multimode fiber, the engineer will of course use multimode fiber in the network. One of the decisions that must be made before purchasing the fiber cable is whether the ends are connectorized or not connectorized. In some cases, only one end can be connectorized because the connectors will not fit through the duct system and therefore must be spliced in after the cable is pulled. Connectorizing cable on site requires special tools that may not be available, making it difficult for the contractor to complete the job under cost; hiring someone with the correct tools and skills to splice on a connector is expensive. The outer jacket of the cable is another decision that has to be made per building, as the cable runs vary dramatically. If a cable run does not have secure conduit to place the fiber into, the engineer may decide to purchase a heavier outer jacket to ward off any intrusions. Special plastic innerduct tubes may also be used to protect the fiber in these situations.

Termination Blocks and Outlets

Termination blocks come in many different styles. The three most commonly deployed are the 66 punch-down block, the 110 punch-down block and the RJ-45 cross-connect panel. The 66 block is used to terminate and cross connect copper wires. The 110 punch-down block, similar to the 66 block, also terminates and cross connects copper cable; and the RJ-45 patch panel is used to terminate and cross connect category 5 or 6 LAN cabling, as shown in Figure 14–49. As with all inside plant scenarios, many different types of termination blocks have been deployed and cobbled together depending on the situation. It is important to understand the purpose of the block; from that, determining how it works is fairly simple: the block connects wire segments together.

Outlets, such as the simple RJ-11 used to terminate the telephone line in a residential house, are used to interconnect the inside plant cable to the termination device, such as a telephone, a fax machine, a workstation, a modem, and so forth. Outlets come in many different sizes and styles. At the workstation, outlets are mounted in the wall or on the desk, as shown in Figure 14–50. The number and variety of outlets available are staggering. Deciding which to use depends on the preference of the building owner, the types of service being offered, and the amount of space available for the outlet. If, for example, a data jack is required, an eight-pin RJ-48 outlet is needed; if the station is only going to have a standard analog telephone set, an RJ-11 outlet is sufficient. In most cases, multiple outlet devices are placed even when all of the wires are not terminated. An outlet with two data jacks and two voice jacks is common as growth and changes are constant.

14.9.2 Distributing Inside Plant Wiring

Distributing cable in a building requires a great deal of coordination between the building owner, the service provider, the telecommunications manager, and the contractor. In addition, the type of building, the age of the building, and the budget allowed for the

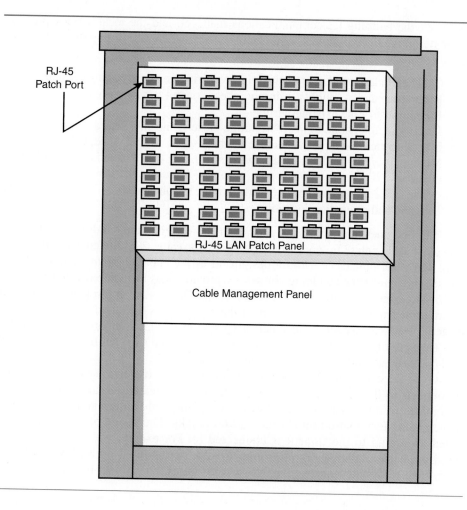

RJ-45
Patch Port

RJ-45 LAN Patch Panel

Cable Management Panel

Figure 14–49
RJ-45 patch panels are used to
connect LAN cabling throughout a
building.

project must all be taken into consideration before beginning the design. If planning is done properly, the future cost of providing communications services to the residents of the building is reduced, as are the headaches and problems.

Older buildings are more difficult to wire as the walls, riser space, and so forth were not built to accommodate modern communications needs. New buildings are designed with communications needs in mind and, as such, are much easier to configure for all types of cabling. Typically, the cabling construction contractor works with the

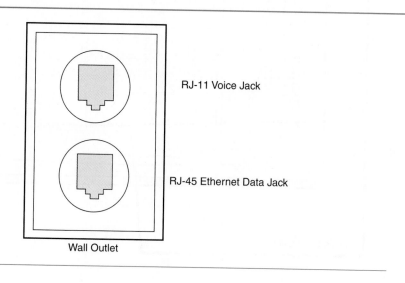

RJ-11 Voice Jack

RJ-45 Ethernet Data Jack

Wall Outlet

Figure 14–50
Wall outlets are often wired for both
voice and data. The RJ-11 jack is
used to interface with a standard
analog voice line, and the RJ-45 jack
is wired for a standard Ethernet
connection.

architect, the building telecommunications manager, and the local service provider when designing the wiring topology. Some concerns addressed during the preliminary meetings are:

1. Cable entrance location and where inside wiring connects to outside plant
2. Riser space and available conduit or where conduit can be placed
3. Running list of the existing wiring
4. Data room, telephone room, distribution closets, and cable entrance room
5. List of services being provided: data, analog voice, digital voice, CATV
6. Medium used to provide services: wireless, CAT6, CAT3, coaxial cable, fiber optic cable
7. Office space layout: number of workstations, telephones
8. Bonding and grounding locations

The first question that must be addressed is, Where in the building does the outside plant cable enter? In large, high-rise buildings, cable typically enters in the basement and is terminated in a room designated as the cable distribution room. If the building is a multitenant building, the service provider may be required—depending on tariffs—to extend the cabling to the tenant's site. There is no set guideline mandating where cable enters a building or whether the cable comes in from a manhole or from an aerial feed. Consequently, technicians spend lots of time wandering around buildings trying to find all of the cable entrances. It may sound silly, but finding the correct cable takes up a good portion of a technician's day.

Once the cable entrance location is found, the next step is to determine how it connects to the existing or future wiring. In most instances, inside plant cable terminates in the same room as the outside plant cable enters and is cross connected using termination blocks such as 66 blocks. The inside plant wiring is labeled according to the floor it is connected to. For example in a fourteen-story building, cables labeled *Flr-1* through *Flr-14* are mounted on a plywood backboard and terminated on twenty-five-pair 66 blocks, as shown in Figure 14–51.

Figure 14–51
Many buildings terminate the inside plant cable onto punch-down blocks mounted on a backboard attached to the wall.

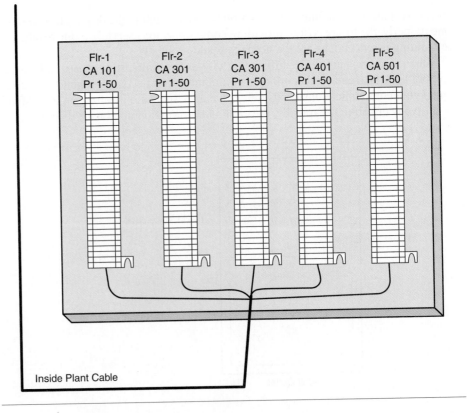

The cable is pulled up through the building in riser space and terminated in a distribution closet on the floor, as shown in Figure 14–52. Riser space may be air ducts, specially designed open passes inside elevator shafts, conduit attached to walls in stairwells, or any other hidden passageway. Riser space is one of the most valuable commodities in a building, as service providers vie for access that may be compared to urban gold.

Providing a central location where inside and outside plant cabling resides helps organize the cabling and provide a point where circuits can be cross connected and disconnected and re-crosses connected to other circuits efficiently. Additionally, cross-connect points are used during troubleshooting and maintenance. Cross-connect wire is used to connect cable pair, as shown in Figure 14–53. Smaller cables run between the distribution closet and the workstations using a fan-out topology, as shown in Figure 14–54, according to the engineer drawings. Within the distribution closet, the wires are terminated on blocks such as 66 blocks or, in the case of category 5 or 6 data wiring, on RJ-45 cross-connect panels. In either case, the cabling coming from the basement connects to the floor wiring at a termination block, as shown in Figure 14–55. Labeling the block with the workstation number is critical, as future moves and changes depend on knowing what wire to move to where. A naming convention should always be formalized before any cable is run in any building, old or new. A cable-running list is the technician's bible, as it provides a view of the entire wiring schematic, which helps reduce time and energy spent on tracking down wires.

At each workstation, an outlet is placed and used to connect the floor wiring to the termination device—telephone, fax machine, workstation, file server, and so forth. Outlets, as mentioned earlier, come in a variety of styles and sizes. Whether the outlet can handle many terminations or just one, it must be labeled using the cable

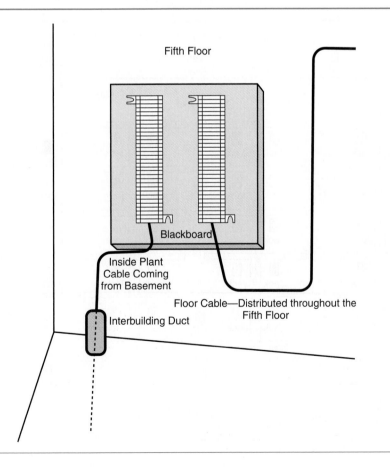

Figure 14–52
Inside building wiring is typically terminated inside a closet located on the floor. A backboard is mounted on the wall and punch-down blocks attached. The punch-down blocks serve as wire termination and connection points. The inside building cable will be cross connected to the floor wiring by attaching jumpers between the blocks.

Figure 14–53

The inside building cable 501 will be cross connected to the floor wiring cable 5001 by attaching jumpers between the blocks. In the diagram, pair 5 of cable 501 is connected to cable 5 of cable 5001. This ties together the cable that extends to the basement and the cable that feeds each office or cube.

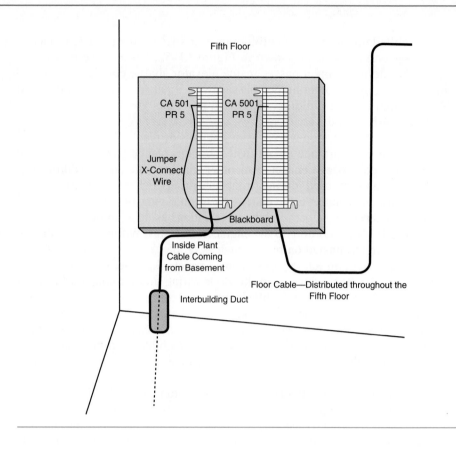

Figure 14–54

Distributing cable throughout an office requires cable runs to be made to each work area. The cabling is tied into the outlet, normally mounted on the wall of the office or cube.

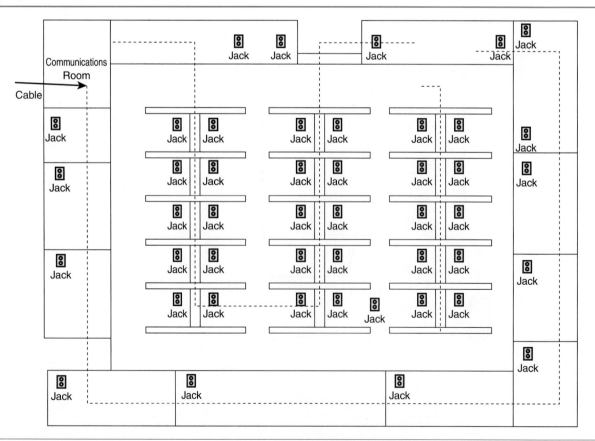

Figure 14–55

Figure (a) shows a telephone connection that requires a point-to-point circuit extending from the central office to the telephone. Inside cabling must accommodate the point-to-point architecture. Figure (b) illustrates how LAN cabling is point to point from the closet to the workstation, a shared cable from the closet to the basement. A hub or switch is used to aggregate the traffic in the closet together onto one medium.

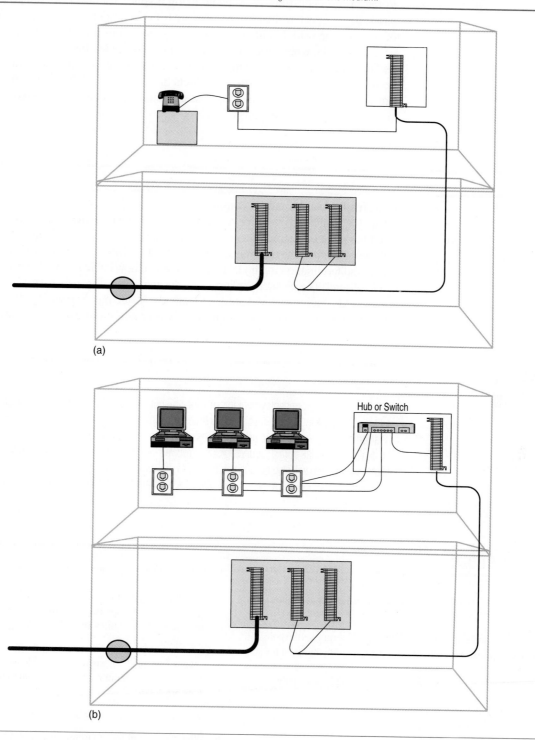

(a)

(b)

running list; in the future, technicians will depend on the marking to direct them to the correct location. Connecting the outlet to the termination device requires a connectorized cable such as an RJ-11 modular cable used to connect the telephone to the jack. Connecting the workstation to the LAN jack, RJ-45, depends on a double-ended RJ-45 cable.

Placing fiber optic cable in a building is similar to placing copper cable. The main differences are how it is cross connected—terminated in the distribution closets—and where the cable enters. Fiber optic cable coming into a building is often directly spliced to the inside plant fiber optic cable using a splice tray or a full splice case. In some cases, the fiber is terminated on a fiber-distribution panel that is mounted in a relay rack or hung on the wall. In either case, the fiber cross connect is made using fiber jumpers. The fiber cable, whether it is multimode or single-mode fiber, is fed up through the riser and terminated in the distribution closet. In many cases, engineers require that the fiber be fed through flexible plastic innerducts in order to avoid the chance the cable will be crushed or pinched. One of the most common problems encountered in inside plant fiber runs is pinched or smashed cable.

SUMMARY

Chapter 14 focused on the communications networks outside plant infrastructure. The chapter was divided into three primary areas that looked at the copper plant infrastructure, the fiber plant infrastructure, and structured access or building wiring infrastructure. The chapter presented the differences in copper cable sizes, noting the types of copper splice cases and termination methods. It also explained how cable is installed both overhead and underground. The chapter explained the differences in fiber cables specifically, cable size termination equipment, and splicing methods. Inside building wiring was discussed as it pertains to structured access designs. Both voice and data wire designs were also discussed.

CASE STUDY

Using the outside plant deployment information you put together in the Chapter 13 case study, build the outside plant network for Green Grass. You will need to determine if the cable should be aerial, underground, buried, or a combination. You will also need to choose what type and size of fiber optic cable to use and how it should be deployed—aerial, buried, or underground. You will need to select splice cases and place them strategically in the network. You will also need to determine how you will terminate the cable at the end premises.

Again, use the information in Appendix A. You should also Google some sites for additional information.

REVIEW QUESTIONS

1. Selecting the type of telephone pole to place in a span requires the engineer to look at the following:

   a. number and size of the cables that will be strung on the pole
   b. whether power or other utility cables will be hung on the pole
   c. whether the pole will be a corner or straightaway pole
   d. a and b
   e. a and c
   f. all of the above

2. All telephone poles require that a guy wire be attached to stabilize the pole in order to keep it standing straight.

   a. true   b. false

3. Why are pole tags important?

4. Two general types of aerial cable are deployed in the network. _____ has the messenger embedded inside the cable, and the second requires the cable to be _____ to the steel messenger.

5. Technicians place twists in aerial cable as they hang it in order to provide a better path for the signal to travel and help reduce cross talk.

   a. true   b. false

6. What element in the environment causes the greatest number of problems to copper cable in the telephone network?

7. What is a manhole chimney?

8. Why should a technician always check for gases in a manhole prior to climbing into the vault?

9. The elements commonly found in manholes include

    a. copper cable.
    b. fiber optic cable.
    c. splice cases.
    d. carrier housings.
    e. remote switch modules.
    f. air tanks.
    g. hooks.
    h. anchors.
    i. a, b, c, d, and f.
    j. a, b, c, d, g, and h.
    k. all of the above.

10. The term *duct* refers to the cable that is placed through the sewer systems in metropolitan areas.

    a. true   b. false

11. Selecting watertight splice cases and carrier housings is critical when designing an underground cable system.

    a. true   b. false

12. Communications cable laid in a trench is referred to as _____ cable.

13. Trenches used to hold the cable are most commonly found in

    a. metropolitan networks.
    b. rural areas.
    c. new developments.
    d. access networks.
    e. b, c, and d.
    f. all of the above.

14. What is meant by the term *cable transition?*

15. A splice case mounted between two cable sections along the cable path is referred to as a(n) _____ splice case.

16. A splice case mounted at a corner pole or junction point is referred to as a(n) _____ splice case.

17. List three types of connectors used to splice copper cable.

18. List two types of terminals used to connect the distribution cable to the drop line.

19. *Demarcation point* is the term used to describe the junction between the feeder cable and the distribution cable.

    a. true   b. false

20. The following are copper termination methods found in multitenant buildings:

    a. 66 block
    b. FDP
    c. PIC closures
    d. ready-access terminals
    e. a, b, and d
    f. a, c, and d
    g. all of the above

21. When sizing a copper cable for a network, the engineer looks at_____, _____, and_____.

22. Which statement or statements are true?

    a. A twenty-six-gauge wire is smaller in diameter than a twenty-four-gauge wire.
    b. A twenty-six-gauge wire has a lower resistance value per foot than a twenty-four-gauge wire.
    c. A twenty-four-gauge wire is smaller in diameter than a twenty-six-gauge wire.
    d. A twenty-four-gauge wire has a lower resistance value per foot than a twenty-six-gauge wire.
    e. A nineteen-gauge wire has a smaller diameter than a twenty-six-gauge wire.
    f. A nineteen-gauge wire has a lower resistance value than a twenty-four and twenty-six-gauge wire.
    g. b and e.
    h. a, f, and d.
    i. a, d, and f.
    j. none of the above.

23. A load coil is used to extend the distance a voice analog signal can travel down a copper pair.

    a. true   b. false

24. The most common color code used to define copper outside plant cable is

    a. orange, blue, white, brown, slate.
    b. green, orange, blue, white, brown, slate.
    c. white, red, black, yellow, violet.
    d. blue, orange, green, brown, slate.

25. How many binder groups are in a superbinder group?

26. If a cable splicer asks you to point out the twenty-fourth pair of a twenty-five-pair cable, what color would you look for in the cable group?

27. When a cable splicer frogs the brown/white pair coming from office 1 to the blue/white pair tied to office 2, what cable pair will you pick up tone on in office 1?_____ What cable pair will you pick up tone on in office 2?_____

28. The CSA refers to the area surrounding an alternate carrier's long distance POP (point of presence).

    a. true   b. false

29. Choose the correct statement as it relates to the terms used to define the local telephone network.

    a. Distribution cable is the large cable that spurs off from the central office. Feeder cable is the cable that feeds off the distribution cable. Branch cables are the cables that spur off the feeder cable.
    b. Drop wire refers to the single-pair wire that services the customer's premises. Branch cable refers to cable that ties the distribution cable to the metropolitan other carrier network cables.
    c. The feeder cable is the large cable that feeds out of the central office. Distribution cables are the cables that spur off the feeder cable and service a given group of streets.
    d. None of the above is correct.

30. T1 Carrier housings are located on average_____ feet apart and _____ feet from the central office and end location.

31. The average loop resistance of a traditional 26-gauge copper loop used to carry analog voice signals ranges between _____ and _____.

32. Explain one reason ribbon fiber cable was introduced.

33. The standard fiber optic cable color code is

    a. green, orange, blue, white, brown, violet, and rose.
    b. blue, orange, green, brown, slate, white, red, black, yellow, violet, rose, and aqua.
    c. blue, orange, green, brown, and slate.
    d. blue, white, orange, black, red, yellow, violet, rose, aqua, and brown.
    e. none of the above.

34. A loose-tube fiber optic cable is seldom deployed in the outside plant network.

    a. true   b. false

35. List the two types of ribbon cable deployed in the outside plant network.

36. Explain the difference between fiber miles and route miles as they pertain to fiber optic systems.

37. Explain why a separate copper wire is buried with the fiber optic cable.

38. A fiber optic splice case is used to connect fiber segments together.

    a. true   b. false

39. List the two methods used to splice fiber optic strands.

40. Improperly grounded components or cable in the network results in

    a. an unsafe condition; stray voltage may endanger anyone exposed to the network.
    b. a reduced chance of outside electrical interferences.
    c. providing a communications order wire that can be used as a test circuit.
    d. improving the reflective characteristics of the signal.
    e. a and b.
    f. a, b, and d.
    g. none of the above.
    h. all of the above.

41. *Plenum* refers to a cable that can be used in the inside plant and the outside plant.

    a. true   b. false

42. List three types of termination blocks used to terminate outside plant copper wire.

KEY TERMS

dead-end pole (362)

push brace pole (363)

messenger (364)

strand (364)

spinner (365)

U-Guard (366)

bell cap (366)

manhole chimney (366)

shoe (367)

pig (367)

gel-filled cable (368)

riser cable (370)

in-line (370)

bridged (370)

curding brush (370)

ready-access terminal (372)

dd (372)

fixed-count terminal (372)

demarcation point (373)

lateral (375)

load coils (376)

binder groups (377)

transpositions (378)

frogs (378)

tone generator (378)

branch feeder cables or F2 cables (381)

digital loop carrier (381)

T-Carrier span (383)

side 1 (384)

side 2 (384)

loop resistance (386)

loose-tube cable (389)

strength member (389)

standard loose-tube ribbon cable (389)

central buffer ribbon cable (390)

plenum cable (390)

NEC (390)

snowshoe (392)

route miles (392)

fiber miles (392)

fiber optics (396)

handhole (396)

segment (400)

span (401)

system (401)

bonding (403)

cable vault (404)

structured wiring (404)

15

Core Networks

Objectives:

After reading this chapter, you should be able to

■ Define the core network.

■ Define network topologies deployed in the core network.

■ Define protocols deployed in the core network.

■ Define equipment found in the core network.

Outline

Introduction

15.1 Physical Structure of a Core Network

15.2 Core Network Architecture

15.3 Core Network Equipment

15.4 Physical Locations of a Core Network

15.5 Mastering the Concepts of a Core Network

■ INTRODUCTION

long (L)
A term used to describe the shortest network span distance in a core or long-haul optical network.

very long (VL)
A term used to describe the medium network span distance in a core or long-haul optical network.

ultra long (U)
A term used to describe the longest network span distance in a core or long-haul optical network.

The core is defined as the long-haul portion of the communications network used specifically to carry long distance traffic—voice, video, and data—between cities, continents, and other regionally diverse areas. The IEC has further segmented the core network into three categories: the *long* or *L*, the *very long* or *VL*, and the *ultra long* or *U*, as shown in Figure 15–1. The difference between the three categories is the level of loss allowed on the span before a regenerator is required to repeat the signal. The IEC standard is used by equipment manufacturers when designing the optical components used to transmit the signal. The guidelines help both the equipment manufacturer and the carrier determine the most cost-effective, viable system to deploy in a long-haul network. The decision to deploy an ultra-long-haul device versus a long-haul device is made by looking at both the distance the signal has to travel and the terrain the signal has to travel across. For example, if a span runs between Pittsburgh and Chicago, the distance is fairly short and the route travels across land, making it unnecessary to deploy the more costly ultra-long-haul equipment. In contrast, a span that connects New

Figure 15–1
Core networks are defined by the distance the link runs between transmitter and receiver. The long-haul, very long-haul and ultra-long-haul loss and distance categories are shown.

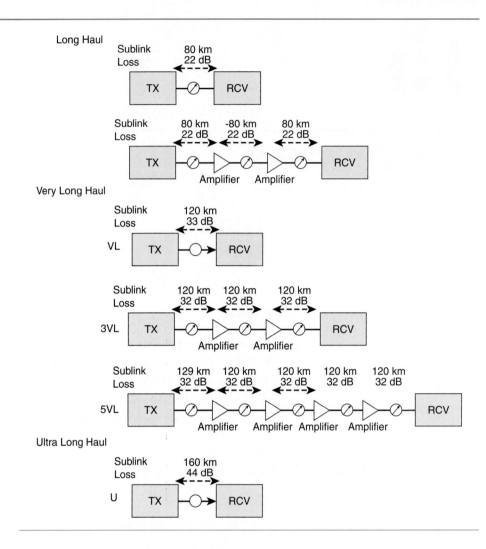

York to London would benefit from ultra-long-haul systems that help reduce the number of repeaters needing to be dropped into the Atlantic Ocean.

Chapter 15 defines the elements that make up the core telecommunications network deployed throughout the world today. The chapter begins by defining all of the physical, architectural, and electronic components that make up the core network. It concludes with a synopsis of real-life situations that take into account the previously defined network components.

■ 15.1 PHYSICAL STRUCTURE OF A CORE NETWORK

The physical structure of the core network consists of all of the elements discussed in the first few chapters of the text—poles, splice cases, manholes, buried cable, underground cable, aerial cable—in addition to equipment used to regenerate or amplify the signal. The primary medium used to carry traffic across the core network is fiber optic cable. Copper cable is typically not used beyond the local metropolitan network; and wireless microwave links, though still in use, are less efficient and less reliable than fiber links. The typical core network is composed of strands of fiber strung along poles or buried under the earth—traversing mile after mile until reaching the far end location. The physical plant is pieced together section by section, as discussed in Chapter 13 and Chapter 14.

The fiber optic cable is sized according to the amount of traffic or needed bandwidth and the cost of cable. Often a *long-haul provider* opts to purchase fiber strands or wavelengths from a wholesale service provider referred to as an alternative distribution company—ADCo—that has an existing network in place. Several large long-haul providers, electric companies, and governments have built extensive coast-to-coast networks that crisscross the Americas, Europe, and Asia and are more than willing to sell or lease portions of their networks to carriers or large companies. It is also common for these companies to swap portions of their networks in order to expand into areas where they lack coverage, thus expanding without having to sink large sums of money into physical plant builds.

It is unlikely that a company would decide to build a long distance or ultra-long-haul network today because of the abundance of available capacity on the existing networks. The costs to build a long-haul network are exorbitant, thus forcing the service provider to justify the need for bandwidth up front even before the design phase is initiated. Today, the fiber glut in the long-haul sector of the communications network is forcing every service provider to analyze each request for network more thoroughly. Additionally, the introduction of low-cost *DWDM*—dense wave division multiplexing—equipment has contributed to the capacity glut in the long-haul and ultra-long-haul arenas. Today, one strand of fiber is able to carry trillions of bits of information. Though the long-haul network builds have essentially ceased to exist, the one area still experiencing growth is networks built to connect smaller cities referred to as Tier 2, 3, and 4 metropolitan areas. The term *regional network* is used to describe the connections between the smaller cities or regional hubs. The equipment deployed in a regional network mirrors that of a long-haul network. The most significant differences between a regional network and a long-haul network are the distance between termination and the person describing the network. Though different terms are used to define the various types of core networks, all deploy the same type of physical plant, network architectures, and transmission equipment. Essentially, each of the network types still depends on fiber optic glass and electronic equipment to carry the information between point A and point B. The differentiations between the networks are the type of fiber and the power of the electronics.

■ 15.2 CORE NETWORK ARCHITECTURE

The *core network* architecture varies depending on the type of information being sent. The four network architectures deployed are the ring, mesh, linear, and point to point. Initially, the core network was built using point-to-point connections that allowed traffic to flow between site A and site B. In the mid-1990s, network designers started to deploy SONET ring architectures that provided full-path protection, meaning that an alternative route was always available. Today, the long distance network is comprised of SONET rings that interconnect with each other allowing traffic to flow seamlessly between multiple locations.

Though many designers felt that they had found the perfect solution to carry long-haul traffic, in the late 1990s a new philosophy was developing. Some of the largest service providers were beginning to deploy the new and somewhat controversial mesh architecture. The main advantage touted by mesh enthusiasts was the ability to carry traffic on all links, working and protection. Unlike the ring architecture that required an idle protection path, the mesh network used protection schemes that deployed traffic across multiple links according to the traffic destination. For example, ring architecture establishes a working and a protection path between sites. The protection path sits idle waiting to be used during a network failure. The mesh network does not designate concrete protection routes as is done in the ring design. Instead, multiple paths are built to each site from different directions providing alternative routes to be used to carry both working and protection traffic. During a failure, there are multiple ways traffic can flow

long-haul provider
A telecommunications carrier that has a long distance network, meaning a network that covers wide areas such as city to city, state to state.

DWDM
Dense wave division multiplexing. Equipment that multiplexes wavelengths across optical networks.

regional network
An optical network that connects smaller cities or regional hubs.

core network
Long-haul network used to carry information between cities, states, and countries.

to the different locations until the fault is repaired. Today, many long-haul providers deploy mesh or partially mesh networks across the globe.

Though commonly industry pundits argue over the value of ring versus mesh or vica versa, the reality is that simpler, less expensive networks such as linear and point to point are still used successfully. The linear network is one that connects multiple locations together in a line, such as from point A to point B, then to point C, and so on. The signal may travel straight through B to C or drop some traffic off at C. Protection on a linear network may be compromised unless alternative routes are established between the locations. Ultimately, a linear design is deployed when the cost to incorporate a fully redundant architecture such as mesh or ring is beyond the scope of the company's budget. Linear networks, though, are still viable, functional designs to consider when building out a connection—especially if the link has minimal traffic.

A fourth design deployed in the long-haul network is a point-to-point design that is used primarily to service high-traffic locations such as between Los Angeles and New York City. Even with the point-to-point design, some type of protection scheme is deployed such as diverse point-to-point paths. In some cases, smaller carriers deploy simple point-to-point networks that do not offer full one for one protection. Instead, the carrier purchases off-net circuits from other carriers to provide backup during disaster situations. *Off-net* is a term used to describe a circuit purchased from another carrier, while *on-net* refers to a circuit riding on the service provider's own network.

The core network architecture is varied and complicated. Most networks are not just one type of architecture but a combination of designs dependent upon the traffic flow; bandwidth needs; and, of course, budget constraints. Figure 15–2 shows four different core network architectures—ring, mesh, linear, and point to point; and Figure 15–3 illustrates the protection schemes used in each network.

off-net
Traffic carried on a network not owned by the carrier.

on-net
Traffic carried on the network owned by the carrier.

Figure 15–2
The four most common network architectures are the ring, mesh, point to point, and linear.

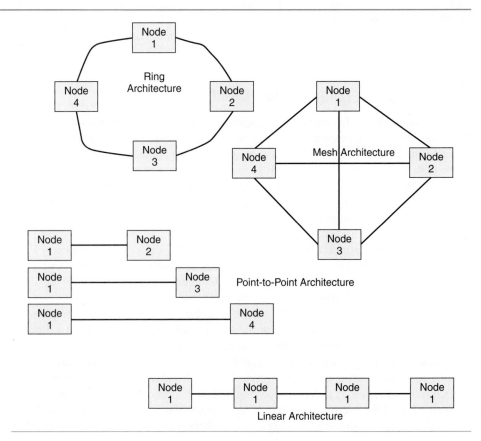

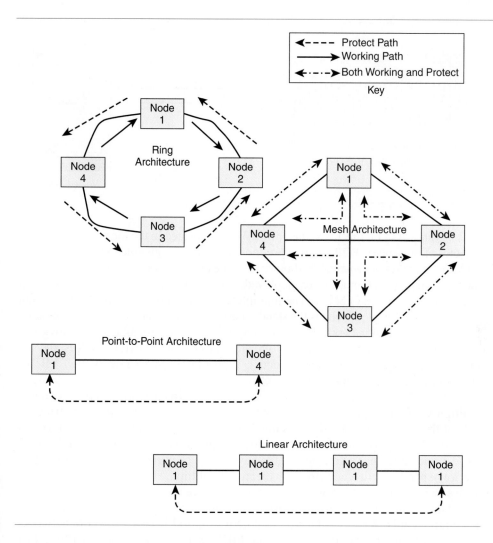

Figure 15–3
Protection schemes used in optical networks vary depending on the network architecture. The ring architecture has a fully redundant protection path that is used when a fiber cut occurs. Mesh networks have multiple routes available between devices. The routes are working routes that absorb the protection traffic when there is an outage. Point-to-point and linear network protection schemes are only available if an alternate path is built.

■ 15.3 CORE NETWORK EQUIPMENT

Along with the core network's physical plant, electronic equipment is required to convert, format, switch, and regenerate the signal. The type of equipment used is dependent upon the type of information being sent. The term *transport* or *transmission* is used to describe the equipment and circuits used to transport information across a network. Early on in the industry, circuits used to connect high-volume sites together carrying multiple signals were referred to as *transmission lines*, similar to the power company's transmission grid. The term has survived even through multiple technology and industry changes. Technicians working in the transmission area of the phone company are responsible for maintaining all types of multiplexers, DWDM systems, amplifiers, cross connects, and any type of equipment responsible for transmitting traffic. In the core network, this definition may be segmented into three key areas: time division multiplexing equipment, wave division multiplexing equipment, and fast-packet equipment.

transport
Term used to describe the transmission systems deployed to carry communications.

transmission lines
Communications circuits used to carry traffic between communications hubs, whether that is between POPs, between central offices, or between distribution hubs.

15.3.1 Time Division Multiplexing Equipment

The TDM portion of the network consists of optical multiplexers, typically SONET add/drop multiplexers—ADMs—and digital cross-connect systems—DCSs. Optical multiplexers are used to add/drop circuits, convert electrical signals into optical signals, package digital signals into different transport protocols, and groom traffic between

Figure 15–4
A common sight in a long-haul switch
site is a SONET MUX connected to the
LD voice switch. The diagram shows a
logical connection between the two
devices. Typically, DS-1 or DS-3 links
connect the two devices.

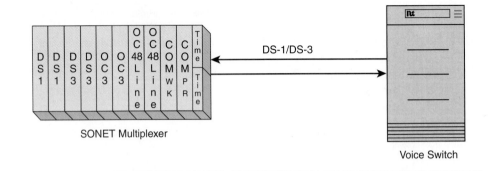

MUX
Shortened version of
multiplexer. Multiplexes,
combines signals onto a
medium.

sites. As shown in Figure 15–4, the optical *MUX* is located in the long-haul switch site or point of presence—POP—and is used to interconnect with voice switches; routers; ATM or frame relay switches; and other multiplexers, both optical and electrical. The most common optical multiplexer deployed in a core network is the Synchronous Optical Network (SONET) MUX. The multiplexer is built around the SONET protocol, as described in Chapter 8, and is the standard used to package and transport information across the long-haul network. Regional, long-haul, very long-haul and ultra-long-haul networks are comprised primarily of OC-48 and OC-192 SONET MUXes and are, therefore, subsequently described in detail. Typically, the lower-rate SONET MUXes are found in metropolitan and access networks.

The MUX is divided into five key areas: tributary, line, cross connect, multiplexor/demultiplexor, and the control or administrative portion. The tributary side of the MUX, also called the channel side, consists of channel cards referred to as *circuit packs* that interface to incoming/outgoing signals such as DS-3, OC-3, OC-12, and so forth. The channel cards have interface ports that connect directly to the low-speed circuit via optical or electrical cabling. The type of interface card varies per vendor and is always specified in the vendor's equipment documentation. One of the most important questions an engineer asks a vendor when evaluating equipment is, What interface types does it support? Typically, in core networks, the lowest tributary rate deployed in the MUX is a DS-3 because most long-haul networks depend on other pieces of equipment to groom and MUX the lower-rate signals. Multiplexers deployed in metropolitan and access networks are built to accommodate lower circuit rates such as DS-1, Fast Ethernet, or DS-0s. The number of ports per card is also a vendor-dependent condition. A DS-3 circuit pack may have twelve DS-3 interface ports per pack, while an OC-48 circuit pack may have only one port per pack. The physical connections also vary per vendor; but, typically, the electronic or copper connections terminate either on the back of the MUX or on a separate panel on the front, while the optical interfaces normally connect directly onto the front of the circuit pack, as shown in Figure 15–5. Many other variations exist between vendors' equipment, such as how the excess cable is stored; the type of connector used; the depth, width, and height of the unit; and so on. Many of these attributes are used as reasons for purchasing a specific type of equipment, especially the way the cabling is distributed on the box. Though seemingly insignificant, many engineers view this as a critical factor in how the equipment will fit in the site, the ease of repair during trouble conditions, and the ability to organize the site.

The circuits are physically connected to the interface ports on the multiplexer via copper cable or fiber optic jumpers. A DS-1 is terminated on wire wrap pins located on the back, front, or side panels of the MUX, as shown in Figure 15–6. Notice the transmit and receive pair that must be connected according to the ins and outs of the box. A second common interface type, especially in core networks, is a DS-3 circuit that connects copper coaxial cable to BNC connectors, one for transmit and one for receive, located on the multiplexer. The multiplexer also may interconnect to

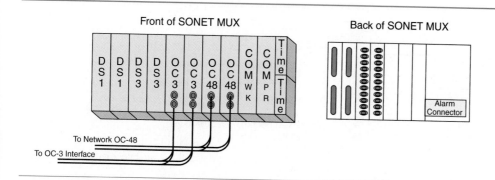

Figure 15–5
Most electrical connections are made from the back of the SONET MUX, and most optical connections are made from the front.

optical interfaces, such as an OC-3 or OC-12, or Ethernet. Optical interfaces typically terminate on the front of the channel card using fiber optic jumpers. Other terminations may be used depending on the type of multiplexer and the type of channel cards deployed.

Beyond connecting the physical interfaces to the MUX, the channel card must also be able to format and package the signal before passing it on to the cross-connect field. The method used to format the signal depends on the protocol used to transport the line side or high-rate side onto the network. If the line side uses the SONET protocol, the channel card must be able to convert the incoming tributary signal into a SONET formatted signal. For example, a DS-1 interfacing into the channel card would be converted into a VT-1, a DS-3 would be converted into an STS-1, and so on. The circuit pack is smart enough to know how to repackage the incoming signal, such as an asynchronous DS-3 into an STS-1 SONET frame. A key concept in the world of transmission is that all information is packaged into frames that are used to define the beginning and end of a bit streams. The structure of the frame is laid out in the protocol standard and is used by equipment vendors when building the MUX. In our example, the circuit packs in a SONET MUX are designed to take whatever type of signal is coming in and structuring it around the SONET standard. The important point to remember is that each multiplexer has a low-speed side referred to as the *tributary side* where the channel cards reside and a line or network side where the high-speed line cards reside. For example, a MUX may have an OC-3 tributary side that is multiplexed into the OC-48 line-side signal. The signals come in on the low-speed circuit packs where they are properly packaged and formatted before moving into the cross-connect field.

tributary side
Low-speed side of the MUX that terminates lower-rate circuits tied to end locations.

A cross connect is established between the tributary side and the line side of the MUX through the back plane or bus of the device. The multiplexer spec lists the back plane speed or cross-connect granularity and is a key parameter to evaluate during the selection process. Commonly, the cross-connect capability is stated as an $N \times N$ cross-connect fabric, also referred to as a switch fabric. For example, an OC-48 multiplexer that has a 1344×1344 cross-connect capability means that 1344 DS-1s can be cross

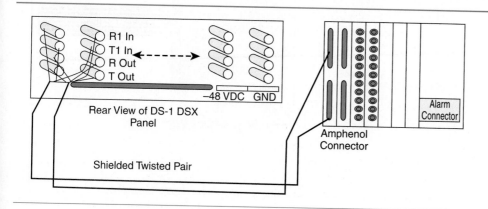

Figure 15–6
Connecting the DS-1 ports from the SONET DS-1 interface card to a DS-1 DSX panel is shown. Note that the connection at the MUX is made using an amphenol connector. At the DSX panel, the individual twisted transmit and receive pairs are wire wrapped into wire wrap pins.

Figure 15–7
Within the multiplexer, the DS-3 is cross connected and groomed into the OC-48 signal after being formatted into an STS-1. The STS-1 is cross connected into an STS-48, which is then converted into an OC-48 signal.

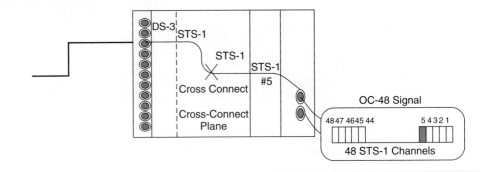

connected at one time across the back plane of the device. In this instance, the multiplexer has full one for one cross-connect capability. Typically, long-haul SONET multiplexers consist of higher-rate channel cards such as DS-3, OC-3, OC-12, or GigE because lower-speed cross connects are performed by equipment sitting in the metropolitan network or a DCS in the POP. The cross-connect field is defined by the number of possible STS-1 cross connects such as 48 by 48 for a one for one OC-48 multiplexer.

The cross-connect portion of the multiplexer is responsible for electronically connecting a signal on one side of the MUX to a signal on the other side of the MUX. A DS-3 arriving on a tributary card may be cross connected to an STS-1 connected to the line card, as shown in Figure 15–7. The electronic cross connect, once established, allows the traffic coming into the MUX from either side to traverse to the respective port, line, or channel. Cross connects may be as granular as DS-0 depending on the switch fabric deployed and the software. Typically, long-haul networks contain MUXes with DS-3/STS-1 granularity and, therefore, are less complicated than lower-rate multiplexers.

The flow of the signal from the tributary card to the line side in an OC-48 MUX with DS-3 channel cards works like this: The signal travels from the channel card through the cross-connect plane to the line card where the lower-rate signals are multiplexed together into one higher-rate signal such as, in our example, 48 STS-1s multiplexed into one STS-48, as shown in Figure 15–8. At this point, the signal has moved from the cross-connect field into the multiplexer/demultiplexer area of the MUX. Some

Figure 15–8
Multiplexing STS-1s together happens in the multiplexing portion of the SONET device. In the example shown, forty-eight individual STS-1s are muxed/demuxed into one electrical STS-48.

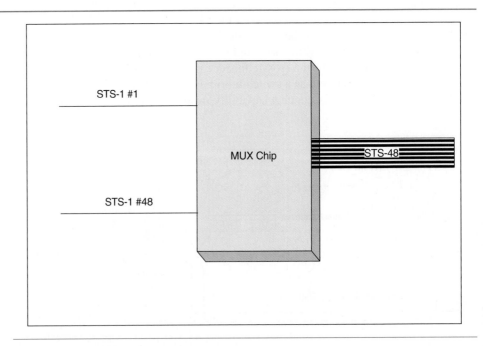

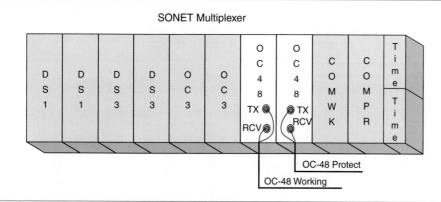

SONET Multiplexer

OC-48 Protect

OC-48 Working

Figure 15–9
The high speed of the SONET MUX
consists of two sets of optical line
cards: one working, one protect. Fiber
optic jumpers are connected to the
front of the card into an optical barrel
connector.

mux/demux chips sit on the line card while others reside in separate cards called trans-mux cards. Again, the layout is vendor specific, but the function is the same: the lower-speed signals are muxed together into one high-speed signal or separated out—demuxed—from a high-speed signal depending on the direction of the signal, that is, whether it is transmitted or received.

The high-speed side of the MUX is where the optical line cards reside and is re-ferred to as the line side or network side of the MUX, as shown in Figure 15–9. The op-tical line circuit packs consist of a transmitter laser or LED and a light-sensitive diode referred to as a *detector* that receives and reads the light signal. The transmitter portion of the card, in addition to providing the light source, has a timing generator that is used to synchronize the bit stream. Once the individual signals have been multiplexed into one high-speed signal, such as forty-eight STS-1s into one STS-48, the transmit card performs an E-to-O conversion—electrical to optical—creating an OC-48 optical sig-nal. The OC-48 laser located on the transmit line card turns on and off according to the bit pattern being sent. In summary, the transmit line card contains a light source, LED or laser, that emits an invisible pulse of light onto the fiber optic strand that forms an information signal. A physical fiber connector located on the front of the card is used to connect to the outside plant fiber optic cable.

In the reverse direction, from the line side to the tributary side, the receive portion of the card accepts the incoming light signal and interprets the pulses using a light-sensitive diode. The optical signal goes through an O-to-E conversion—optical to electrical—creating an STS-48. Each high-speed line card has a duplicate card used for protection in case the working OC-48 card fails. The term used to describe this type of protection is *1 + 1,* meaning that the signal travels through both the working and pro-tection sides of the network simultaneously. If the working card or path fails, the switch overtime is minimal—under 50 ms—due to the 1 + 1 nature of the protection scheme.

After the O-to-E conversion, the signal moves on to the demux section of the device where the high-speed signal—STS-48—is demultiplexed down into individual STS-1s. From the demux portion of the MUX, the signals are sent to the cross-connect field where they are distributed to the appropriate channel card. The channel card reformats the STS-1 into a framed DS-3 and sends it out onto the coaxial cable connecting to the terminating equipment. Figure 15–10 follows the entire process from the ins to the outs.

The last area of the MUX to be described is the control or administrative portion. The control section of the device is responsible for holding software code, gathering and distributing alarms, providing telnet or local access into the device and other vari-ous administrative tasks. Some manufacturers' devices store the cross-connect infor-mation and software code on the channel or line cards, leaving the control portion of the device for alarm functions and remote access. When this is the case, the control por-tion of the device may fail without affecting traffic. When the control cards do hold memory, a backup card is always deployed to guarantee traffic will not be affected if the card fails.

Figure 15–10
The figure shows a DS-3 as it travels between a SONET MUX and a DS-3 terminating device such as an M13 multiplexer. Note how the two devices meet at the DS-3 DSX and the cross connect between them has a reversal or cross: transmit to receive and receive to transmit.

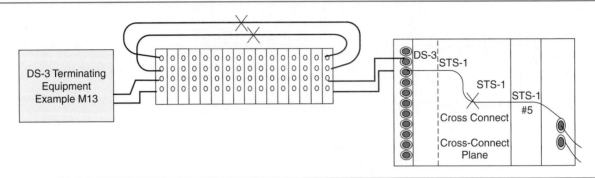

Synchronization circuit packs are also found in a SONET MUX. The packs are typically considered to be part of the administrative area because they do not carry live traffic. Each SONET MUX must have some sort of a timing generator, function of which is to keep the bits synchronized. The timing card receives a timing reference from a BITS—Building Integrated Timing Supply—source that pulls in a Stratum 1 level clock from the GPS—Global Positioning System—satellite and redistributes it to the MUX as a Stratum 2 level clock. If the timing packs were not installed, the bit stream would lose synchronization relatively quickly causing the signals to become confused and corrupted. In the case of a SONET, only one external timing supply is needed to provide clocks for multiple devices on the network. The clock is fed into the MUX, which then distributes it across the network to the other network nodes. Each device, though, must have timing cards installed in order to accept the incoming timing signal from the network and distribute it to the circuit packs. A working and a protect timing circuit pack are deployed to ensure proper synchronization of devices when either a card or incoming timing source failure occurs. Due to the complexity of timing a network, a detailed discussion is presented later in the text.

A DCS—digital cross-connect system—is a device placed in the long-haul carrier's switch site to groom circuits across a network. DCSs are sized according to the expected number of incoming and outgoing circuits. The three common DCS sizes are 128 DS-3s, 256 DS-3s, and 1034 DS-3s. The smaller units may grow up to the 1034, though a processor upgrade is usually required in addition to adjacent space. Though the units are sized according to the number of DS-3 cross connects, they are capable of cross connecting lower- and higher-speed circuits such as DS-1s and OC-3s or OC-12s. Due to this wide range of functionality the DCS is referred to as a *broadband DCS* or a *3/1*. The *narrowband DCS,* called a *1/0 DCS,* cross connects DS-0 channels within T1 circuits. Since the long-haul switch site does not require extensive DS-0 grooming, relatively small 1/0 DCSs are deployed. It is common to see a small, one-shelf 1/0 DCS handle the entire switch site's traffic. Figure 15–11 illustrates how 3/1 and 1/0 DCSs are used in a long-haul network.

The structure of a DCS is similar to a multiplexer but on a bigger scale. The DCS is divided into three main sections: the interface, the cross connect, and the administration. The interface cards are similar to a SONET MUX's interface cards. They have physical connectors that terminate the circuit such as wire wrap pins for T1 interfaces, BNC coaxial cable connectors for DS-3 and STS-1 circuits, and optical connectors for optical interfaces. The interface cards also format the signal into a SONET-friendly frame such as a VT-1 frame for DS-1 circuits and STS-1 frames for DS-3 circuits before passing it on to the cross-connect field.

The cross-connect field is responsible for forming an electronic cross connect between the two circuit locations. For example, a circuit such as a DS-1 traveling from New York City through the Indianapolis switch site to Bismarck, North Dakota, depends on the DSC to make a connection between the two legs of the route. The DS-1 rides on

broadband DCS
Digital cross-connect system used to electronically cross connect high-speed DS-1 channels into DS-3s or STSs, DS-3 or STS channels within OC circuits.

narrowband DCS
Digital cross-connect system used to electronically cross connect DS-0 channels within a DS-1 circuit to other DS-0 channels.

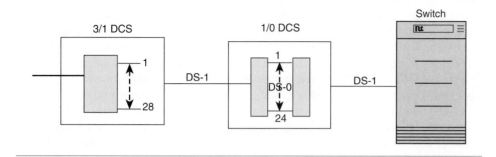

Figure 15–11
Circuits are fed into a DCS where they are groomed and multiplexed. The example shows how a DS-1 is groomed out of a DS-3. The DS-1 feeds into a 1/0, where it is groomed down to DS-0 and cross connected into a DS-1 feeding into the switch.

DS-3 circuit 102 from New York City to the DS-3 interface port on the Indianapolis 3/1 DCS. The DS-1 within the DS-3 is groomed or cross connected to DS-3 306 traveling to Bismarck, North Dakota. The other twenty-seven DS-1s on DS-3 102 may be groomed onto different DS-3 circuits traveling to other locations or may be terminated into the long distance switch in Indianapolis. The DCS is similar to a railway hub where boxcars are moved from one train to another depending on their destination. The advantage the DCS provides is the elimination of physical cross connects between passive termination devices such as DSX—digital signal cross-connect—panels. Instead, the cross connects are programmed into the DCS using a terminal and may be easily changed when needed. The cross-connect portion of the DCS has 1 + 1 redundancy, meaning that the signal is bridged through both the working and protect cross-connect fields, thus providing millisecond switchover times and ensuring that no data will be lost during hardware failures.

The administrative or control section of the DCS may be looked at as the big brain of the unit. All of the circuit configurations, cross connects, alarm parameters, timing functions, and software code are stored in the administrative portion of the DCS. The administrative portion has a full 1+1 redundancy in addition to a tape or disk drive that is routinely backed up to ensure that no information is lost during a catastrophic failure. A timing source is also fed into the administrative section where it is distributed to the interface cards and used to provide synchronization. Figure 15–12 shows the structure of the DCS and defines the function of each section.

The DCS used in the international network is different from those deployed domestically in that it has the ability to convert the North American standard circuit format into the European standard: T1 to E1 or T3 to E3, SONET to *SDH* (Synchronous Digital Hierarchy). The international DCS is usually found only in gateway switch sites such as in New York City, London, Los Angeles, Mexico City, and so forth. The international conversions are performed within the DCS eliminating the need for additional conversion equipment.

SDH
Synchronous Digital Hierarchy. The digital signal rates and protocol used outside North America.

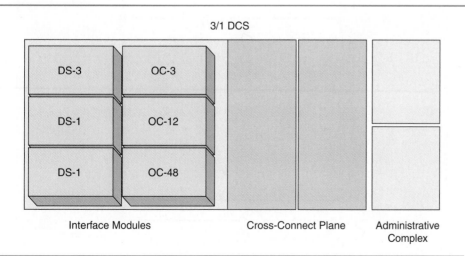

Figure 15–12
A 3/1 DCS consists of interface modules where circuits terminate, the cross-connect plane, and the administrative complex.

15.3.2 Wave Division Multiplexing Equipment

Wave division multiplexing equipment is also commonly found in core networks. The most common type of WDM equipment is the DWDM—dense wave division multiplexer. The DWDM's role in the network is to multiplex multiple wavelengths that each carry a signal such as an OC-48 onto one fiber optic strand. For example, eight OC-48s may be fed into a DWDM device and multiplexed onto one transmit and one receive fiber strand. If the DWDM were not used, sixteen strands of fiber would be needed between the two locations to carry the eight OC-48s, as shown in Figure 15–13. Determining the number of channels or wavelengths to deploy between locations is accomplished by looking at the traffic requirements and expected growth rate. Today, the number of channels multiplexed onto a fiber pair ranges from four to in the hundreds. The limit has yet to be reached. Thanks to the DWDM, it is no longer necessary to construct additional fiber cable when additional capacity is needed between sites.

The common DWDM architectures found in the core network consist of *protected lambda*, a term also used to define a wavelength, or *unprotected lambda*. Protected DWDM is designed by building a protection wavelength on a separate path ensuring path diversity between points. The second type of design is an unprotected lambda, meaning that a single point-to-point path is available to the traffic and if the lambda fails, the signal is lost—thus, the data flow is disrupted. Cost is the main reason carriers opt for unprotected lambda architecture. Some vendors have implemented protection schemes similar to the SONET ring architecture standard where signals are bridged both on the working and protect wavelength helping to guarantee automatic protection switching, APS, thus reducing or eliminating any down time during disasters. As you may imagine, it is very unnerving to see a network with multiple high-speed signals built without some form of protection available.

The DWDM has four functional areas: transponder, multiplexer, transmitter/receiver, control or administration. If you look at a DWDM system from the low-speed side in, the first section encountered is the transponder. The transponder performs a frequency conversion by accepting the incoming light signal from an optical source such as an OC-48 MUX and converting the incoming light, typically a 1310 or 1550 nm wavelength, into a standard frequency as defined by the ITU grid.

Before we continue explaining the DWDM equipment, we first need to detail the ITU spectral bands and how they are used by DWDM systems in the long-haul network. The ITU standard has been organized into several frequency bands or windows. The first band or window, called the O-band or original band, consists of wavelengths ranging from 1260 to 1360 nm. The second band is the E-band or extended band and includes wavelengths from 1360 to 1460 nm. The extended band contains the water peak

protected lambda
An optical wavelength deployed with a protection wavelength built in a diverse network providing redundancy.

unprotected lambda
An optical wavelength used to carry information in an optical system. No redundancy is available for the wavelength.

Figure 15–13
A DWDM system would eliminate the need to put in multiple strands of fiber between locations. Instead, the signals are multiplexed onto a strand.

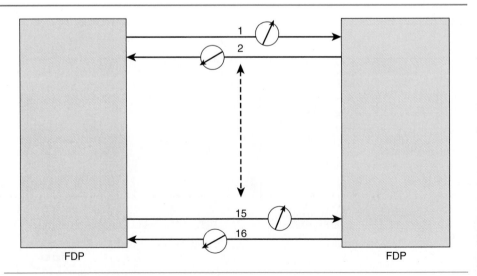

area, as described in an earlier chapter, and is therefore not used in a long-haul network. The S-band or short wavelength band includes the 1460 to 1530 nm wavelengths and is the third window defined by the ITU. The O, E, and S bands are not used in long-haul DWDM systems today. The C-band or conventional band is the most popular band deployed in DWDM systems and consists of frequencies between 1530 and 1565 nm. The optical amplifiers, *EDFAs* (erbium-doped fiber amplifiers), function best within this frequency range and are the main reason for the C-band's popularity in the long-haul network. The L-band or long wavelength band is also used in DWDM long-haul systems and covers the 1565 to 1625 nm wavelengths. The final band defined by the ITU is the U-band or ultra band and is not used other than for supervisory channels in the long-haul DWDM systems. The wavelengths included in this band range from 1625 to 1675 nm.

The next portion of the DWDM system is the multiplexer or coupler, which is responsible for combining all of the newly converted frequencies onto one transmit strand of fiber. The multiplexer is a fairly simple device that passively couples and filters the optical signals together. Once the signals are multiplexed together, the light source or transmitter pulses out the bits of information from the multiple wavelengths onto the fiber strand. At the far end, a detector receives the signal coming in and passes it on to the demultiplexer device that uncouples the signals into their respective wavelengths. The transponder then converts each of the ITU-defined wavelengths back into the correct terminating wavelength and sends the signal onto the fiber jumper connected to the terminating multiplexer—such as a 1550 nm wavelength to the OC-48 multiplexer.

The best way to explain how a DWDM works is to use an example relevant to a long-haul network. Our example includes four optical systems comprised of one OC-192, one OC-48, and one GigE and one OC-12 connected to the DWDM transponder through fiber optic jumpers, as shown in Figure 15–14. The DWDM transponder

EDFA
Erbium-doped fiber amplifier. Used in long-haul networks to boost the strength of the optical signal.

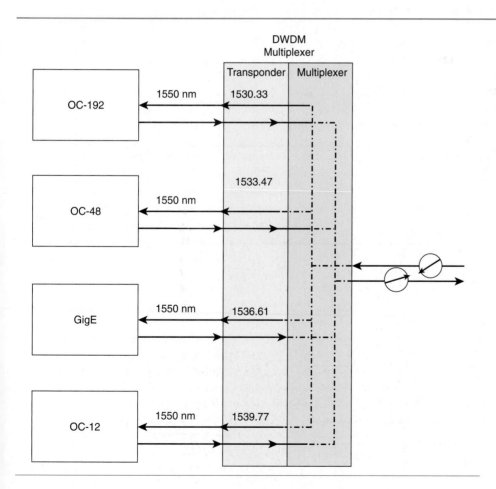

Figure 15–14
DWDM systems combine multiple light signals into one transmission link.

takes each of the incoming wavelengths and converts them into four ITU-specific C-band lambdas, as shown on the diagram. After the conversion process, the newly formed signals travel across the backplane of the MUX to the coupler where they are multiplexed together into one optical signal. The optical signal is pulsed out from a laser onto the network. The far end DWDM system receives the signal riding on the fiber and passes it on to be demultiplexed or filtered into the individual wavelengths. The wavelengths are then fed to the transponders where each is converted back into its original wavelength: 1550, 1310, or 850 nm. Once the conversion is complete, each individual signal is outpulsed onto the fiber optic jumper that connects the DWDM equipment to the optical MUX.

DWDM systems may also have a dispersion-compensation device built in to help offset the effects of chromatic dispersion. Due to the great distances between signal regeneration, chromatic dispersion causes the signal pulse to spread and lose shape as it travels down the fiber strand, as shown in Figure 15–15. Once the signal spreads or distorts to a certain degree, the receiving equipment no longer is able to differentiate the pulses and thus errors occur causing corrupted data. During the initial system integration, each span in the long-haul network is characterized for chromatic dispersion using special test devices that measure the amount of dispersion and its probable effect on the signal. Dispersion compensators nullify or cancel out some of the dispersion on the link helping to increase the distance the signal can travel before it requires an O-to-E-to-O conversion performed by a regenerator.

Figure 15–15
Chromatic dispersion causes the pulses to spread as the signal travels down the fiber strand. Pulse spreading can cause the one and zero bits to meld into each other's time periods, as shown in the first diagram. Pulse spreading causes errors to occur on the circuit, thus corrupting the data. Chromatic dispersion creates real problems on high-speed, long distance circuits because the higher the speed, the closer the pulses, as shown in the second diagram. Therefore, the bit-per-second rate and the distance are the two key factors when determining how chromatic dispersion will affect the signal.

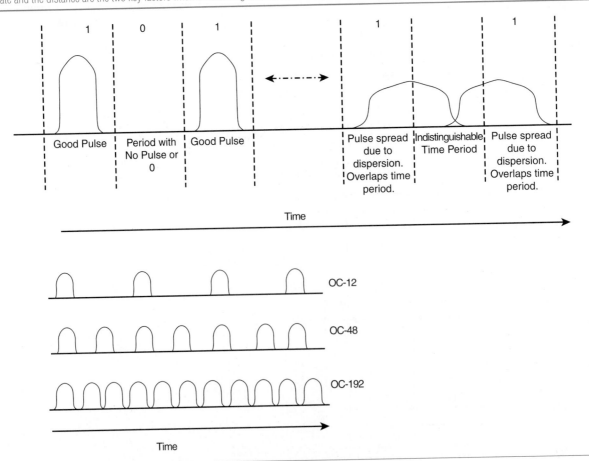

Though dispersion-compensation devices are very expensive, the savings are often substantial when compared to deploying additional regenerators. Each time a regenerator is placed in a network, a hut, power, AC, and so forth are required. Today, many newer networks handle dispersion compensation by deploying a device developed in early 2000 called a *soliton*. Technology continues to solve the constant issue of "how far will my signal travel before I have to do something?" The soliton is one of the most significant breakthroughs in the long-haul network, thanks to its ability to extend the distance the signal can to travel before requiring regeneration. The DWDM system may or may not contain one of these dispersion-compensating devices.

In addition to the transponder, the coupler, and some type of dispersion compensator, some DWDM systems also contain an optical amplifier that helps extend the distance the signal is able to travel. Again, the addition of an amplifier is determined during system integration.

Once the signal peters out to a point where the pulses are barely distinguishable, a regenerator is required to re-create the pulse. Separate from the DWDM system, the purpose of the regenerator is to repeat or re-create the incoming signal to the same specifications of the original signal leaving the transmitter at the originating end. The regenerator performs an optical-to-electrical-to-optical conversion that allows it to completely regenerate a new signal with the same amplitude or strength of the original. As discussed earlier in the chapter, regenerator huts are placed strategically in the network according to the manufacturer's loss budget specifications. A regenerator is nothing more than a MUX without the channel cards and, as such, requires all of the miscellaneous items needed to hold the device, power it, and keep its temperature just right. Therefore, an enclosure called a hut is deployed if a building is not available. Huts are small buildings usually constructed of fiberglass that are placed on a cement pad in the outside plant, typically in a field, along a railroad track, or in some other desolate inexpensive location. The engineer must obtain right-of-way from the landowner; have power run to the hut by the local power supplier; install generators and batteries for backup power; and build an environmentally safe site with AC, security, and fire suppression.

The structure of a regenerator is simple. The chassis holds two types of cards: administrative cards and line cards. The optical fiber is connected to the line cards both on the in and out portion of the MUX, as shown in Figure 15–16. Note that there are two sets of optics because the signal has to be received and transmitted on each side—from and to the originating side and from and to the terminating side. The signal comes into the multiplexer's receiver where it is converted into an electrical signal—such as an OC-48 into an STS-48. The STS-48 is then mapped intact to the transmit card that performs an electrical-to-optical conversion. The newly created optical signal is transmitted out onto the fiber optic network in the opposite direction from which it arrived. The signal continues to travel down the span, healthy and ready to travel. The

soliton
Device used in fiber optic networks that reduces chromatic dispersion in the network.

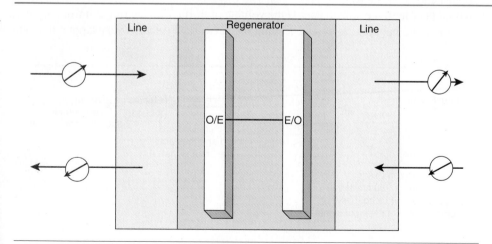

Figure 15–16
Regenerators accept an incoming list signal from the network, convert it into an electrical signal that mirrors the original, and convert it back into an optical signal before shipping it out into the line. Regenerators are placed strategically along the fiber route according to the link loss and signal strength.

administrative cards perform the same function as they do in an add/drop multiplexer, described earlier in the chapter. The unit also contains timing cards that are used to distribute timing across the network. The regenerator is a fairly simple device that should require little maintenance once in place.

The last piece of equipment used in conjunction with WDM systems is the optical amplifier. The EDFAs—erbium-doped fiber amplifiers—are used to extend the reach of the signal without performing an O-to-E-to-O conversion. The difference between a fiber amplifier and a regenerator is that the amplifier boosts the incoming signal optically and the regenerator recreates the new signal electrically. An optical amplifier (op amp) is small, low powered, and much less expensive than the regenerator. You may ask why op amps are not used exclusively because they are less expensive than regenerators. Unfortunately, the signal may be amplified only so many times before it requires regeneration because of signal distortion caused by noise and interference inherent in the system.

The EDFA consists of a piece of fiber doped with erbium, a pump laser, a coupler, and an isolator. The erbium-doped fiber is able to convert energy from the fiber by coupling the signal from a pump laser into the wavelength needing to be amplified. The laser is either a 980 or 1480 nm laser that injects itself into the doped fiber and is coupled with the core wavelength to produce a stronger, more powerful signal. Optical isolators are placed at the beginning and ending of the doped fiber to help filter out spurious noise and interference. Figure 15–17 details the structure of an EDFA. The beauty of the EDFA is its ability to amplify all of the wavelengths across the C-band within the optical signal. Newer amplifier types are starting to emerge, though the EDFA still remains the most popular in DWDM systems.

Noise caused by the amplification process is referred to as ASE (amplified spontaneous emission) noise and will affect the signal-to-noise ratio adversely after being amplified multiple times. Therefore, the number of op amps placed in sequence is limited due to the negative effect of ASE noise. ASE noise is discussed in more detail in a later chapter.

15.3.3 Fast-Packet Equipment

Core networks also depend on fast-packet switches and aggregation devices to package and carry traffic across the network. The three most commonly deployed fast-packet devices in the core network are the IP core router, the ATM switch or aggregation device, and the frame relay switch or aggregation device. The IP core router has become a favorite over the past few years. The core router's role in the network is to aggregate incoming streams of IP data, analyze the packets by cracking them open and reading the routing instructions inside, and route them to the correct destination according to the level of priority assigned to the packet. The main difference between a core router and a host or network router is the size of the processor, the size of the interfaces, and the sophistication of the routing protocols. The processor on the core router must be substantial in order to be able to handle the volume of traffic flowing through from ingress, going into the network, to egress, leaving the network. Typically, very large redundant

Figure 15–17
EDFAs are used in long-haul networks to amplify the light signal every so many miles depending on the loss budget of the span. An EDFA consists of a piece of fiber doped with erbium. A pump laser injects the doped fiber with a wavelength that excites the ions and thus boosts the input signal.

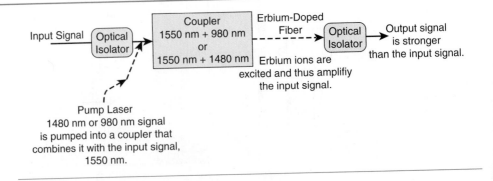

processors are found in the core router giving it the distinctive title of carrier class—fully redundant and fault resistant. The carrier class core router is built with large processors, memory, and system redundancy, helping guarantee uptime performance.

The second role of the IP core router is to crack open the packets in order to determine where they are going and the priority level of the packets. For example, an IP packet that is carrying VoIP requires a higher priority level than that of just plain data. The VoIP packet must reach its destination within a specific time in order for the transmission to be coherent to the user. The most common routing protocol used in core routers is the Border Gateway Protocol. The BGP is used to determine which link the data should be sent out on and the alternate route or routes to use if needed. The BGP is explained in detail later in the text.

The core router also is responsible for interfacing into the network through network interface ports. The size of the port is dependent on the bandwidth requirements and the existing network connections. For example, a core router may connect directly into a DWDM system through an optical interface such as a 2.5 Gig or 10 Gig port. The core router may also interface into a SONET MUX, OC-N interface, and traverse the network encapsulated in a SONET frame. On the opposite side of the router, the ingress sides, tributary circuits interface into the core router from various network routers and access devices. The core router aggregates all of this traffic similar to a central post office and routes the traffic according to the IP addresses within the packet, similar to a postal address. Routing and IP addressing are explained in detail elsewhere in the book. For now, understand the IP core router's role in the core network is to aggregate IP traffic, route the packets, and interface into both the ingress network and the egress network.

The core router consists of four primary areas: ingress ports, egress ports, central processing unit, and switching or routing fabric. The number and size of ingress/egress ports available vary dependent on the vendor but typically include DS-3, OC-3, OC-12, OC-48, OC-192, and GigE. The physical ports reside on circuit packs referred to as *blades*, as shown in Figure 15–18. The DS-3 circuit terminates on two BNC connectors, transmit and receive; the optical connections terminate on standard fiber optic connectors such as SC or ST, and the FastE connects into a standard RJ-45 style port. The egress ports on the router reside on redundant circuit blades. Beyond interface cards, the core router has redundant CPU circuit packs that hold the software code, routing information, processors, and so forth.

The routing or switching fabric of the router is responsible for moving traffic between interface ports such as from the ingress side of the router to the egress side. The size of the switching fabric must be analyzed before deployment to ensure that no traffic will be blocked or discarded. Additional buffers may be added to a router to hold overflow traffic when the network is congested. The insides of the router—the CPU,

blades
Term used to describe a circuit pack deployed in a system.

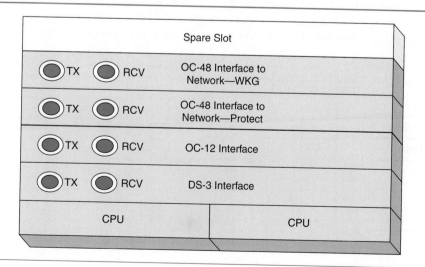

Figure 15–18
Core network routers interface into the network at high optical rates such as OC-48. Due to the high data rate, a protection card is often configured to ensure against any hardware failures.

memory, buffers, and software code—are much more impressive than the actual hardware, which is fairly simple. Overall, a core router is a very expensive, complicated device that requires a great deal of experience to operate.

A second type of fast-packet device found in core networks is the ATM switch or aggregation box. The ATM switch is used to assemble, dissemble, switch, and prioritize ATM cells. The ATM switch is a Layer 2 switch that establishes nailed up connections between locations and routes ATM framed cells onto the network. In most core networks, an ATM switch is used to carry data and voice across the long-haul network interfacing directly to the metropolitan networks. The physical structure of an ATM core switch is comprised of interface ports, CPU, and a switching fabric, as shown in Figure 15–19. The size of an ATM switch deployed in a core network is determined by many factors, such as the types of interfaces coming into the device on the low-speed or ingress side and the size of the connection leaving the switch on the high-speed or egress side. Large service providers typically have fairly high-speed circuits, DS-3 and up, feeding into the core ATM switch. BNC connectors are used to terminate the DS-3 circuit and optical connectors such as SC or ST are used to terminate fiber jumpers connecting to the lower-speed access devices. The egress side of the switch has high-speed interface ports, such as OC-12c, which are terminated similar to the ingress side of the box by optical connectors such as SC or ST. The ATM switch may be directly connected onto the network or be multiplexed into a DWDM system depending on the design.

The ATM processor is responsible for holding the software code, assembling and dissembling cells, providing access into the box, analyzing and distributing alarms, and carrying a customer profile. The size of the processor is dependent on the vendor's specs and the carrier's needs. The processors are deployed in a 1+1 protection scheme, meaning that the system is fully redundant. The size of the switch fabric of the ATM switch is critical to the switch's performance. The amount of buffer space and the type of traffic policing, flow control, and so forth are defined by the vendor and should always be reviewed before purchasing a box. The vendor's specs list the size of the switch fabric, the methods used for traffic flow and traffic policing, and the cell rate allowed through the system.

The final type of fast-packet device found in the network is a frame relay switch or aggregation box, as shown in Figure 15–20. The frame switch, similar to the ATM switch, is placed in the POP or switch site to interface with the long-haul network. The frame switch has interface ports, CPU, and a switch fabric. The ingress interface ports on the frame switch are typically lower speeds than that of the ATM or core router. DS-1 and DS-3 ports are commonly found in core network frame switches. The DS-1 circuit terminates onto wire wrap pins or an RJ-48 connector; and the DS-3 connects to BNC connectors, one for transmit and one for receive. The egress side of the switch is DS-3, OC-3, or OC-12, depending on the requirements of the network and the type of equipment deployed. The DS-3 circuit is connected to BNC connectors, one transmit and one receive, and the OC-N circuits are connected to fiber jumpers such as SC or ST.

Figure 15–19
ATM switches typically have line cards used for both ingress and egress, depending on how they are programmed; a switch matrix; a CPU; and an administrative pack.

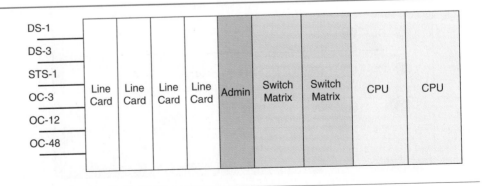

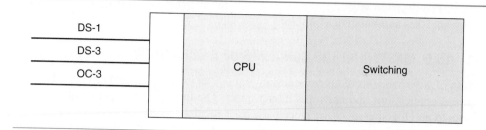

The processor is used to hold the software code, package the information into frames, prioritize the frames, analyze and distribute alarms, and provide access into the box. Each processor is deployed in duplicate to ensure redundancy during hardware outages. The switch fabric is defined by the vendor and refers to the number of frames switched at one time through the unit. Frame switches are less complicated than core routers because they are strictly Layer 2 devices and not concerned with routing protocols or cracking the packet open. Buffering is also performed within the frame switch and is often used during high-volume periods to help avoid lost frames. The frame switch has been around for some time, though today is less prevalent in the core long-haul network than an ATM or core router.

15.4 PHYSICAL LOCATIONS OF A CORE NETWORK

We have discussed the various types of equipment found in a long-haul network in detail and now need to describe where all of this equipment is physically located. Depending on whether the long-haul provider carries voice, data, or a combination determines the type and location of the switch site in addition to how the site is constructed. Long distance voice providers select the number of long distance switches required to handle the traffic load across their footprint. In doing so, they must determine which cities warrant full switch capability and which require a smaller hub housing only transmission equipment, no switch. The same is true with data service providers. The core routers must only be placed in very high-volume sites because of the high cost of the equipment and the high cost for ongoing maintenance.

Therefore, the designer must determine where best to place the routers and how best to back haul the traffic to and from the site. Traffic is back hauled from the small hub site to a switch site where the information is processed and switched accordingly. The term POP is used to represent the interconnection location, whether that is a full-blown switch site or a small hub. Each long-haul provider establishes a POP in an area that it wants to interconnect into. The POP is the location where other carriers, both local and long haul, meet the service provider and exchange traffic. Typically, a long-haul provider has two types of POPs, a fully blown switch and core router site and a smaller transmission-equipment-only hub site.

A switch site houses all of the communications equipment. The most common devices found in a switch site are the long distance switch, transmission equipment such as a SONET MUX, DWDM equipment, a DCS, fast-packet devices, and miscellaneous servers used for maintenance and remote access. The switch site must be *carrier class*, meaning that it must have full power backup supplied by both batteries and a gas or diesel generator. The floor space should have sufficient *HVAC* available and a functioning fire-suppressant system. The space should be kept clean and have a good ventilation system to help cut down on dust and mildew. In addition, the site should have a security system to keep unwanted visitors from entering. The cost to build a carrier class site is substantial and is often the reason a service provider opts to back haul the traffic instead of building out a new switch site. The cost to deploy a hub site is much less expensive because many times the site is colocated within a carrier-class site owned by another carrier or a *colocation* company that specializes in leasing carrier-class

carrier class
Term used to describe site that has redundant power, redundant entrances for circuits, redundant circuits, security, and other features related to maintaining a 99.999% uptime location.

HVAC
Heating, ventilation, and air conditioning.

colocation
Several service providers lease space in a carrier-class site and place communications equipment in the colocation site and connect to the carrier-class site's network.

space. The decision whether to deploy a full site or a hub is made after hours of network analyses and budget calculations.

15.5 MASTERING THE CONCEPTS OF A CORE NETWORK

Mastering the concepts of network design requires an in-depth understanding of how the various applications and networks interconnect. The following few paragraphs will attempt to pull together all of the piece parts associated with core networks by using real-life scenarios that depict commonly deployed applications. Scenario 1 involves a long distance telephone company, TelTel, that has established a fiber optic long-haul network across the United States. The carrier has ten main switch sites, twenty-five smaller hub sites, and a fully deployed optical network using DWDM technology, as shown in Figure 15–21. Notice how the links crisscross the country forming multiple paths to the different switching centers. As we move deeper into the network, Figure 15–22 shows how the DWDM equipment, the EDFAs and the regenerators are placed between the switch and POP sites. The SONET MUXes are used to aggregate the traffic from the switch or local handoff and package it into SONET frames that are transported directly onto the DWDM wavelength. The final diagram illustrates how the switch interconnects into the transport network and how the local exchange traffic interconnects into the switch or POP site.

The second scenario illustrates a Tier 1 ISP called IPIP. IPIP accepts traffic from multiple local ISPs, CLECs, ILECs, and large companies. The destination of the traffic is the World Wide Web and, as such, requires IPIP to interconnect to the network access points called *NAPs*. As shown in Figure 15–23, IPIP's network consists of both leased and owned fiber infrastructure and requires that portions of the network contain DWDM equipment while other sections, low traffic load areas, consist of leased OC-N rate circuits. Moving deeper into the network, as shown in Figure 15–24, shows how the core router interconnects into the fiber long-haul network at the various OC-N rates. Figure 15–25 illustrates the method used to interconnect the local providers into IPIP's core router. As you may note, the IP traffic comes in from ISP-1 and interconnects into the core router DS-3 port. The IP packets flow through the router where they are cracked, the IP address is examined, and a destination address is determined. The routing protocol in the core router points the traffic to the correct interface port where the information exits and heads across the network. The information sent by the ISP may travel multiple routes, traveling through many core routers before reaching a peering location, private or public, where it traverses onto another network such as a second ISP to the final Web site.

NAPs
Network access points. Interconnection points where Tier 1 ISPs terminate circuits and interconnect to other Tier 1 ISPs.

Figure 15–21
TelTel's network consists of three SONET rings with ten switch sites, twenty-five POPs, and the deployment of DWDM. The cities shown have long distance switches and data switching equipment.

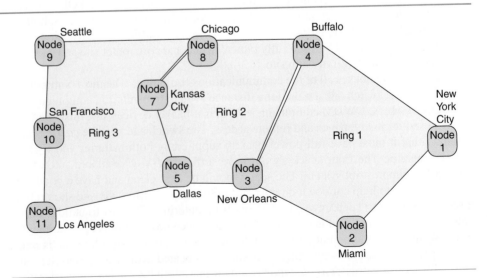

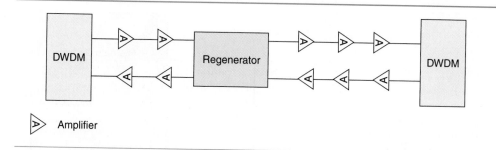

Figure 15–22
Between switch sites, fiber optic networks have been deployed using DWDM equipment. EDFAs and regenerators are used to boost or repeat the signal according to the loss budget design.

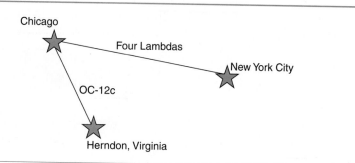

Figure 15–23
IPIP's long-haul network architecture consists of a point-to-point DWDM system backed up by one diverse OC-12.

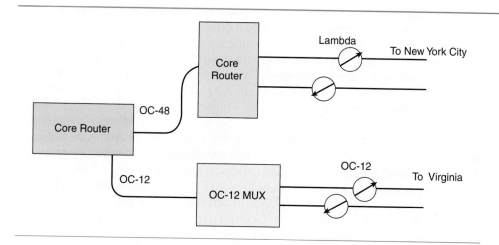

Figure 15–24
The equipment configuration in the Chicago switch site consists of a core router tied to both the DWDM and an OC-12 MUX.

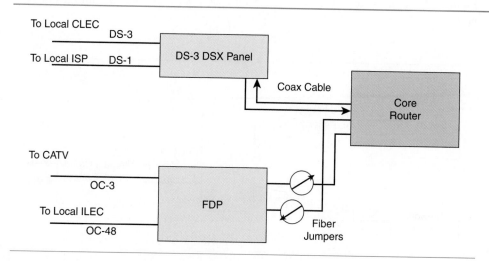

Figure 15–25
IPIP interfaces with circuits from the local network at its core router. DS-3s, OC-3s, and OC-48s feed into the ingress side of the router.

Figure 15–26
LittleTel interconnects with alternate
carriers' long-haul networks in order
to achieve the benefits of least-cost
routing and redundant networks.
LittleTel does not own its own long-
haul network but instead interfaces
into other providers' networks via
OC-N connections.

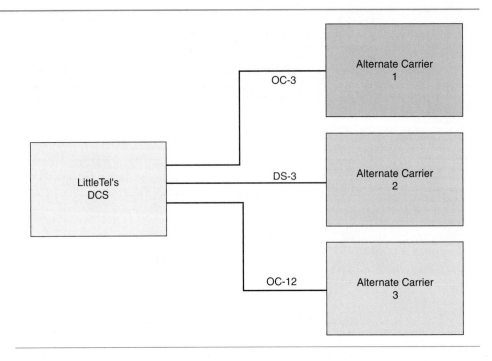

The third scenario involves a smaller voice and data carrier that depends exclusively on other carriers' networks to transport their traffic across the country. The carrier, LittleTel, has built a small network using long distance switches to process and switch voice traffic and network routers to route packets to their destinations. LittleTel buys bandwidth from carriers and alternative distribution companies it does not have to build its own physical plant. The traffic it ships may ride on multiple carriers' networks depending on the cost per bit and the availability of links. For example, LittleTel's voice switch is configured to point traffic to the least-cost route according to traffic tables established each morning by the traffic engineer. Updating the traffic tables daily for time-of-day routing or least-cost carrier routing is handled by groups of analysts who work on negotiating bandwidth costs from carriers guaranteeing LittleTel the lowest per minute costs. As shown in Figure 15–26, the LittleTel network consists of optical interfaces from SONET MUXes in addition to DS-1 and DS-3 connections into other carriers' SONET equipment. LittleTel does not own any DWDM equipment but does interface into other carriers' DWDM equipment. Figure 15–27 illustrates how the voice switch and network router interconnect into the off-net carrier's network. Figure 15–28 illustrates the local telephone company connection into LittleTel's network.

Figure 15–27
A more granular view of how LittleTel
interfaces its switch and router to an
alternate carrier is shown in the
diagram. Note how the DCS provides
a way to aggregate and groom traffic
before interfacing with the carrier.

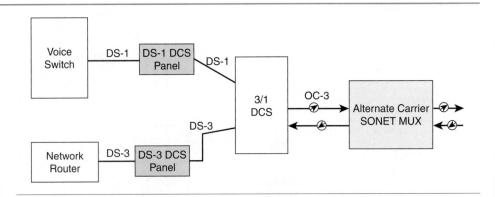

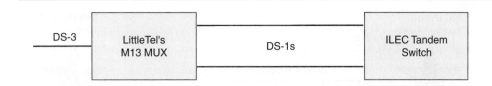

Figure 15–28
LittleTel interconnects into the local ILEC switch by dropping out DS-1 circuits using an M13 MUX.

Scenario 4 looks at a regional carrier, RegTel, that is carrying traffic between Cleveland, Ohio, and Columbus, Ohio. The network consists of two strands of fiber terminating into the colocation space in a carrier hotel. A DS-3 has been purchased from a different carrier to provide a diverse path between the two cities in case the optical fiber is disrupted. RegTel has one OC-48 MUX deployed at each end that is used to carry private line services such as DS-1s, DS-3s, and Fast Ethernet between the two cities for service providers such as CLECs, small ISPs, and large businesses. Figure 15–29 depicts the physical infrastructure of the point-to-point network; and Figure 15–30 depicts the network equipment, the OC-48 MUX, and the terminating circuits.

In scenario 5, an international communications carrier, EuroTel, has deployed an international network that connects the North American continent to Europe. Figure 15–31 provides a high-level view of the network as it connects the two countries, Canada and Ireland. From the figure, you should note the two paths available for traffic to travel. EuroTel has purchased twelve strands of fiber and utilizes DWDM equipment in order to increase the capacity of the links. As shown in Figure 15–32, amplifiers and regenerators are required to carry the signal across the ocean. The amplifiers and regenerators are built specifically to endure conditions at sea. The form factors encasing the units is of course watertight and the equipment is built to withstand heavy storms, shark attacks, and other mishaps associated with being submerged

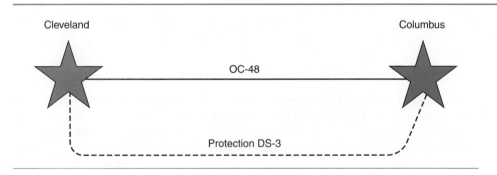

Figure 15–29
RegTel's network topology between Cleveland and Columbus consists of one OC-48 connection with a backup DS-3 running on a diverse path.

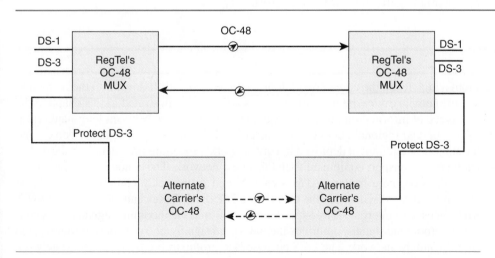

Figure 15–30
RagTel's network equipment includes an OC-48 MUX tied to its fiber network. The diverse DS-3 also feeds from the OC-48 MUX to the alternate carrier's network.

Figure 15–31
The network topology of a link between St. John's, Newfoundland, and Shannon, Ireland, consists of two routes, each consisting of twelve strands of fiber.

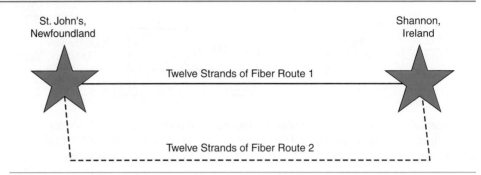

Figure 15–32
EuroTel's network consists of underwater submarine cable with specially designed amplifiers and regenerators. The number of amplifiers and regenerators depends on the loss budget of the link.

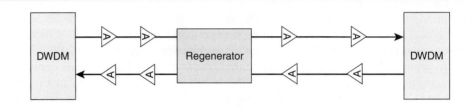

Figure 15–33
EuroTel's network consists of the North American T1, SONET standard, and the European E1, an SDH standard. In order for the two networks to speak, a conversion must be performed at one end of the network. EuroTel has placed its conversion device, a function handled with the DCS, in its New York City office.

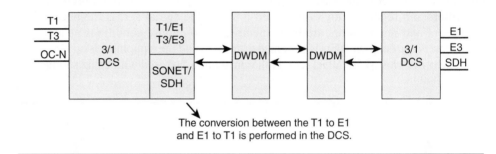

under water. The final illustration, Figure 15–33, shows the transport equipment connecting into the DWDM network. Note that the DCS used to terminate the cables at both ends. The DCS in the New York office converts the outgoing signal into the European standard E1s before interfacing into the DWDM equipment. In the diagram, note how the DCS has an optical interface configured as an SDH standard, not a SONET standard. One end has to perform a conversion; in our scenario, the conversion is performed at the New York City end.

SUMMARY

From a high level, the core network looks simple. As we drill down into the many layers, the complexity of the network becomes apparent as illustrated in this chapter. The first piece of information presented explained the difference between long-haul, ultra-long-haul, and regional core networks as defined by the ITU. The chapter delved into the subject of equipment deployed in core networks, spending time detailing the different types of transport equipment found in a core network. The chapter looked at TDM, wave division, core routers, ATM switches, and frame relay equipment. The TDM equipment types discussed in the chapter are SONET MUXes and DCSs. The SONET MUX is used to interface lower-speed circuits, format the incoming signals, cross connect or groom the signals, multiplex the low-speed signals, and transport the aggregated signals onto the network. The DCS purpose is to groom traffic from and to the network similar to a train relay station where the boxcars are moved from one train to another. The DCS and the SONET MUX may be found in both switch sites and hub sites.

The second type of equipment category discussed is the DWDM equipment, or dense wave division multiplexing equipment. DWDM is used to increase the carrying capacity of a strand of fiber by dividing the light spectrum into specific frequencies that are able to ride together across the fiber pipe. DWDM systems are sold by the number of wavelengths such as 32 lambdas, up to as many as 160. Along with DWDM systems, regenerators and amplifiers are used in the long-haul network to increase the distance the signal is able to travel. The amplifier boosts the signal, and the regenerator completely re-creates a new identical signal. Amplifiers are less expensive, smaller, and easier to place than regenerators but are subject to signal/noise limitations: the noise is amplified with the signal, eventually overpowering the signal. Regenerators perform a full optical-to-electrical-to-optical conversion in order to regenerate the signal to its original level. Building a small hut to house a regenerator contributes substantially to the cost of regeneration.

If this were five years ago, the fast-packet equipment description would be placed in the future network section of the text. Today, all carriers deploy some sort of fast-packet technology to carry their data and VON, voice over network, information. For example, as described earlier, the core router is responsible for aggregating IP traffic and routing it to its destination using sophisticated networking protocols such as BGP. The IP router is commonly used by all Tier 1 ISPs to ensure always uptime thanks to the redundant design of the core router. Unlike older network routers, "reboot" is a word of the past.

ATM switches are used to carry voice, video, and data across the core network. The ATM switch packages the information into fixed-length cells that are transported onto virtual circuits. The ATM cells are assembled according to the type of information being transported ensuring that latency-sensitive traffic such as voice and video will reach their destination uncorrupted. The ATM core switch is large enough to interface directly into DWDM systems yet small enough to interface with lower-rate signals from small service providers or ISPs. ATM, though often scorned and condemned to obsolescence, continues to plug along as a transport method for core networks.

The final fast-packet technology discussed was the frame relay switch or aggregation box. Frame relay has endured multiple attempts on its life throughout the past few years. The technology is so simple and efficient that the actual users of the technology refuse to give it up. The long-haul network rarely ships bare frame across a pipe. In most situations, the frame is encapsulated into ATM or into SONET frames and multiplexed up to a higher-rate signal. The frame switch interfaces at rates as low as DS-1 and aggregates the traffic onto circuits with bit rates as high as OC-3. The frame switch is a common sight in most long distance providers' switch sites and will continue to provide service into the future.

The location where all of the equipment in the long-haul network is housed was discussed. Carriers build out two different size sites: a smaller hub site and a full-blown switch site. The hub site is used to aggregate traffic onto the core network using transport equipment and DWDM. The hub site does not house a switch and therefore does not perform any type of switching. It may house a smaller core router depending on the amount of traffic traversing the site from the local market. The switch site must be built with full backup power, both batteries and a generator, HVAC, fire suppression, and security in order to be called carrier class. The hub site is placed at a carrier hotel colocation cage or some other carrier class facility.

The purpose of the final section of the chapter was to assemble many of the concepts presented earlier and apply them to several real-life scenarios in order to help organize the information. The scenarios depicted a long distance voice provider, a data provider, a small resale carrier, a regional carrier, and an international carrier.

The information presented in this chapter should be used in conjunction with Chapter 13 and Chapter 15 as they go into more detail and provide additional insight into the way long-haul networks function.

CASE STUDY

You need to determine which Tier 1 ISP you will interconnect to in order to provide access to the outside world. You will need to decide which of the two carriers detailed in Appendix A will win your business. The goal of this exercise is to look at the piece parts of the ISP's core networks in order to determine which of the two provides the best solution for your needs. Remember, cost is not the only parameter that should be considered in the selection process.

REVIEW QUESTIONS

1. What is the difference between a long and a very long core network as defined by the IEC?

2. What is the primary transmission medium used to carry traffic across the core network?

3. The primary reason DWDM systems are used in long-haul networks is to increase the number of circuits that can be carried on a single fiber strand.

   a. true   b. false

4. The core network architecture is

   a. point to point, mesh.
   b. point to point, mesh, ring, linear.
   c. point to point, ring.
   d. mesh.
   e. none of the above.

5. Explain why protection paths are critical components in a core network.

6. An ADM, add/drop multiplexer, is deployed in a core network to do all the following:

   a. add drop circuits
   b. regenerate the signal
   c. pass traffic through the multiplexer
   d. groom channels from one circuit to another
   e. amplify the signal
   f. a, b, c, and d
   g. a, c, d, and e
   h. all of the above

7. Optical multiplexers are divided into five areas:

   a. administrative or control, mux/demux, tributary, access, line
   b. administrative or control, mux/demux, tributary, line, and cross connect
   c. cross connect, access, mux/demux, tributary, line
   d. line, access, tributary, amplification, regeneration, circuit packs
   e. none of the above

8. A SONET MUX converts a framed DS-1 into a(n) _____ and a framed DS-3 into a(n) _____.

9. Explain what is meant by the term *tributary* as it relates to a SONET MUX.

10. *1 + 1 redundancy* refers to a protection method that

   a. provides two paths, one for the working and one for the protection circuit. When one fails, the second path immediately takes over.
   b. provides a standby protection path. A switchover is initiated when the working path fails.
   c. requires a manual switchover to the protection path when the working path fails.
   d. all of the above.
   e. none of the above.

11. Synchronization circuit packs are used to carry information plus provide timing.

   a. true   b. false

12. A digital cross-connect system is used to groom channels within one circuit to another circuit through an electronic cross connect.

   a. true   b. false

13. Define the term *lambda*.

14. Wavelengths used to carry communications circuits are divided into windows. The windows defined by the ITU are

   a. original, extended, short wavelength, conventional, and long wavelength.
   b. 1260 to 1360 nm, 1360 to 1460 nm, 1460 to 1530 nm, 1530 to 1565 nm, and 1565 to 1625 nm.
   c. conventional and long wavelength.
   d. 1530 to 1565 nm and 1565 to 1625 nm.
   e. original, extended, and short.
   f. 1260 to 1360 nm, 1360 to 1460 nm, and 1460 to 1530 nm.
   g. a and b.
   h. b and c.
   i. e and f.

15. What does EDFA stand for? How is it used in an optical network?

16. List three functions a core router performs in the core network.

17. Define the term *POP*.

KEY TERMS

long (L) (415)

very long (VL) (415)

ultra long (U) (415)

long-haul provider (417)

DWDM (417)

regional network (417)

core network (417)

off-net (418)

on-net (418)

transport (419)

transmission lines (419)

MUX (420)

tributary side (421)

broadband DCS (424)

narrowband DCS (424)

SDH (425)

protected lambda (426)

unprotected lambda (426)

EDFA (427)

soliton (429)

blades (431)

carrier class (433)

HVAC (433)

colocation (433)

NAPs (434)

16

Metropolitan Networks

Objectives

After reading this chapter, you should be able to

- ■ Define network topologies deployed in the metropolitan network.
- ■ Define protocols deployed in the metropolitan network.
- ■ Define equipment found in the metropolitan network.

Outline

■ INTRODUCTION

Metropolitan telecommunications networks are the critical link between the long-haul network and the "to-the-customer" access network. Though metropolitan networks resemble long-haul networks in many ways, there are distinct differences between the two. *Metropolitan network* is a very loose term used to define the network that connects large aggregation points such as central offices, *ISPs*, *carrier hotels*, and very large businesses. The metropolitan network is in itself a small, fully functional network that is capable of handling all types of information across all types of meda. The highway system is often used as an example to help explain the dividing line between long-haul and metropolitan networks. The interstate highway system, such as Route 95 running between New England and Florida, may be compared to the long-haul network; and the local expressways or beltways around the cities and the main arteries through the cities compare to the telecommunications infrastructure in a metropolitan area. The metropolitan network winds its way around connecting phone company offices, carrier hotels, CATV MSOs, and long distance entrance hubs and on and on forming a maze of circuits, connections, and aggregation points.

metropolitan network
Communications network that connects aggregation points such as carrier hotels, central offices, ISPs, and large businesses in the metropolitan area.

ISP
Internet service provider. A communications company that provides Internet services to subscribers. AOL, Earth Link, and Netscape are ISPs.

carrier hotel
A physical location—typically an office—in the downtown area used to house communications equipment. Large service providers lease a floor in the building to house their communications equipment and connect their networks with other carriers' networks.

Similar to the long-haul network, transmission equipment is responsible for assembling and packaging the information that is passing around the network and carrying it to its destination. Observing a difference between equipment deployed in a metropolitan network and that used in a long-haul network is often difficult, because many of the same manufacturers produce both; and often the same models are deployed in both networks. The significant difference is the size of the equipment, that is, the number of circuits it can terminate and the type of software installed.

The purpose of this chapter is to present a view of the metropolitan network that shows the types of transmission equipment deployed and the methods used to package the information. The chapter has been divided into three primary areas: network architecture, protocols, and equipment.

■ 16.1 COMMON METROPOLITAN NETWORK ARCHITECTURES

The four most commonly deployed network architectures in the metropolitan space are ring, mesh, linear, and point to point. Determining which of the four is the most widely deployed is difficult. The ring architecture is the dominant method deployed by ILECs and CLECs. ISPs and ICPs—Internet Communications Protocols—often opt for mesh architecture to accommodate their focus on IP as a transport method. Large enterprises and other individual carriers depend on point-to-point systems as they are much more cost-effective, though less reliable. Overall, a conglomeration of network types, even pieces of networks, are found in the metropolitan area.

The ring architecture that uses the SONET protocol over fiber optic strands is one of the most widely accepted architectures. It provides an extremely reliable, flexible way of carrying traffic to the many locations. Connecting individual nodes to an adjacent node forms the ring topology providing full 1 + 1 protection. When the fiber is cut, traffic immediately—within 50 ms—reverses its direction and travels around the opposite side of the ring: clockwise, counterclockwise. ILECs have built hundreds of various-size SONET rings. Some are very small, such as a three-*node* OC-3, and some are huge, such as three interconnected rings, each with sixteen nodes, as shown in Figure 16–1. Ring architecture allows for traffic to be added and dropped at any of the locations—thus, the term used to describe the SONET MUX, *ADM*, add/drop multiplexer.

The second topology, mesh, has gained in popularity mainly because of the push by the IP crowd to IP the world. The value of a mesh network is that all links may be used fully, unlike the ring, which reserves the protect path. The downside to the mesh architecture is the slow switchover time during outages. Additionally, often it is less expensive to add a node to a ring in the metropolitan space than building point-to-point mesh connections between locations, as shown in Figure 16–2.

The third network type is the linear or pass through. The linear network allows traffic coming in from one location to be either dropped or passed through to a third location and so forth. Linear networks are commonly used to connect central offices together allowing for traffic to be added/dropped or shipped straight through. The ability to pass traffic through a location without having to MUX it down and then MUX it back up lowers the cost for equipment and maintenance. Figure 16–3 depicts a typical linear design in a metropolitan area. Protection is provided only if diverse routes are established between the devices.

The final architecture to be discussed is the good-old standby, the point to point. Though the point-to-point topology offers no real redundancy during network outages unless a diverse path has been provisioned, it still is one of the most popular designs deployed in the local space. The main reason is that many of the circuits in the metropolitan space are low-bandwidth extensions that are used to tie in remote sites. Not all

node
A common term used to describe the SONET multiplexer.

ADM
Add/drop multiplexer. Refers to a SONET multiplexer that can add and drop circuits such as adding DS-1s at one location and dropping a DS-3 with the DS-1 embedded inside at a different location.

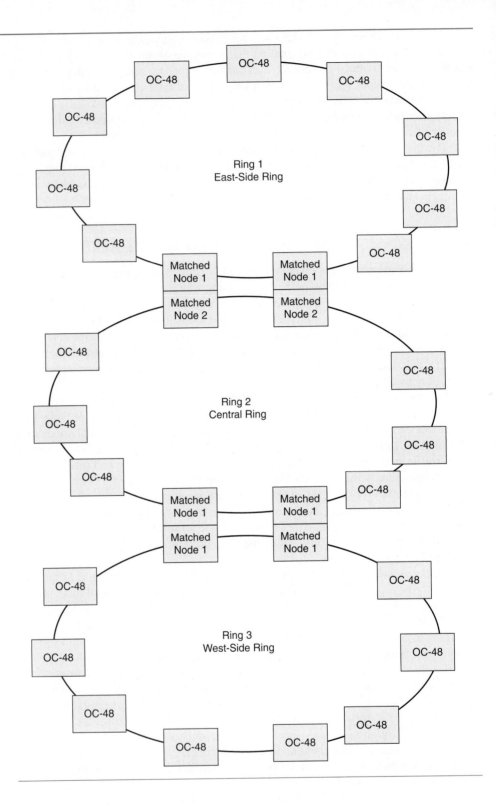

Figure 16–1
Interconnected SONET rings used to
network different regions within the
metropolitan area.

OC-48

OC-48

OC-48

OC-48

OC-48

OC-48

OC-48

OC-48

Ring 1
East-Side Ring

OC-48

OC-48

Matched
Node 1

Matched
Node 1

Matched
Node 2

Matched
Node 2

OC-48

OC-48

OC-48

Ring 2
Central Ring

OC-48

OC-48

OC-48

Matched
Node 1

Matched
Node 1

Matched
Node 1

Matched
Node 1

OC-48

OC-48

OC-48

OC-48

Ring 3
West-Side Ring

OC-48

OC-48

OC-48

OC-48

OC-48

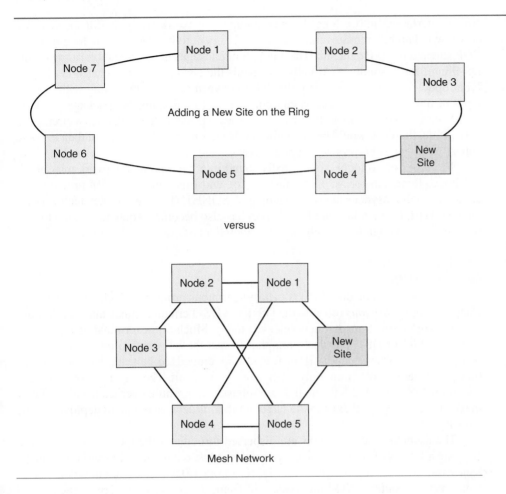

Figure 16–2
Adding a new site to a ring network requires two connections to the existing network, as shown in the first diagram. Adding a new site to a partially meshed network or fully meshed network requires multiple fiber connections to multiple nodes.

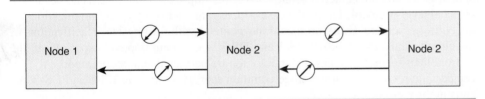

Figure 16–3
The diagram depicts a typical optical linear network topology.

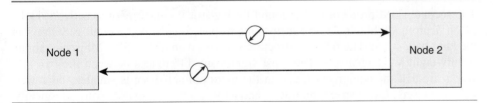

Figure 16–4
The diagram depicts a typical optical point-to-point network topology.

circuits in the metropolitan space require full OC-48 SONET rings. Figure 16–4 shows an example of a point-to-point network.

■ 16.2 PROTOCOLS DEPLOYED IN METROPOLITAN NETWORKS

Similar to the core network, the metropolitan network relies on the standard transport protocols—TDM, ATM, Ethernet, and IP. Embedding protocols and framing methods in order to aggregate traffic into a higher-rate signal is a common practice in a metropolitan network and the key function of transmission equipment. For example, a DS-3

POP
Point of presence. Refers to the communications hub or central location where the service provider hands traffic to other service providers and the long distance network. Often, the POP resides in a carrier hotel.

asynchronous
Term used to define an untimed digital circuit.

all optical
Term used to describe a network that carries information directly on fiber optic strands without needing a physical layer protocol.

IP only
Refers to a transmission scheme that places IP onto fiber directly without encapsulating it into a Layer 2 and Layer 1 protocol.

circuit interfacing into a long distance company network from an ISP located in a downtown data hub requires transmission equipment to be placed at the long distance *POP* site and the ISP data hub. The equipment packages the information riding on the DS-3 into, for example, ATM cells and hands the signal off to a higher-order MUX. Fiber optic cable is used to carry the DS-3 between the two sites along with various types of traffic. The transmission equipment used to multiplex and package the ATM cells is part of the metropolitan network. If it were possible to break open a circuit and peer inside, the view would be intriguing in that the protocols, frames, and data would appear as layers, sublayers, and sub-sublayers.

The types of protocols traditionally found in a metropolitan network may be divided into three categories: TDM, fast packet, and Ethernet. The TDM protocols include the older asynchronous standard and SONET. The fast-packet technologies include ATM, frame relay, and IP. Ethernet has also become a contender in the metropolitan space, though it is usually not considered a fast-packet technology.

16.2.1 TDM Protocols
Asynchronous Protocol
The first protocol to be discussed is the asynchronous protocol, which was detailed in Chapter 8. *Asynchronous* refers to the DS-1 and DS-3 circuit formats that form the underlying core of the transport network even today. Much of the terminating equipment still uses DS-1 or DS-3 physical ports that format the information into either a DS-1 or DS-3 frame. There may be IP or frame relay embedded beneath, but the physical transport layer is most commonly a DS-1 or DS-3 circuit. The Layer 1 protocols, into which the DS-1 and DS-3 frame fall, are robust, require little overhead, and are fairly inexpensive to deploy thanks to the large number of devices built and deployed in the industry.

The asynchronous protocols are dispersed throughout the metropolitan network carrying all types of traffic—voice, data, video. Most commonly, they are found interfacing into SONET multiplexers; into DCSs—both 1/0 and 3/1s; and into data equipment such as routers, ATM switches, and frame relay switches. Terms such as *all optical* or *IP only* are bantered about, leaving the impression that there is no need for a physical layer protocol. This is the case with as many terms in telecommunications, is not entirely accurate. Every protocol must contend with the physical specifications of the circuit. Thus, the standard DS-1 and DS-3 will remain popular, especially in the metropolitan space, until new methodologies are perfected and, more importantly, accepted. Figure 16–5 illustrates three common scenarios showing DS-1 and DS-3 circuits in the metropolitan space.

SONET Protocol
The second TDM protocol that is found throughout the metropolitan network is the SONET protocol. SONET in the metropolitan space is exactly the same as SONET in the long-haul space. The framing structure is built on an STS-1 52 5 Mbps containing twenty-eight VT-1 channels. The most significant differences between the two networks are that the equipment found in the metropolitan space is smaller, has lower-powered lasers, lower-speed interfaces, and a more granular switch matrix. Logically, the metropolitan space requires that smaller-rate circuits interface into the network at varying points such as the central office or a regional ISP. Also, the distances between interconnection points are much shorter—typically, less than fifty miles—eliminating the need for expensive high-powered lasers.

Today, SONET is the most widely deployed transport method in the metro space. Even though a business may send their traffic to the ISP via an Ethernet pipe, more than likely that circuit is embedded within a higher rate signal that invariably interfaces into a SONET MUX. The SONET MUX must be able to accept the Ethernet signal along with many other types of signals such as the DS-1 and DS-3. Therefore, it is important to recognize the significance of the SONET protocol in the metro space.

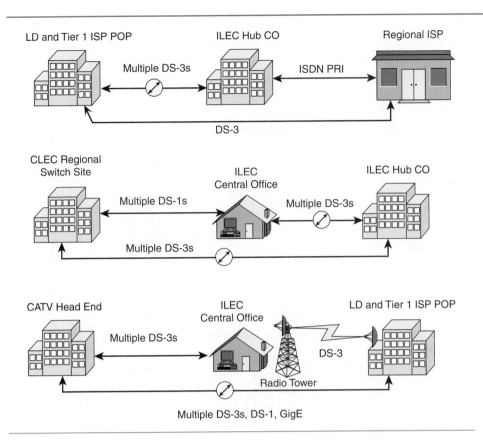

Figure 16–5
DS-1 and DS-3 circuits are used throughout the metropolitan network to connect different carriers together.

Figure 16–6 illustrates several situations where SONET is used in the metropolitan space.

16.2.2 Fast-Packet Protocols
ATM

The third type of protocol deployed in the metropolitan network is ATM—Asynchronous Transfer Mode—which is a common player in this space. It is seen as a viable way to carry all types of traffic including voice, video, and data in a more efficient way than that of TDM—statistically multiplexed versus time multiplexed. ATM evolved from a standard established in the mid-1990s. The goal of the participants in its evolution was to produce a protocol that could carry all types of traffic reliably. ATM's acceptance, though, has not been as significant as hoped. Today, several upstarts such as *MPLS*—multiprotocol label switching—are challenging ATM and succeeding in certain areas.

The most significant accomplishment ATM provides over the other protocols is its ability to guarantee service levels for time-sensitive and non-time-sensitive information such as voice and mission-critical data. In addition, ATM allows information to statistically use the bandwidth, similar to IP, thus improving the efficiency of the circuit. The protocol assembles information into cells according to the traffic type, as explained in detail in Chapter 7.

Multiservice carriers that transport all types of information and require carrier-class networks have become the biggest proponents of ATM. When the providers carry voice, they must meet certain expected standards comparable to the traditional TDM network. The problem many of the new competitive carriers faced was the cost to lease facilities from the ILEC. Therefore, the ability to carry voice in a carrier-class manner and use bandwidth more efficiently resulted in the adoption of ATM as their transport method.

MPLS
Multiprotocol label switching. A transmission method used to carry information across the network. MPLS is a Layer 2/3 protocol deployed by carriers.

Figure 16–6
SONETs are extensively deployed in the metropolitan network. They form the interface between the core network and the access network.

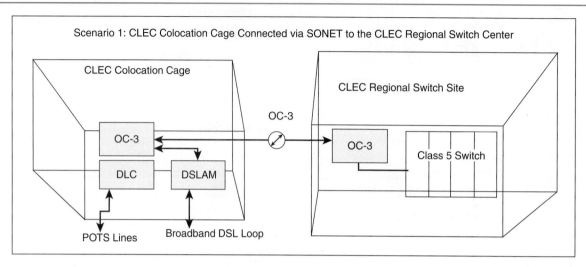

Scenario 1: CLEC Colocation Cage Connected via SONET to the CLEC Regional Switch Center

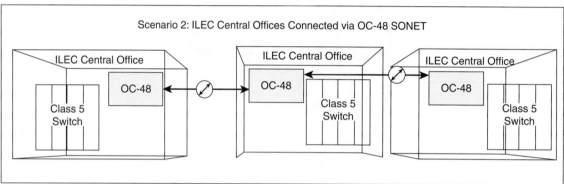

Scenario 2: ILEC Central Offices Connected via OC-48 SONET

Scenario 3: Long Distance Communications Provider Interconnecting with ILEC and MTSO

CLEC
Competitive local exchange carrier. Service provider that offers local telephone service competing with the incumbent local exchange carrier or ILEC.

ICP
Integrated communications provider. Similar to a CLEC, the ICP offers communications services to end users. The ICP specifically offers multiple types of service. *CLEC* and *ICP* are essentially the same.

Therefore, the *CLECs* or *ICPs* that evolved since the 1996 Telecom Act are the most notable users of ATM in the metropolitan space. ATM is used in the *colocation* sites to aggregate all types of traffic: voice, data, DSL, and, most notably, VoDSL— Voice over Digital Subscriber Line. The equipment vendors that produced *DSLAMs* saw the chance to increase their market share by pushing the value of VoDSL, which, in turn, strengthened the need for ATM. The DSLAMS of the late 1990s changed from frame relay on the upstream side to ATM specifically to accommodate the delivery of Voice over DSL. Today, the surviving CLECs and ICPs continue to depend on ATM to help lessen their facilities' cost to the ILEC while still being able to offer high-quality voice services comparable to that offered by the ILEC. In addition, ILECs also jumped on the bandwagon and decided to implement ATM, allowing them to deploy VoDSL similar to the CLECs. Unfortunately, the crash experienced in telecommunications in the early 2000s halted the expansion of large VoDSL rollouts by ILECs and CLECs causing the spread of ATM in the metropolitan space to slow. ATM remains a player in the metropolitan space, even though many critics have predicted its death. So far, none have been proven right. The most likely scenario is that the metropolitan space will always contain a certain amount of ATM traffic, similar to TDM.

Frame Relay

Frame relay, developed in the early 1990s, is one of the most resilient protocols ever and continues to baffle analysts as it plugs along gaining favor and market space. The frame relay protocol deployed in the metropolitan space is exactly the same as that found in the long-haul space. It, like ATM, falls into the fast-packet group of protocols meaning that it statistically multiplexes the variable-length frames onto the pipe, as described in detail in Chapter 7. Frame relay's simplicity is the main reason for its continued popularity. In addition, a frame relay interface, which typically depends on a TDM DS-1 pipe, has fewer overhead bits and thus eats up less bandwidth than ATM.

Within the metropolitan area, frame relay is used to connect offices together, carry DSL—Digital Subscriber Line—traffic back to the switch site, and aggregate multiple lower-speed circuits into one. The most common use of frame relay circuits is to provide a *WAN*—wide area network—connection between routers. The odds are that if you walked into a small business today and pulled up the config file on its router, you would be able to view the frame relay port configuration. Figure 16–7 shows a common use for frame relay in the metropolitan space.

IP

Internet Protocol (IP) is one of the most widely used protocols in the network today. IP is used to carry traffic—today, both data and voice—between sites, to the Internet, and between service providers. IP has always been very popular with the data-applications folks. They have steadily worked on perfecting the protocol into a very sophisticated means of carrying all types of traffic. As we mentioned earlier, all protocols depend on physical or Layer 1 protocol to ensure that the circuit parameters are met. The goal of IP designers is to eliminate this need. They may be viewed as purists, believing that only IP should be allowed in a network. Wars between protocol designers and engineers are as volatile and often as uncompromising as wars fought over religion. IP proponents tend to stimulate very heated discussions between ATM, TDM, and frame advocates.

colocation
Term used to describe how the CLEC or ICP leases space in the ILEC's central office. Cages are built in the CO, and the CLEC's communications equipment is installed.

DSLAM
Digital subscriber line access multiplexer. A multiplexer used to aggregate DSL circuits.

WAN
Wide area network. The network that connects physically separated locations.

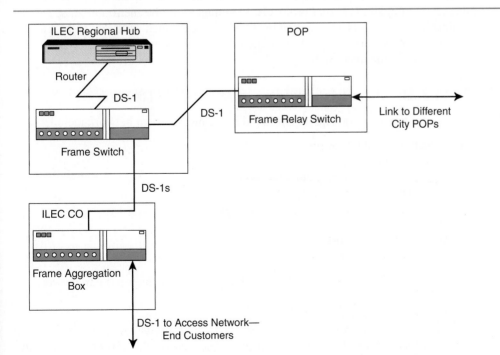

Figure 16–7
Frame relay networks are common in metropolitan networks. One of the simplest scenarios is connecting customers to their local CO via DS-1 circuits that interface into a frame aggregation box. From there, the frame traffic is routed to the main distribution hub where it is directed back into the local metropolitan network or pushed out to the POP where it rides on the regional and core networks to its destination.

IP is used to direct traffic to a final destination using an address, as described in Chapter 12. Being a higher-layer protocol, IP is much more complicated than TDM, frame, and ATM. It uses complex algorithms to determine what should be done, and when, to each packet it receives. Today, the IP packet may be configured similar to an ATM cell with a high-priority status for time-sensitive traffic such as voice. In a closed network, that is, one that does not allow public traffic on it, using IP for time-sensitive traffic works well. The problems start to arise when placed on a public network that has many more variables, creating a greater chance for delays and contention. Still, in the metropolitan space, there are several new startup carriers that are deploying IP-only networks.

IP is more difficult to categorize in the world of transmission protocols, because it does more than just assemble data for transport. It also must think about what to do with the data and look ahead to see which route is best. The interface port for IP is still tied to the DS-1, DS-3, or other TDM interface speeds and, as such, is often viewed not as a transport protocol but as a network protocol—hence, its position in the OSI stack. Today, though, it must be considered a contender in the transport technologies as designers continue to improve its ability to carry all types of traffic directly on the physical medium.

16.2.3 Ethernet

Ethernet, the old standard, has surprised even the diehard frame relay gurus as being a very versatile and inexpensive—especially inexpensive—means of transporting traffic across the WAN. Fast Ethernet, 100 Mbps, is one of the fastest-growing interfaces on the WAN side of routers. Companies have seen that they are able to connect their many sites using the standard Ethernet interface, eliminating the need for frame relay or ATM protocols. Providers spouting "Ethernet Only" mantras sprang up in the early 2000s offering WAN extensions to customers for flat fees. Beyond liking the transport method due to its cost, many network managers eagerly embraced Ethernet on the WAN interface. Network managers are familiar with Ethernet on the *LAN* side of routers, thus making them comfortable deploying it on the WAN side. Though in this section of the text we are not looking at the service, we are looking at what the metropolitan network is carrying. The Ethernet signal coming from the customer must interface into another Ethernet interface in the metropolitan equipment, oftentimes a SONET MUX. The SONET equipment formats the Ethernet frame into STS frames and carries them across the network. Therefore, as with IP, the metropolitan equipment deployed in the metropolitan space must be able to interface with and carry in an efficient manner the Fast Ethernet and Gigabit Ethernet protocols. In summary, in the metropolitan space, the physical layer protocols are handled by TDM protocols such as SONET; the Layer 2 functionality is performed by ATM, frame relay, or Ethernet; and, typically, all Layer 3 functionality is handled by IP.

LAN
Local area network. Refers to the network that connects devices within a small area such as between floors of a building or between buildings on a campus.

■ 16.3 EQUIPMENT FOUND IN METROPOLITAN NETWORKS

The types of transmission equipment found in the metropolitan network include:

- Asynchronous multiplexes
- SONET multiplexes
- Digital cross-connect systems (DCSs)
- ATM switches and aggregation boxes
- Frame relay switches and aggregation boxes
- Routers
- Passive optical networks
- Wave division multiplexers

A description follows of how each of these fits into the metropolitan network.

16.3.1 Asynchronous Multiplexers

First on our list, asynchronous multiplexers are found everywhere in the metropolitan space. ILECs, CLECs, ICPs, ISPs, and others depend on these small, inexpensive

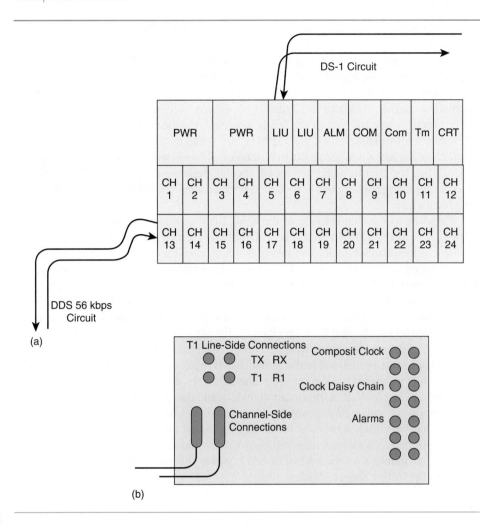

DS-1 Circuit

PWR | PWR | LIU | LIU | ALM | COM | Com | Tm | CRT

| CH 1 | CH 2 | CH 3 | CH 4 | CH 5 | CH 6 | CH 7 | CH 8 | CH 9 | CH 10 | CH 11 | CH 12 |
| CH 13 | CH 14 | CH 15 | CH 16 | CH 17 | CH 18 | CH 19 | CH 20 | CH 21 | CH 22 | CH 23 | CH 24 |

DDS 56 kbps Circuit

(a)

T1 Line-Side Connections
TX RX
T1 R1
Channel-Side Connections

Composit Clock
Clock Daisy Chain
Alarms

(b)

Figure 16–8
A channel bank consists of a line side and drop side, as shown in (a). Wire connections for each area are typically located on the back, as shown in (b). Newer models are very small—the size of a pizza box.

multiplexers to aggregate multiple DS-0s into a DS-1, or DS-1s into a DS-3, or DS-3s into a higher 560 Mbps asynchronous signal. The local phone companies have turned up thousands of these units over the years and today still continue to depend on them to move traffic around the network.

Channel bank is the term used to describe the box that multiplexes DS-0s into a DS-1. The channel bank has been around since the birth of the digital network. It was the first method used to format analog and digital DS-0s into a DS-1 circuit. Figure 16–8 depicts a typical channel bank as found in an ILEC's central office. The current models of channel banks are very small and often software programmable, eliminating the need to swap out channel cards when a new service is implemented.

The most common use for these channel banks is to aggregate incoming special service circuits such as digital 56 kbps, ISDN, *foreign exchange (FX)*, and analog data into a DS-1. The channel bank accepts the incoming signal on a channel card; converts the signal into a digital format if needed, such as with analog data; and multiplexes the signal into a channel on the T1 circuit. The SS-7 network, used to control the voice network, runs through channel banks as digital 56 kbps circuits linking COs. Extending ISDN service outside the serving wire center is accomplished by feeding the ISDN signal into a channel bank. FX—foreign exchange—circuits, which are discussed in detail in the access network section, are fed into channel banks and carried to a remote office. As is evident, the channel bank serves a very important function in the metropolitan network. It allows the carriers access to the lower-rate signals, in addition to multiplexing them into a higher-rate signal.

channel bank
Multiplexer used to aggregate DS-0 channels into DS-1s.

foreign exchange (FX)
Circuit used to extend a voice line between serving wire centers.

In the metropolitan network, the typical location for a channel bank is in the special services area in the central office or switch site of a competitive carrier. The channel bank's channel side is wired to the main distribution frame through amphenol connectors and termination blocks. The outside wiring of the individual circuits, such as ISDN or digital 56 kbps, is connected to the termination block located on the main distribution frame, as shown in Figure 16–8. Frame wire, usually a red/white pair, is run between the vertical side of the frame where all outside cables terminate to the special services termination block that is physically wired to the channel bank. All *special service circuits* are wired using a different color of frame wire from POTS lines specifically to flag the circuits as higher-speed, special, dedicated circuits. ISDN is typically wired using orange/white, while special service circuits are red/white, and POTS lines are green/white.

Opposite to the channel side of the bank, the line-side or high-speed circuit side, a DS-1, is wired to a DSX panel. The transmit and receive pair are wire wrapped onto the back of the channel bank as shown in the diagram. The opposite end of the cable is wire wrapped on the back of the DSX panel. In addition to the transmit and receive pairs that carry the DS-1, each channel bank requires a timing feed from a stable synchronous clock. The timing feed is also wire wrapped on the back of the channel bank. In most instances, several channel banks reside adjacent to one another and share one timing supply. One channel bank receives the timing signal and acts as a source for the adjacent bank through a wire wrapped connection, as shown in Figure 16–9. Each channel bank is daisy chained to the previous and receives its clock in this manner. The clock feeding a channel bank is referred to as a 56 kbps composite clock source, as was explained in Chapter 8. It is important to understand that a channel bank will not function properly without a timing source. If the location where the channel bank resides does not have a timing source, as is the case in almost all business locations, the unit can be optioned as a slave to the upstream master and use the incoming signal as its clock. All channel banks must be optioned as either the master or the slave depending on where the timing source resides.

special service circuits
Circuits other than POTS lines. Carry services such as DDS, DS-1, analog data, FX, OPX, and so forth.

Figure 16–9
An external clock source is fed to one channel bank in the relay rack. The clock signal is daisy chained from one bank to the next.

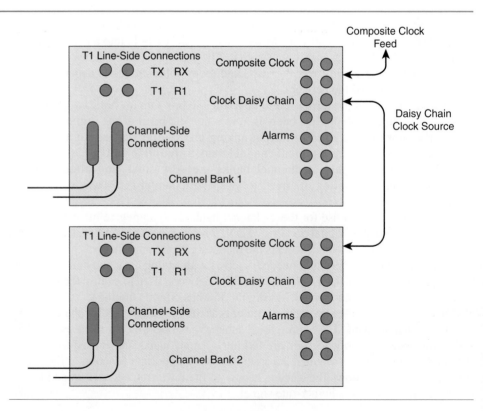

Several options must be set on a channel bank before service can be initiated. The first, as just mentioned, is optioning the bank for timing, either master or slave. This is accomplished either by moving a dipswitch or jumper in the older style multiplexers or through software commands on newer channel banks. The second option that must be set is the type of framing and line code associated with the T1 signal. The options are either ESF or SF and B8ZS or AMI, as explained fully in Chapter 8. Additional options may be required depending on the type of channel bank and the design requirements. Overall, the channel bank is a fairly simple device that continues to remain an integral part of the metropolitan network configuration.

The second type of asynchronous multiplexer found in the metropolitan network is the *M13*. The *M13* MUX multiplexes twenty-eight DS-1s into one DS-3. This multiplexer is one of the most widely deployed in metropolitan networks by ILECs, CLECs, ICPs, ISPs, and large business customers. Today, the multiplexer is amazingly small, about 1.25 in. high, and fits into 2 U of rack space in a standard relay rack (1 U = 1.75 in. or 4.44 cm). A craft interface port is available to the technician for programming the box, and a TCP/IP telemetry interface port is used for remote access. The advancement of this one box alone is amazing and shows how innovation not only helps produce new devices but also improves the old.

The structure of an M13 multiplexer is shown in Figure 16–10. A larger style MUX is shown to help illustrate the ins/outs and circuit. Newer boxes resemble a large pizza box and are therefore referred to as *pizza box MUXes*. Circuit boards in pizza boxes are no longer visible. Instead, they are placed inside the device, similar to circuit boards mounted inside a PC. In either case, the circuit packs perform the same function. Similar to the channel bank, the M13 has two sides: the line side and the channel side. The difference is that the channel side, also called the low-speed side, is terminating DS-1s and the line or high-speed side is terminating DS-3s. As you may note from the diagram, the DS-1 circuit is wire wrapped onto the back of the MUX with two pairs of wires—a transmit pair and a receive pair. From the MUX, the two pairs of wires are connected to the back of a DS-1 digital signal cross-connect panel (DSX panel). Also shown in the diagram are the remaining twenty-seven circuits that are fed into the MUX, each wire wrapped to the back of the chassis. Some manufacturers build *amphenol cables* to connect the MUX to the DSX. One end of the cable has an amphenol plug that presses on to an amphenol jack on the MUX; and the out end of the cable is open—meaning not connectorized, allowing it to be wire wrapped onto the DSX-1 panel. The ease by which the wiring is accomplished is the main reason many vendors have migrated to amphenol connectors. The downside to this wiring method is that if one of the wire pairs has a problem, all of the circuits must be disconnected in order to repair the cable. That means twenty-seven other T1 circuits must be taken out of service—not a pretty sight.

Coaxial cable is used to wire the line side or DS-3 side of the MUX and is terminated using *BNC connectors* on the back or side of the MUX. The two strands of cable—transmit and receive—are run between the MUX and a port on the DS-3 DSX panel. Again, BNC connectors are used to terminate the coaxial cable at the DS-3 DSX panel. Unlike the channel bank, the M13 MUX does not require a timing source because the signal is asynchronous, as explained in Chapter 8.

M13
Multiplexer that multiplexes DS-1 circuits into a DS-3 circuit using the M13 DS-3 frame structure.

amphenol cable
Connector used to terminate multiple copper pairs.

BNC connector
A coaxial cable connector used to terminate coaxial cable to a piece of hardware.

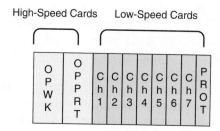

High-Speed Cards   Low-Speed Cards

Figure 16–10
Typical structure of an M13. The high-speed side of the MUX may be connected via fiber optic jumpers or coaxial cable.

Similar to a channel bank, several options must be configured both on the channel side and on the line side of the MUX. Either toggling a DIP switch (dual in-line package switch) or selecting the parameter using software commands accessed through the craft interface port from a PC is used to configure each channel card. For example, if the circuit requires ESF framing, B8ZS line code, the DIP switch must be toggled toward the ESF position or selected from the menu screen. Configuring a MUX via the craft interface using a PC is much simpler and much less intrusive than having to pull out the circuit pack and fiddle with a DIP switch. The options found on the line-side DS-3 circuit include framing—M13, C-bit parity, or unchannelized; alarming options such as X-bit designations; and line build out (LBO) as defined by the engineering work orders. The options found on the line side or DS-1 side of the MUX include framing—ESF or SF; line code—B8ZS or alternate mark inversion (AMI); and alarming designations.

16.3.2 SONET Multiplexers

A detailed description of the SONET MUX is explained in Chapter 15. The purpose of revisiting some of the information already presented is to explain how SONET is used in the metropolitan space. In order to do this, it important to detail the physical structure of a typical metropolitan MUX and explain how it interconnects with other pieces of equipment.

One of the primary differences between the SONET equipment found in the core network and that found in the metropolitan network is the line speed, that is, the OC-N rate of the multiplexers deployed. An OC-1 MUX would never be used in a long-haul network, and it is very unlikely that an OC-3 or OC-12 would either. An OC-192, before the early 2000s, would never be used in the metropolitan area due to the higher cost optics. Today, OC-192s are deployed in the metropolitan network ever since vendors reduced prices in the early 2000s after CLECs and long distance carriers stopped buying equipment due to the telecom crash. Regardless of the economics of the industry, the important point to realize is that the structure of the MUX, whether large or small, is the same. Every SONET MUX, as explained in Chapter 15, is divided into five sections: the tributary side, the line side, the cross-connect or switch matrix, the multiplexer/demultiplexer, and common cards. There is a tributary or channel side and a high-speed line side. If the MUX is an OC-1, the tributary side may be multiple DS-1s or Ethernet interfaces and the line side an OC-1. If the MUX is an OC-48, the tributary side may be a combination of DS-1s, DS-3s, Fast Ethernet, and OC-12s while the line side is an OC-48. Because the long-haul spans cover long distances before adding or dropping traffic—high-speed line-side cards such as OC-192s with high-speed tributary cards such as DS-3—OC-3s prove to be the most cost-effective. Within the metropolitan space, the spans are much shorter carrying many more types of information to a variety of aggregation points such as ISPs, IXCs, CLECs, and ILECs. Therefore, a SONET MUX capable of handling very low speed tributary cards is essential. On the line side, often a network is built either in a low traffic area requiring a very low speed line card such as an OC-1 or an OC-3. The metropolitan network demands SONET equipment capable of handling both very low speed tributary interfaces and very low and high speed line-side optics. As may be expected, the granularity of the switch matrix or cross-connect field must support the variety of low- and high-speed interfaces.

To explain how a SONET MUX is used in the metropolitan space, we will walk through two scenarios that should help organize the concepts in a more meaningful way. The first looks at an OC-3 SONET point-to-point network deployed by an ILEC for circuit backhaul, and the second looks at an OC-48 SONET used to connect a CLEC's switch site to its colocation cage.

The OC-3 is one of the most widely deployed smaller multiplexers in the metropolitan space. You will find the MUX in central offices, in *digital loop carrier (DLC)* sites, in large business complexes, and in many other locations. In our example, we look

digital loop carrier (DLC)
A multiplexing device used to aggregate end subscriber services and transport them back to the central office or CLEC regional switching center.

at an OC-3 placed in a DLC cabinet in the field. A DLC is a multiplexer used to aggregate subscriber services such as POTS lines and transport the circuits to a central office or CLEC regional switching center. In the ILEC network, a DLC is placed strategically near the end subscribers such as in a cabinet next to a housing development or shopping complex. The multiplexer is installed inside the environmentally controlled cabinet or hut. In the central office, the receiving MUX is mounted in a 7 ft. relay rack in the carrier/special services area. In our example, the multiplexers placed at the remote location and in the central office are OC-3 SONET multiplexers. DS-1s from the DLC are aggregated in the field and transported to the MUX in the CO or, in the reverse, transported from the CO to the remote site.

Typically, an OC-3 DS-1 channel card terminates between four and eight DS-1 circuits, as shown in Figure 16–11. A wire wrap panel located in the cabinet near the MUX terminates the DS-1 circuit coming out of the DLC's line card and is used as a cross-connection point into the OC-3 DS-1 channel card. The DS-1 cards, similar to those in an M13, must be optioned according to the engineering spec for framing, line code, and alarming. Looking at the signal path from the DLC to the MUX shows the signal travels from the DLC DS-1 line card to the cross-connect panel, then into the DS-3 tributary card through wire wrap pins or an amphenol connection on the back of the OC-3 MUX. Once in the MUX, the signal is formatted within the line card and converted into a VT-1 signal. The VT-1 is groomed into the appropriate STS-1 channel in the cross-connect field and is then multiplexed up into an STS-3. The electrical STS-3 travels through an electrical-to-optical converter where it becomes an OC-3 signal. The laser outpulses the OC-3 signal onto the fiber optic strand connected to the optical connector on the front of the optical line card. Optical jumpers connect the line card to a small *fiber distribution panel (FDP)* that is used as the interconnection point between the outside plant fiber cable and the equipment.

fiber distribution panel (FDP)
A piece of equipment used to terminate optical cables. It is used to connect the outside plant fiber to the optical equipment.

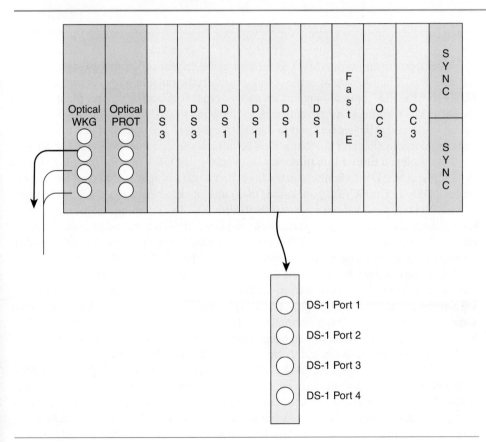

Figure 16–11
Optical interface into SONET MUX.

Figure 16–12
SONET MUX to FDP to outside plant cable.

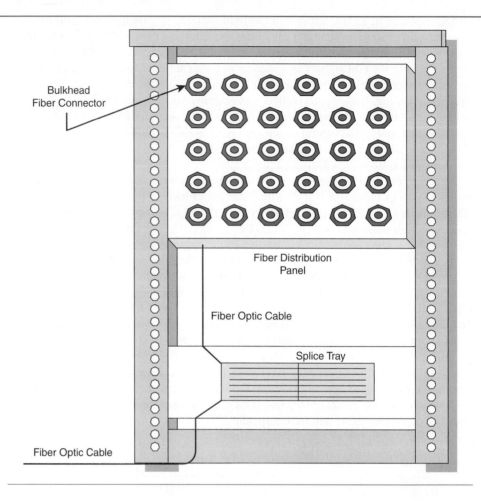

Bulkhead
Fiber Connector

Fiber Distribution
Panel

Fiber Optic Cable

Splice Tray

Fiber Optic Cable

The opposite end of the MUX is located in the central office in the carrier/special services area. The fiber optic cable coming in from the outside plant terminates on an FDP. From the FDP, optical jumpers are used to cross connect the outside plant cable to the OC-3 MUX sitting in a 7 ft. relay rack, as shown in Figure 16–12. The OC-3 receives the optical signal at the receive port of the OC-3 line card. The optical signal is then converted into an electrical STS-3 that is then demultiplexed into three STS-1s. The STS-1s are demultiplexed further into three sets of twenty-eight VT-1 signals that are groomed to their respective DS-1 channel cards where the VT-1 framing is stripped off leaving the original DSF-1. The OC-3 channel-side DS-1 cards are hardwired to a DS-1 DSX panel. As with all DSX panels, there are two sides to a circuit—the incoming and the outgoing—respective of the direction. In this example, the DS-1 circuits terminated at the DSX are physically cross connected to a DSX port hardwired to the voice switch, as shown in Figure 16–13. By making a hardwire cross connect, the DS-1 from the switch travels to the MUX and out onto the fiber optic cable to the end MUX at the remote site. The DS-1 travels in the opposite direction from the remote site through the OC-3 network to the DSX panel and to the switch as previously explained. The importance of knowing which direction is which cannot be overemphasized. A circuit always has two paths, from A to B and B to A. Circuit troubles often happen on one path only, and knowing where to connect the test equipment is critical to the success of finding the problem.

In the metropolitan space, an OC-48 multiplexer is the favorite of all SONET multiplexers. The OC-48 is found in central offices, in competitive carriers, in long-haul carriers, in ISPs, and on and on and on. Our scenario will focus on an OC-48 connecting a CLEC's switching center to its colocation site. The OC-48 MUX at the switching site takes up about half of a 7 ft. relay rack in the transmission area. In our second scenario, the CLEC is terminating ten DS-3s into the DCS also located at the switch site. At the

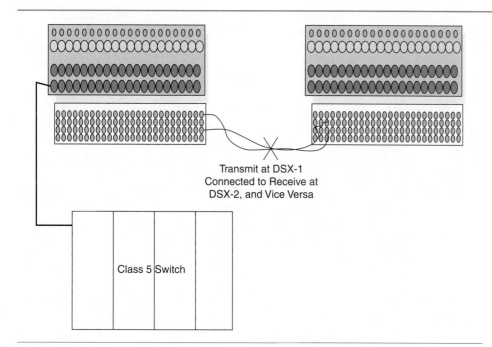

Figure 16–13
T1 circuit is cross connected at the DSX-1 panel. Circuit coming from the outside span is cross connected to the class 5 switch.

colocation site, the MUX terminates 240 DS-1s and 5 DS-3s. Connecting the DS-3 circuits at the switch site to the DCS requires coaxial cable connections between the DS-3 DSX panel and the OC-48 MUX. The connection between the DCS and the OC-48 is made at the cross-connect panel. Each coaxial cable is terminated on BNC connectors on the back or side of the MUX. In our example, the DS-3 line card in the MUX terminates twelve DS-3s. Because the OC-48 we are dealing with has a full one-for-one switch matrix, the DS-1s are converted into VT-1 signals and groomed across the switch matrix individually to the appropriate STS-1 channel. The ten STS-1s are converted into an STS-48 that then goes through an electrical-to-optical conversion changing it into an OC-48 signal. The OC-48 signal is shipped out onto the fiber optic jumper that connects the MUX to the FDP. The FDP performs the job of a cross connect between the MUX and the outside plant cable. Figure 16–14 shows the incoming and outgoing connections.

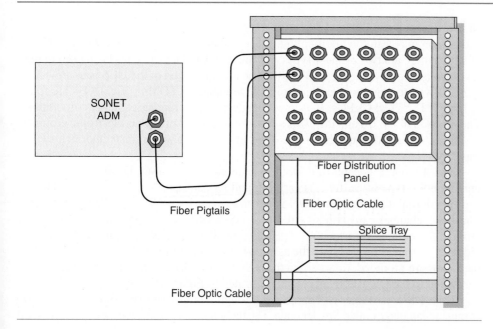

Figure 16–14
Fiber distribution panels are used to tie fiber optic equipment such as a SONET MUX to the outside world or to other devices. The diagram depicts a MUX tied to an outside plant fiber pair through jumpers.

The signal arrives at the colocation site and is immediately converted from an OC-48 signal into an electrical STS-48. The STS-48 is demultiplexed into the individual STS-1s, then into the individual VT-1s. The VT-1s are switched or cross connected to the correct DS-1 tributary ports where the VT-1 framing is stripped off revealing the original DS-1. The DS-1 tributary side is hardwired to the DS-1 DSX panel. DS-3 circuits coming out of the MUX terminate on a DS-3 port that is hardwired with coaxial cable to the DS-3 DSX panel. In most instances, the OC-48 is deployed not as a point-to-point network as in our scenario but in a ring configuration allowing traffic to be added and dropped at multiple points in the network. Our scenario showing how two OC-48 multiplexers connect two sites is much simpler than multiple OC-48 multiplexers sitting in many different colocation sites all sharing the same two pairs of fiber strands.

Both the OC-3 MUX and the OC-48 MUX require a timing source to be terminated to supply timing to the network. Similar to the channel bank, only one MUX has to be fed with an external timing source. The other multiplexers in the network may be optioned as line timed and pull timing off the line. All SONET multiplexers have redundant timing cards and allow for redundant timing supplies, as explained in a previous chapter. The timing source—in this case, a DS-1—is hardwired from the DSX panel where it is cross connected to the timing supply and the back of the multiplexer. The timing cards seize the timing signal coming in, locking on to it before passing it on to the line cards that use it as a reference source. Most multiplexers have the capability to sustain timing without an external source for on average a 72 hr. period. The MUX has an internal Stratum 3E clock that is used as a backup timing source when the external clock source is compromised. The 72 hr. period gives the network technician time to isolate and repair the timing network and, hopefully, avoid network degradation.

Beyond timing, all SONET multiplexers are usually wired for remote access either through TCP/IP ports or RS-232 ports. TCP/IP is the preferred method for remote access. Many times, engineers forget why it is so important to spend the money on a separate network just for remote access. However, they tend to remember after a major disaster when access through the optical *data communications channel (DCC)* is disrupted causing the MUX to be isolated or not accessible except by a physical visit. Connecting a MUX via TCP/IP requires a small end router to be used to interface the TCP/IP Ethernet port to a WAN interface feeding the upstream network. Figure 16–15 illustrates what is needed when connecting the MUX to a separate frame relay network. The advantage of such network is the ability to access the MUX even when the optical network is down.

data communications channel (DCC)
A channel within the SONET frame that is used to carry overhead and command information to the SONET multiplexers.

16.3.3 Digital Cross-Connect Systems

A common player in the metropolitan space is the digital cross-connect switch (DCS) of which there are two primary types: 1/0 and the 3/1. The DCS has been used in metropolitan networks since the mid-1990s. The main function of the DCS is to provide an efficient way to aggregate and groom large amounts of TDM traffic. DCSs are found in large ILEC COs; CLECs; ICPs; switch sites; and, in rare cases, large corporations. The function of the 1/0 and the 3/1 is similar in that they both act as electronic groomers of channels, the difference being the types of circuits they terminate and their overall capacity.

The 1/0 DCS accepts the DS-1 containing twenty-four DS-0 channels, each unique in the type of traffic it is carrying and its configuration. One channel may be carrying digital data, another ISDN voice and data, another SS-7 packet, and so forth. The DS-1 channel card is hardwired to a DS-1 DSX panel. As the signal traverses through the card, it is formatted, timed, and readied to be cross connected. The individual DS-0s are groomed onto the assigned DS-1 according to their destination. Once the DS-0s are groomed, they are fed through the new DS-1 channel card, which is also hardwired to the DS-1 DSX panel, as illustrated in Figure 16–16.

Each DS-1 card may be configured through a craft terminal using a standard 1/0 command language called Snyder code. The technician programs the card for the correct

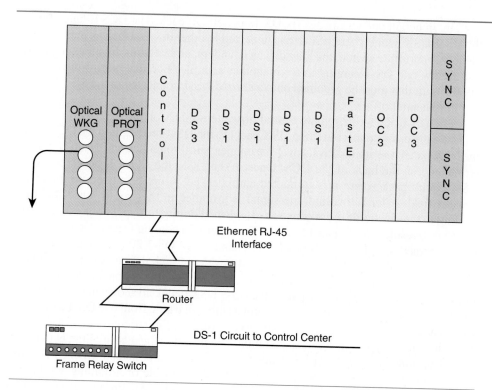

Figure 16–15
Most optical networks require separate management networks that are used to remotely access the devices, receive alarms, and so forth. This configuration illustrates a common method used for management control.

framing—again, ESF or SF, the correct line code, and alarming. The technician also provisions each DS-0 channel according to the type of traffic it is carrying. For example, a digital data circuit requires the technician to set the bit rate of the channel and signaling method. An FX circuit must be configured for the correct signaling and so forth.

1/0 DCSs range in size from one shelf terminating as few as twelve DS-1s to multiple shelves terminating many DS-1s, both referred to as *narrowband DCSs*. The shelf typically contains the DS-1 circuit cards that terminate multiple DS-1s, redundant timing cards, and administrative common cards sometimes containing the cross-connect field. A timing source is fed into the DCS, providing a stable timing supply and a remote access port; either TCP/IP or an RS-232 is available for remote access. In addition, a craft interface port is available to connect a dumb terminal for programming the device.

Again, explaining how the DCS plays in the metropolitan network is best accomplished by using a scenario showing how a DCS in an ILEC CO is used to groom traffic between offices. In our example, the DCS terminates 360 DS-1 circuits and consists of three shelves sitting in relay racks in the special services area of the CO. Each shelf

narrowband DCS
The narrowband digital cross-connect switch electronically grooms DS-0 circuits between DS-1 circuits.

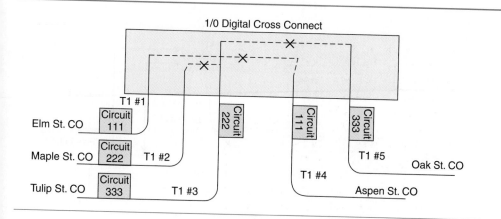

Figure 16–16
1/0 digital cross connects are used to groom individual 64 kbps DS-0 channels from one T1 to a second T1 electronically.

is 12 in. in height and terminates six DS-1s per circuit pack. The DCS is hardwired to the DS-1 DSX panel where it is cross connected to other devices such as multiplexers, the local end-office switch, the long distance tandem, and so forth. In our example, we will follow one DS-1 coming in from another central office that is carrying an SS-7 DS-0 circuit that must be groomed onto a second DS-1 connected to the SS-7 hub site in a third central office.

The DS-1 tied to the remote central office terminates on the DS-1 DSX panel. It is physically cross connected at the DSX panel to the DS-1 feeding into the DCS. From the DSX panel, the signal travels over the copper pair through the amphenol connector terminated on the back of the DCS. Once in the DCS, the signal is accepted by the DS-1 circuit pack where it is formatted and checked before being sent to the cross-connect field. After formatting, the signal is groomed in the cross-connect field according to the electronic cross connects established during turnup. In our example, the SS-7 DS-0 resides in channel 12 on DS-1 number 106 coming from the remote office (Maple Street) and must be groomed into channel 5 of DS-1 206 heading to the third CO where the SS-7 STP resides. The signal travels through the cross-connect field to the channel card that terminates DS-1 206. After leaving the channel card, the signal rides on the twisted copper wire pair to the DS-1 DSX panel, all shown in Figure 16–17. The benefits the DCS provides are apparent as moving traffic from one DS-1 to another using channel banks would require extensive wiring configurations.

The second type of DCS most commonly deployed in the metropolitan network is the 3/1 or broadband DCS. The *3* stands for DS-3, and the *1* stands for DS-1. This today is a bit of a misnomer, because the 3/1 DCS is capable of grooming VT-1s and DS-1s onto DS-3s, OC-3s, OC-12s, and OC-48s. It is also able to groom DS-3s or STS-1s onto OC-3s, OC-12s, and OC-48s. It is truly a broadband-grooming device. Because the function of the 3/1 is the same as the 1/0, we will not bore you by going through the train analogy again. However, you should realize that the 3/1's purpose in life is to move channels between circuits electronically; the only difference is the size of the channels.

The 3/1 is a bit more complex than the 1/0 as it terminates DS-1s, DS-3s, STS-1s, OC-3s, OC-12s, and OC-48s. It is capable of grooming DS-1s, DS-3s, and STS-1s between any of these interfaces electronically. The structure of the 3:1 consists of three primary areas: tributary circuit packs, the cross-connect plane, and the administrative complex. The tributary side of the DCS is responsible for physically terminating all of the circuits such as DS-1s, DS-3s, and OC-Ns. It is also responsible for formatting the incoming signals into a SONET standard frame such as a VT-1 or STS-1. Options similar to those found in a MUX must be configured for each circuit terminating into the DCS.

The second area of a DCS and the most critical is the cross-connect plane. The cross-connect plane forms the electronic cross-connect field that establishes electrical paths between the various tributary ports. The redundancy of the cross-connect plane is critical and is therefore always a 1+1 protection design, meaning that the signal travels through both the *A* and *B* sides simultaneously. At the output of the plane, the best signal is selected according to error-checking schemes built into the unit. The

Figure 16–17
A digital cross-connect system is responsible for grooming traffic around the metropolitan network.

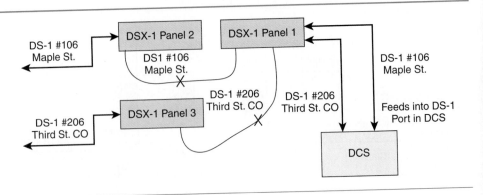

administrative portion of the DCS, also called the brains of the device, like the cross-connect plane, consists of two complete copies—an *A* and a *B*. The administrative section performs several functions, one of which is to house the cross-connect information as it is entered manually by the technician. Both the working and protect shelves of the administrative complex have hard drives that are used to store the cross-connect and circuit configuration parameters. In addition, a backup tape or disk drive is used to provide additional protection in case the unit crashes. The *A* side and *B* side work together in a 1 + 1 protection scheme, meaning that one side works while the other stands in the wings waiting for the working side to fail. The hard drives are backed up every so many minutes, and the tape or disk drive is backed up once a day.

A three-part example showing how circuits are groomed in a broadband DCS is presented next. The first part illustrates how a DS-1 circuit is groomed into a DS-3; the second part shows how a DS-3 is groomed into an OC-12; and the third shows how a DS-1 is groomed into an OC-48.

A DS-1 being fed out of a channel bank sitting in a relay rack in the special service area of the central office terminates at the DS-1 DSX panel. The DS-1 leaving the channel bank is destined for a long distance company located across town. At the DSX panel, the DS-1 circuit connected to the channel bank is cross connected to the DS-1 circuit feeding into the tributary side of a 3/1 DCS, as shown in Figure 16–18. Each DCS DS-1 port shelf is capable of terminating fifty-six DS-1 circuits requiring that a 128-pair cable is run between the port shelf and the DSX panel. The wires are spun down on wire wrap pins at the DSX panel and terminated using amphenol connectors prewired at the factory at the tributary shelf.

The signal leaves the channel bank across the T1 shielded twisted pair cable and into the port on the DSX panel. From the DSX port, the signal jumps onto the cross-connect wire used to connect the channel bank to the DCS. After leaving the second DSX port, the signal travels on the twisted copper wire pair connected to the DCS tributary shelf. The port card looks at the incoming signal, taking note of any problems before packaging it into a VT-1, 1.728 Mbps frame. Once in VT-1 form, the signal travels into the cross-connect plane where it is routed electronically to the destination address, which in our example is a DS-3 circuit destined for the long distance telephone company across town. The VT-1 enters the DS-3 port card where the VT-1 framing is stripped off and the DS-1 is revealed. The DS-1, now riding with twenty-seven other DS-1s, leaves the DS-3 port card on coaxial cable. The DS-3 DCS port is hardwired

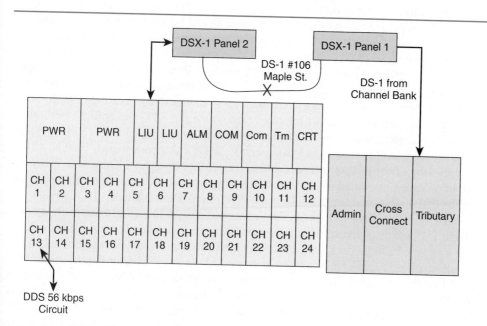

Figure 16–18
Common connection in the metropolitan space between a DS-0 channel bank and a DS-1 port on the 3/1 DCS.

Figure 16–19
A common practice in the
metropolitan network is to use the 3/1
DCS to groom DS-3 circuits from one
network to another.

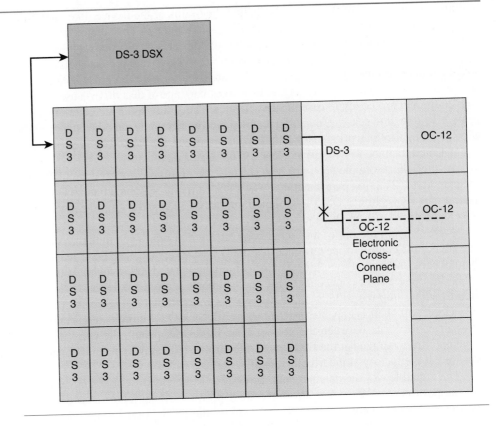

using coaxial cable to a DS-3 DSX panel. From the DSX panel, the signal will travel across coaxial cable jumpers used to create a physical cross connect between the DCS DSX port and an optical multiplexer port also tied to the DS-3 DSX. Cross connecting the two DS-3 DSX ports requires two coaxial cables connected to the transmit-and-receive BNC connectors making sure to form a cross of the ins and outs. Simply stated, the transmit of the DCS should be connected to the receive of the SONET MUX and the transmit of the SONET MUX should be connected to the receive of the DCS. All DSX cross connects require the jumpers to be flipped in order to feed the transmit signal of one device to the receive port of the other, and vica versa. The exception to this rule is when a DSX panel has a built-in cross connect placed between two ports internally.

The second example follows a DS-3 as it flows through the DCS and is multiplexed or groomed onto an OC-12 signal. A DS-3 arrives at the DCS on coaxial cable connecting the DS-3 tributary port to the DS-3 DSX panel, as shown in Figure 16–19. The signal is formatted into an STS-1 frame and directed into the cross-connect field where it is groomed onto the assigned STS-1 channel. From the cross-connect field, the STS-1 is electronically fed to an OC-12 tributary card where the STS-1 is multiplexed into an STS-12. The electronic STS-12 is converted into an optical signal that is then outpulsed onto the fiber jumper connected to the front of the OC-12 tributary card. The OC-12 signal rides on a fiber jumper to a port on the fiber distribution panel. Once at the fiber distribution panel, the signal moves onto fiber jumpers used to connect the equipment side of the FDP to the outside plant side of the FDP. An FDP similar to the DSX panel is the connection point between the two sides of the circuit—in this case, equipment and network. The outside plant cable terminates on the back of the FDP. The inside fiber cable is fed down the building riser to the location where the outside plant fiber is terminated and is either spliced together at that point or connected via a second fiber jumper at an FDP.

Our third example follows a signal as it moves through a broadband DCS from the DS-1 port to the OC-48 port. As explained before, the DS-1 port of the DCS is hardwired to a port on the DS-1 DSX panel. The signal travels across this connection through a line interface module—LIM—used to physically balance the signal before sending it into or out of the circuit pack. LIMs are small circuit boards around the size of a poker chip. LIMs may go bad, and therefore, should always be checked when a signal is experiencing problems. Once the signal enters the circuit pack, the signal characteristics are checked before being formatted into a VT-1 frame. Each circuit pack terminates multiple DS-1s, and each shelf terminates several circuit packs. As with all transmission equipment in telecommunications, the shelves are built with exact multiples of DS-1s that are multiplexed together into DS-3s or STS-1s. In our example, the DS-1 has been converted into a VT-1 that is then multiplexed together with twenty-seven other VT-1s to make one STS-1. Fiber optic cable is used to connect the interface port shelves of the DCS to the switch matrix in order to reduce the chance of cross talk and to provide a high-bandwidth pipe for the hundreds of signals traveling between the two functional areas. The STS-1 rides across the fiber optic connection to the switch matrix where it is groomed into a channel according to the cross-connect information stored in memory.

For example, our DS-1 is destined for a site being fed by an OC-48 system terminating directly into the DCS. Therefore, the VT-1 riding in the STS-1 is cross connected to a VT-1 tied to the OC-48 circuit pack as shown in Figure 16–20. Note that the diagram shows the *TL1*—Transaction Language 1—command used to establish the cross connect between the two line cards—the DS-1 port and the OC-48.

After moving through the cross-connect plane, the VT-1 signal is groomed into an STS-1 destined for the OC-48 card. At the OC-48 line card, the STS-1 is multiplexed together with forty-seven other STS-1s, forming an STS-48. As with the STS-12 in the previous example, the STS-48 is converted into an OC-48 signal and is outpulsed onto the fiber jumper that connects the DCS to the FDP. As shown in both examples, a DCS efficiently moves traffic electronically from one circuit to another and multiplexes/demultiplexes lower and higher signals together.

TL1
Transaction Language 1. Interface command language used to send commands to SONET equipment including SONET multiplexers, broadband DCS, and so forth.

16.3.4 ATM Switches and Aggregation Boxes

The ATM switch or aggregation box is used in the metropolitan space primarily to aggregate many lower-speed signals into a higher-speed pipe and back haul the traffic to an interconnection point. The interconnection point may be a regional or Tier 1 ISP; an ILEC, CLEC, or ICP; or a private business such as a large corporation. The equipment is structurally the same as that described previously in the section on the core network. The main difference between the two is the level of processing performed by the metropolitan ATM switch and aggregation device when compared to the long-haul switch. In the metropolitan space, there are many different types of traffic being fed into the ATM switch. These include various types of voice traffic, data traffic, and video traffic. The box has to be able to precisely assemble each cell according to the traffic type. Additionally, the box has to police the ingress and egress portions of the pipe to guarantee that there is no blockage of cells as they flow through the box. The long-haul ATM switch is typically just aggregating already assembled ATM cells and watching the flow

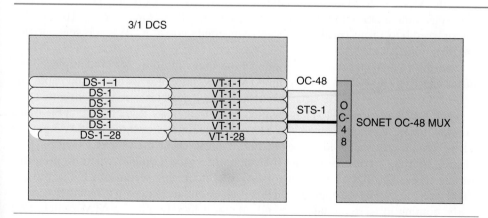

Figure 16–20
Digital cross connects groom DS-1 circuits into VT-1 circuits that ride inside an STS-1. Often, an STS-1 interfaces to a SONET MUX via an OC-48 card.

of the cells to ensure that congestion is kept to a minimum. Also, the long-haul ATM usually has very high speed interface circuits such as OC-48 and OC-192, lessening the issues of traffic shaping and traffic policing. When a device is trying to aggregate multiple DS-3 circuits and shove them out on a DS-3 pipe, buffering, shaping, and policing become much more of a headache. Why, you may ask, doesn't the metropolitan provider buy a bigger box with higher-rate interfaces? The reason of course is cost.

The ATM device in the metropolitan space is structured the same as in the long-haul space except that it may not have the same level of redundancy because it is responsible for carrying fewer bits; therefore, protection is less critical than for a box carrying a full OC-48. Typically, in the metropolitan space, an OC-3 or OC-12 is sufficient on the upstream side and circuits as small as DS-1s may feed into the unit on the tributary side. The ATM switch is divided into four primary sections: the tributary side, the line or upstream side, common cards, and the processor. Where each of these physically resides is irrelevant. They perform the same functions whether they are contained on the tributary circuit packs or on their very own circuit packs. This is especially true of the processor. Some vendors have integrated the processors on the tributary cards trying to lessen the need for dual processor cards. Others have 1+1 processor cards, and some have only one. In many cases, metropolitan ATM switches and aggregation devices, which today are fairly synonymous with each other, do not provide redundancy on the tributary cards at the hardware level. Instead, they allow the traffic to share other virtual channels if there is a port failure. This works fine if the pipes are not heavily oversubscribed and if there is more than one circuit terminating at the same location. Aggregation boxes are often much smaller and do not have switching fabrics as do the ATM switches. Today, though, many vendors have included at least a small switch fabric in many of the aggregation boxes. The types of interfaces found in an ATM switch and aggregation box range from DS-1 up to OC-48. In the metropolitan space, DS-1 interfaces have become much more prevalent while OC-48s are cautiously deployed.

The following example illustrates how an aggregation box and an ATM switch are deployed by a CLEC that sells voice, DSL, and VoDSL. The provider has placed an ATM aggregation device in its colocation site. The device has DS-1 TDM circuits carrying voice from a digital loop carrier—DLC, DS-1 frame relay circuits carrying DSL traffic from a DSLAM, and a DS-3 ATM circuit carrying VoDSL traffic from a second DSLAM. The aggregation device accepts the TDM DS-1 and converts the signal into an ATM circuit emulation pipe—also called *CBR*, constant bit rate, or circuit emulation, as explained in detail in Chapter 7. CBR packaging is similar to TDM in that it nails up a path specifically for that traffic, in our case the DS-1 TDM. Next, the box takes the frame relay DS-1 and converts it into AAL5—ATM adaptation layer 5—data traffic or *UBR*—unspecified bit rate, also explained in Chapter 7. Finally, the ATM DS-3 that is made up of partially VoDSL traffic and partially data traffic is packaged accordingly. Variable bit rate (*VBR*) or AAL2 segmentation assembly is used to format the VoDSL traffic; and UBR, AAL5 SAR—segmentation and reassembly—process is used to format the data traffic. On the upstream side of the box, a DS-3 interface connects the ATM box to a DS-3 DSX panel. From the DS-3 DSX panel, the signal is cross connected onto a DS-3 port connected to an OC-48 MUX tributary card. The DS-3 is multiplexed into an electrical STS-48 and converted into an OC-48 that is sent out onto the fiber optic cable.

At the switch site, the DS-3 coming from the colocation site is multiplexed out of the SONET OC-48 and fed into an ATM aggregation box using coaxial cable and DS-3 DSX panels. At the DSX port, a coaxial cross connect is used to connect that port to a port hardwired to an ATM aggregation box sitting in a 7 ft. relay rake in the transmission lineup. The signal enters the ATM aggregation device through a DS-3 port on a DS-3 circuit pack that holds three other DS-3 ports. Using the VPI/VCI—virtual path identifier/virtual circuit identifier—indicators the box determines where to point the various types of traffic. As most vendors' ATM switches do not convert CBR traffic into TDM, and vice versa, the aggregation box is an essential element in this network. If the traffic coming in were all data, it could interface directly into the ATM switch. One

CBR
Constant bit rate refers to the format an ATM cell is given when it carries real-time traffic such as voice or video. CBR nails up the path between the devices in the same way a TDM channel is nailed up. CBR is the same as AAL1.

UBR
Unspecified bit rate refers to the format an ATM cell is given when the traffic is not time sensitive and does not require prioritization. AAL5 is the same as UBR.

VBR
Variable bit rate refers to the format an ATM cell is given when it carries traffic that is time sensitive, such as voice, but is multiplexed statistically with priority stamps used to maintain quality. VBR is the same as AAL2.

thing to keep in mind is that the ATM switch ports are normally much more expensive than the aggregation box ports and, as such, are engineered precisely in order to not waste bandwidth. The CBR traffic is terminated onto a DS-1 port that converts the ATM cells into TDM channels that mirror the original. From the DS-1 port, the signal rides on twisted copper wire that is wire wrapped to a DS-1 DSX panel. At the DSX panel, a cross-connect jumper ties the ATM aggregation box DS-1 to a *DFI*—Digital Facility Interface—switch port where it is terminated.

Next, the aggregation box grabs the AAL2 traffic and points it to an output DS-3 port connected to a DS-3 DSX panel via coaxial cable. A VoDSL gateway device, used to convert the AAL2 traffic into standard TDM voice, sits between the aggregation box and the circuit switch. The aggregation box and the VoDSL gateway are connected at the DS-3 DSX through a coaxial cable jumper. The VoDSL gateway is connected to the circuit switch in several ways. DS-3 ports or DFI ports are used to connect the switch to the gateway. In some cases, the switch and the gateway are connected through OC-N interfaces, though less often than STS-1s or DS-3s. The VoDSL conversion box is responsible for converting AAL2 VoDSL traffic into TDM channels and handing it on to the voice switch in a format it likes.

Finally, the ATM aggregation box directs the AAL5 data traffic to a port connected to the ATM switch. Figure 16–21 shows the placement and connections associated with an ATM switch located in the metropolitan space as we have described. Typically, an OC-3 or an OC-12 connection is used to connect the aggregation box to the switch, mainly because of the link's ability to carry large amounts of traffic and, of course, because of the lower cost associated with the optical port when compared to the electrical DS-3 port. The main objective of any ATM engineer is to reduce the cost per port on an ATM switch; and, because ATM switches are high-end devices, OC-N interfaces—specifically, multimode—are less costly than DS-3 interfaces. Many service providers deploying DSL traffic have adopted the use of a device called an *SMS*—Subscriber Management System. The SMS is used to aggregate the VPI/VCI and distribute the traffic efficiently before handing it off to the ATM switch sitting at the edge of the network, as shown in Figure 16–22. The SMS and the aggregation box interconnect through DS-3 or OC-N interfaces and share path information accordingly.

Whether the AAL5 traffic is first handed to an SMS device or directly to the ATM switch, its final destination remains the same. The ATM switch is responsible for directing the cells to the output or egress ports toward its final destination according to the VPI/VCI indicators. In some cases, if the switch site is large and used as a data hub, a carrier-class router is used to direct traffic at a Layer 3 level using source and

DFI
The Digital Facility Interface is the digital interface port on a voice switch.

SMS
A Subscriber Management System is a piece of equipment used to aggregate traffic coming from multiple DSLAMs before it is passed to the router; or in the reverse, the router passes the traffic to the SMS that then distributes it to the DSLAMs.

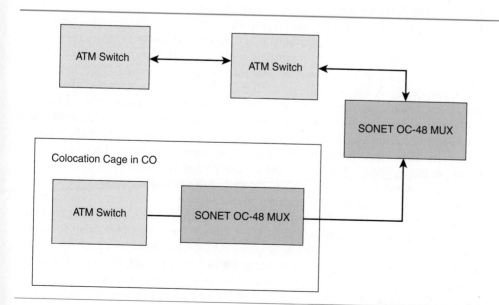

Figure 16–21
ATM switches are typically located in a centralized location and used to switch ATM cells between multiple ATM end points. The diagram shows a typical ATM metropolitan network topology.

Figure 16–22
SMSs, subscriber management systems, are needed in the metropolitan space to manage traffic flows from broadband customers. Typically, an SMS interfaces with the aggregation box that handles the ins and outs to and from the ATM switch.

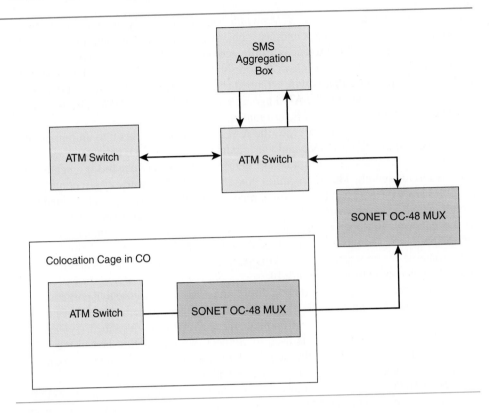

destination addresses. In this case, the ATM switch is connected to the router via a DS-3 or OC-N interface. The router cracks the packets, looks at the source and destination addresses, and either sends them back to the ATM switch on the appropriate path or sends them out directly to the core network. This may sound confusing—and of course it is—but it is important to understand that traffic crossing the metropolitan network is not destined for just one site and, in addition, traffic is not all the same, as shown in this example—VoDSL, voice, data. All of the devices just described are commonly deployed in the metropolitan space and are networked together in a way that allows information to traverse between metropolitan and core, and core and metropolitan, seamlessly.

16.3.5 Frame Relay Switches and Aggregation Boxes

The use of frame relay switches and aggregation boxes in the metropolitan network is very common. Not only do data-only users depend on frame relay circuits, many DSLAMs use frame as their upstream link to the hub site. The frame relay switch is much smaller than the ATM switch. The interfaces range from a DS-0 to a DS-1, DS-3, and OC-3/12. The structure of the frame switch is similar to the ATM switch in that it consists of interface tributary ports, a switch matrix, common administrative cards, and upstream or line cards. The internal workings of a frame relay switch also use simple buffering and traffic policing methodologies to help ensure traffic delivery. The more sophisticated the frame switch, the more expensive. Therefore, many providers play a game of, How far can we push it before traffic is blocked and customers are angry? Usually, less-expensive units suffice, making frame relay a viable means for carrying traffic across the metropolitan space.

A frame relay switch often resides at a large central office or an ISPs regional hub. The device is used to accept incoming DS-1 circuits carrying frame relay formatted data. The switch is responsible for setting up DLCIs—data link connection identifiers, as explained in Chapter 6, and distributing them to the correct location. The frame switch commonly interfaces with a DSLAM or router that contains frame relay capable interface ports. Our example of a typical frame connection in the metropolitan space shows how a frame switch located at a regional ISP interconnects to a frame interface in the ISP's router located at a remote hub site.

The router at the remote hub site aggregates dial-up traffic for the ISP and transports it via frame relay across a DS-1 to the ISP's regional site. The frame relay interface in the router formats the frame relay circuit according to the information given by the engineer. For example, a DLCI number is provisioned from the router to the frame switch and another from the frame switch to the router. The interface in the router must also be configured for the type of DS-1 framing and line code. The DS-1 connects to the router through an RJ-48 port on a *WIC*, wide area network interface card—in the router and is wire wrapped onto the back of a DS-1 DSX panel. From the DSX panel, the DS-1 is cross connected over to the T1 provider's network through a hardwired connection from the DSX panel to a Smart Jack, also called a NIU—network interface unit, sitting in a shelf on the wall.

At the opposite end of the circuit, the DS-1 from the T1 provider terminates on a DS-1 DSX panel at the ISPs regional site. The DS-1 is cross connected to the DSX port and then hardwired into the frame relay switch. The wire connection terminates on the back of the frame switch either through wire wrap pins or amphenol connectors. The signal travels into the DS-1 interface pack where it is switched according to the circuit configuration and pointed to the correct upstream interface, which in this case is a DS-3 circuit connecting the ISP to a Tier 1 Internet provider.

16.3.6 Routers

Routers are used throughout the metropolitan network as aggregation and routing devices, as explained in Chapter 12. The objective in this chapter is to show where routers reside in the metropolitan network and how they interface with other transport devices. Routers may be categorized as either *host* or *network* in the metropolitan space. In the long-haul network, most routers are called *core*, meaning that their size is impressive and their capability to move traffic is huge. In our case, most routers are used to route packets, not to transport large volumes of traffic. Typically, a router interfaces with a transport MUX of some kind and depends on it to aggregate up to the OC-48 level. A router has an interface that connects to what is called the WAN—wide area network—side of the network. The WAN side of the network may consist of a DS-1, as just described in the frame relay section, or a DS-3, an OC-3, or even an OC-12.

In our example, we will look at a DS-3 connection with a medium-size network router sitting in a data hub site deployed by an Ethernet provider. The network router is tied to a core router that is tied to a SONET OC-3 pipe feeding into a Tier 1 Internet provider network such as MCI, AT&T, Global Crossing, or Sprint. On the ingress side of the network router, Fast Ethernet interfaces carry the data into the device. The processor cracks the packets to see how it should route the traffic according to the routing protocol. It notes the IP addresses—source and destination—and sends the information accordingly. The traffic is moved across the router to the OC-3 port where it is formatted into an ATM cell and shipped out riding on multimode fiber that connects to a multimode optical port on the carrier's SONET MUX. The carrier feeds the OC-3 signal into its optical MUX and transports it to the Internet peering point, such as an NAP. The signal rides on the fiber pipe to the distant location where it is demultiplexed and connected to the core router as an ATM DS-3. The core router strips off—disassembles—the ATM cells revealing the IP packets. Using the IP source and destination address, the router directs the information accordingly.

Most Tier 1 and Tier 2 ISPs sell bandwidth in increments of DS-3s such as one up to forty-eight. If the end user—in this case, the Ethernet provider—wants to increase the bandwidth, it calls its carrier, which increases the bandwidth on the pipe by logging into its system and opening up a larger path. For example, if our provider, who initially required one 52 Mbps worth of bandwidth, needed an additional 52 Mbps worth of bandwidth, it would call up the Tier 1 carrier and ask for 104 Mbps. The carrier would open up the OC-3 connecting its two devices to 104 Mbps and charge the customer accordingly.

WIC
The wide area network interface card is the DS-1 interface card in a DSLAM.

passive optical network (PON)
A fiber optic network with passive filters deployed to carry multiple wavelengths across the fiber optic strand.

16.3.7 Passive Optical Networks

Passive optical networks (PONS) are commonly used in the metropolitan network to increase the capacity of the fiber optic strands. A PON—similar to a WDM system—separates the light signals into individual channels or frequencies. The difference is that a PON, as it says, is passive, meaning that there are no electronics, just passive filters, similar to a prism that separates out wavelengths or colors. This method of deploying multiple wavelengths on a fiber strand is very, very cost-effective, especially when compared to WDM systems.

Though passive optics are small, simple to implement, and resilient, they do have several limitations, the most important of which is the short distance the signal can travel between PON sites. The PON is not responsible for amplifying or formatting the signal. It is only concerned with providing a way to expand the capacity of the fiber by providing multiple paths on one strand for the light to travel across, almost like dividing up the fiber pipe into multiple pipes. As such, PON deployments have become important additions to congested metropolitan fiber networks. The key point to draw from this is that a PON does not fit every application and must be carefully engineered according to the loss budget of the system in order to work properly.

A PON is a very small device that is installed in shelves in relay racks or mounted on a wall near a termination box, in cases in manholes, and so forth. The versions differ, and some are very sophisticated in that they are able to filter many wavelengths across one fiber, as commonly seen in the CATV networks. Though the devices range from very inexpensive two-filter devices to larger, more sophisticated devices, passive optical networking is an important component in the metropolitan network.

16.3.8 Wave Division Multiplexers

During the early 2000s, WDM reached into the metropolitan space as the designs of smaller, lower-powered, less-expensive systems were produced by equipment vendors. The need for hundreds of channels traveling on one fiber strand between two distinct points is not the critical criterion used for systems residing in the metropolitan space. The metropolitan network has many from and to locations, in addition to shorter distances between locations. Another factor that limited the deployment of WDM in metropolitan areas was the lower bandwidth requirement for individual channels. As seen from the discussion earlier in the chapter, the metropolitan space is a conglomeration of smaller circuits aggregated into larger circuits that are distributed at different points in the network, as well as containing all types of traffic. The traffic enters and exits the network at many locations, forcing the equipment to add and drop signals frequently, unlike the two-point core network scenario. The core network required large pipes carrying high-bandwidth circuits over long distances adding and dropping circuits at the two ends. The WDM systems proved to be more cost-effective than constructing new fiber runs over such long distances. In contrast, in the metropolitan space, building additional fiber runs or using PONs proved to be a better solution to alleviate fiber exhaust.

After the telecom crash in early 2000, vendors quickly realized that in order to sell their WDM product they would need to do two things: reduce the size and dramatically reduce the price. Both were accomplished, and a new wave of WDM systems entered the metropolitan space—*CWDM*, coarse wave division multiplexing. CWDM systems carried fewer channels and used lower-powered lasers, making them less expensive to engineer and build and, in turn, making them more useful to the local service provider.

Similar to those deployed in the long-haul network, the CWDM system is divided into four sections: transponders, multiplexer or coupler, laser, and common cards. The transponders used in the metropolitan space may like the long-haul network, be designed using the ITU C-band. The differences are that, typically, there are fewer channels and the spacing between the channels is greater. The rule of thumb is, the greater the spacing between channels, the less expensive the equipment. The extended-band fiber, such as Lucent's AllWave™ and Corning's SMF-E, were designed specifically for the metropolitan space for this exact reason. The increased spectrum available— 1310 to 1625 nm—gives vendors the ability to design systems with wider channel

CWDM
Coarse wave division multiplexing is a WDM technology that multiplexes wavelengths together on one fiber optic strand. CWDM differs from WDM primarily in the number of channels or wavelengths it can multiplex.

Figure 16–23

Coarse wave division multiplexing is deployed in the metropolitan network to reduce the number of fiber strands needed to carry multiple optical systems.

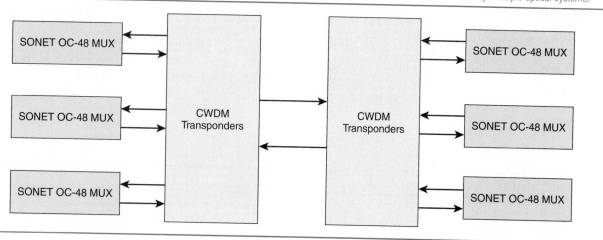

spacing across the entire spectrum. CWDM is built using wider channel spacing, thus making it less expensive to deploy.

A CWDM system in the metropolitan space can be found in many diverse locations: carrier hotels, ILEC COs, CLECs and ICPs, switch and colocation sites, and large enterprise customers. The most common deployment of CWDM can be found in locations where the cost to place new fiber in the ground exceeds the cost of deploying the CWDM system. For example, several years ago, a large enterprise customer built a six-strand ten-mile fiber network connecting its two office buildings together. Within the past few months, the network manager observed that the bandwidth on the six strands of fiber is at 70% capacity, meaning that only 30% of the fiber is available for new circuits. Additionally, she has been informed that the sites have been slated to become the disaster recovery center for the company, meaning that the two connected buildings must be able to transfer huge amounts of data instantly as the data-storage units will need to continuously mirror each other. The cost to run a new fiber cable is exorbitant—around $300 per foot per fiber duct, thanks in part to the municipality that is now requiring reconstruction costs in full for all road disruptions and the like. The network manager, after running through a cost analysis, has decided to place a CWDM system in the network that will increase the capacity of two fiber strands substantially. The system deployed is equipped with six wavelengths, each capable of carrying gigabits worth of information. The system takes up half of a 7 ft. relay rack and requires two feeds of DC power feeds, one working, one protect. Fiber jumpers connect each of the transponders to an FDP where they interconnect to the terminating MUX or optical equipment, such as the data-storage device. The optical line card connects through fiber jumpers to the FDP where it is cross connected to the outside plant fiber. At the opposite location, the same setup exists, all shown in Figure 16–23. Because the CWDM device is capable of combining sixteen wavelengths onto one transmit and one receive fiber strand, the network manager is sure she will have enough capacity for a very long time without having to dig up the streets and disrupt the existing plant.

SUMMARY

Metropolitan networks encompass all types of network architectures, networking protocols, and networking equipment. Metropolitan networks differs from the long-haul or core networks primarily in size and the number of locations where circuits are added or dropped. A metropolitan network requires that the devices groom lower-level circuits and interface with lower-rate devices. The chapter described the different network architectures, the protocols, and the types of equipment found in the metropolitan space.

CASE STUDY

Miranda Rogers, Big Bob's youngest daughter, has been assigned the task of designing a metropolitan network that provides connectivity to both central offices, the cellular switching center, the new CATV switching center, and the new VoIP company being built by Hector and his sister Angel.

Use the information in Appendix A to develop a preliminary design that connects all of the locations together and provides each location the ability to add/drop different circuit types.

Appendix A also provides a list of services each site requires.

REVIEW QUESTIONS

1. A metropolitan network carries information between locations within the metropolitan area such as between central offices, ISPs, POP and ISP, and large businesses.

   a. true   b. false

2. List the four most common network topologies deployed in the metropolitan space.

3. The common protocols deployed in the metropolitan network are

   a. ATM, IP, MPLS, and frame relay.
   b. ATM, IP, MPLS, frame relay, and Ethernet.
   c. IP, MPLS, and Ethernet.
   d. SONET and asynchronous TDM.
   e. a and d.
   f. b and d.
   g. all of the above.

4. A VT-1 circuit falls under the asynchronous TDM category.

   a. true   b. false

5. A channel bank aggregates subscriber circuits such as FX, DDS, and ISDN into SONET OC-N circuits.

   a. true   b. false

6. The M13 is used to

   a. multiplex twenty-four DS-1 circuits into one M13 framed DS-3.
   b. multiplex 672 DS-0s into one M13 framed DS-3.
   c. multiplex twenty-eight DS-1s into one M13 framed DS-3.
   d. multiplex twenty-four 64 kbps circuits into one M13 framed DS-3.
   e. multiplex twenty-four 64 kbps circuits into one DS-1.
   f. a and b.
   g. b and c.
   h. b, c, and e.
   i. all of the above.

7. What is the difference between a SONET multiplexer deployed in the core network and a SONET multiplexer deployed in the metropolitan network?

8. Ethernet has become one of the most popular transport methods used in the metropolitan space due to the following:

   a. Ethernet provides an easy method for handling Layer 3 functionality.
   b. Ethernet is already used in the LAN space making it less "scary" to IT managers needing to deploy a WAN solution.
   c. Ethernet equipment is inexpensive.
   d. Newer Ethernet interfaces, FastE and GigE, provide a high-bandwidth WAN pipe.
   e. Service providers started offering low-cost, high-bandwidth Ethernet pipes.
   f. A, b, and d.
   g. B, c, d, and e.
   h. All of the above.

9. Frame relay continues to be a popular protocol in the metropolitan space.

   a. true   b. false

10. ATM is used in the metropolitan space to carry both time-sensitive and non-time-sensitive traffic.

    a. true   b. false

11. A narrowband DCS is also called a(n) _____ DCS.

12. A broadband DCS is also called a(n) _____ DCS.

13. A narrowband DCS aggregates lower-speed channels into higher-speed channels by grooming _____ into _____.

    a. DS-1s, DS-3s
    b. DS-3s, STS-1s
    c. DS-1s, VT-1s
    d. DS-0s, VT-1s
    e. DS-0s, DS-1s
    f. none of the above

14. A broadband DCS is used to groom lower-speed circuits into higher-speed circuits by grooming _____ into _____.

    a. DS-1s, DS-3s
    b. DS-3s, STS-1s
    c. DS-1s, VT-1s
    d. DS-3s, OC-3s
    e. STS-1s, OC-3s or OC-12s
    f. OC-3s or OC-12s, OC-48s
    g. STS-1s, OC-48s
    h. all of the above

15. Draw the line-side connection between a SONET MUX and the outside plant.

16. Diagram a network that has three OC-3 multiplexers connected in a ring topology.

17. Diagram a network that has an OC-48 multiplexer connected to three OC-3 multiplexers located in three different central offices.

18. Diagram an ATM switch connected to an ATM aggregation box sitting in a colocation cage in a central office. Include the connection between the ATM aggregation box and a DSLAM.

19. What does PON stand for? Why is it used in the metropolitan network?

20. Explain the difference between a CWDM system and a DWDM system.

KEY TERMS

metropolitan network (442)
ISP (442)
carrier hotel (442)
node (443)
ADM (443)
POP (446)
asynchronous (446)
all optical (446)
IP only (446)
MPLS (447)

CLEC (448)
ICP (448)
colocation (449)
DSLAM (449)
WAN (449)
LAN (450)
channel bank (451)
foreign exchange (FX) (451)
special service circuits (452)
M13 (453)

amphenol cable (453)
BNC connector (453)
digital loop carrier
    (DLC) (454)
fiber distribution panel
    (FDP) (455)
data communications channel
    (DCC) (458)
narrowband DCS (459)
TL1 (463)

CBR (464)
UBR (464)
VBR (464)
DFI (465)
SMS (465)
WIC (467)
passive optical network
    PON (468)
CWDM (468)

17

Access Networks

Objectives

After reading this chapter, you should be able to

- Define the access network.
- Identify the different types of access circuits.
- Identify the different types of equipment used in the access network.
- Identify the network architecture deployed in the access network.

Outline

■ INTRODUCTION

Access network is a term used to describe the types of circuits used to service the end user in the metropolitan area. The telephone line connecting your home to a central office is part of the access network as is the fiber optic GigE service feeding a large enterprise customer. Of all the network segments—core, metropolitan, access—the access network encompasses the greatest diversity of facilities and varied types of services. As such, the job of designing, building, and maintaining the network is complicated and requires a great deal of skill and telecommunications savvy. The goal of this chapter is to present a view of the most popular access technologies, focusing on the circuit type, the equipment, and the network architecture. Included in the chapter are sections detailing the following access technologies: POTS, ISDN, special services circuits, broadband services, T-Carrier circuits, GigE and FastE services, and optical services.

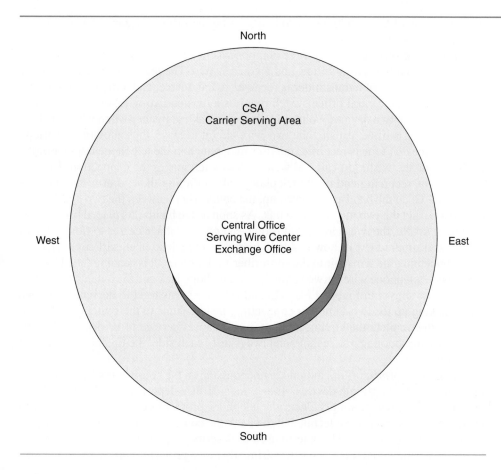

North

CSA
Carrier Serving Area

West

Central Office
Serving Wire Center
Exchange Office

East

South

Figure 17–1
Central offices located strategically throughout a local telephone network are designed to service an area referred to as a CSA. All locations residing within the CSA are served by the central office. The services connecting to the locations within the CSA are referred to as access services.

Before we discuss individual access technologies, we need to explain the underlying structure of the network that carries these services. From the beginning, the local network consisted of wire-line connections to each individual home and business. Buildings called central offices were constructed throughout the metropolitan area as aggregation points for the hundreds of customer connections. Areas surrounding each central office were defined as CSAs—carrier serving areas—encompassing a defined geographic region fed by the central office; engineers drew a circle around the central office, as shown in Figure 17–1, designating the CSA. It is this connection between the central office and the end customer that constitutes the access portion of the network. Today, the central office may be bypassed in certain situations such as when a fiber optic cable is constructed specifically to connect two locations. In this instance, the fiber optic cable is the access network carrying traffic between the two end points. In most cases, though, the local telephone company, ILEC, continues to provide the access network's infrastructure. Today, new entrants into the access space include CATV, cellular, Wi-Fi, and other wireless technologies. The newest service providers entering this area are power companies that offer broadband services across the power lines, a subject discussed in the emerging technologies chapter. Access network services may ride on all types of physical media—copper, fiber, or wireless—depending on the design of the network. The medium is irrelevant to the service being offered as a one travels across a copper pipe in the same way it travels across a fiber optic strand: a one is a one is a one.

■ 17.1 PLAIN OLD TELEPHONE SERVICE

17.1.1 From the Central Office to the Subscriber

Plain old telephone service (POTS) is the oldest access technology in the network. In fact, it is why there is an access network. The first wires strung by Alexander Graham Bell's company in the late 1800s introduced the public to dial tone in their homes or

plain old telephone service (POTS)
A term used to describe the analog voice service deployed in the local telephone network.

businesses, and POTS continues to this day to be the primary means of communications around the world.

Telephone service requires three components: a telephone, a connection between the end user and the central office, and a voice switch. The telephone is a very simple device that consists of a transmitter, a receiver, a dial device, and a ringer. Connecting a telephone to the central office switch requires a twisted two-wire, one-pair connection between the two devices. Voltage is placed on the line by the switch when the telephone receiver is lifted, causing the two-wire circuit to close, thus allowing electricity to flow. Once dial tone is received, the user may punch in the telephone number of the person he or she wishes to call and wait as the switch makes the connection. Ringing occurs at the receiving end of the telephone call, notifying the end user that a call is coming in. Once the handset is picked up, the connection between the two end users is established and the two may begin to talk. As soon as the handset is placed back on the telephone cradle, the two-wire circuit is opened and the voltage ceases to flow. This, of course, is a simple view of how dial tone works, as explained in an earlier chapter.

Connecting the wire pair to the subscriber's telephone set is accomplished by running inside telephone wire between the RJ-11 telephone jack and the outside telephone protector, as shown in Figure 17–2. *Tip* and *ring* are terms used to designate the two wires making up the two-wire pair connecting the telephone to the switch. As the two are used to complete the circuit, a break in one causes service to be disrupted; thus, during troubleshooting, the technician always tests each side, the tip and the ring, for continuity.

Telephone protectors are mounted either outside on the house or inside in the basement and are fairly simple devices. Today, most ILECs deploy *network interface devices (NIDs)*, which are simply electronic telephone protectors capable of being looped back remotely allowing the technician to check the continuity of the outside wire pair before initiating a truck roll—a term used to describe dispatching a technician to the location. Wiring an NID is similar to wiring a passive protector and, as such, is fairly

tip
One conductor of the two-pair copper conductor used to carry electrical signals.

ring
One conductor of the two-pair copper conductor used to carry electrical signals.

network interface device (NID)
The unit mounted at the subscriber and used to terminate the outside plant copper pair to the inside plant copper wiring. The NID is a passive device used to terminate wire pairs; often referred to as the *protector*.

Figure 17–2
Connecting the telephone set to the central office requires a protector or NID—network interface device—where the inside and outside wires are tied together.

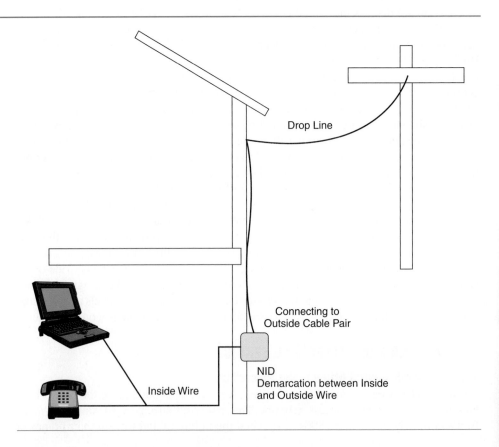

Drop Line

Connecting to
Outside Cable Pair

Inside Wire

NID
Demarcation between Inside
and Outside Wire

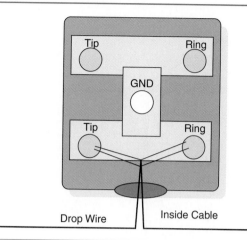

Tip

Ring

GND

Tip

Ring

Drop Wire

Inside Cable

Figure 17–3
Protectors placed on the side of
buildings or inside a basement or
other termination room consist of tip,
ring, and ground lugs.

simple and straightforward, as shown in Figure 17–3. One of the most critical tasks—
often overlooked—is ensuring that a good ground connection is made at the termination point. The lack of a good ground may cause intermittent noise on the line, which
can cause a difficult-to-locate trouble. Additionally, stray voltage coming from lightning strikes, electrical lines, or other electrical sources is diverted to ground helping to
reduce the chance of entering the inside telephone line. Grounds are critical to the
safety of both the resident and the technician.

Connecting the protector to the cable infrastructure requires a drop wire connection between terminals mounted on the telephone pole or in a pedestal, as explained in
Chapter 14. The *drop line* also has a tip and ring pair that is connected to the distribution cable tip and ring pair, as shown in Figure 17–4. Drop wires are often trouble
points in the network.

At the opposite end of the circuit, the telephone pair riding inside a large outside
plant cable with hundreds of other subscribers' wire pairs, snakes its way to the vertical side of the *MDF*—main distribution frame—through the cable vault typically located in the basement of the central office. Every central office has an MDF that is used
as the wiring hub for incoming access circuits. The MDF is divided into a vertical side
and a horizontal side and is explained in detail in Chapter 18. Wires from the outside
world are terminated on the vertical side of the frame either by wire wrapping them
down on wire wrap pins or soldering them onto connection points. Frame wire is used
to connect the vertical side to the horizontal side of the MDF, as the horizontal side is
where all terminating devices such as the voice switch are wired. Frame wire color corresponds to the type of circuit being connected and typically follows a standard color
code as set forth by the engineering department. In most central offices across North
America, green/white wire is used to represent POTS circuits, orange/white represents
ISDN, and red/white represents special service circuits—white being tip and the other
color ring.

Terminating equipment located in the CO is wired to blocks on the horizontal side
of the MDF and labeled accordingly. For example, a switch is wired to termination
blocks on the horizontal side of the MDF during the installation phase of the switch.
The dial tone heard by the end user travels from the switch through the wires connecting the switch to the termination block on the horizontal side of the frame and through
the green/white frame wire to the wire wrap pins on the vertical side of the frame. After leaving the MDF, the dial tone travels across the inside CO wiring that snakes its
way to the cable vault where it is spliced into the large outside plant cable leaving the
building. During troubleshooting, the technician must know how the pair travels
through the network in order to find key connection points where dial tone can be accessed. The first location tested is the vertical connection point on the frame. This point
immediately tells the technician whether the trouble is on the inside or outside portion

drop line
Copper cable used to
connect the subscriber's
cable pair in the distribution
cable to the customer's
demarcation point, typically
an NID or protector.

MDF
Main distribution frame.
Found in a central office and
used to terminate and cross
connect outside plant copper
wire to inside cables.

Figure 17–4
The two wire pair from the drop wire
is terminated at the terminal. The
terminal ties the drop line to the
distribution cable.

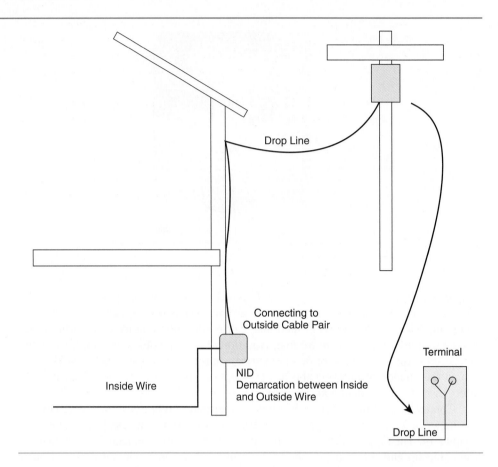

Drop Line

Connecting to
Outside Cable Pair

Terminal

Inside Wire

NID
Demarcation between Inside
and Outside Wire

Drop Line

of the circuit. If dial tone is leaving the office, the technician will move outside and listen for dial tone at a cross box, the terminal, or the protector. Refer to Chapter 14 for details on the outside plant cabling infrastructure.

17.1.2 Digital Loop Carrier Technologies

Though individual POTS lines fed from the CO to the subscriber's location are still very common in the access network, newer technologies have changed how dial tone is carried to the end location. Digital loop carriers (DLCs) have been deployed in the access network since the advent of T1 circuits in the 1970s. The purpose of a DLC is to help reduce the huge number of wires strung between the CO and the hundreds of end customers fed by the serving wire center—another term used to describe the central office. DLCs are electronic devices placed either in cabinets sitting on cement pads, inside buried vaults, or inside small remote offices. One side of the DLC connects to the local subscribers in the area using the same twisted wire pair used throughout the access network. The POTS lines, similar to those already described, connect to the DLC channel cards that terminate POTS and other services such as ISDN, DSL, and special services. Each of the signals from the various channel cards is multiplexed into T1 circuits that are fed back to the central office on T1 links. As a T1 circuit is capable of carrying twenty-four DS-0s, the value of the DLC is apparent—fewer outside plant wires to contend with. In addition traffic engineering studies have shown that concentrating subscribers such as 2:1, that is, supplying one DS-0 channel for every two customers still provides acceptable grade of service levels to the subscribers. In fact the newer GR-303 interconnection standard used in the current DLCs allows for concentration ratios as high as 45:1, though few service providers would be foolish enough to use such high levels. A 4:1 ratio seems to work well in most areas, though traffic engineers must be vigilant and watch for blocked calls.

Today, next-generation DLCs are very sophisticated in that they terminate all types of circuits such as ISDN, special service circuits, DSL, and so forth. Also, they use SONET systems that not only provide fully redundant paths but also allow for traffic to be added and dropped at each of the nodes on the ring. The benefits gained by deploying DLCs are substantial and their use often is very easy to cost justify. The main point to be made in this section on POTS is that the DLC sends the same dial tone service to the customer as that coming from the switch. The switch is still in control, as the DLC is a device hanging off of the switch. As already mentioned, a protocol standard, GR-303, is used to define how the switch talks to the DLC, such as call setup, call teardown, and so forth. T1 circuits connect the DLC to the switch and are formatted using the GR-303 framing structure. The previous standard, called GR-08 or TR-08, is still prevalent in the network though the additional benefits offered by GR-303 have persuaded many service providers to convert when possible. However, it should be noted that there are many older DLCs, such as Lucent's SLC-96 and Nortel's DMS-1 Urban, that continue to function just fine; because the subscriber base in the areas in which they are located have not grown, engineers are reluctant to change them out.

DLCs connect to the switch in two ways. *Universal DLC* connections refer to a configuration in which a remote DLC in the field connects to a second DLC in the central office called the COT—central office terminal. COTs terminate the individual POTS lines at the MDF. The second type of DLC, called the *integrated DLC*, connects directly to the switch at the T1 or higher level interface. Integrating the DLC into the switch eliminates hundreds of wired pair connections at the MDF. Obviously, this not only saves in installation costs— less wiring required—but also reduces maintenance costs— fewer wires to terminate. Figure 17–5 shows both the COT and integrated DLCs.

DLCs have grown from dial tone only devices to sophisticated access devices that allow for multiple services to be terminated. Because the number of subscribers fed by DLCs directly has increased as much as 30% in some areas, the FCC mandated that

universal DLC
Two DLCs back to back, one in the field and one in the central office. The CO DLC interfaces to the switch through line-side connections. Also referred to as the *central office terminal (COT)*.

integrated DLC
One DLC in the field interfaces to the switch through DS-1 Digital Facility Interfaces (DFIs). The DLC is integrated into the switch, eliminating the need for a second DLC in the CO.

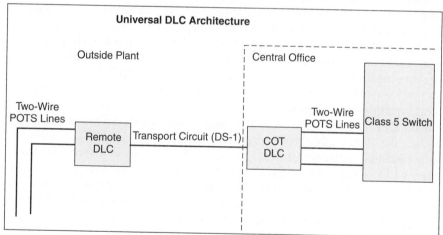

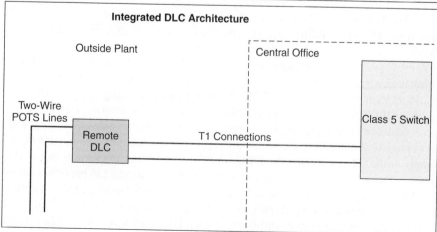

Figure 17–5
Digital loop carriers are placed strategically throughout the access network. Two common architectures found in the network are the universal and the integrated.

unbundled network element (UNE)
Term defined in the 1996 Telecom Act to describe the physical loop available to competitive local exchange carriers that feeds from the central office to the subscriber's premises.

SLC-96
Lucent's DLC that is capable of aggregating ninety-six subscriber lines onto several DS-1 circuits. One of the first DLCs deployed in the network. Often, the term *SLC* is used in place of *DLC*. DLC is the correct term.

TR-08 protocol
Protocol used to define how the remote DLC talks to the switch. TR-08 allows a 2 to 1 concentration ratio, meaning that two subscribers may share access to one channel, thus doubling the number of subscribers that can be serviced by the DLC.

ILECs must offer unbundling services to CLECs—competitive local exchange carriers—at the DLC. The CLEC may purchase an *unbundled network element (UNE),* which is simply a wire pair connected to the subscriber's location from a CLEC-owned DLC colocated in the serving central office. Many ILECs offer transport services such as ATM virtual circuits from the DLC back to the central office. The VCs carry the CLEC's DSL traffic coming from either their own remotely located DSLAM or from a DSL channel card in the ILEC's DLC. At the CO, the CLEC's VCs are groomed out and interfaced into the CLEC's equipment in the CO. As is obvious, the access network involves many different network topologies though each uses the same types of equipment.

Digital loop carrier equipment includes many types of configurations that are deployed depending on size, location, and services to be offered. As mentioned earlier, older DLCs have been deployed for years and continue to exist in all ILEC networks. The difference between the older and newer versions is fairly substantial, as the older are less software dependent in addition to allowing for fewer terminations per device. A common configuration in many ILEC networks, the Lucent *SLC-96* terminates ninety-six subscriber lines at the DLC. Four copper T1 circuits are used to connect the DLC to the switch and carry the ninety-six DS-0 channels filled with the subscribers' voice conversations. A protection T1 is also often turned up to provide a 4:1 protection scheme, meaning that if any of the four T1s were to fail, the traffic would move onto the protection path. Often, a 2:1 concentration ratio is used and, as a result, only two T1 circuits are turned up between the switch and the DLC. The SLC-96, using the *TR-08 protocol,* allowed for a 2:1 concentration ratio providing forty-eight DS-0s for ninety-six subscribers. The subscriber lines terminate into a cross box, and the cross box connects into the DLC.

New DLCs were deployed throughout the 1980s. Through the mid-1990s, copper T1 circuits continued to be used to connect DLCs to the central office. Once fiber optic systems were deployed, engineers began to run fiber optic cable to the DLC cabinets and place asynchronous fiber optic multiplexers inside. The DLC then connected via copper to the back of the FOT—fiber optic terminal—where the signals were groomed into DS-3s, which were fed out onto the fiber strands. Eventually, DLC manufacturers began to integrate the fiber optic MUX into the DLC, creating a SONET-fed device that could be part of a SONET ring. The devices were capable of terminating over a thousand subscribers in addition to offering many other services such as ISDN, DSL, and so forth. The traditional DLC and new-generation DLC architectures are shown in Figure 17–6.

Once the 1996 telecom bill was passed, small, copper-fed DLCs capable of terminating hundreds of subscriber lines in a small 15 in. chassis were introduced to serve the new CLECs. The value of the new DLC included its size, taking up about one-fourth of a relay rack; the large number of terminations; its use of GR-303 protocol, allowing for high concentration ratios; and low power. Lucent's AnyMedia® FAST terminates not only POTS but also DSL signals.

The following two scenarios should help illustrate how dial tone travels through a DLC. The first depicts the traditional ILEC DLC sitting out in a field near its subscriber base. The second illustrates how the DLC is used by a CLEC in a metropolitan network.

A DLC sitting in a cornfield on a cement pad near a small city feeds several new housing developments and a new strip mall. The DLC has the capability to serve 1094 subscribers within the defined serving area. A SONET OC-3 ring connects the DLC to two other DLCs and the local central office switch. When a subscriber picks up the handset of his or her telephone, the channel card senses the off-hook condition and applies voltage to the line. The channel card in parallel sends a message via the signaling channels in the T1 connecting the DLC to the switch telling the switch that such and such a line has gone off-hook. The subscriber holding the handset to his or her ear hears the dial tone and punches in the telephone number of the person he or she is calling— calling party to called party. The digits travel across the two-wire pair from the subscriber's home to the cross box sitting next to the DLC in the cornfield. They continue through the cross box and enter the channel card in the DLC. The digits are multiplexed

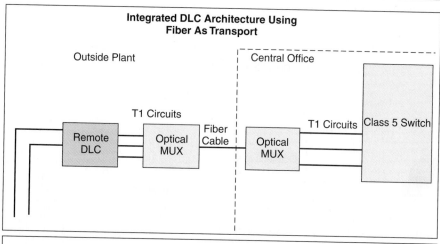

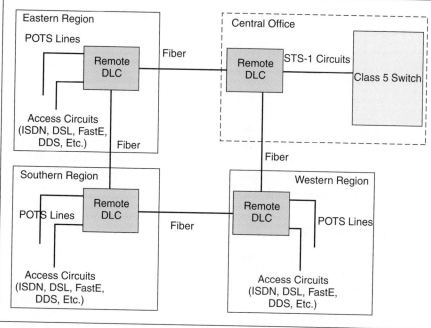

Figure 17–6
Digital loop carrier architectures have
evolved over the past twenty years.
Today, many carriers tie DLCs
together in a region using SONET
ring architectures.

into a DS-0 channel on the T1 circuit that carries it to the switch where it is routed to
the appropriate destination. When the subscriber receives a call, the telephone rings notifying him or her of the incoming call. Each DLC supplies ringing through a ringing
generator located within the DLC's cabinet. Often troubles occur in ringing generators
causing technicians to scratch their heads for a few hours before figuring out the cause
of the problem. As shown in this example, the DLC does not switch traffic but only
multiplexes and carries it to the switch at the central office.

The second scenario involves a DLC placed in a colocation cage in an ILEC's central office by a CLEC. The DLC is mounted in a standard 7 ft. relay rack and powered
using central office DC power. The subscriber wiring either connects onto a termination frame within the cage or is directly connected to a POT—point of termination—
frame owned by the ILEC. From the POT frame, cable is run to blocks on the horizontal
side of the MDF. When an order is issued by the CLEC to the ILEC for service to such
and such an address, frame wire is used to connect the block on the horizontal side of
the frame to the outside cable location on the vertical side of the frame. Dial tone travels from the CLEC's switch sitting at a remote switching center across interoffice
facilities—called IOFs—to the T1 high-speed card in the DLC sitting in the cage in the
ILEC's CO. The DLC demuxes the T1 signal and directs each channel to its assigned

Figure 17–7
The telephone access circuit travels from the switch across the MDF where it is connected to the outside plant cable. The outside plant cable terminates at the customer premises on a network interface device. Inside wire is the final leg of the circuit.

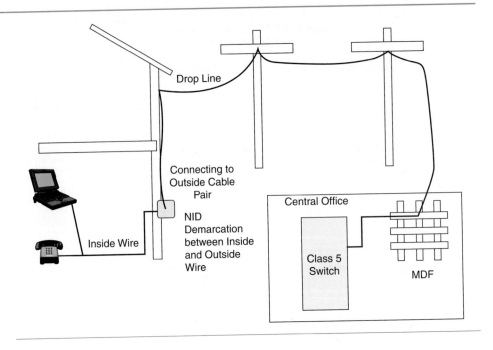

channel card where it is fed onto the twisted pair wire connected to the CLEC frame in the cage. From the cage, the signal travels to the ILEC MDF where it is tied via frame wire to the correct outside plant pair. From the MDF, the signal travels through the CO and across the feeder/distribution plant to the subscriber's telephone. Figure 17–7 illustrates the end-to-end connections we have described.

■ 17.2 ISDN SERVICE

Integrated services digital network has been around since the early 1990s. ISDN is the first and only truly digital switched service offered by the telephone companies. The most important reason ISDN was developed was to extend the digital network and out-of-band signaling capabilities to the end subscriber. Providing a switched digital circuit all the way to the customer premises provides several advantages to the subscriber. One of course is additional bandwidth as an ISDN BRI—Basic Rate Interface—line carries two 64 kbps channels called *B* or *bearer* and is capable of carrying either voice or data. Within the digital ISDN signal, a *D* or *data* channel is configured to carry signaling information back to the switch making the ISDN telephone a manageable device on the network, a real plus for telephone companies. Therefore, an ISDN BRI signal consists of two 64 kbps bearer channels and one 16 kbps data channel equaling a line speed of 160 kbps of bandwidth.

As traditional POTS was a switched analog line technology, ISDN was posed to move in and completely digitize the local loop. Unfortunately, it did not make the inroads expected of it and, consequently, became a technology used by specific applications only. Not that ISDN faded into oblivion—it just did not take over the access network as predicted. Even though it does not dominate the access network, today ISDN BRI plays a substantial role as it is used to connect small businesses giving them special features such as four-digit dialing, multiple call appearances, and so forth.

Switched digital services is a term used often to describe ISDN services. Switched refers to a service that feeds into the PSTN—Public Switched Telephone Network—where it is routed according to a telephone number. ISDN was not the first digital service offered to the end customer. AT&T's 56 kbps private line digital data service—DDS—was the first digital circuit sold to end customers and offered a point-to-point or point-to-multipoint connection that did not interface into the voice switch.

bearer
Term used to describe the B channels of the ISDN BRI circuit. The channels bear, or carry, information signals such as voice or data.

data
Term used to describe the D channels of the ISDN BRI circuit. The D channel carries both signaling and low-speed data.

switched digital services
Term used to describe ISDN services.

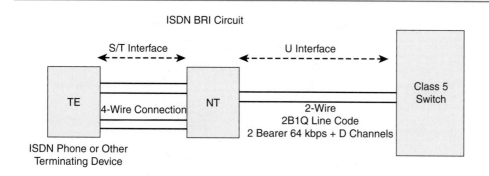

Figure 17–8
Standard ISDN BRI interface includes a two-wire connection between the switch and the customer premises and a four-wire connection between the NT and the end terminating device.

17.2.1 ISDN Basic Rate Interface

ISDN BRI services include an ISDN port on the switch, a two-wire connection from the central office to the subscriber, an ISDN interface device called an *NT-1*, and an ISDN terminating device such as an ISDN telephone. The standards bodies have defined interfaces across the BRI connection, as shown in Figure 17–8. As shown in the diagram, the connection from the switch to the NT-1 at the customer premises is referred to as the *U interface* and requires a two-wire handoff at the NT-1. This, of course, is the same copper pair used by POTS minus bridged taps and load coils, as explained in Chapter 1. The NT-1 receives the 160 kbps signal from the line, processes it, and passes it on to the ISDN device via a four-wire connection. A two- to four-wire conversion happens inside the NT-1. Connecting the NT-1 and the ISDN device called a TE1—Terminal Equipment Type 1—is referred to as the S/T interface and is a 192 kbps four-wire connection. When an end device is not ISDN capable, a unit called a *terminal adapter (TA)* is used to convert the S/T interface into the appropriate format. The term TE2 is used to describe a non-ISDN-capable device. The RJ-45 jack is used to terminate the ISDN four-wire and two-wire interface, as shown in Figure 17–9.

Though the U interface carries a 160 kbps signal, the available bandwidth to the user is only 144 kbps due to a 16 kbps overhead tax. Similar to the U interface, the 192 kbps line speed of the S/T interface includes 144 kbps of bandwidth and 48 kbps of overhead. The purpose of each interface definition is to help technicians sectionalize a trouble as all ISDN test sets have both U and S/T test capability.

NT-1
Network termination 1. Used to connect the ISDN device to the network.

terminal adapter (TA)
A device used to convert an ISDN digital signal into an analog signal.

Equipment

Equipment used in an ISDN BRI circuit includes the devices we have mentioned: the NT-1, the TE1, and the TA. Today, many telephone sets are built with integrated NT-1s. A TA is also often integrated into the device or into a special card in the device. ISDN telephone sets vary dramatically and offer many features such as caller ID, company rolodex, stored profiles, selective ringing, call blocking, and so forth. Though features are numerous, the cost for an ISDN telephone set is still higher than that of a traditional analog telephone set. The savings though, comes from being able to use the call features to help improve the efficiency of the office.

Routers have ISDN WAN cards that are used for circuit redundancy. If the T1 circuit fails, the router initiates a call across the network through the ISDN WAN card and establishes a connection on the PSTN as configured by the routing tables in the router. Traffic is shipped across the ISDN line helping to keep the network up during a network outage. Other devices used to connect multiple data users to ISDN lines are available and are used primarily by small businesses employing a handful of workers.

Physical Plant Infrastructure

The two-wire pair connecting the central office and the customer premises requires that the line is conditioned for digital services, meaning that the bridged taps and load coils are removed as mentioned earlier. In addition, it is important to remember that because

Figure 17–9
Wiring the T interface and U interface to standard RJ-45 jacks (may be terminated on a standard RJ-11 jack).

T Interface: Four-Wire Connection Plus Power

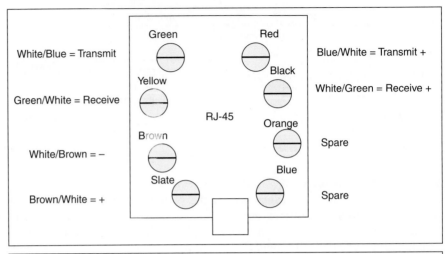

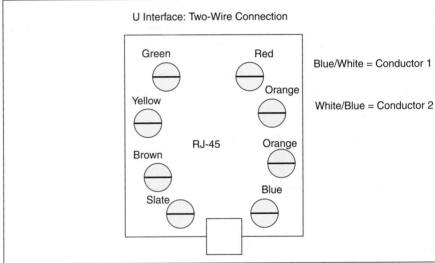

ISDN runs on a higher frequency than traditional voice services, the signal can travel only 18,000 ft. on twenty-six-gauge wire before having to be repeated. A repeater extends the distance the signal travels by about 16,000 ft. Therefore, even though the ISDN line is able to use the existing outside plant copper cable infrastructure, special consideration has to be made before turning up service. In fact, due to the number of load coils and bridged taps in the network, many ILECs assign the task of conditioning whole groups of lines to special crews. Even with the push to condition multiple cables for digital services, turning up an ISDN line typically takes longer than turning up a POTS line.

Line Coding Method: 2B1Q

In order for the two-wire circuit to carry the 160 kbps of information—remember, the voice line was designed to carry frequencies between 300 and 3000 Hz—a new line-coding scheme was devised called 2 Binary 1 Quaternary or 2B1Q. Unlike AMI line coding that carries one bit per signal change, 2B1Q carries two bits per amplitude and polarity change, as discussed in Chapter 4. A result is an increase in the bit-per-second rate made possible thanks to the ingenious method of manipulating the amplitude and polarity of the signal. Additionally, the 2B1Q format uses frequencies from 0 to 392 kHz, a much wider spectrum than the traditional POTS line.

Powering the Service

Powering an ISDN terminating device differs from powering the standard telephone set that is energized whenever the handset is picked up, that is, energized by the switch. ISDN devices must be plugged into the AC outlets at the customer premises. If power fails and there is no battery backup, telephone service is lost. Though designers saw this as a minor concern, consumers felt differently. Today, small businesses have battery backup for their services, thus reducing the problem of having to provide backup power for ISDN equipment. Residential subscribers have yet to migrate in large numbers to ISDN services; therefore, the power issue is less of a problem.

Inside Building Wiring

Connecting the outside protector to the inside ISDN jack is simple in that it is a standard, two-wire RJ-11 connection. Once the signal passes through the NT-1 and is converted into a four-wire circuit, jack wiring becomes trickier. Often, when trouble occurs during turnup, improper wiring of the four-wire connection at the customer premises is the cause. A standard RJ-45 connector is used to terminate the four-wire connection between the NT and the TE. The RJ-45 connector is the standard used on all ISDN devices.

When multiple ISDN circuits are terminated at one site, passive connection blocks such as 66 blocks or 110 blocks, as discussed in an earlier chapter, are used. Note that the NT-1s are grouped together in a shelf in the telephone room and the four-wire S/T connections are distributed to the termination blocks that are used as cross-connection points to the various offices or workstations. Category 3 or 5 cable is run to each ISDN jack. The ISDN device connects to the termination jack using an RJ-45 connector, as already mentioned. The variety of ISDN jacks is endless, some using punch-down connectors, some screws. Fortunately, though, all use the same color code and wiring schematic. Companies find it much simpler to distribute power from the telephone room using the brown/white pair.

Interoffice Connection

As ISDN BRI service requires the class 5 switch to be ISDN capable, offering ISDN service to entire metropolitan areas requires each switch to be upgraded to handle ISDN circuits. Unfortunately, this involves not only a software upgrade but also a hardware service module upgrade. Consequently, because of the cost, many smaller central office switches are not upgraded. Customer circuits fed by the smaller switch must be trunked back to an ISDN-capable switch in the network via T1 lines. In order to accomplish this, the customer's ISDN line is fed into BRITE—Basic Rate Interface Transmission Extension—cards residing in a channel bank and multiplexed into a T1 circuit. The ISDN line feeding the end customer takes up three of the twenty-four slots in the channel bank, though only one channel card is required. Dummy cards are placed in the adjacent two slots to prevent a technician from accidentally populating that slot with another service card. The ISDN BRI signal is multiplexed into the T1 circuit that connects the two offices together. At the ISDN-capable office, the T1 again terminates into a BRITE card in a channel bank in that office. From the channel card, the ISDN signal travels on inside plant wires that terminate on a block on the MDF. An orange/white frame wire connects the channel bank block on the MDF to wire wrap pins, tip and ring, on the switch block also mounted on the frame. From the switch block, the signal travels across wiring that connects the block to an ISDN port in the switch. It is important to point out that the BRI between the channel bank and the switch is a two-wire connection using the tip and ring designations. Throughout the local network, many channel banks are populated with BRITE cards used to carry ISDN signals to a particular switch. In our example, we followed an ISDN signal from a customer through a non-ISDN-capable central office to an office with an ISDN-configured switch. A second example of using trunking between switches to carry ISDN BRI signals is when ISDN Centrex is deployed and the business wants to offer four-digit dialing similar to a PBX at all of its locations. This requires that locations fed out of a different switch than that of the main headquarters be tied via trunking circuits as just explained in order to receive ISDN lines with compatible telephone NXXs.

Services

ISDN BRI circuits are used for multiple services including data, voice, video conferencing, and backup WAN links. Small business customers are able to connect several phones to one outgoing ISDN BRI line, thus reducing the cost for telecommunications services. *Multiple call appearance* is a feature of ISDN that allows multiple numbers per BRI line, though technically only two users may talk to the outside world at any one time because there are only two bearer channels available. Business owners calculate the number of incoming and outgoing calls to determine the number of users they dare place on one ISDN circuit.

A second service offered by ISDN BRI lines is the ability to carry both voice and data on one connection. A real estate agent provides a good example of a user that takes advantage of this service. The real estate agent may speak to a client using the ISDN telephone set at the same time as accessing information from a file server located in a different office using the data channel. The ability to talk on the telephone and network using the PC is one of the key advantages of ISDN, especially for people working at home. The advent of DSL and cable modems has slowed the growth of ISDN in this arena as DSL and cable provide a good alternative to data access without having to tie up the voice line.

Video conferencing is one of the most popular uses of ISDN BRI circuits and, thanks to its switched capability, will continue to be used to provide these services. In addition, WAN backup has become one of the most common deployments of ISDN BRI as an alternative to purchasing a second T1 between sites. The customer pays the monthly cost for the ISDN circuit and a minute-per-use charge only when the circuit is initiated.

17.2.2 ISDN Primary Rate Interface

The second type of ISDN circuit found in the access network is the ISDN PRI—Primary Rate Interface—circuit that carries twenty-three 64 kbps channels and one 64 kbps D channel. As with BRI circuits, the D channel is used to carry out-of-band signaling information, thus giving it many more capabilities than a traditional T1. Today, ISDN PRI circuits are used to connect PBXs to the central office switch, ISPs to the central office switch, and intracompany networks to the central office switch. ISDN PRI services are explained in detail later in the chapter.

■ 17.3 SPECIAL SERVICES CIRCUITS

An explanation of the access network would not be complete without talking about the various types of special services circuits. Though today seldom sold, several continue to be used as they provide cost-effective methods for carrying information. All special services circuits are used by business customers and for many years have been the bread and butter of the local telephone companies. In fact, special services circuits subsidized residential services and, as such, became the core moneymakers for the phone companies. Today, the most common special service circuits in the access network are 56 kbps digital data; OPX—Off-Premise Extension; FX—foreign exchange; and analog trunks—DID and E&M, tie lines. In the past, analog data circuits were popular and very prevalent in the business community. Today, with the advent of digital technologies, analog data circuits, though still found in the local network, are being replaced by the newer technologies and therefore are not discussed.

17.3.1 56 kbps Digital Data Circuits

AT&T introduced DDS—digital data service—in the late 1970s as a private line service capable of carrying 56 kbps of information. As a private line service, DDS was offered to business customers that wanted a connection between two or more of their

multiple call appearance
ISDN offers the subscriber multiple numbers per ISDN circuit.

OPX
Off-Premise Extension. A circuit that extends a line from a PBX to a remote location.

locations. Two types of 56 kbps circuits were available: point to point or multipoint. Today, 56 kbps service is still used and fairly popular with business customers.

A 56 kbps private line service requires a terminating device at each of the customer locations called a *DSU/CSU*—data service unit/channel service unit. A 56 kbps line is a four-wire circuit that rides on standard POTS lines between the central office switch and the customer's premises. Though "56 kbps digital circuit" is used to describe the service, lower speeds are also sold under the same name using the same devices. The traditional speeds sold by most service providers are 2.4, 4.8, 9.6, 19.2, and 56 kbps.

Digital Data Network Architecture

As shown in Figure 17–10, the point-to-point service does not travel through the switch at either of the central offices—thus the term *private line;* only the customer's traffic is allowed on the circuit. The four-wire connection at the customer location is designated by a transmit pair and a receive pair that carry the AMI line-coded signal to the central office. As with all higher-speed signals, the distance allowed between sites is limited. In the case of digital data circuits, the signal distance relates directly to the bit-per-second rate.

As with ISDN BRI circuits, the digital 56 kbps circuit often feeds into a channel bank in the central office and is multiplexed into a T1 circuit that connects to a channel bank in a different central office. The network architecture must include this step if the two locations are fed out of two different offices. Therefore, the most common digital private line service rides in a T1 channel through the local telephone company's network. For example, a four-wire digital 56 kbps line attached to the subscriber's location terminates on the vertical side of the MDF in the central office. A red/white frame jumper is used to connect the termination on the vertical side to a special service block on the horizontal side of the MDF. The special services block is hardwired directly to a channel bank sitting in the special services area of the central office using wire wrap pins at the block on the MDF and an amphenol connector on the back of the channel bank. The signal enters into a digital data channel card that formats it and passes it on to the multiplexer that grooms it into one of the twenty-four channels in the T1. In most instances, the T1 is fed into a higher-order multiplexer—probably a fiber optic MUX—that carries it to the far end central office where it demultiplexes it and feeds it to the channel bank. Thus, two back-to-back channel banks are connected via the T1. The digital data circuit is demultiplexed in the channel bank and passed on to a channel card that sends it out across a four-wire circuit connected to a block on the

DSU/CSU
Data service unit/channel service unit is the device that interfaces with the network and the communications equipment.

private line
Refers to a local special services circuit that does not go through the switch.

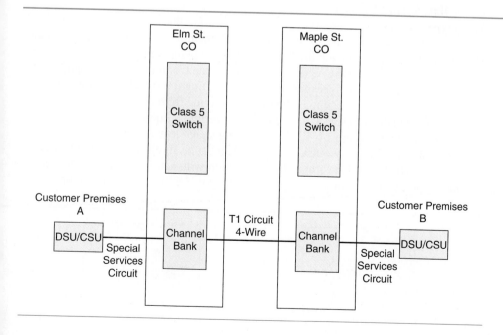

Figure 17–10
Special services circuits travel through the serving central office where they are fed into devices called channel banks. The circuits do not interface into the class 5 switch.

Figure 17–11
Often banks purchase special services circuits for each branch. In our example, the 56 kbps circuits feed back to a central hub creating a hub and spoke architecture.

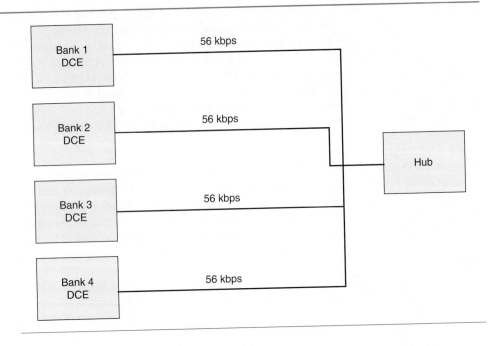

multipoint topology
Several locations are fed by the same circuit. Each location is tied back to a central hub through a shared circuit. Banks connecting to a main branch site is a common multipoint circuit application.

DCE
Data Communications Equipment. The generic term used in the industry to describe the interface device used to interface the network to the end termination equipment or DTE.

DTE
Data Terminal Equipment. The generic term used in the industry to describe the end computing equipment such as a server or a computer.

horizontal side of the MDF. Again, red/white frame wire is wire wrapped to the pins on the block on the horizontal side and the termination point on the vertical side of the frame. The inside plant cable terminated on the vertical side of the frame snakes its way to the cable vault where it is spliced into outside plant cable that traverses the outside plant and terminates the four-wire connection at the customer's location.

Many digital data circuits are configured in a *multipoint topology*, which means that several locations are fed by the same circuit. Two types of multipoint circuits are found in the local network. The older of the two is a hub-and-spoke architecture, in which one location is the hub that feeds many smaller locations, as shown in Figure 17–11. Banks often deploy multipoint hub networks that are used to connect the headquarters' mainframe to ATM or branch offices. A second type of multipoint network deploys a fully meshed architecture allowing two-way communications between different sites, not just the hub and the remote.

Digital Data Service Equipment

Many types of equipment are found in a digital private line service, as well as a wide variety of interface types that are used to connect the devices. As mentioned earlier, a DSU/CSU is used to terminate the four-wire outside plant circuit at the customer site. In the world of data, the DSU/CSU is the *DCE*—Data Communications Equipment— of the circuit and, as such, is responsible for formatting the signal according to the direction it is traveling. For example, the incoming signal from the central office must be passed from the DSU or DCE to the *DTE*—Data Terminal Equipment—via a V.35 connection, as shown in Figure 17–12. The outgoing signal leaving the DSU/CSU and interfacing onto the four-wire outside plant cable must be shipped at a specified decibel level in an AMI line-coded format. In conjunction with acting as liaison between the outside world and the terminating equipment, the DSU/CSU may also be looped back

Figure 17–12
A DCE, which may be a DSU/CSU, typically connects to DTE via V.35 interface.

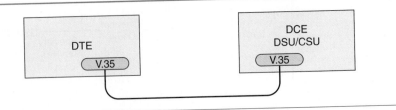

offering a troubleshooting point in the network. Sophisticated DSUs/CSUs are capable of dropping/adding DS-0 circuits, making them very useful to the end user.

As mentioned earlier, a channel bank is often used when connecting two remote locations using a digital private line service. Digital data channel cards are placed in the channel bank and configured according to the speed of the circuit. Beyond a channel bank, many local networks include a 1/0 DCS that may terminate a 56 kbps private line circuit using a 56 kbps channel card or, as in most cases, groom the DS-0 carrying the private line service to the appropriate T1 as explained in an earlier chapter. Many DCSs are able to perform bridging functions and are often used for video conferences or other services requiring bridging.

Commonly, digital private line services are sold to business customers who want to connect two or more locations using a private link that offers security and cost savings. As the charges for minutes of use compute on dial-up lines, many business customers find it easy to cost justify implementing digital private line services between locations. It should be noted that private line services are not restricted to the local areas, as long distance companies sell hundreds each year; however, each still rides that last mile on the local provider's access network.

17.3.2 Off-Premise Extension Circuits

Amazingly, through all of the network upgrades, the introduction of DSL, Ethernet, and all of the other cutting-edge technologies, the simple analog OPX—Off-Premise Extension—circuit is still being deployed in the local network. The OPX offers a way for customers to extend the reach of their PBX without having to deploy an additional PBX or key system at the remote site. One of the most common examples is a school with a PBX sitting in the main office and an OPX line extending a station connection to a bus garage sitting some distance from the school. An OPX line simply connects a PBX station port—the port the telephone hangs off—to a two-wire outside plant connection tied to a telephone sitting in a remote location.

The OPX line consists of a two-wire connection from the cable termination blocks, such as a 66 block, used as a cross-connect point between the PBX and the outside plant cable, as shown in Figure 17–13. The two-wire connection feeds back to the central office via the outside plant infrastructure and terminates on the vertical side of the frame. A red/white frame wire is used to connect the point on the vertical side of the frame to a special service block mounted on the horizontal side of the frame. The block on the horizontal side of the frame is connected via inside cabling to a shelf sitting in the special service area of the central office. An OPX card in the shelf amplifies the signal if needed and passes it back out onto a two-wire pair that again is connected to the same special services block on the MDF. A red/white frame wire connects to the vertical side of the frame where the outside plant pair feeding the remote site terminates.

Troubleshooting an OPX circuit requires an understanding of where the dial tone is coming from and in what direction the signal is traveling. For example, dial tone at

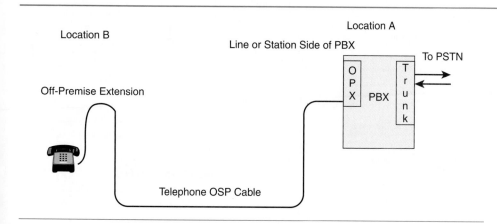

Location B

Location A

Line or Station Side of PBX

Off-Premise Extension

To PSTN

OPX PBX Trunk

Telephone OSP Cable

Figure 17–13
An OPX line is fed out of an OPX or station side port and fed back to the OSP—outside plant—cable feeding a location down the street or in the next lot or so forth. If the distance between the PBX and the remote location is too great, the OPX line will ride back into the CO where it is amplified and tied to the end station.

the remote station is generated by the PBX. Often, technicians become confused and turned around when trying to troubleshoot an OPX circuit. Technically, they are fairly simple in that they are POTS lines fed out of a PBX, not a central office switch. The signal carried across the line is the same as that carried on a traditional POTS circuit, and the loop resistance is typically much lower than that of a central office switch, around 850 ohms.

17.3.3 Foreign Exchange Lines

If this text were being written twenty years ago, the section on FX—foreign exchange—lines would be at least five times longer. FX lines are deployed much less often thanks to the call-forwarding feature on the switch. An FX line is used to extend a customer's telephone service from a different switch. For example, a customer that is served out of the Maple Avenue switch with an exchange code of 232 wants a telephone line terminated with the exchange code from a different central office such as the Walnut Street office with an exchange code of 545. The subscriber wants its customers to be able to dial the 545 number and reach the location fed by the 232 Maple Avenue exchange. In order to do this, the FX circuit is nailed up between the two offices and connected to the wire pair feeding the customer's Maple Avenue location. When the handset is picked up, dial tone is being generated by the Walnut Street CO, not the Maple Avenue CO.

Connecting the circuit between the two sites requires a channel bank at each office with an FXO (office) channel card at the office end and an FXS (subscriber) channel card at the station end—the office serving the customer. The line feeding out of the switch terminates as usual on wire-wrap pins on the block mounted on the horizontal side of the MDF. The block is connected via frame wire to the special service block mounted on the MDF, as shown in Figure 17–14. Cabling connects the special service block on the MDF to the channel bank located in the special service section of the CO. An FXO channel card is placed in the channel bank and is used to connect to the switch on one side and pass the signal on to the multiplexer in the channel bank on the other. The multiplexer grooms the FX DS-O into the T1 circuit that connects to the channel bank in the Maple Street central office where the T1 signal demultiplexes out the DS-0 channel and feeds it to the FXS channel cards in the channel bank. A two-wire connection extends up from the channel card to a special service block on the MDF in the Maple Street office. A frame wire is used to connect the wires on the special service

Figure 17–14
Special services circuits terminate on the vertical side of the MDF and are cross connected via frame wire to a channel bank or other transmission equipment.

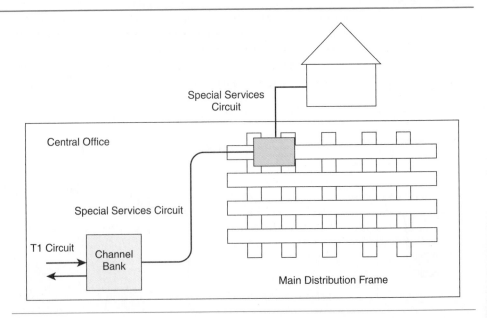

block to the customer's wire pair on the vertical side of the frame. From the vertical side of the frame, the signal travels across a two-wire voice grade line to the customer's location.

17.3.4 Analog Trunks
Signaling Methods

Before discussing analog trunks, it is important that we first explain the different types of signaling used on copper analog circuits to establish service between the switch and the end device. The three signaling methods to be discussed are loop start, ground start, and E&M signaling. Loop start signaling is the most commonly deployed method in the access network as it is used on all voice grade POTS lines. *Loop start* refers to a signaling method that interprets the closed loop condition caused when a telephone handset is lifted, or a PBX places a 600-ohm load, or a telephone set provides a 600-ohm load across the tip and ring conductors. Once the switch senses the loop closure, it sends dial tone and marks the line as busy. An idle loop start line has 0 V on the tip side of the line and –48 V on the ring. When a call is coming in, the ring side of the circuit carries a 100 V, 20 Hz ringing signal. Therefore, if you are a cable splicer sitting in a nice wet manhole splicing a bunch of wires together, you will instantly feel the shock of the 100 V, 20 Hz signal if you accidentally touch the copper ring conductor at the same time ringing occurs on that line.

Ground start signaling is the second most popular signaling method found in the access portion of the network. Many PBXs, coin phones, and channel banks use ground start signaling to establish calls. Ground start signaling requires the PBX or end device to place a ground on the ring lead for an instant to signal to the switch that it wants to receive dial tone. The switch immediately returns a ground on the tip lead letting the PBX know that the request was received. Once the PBX has received the ground on the tip lead, it completes the loop by placing a 600-ohm load between the tip and ring similar to loop start signaling. Often, technicians troubleshooting lines in the access network forget to check the line for ground start signaling. They attach their test sets to the line and listen for dial tone. If they fail to receive dial tone, they may assume the line is bad between the point that they are testing and the central office. The result is they waste time troubleshooting the wrong section of the line. Good records help eliminate this problem, though it still occurs with some regularity as good records are often not available.

The E&M signaling method is used to establish service between two switches, primarily two PBXs connected by tie lines. A *tie line* is nothing more than a private line circuit used to connect two PBXs together. *E&M* is the signaling method used by the two switches to establish communications such as setting up and tearing down calls. E&M is a much more complicated signaling method than ground start signaling, for example, as it requires an additional two-wire pair to carry signaling information between the switches in addition to the wire pair or pairs—there are two-wire and four-wire E&M tie lines—used as voice trunks. Similar to ground start and loop start signaling, ground, battery, and current flow are the signaling variables used to establish a connection between the two devices. If, for example, a call is made from a station attached to PBX A to a station attached to PBX B, the following steps would occur: When station A goes off-hook, it receives dial tone from PBX A. The user dials the outgoing call code—such as 9—followed by the four-digit station number of station B—9-2345. PBX A's immediate reaction once it receives a 9 is to ask for a free trunk between itself and PBX B. Since the M lead sends signals and the E lead receives signals, PBX A sends the signal requesting service on the M lead, which is received on PBX B's E lead. PBX B sends a –48 V signal on its M lead to PBX A's E lead notifying it that it has received its request and that it is available to carry the call, as discussed in Chapter 10. Dial tone and digits are carried across the voice pair, not the E&M signaling leads. Only battery and ground are carried across the signaling leads for the purpose of establishing a connection. In the past, rotary telephone pulses did travel across

loop start
Loop start signaling happens when the loop changes state: loop is open, no dial tone. Handset is picked up; the circuit closes and initiates dial tone.

tie line
A circuit that connects two switches together—typically, two PBX switches.

E&M
Signaling method used between two switches.

the M lead as they contain only two states, on or off. Several types of E&M signaling are available. Before troubleshooting or turning up a new service using E&M signaling, it is wise to refer to the equipment manual for insight into how the leads are used and connected.

DID Trunks

DID trunks
Direct inward dialing trunks are used for incoming calls to a PBX, key system, or channel bank from the serving switch.

Direct inward dialing (DID) trunks are very common in the access network. *DID trunks* allow calls to flow from the PSTN to customer premises—inward to the customer. If calls coming into the business from the outside world are fewer than the number of employee telephone stations, then DID trunks provide an economical way to extend voice services in a business. Using traffic analysis, the end subscriber determines how many actual lines are needed to service the incoming call volume and purchases that many lines. Telephone companies sell blocks of numbers such as 344-1000 to 1200 to DID trunk subscribers. Typically, they are sold in blocks of 100 numbers that are then assigned by the PBX to one of the end stations. For example, if a company has 1000 employees, each with his or her own telephone set, the business owner could purchase 100 DID trunks and ten blocks of 100 numbers equaling 1000 telephone numbers. The end user has determined that at any one time only 100 incoming calls occur and therefore he only needs 100 DID trunks.

A DID trunk rides on the same copper two-wire connection as a traditional voice grade line. Different though from a voice grade line, a DID trunk is configured through software in the switch as a trunk, a switch-to-switch connection. A DID connection is initiated when an incoming call is received by the central office switch for one of the numbers assigned to the DID trunk. The switch immediately completes the loop by placing a conductor across the pair; and, as with loop start signaling, current flows through the loop to the PBX notifying it that a call is coming in. The PBX responds to the central office switch with a "wink," which is nothing more than a quarter-second reversal in polarity used to tell the switch that it has registers available and is ready to accept the four-digit number. The PBX directs the call to the correct end station and rings the telephone set. Once the station is picked up, the PBX again reverses the polarity signaling the switch that the call has gone through. The central office switch uses this polarity reversal as notification to start billing the line for usage.

DOD Trunks

DOD trunks
Direct outward dialing trunks are used for outgoing calls from a PBX, key system, or channel bank to the serving switch.

Similar to DID trunks, *DOD*—direct outward dialing—*trunks* are used to connect a customer's PBX directly to the end-office switch. The number of DOD trunks deployed seldom equals the actual number of users depending on the volume of outgoing calls. The value of having fewer trunks than users is obvious: the fewer trunks, the less cost. A switch allows a trunk to be configured as either a DID, a DOD, or two way. If the trunk is optioned as a DOD, incoming calls are not allowed on the trunk as is true for DID trunks in the reverse.

■ 17.4 BROADBAND SERVICES

broadband services
A high-speed circuit used to carry traffic coming from and going to the Internet. DSL, cable modem, and so forth are considered broadband services.

The ILEC and the CATV provider offer *broadband services* to both the residential and business customers. CATV providers carry the broadband signal on the coaxial cable feeding the customer's CATV box. The ILEC uses the twisted copper pair feeding the analog telephone set to carry the DSL—digital subscriber line—broadband signal. In either case, the end user is given a pipe that carries megabits per second of information into the home. Today, many ILECs are replacing the twisted copper pair into the home with fiber connections. Regardless of the type of pipe connected to the end user, the signal carrying the information remains the same; it is only the bandwidth or information carrying capacity that changes.

The third provider of broadband access technology is the cellular service provider. Cell companies or PCSs, depending on how they define themselves, have deployed broadband wireless technologies, a topic discussed in a later chapter, capable of providing broadband connections to the device, not just the residence, across the providers' footprint. The advantage is obvious. A businessperson on the road may have access to a broadband pipe allowing him or her to connect at high speeds using the wireless adapter card in the PC.

The final broadband technology in the market is the *Wi-Fi* or *WiMAX* service provider. Wi-Fi, a wireless technology, offers the consumer a broadband pipe to connect to the Internet—similar to CATV and the ILEC. The difference is that the Wi-Fi provider's signal rides on airwaves to the customer's location, eliminating wire-line access into the premises.

17.4.1 Digital Subscriber Line Services
DSL Beginnings

From the beginning, service providers and equipment manufacturers have been driven to develop a high-speed signal capable of traveling across the existing copper infrastructure. Finally, in the early 1990s, HDSL—high bit-rate digital subscriber line—was introduced as a way to carry the 1.544 Mbps T1 signal over the traditional voice grade infrastructure. HDSL, using the 2B1Q line-coding format, was able to transmit a full T1 across a four-wire twenty-six-gauge cable 12,000 ft. Once HDSL was introduced and proved to be successful, engineers and entrepreneurs started to work on other line-coding schemes that would open up new high-speed services, coined broadband services.

As technologists studied the effects of signal transmission from a central office to the end customer over copper wire, it was noted that a signal was stronger traveling from the central office to the end location than in the reverse, the end location to the central office. From this data, it was surmised that a higher bit rate signal could travel from the CO to the subscriber and a lower bit rate signal from the subscriber to the CO, resulting in an asymmetrical service—soon named ADSL—Asymmetric DSL. Luckily, traffic flow on Internet circuits fit this architecture—faster on the download, slower on the upload, thus resulting in a product that accommodated the residential Internet user. The maximum downstream rate was set at 6 Mbps, while the upstream speed maxed out at 512 kbps. Once ADSL was established as a viable service offering, an additional feature was added that allowed for the signal speed to vary or rate adapt to the conditions of the line. As we all are aware, the outside plant copper infrastructure is comprised of various cable gauges and other anomalies that cause the signal to lose strength or attenuate at different rates depending on the conditions of the connection. Therefore, the introduction of RADSL—Rate Adaptive DSL—allowed this new service to be offered to a much larger population.

The second type of DSL service developed from the need for symmetrical services—the same speed on the upstream and the downstream path—for business customers. The result was the introduction of SDSL—Symmetric DSL—which became the DSL product offered by CLECs and DLECs in the late 1990s and early 2000s. SDSL, similar to a T1 circuit, offers a maximum line rate of 1.544 Mbps on both the upstream and the downstream circuit. Similar to RADSL, SDSL was also deployed as a rate-adaptive service by most vendors, allowing the line rate to adjust to the condition of the circuit. In addition, service vendors built in specific rate levels that could be sold accordingly. For example, many service providers offered rates of 64, 128, 256, and 384 kbps and 1.544 Mbps. Of course, if the distance between the CO and the end user exceeds the allowed loss budget, the speed of the circuit must be adjusted accordingly: lower speed, less attenuation, greater distance.

Several variants of SDSL and ADSL have been introduced into the marketplace. DSL's competitor is the cable modem offered by CATV companies. Though in many instances business customers, due to security issues, prefer DSL, residential users have readily chosen cable modem service for Internet access even though DSL continues to be a good service offering and, as such, must be included in local service providers' portfolios.

Wi-Fi
Wireless Fidelity is a wireless technology that allows the subscriber to surf the Internet from anywhere within the Wi-Fi footprint, also called *hot spot.*

WiMAX
Wireless technology that is used to tie Wi-Fi hot spots to the network.

line code
Term used to describe how
the bits are formatted on the
line.

Line Code

In the beginning if the words *line code* were uttered in the presence of DSL advocates, it was certain that a heated argument would ensue. Two line-code standards were put forth during the initial phase of DSL deployment—DMT and CAP, segregating the DSL advocates into two very distinct camps. As DMT—discrete multitone—was the first technology to prove that it could carry the required 6 Mbps signal on the downstream side of the circuit, it won the line-code battle and became the standard in the industry. Subsequently, the CAP—carrierless amplitude phase—line code excelled in several key areas, making it a de facto standard accepted in conjunction with DMT.

As discussed in Chapter 4, DMT and CAP line-code modulation techniques manipulate bits across the frequency spectrum. CAP's line code sends two to nine bits per baud by manipulating the amplitude and phase of the signal, similar to the method used in QAM—quadrature amplitude modulation. Constellations surrounding the core frequencies are what allow CAP to modulate so many bits per baud, as discussed in Chapter 4. DMT uses discreet frequencies divided into evenly dispersed bands, each capable of carrying multiple bits per signaling rate or baud. Most equipment manufacturers, exhausted by the line-code argument, have forfeited the war and are producing DMT- and CAP-capable circuit packs. A third line code called MVL—Multiple Virtual Line—developed by the Paradyne Corporation is also deployed in the network even though it is a proprietary line code owned by the vendor. MVL is able to carry a signal much greater distances.

Equipment and Connections

Though line code makes a nice story to tell your grandchildren, understanding how a DSL connects into the access network is more important if you are an engineer or technician working for a service provider. A DSLAM—digital subscriber line access multiplexer—is the equipment used to aggregate DSL circuits and multiplex them into a higher-speed transport pipe such as DS-1s, DS-3s, or OC-Ns. The DSLAM may be segregated into two sides; one aggregates customer DSL circuits, and the other multiplexes the DSL into a transport signal that is sent out onto the high-speed circuit interfacing the network. Therefore, a DSLAM has a line side—DSL circuits—and a network side—transport circuit.

DSL circuit packs are built to modulate the signal according to the specified line code. For example, a DSLAM may feed fifty residential customers using ADSL circuit packs and 100 business customers using SDSL circuit packs. The first circuit pack may provide SDSL service and the second ADSL, and the third HDSL, and so forth. The DSLAM has to be very flexible in both the types of DSL it supports and the number of ports available per chassis, as competition between DSL vendors is fierce.

The opposite side of the DSLAM, the network side, is responsible for packaging the hundreds of DSL signals into a transport protocol such as frame relay or ATM. In the beginning of DSL deployment, the transport methodology wars were as fierce as the line-code wars. Each vendor struggled to prove that its design was superior to that of its competitors. The result was a combination of strategies that merged into a fairly standard architecture. ATM proved to be the most popular aggregation method due to its ability to carry VoDSL.

The size of the pipe used to connect the DSLAM to the main routing center also caused vendor disputes. Initially DSLAMs were built with DS-3 and OC-3c interfaces. Soon it was realized that *DLECs* and CLECs had to lease the IOFs—interoffice facilities—used to connect the DSLAM to the switching center. Because most service providers oversubscribed their lines, it was unnecessary to purchase big pipes such as DS-3s or OC-Ns to haul the traffic back to the RSC—Regional Switching Center. Some DSLAM vendors were lucky enough to have built a T1 interface that helped lower the transport costs, especially when the equipment sat in a small office. Today, all DSLAMs offer multiple physical interfaces that allow the service provider to provision a lower-speed circuit initially and grow as its customer base grows.

Because DSL service is statistically multiplexed onto the upstream transport circuit, oversubscription must be balanced with enough bandwidth to handle the traffic flows. Once congestion happens, the service provider must upgrade the network link

DLEC
Data local exchange carrier.
A data-only competitive
carrier in the local network.

to accommodate the increase in traffic flow. Logically, traffic monitoring is essential, as slow service is a sure way to lose customers. Most DSLAMs have built-in traffic monitors using the ATM protocol QoS features to notify them when cells are discarded and congestion occurs.

Most DSLAMs are found in the central office close to the subscriber's access circuit. Though the size of the DSLAMs varies according to the vendor, most fit easily in half of a 7 ft. relay rack and are able to terminate as many as 400 subscribers. What is the power drain? That is one of the first questions a designer should ask a DSLAM vendor, as some units require high-amperage feeds in order to support the high volume of DSLs. As DSLAMs provide voice services, it is also important to make sure the device works with A and B load sharing power.

Subscriber Management System

Equipment placed in the RSC or main data hub site is responsible for performing Layer 3 routing of DSL traffic to the appropriate destination. A device built specifically to terminate DSL traffic, the SMS—Subscriber Management System—sits in the data-switching center and is connected to all of the DSLAMs in the network. The job of the SMS is to efficiently aggregate all of the DSL signals and direct them to the appropriate routing or switching equipment similar to modem pools used on dial-up Internet devices.

Connecting the DSLAM to the Network

As DSL circuits are used for three primary purposes—Internet access, VPNs (Virtual Private Networks), and VoDSL—equipment in the main switching center must be able to direct the traffic accordingly. If the signal is destined for the Internet, the SMS sends the traffic to the core router for processing. If the signal is destined for a secondary location on a VPN, the SMS sends the traffic to the ATM circuit connected to the DSLAM feeding the secondary location. If the signal is destined for the circuit switch, the traffic is sent to the VoDSL gateway tied directly to the switch. Figure 17–15 shows how a DSL circuit connects through the access network.

The physical connection between the SMS device and the routing/switching equipment is as follows: The SMS connects to a core router either through a DS-3 or

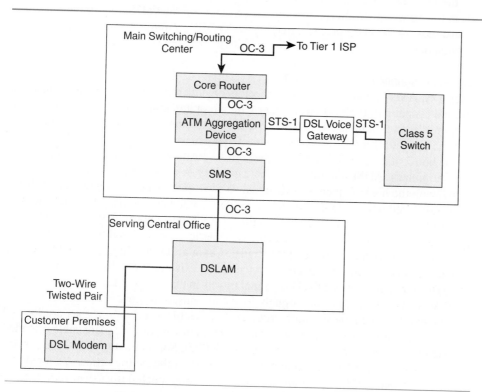

Figure 17–15
A DSL circuit connects to the serving central office by a two-wire connection in the same way as standard telephone service. The circuit is aggregated with other DSL circuits and connected to the central switching/routing center where the traffic is groomed according to type.

OC-N circuit. If the circuit is a DS-3, coaxial cable is used to connect the SMS box to the DS-3 DSX panel. A coaxial cable cross connect is placed between the DSX port tied to the router and the DSX port tied to the SMS box. If the circuit connection is an OC-N, fiber jumpers are used to connect the SMS device to a port on the fiber distribution panel—FDP—where it is cross connected to a port tied to the router. Both the router and the SMS devices have optical line cards where the fiber optic jumpers are terminated using standard optical connectors: FC, ST, SC. Because the two devices reside in the same building, multimode fiber optic jumpers and a low-powered laser are typically deployed as their cost is much less than high-powered lasers. Traffic coming into or leaving the router on the egress side, leaving the local network, typically rides on metropolitan circuits that feed into Tier 1 ISPs.

Connecting a DSL signal to a circuit feeding a second DSL, thus creating a Virtual Private Network, requires the SMS device to route the cells to the correct outgoing IOF circuit. As is common, most IOF circuits are DS-3 and connect to the SMS through DS-3 DSX locations. The incoming DS-3 connected to a remotely located DSLAM terminates at the DS-3 DSX port. Coaxial cable is run between the SMS device and a DS-3 DSX port. A cross connect is placed between the incoming line and the equipment by running a physical coaxial cross connect. If the DSLAM ties to the central switching center via an OC-N circuit, fiber jumpers are run from the OC-N multiplexer to a port on the FDP. Fiber jumpers also are used to connect the SMS to the port on the FDP creating the connection between the DSLAM and the SMS.

VoDSL traffic travels through the SMS device to a VoDSL gateway device that converts the ATM cells into TDM channels configured in the GR-303 format. The connection between the two devices is most often a DS-3. Coaxial cable is run between the SMS device to a DS-3 DSX port. The VoDSL device is tied to the SMS by running coaxial cable to a port on the DS-3 DSX, and then a cross connect is made between the two ports to tie them together. The coaxial cable terminates on BNC connectors that tie to a DS-3 port on the box. Similarly, the VoDSL gateway terminates the DS-3 on a BNC DS-3 port. From the VoDSL device to the switch, an STS-1 connection is common. This involves running coaxial cable from the VoDSL gateway again to a port on the DS-3 DSX panel and connecting an STS-1 BNC port to a port on the DS-3 DSX panel. Again, a cross connect is placed between the two ports to tie them together. Traffic interfacing into the switch from the VoDSL equipment has been converted into the TDM protocol making the VoDSL equipment a cell to time slot conversion device.

Wire-Side Connections on a DSLAM

Most DSLAMs provide amphenol connectors on either the back or front to connect the twisted copper pair to the device. Either wire wrapping, punching down, or connectorized ends such as amphenol connectors are used to connect the cable to a termination block. Frame jumper wire is used to connect the wire pair terminated on the frame block on the horizontal side of the frame to the vertical position where the outside plant cable pair terminates. ADSL circuits have the capability to carry a VF—voice frequency— signal with the high-frequency DSL signal. A filter shelf is placed next to the DSLAM or incorporated into the DSLAM chassis and is used to split out the VF signal and pass it on to the voice switch, as shown in Figure 17–16. Filters allow the 300 to 3400 Hz frequencies to remain intact, thus allowing the line to carry the traditional voice signal along with the higher frequency DSL signal. Using a POTS splitter allows the service provider to give the subscriber DSL service on the existing POTS line.

The wire pair carrying the DSL signal travels in the same cables as a regular POTS line. As with all high-frequency signals, load coils must be removed on all DSL circuits; in addition, bridged taps should be removed if possible, especially if they are located close to either the CO or termination equipment. In theory, outside plant cable is loaded to extend the range of the voice signal after 18,000 ft. Subsequently, most DSL service is restricted to subscribers residing within 18,000 ft. of the central office. As noted earlier in the chapter, DSL signals are capable of traveling much farther than 18,000 ft. up to 29,000 ft. on lower-gauge cable when transmitting a slower-speed signal.

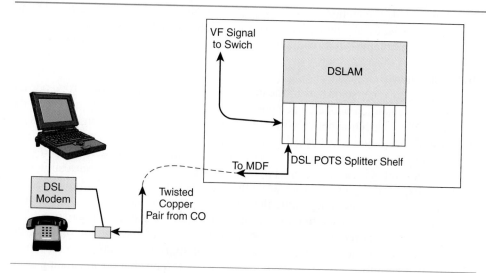

Figure 17–16
ADSL circuits can carry both the high-speed broadband DSL signal and the traditional POTS signal. A splitter is used at each end of the circuit to filter out the voice frequency (VF) signal directing it to the switch. The DSL signal is fed into the DSLAM.

Hence, the distance limitation is the result of the ILEC's decision not to deload the outside plant for DSL service, as the task would be daunting. As a bridged tap causes a reflection of the signal that may cause distortion of the original signal, removing taps close to either end of the circuit (CO end or subscriber end) is critical. ILECs have committed to remove so many feet of bridged taps and all load coils if the circuit is less than 18,000 ft. Most ILEC tariffs detail what will be performed on a DSL in order to condition it properly for service.

A second hurdle to overcome when deploying DSL service is the large number of lines fed through DLCs. The only DSL service offered on DLC lines is *IDSL*—ISDN Digital Subscriber Line—and offers a maximum line rate of 144 kbps, the same as ISDN BRI. Older DLCs are not capable of terminating DSL, causing provisioning and outside plant engineers headaches when sales sells DSL to a subscriber fed by an old DLC. If possible, the engineer pulls the customer wire pair out of the DLC and connects it directly to the distribution cable, bypassing the DLC.

IDSL
ISDN Digital Subscriber Line. A circuit type that offers ISDN over DSL loops.

Terminating a DSL at a Customer Premises

Terminating a DSL circuit at customer premises is similar to terminating a POTS line in that the DSL terminates at the protector whether that is an outside line protector or a termination block in the telephone closet. From the demarcation point, inside wire is fed from the demarcation block through the building to the location where the DSL service is to be terminated. A standard RJ-11 jack is used to connect the inside building wiring to the DSL modem. Once the connection is completed, LEDs on the modem light indicating the line is synced up to the DSLAM and is ready to carry traffic. On the opposite side or LAN side of the DSL modem, the computing device connects via an RJ-45 connector to the modem, as shown in Figure 17–17. As DSL is a finicky, high-speed service, poorly made terminations at any of the connection points cause the DSL

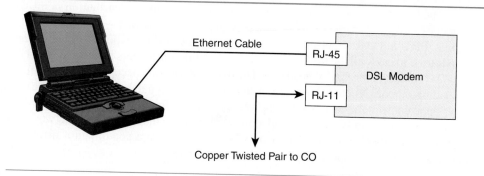

Figure 17–17
The DSL modem connects to the terminating device, in our example a laptop computer, via an Ethernet RJ-45 connection. The DSL circuit connects into the DSL modem by terminating on a standard RJ-11 jack.

signal to degrade thus inhibiting the speed of the signal. In some cases, it prevents the line from syncing up altogether. Therefore, connections should always be checked and rechecked if problems occur on the line or if the line speed is less than expected.

Traffic Monitoring

It is important to understand how traffic traverses the DSL network, as the wrong portion of the network is often blamed when congestion occurs on the line. In many cases, the two-wire DSL connection between the subscriber and the DSLAM is singled out as the reason traffic is sluggish. Congestion is more likely to occur on the network side of the DSLAM as oversubscription or transmission problems cause traffic to slow. In addition, slowness often is caused by that big network in the sky, the WWW, especially when the ISP is experiencing problems or when the network is underengineered and oversubscribed. Therefore, monitoring traffic flow on DSLs must be done in order to maintain good customer service in addition to helping reduce time spent troubleshooting.

Most DSLAMs provide views of network performance on both the line and the network sides of the device. In addition, special auxiliary traffic programs have been released that help to isolate trouble areas. Engineers responsible for mandating oversubscription ratios often find themselves on the firing line when suddenly a flurry of users decide to download the latest version of some silly game. The difficulty in predicting usage is apparent as the customers call in screaming for more bandwidth. On the flip side, the engineer is faced with angry e-mails from the budget department when he or she overengineers the network and gives those Internet-crazed customers unlimited bandwidth. As DSL is still in its infancy compared with other technologies, most engineers find it necessary to tweak the system frequently to ensure that they do not tick off too many people on either side of the fence.

17.4.2 Cable Modem

Cable television providers have aggressively entered the broadband market. Today, CATV carries a good portion of broadband Internet traffic and continues to expand its footprint with bundled services such as VoIP. The *cable modem*, based on the DOCSIS protocol, a subject explained in detail in Chapter 21, typically is part of the cable television device placed at the customers home.

cable modem
Modem used to terminate broadband services offered by the CATV provider.

The cable modem receives its network feed from the coaxial cable connection terminated on the back of the cable box. Specific frequencies have been carved out to carry the broadband signal across the CATV network. From the cable box, an Ethernet connection is used to connect the cable modem to the end device such as a PC or router. The CATV broadband service uses a broadcast network technology in the access space to carry traffic to the end device. This means that all customers fed by the same coaxial cable television feed reside on the same broadcast segment, similar to Ethernet. When a packet destined for April's PC sitting in her office on the second floor is shipped out on the coaxial cable, all of April's neighbors will receive the packet (packet is broadcast to everyone) but only April's device will recognize it and grab it off the line. The device accepts the traffic if it is addressed to MAC and IP addresses, which are used to differentiate end devices on a network segment similar to Ethernet technology used in an internal LAN. Therefore, you may hear people say CATV uses a bus technology or a broadcast technology.

The CATV access network looks like a tree, one trunk or distribution point with branches spreading out feeding customers within a specified area—everyone living on Maple, Elm, and Oak streets, for example. From the distribution point, a fiber connection is used to carry all the traffic from all the branches—segments feeding Maple, Elm, and Oak—back to the CATV head end. The distribution point accepts the electrical signal riding on coaxial cable, converts it into an optical signal, and transports it across fiber optic cable to the main switching/routing center referred to as a head end (similar to the POP or tandem switch in the telecom world). In the industry, CATV networks are called hybrid fiber/coax (HFC) networks because they are a combination fiber optic and coaxial cable transport medium in the access space.

CATV networks are explained in greater detail in Chapter 21 and Chapter 22. For now, it is important to know that cable modem falls into the access space as a broadband service offering to both the residential and business customer base.

17.4.3 Cellular Broadband Service

Cellular or PCS companies have also entered the broadband space. New technologies have evolved allowing the service providers to offer high-speed connections anywhere within their footprints. Because today most of the cellular companies provide national coverage, the access to broadband service almost anywhere in the United States is possible. The beauty of this offering is that the customers are no longer tethered to their physical addresses. They may sit in the middle of a lush green pasture with their laptops and surf the Web. The technology behind the cellular broadband offering is explained in detail in Chapter 23 and Chapter 24. At this point, the main point to grasp is that cellular service providers are contenders in the broadband access space.

17.4.4 Wi-Fi

Wi-Fi broadband services are gaining in popularity, specifically in areas not serviced by CATV or ILEC broadband services. Towns and small cities have entered the broadband space in order to "wire" their communities. Business today and new residents demand easy access to the Internet at high speeds. Towns not wired for broadband services are at a disadvantage. Wi-Fi connects the residents in an area through wireless connections terminating at a central distribution point. The distribution point feeds back to the central hub that connects to the ISP or other service provider. A detailed explanation of the Wi-Fi network and protocols is given in Chapter 24.

■ 17.5 T-CARRIER CIRCUITS

17.5.1 T1s in the Access Network

T1 high-capacity circuits are used to carry all types of services in the access network. Service providers sell T1s to all types of business customers for all types of applications: LAN connections, PBX tie lines, aggregating slower-speed circuits, and on and on and on. End users have come to depend on T1 services to carry their traffic anywhere in the network, including metropolitan and long-haul networks. For example, a T1 may connect two locations in a city and carry the company's voice and data traffic, eliminating the need for multiple measured business lines. The customer pays for the circuit, not the minutes of use. As such, it has become very easy to cost justify purchasing a T1 once the cost for minutes of use has been calculated for equivalent dial-up lines. Consequently, the access network has realized an exponential growth in T1 services to end users for several years. Service providers also deploy T1 circuits as a means to aggregate smaller-speed lines onto one facility helping to reduce the amount of copper needed in the network. If an aerial picture were taken of a typical metropolitan access network, the spiderweb of T1 circuits would be amazing, as the T1 has become the foundation of transmission systems.

As explained in Chapter 8, the T1 signal consists of twenty-four 64 kbps DS-0 channels equaling a transmission rate of 1.544 Mbps. The signal travels on a four-wire connection, one pair for transmit and one for receive. T1 circuits are often groomed into higher-rate circuits such as DS-3s or OC-Ns before being carried across the network on fiber optic systems. Even when multiplexed into a higher-rate signal, the T1 remains intact.

17.5.2 T1 Equipment

Equipment used to connect T1s varies as dramatically as the number of different applications dependent on T1s. Equipment may be divided into two categories, terminating equipment and network equipment. The network equipment is simply the device used to interface the line to the terminating equipment. The NIU—network interface unit, also called a Smart Jack—is placed at the end of the circuit and is used as an interface between the network and the terminating equipment. An NIU, if placed at the

Figure 17–18
An RJ-48 termination is made
between the NIU and the DCE. Four
wires coming from the CO are
terminated onto the Smart Jack.

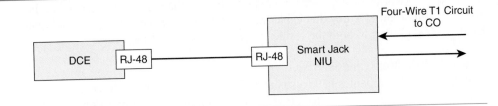

Figure 17–18
An RJ-48 termination is made
between the NIU and the DCE. Four
wires coming from the CO are
terminated onto the Smart Jack.

NIU
Network interface unit. The
interface device used to
terminate T1 circuits coming
from the network to the
termination device, such as a
CSU.

end of a line-powered copper circuit, is responsible for filtering any power spikes or other electrical problems coming off that line. In the opposite direction, the NIU is used as a repeater making sure the signal meets the correct physical characteristics such as the correct transmit decibel level before heading out into the network. One of the features a Smart Jack offers is remote and local loop-back capability, giving the technician a way to remotely test the line. The T1 line coming into the site is physically tied to the Smart Jack using terminating screws. An RJ-48 modular jack is used to connect the opposite of the NIU to a CSU, as shown in Figure 17–18.

Inside the customer's premises, the *NIU* is the service provider's demarcation point. The customer does not own or maintain the NIU. After the NIU, the T1 circuit is connected into a CSU—channel service unit. The CSU, similar to the NIU, is an interface device placed before the terminating equipment. Often, as with most routers, the CSU is built into the WAN circuit pack. Even if the CSU is built into the circuit pack, it still performs the same functions, which include preparing the signal either for the terminating equipment or, in the opposite direction, for the network. Any voltages or other electrical occurrences outside the standard values are eliminated, and many CSUs are sophisticated enough to perform error checking. In addition, the CSU, like the NIU, can be looped back remotely to help eliminate the need for the technician to drive to the site.

The CSU is connected to the NIU using a four-wire circuit with an RJ-48 connector. A V.35 connection is made between the CSU and the terminating equipment, as shown in Figure 17–19. Some CSUs are very sophisticated in that they are able to multiplex/demultiplex individual DS-0 channels and direct them to different terminating devices. The signal leaving the CSU, though, is still a T1, just with some empty channels.

A channel bank is a device that is commonly found in the access network both at the central office and at the customer's premises. The role of the channel bank is to multiplex/demultiplex DS-0 channels onto or off a T1 circuit. Most channel banks today are small, pizza-size boxes that require a minimum amount of power and are configured through a craft interface port using software commands. Older channel banks, still making up a large amount of those currently deployed, are configured the old-fashioned way by moving dip switches and setting jumpers. Either way, both perform the same function. Twenty-four channel cards are present on the line side of the channel bank consisting of all types of DS-0 circuits including ISDN BRITE cards, DDS 56 kbps cards, FX, OPX, DID, and so forth. Service providers use channel banks to groom DS-0 channels out at the central office and feed them to the subscriber across the copper infrastructure. If you were to tour a central office, you would see a large number of channel banks mounted in relay racks in the carrier room area of the office.

A second type of T1 device found in a central office is the 1/0 DCS that was discussed in a previous chapter. The 1:0 DCS was built to replace back-to-back channel

Figure 17–19
The DCE and DTE tie together
typically using a V.35 interface.

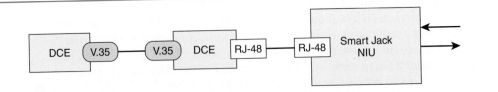

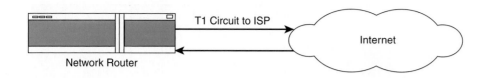

Figure 17–20
An egress connection is shown
connecting the network router to the
Internet via the ISP.

banks by eliminating the need to demultiplex a DS-0 before feeding it into a second channel bank. The 1:0 DCS acts as the train station moving DS-0s from one T1 to another electronically, eliminating the need to physically wire the devices out. Not every central office has a 1:0 DCS, as its cost is much higher than that of a channel bank. Medium and large central offices may have DCSs deployed depending on the number of circuits feeding into the office.

Many small businesses have routers that interface into the Wide Area Network using T1 circuits. The *WAN interface card (WIC)* resides in the router and is configured as a physical interface. Most WAN connections are established using frame relay as the packaging method and, as such, must feed into a frame relay switch or interface on the opposite end. Several types of connections are made by T1s interfacing into a router. The first is connecting into the Internet by establishing a T1 circuit between the router and the ISP's network, as shown in Figure 17–20. The router sends all Internet traffic out on the T1 that connects to the ISP. A second type of connection, often referred to as *VLAN or virtual LAN*, is used to connect two remote sites in order for them to communicate locally. Each of the locations looks as if it belongs to the same LAN and is able to send traffic between servers seamlessly. Though the T1 circuit is technically configured as a WAN, it is carrying LAN traffic across the outside network creating a LAN-to-LAN internetwork.

ISPs purchase ISDN PRI circuits that connect into the local circuit switch and provide trunks similar to DID trunks between a PBX and the CO switch. The ISP is able to configure circuits using standard concentration ratios to help improve on the efficiency of the connection. PRI, as explained earlier, offers signaling capability to the end device and, as such, provides numerous features that help to reduce the number of circuits needed between the devices. In addition, PRI trunks help reduce the cost of access making the service very desirable to the ISP who is always counting pennies in order to keep afloat.

A PBX—private branch exchange—also terminates T1 signals into integrated T1 cards called DTIs—digital trunk interfaces. A PBX terminates T1s for several reasons. One is to connect directly into the service provider's end-office switch allowing the PBX to establish trunking circuits between them. Some trunks are assigned as incoming and some as outgoing—DID/DOD—depending on the traffic patterns. A second use for a T1 by a PBX is to bypass the local telephone company and directly connect into the long distance network, thus eliminating any local access charges. A PBX, as it is a small switch, routes all long distance calls to the T1 connected to the long distance switch. A third use of a T1 feeding a PBX is as a tie line used to connect two PBXs. All calls destined for a station hanging off the remote PBX are routed onto the tie line and, as such, do not incur any usage charges as they would if the calls traversed the PSTN.

Within the access network, T1 circuits often terminate into fiber optic systems, as discussed in an earlier chapter. A SONET MUX contains a T1 interface card that accepts a copper four-wire circuit, as shown in Figure 17–21. Microwave systems also carry T1 circuits within a DS-3. There are many M13 multiplexers in the access network, both in the switching offices and at the customer premises. The T1 circuit feeds into interface ports on the M13 as four-wire copper circuits. Finally, T1 circuits are used to connect DLCs to the central office, as discussed earlier in the chapter. Old and new DLCs have copper four-wire T1 interfaces.

WAN interface card (WIC)
A term used to describe a T1
interface port in a router or
other terminating device.

VLAN or virtual LAN
A virtual path or channel
established between devices
in a network. One circuit may
have several VLANs. A VLAN
may service many locations.

17.5.3 T1 Network Architecture

A copper-fed T1 comes in two varieties: the standard AMI T1 and the 2B1Q HDSL T1. Both require a four-wire circuit, transmit and receive; and both require a device at the

Figure 17–21
A SONET MUX has the ability to accept copper T1 circuits and groom them into VT-1.5. Typically, multiplexers deployed in the access network contain DS-1 interface cards.

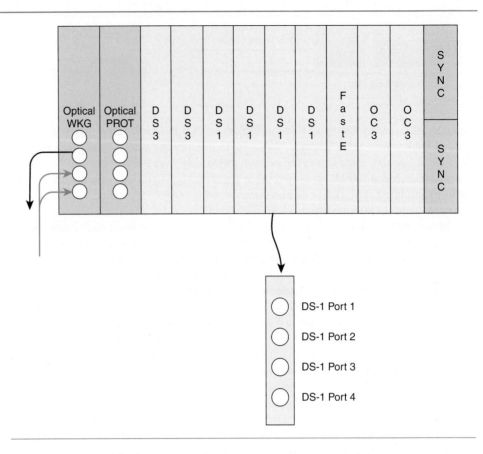

end such as a Smart Jack. The difference between the two is the distance the signal can travel before needing a repeater and the type of cable it rides on. A traditional copper T1 circuit rides on special carrier cables separated from the hundreds of POTS lines feeding subscribers. The T1 circuit feeds into carrier housing every 6000 ft. that holds in-line repeaters. Within the central office, the circuit terminates into a CO repeater responsible for repeating the outgoing signal and powering the circuit. At the customer premises, the circuit is demarced at an NIU, as discussed earlier. An AMI line code is used to format the signal, meaning that every other pulse is of opposite polarity, as discussed in Chapter 4; and each signaling state carries one bit of information requiring a frequency of 756 kHz in order to carry the 1.544 Mbps. One wire pair carries twenty-four DS-0s from direction *A* to *B,* and the other pair carries twenty-four DS-0s from direction *B* to *A*. The transmit and receive pair must be separated by a minimum of twenty-five pairs.

HDSL T1 circuits ride on the same wires as the POTS circuits. A separation between the transmit and receive pair is not necessary as the signal structure is divided as transmit and receive on each pair. Additionally, the HDSL T1 signal is able to travel a full 12,000 ft. before needing to be repeated. 2B1Q line coding reduces the maximum frequency to 356 kHz, as four bits are transported for every baud. The HDSL circuit terminates into an HDSL NIU at the customer premises and into an HDSL device similar to a CO repeater at the central office. The CO side of the HDSL circuit is responsible for repeating the outgoing signal and powering the line. The repeater housings used to hold HDSL T1 repeaters are smaller than T-Carrier housings as they only hold up to twelve repeaters whereas T-Carrier housings hold as many as fifty repeaters.

Though the differences between the two copper T1s are significant, the signal is the same. An end customer would not know the difference between the two as the interfaces between the NIU and the CSU or the HDSL NIU and the CSU are exactly the same. The purpose of deploying HDSL should be evident. First, it reduces the cost in that it is able to ride in the existing cable plant and does not need to be separated from other circuits making the provisioning much simpler. Additionally, the signal is able to travel a full 6000 ft. farther before needing to be repeated, thus reducing the cost of in-line repeaters. One additional difference between the two types of copper T1 circuits is the way they are terminated in the CO. An HDSL T1 is terminated on the MDF in the same way a POTS circuit is. In contrast, T1 carrier circuits are tied together on a cable head that is mounted typically in the carrier room. Both types, though, end up tied to a DS-1 DSX panel in the carrier room, as described earlier. HDSL is the most common way to deploy T1 circuits to end customers today. Almost all cell sites are connected via HDSL T1 circuits, as are small businesses. Engineers have grasped the technology in full.

Fiber optic systems are used to carry T1 circuits across the access network. Many large metropolitan buildings have been wired with fiber cable. Fiber multiplexers are installed in the building and are used to feed T1 circuits to the various customers in the building. The T1 signal riding in the fiber system is the same as that riding on a copper T-Carrier span. The main difference is that the electrical signal is multiplexed into higher-rate signals such as an OC-12 and converted from an electrical signal to an optical. Terminating the T1 at the customer premises still requires an NIU that is powered via an AC outlet in the room in which it is installed. This is the only real difference in that the customer is now responsible for providing backup power in case the building power fails.

17.5.4 Connecting T1 in the Central Office

Though the method used to carry the T1 circuit across the access network varies, as already explained, the connection made in the central office is the same. The T1 circuit is tied to a port on the DS-1 DSX panel and cross connected according to its destination as described earlier in the chpater. Most phone companies designate the type of connection made at the DSX by using different colored LEDs on the DSX panel. For example red LEDs represent copper T1 spans—either HDSL or T-Carrier. Orange represents a T1 interfacing into a fiber optic system. Yellow represents a T1 connected to a T1 port in the switch called a DLCU—digital line carrier unit. Green represents a T1 connected to a channel bank or other terminating equipment such as the DCS. A T1 coming in from a copper T1 span may be cross connected to a T1 interfacing into a fiber optic system used to connect two central offices. This connection is referred to as a network-to-network connection. If the technician were to plug his or her test set into the monitor port of the copper T1, he or she would observe the signal coming into the central office on the receive pair. If the technician moved his or her test set to the monitor port of the fiber system T1, he or she would be observing the signal coming off the fiber optic network. Similarly, if a T1 circuit coming in from a customer site on a copper span is cross connected to a switch port, the T1 signal coming out of the switch is monitored when the technician plugs his or her test set into the monitor jack of the T1 tied to the switch port. Whether the color codes we have described are adhered to or not, it is important for anyone working in telecommunications networks to understand how to sectionalize a trouble through the use of DSX positions. Misinterpreting the direction the trouble is coming in from causes hours of lost time and very severe headaches.

The physical connection made at the DSX panel requires the shielded twisted wire pair to be wire wrapped on the back of the DSX. There are two in pins and two out pins used to attach the twisted wires to. Most DSX panels are divided into twenty-eight port sections that correspond to the twenty-eight DS-1s that make up the DS-3 circuit. Today, large DSX panels have been deployed that require less space and allow for many more port terminations. The two most common sizes found in the network are a twenty-eight position DSX and an eighty-four position DSX.

The front of the DSX has three jack ports, the top used as a monitor port, the second as a transmit port, and the third as a receive port. The transmit and receive ports are used to attach a test box and perform an intrusive test on the line, terminating the signal into the test box. The two ports also provide a loop back and patch point in the network, as shown in Figure 17–22. A loop-back plug is placed into the jacks causing the signal to loop around and head back to its origin. A patch cord is used to connect the DSX port to another DSX port, thus completing a circuit connection such as from a

Figure 17–22

A loop-back is used to loop the circuit in one direction. The diagram shows a loop-back plug being placed in the DSX port tied to the class 5 switch. The signal will travel from the switch to the DSX panel and back to the switch. A test box can be connected to the monitor port on the DSX panel to determine if the signal traveling out of the switch is good or bad. Placing a loop toward the switch should make the switch port look up and good.

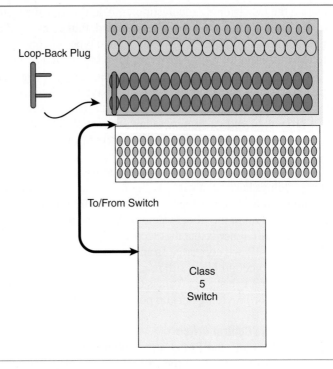

Loop-Back Plug

To/From Switch

Class 5 Switch

copper T1 to a switch port, as shown in Figure 17–23. Patching a circuit is a temporary way to cross connect the two ports during turnup or troubleshooting of the network.

Below the jack ports, wire wrap pins are used as the cross-connect point between the two DSX ports. A cross-connect wire is wrapped onto the five pins according to a color code that is used primarily to keep the cross straight. The top pin is used as a ground and is what causes the LED of the connected DSX port to flash when a bantam jack plug is inserted into the monitor port. This sounds trivial in the big scheme of things, but if you are a technician trying to figure out where a circuit is cross connected to and it is 2:00 in the morning, you very much appreciate the flashing LED. The second and third pins from the top are the *in* of the circuit, meaning that it is where the receive of the incoming signal from the opposite DSX port is tied. The fourth and fifth pins are the *out,* meaning the transmit out of that DSX jack to the opposite DSX port. When a cross connect is made between the two DSX ports, a cross is placed between the ins and outs as the one circuit's in is the other circuit's out, and who's on first. Figure 17–24 helps

Figure 17–23

A patch cord is used to temporarily connect two ports on a DSX panel.

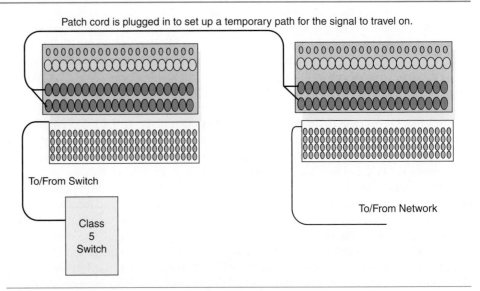

Patch cord is plugged in to set up a temporary path for the signal to travel on.

To/From Switch

Class 5 Switch

To/From Network

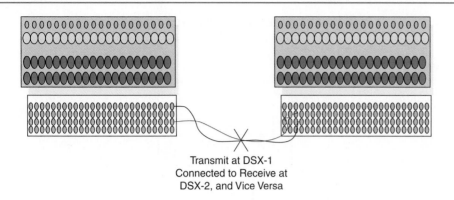

Transmit at DSX-1
Connected to Receive at
DSX-2, and Vice Versa

explain this very important concept. If the blue/white pair is wire wrapped onto pins 2 and 3 of DSX position 1, at the opposite end the blue/white pair is wire wrapped onto pins 4 and 5 of DSX position 2. In conjunction, if the orange/white pair is wire wrapped onto pins 4 and 5 on DSX port 1, it must be wire wrapped onto pins 2 and 3 of DSX position 2—thus, the cross. Luckily, the T1 wiring scheme is not tip-ring sensitive, meaning that the tip does not have to be terminated on pin 1 and ring on pin 2. Instead, the tip and ring of the pair may be wire wrapped onto either the first or second pin. An important point to grasp is whether the termination of the T1 is from fiber, a switch, or equipment, the DSX panel is always wired the same.

17.5.5 Inside T1 Wiring

Wiring the T1 to the NIU at customer premises varies as much as the applications riding on the T1. Telephone companies may decide to terminate the NIUs in an NIU shelf if the customer has multiple T1s. Small customers may receive an NIU mounted in a closet behind the receptionist desk. It would be impossible to discuss all of the methods used to install an NIU. Inside wiring also varies per customer and, as such, is often one of the biggest trouble areas in the network.

As T1s are high-speed and, consequently, high-frequency signals, inside wiring should comply with the standards established in the industry. Most service providers require T-screen cable, meaning cable that is shielded twisted pair specced for high-frequency signals. Often, existing inside wire is used between the basement, where the outside plant cable terminates, and the end termination site, where the NIU is placed. The reason this happens is that the cost to run special T-Carrier wire is too high or duct is not available. Either way, a four-wire connection is needed between where the cable comes in and the location where the NIU is to be mounted. This is true also when an FOT system is deployed and the multiplexer sits in the basement feeding all of the offices in the building. Connecting the T1 still requires a copper four-wire connection.

■ 17.6 GIGABIT AND FAST ETHERNET SERVICES

One of the fastest growing access technologies is Gigabit Ethernet services offered to business customers. For example, a strip mall may purchase a GigE or FastE pipe and segment the bandwidth to the tenants for a fee. Similar to T1s, GigE and FastE are providing high-capacity circuits from the customer's premises to a central distribution location such as a POP or distribution point in the access network. Ethernet services will continue to grow as fiber deployment extends through the access network.

■ 17.7 OPTICAL SERVICES IN THE ACCESS SPACE

As found in the core and metropolitan networks, SONET and DS-3 circuits are deployed in large numbers in the access network. SONET services are sold primarily to large companies that require very reliable high-bandwidth connections to multiple locations in the network. Additionally, SONET and DS-3 services are sold to CLECs,

Figure 17–25
IOF—interoffice facilities—are sold
by ILECs to other service providers.
DS-1, DS-3, and SONET circuits are
sold at a reduced cost to the provider.
The access network circuits are
groomed onto IOF circuits that tie to a
regional switching center.

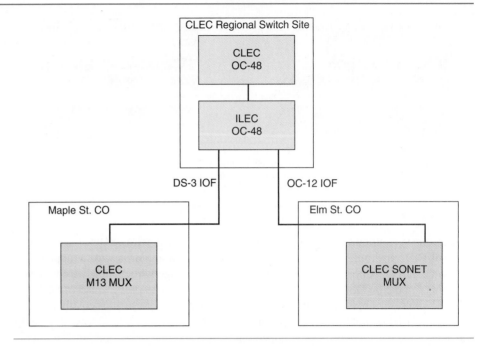

DLECs, and ICPs for IOF—interoffice facilities—and UNEs—unbundled network elements, as shown in Figure 17–25. The structure of the network is the same as that of the metropolitan network; because of this, we do not spend much time discussing these services. The systems are sold in ring or point-to-point configurations and are groomed down to the circuit level needed by the customer. For example, a large company may purchase an OC-3 SONET ring to connect three of its locations around the city. Today, all of the large ILECs offer SONET ring services to business customers and competitive carriers.

SUMMARY

The access network is one of the most complicated portions of the telecom infrastructure, primarily because of the wide variety of services available and deployed. Today, the extension of fiber to the access network and the aggressive deployment of broadband services, including DSL, make the access portion of the network one of the most changeable segments of the industry. Later in the text, newer technologies such as Wi-Max and cellular broadband are discussed as they are the newest entrants in the access space. A good rule of thumb to follow when discussing access network technologies is that all technologies deployed in the core network will migrate to the metropolitan network and all technologies that are deployed in the metropolitan network will eventually migrate to the access network.

CASE STUDY

Miranda completed the metropolitan network design as her father requested. He now has asked her to provide a list of services that should be offered to the end subscribers. In addition to the list of services, Miranda will need to explain how the services will be deployed across the access network and tie into the metropolitan network.

Use the information in Appendix A to complete this exercise. Remember, there is no one right answer.

REVIEW QUESTIONS

1. List three types of carriers that offer services over the access network.

2. Define the access network as it pertains to the communications network.

3. The following are considered part of the access network:
   a. central office
   b. ISDN circuit
   c. POTS line
   d. FX circuit
   e. broadband pipe
   f. T1
   g. GigE
   h. all of the above.

4. The ISDN BRI circuit D channel is used to carry Voice over IP calls across the Internet.
   a. true
   b. false

5. Why do businesses purchase DID/DOD trunks?

6. The following are considered traditional special services circuits:
   a. ISDN BRI
   b. ISDN PRI
   c. digital data service
   d. Off-Premise Extension
   e. analog trunks
   f. DS-1
   g. GigE
   h. FastE
   i. c, d, e, and f
   j. a, b, c, d, e, and f
   k. g and h
   l. all of the above

7. What is the term used to describe traditional analog telephone service?

8. The conductors that make up the copper pair are referred to as _____ and _____.

9. A protector used to terminate the copper loop at the subscriber's location is referred to as a(n) _____.

10. The term *drop line* is used to refer to the cable that is dropped off the back of the DSL modem and used to connect to the terminating equipment.
    a. true
    b. false

11. The outside plant cables are terminated on _____ in the central office.

12. An integrated DLC eliminates the need for a DLC in the central office. The DLC interfaces into the switch through what port?

13. Next generation DLCs have circuit packs that terminate _____ services.
    a. ISDN
    b. POTS
    c. FX
    d. DSL
    e. DS-1
    f. DS-3
    g. FastE
    h. OPX
    i. all of the above

14. An ISDN BRI circuit requires two copper pairs from the central office to the NT-1 sitting at the subscriber's premises.
    a. true
    b. false

15. ISDN PRI circuits are used to
    a. tie a PBX to the serving central office switch.
    b. provide trunk connections from an ISP and the central office switch.
    c. provide a connection between routers.
    d. carry Fast Ethernet from one subscriber to a second.
    e. a, b, c, e, and f.
    f. all of the above.

16. The 56 kbps digital circuit was the first digital circuit offered to the subscriber by the telephone company.
    a. true
    b. false

17. A subscriber that wants to add a telephone service to an office located three blocks away may order a(n) _____ circuit.

18. The terms *loop start* and *E&M* refer to _____ methods.

19. Name three broadband services.

20. What advantage does a broadband pipe provide the end user?

21. Terminating a DSL or CATV broadband service at the subscriber's premises requires the service provider to install a router.
    a. true
    b. false

22. Name three applications that require the customer to purchase a T1 circuit.

23. An NIU is used as an interface between the outside plant circuit and the terminating equipment at the site.
    a. true
    b. false

24. Why do service providers deploy HDSL T1 circuits?

25. Name three reasons the end customer purchases a FastE or GigE service.

KEY TERMS

plain old telephone service
   (POTS) (473)

tip (474)

ring (474)

network interface device
   (NID) (474)

drop line (475)

MDF (475)

universal DLC (477)

integrated DLC (477)

unbundled network element
   (UNE) (478)

SLC-96 (478)

TR-08 protocol (478)

bearer (480)

data (480)

switched digital services (480)

NT-1 (481)

terminal adapter (TA) (481)

multiple call appearance (484)

OPX (484)

DSU/CSU (485)

private line (485)

multipoint topology (486)

DCE (486)

DTE (486)

loop start (489)

tie line (489)

E&M (489)

DID trunks (490)

DOD trunks (490)

broadband services (490)

Wi-Fi (491)

WiMAX (491)

line code (492)

DLEC (492)

IDSL (495)

cable modem (496)

NIU (498)

WAN interface card
   (WIC) (499)

VLAN or virtual LAN (499)

Communications Services Providers

Photo courtesy of USDA/NRCS/NCGC National
Cartography and Geospacial Center

18

The Public Switched Telephone Network Central Office

Objectives

After reading this chapter, you should be able to

- Discuss the physical structure of the switching center.
- Discuss the components found in the switching center.
- Discuss the function of the switching center.

Outline

Introduction

■ INTRODUCTION

Building a new switching center from the ground up is not an easy task. A properly functioning switching center requires that the details be well planned and the building well organized. This chapter discusses in detail each component found in an ILEC's central office, a CLEC's regional switching office, and a long distance switch site.

■ 18.1 SWITCHING EQUIPMENT

18.1.1 Local Telephone Network Central Office Switches
Switch Layout
Our tour of the local telephone company's central office starts with the serving end office class 5 switch. Chapter 7 discussed how the switch functions. We now are ready to describe what the switch looks like, where it is located in the central office, what it is connected to, and how it is maintained and provisioned.

The central office class 5 switch is housed in an environmentally controlled room. The temperature of the room must be kept fairly constant to avoid overheating the electronics. Because a switch is a large computer, it generates a great amount of heat as it computes continuously. Without sufficient temperature control, the switch heats up until the processors start to break down. At that point, the switch shuts itself down to avoid permanent damage. One of the costliest items in a switch center is the air conditioning unit that keeps the switch cool.

In addition to temperature control, most central office switch rooms are built with special air filters to reduce the amount of airborne dust and contaminants. Today, switches can be placed in less environmentally clean areas because of the improved design of cabinets, placement of circulating fans, and overall resiliency of the electronics. Older central offices in ILECs are normally found in a separate closed-off room.

A switch's *footprint*, the term used to define the square footage of the area, varies depending on the size of the office. Each office that houses a host switch requires a minimum of two processor bays, an administrative bay, and a line and trunk bay. The size of the switch varies depending on the number of lines served by that office, the number of trunk connections required, and the make of the switch. The switch engineer is responsible for sizing the switch by using demographic information, traffic studies, and switch specifications. Once the switch is sized, the engineer puts together a detailed layout showing the space the switch will occupy and the air conditioning and electrical power needed.

footprint
Amount of physical space a piece of equipment requires.

Once the dimensions of the switch are determined, it is time to determine how to place each of the equipment bays within the lineup. Figure 18–1 shows a typical 7 ft. by 23 in. equipment bay. Manufacturers provide detailed rules concerning the positioning of each module, the number of bays allowed in one lineup, and where each piece of the switch can reside.

Once the dimensions and position of the switch have been decided, the design of the overhead ironwork begins. *Ironwork* is used in telecommunications to describe ladder racking, *relay racks*, and other miscellaneous support structures. As shown in Figure 18–2, ironwork is used to hold all the circuit cables and power cable that interfaces into the switch. An alternative to placing ironwork above the switch is to use a false floor beneath the switch. Normally, though, ironwork to hold cables above the

ironwork
Iron support structure used to secure relay racks and hold cable.
relay rack
Rack used to hold equipment in communications offices.

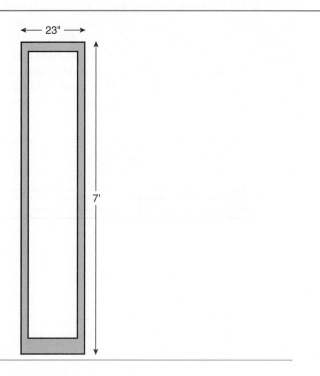

Figure 18–1
Standard 7 ft. by 23 in. relay rack used to mount equipment in the central office.

Figure 18–2
Ironwork is placed over the top of the relay racks to secure the relay racks and carry the many cables feeding the various pieces of equipment.

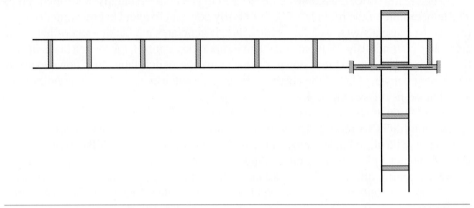

switch is preferred to false floors. A ladder rack is, as it sounds, a rack that looks like a ladder. The iron rack is hung above the switch from the ceiling or perched on poles mounted in the floor. Cables are strung across the ladder rack, as shown in Figure 18–3. The racking extends past the switch toward the cable distribution portion of the central office and holds the cables coming into and leaving the switch room. The switch room is fairly sparse, holding the switch, ironwork, and an air conditioning unit.

Connecting the Switch to the Outside World

A switch is useless without connections to the outside network. It must connect to multiple long distance networks, to other local switches, to operator services switches, to 911 centers, to CLEC switches, to ISPs, and to customer locations. A large number of cables are needed to connect the switch to the outside world, which means that a detailed cable plan must be produced to ensure proper, efficient cabling of the central office. A poorly planned switch site is messy and difficult to work with because of the jumble of cables that are almost impossible to trace. One of the most important tasks a switch engineer performs is planning for the future—within that plan, cable layout is critical.

Connecting to Other PSTN Switches   The central office switch connects to other PSTN switches, such as IXCs, CLECs, and other ILECs, using digital circuits. The size of the pipe established between the switches depends on the amount of traffic that travels between the two companies. DS-1 and DS-3 circuits are the most common ways to connect two switches. Switch-to-switch connections are always digital circuits whether they are DS-1, DS-3, or higher rate interface such as OC-3.

Figure 18–4 illustrates the cable layout of a DS-1 interface into the switch. The cables run on the ladder racking above the switch, then drop down like a waterfall to the

Figure 18–3
Cable rides along the ironwork.

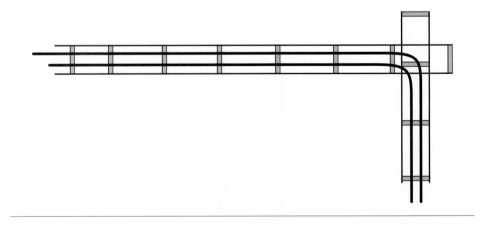

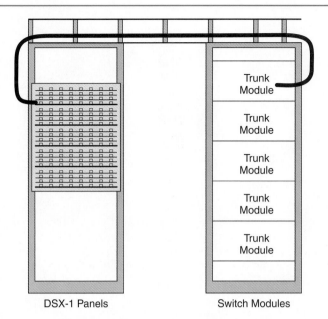

Figure 18–4
Connection between the switch modules and the outside world. The cable feeding the switch modules connects to an interconnection point for distribution in the central office.

DSX-1 Panels   Switch Modules

trunk interface ports on the switch. There, each transmit and receive wire is spun down on *wire wrap* pins attached to the back of the switch module. A standard wire wrap termination is shown in Figure 18–5, which shows how the wire is spun down—a term used to define a wire wrap connection on a pin conductor. T1 cable used in switch sites must meet certain electrical specifications. The wires are normally grouped into twenty-five-pair cables and encased in a special shielded jacket that helps protect the signals from electromagnetic frequencies emitted from adjacent circuits, switching equipment, and environmental influences like fluorescent lights. One end of the cable will be attached to ground in order to isolate the cable from stray voltages that may be induced by touching the ironwork or other conductors. The twenty-five-pair grouping of wire follows the standard color code shown in Table 18–1. The first pair to be terminated is

wire wrap
Termination pin used to physically wrap wire around.

Figure 18–5
Wire wrap pins are common termination connectors for copper wire conductors. (a) Shows the back of the panels with pins. (b) A close-up view of one pin with wire wrapped around it forming the connection.

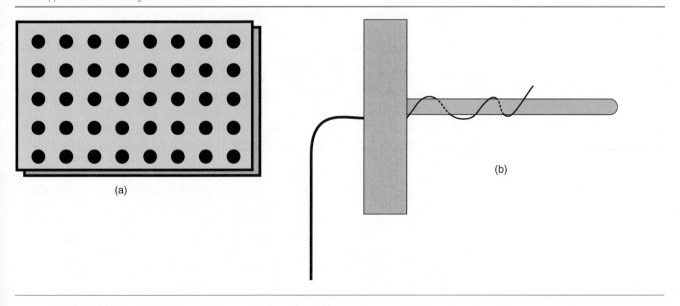

(a)

(b)

Table 18–1
Standard Copper Cable Color Code or
Solid Colors.

Pair Number	Tip Color	Ring Color
1	White	Blue
2	White	Orange
3	White	Green
4	White	Brown
5	White	Slate
6	Red	Blue
7	Red	Orange
8	Red	Green
9	Red	Brown
10	Red	Slate
11	Black	Blue
12	Black	Orange
13	Black	Green
14	Black	Brown
15	Black	Slate
16	Yellow	Blue
17	Yellow	Orange
18	Yellow	Green
19	Yellow	Brown
20	Yellow	Slate
21	Purple	Blue
22	Purple	Orange
23	Purple	Green
24	Purple	Brown
25	Purple	Slate

the white/blue pair, the second is the white/orange, the third is white/green, and so on. Figure 18–6 shows test tools used to troubleshoot the pairs of wires.

The DS-3 circuit is the second most common way to connect two switches together. A DS-3 connection requires coaxial cable to terminate at the digital interface port on the switch. Similar to the twisted wire cable, the coaxial cable is dropped from the ladder racking in a waterfall fashion, as shown in Figure 18–7, and terminated on BNC connectors located on the back of the switch module. A BNC connector is similar to the connector attached to a cable television box. The opposite end of the coaxial cable is attached to a multiplexer located in the transmission area of the switch site. The distance between the switch and transmission room must not exceed 450 ft., due to signal attenuation. If the distance is greater, in-line repeaters must be placed to compensate for the signal loss. The gauge of the coaxial cable determines the distance the signal will travel before it loses strength. The range listed by most manufacturers is 350 to 650 ft.

Another method used to interconnect the switch to the outside world is fiber optic cable. Similar to the DS-1 twisted pair and the DS-3 coaxial cable, the fiber optic cable follows the structure of the ladder rack. Unlike the copper cables, it normally rides inside a fiber optic trough attached to ironwork. The distance the fiber has to run determines whether fiber jumpers or fiber cable will be used. Fiber jumpers are individual fiber strands encased in a flexible plastic jacket. The fiber jumpers are used to connect two pieces of equipment together and range in length from a few feet up to more than 100 ft. If numerous connections have to be terminated and the distance between the two devices is great, a fiber optic cable that holds many strands of fiber is used. The fiber optic cable encases multiple strands with one plastic jacket. The two termination methods are shown in Figure 18–8. Fiber optic cables or jumpers normally terminate in the transmission room or transmission area where all of the fiber optic multiplexers reside.

These three cable types are used throughout the industry to connect switches. The initial switch-to-switch connection used copper wire to connect switchboards together.

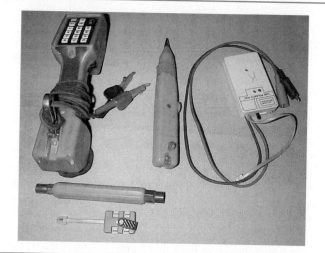

Figure 18–6
These tools are used to troubleshoot copper cable. Clockwise from bottom: connector clip, cabinet wrench, test set, wand to detect tone, and toner. (Photo courtesy of M. A. Rosengrant)

Figure 18–7
Connection between the switch modules and the outside world. The cable feeding the switch modules connects to an interconnection point for distribution in the central office at a DS-3 rate using coaxial cable.

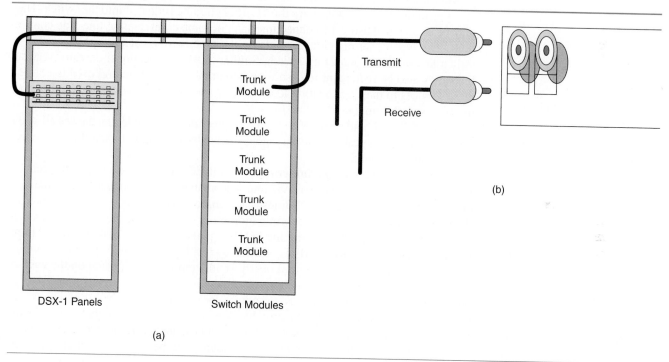

Figure 18–8
Fiber jumpers that connect the outside fiber cable to the optical equipment. The fiber distribution panel is the interface point.

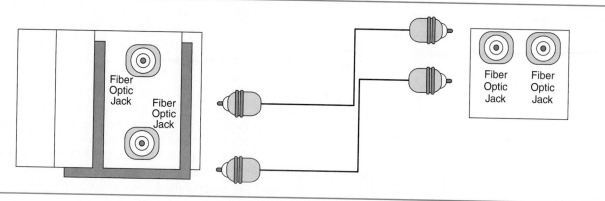

The only difference today is that switchboards have been replaced with electronic switches and signals are multiplexed together onto copper wire, coaxial cable, or fiber optic strands.

Interconnecting the Switch to Devices Other than a Switch   A switch requires more than just connections to other switches. It has multiple tasks to perform and must be able to talk to supplemental devices that work with the switch to perform those tasks. The tasks include operator services, 911, SS-7, voice mail, and many others. For the switch to be able to switch calls, it must have a connection into the SS-7 network, which consists of a link between the switch and an STP. Most switches connect to the STP using either a T1 signal or a 56 kbps signal. Diverse links into the SS-7 network are essential to ensure that the switch will continue to function even when one of the links fails. Without signaling, calls will not complete. The termination is similar to DS-1 feeds.

Local or tandem switches have links that connect them to the 911 centers. The 911 interface requires a T1 digital circuit that connects the switch at a digital circuit interface unit. The software in the switch defines the 911 port and routes all 911 calls onto that physical circuit. Each switch has duplicate 911 circuits riding diverse paths that provide alternate routes to ensure against downtime.

Operator services trunks are a third type of circuit terminated into the switch. They may be established using T1 connections and, in some cases, analog line-side connections. The type of termination depends on the type and age of the operator services switch and the interconnecting switch. Similar to 911 connections, operator services links may also be redundant to ensure service reliability. The method used to connect these services to their terminating locations is very similar to the method used to connect two switches together.

The SS-7, 911, and operator services circuits ride in the cable that carries all of the trunking traffic. The cable is strung along the ladder racking between the switch and the transmission room.

Connecting to the Local Subscriber   The ILEC's end-office switch connects directly to the local subscriber pair via line-side switch ports. Connecting the line-side of the switch to the outside POTS lines is accomplished by running copper wires from the line-side switch ports to the termination block mounted on the *main distribution frame* (MDF). The cable carrying the POTS signals is strung across the ladder racking in the switch room to the MDF, which may reside in the same room but, more often, in a separate area of the central office. Wire wrap pins are located on the back of the line-side modules of the switch where the color-coded wires are spun on. The opposite end of the cable is connected to switch blocks mounted on the MDF. The switch blocks consist of a passive termination block with pins on one side that terminate the switch ports, and pins on the opposite side that terminate the frame wire that connects to the outside pair. Figure 18–9 depicts the connection between the switch line units and the MDF.

Switch Alarms   A switch is responsible for switching thousands of calls every second. It goes into alarm when a circuit pack fails, a circuit goes down (referred to as a loss of a facility), errors are received on a facility, one of the processors fails, or other problems occur. Switch alarms notify the *network operations center* (NOC) that monitors the network for problems or the local switch technician. Building a switching office requires that office alarms be configured to ensure that the site functions. When it does not function, help is dispatched to fix the trouble.

Every switch has alarm ports defined as critical alarms, major alarms, and minor alarms. Copper wire cables are run from the switch's alarm ports to an alarm block on the MDF. The alarm block bridges the alarm leads to a contact closure device that interprets the incoming alarm signals, then passes them on to the technicians via a computer console tied directly to the switch. In addition to sending the alarms to the technicians' workstation, the alarms are fed back to the NOC where desk technicians monitor the network remotely twenty-four hours a day. The alarms in the office are often attached to visual and audible alarm devices such as lights and gongs.

main distribution frame
A wiring arrangement that connects the telephone lines coming from outside on one side and the internal lines on the other.

network operations center
The central place that monitors the status of a corporate network and sends out instructions to repair bits and pieces of the network as they break.

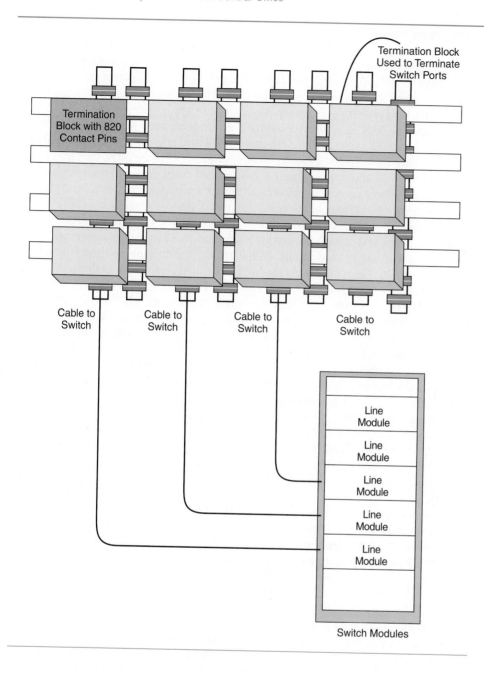

Figure 18–9
Horizontal side of the main distribution frame. Line-side terminations into the local class 5 switch.

When an alarm comes in, the light at the head of the row of relay racks lights, indicating that the trouble can be found in that particular row of equipment. Gongs also go off when an alarm is received. Most switch offices have several variations of audible alarms, each representing either a type of alarm or the severity of the alarm. The alarm remains in the *alarm state* until someone manually acknowledges it and turns it off, even if the trouble has not been cleared.

In addition to local alarms, a remote network operations center (NOC) also receives the alarms from the switch site. The NOC's personnel log in to that particular switch to determine the problem. The NOC may see that the switch site's air conditioner has failed and that the room temperature is rising above the allowed limit. The alarms on a switch give the local and remote technicians switch alarm information in addition to site information, such as environmental alarms. In order to maintain the 99.999% uptime standard that all telephone companies strive for, alarms must be fed locally and remotely to provide the status of the switch site.

alarm state
Notification state when trouble occurs, defined as critical, major, minor, and so on.

Switch Provisioning and Maintenance

maintenance control center
A central place in a stored program control central office from which system configuration and trouble testing are controlled.

A local *maintenance control center* (MCC) and a remote NOC are the control centers for switch provisioning and maintenance. The structure within the company determines the amount of authority each center holds. The traditional ILEC grants the switch technician complete autonomy over the switch site. The switch technician makes sure the switch is running properly—that the power, air conditioning systems, and building were maintained. The NOC acted as the intermediate point between the customer calling in and the switch technician that fixed the problem. Newer telephone companies and even the ILECs have now granted a greater amount of authority to the NOC than in previous years, thanks to its ability to remotely access almost every device in the switch center. Many central offices do not have dedicated switch technicians. Instead, the NOC now monitors the site and dispatches a roving technician when a problem occurs.

The MCC is the hub of activity at most switch sites. It is typically located next to the switch room and houses a computer terminal that is directly connected to the switch. The terminal is used to talk to the switch. Technicians also use it to tell the switch what to do. The main reason the MCC is located close to the switch room is that often a circuit pack needs to be reset or changed. The technician is able to easily move between the terminal in the MCC and the switch. Because the MCC is the technician's office, the switch technician spends a good deal of time in the MCC room troubleshooting switch issues, provisioning circuits, performing routine maintenance on the switch, and working with other telephone company departments. Most MCCs are set up to accommodate at least two computer terminals, a printer, several telephones, bookcases for manuals, file cabinets for circuit designs, and at least one coffeepot. A central office would not be complete without an MCC room.

Remote NOCs are used to monitor the state of the switch. The NOC is connected to the switch via internal LAN connections that allow the NOC technician to log in to the switch remotely and perform the same functions that the switch technician located in the MCC performs. The other devices at a switch site—such as the transmission equipment, the generator, and the air conditioner—are often attached to the NOC through an internal LAN connection. Most telephone companies have now deployed sophisticated operations support systems (OSSs) that are connected logically to each device in the switch site. The goal of an OSS is to consolidate the provisioning and maintenance of all the devices in the switch office. The OSS is also expected to allow for flow-through provisioning that automatically configures a circuit when a request is passed down from the order entry software program. Automatic provisioning is now a reality, but the details are still being refined.

18.1.2 CLEC Regional Switch Sites

Regional Switching Center
A control center connecting sectional centers of the telephone system together.

The switch located in the CLEC's *Regional Switching Center* (RSC) is a class 5 digital switch. It is able to switch calls the same way the ILEC class 5 switch switches calls. The CLEC purchases its class 5 switch from the same vendors that supply the ILEC with its class 5 switch. The switches have trunk and line-side terminations, along with auxiliary terminations for operator services. The CLEC regional switch site is very similar to the ILEC central office switch site. The CLEC, like the ILEC, places its regional switch in a controlled environment making sure the temperature is maintained according to manufacturer's recommendations. The CLEC switch also connects with IXCs, ISPs, ILECs, cellular providers, and other CLECs using DS-1, DS-3, and optical interconnection facilities. The facility's interconnecting switches are laid out in the same fashion as the ILEC's central office and cable is placed in ladder racking mounted above the switch. The copper and fiber optic cable is run into the transmission room or transmission area where the circuits interface transmission equipment such as multiplexers and DCSs. The CLEC switch connects to other switches by way of DS-1 circuits, DS-3 circuits, OC-3 circuits, or OC-12 circuits and are exactly the same as the ILEC's DS-1, DS-3, or OC connections. The CLEC also must connect with the 911 centers and the operator services switch and have at least two A-link circuits diversely fed into the signaling network.

Although the CLEC RSC appears to be an exact replica of the ILEC central office, there are some important differences that need to be explained to understand the total

Figure 18–10
The relay racks inside the dotted line are placed inside the ILEC's central office by the CLEC. The CLEC depends on the ILEC's outside plant facilities and must interconnect to the ILEC's MDF. The POI—point of interconnection—is where the ILEC allows the CLEC to connect to the ILEC's pair.

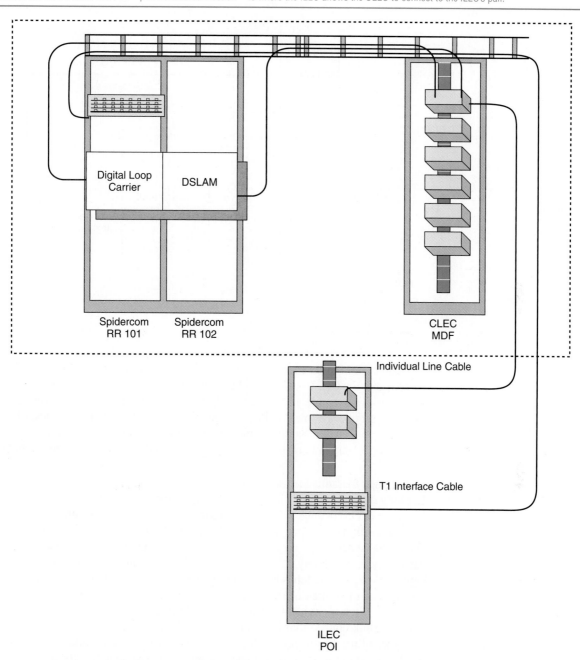

PSTN network. The CLEC switch, unlike the ILEC's central office switch, does not have direct line-side terminations to the subscriber. Instead, the CLEC connects to special equipment located in a cage in the ILEC's serving central office. Figure 18–10 illustrates a typical connection between the CLEC RSC and the ILEC's central office. The special equipment connects directly to the ILEC's MDF where it is cross connected to the POTS pair feeding the customer's residence. This termination may be viewed as the line side of the switch. The difference is that the line-side port modules are not sitting next to the serving switch but are located remotely in a cage in an ILEC's central office. The CLEC owns its own switching equipment but does not own the wire pairs feeding the customer's residence. Therefore, it must colocate its equipment in the serving central office to gain access to the subscribers served out of that office.

Figure 18–11
A backhaul scenario in which the traffic initiating in Plattsburgh is carried to the class 5 switch sitting in Burlington, switched, then returned to ILEC's tandem to be switched to the end user in Plattsburgh.

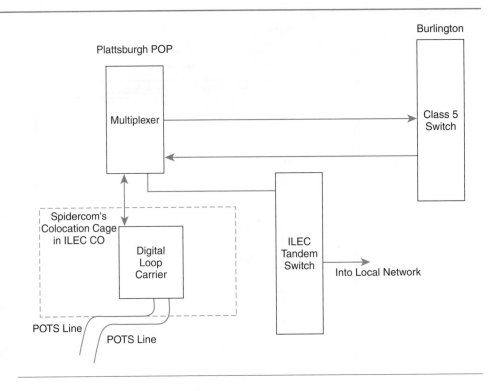

The second difference between the CLEC's RSC and the ILEC's serving central office is that the CLEC switch might not be located in the same city as the customers it is serving. For example, in Figure 18–11, the CLEC switch is located in Burlington, Vermont, but serves customers in Plattsburgh, New York. The term used to describe the connections between the switch and the remote colocation sites is *backhauling*. It describes how traffic is hauled back to a central switch located in a different geographic region.

18.1.3 Long Distance Switch Sites

The long distance (LD) switch site is similar to the class 5 switch site in that it also requires an environmentally controlled room. The LD switch terminates digital circuits from other digital switches such as ILECs, CLECs, ISPs, cellular providers, and other IXCs. The switch looks very much like a class 5 switch in that it consists of an administrative bay, a switch bay, and trunk modules arranged in a similar fashion, as shown in Figure 18–12. The switch is less complex, however, because it does not have to switch local calls and therefore does not require thousands of line-side terminations. The purpose of an LD switch is obviously to switch long distance calls. The LD switch only switches trunk-to-trunk connections by using the dialed number and the customer's record. The two types of trunk connections found on an LD switch are intermachine trunks (IMTs) and Feature Group A, B, or D. They connect two LD switches together. The Feature Group trunks connect the LD switch to the ILEC's tandem switch or directly to the ILEC's end-office switch, the CLEC's regional switch, or a cellular provider's switch.

The long distance switch site is similar to the ILEC and CLEC switch sites but differs in several ways. The most significant difference is that the long distance switch site does not have an MDF because it does not have line-side terminations. Instead, a smaller facility termination bay connects miscellaneous circuit connections such as SS-7 links, telephone circuits used at the site, alarm connections, and interdevice connections. The main portion of the long distance switch site is composed of the switch and the digital cross-connect panels that terminate the hundreds of trunk connections.

The cable from the LD switch is run along ladder racking to the transmission room or area. The wires are spun down on wire wrap pins located on the back of the trunkside interconnection modules. The number of cables required depends on the size of

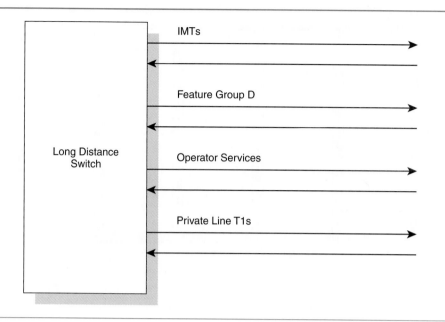

Figure 18–12
Long distance switch trunks are used to connect to different types of switches. The IMT connects two long distance switches together. The Feature Group D connects the long distance switch to a local tandem or end office. The operator services trunks connect to an operator services TOPS trunk. Private line T1s connect to PBXs or other customer premises equipment.

the switch and the number of directly connected fiber optic terminations. Many LD companies now interconnect with other carriers using fiber optic terminations. Using fiber eliminates a majority of copper T1 and T3 connections. Overall, the structure of a long distance switch site is the same as the local switch site, the cellular switch site, and the CLEC's regional switching center.

■ 18.2 TRANSMISSION EQUIPMENT

Obviously, every switch site has to have a switch, but every switch site must also have transmission equipment to connect the switching equipment to the network. The equipment placed in the transmission room is used to transport voice and data traffic across the PSTN.

18.2.1 Transmission Space

Every switch site, whether it is an ILEC, a CLEC, or an IXC, needs a portion of the office to house transmission equipment. The transmission space size depends on the size of the office. It can be as small as a 10 ft. by 10 ft. area and as large as multiple floors in a building. Like the switch site, the transmission equipment requires a temperature-controlled area.

Relay racks are mounted next to one another, as shown in Figure 18–13. They are normally 7 ft. by 23 in. and made of either aluminum or steel. Various pieces of equipment are mounted in the relay racks. The ironwork is mounted above the relay racks forming cable trays for the cable that has to be run in the transmission room. Figure 18–14 depicts a typical transmission area in a switch site.

18.2.2 Cross-Connect Panels
Defining DSX Panels

Every telecommunications company uses cross-connection panels. Several manufacturers, including ADC and Telect, manufacture different types of digital cross-connect panels. The purpose of a cross-connect panel is to provide an intermediate meeting point for the circuits coming into and leaving the site. The panel provides a point of interconnection. DSX stands for *digital signal cross connect*. The DSX panel is a passive fixed device mounted in a relay rack in the transmission room that physically cross connects two circuits together. Many different styles of DSX panels terminate a variety of cables. The two most common panels terminate DS-1s and DS-3s. Long distance switch sites contain numerous relay racks filled with DSX panels.

Figure 18–13
Five relay racks lined up in a row.

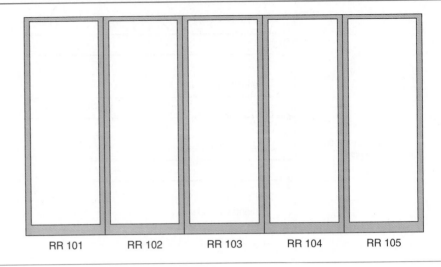

RR 101   RR 102   RR 103   RR 104   RR 105

Figure 18–14
Typical transmission area used to hold multiplexers, cross-connect panels, and numerous cables.

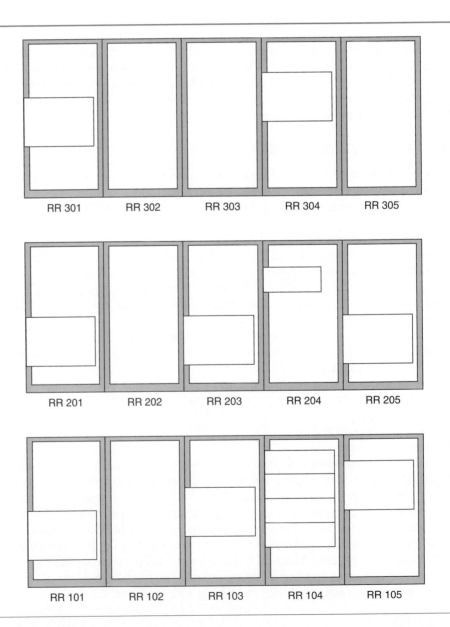

RR 301   RR 302   RR 303   RR 304   RR 305

RR 201   RR 202   RR 203   RR 204   RR 205

RR 101   RR 102   RR 103   RR 104   RR 105

Figure 18–15
DSX-1 cross-connect panel that terminates twenty-eight T1 and circuits. T1 and R1 are the termination points for the transmit pair of the four-wire T1 circuit. TR is the termination point for the receive pair of the four-wire T1 circuit. The ground pin on the LED above the monitor jack is used to identify the opposite end of the cross connect.

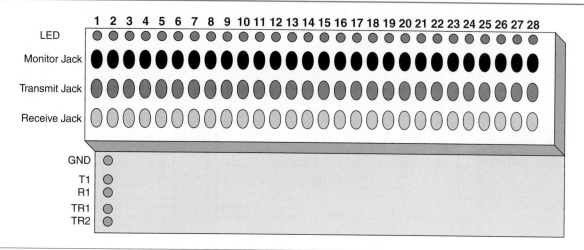

DS-1 DSX Panels

DS-1 DSX panels vary in size, in the number of DS-1 terminations, and in the physical layout of the wire wrap pins. Choosing the correct DSX panel depends on the application, the size of the space, and the buyer's personal preference. The function of the panel is not affected if the DSX wire wrap pins are located on the back of the panel instead of on the front of the panel. The wide range of styles exists because different carriers have different needs. The most common DSX panel has wire wrap pins on the back of the unit that connect to the network or to a piece of equipment and wire wrap pins on the front for the cross connect. Figure 18–15 illustrates a typical DS-1 DSX panel. A wire wrap tool is used to wrap the wire onto the cross-connect pins, as shown in Figure 18–16. The cross-connect feature is performed by flipping the transmit and receive pair on the front of the panel, as shown in Figure 18–17. Twenty-eight individual wire wrap positions with bantam jacks are located on the front of the DSX panel. Each of the positions corresponds to one DS-1 port. Because a DS-3 is made up of twenty-eight DS-1s, many DSX panels are built in twenty-eight DS-1 increments and many have more than twenty-eight positions. A commonly used DSX panel has eighty-four DS-1 positions, equating to three DS-3 circuits, each terminating twenty-eight DS-1s.

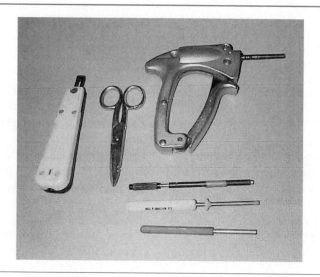

Figure 18–16
These tools are used to terminate copper wire onto wire wrap and to block connectors. Clockwise from left: punch-down tool, snips, wire wrap gun, three unwrappers. (Photo courtesy of M. A. Rosengrant)

Figure 18–17
Cross-connect panels are used to connect two separate ends of a circuit and to provide a point for testing the circuit. (a) The back of the DSX is shown connecting a piece of equipment to the network, then to a switch. (b) The panel functions as a test and cross-connect point.

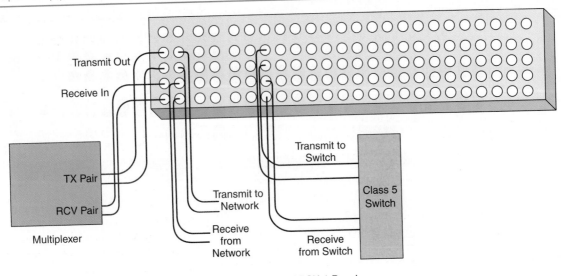

(a) Rear View of DSX-1 Panel

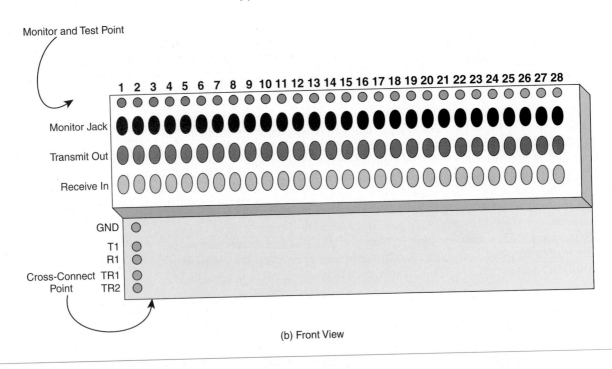

(b) Front View

Additionally, on the front of the panel, each DS-1 termination has three jacks that are used for maintenance. The top jack is the monitor jack, the second jack is bridged to the receive pair on the back of the panel, and the third jack is bridged to the transmit pair on the back of the panel. The jacks are referred to as *bantam jacks* and accept barrel bantam plugs, as shown in Figure 18–18. Test equipment is attached to the maintenance jacks using bantam jack cords. For example, to monitor the signal, we would plug one end of the cord into the monitor jack on the DSX panel and the other end into the bantam jack located on the front panel of the test equipment. The purpose of this test would be to monitor the receive signal—the signal coming in from whatever is terminated on the back of the DSX panel—for errors or a trouble condition. For example, if one of the DS-1 connections from the switch is terminated on the back of the DSX

Figure 18–18

(a) The small bantam jack is used to connect the test jacks in the DSX panel to test equipment or to patch a circuit quickly without having to wire the connection down. (b) The loop-back plug is used to loop the circuit back at that point in the network. It is plugged into the jacks on the DSX panel.

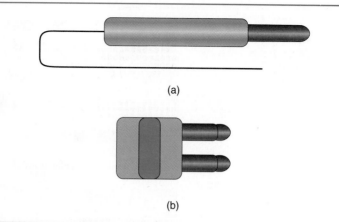

(a)

(b)

panel, the test equipment plugged into the front monitor jack views the signal being transmitted out of the switch to the DSX panel. The monitor jack is bridged onto the receive pair of the wires. The test is a *nonintrusive test*, meaning that it will not interrupt service. It simply monitors the signal for errors. A DSX panel is used continuously by technicians who are trying to determine the condition of a circuit.

    Intrusive tests break the circuit and are only done when the circuit is already down or a maintenance window has been scheduled. A common intrusive test is a loop-back test. For example, in the previous scenario, we monitored the signal coming in from the switch. To prove that the switch was not the cause of the trouble on the line, we could place a loop-back plug in the receive and transmit jacks back toward the switch, as shown in Figure 18–19. By placing a loop plug in the transmit and receive jacks, we have looped the signal coming from the switch back to the switch. The switch sees the

nonintrusive test
Circuit test that does not disrupt service.

intrusive test
Circuit test that disrupts service.

Figure 18–19

The transmit and receive jacks on the DSX panel are connected into the test box. The test box sends out a test pattern and analyzes the incoming signal pattern.

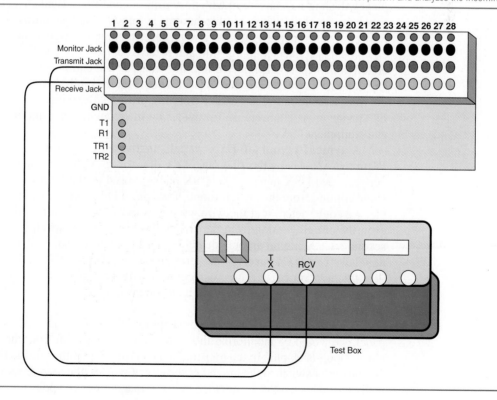

Figure 18–20
A typical DSX layout found in a telephone center such as a central office or long distance switch site. The DSX panels are placed together in relay racks.

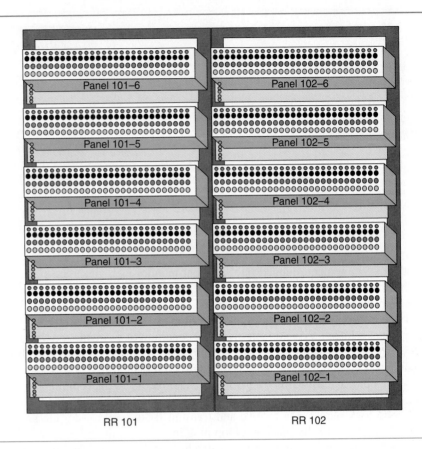

Panel 101–6   Panel 102–6
Panel 101–5   Panel 102–5
Panel 101–4   Panel 102–4
Panel 101–3   Panel 102–3
Panel 101–2   Panel 102–2
Panel 101–1   Panel 102–1
RR 101   RR 102

incoming signal and determines whether it is a good signal or a corrupted signal. If the switch port sees a good, clean signal, the switch port will reset and the test equipment will show a clean circuit with no alarms or conditions. Bantam jacks on the DSX panel are essential when troubleshooting a circuit. Another intrusive test involves feeding a stress pattern into the circuit and watching the signal for errors. The test equipment is connected to transmit and receive bantam jacks, and a stress signal generated in the test box is injected into the line. The test equipment monitors the signal and determines whether the circuit is clean or is errored and corrupted. The DSX panel provides a location to sectionalize troubles, provides flexibility by being able to easily change cross connects, and provides an interconnection point between facilities and equipment.

A typical layout of DSX panels in the transmission room is shown in Figure 18–20. Using this diagram, we will walk through a typical cross connection between two DSX panels. The DSX panel is used as the termination point for the cable coming from the switch room. If we placed tone on one of the pairs in the cable, we could detect it at the wire wrap pins of the digital port on the switch. However, the circuit coming from the switch has to leave the switch site and travel to a second switch site on an outside cable that is also terminated on a DSX panel. We now have two DSX panels—one terminating the switch cable and one terminating the outside plant cable. The cross-connect wire wrap pins on the front of the DSX panels connect the two DS-1 circuits terminated on the back of the two DSX panels, as shown in Figure 18–21.

If we were to plug our test equipment into the monitor jack of the outside DSX panel, we would be monitoring the receive signal coming in from the outside. If we place a loop-back plug in the transmit and receive jacks of the outside DSX panel, the equipment or switch at the far end of the circuit will see the loop. DSX panels are typically labeled using the T1 number, the span that defines end points of the circuit, and

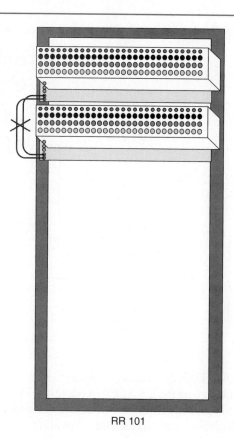

RR 101

Figure 18–21
Cross-connect jumper between two DSX positions. The transmit from panel 1 is connected to the receive on panel 2. The receive of panel 1 is connected to the transmit of panel 2—thus, the term *cross connect*.

a circuit ID. Above the monitor jack on the DSX panel, small LEDs identify which DSX panel and port the cross connect is tied to. In most ILECs, the LEDs are different colors and signify different termination types. A red LED signifies a copper T1 span, an orange LED a fiber optic span, a yellow LED a switch port, and a green LED a channel bank or some other piece of carrier equipment. The use of DSX-1 panels is essential for any type of carrier or communications provider.

DS-3 DSX Panel

The DS-3 DSX panel serves the same purpose as the DS-1 DSX panel, in that it is an interconnect point for DS-3 circuits. The main difference between a DSX-1 and a DSX-3 is that a DSX-3 panel terminates coaxial cable connections and the DSX-1 panel terminates twisted copper pair connections. The two most common types of DS-3 DSX panels are the external cross connect and the internal cross connect. The external cross-connect panel has DS-3 module cards that fit into a metal chassis, and BNC connection ports protrude from the back of the module, as shown in Figure 18–22. The two top connectors terminate the incoming signal. The third and fourth connectors are used for the cross-connect cable that links the DSX module to a second DSX module where the other end of the circuit terminates. As shown in Figure Figure 18–23, the two modules provide a way to connect the two ends of the circuit together. The transmit from module 1 is connected to the receive on module 2, and the receive of module 1 is connected to the transmit of module 2. This is a very important concept in the world of transmission—each signal leaving one piece of equipment is that equipment's transmit, but that same signal arriving at the opposite end's equipment is that equipment's receive.

The second type of DS-3 DSX panel has an internal cross connect eliminating the need for the physical cross connects just described. The DS-3 cable from one end is

Figure 18–22
Coaxial connection from a multiplexer
to a DSX-3 module.

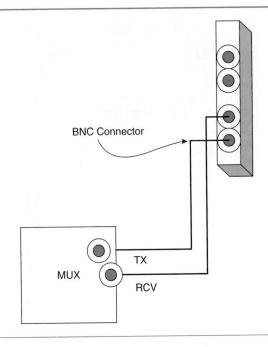

terminated onto the DS-3 DSX module, as shown in Figure 18–24. The opposite end of the circuit is also terminated on the DS-3 DSX module but on the bottom two BNC connectors, also shown in Figure 18–22. The DS-3 module has an internal cross connect that connects the two ends of the circuit together, thus providing a path for the signal to travel. The convenience afforded by the internal cross-connect module is obvious. The technician no longer has to make up short coaxial cables to cross connect DS-3 modules.

Figure 18–23
Two DSX-3 modules connected to two separate multiplexers. The DS-3 circuit from one MUX is connected to the DS-3 jack of another MUX through the DSX-3 cross-connect modules. The cross-connect modules are connected by coaxial cable. The transmit from one goes to the receive of the other, and vice versa.

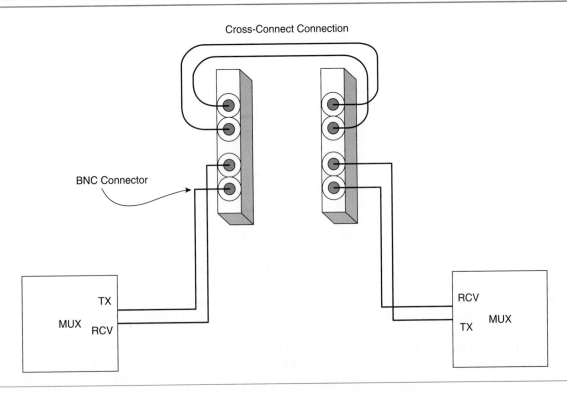

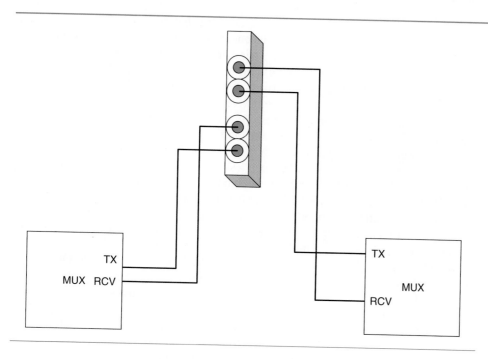

Figure 18–24
DSX-3 module with an internal cross connect that eliminates the need for a second DSX-3 module and physical cables. The DS-3 circuit from one MUX is connected to the DS-3 jack of the other MUX through the DSX-3 cross-connect modules.

18.2.3 Transmission Equipment

We discussed transmission equipment in detail in an earlier chapter. Here we explain where the transmission equipment resides in the switch site, how the equipment is connected to the network and the switch, and why transmission equipment layout is critical to building a reliable, efficient communications network.

Multiplexers in the Transmission Room

The concept of the multiplexer (MUX) was explained in detail earlier. Now we are interested in where a MUX is placed in the switch site, how it is used, and the various types of interfaces commonly found in the equipment placed in the switch site of local and long distance carriers. Multiplexers are mounted in relay racks in the transmission room or transmission area of a switch site and are used to combine lower-speed signals into one high-speed link. The multiplexer can be divided into three generic areas: common cards, line cards, and tributary or interface cards. The common cards may be viewed as the brains of the unit; their number varies depending on the type of multiplexer. The term *common card* refers to the cards that work for the MUX and all other cards in the unit. They are common to all other card types and are not used to interface lines going in or coming out of the box. An example of a common card is the processor card of the MUX. The processor card holds the latest version of code, it may have a database of circuit information, and it communicates alarms to the upstream network. Multiplexers placed in service provider networks normally require redundant processor cards to ensure that the box will function even if one of the processors breaks down. A second type of common card is the craft interface terminal (CIT) card that allows the technician access into the box. The CIT card may also be used to perform tests or handle remote hookups. A third common card that is always found in fiber optic multiplexers is that timing or synchronization card that provides an interface to the external timing supply and an internal stratum clock signal. Generally, the processor card, the CIT card, and the timing cards make up the common equipment in the MUX. As with everything else in telecommunications, there are variations to this rule.

The second functional group of the MUX is the line cards, which interface the large multiplexed pipe. In some instances, they are T1 interface cards that carry multiple DS-0s. In other cases, they are OC-48 cards that carry forty-eight DS-3s or thousands of DS-0s. The line card is the one that interfaces with the big pipe traversing the network.

The third functional group is the interface or tributary cards. The tributary cards may vary in speed and interface type depending on the multiplexer. For example, an OC-48 fiber optic MUX may have DS-3 tributary cards that interface twelve DS-3 circuits, an OC-3 interface card that interfaces three OC-3 circuits, and a GigE card that interfaces a Gigabit Ethernet signal. Each of the interface or tributary cards is aggregated together onto the line by being multiplexed into the line card. Small multiplexers, such as a channel bank, may have ten foreign exchange (FX) analog interface cards and fourteen DDS port cards multiplexed into one T1 line signal.

The variety of multiplexers depends on the interconnection speeds needed between sites. A full OC-48 pipe required between a switch site and an adjacent switch site requires that an OC-48 multiplexer be located in the transmission room. A channel bank may also be needed to multiplex DS-0s into a DS-1 pipe feeding a distant switch location. Lower-rate multiplexers often feed into higher-rate multiplexers, creating a long line of smaller signals being "eaten" by bigger signals, similar to a chain of fish each eating the smaller fish. For example, a channel bank that is multiplexing two SS-7 links into one T1 signal will interface into the OC-12 multiplexer and, in turn, the OC-12 multiplexer will interface into the OC-48 multiplexer. Today's large networks often include DWDMs that take OC-48 signals from the OC-48 multiplexer and wave division MUX them into one optical signal made up of as many as eighty individual wavelengths. So, every multiplexer has lower-speed signals feeding into interface packs that are combined into high-speed signals that are fed to the outside through line cards. Even the largest signal can become an interface signal to some higher-rate multiplexer. A high-speed multiplexer such as the OC-48 carries layers of multiplexed traffic. It is important to note that transmission happens in both directions. The higher-speed signal drops out lower-speed signals that are fed into the interface circuit packs of the multiplexer.

A typical transmission room holds a variety of multiplexers, from DS-1 upstream channel banks to high-end DWDMs. The layout of the equipment and the interconnection of the cable are critical parts of the overall success of the switch site's ability to handle large volumes of traffic. We will begin by looking at a DS-1–fed channel bank that is carrying an SS-7 DS-0 circuit. A switch has two options when connecting SS-7 circuits. The first is a full DS-1 circuit that interfaces directly into an SS-7 port on the switch and terminates as a full DS-1 into the STP wherever that is in the network. The second SS-7 circuit type is a 56 kbps DS-0. The four-wire connection at the switch is normally connected to a higher-order multiplexer before being transported out onto the network. In our example, the SS-7 copper four-wire circuit travels between the switch room and the transmission room by being placed in the ironwork established between the locations. The DS-0 four-wire circuit normally terminates on a block located on the MDF before traveling to the transmission area. The block on the MDF is where all DS-0s needing to interface into the channel bank meet. From the MDF block, a twenty-five-pair cable is extended into the transmission area and terminated on the back of the channel bank using a prebuilt amphenol connector, as shown in Figure 18–25. The SS-7 circuit enters the channel bank's DS-0 56 kbps circuit pack, and the DS-0 is multiplexed up into the T1 signal. The T1 wire from the channel bank is terminated on one port on one of the DSX panels. In most cases, individual T1s do not exit the site as just one T1 circuit. Instead they are fed into higher-order multiplexers that leave the site via fiber or microwave technologies. In our example, the T1 circuit we wire wrapped down on the back of the DSX panel must be multiplexed into a DS-3 circuit that is destined for the location where our SS-7 circuit terminates.

The twenty-eight DS-1 interfaces feeding an M13 multiplexer are wire wrapped down on the back of the DS-1 DSX. The high-speed DS-3 leaving the multiplexer rides on coaxial cable that terminates on a DS-3 DSX panel. Our SS-7 T1 must somehow feed into the M13 multiplexer as a T1 signal. A physical cross connect is made between the M13's DS-1 DSX panel port 12 and the port on the DSX where the channel bank's T1 is terminated. Once the cross connect is in, the circuit is complete from

1. The switch to the channel bank
2. The channel bank to the DSX panel

Figure 18–25

Typical connection between a channel bank's individual channels and the T1 interface on the DSX panel.

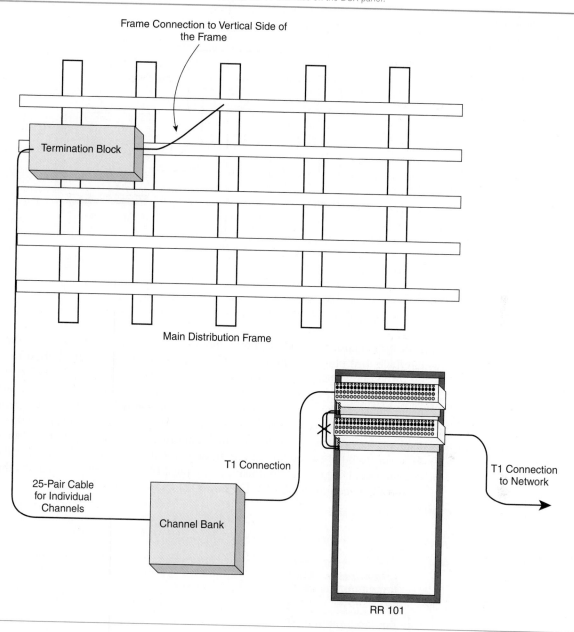

3. The DSX panel to the second DSX panel
4. The second DSX panel to the M13 multiplexer
5. The M13 multiplexer to the DS-3 DSX panel
6. The DS-3 DSX panel to the fiber optic OC-12
7. The fiber multiplexer's laser out to the far end optical receiver
8. The optical receiver to the DS-3 interface circuit pack
9. The interface circuit pack to the internally cross-connected DS-3 DSX panel
10. The DS-3 DSX panel to an M13 multiplexer
11. The M13 multiplexer to a DS-1 DSX panel
12. The DS-1 DSX panel to a second DS-1 DSX panel
13. The DS-1 DSX panel to the SS-7 STP

The one step we left out is how the signal leaves on the fiber optic cable. The OC-12 also has to be multiplexed into a higher upstream signal. The OC-12 upstream fiber jumper is connected directly to an OC-12 interface pack plugged into an OC-192

Figure 18–26
Fiber distribution panels are used to cross connect fiber multiplexers to other fiber multiplexers or to the outside network.

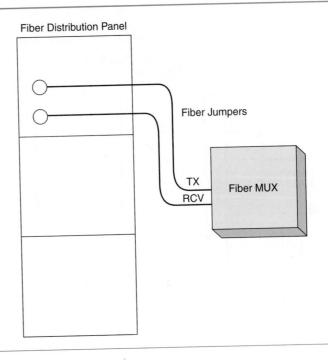

fiber optic multiplexer. At this point, the SS-7 circuit has made it as far as the OC-12 interface card plugged into the OC-192 multiplexer, which is the highest rate multiplexer in our scenario. The upstream fiber optic jumper from the OC-192 multiplexer is fed into a fiber optic distribution panel. The connection made at the distribution panel connects the fiber from the multiplexer to the outside plant fiber feeding the transmission room. An internal fiber connects the outside plant at manhole 0—the first manhole leaving the switch site—to the fiber distribution panel. The signal is now able to travel from the switch all the way through the transmission room and finally to the outside world. Figure 18–26 illustrates a typical fiber distribution panel and the many types of connections feeding it. Figure 18–27 shows the SS-7 circuit.

Figure 18–27
The connection required to carry SS-7 messages. A T1 connects the switch to an STP. From the STP, the T1 is connected to another STP's resident in the SS-7 network.

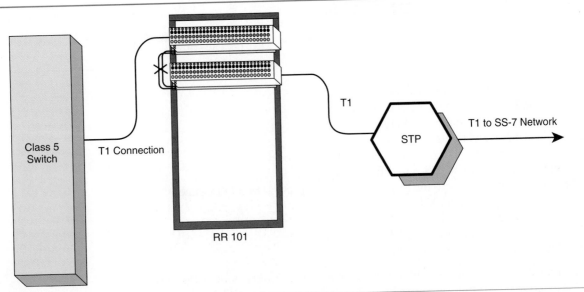

Most networks have now ...
performed remotely. The next s...
states and produce a simplified...
not easy because the "what if" ...
lems is huge. The advantage of ...
cians would not need to spend...
them more quickly. In addition...
vices, as well as the number of ...
more sophisticated method of r...

18.5.2 Network Operations Cen...
The network operations center...
the switch site, typically by w...
work. An alarm condition appe...
and the technician can then log...

Smaller carriers have speci...
alarm condition occurs. The net...
and pages the technician on cal...
another off-site location and fix...

■ 18.6 CENTRALIZED PO...

An area of the switch site that...
transmission equipment, and e...
office power plant is located ir...
power source for the entire cent...
ing AC into the −48 V DC feed...

For the switch site to rem...
provided along with a backup...
The first source of power is alw...
backup, and the third is the d...

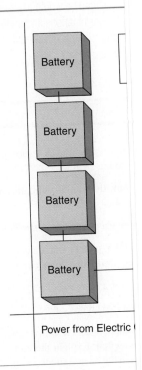

Power from Electric ...

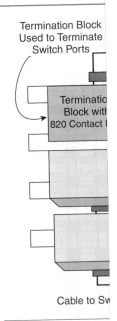

Termination Block
Used to Terminate
Switch Ports

Terminatio...
Block with
820 Contact ...

Cable to Sw...

vertical side of the f...
proper line number ...

Horizontal Side of th...
The *horizontal side*...
width of the frame. ...
posite the vertical si...
in Figure 18–31. Nu...
dreds of twisted cop...
in the switch room. ...
wire wrap pins on tl...
so when a wire is w...
a signal can travel t...

The cables lea...
mounted on the hor...
defined earlier in th...
horizontal side of tl...
switch termination...
mination on the ve...

An example o...
Amherst Street and...
central office, the f...
service needs to be...
scans down the ord...
telephone service. ...
mination point. Thc...
point on the switch...
teenth pin on termi...
runs a frame jumpe...
nation block 5 pin...
nected from his ho...
office. Henry is ve...
miliar dial tone.

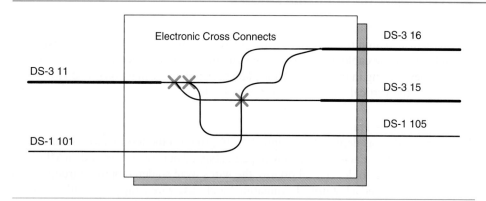

Electronic Cross Connects

DS-3 16

DS-3 11

DS-3 15

DS-1 105

DS-1 101

Figure 18–28
The DAC—digital access cross
connect—is used to cross connect
circuits, electronically eliminating the
need for physical DSXs and
multiplexers.

Digital Signal Cross Connect
The use of electronic digital signal cross-connect (DSX) equipment has spread through the network at an amazing pace. The DSX was introduced in the transmission network about twelve years ago with the DS-1 cross connect, which is used to help route individual DS-0s onto different DS-1s without physically wiring out every connection. From the DS-1 cross-connect system, the 3/1 DSX evolved allowing DS-1s riding in DS-3s to be moved electronically, again eliminating the need to physically wire circuits together. Only the largest switch sites have DSXs due to their cost, maintenance requirements, and size.

To show how a DSX improves the efficiency of the network, we will walk through the previous example and add a DSX to the switch site. Our SS-7 DS-0 still leaves the switch as a four-wire DS-0 interface and travels to the channel bank, where it is multiplexed into a DS-1. The DS-1 is still connected to the DSX port. The channel bank side of the connection is now cross connected to a second DSX panel that interfaces the DSX. The signal travels into the 3/1 DSX through a DS-1 port. The DSX has DS-1, DS-3, and OC-12 interface circuit packs. The DS-1 carrying the SS-7 DS-0 is cross connected to a VT-1 riding inside a VT-1, which is, in turn, riding inside the OC-12 signal. The OC-12 fiber jumper connects the OC-12 interface circuit pack in the DSX to the OC-12 interface pack of the OC-192, which carries the signal through the fiber distribution panel to the outside world.

The DSX was able to eliminate additional physical connections, such as the DS-1 interface into the M13, the DS-3 between the M13 and the OC-12, the actual OC-12, and the fiber connection to the OC-192. The DSX helps eliminate physical connections between multiplexers, reduces the number of physical interface ports a signal travels through, and improves the efficiency of the provisioning process. Figure 18–28 illustrates this example.

18.2.4 Power in the Transmission Room
The power requirements for multiplexers, DSXs, and channel banks are found in manufacturer specs. Most telecommunications equipment, including the switch, requires a −48 V feed. To guarantee a 99.999% network, two separate power feeds are deployed in the transmission room—an A feed and a B feed. If the A feed is lost, the equipment will switch to the B feed power source. The formula to determine the fuse size for the equipment varies according to the equipment vendors' defined specs for total power drain. Many power engineers use the two-and-one-half times drain rule to determine the rate to fuse a piece of equipment and the amount of total power required per transmission room. The source used to power the transmission room is backed up by on-site generators and backup batteries—the same used by the switch. Very seldom does a power outage cause a switch site to lose power but when it does, it is a serious problem.

Figure 18–34
Fiber distribution splice box mounted on the wall of the cable vault. The outside fiber cable is fed into the box where it is connected to the inside fiber cable that is fed to the transmission room.

vertical side of the
Termination of all plant lines.

Figure 18–29
The vertical side of terminates cable fro plant.

operations support system
Methods and procedures that directly support the daily operations of the telecommunications structure.

batteries immediately kick in and provide power to the site. Within a few seconds, the backup generator starts up and takes over from the batteries as the power source. The generator runs until the electric company restores commercial power. The only downside to depending on generators is that they, too, require a source of energy. Diesel fuel or natural gas is commonly used to power the generators, and diesel-fed generators require refueling if the power stays out for an extended period of time.

SUMMARY

The purpose of this chapter was to explain each of the elements found in a switch site. The environmentally controlled switch room, the layout of the switch in the switch room, the method used to connect the switch room to the transmission room and MDF, and the different pieces of equipment found in the transmission room were discussed. How the DSX helps reduce the number of physical connections was explained, as was how the DSX works. The MDF found in a local end-office switch was presented and compared to the MDF found in the CLEC and IXC switch site. The method used to connect circuits together using DSX panels was outlined and the interfaces between the fiber optic multiplexers were touched upon.

The cable vault was detailed as the location where outside cable enters the switch site, and splicing points were noted for both copper and fiber optic cable. Finally, the network management of the devices in the switch site was discussed and the need for sophisticated network management systems was explained. The final topic of discussion was the power plant, including the backup generator, backup batteries, and −48 VDC power supply.

CASE STUDY

As you may have guessed after reading this chapter, your assignment is to lay out the central offices for Green Grass Telephone Company. You will need to draw up floor plans showing the switch room, transmission room, cable vault, and Blue MacGillian's MCC room. Blue was just hired by Big Bob as the first employee of Green Grass. Blue is Big Bob's brother-in-law and has decided he would like to be a switch technician. You will also need to size the MDF for the number of cable terminations and include the correct number of DSX panels for DS-1 and DS-3 termination. Fiber distribution panels should also be sized correctly as should the power plant and cable layout for the building.

REVIEW QUESTIONS

1. What does the term *environmentally secure* mean when referring to a switch site?

2. What is the difference between a central office and a long distance switch site?

3. What is the difference between an ILEC's end office and a CLEC's regional switch site?

4. Why does a switch require so much air conditioning?

5. Draw a main distribution frame and label the vertical and horizontal sides.

6. Explain how the outside cables are terminated on the vertical side of the frame.

7. Explain how the switch is connected to the horizontal side of the frame.

8. Draw a picture showing how typical POTS lines enter the central office and connect into the switch.

9. Draw a picture showing how the trunking circuits connect the switch to the outside world.

10. Why does a switch site have a generator on site?

11. Explain the purpose of a DSX panel in the transmission room and why two ports are required when connecting facilities.

12. What is the standard size for a relay rack?

13. Why are alarms important in a switch site?

14. In a transmission room, other carriers terminate circuits in order to gain access to the switch. Explain the path of a DS-3 circuit that is terminating in the switch site from an alternative carrier.

15. Explain why a channel bank is often found in the relay rack lineup of a transmission room.

16. What type of multiplexing equipment is found in the transmission room?

17. What are DACs used for and how do they improve the efficiency of the switch site?

18. What does the term *ironwork* mean?

19. Trace the path of a long distance call starting from a telephone served out of the central office through the switch site to an alternate long distance carrier. List each step the call takes as it travels through the serving end office to the long distance switch site.

20. Trace the path of a long distance call starting from a telephone served out of a DLC owned by a CLEC. List each step of the call as it travels between the ILECs and CLEC and between the CLEC and IXC.

TROUBLESHOOTING

Problem 1: You have just received a call from the NOC that the switch is shutting down and workers at the site do not know what to do. You get ready as quickly as possible and head out to the switch site. As you step through the door, you instantly realize the HVAC has quit and the switch is ready to shut down. Explain what you would do next.

Problem 2: You have been asked to rearrange two circuits between two central offices. Using Diagram 18–1 in Appendix C, diagram out how you would connect T1 56 riding in span 10 to T1 20 riding in span 30.

Once you have moved the cross connect on the DSX panel, you will need to stress test the circuit. Again using Diagram 18–1,

show how you would loop back the circuit in both directions. Once the loop back is in place, you will need to call Blue at the other central office to confirm that the multiplexer there sees the loop back and is operating.

Next you will need to perform an intrusive test using test equipment and bantam cords. Show on the diagram which plug holes you will place the transmit and receive cords in. Once you have placed the cords, you will need to run several stress patterns. Using Appendix B, list the stress patterns you are going to run and explain the purpose of each.

If you were to receive CRC errors, what would this indicate? What would BPV—bipolar violation—errors indicate? What would framing errors indicate? Finally, what would timing slips indicate?

KEY TERMS

footprint (509)

ironwork (509)

relay rack (509)

wire wrap (511)

main distribution frame (514)

network operations
 center (514)

alarm state (515)

maintenance control
 center (516)

Regional Switching Center
 (516)

nonintrusive test (523)

intrusive test (523)

vertical side of the MDF (532)

horizontal side of the
 MDF (533)

line equipment number (533)

operations support system
 (536)

19

The Public Switched Telephone Network Topology

Objectives

After reading this chapter, you should be able to

- Describe the outside plant of the local PSTN.
- Describe the outside plant of the long distance PSTN.

Outline

Introduction

19.1 Outside Cable Layout in the Local PSTN

19.2 Equipment in the Outside Plant of the Local PSTN

19.3 Outside Plant Layout for the Long Distance Network

19.4 Carriers' Carriers

19.5 Right-of-Way for Terminals

◼ INTRODUCTION

Most of us do not notice the parts that make up the outside telephone plant. We drive past antennas, walk around open manholes, and become annoyed if a telephone truck is blocking the roadway. The expanse of the telephone network is hard to fathom. Cables run from telephone pole to telephone pole, and miles and miles of cable run in a labyrinth of canals under streets and buildings.

Chapter 19 provides an overview of the outside plant network. It looks at the way cable is laid out in the local telephone network, at the equipment placed in the outside plant, and at the right-of-way issues facing every telecommunications carrier from wire line to wireless. Although the local plant and the long distance plant have many similarities, the chapter differentiates between the two, providing an insight into how their networks are laid out, how they connect, and the consequences of sharing capacity.

■ 19.1 OUTSIDE CABLE LAYOUT IN THE LOCAL PSTN

19.1.1 Copper Cable Layout

Copper cable has a very long life. In many cities, copper cable placed in the ground at the turn of the twentieth century is still being used today. The architecture of the plant is critical to ensuring cable life, plant flexibility, and circuit capacity. As shown in Figure 19–1, the cable coming from the central office meets the outside plant cable in *manhole 0* (or *telephone pole 0*) located just outside the central office. Manhole 0 is a common term used to represent the last manhole before the central office. This manhole is often in the middle of the street or in the yard in front of the central office. Telephone pole 0 is the last telephone pole before the central office. Manhole 0 or telephone pole 0 is also the meeting point for other service providers' cable. For instance, if a carrier wishes to connect its fiber cable to equipment located in the ILEC's central office, it will need to run fiber cable to manhole 0, where it is spliced with the cable entering the central office. In many situations, manhole 0 becomes the entrance point for the non-ILEC carrier cabling.

The cables terminating at manhole 0 or telephone pole 0 are very large and black and carry as many as 6000 subscriber pairs. The cable is dispersed throughout the *carrier serving area* (CSA), running between poles or snaking underground in ductwork. The CSA is the area served by the central office, as illustrated by Figure 19–2.

manhole 0
First manhole outside the central office.

telephone pole 0
First telephone pole outside the central office.

carrier serving area
The area served by the central office.

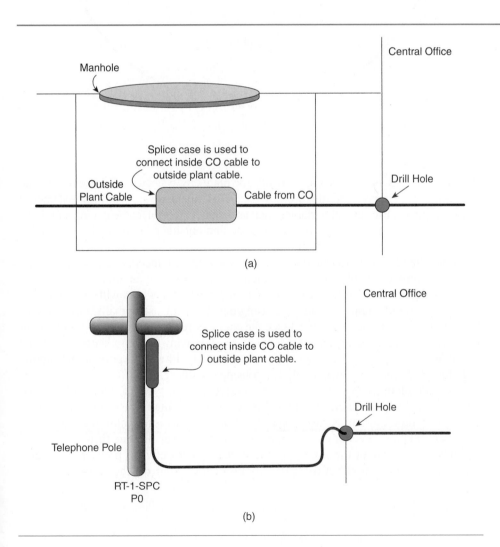

(a)

(b)

Figure 19–1
The interface between the central office and outside plant occurs at either (a) manhole 0 or (b) telephone pole 0.

Figure 19–2
A typical carrier serving area defined
around the central office. The CSA
encompasses all locations that will be
served by the central office.

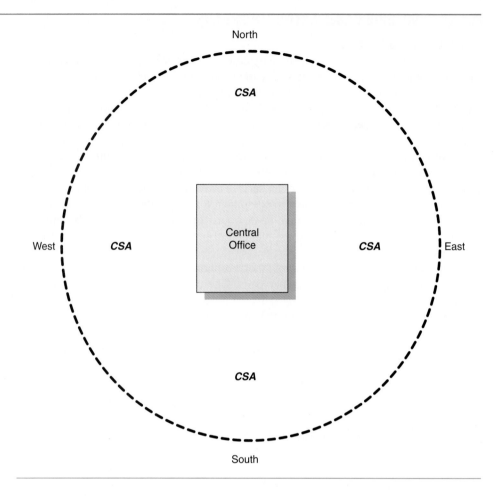

The term *CSA* has always been used by outside plant engineers to define the region that the central office serves and, therefore, which subscribers are served from that central office. Engineers use several variables such as distance, terrain, cable gauge, and right-of-way issues when designing an office's CSA, and outside plant drawings show large circles around the different central offices.

Once the CSA has been defined, the outside plant engineers begin to lay out the cable. The number of subscribers and potential subscribers in an area determines the size of the feeder and distribution cable. The engineer often works with area planning boards that provide input into future housing developments, strip malls, or other sites that will require telephone service. Once the engineer feels comfortable with the number of subscribers and their location, a mechanical drawing program is used to lay out the cable runs. Figure 19–3 shows a typical cable run with many spurs of distribution cable feeding each street, strip mall, and housing complex.

Outside plant cable is categorized into three general types: feeder cable, distribution cable, and drop cable. Feeder cable is the main cable connecting to telephone pole 0 or manhole 0 and varies in size from as small as a few hundred pairs to as large as 6000 pairs. Feeder cable is similar to a large river flowing through an area. It winds its way along the main thoroughfares, either between poles or in underground ductwork. As with a river, distribution cables spur off the feeder cable the

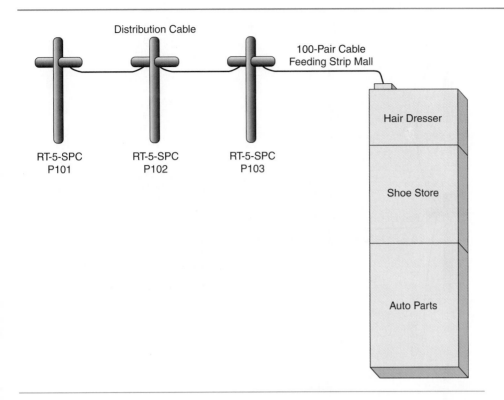

Figure 19–3
Distribution cable feeds into the strip mall, and each store in the strip mall is fed from the distribution cable. The telephone poles are named as shown. RT = route, 5 = route number, SPC = company name, P10X = pole number.

same way creeks spur off a river. The distribution cable hangs between the telephone poles running down individual streets. It may be as small as twenty-five pairs or as large as a feeder cable, depending upon the size of the customer base in the area. Drop cable spurs off the distribution cable and extends to a subscriber's home or business. Termination boxes hang off the distribution cable or pole where the drop cable connects. Drop cable may carry only one pair or as many as twenty-five pairs to the subscriber. The typical drop pair feeding residential customers consists of two copper pairs, one blue and white and the second orange and white, surrounded by a black plastic jacket.

Figure 19–4 illustrates how one wire pair from the central office travels through the feeder cable, then the distribution cable, and finally to drop cable attached to the subscriber's home. For example, feeder cable 800 carries 3000 subscriber pairs down Main Street. A 200-pair distribution cable spurs off the feeder cable at Oak Street, running past each of the homes. At 230 Oak Street, a drop wire is connected to the terminal block mounted on the telephone pole. Pair 200 in the distribution cable is connected to the third lug on the termination box. A drop wire is run from the protector mounted on the house to the termination block on the pole. The wire connecting to the distribution cable and the drop wire are joined in the terminal block when the drop wire is connected to the third lug. The outside plant records show that pair 200 of cable 800 is being used by the subscriber living at 230 Oak Street. If we were to go to the vertical side of the MDF and connect our test set to pair 200 in cable 800 and then go off-hook, we would hear a dial tone from the switch. When we hear the dial tone, we can dial the line verification, usually 511 or 311, to hear the line number of that pair. We could also listen in on the subscriber's conversation because cable

Figure 19–4
Distribution cables, feeder cables, and cable numbering scales.

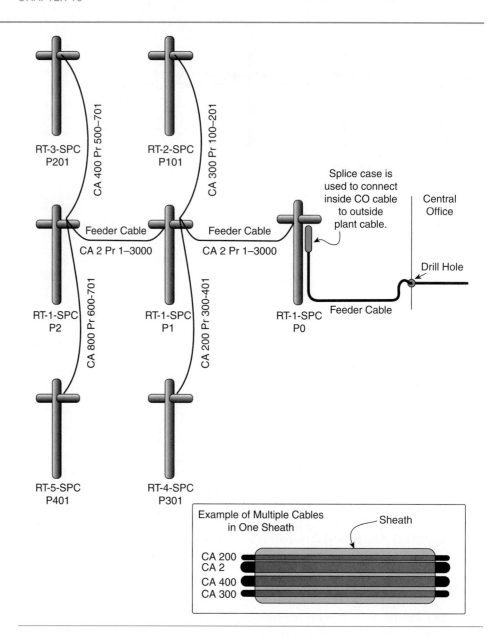

800, pair 200 is connected end to end, switch to subscriber. Therefore, the wire pair rides in the feeder cable, the distribution cable, and finally the drop wire, as shown in Figure 19–5.

19.1.2 Fiber Optic Cable Layout

The main difference between the layout of fiber optic cable and copper cable is that fiber optic cable normally does not terminate at the subscriber's home. Fiber optic cable connects the central office to large buildings, other central offices, remote switch sites, digital loop carrier systems, and POPs in a local telephone region. It is also used to carry large amounts of traffic between two points. The fiber optic cable shares the same ductways and telephone poles as the copper cable, and the same technicians maintain both the copper cable and the fiber optic cable. Fiber optic cable, like copper cable, terminates in manhole 0 or on telephone pole 0 before entering the

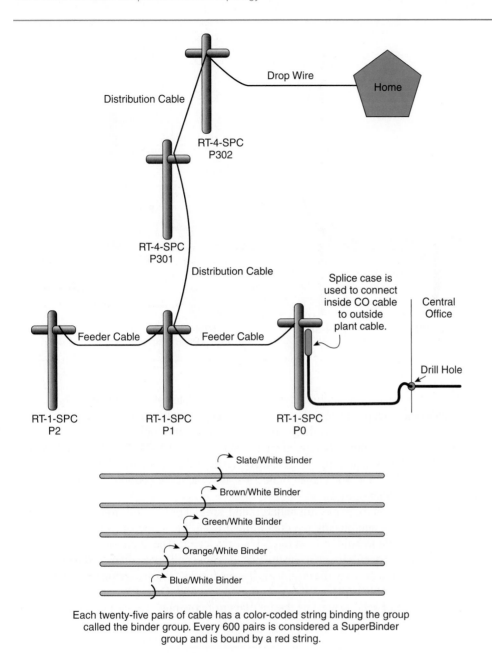

Figure 19–5
The individual pair from the CO feeds from the feeder cable to the distribution cable and through the drop cable.

Each twenty-five pairs of cable has a color-coded string binding the group called the binder group. Every 600 pairs is considered a SuperBinder group and is bound by a red string.

central office. Figure 19–6 illustrates the typical layout of fiber optic cable in a local network.

Fiber optic cable is much smaller in diameter and weighs less than copper cable. It is housed in strong polyurethane jackets that look identical to the black outer casings that enclose copper wire pairs. A small black cable runs next to the large traditional telephone cables between telephone poles. A loop extends from the cable at intervals, as shown in Figure 19–7. The loop is called slack and is designed in every span to provide extra cable in case a section needs to be cut away or a disaster happens, such as a digger hitting the cable and pulling it up or out. Whenever the fiber optic cable is distributed to an underground cable or terminated at a carrier box, orange plastic tubing is used to enclose the cable being spurred off the main feeder. The orange tubing signifies that there

Figure 19–6
Fiber cable is distributed throughout the CSA along the cable runs. The fiber cable runs along the feeder cable to the distribution cable and to the strip mall.

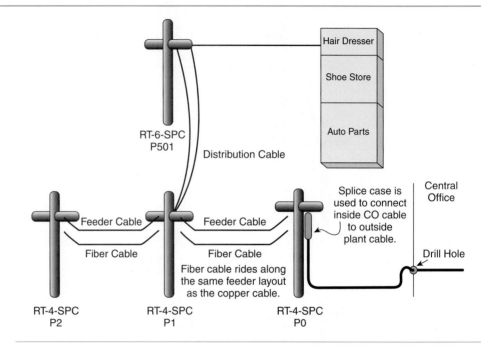

is fiber in that location, in addition to providing extra protection to the fibers. Figure 19–8 illustrates innerduct protection of fibers. As with copper cable, fiber optic cable also contains multiple fiber optic strands in one cable. The number of strands per cable ranges from 6 to 864. Commonly deployed cable counts are 6, 12, 24, 72, 96, 144, 288, 432, 576, and 864 fibers. The type of cable also varies in that it is either loose tube or tight tube, ribbon cable or strand cable. Table 19–1 lists the color code used to identify fiber optic strands in fiber optic cable.

When fiber optic cable is run underground, special precautions must be taken when digging the trench and placing the duct in the trench. Most ducts contain four separate sections, each able to hold at least one fiber optic cable. The depth of the trench is important. It must be deep enough to ensure that the cable will not be disturbed when a backhoe digs in the area or when the ground pushes up the duct during spring thaw. Therefore, most fiber optic cable is placed at least 6 ft. underground. Once the duct is placed in the trench, soft dirt or sand is used as the first layer of coverage. Using the dirt taken from

Figure 19–7
Fiber slack loops are placed in the cable during construction to allow for repairs to be made in case of a fiber cut. (Photo courtesy of M. A. Rosengrant)

Figure 19–8

Shown is a piece of conduit and four ducts used to hold fiber optic cable. (Photo courtesy of M. A. Rosengrant)

the trench as the first coverage can cause the duct to collapse from the boulders or rocks found there, making it very difficult to pull fiber optic cable through the duct.

Two methods are used to pull fiber through the duct. The first method requires that the fiber be attached to mule tape that is placed in the cable during construction. The tape is pulled from the opposite end and, consequently, so is the fiber. The second method shoots a small metal torpedo through the duct. The torpedo has a string attached that is used like the mule tape as a pull cord for the fiber cable. The importance of having a solid duct is obvious. If the duct is collapsed, you must dig up the trench and replace the collapsed portion. Solid ducts protect the fiber cable inside. Tracer wire is used to identify where the fiber duct is located. It is simply a copper wire placed a certain number of inches above the duct. When the duct has to be located, metal detectors are used to find the tracer wire, which is also used as a reference for diggers. They know the duct will be found a certain number of inches under the tracer wire and that they need to either dig by hand or carefully dig using the backhoe. Some companies have been known to place the tracer wire in the duct or right on top of it. The consequences of doing this can include cut duct and cable because the depth of the cable is unknown.

Fiber cable is organized into three unique sections referred to as the system, the span, and the segment. Each end-to-end fiber system is called a fiber optic system and is designated by a system number. Within a fiber optic system, spans are sectioned off to differentiate sections between two points, such as central office A and central office B. The span is given a unique number and a name that is used as a reference during engineering, provisioning, and troubleshooting. Each span contains multiple fiber optic sections called segments. A fiber segment refers to the length of fiber between two splice points.

Splicing the segments requires skilled technicians, a clean work environment, and special splicing equipment. The two types of fiber splicing used today are mechanical

Fiber Strand	Color Code
1	Blue
2	Orange
3	Green
4	Brown
5	Slate
6	White
7	Red
8	Black
9	Yellow
10	Violet
11	Rose
12	Aqua

Table 19–1

Fiber Cable Color Code.

and fusion. Mechanical splicing joins the two fiber ends together using a mechanical clamp. Very few companies use mechanical splicing due to the high splice loss, but it can be used as a temporary fix during fiber cuts.

Fusion splicing is the most commonly used technique today. A machine is used to join the two ends of the fiber and fuse the glass, similar to melting metal rods together. Cleaning the fiber properly and cleaving the ends to make sure they are even and flush are critical factors in the success of the splice. Dust and dirt ruin many fiber splices. One small speck of dirt on one of the fiber ends attenuates the signal dramatically and can even block it from passing through the splice. *Cleaving* the fiber to make sure the ends are flush is also a factor in reducing loss in the splice. A poorly aligned splice causes a higher signal loss through the splice and, thus, higher signal loss along the span.

Splicing two different types of fiber produces additional splice loss caused by the misalignment of the fiber cores due to the difference in core diameter. A splice between two dissimilar types of fiber is referred to as a transition splice. Transition splice loss averages around 0.1 dB.

Fusion splicing is accomplished by placing the two ends of fiber in the fusion splice tray where ends are lined up by using a magnifying eye that zooms in on the fiber ends. Once the ends are lined up and the view shows there is no debris, the fusion button on the machine is pushed, causing the two ends to fuse together. The splicer then uses the *optical time domain reflectometer* (OTDR) to measure the loss through the splice. If the loss is too great, the fiber is broken and spliced again. After each segment is spliced, an end-to-end OTDR trace is taken, along with a power level reading to ensure that the total span loss is less than the maximum allowed level. OTDR tests are performed by connecting the fiber under test to the OTDR laser using a fiber optic jumper. The OTDR is configured according to the type of fiber, the refractive index coefficient of the fiber, the pulse width, the acquisition time, the distance, and the wavelength. The OTDR sends out a light pulse that reflects back into the OTDR detector where the signals are used to determine the characteristics of the fiber. In addition to the trace, the OTDR also provides a summary of the fiber. The loss, the loss per splice, the signal-to-noise ratio, and the length of the fiber appear in easy-to-read tables stored in the OTDR. An OTDR test should be performed in both directions in order to ensure correct loss readings. Table 19–2 lists the acceptable loss characteristics of a fiber splice. Ribbon cable combines twelve to twenty-four fiber strands in one unit that, using a fusion splicer, are all spliced at one time. Like any fiber optic cable, ribbon cable is cleaned, cleaved, and placed in a tray to be fused. The advantage of using ribbon cable rather than single strands is the time saved by splicing all twelve strands at one time. Newer OTDRs are also able to terminate twelve strands and switch between them when testing. The advantage is obvious.

19.1.3 Splice Cases and Cross Boxes

Running one continuous cable pair from the central office to a subscriber located several blocks or miles away is impossible. Connecting the feeder cables to the distribution cables and to other feeder cables requires special cross-connection points and termination blocks that function as interconnection points for the many cable segments. The two devices used for this are splice cases and cross boxes.

Splice cases are used to connect segments of cable along the cable path. Figure 19–9 shows a typical splice case hanging from an aerial cable. Cables are fed into the

---

cleaving
Cutting the fiber strand using a cleaving tool.

fusion splicing
Splicing method for fiber optic cable.

optical time domain reflectometer
Test equipment used to test fiber.

splice case
Case used to splice cable ends together and access them easily.

Table 19–2
Loss Calculations for Fiber Span.

Item	Loss
Fiber optic connector	0.3 dB
Splice loss	0.1 dB
Transition splice loss	0.25 dB
Single-mode fiber	0.25 dB/km @ 1550

Figure 19–9
Splice case. (Photo courtesy of M. A. Rosengrant)

splice case on each end and spliced together to connect the sections. Table 19–3 lists the color code used to differentiate the many pair of copper wires in the cable. Numerous variations of splice cases are found in outside plant networks. Technicians become accustomed to certain models and often complain if the outside plant engineer introduces a new type of splice case. The splice case must be water resistant, rodent resistant, and wind resistant; therefore, manufacturers are always trying to improve the layout of the case. The splices inside the case are made using Scotchloks, illustrated in Figure 19–10, that connect the two wire segments without stripping the wire beforehand. The end of

Table 19–3
Standard Copper Cable Color Code for Solid Colors.

Pair Number	Tip Color	Ring Color
1	White	Blue
2	White	Orange
3	White	Green
4	White	Brown
5	White	Slate
6	Red	Blue
7	Red	Orange
8	Red	Green
9	Red	Brown
10	Red	Slate
11	Black	Blue
12	Black	Orange
13	Black	Green
14	Black	Brown
15	Black	Slate
16	Yellow	Blue
17	Yellow	Orange
18	Yellow	Green
19	Yellow	Brown
20	Yellow	Slate
21	Purple	Blue
22	Purple	Orange
23	Purple	Green
24	Purple	Brown
25	Purple	Slate

Figure 19–10
Connection devices are used to splice two ends of copper wire together. Scotchloks (shown) are one of the most commonly used connection devices. (Photo courtesy of M. A. Rosengrant)

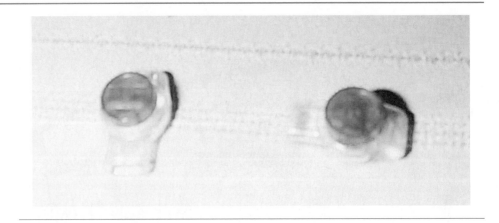

each segment is pushed into each side of the Scotchlok, which is then pressed together using a special tool that looks like a pair of pliers but has flat prongs, as shown in Figure 19–11. The Scotchlok has small metal teeth that penetrate the wire inside the button, thus making a contact between the two wires. Splice cases also vary depending on the manufacturer and the application. Most splice cases hang from the cable, as shown in Figure 19–12, and open from the top. The cable splicer opens the top and pulls open the outer plastic cover to expose the wires for splicing. The type of closure varies in the way it holds the wire, the cover that closes the case, and the way it hangs from the wire.

Figure 19–13 shows a typical *cross box*. Cross boxes are placed in the outside plant to help organize feeder and distribution cables by providing a point-for-cross connection. For instance, feeder cable 700, called F1 cable or feeder 1 cable, comes into cross box 50. Feeder 700 is divided into three smaller cables now referred to as F2 cables or feeder 2 cables. The three cables feed three different areas, as shown in Figure 19–14. The cross box can take one pair from F1 and cross connect it to one pair

cross box
Metal unit placed in the outside plant where cables are distributed to various locations.

Figure 19–11
The splicer uses special tools (shown) when using Scotchlok connectors. These tools help speed up the splicing process. (Photo courtesy of M. A. Rosengrant)

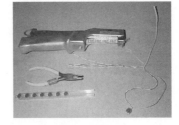

Figure 19–12
Fiber splice cases are similar to splice cases used for copper cable in that they enclose the splice point of the cable.

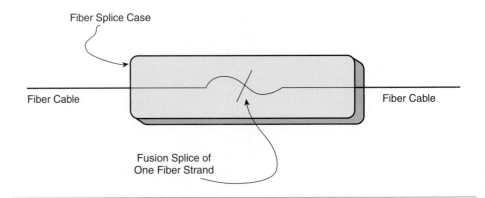

Fiber Splice Case

Fiber Cable

Fiber Cable

Fusion Splice of
One Fiber Strand

Figure 19–13
Small cross box. (Photo courtesy of
M. A. Rosengrant)

in F3. If, in the future, the pair in F1 needs to terminate at a different location fed by F2 cable, the cross connect is removed between F1 and F3 and placed between F1 and F2, as shown in Figure 19–15. The cross box comes in many different sizes and styles and is often messy. A label sheet is located on the inside door for the cable wire number. Without proper labeling, the technician would have an almost impossible job when trying to find a particular pair. A red cap indicates special service circuits such as T1s, ISDNs, or DDSs, indicating that they should not be disturbed. Unfortunately, not all technicians abide by the red tag rule, so on occasion a special circuit pair is stolen for someone's dial tone. Many special service circuits such as T1s, DDSs, or even ISDNs do not have dial tone on the pair and therefore appear as if they are dead and available for use. Technicians listen for dial tone on test sets clipped to the wire pair when determining if the pair is available for use.

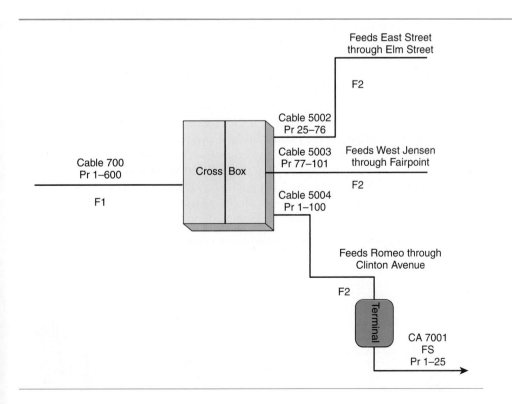

Figure 19–14
The cross box distributes F2 cable to three different destinations. F2 cable also feeds into a second terminal and exits as F3 cable.

Figure 19–15
Distributing the cable pair from the F1 cable to the many other F cables is often handled at a cross box. The cross box provides a physical cross-connect point.

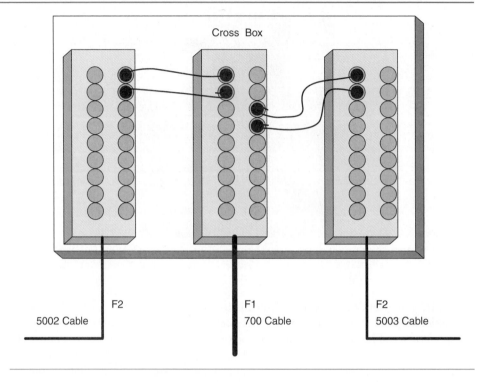

Cross Box

F2
5002 Cable

F1
700 Cable

F2
5003 Cable

19.1.4 Bridged Taps and Load Coils

bridged tap
Length of wire extending beyond the termination point.

A *bridged tap* is an unused section of wire that is still connected to a working wire pair. Figure 19–16 illustrates a typical bridged tap found in a local outside plant network. It serves no purpose and can cause interference on the line they are connected to, especially if the circuit is carrying special service circuits. Today's digital circuits, like ISDN and DSL, are especially sensitive to bridged taps on the line. A bridged tap acts like an antenna connected to the circuit, so the signal traveling down the wire pair to the destination splits in two and travels down the bridged tap. Once the signal traveling down the bridged tap hits the end of the cable, it boomerangs back, causing a reflection that creates noise and disturbs the original signal. The information signal may then be corrupted. It is interesting to note that the shorter the bridged tap and the closer it is to the originating or terminating ends, the worse the interference. When the bridged tap is long and located in the middle of the circuit, the reflected signal is lost before it hits either the originating or terminating ends. Therefore, the position and length of the bridged tap determine how it will affect the signal. Most ILECs define the allowable bridged tap length and location on the circuit in their tariff. Bridged taps may be removed but they are often difficult to locate. Using a time domain reflectometer is similar to an OTDR by sending a signal out onto the line and reading the reflected signal coming back. The TDR sends an electrical signal, not an optical signal. The TDR trace that draws on the screen provides information on the length of the cable, the loss of the cable, the number and location of load coils, and the number and location of bridged taps.

Figure 19–16
Bridged taps are spurs of wire that hang off the working cable pair. The bridged tap does not provide any function. In fact, it adds resistance to the line and can cause unbalanced circuits.

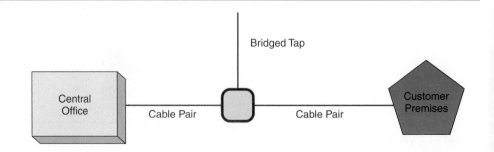

Bridged Tap

Central Office

Cable Pair

Cable Pair

Customer Premises

Load coils, unlike bridged taps, are not always bad. In fact, load coils are placed in the cable to improve voice transmission levels for circuits that travel more than 18,000 ft. A load coil is simply an inductor wire wrapped around a core. It is placed in-line on the cable pair to increase the distance that the voice signal can travel. Many believe the load coil is an amplifier, but it is not. It is a passive device that works with the electrical characteristics of the copper line to improve the wire's ability to carry frequencies between 300 and 3800 Hz. The purpose of the load coil is to cancel out the inherent capacitance that builds up on a wire pair as it travels farther away from the CO. The strength of the signal is adversely affected by the increased capacitance on the line and, consequently, reduces the distance the signal can travel. The load coil, being an inductor, cancels out the capacitive effect, thus zeroing out the impedance and, hence, reducing the attenuation of the signal. The longer the wire, the greater the capacitance. Load coils, for this reason, are placed in circuits that are greater than 18,000 ft. Very rarely is a load coil placed in a circuit that is less than 18,000 ft. The TDR is used to locate load coils in a cable. Smaller, less expensive load coil testers may also be used.

load coil
Inductor placed along the wire path to help extend the distance the voice signal can travel.

For years, load coils solved many outside plant engineers' problems. They were able to build farther out from the central office, thus increasing the total CSA for the office to distances of as much as 28,000 ft. Unfortunately, digital signals and signals that have frequencies above 4000 Hz are blocked by load coils, so they must be removed when ISDNs, T1s, DDSs, or DSLs are deployed.

If a circuit has a loaded bridged tap, or if the load coils are not spaced correctly on the line, the signal will be corrupted. The first load coil is placed 3000 ft. from the central office and every load coil after that is placed 6000 ft. apart. Load coils vary in size from small units that look like hockey pucks to many small plugs, placed in large load coil cases, attached to poles, or placed in manholes. They are spliced into the wire, as shown in Figure 19–17, and are often left on a circuit even after the circuit is rearranged and given to a different customer. If they are left on a bridged tap or the load is misplaced on the cable, the circuit will experience elusive problems such as intermittent noise, high loss, sudden dropouts, and other hard-to-pinpoint troubles. If the circuit is going to carry a digital signal, it will be deemed bad and will require further troubleshooting. When a bridged tap has load coils, problems occur that require many troubleshooting hours to resolve. Once the technician finds the loaded bridged tap, however, it is easily removed and the problem is solved.

19.1.5 Terminal Boxes

A *terminal box* is a device that connects a subscriber's location to the distribution cable. As shown in Figure 19–18, a terminal box is mounted on a pole near the distribution cable. Drop wire is fed from the resident's home to the terminal box. Inside the terminal box, the distribution end and the drop wire are terminated on one lug front and back completing the connection. Figure 19–19 illustrates the connection between the drop wire and the distribution cable.

terminal box
Unit used to connect drop wire to the cable.

Numerous termination boxes are available on the market. The boxes are categorized as underground, aerial fixed count, ready access, pedestal mount, and pole mount. Technicians open up terminal boxes using special access keys that look like Allen wrenches. Troubleshooting, installation, and disconnecting service may be done in the terminal box. The technician uses a telephone test set to check for dial tone, query the line number, and test for cable troubles. The test set is connected to the lugs, and the switch hook is released, causing an off-hook condition. If the distribution cable pair is bad, the technician will not hear dial tone on that pair coming from the central office. If dial tone is present, the problem must be between the termination box and the customer premises.

When drop cable is buried, a ground terminal enclosure connects the subscriber to the distribution cable. Other ground enclosures are used as termination points as well as splice points for multiple cables in an area. For example, the enclosure in Figure 19–20 houses multiple wires that connect to local homes, and pedestal is one example of a termination method used for buried cable. Aerial fixed-count pole-mounted termination boxes are constructed of metal casings and fixed metal termination lugs. Aerial ready-access terminals are hung on the cable. The ready-access

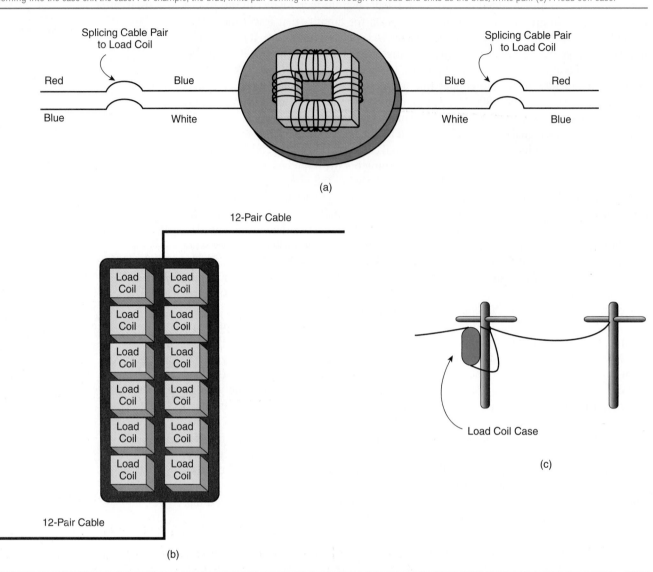

Figure 19–18
Customer premises equipment termination pail and protector. (Photo courtesy of M. A. Rosengrant)

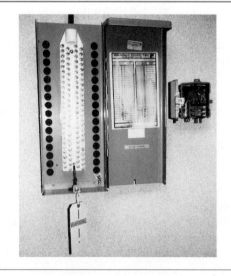

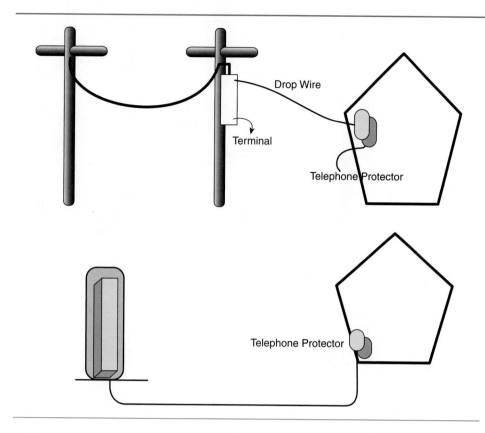

Figure 19–19
Fixed-count terminals are mounted on poles or inside pedestals. Both are used to interconnect the drop wire to the cable pair.

terminal does not have fixed lugs where the wires are terminated. Instead, the distribution pair and the drop pair are Scotchlok'ed together using a Scotchlok splicing button. Aerial terminal boxes may be spliced right into the cable and hang independent of the pole, but they are usually mounted near the pole. When the front cover is removed, the connection points are exposed. Terminal boxes' connection points vary depending on the manufacturer; some have screw-down lugs while others use Scotchloks to secure the two wires together. One terminal box may feed multiple subscribers; and apartment buildings, malls, and businesses often have large terminal blocks that can interface larger cables into their sites. Figure 19–21 depicts a typical apartment complex being fed with a twenty-five-pair termination cable.

Figure 19–20
Buried drop cable feeding a customer's premises is connected to a termination pedestal (shown). The distribution cable and the drop cable cross connect inside the pedestal case, similar to aerial terminal cases. (Photo courtesy of M. A. Rosengrant)

Figure 19–21
Apartment buildings require that multiple pairs of wires terminate in the telephone room. Cabling is distributed through the rest of the complex via house wire.

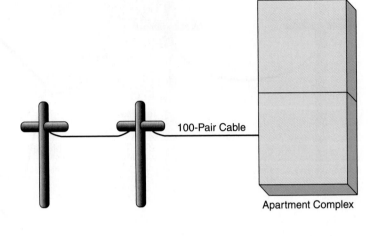

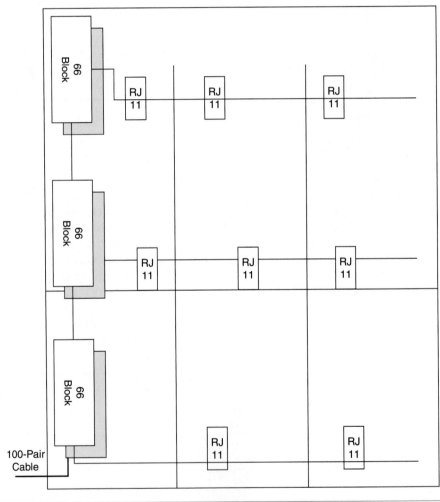

Hanging terminals, splice cases, drop wires, and pole attachments vary depending on the manufacturer of the product. Outside plant engineers must decide which of these connectors will be used before laying out the outside plant. The type of terminal chosen will determine the method used to mount the unit on the pole. For example, if the terminal is a fixed-count terminal similar to the one shown in Figure 19–22, the mounting scheme requires four bolts to be placed near the cable at the top of the pole. The terminal block is then anchored on the bolts and the cable is fed in from the bottom of

Figure 19–22
A fixed-count terminal mounts on a telephone pole and is used to connect the distribution cable to the drop wire. A fixed-count terminal has a fixed number of connections that are used as cross-connect points between the two cables. (Photo courtesy of M. A. Rosengrant)

the terminal. The drop wire is attached to the terminal by feeding it through the cable hole and establishing a drop loop, as shown in Figure 19–23. A second commonly used terminal is the ready-access terminal that is spliced into the cable, as shown in Figure 19–24. This terminal requires that the technician connect the drop wire to the appropriate cable pair by Scotchloking the wire pair from the CO to the end of the drop wire, as shown in Figure 19–25.

19.1.6 Underground Networks

Beneath cities, towns, and boroughs, a maze of cable winds its way through an underground network of manholes and ductwork, all documented in elaborate drawings. In the past, draftspeople made diagrams on large sheets of drawing paper showing the cables, poles, manholes, and various other objects in the outside telephone plant. These diagrams are now generated by sophisticated drawing programs linked to huge databases containing information in streets and roads in the area. Whether the cable inventory is on

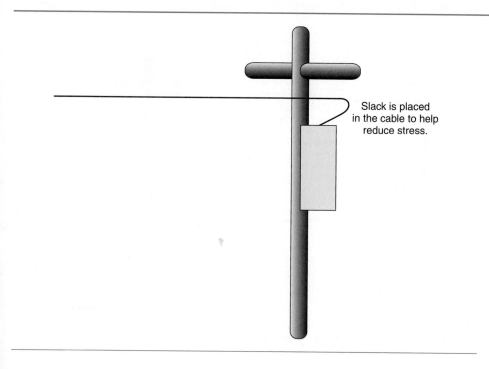

Slack is placed in the cable to help reduce stress.

Figure 19–23
The cable feeding the terminal is looped back to allow for slack.

Figure 19–24
A ready-access terminal is placed along the cable strand and depends on connection crimps, such as those shown in Figure 19–10, to connect the drop wire to the distribution cable. (Photo courtesy of M. A. Rosengrant)

paper or in a PC, the cable layout remains the same. Drawings, referred to as As Builts, are developed by the outside plant construction crew. These include the location of the manholes, the poles, and the pedestals; the physical location of the cable; and the depth it is buried. The drawings are then given to the outside plant engineer who loads them into the mapping database. Technicians, engineers, and the provisioning technicians depend on the drawings to build cable spans, to splice cable, to troubleshoot, and to plan for the future. Our description of the outside plant underground network begins by exploring the manhole.

A manhole is just a hole in the ground that varies in size, depending on its purpose. Cement walls and floors keep the earth from caving in. Before entering the manhole, water that seeps up from the ground and through the duct system may have to be pumped out. A gas test must also be performed to ensure that there is enough oxygen in the air and that no toxic gases are present. There have been more deaths from manhole gases than from pole-climbing accidents; therefore, manhole safety procedures must be followed. Large black cables are pulled into the manhole through holes in the cement walls. Copper cables and fiber optic cables run side by side in the underground

Figure 19–25
A ready-access terminal hangs directly in the span of cable. The POTS line is stripped out of the cable and spliced to the drop wire.

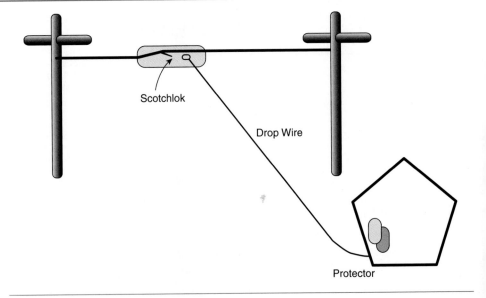

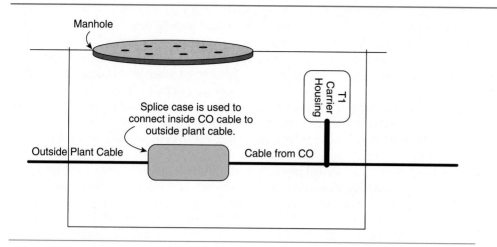

Manhole

Splice case is used to
connect inside CO cable to
outside plant cable.

T1
Carrier
Housing

Outside Plant Cable   Cable from CO

Figure 19–26
Manholes house many items. Splice
cases are used to splice cable
sections together. Carrier housings
hold repeaters.

network, coming into the manhole on one side and exiting on the opposite side. Sometimes the only way to differentiate between the two is to note the size difference; however, some companies mark their fiber optic cable with orange or yellow coatings. Both fiber optic cable and copper have lead tags attached, marking the manhole number, cable number, and pair counts. Cable splicers use these labels to find the correct splice terminal and cable in the manhole, because large manholes have several splice cases and cables. The splice case may hang from the cables in the manhole, as shown in Figure 19–26. A manhole may also have a repeater housing, which is used to hold T1 or ISDN repeaters. The housings vary in shape and size but are similar to those that hang on poles. The T1 span cable is spliced into the repeater housing in twenty-five-pair increments. Repeater housing normally holds twenty-five repeaters, an order wire for communications, and a test pair for testing.

The cables that run through a manhole are encased in heavy, black, waterproof, plastic tubing. When new cable is laid between two manholes, it is shoved into one duct of the innerduct holes and then pulled through until it reaches the next manhole. A splice case is then used to connect the cables to other cables in the manhole, as shown in Figure 19–27. Once the cable is secured in the manhole, a splice case is used to connect the two cable ends by opening each and exposing the wire pair, then securing them into the case. The cable splicer splices each wire pair together, creating a complete cable segment between two manholes. The outer jacket of the cable must be water resistant and tough enough to withstand rodents. Rats and squirrels love to gnaw on cable and are responsible for a fairly high percentage of cable troubles.

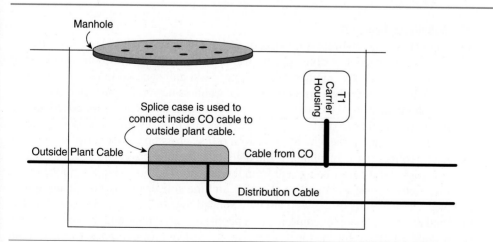

Manhole

Splice case is used to
connect inside CO cable to
outside plant cable.

T1
Carrier
Housing

Outside Plant Cable   Cable from CO

Distribution Cable

Figure 19–27
Manholes not only hold splice cases;
they are also used to distribute cable.

Because of water running through and seeping into cable ducts, air must be injected into the cables to keep water out. One small pinhole in a cable can allow water to seep in and break down the insulating jacket surrounding the bare wire pair. The plastic insulation surrounding the wires inside the cable slowly deteriorates when exposed to water, and paper-insulated wire is instantly affected when water seeps into a cable. The paper disintegrates and exposes bare copper wire to bare copper wire, shorting out the circuit. Cable outages from water seeping into the cable are common. Pushing air into the cable and maintaining a required air pressure keep the water out. The air compressor is housed in the central office and is monitored daily to make sure the pressure in the cables remains static. The first warning sign that cable trouble is likely often comes from lower-than-normal pressure readings. Every local telephone company has a special crew of splice technicians who are extensively trained to ensure that the cable maintains air pressure. When a splice case is opened, the air pressure must be maintained by capping off the exposed end. A special silicone gel may also be used to help keep water out of the cable. Gel-filled cable is commonly used in marshy wet areas, primarily in buried cable.

19.1.7 Aerial Networks
The aerial network is comprised of telephone poles and cable encased in a waterproof, rodent-proof plastic jacket strung between the poles. The outside plant engineer determines the tensile strength required to keep the cable at a certain height above the ground and a maximum weight to ensure that the poles will not snap. The engineer also determines how the cable will be connected to the telephone pole, where the splice cases will be placed, the type and placement of the terminal blocks, how far apart the telephone poles should be placed, and the physical location of each pole. Telephone cable, electric wire, and cable television cable commonly share the same telephone poles. Industry standards dictate that the electric wire be placed at the top of the pole. The telephone cable is placed below the electric wire and the cable television cable is placed below the telephone cable. The ownership of the pole depends on who established the right-of-way in the municipality. The electric company or phone company often owns the pole and charges a pole tax to anyone wishing to attach their cable to the pole. Other communication providers may need to attach their cable to the telephone poles. The new providers must establish interconnection agreements with the pole owner and obtain right-of-way agreements from each municipality and state government in the area. Both the owner of the pole and the municipality often charge the provider a per-pole fee.

Like the cable in the manhole, aerial cable is adversely affected when water seeps through the protective jacket. Air pressure is used here, too, to keep water from seeping in and breaking down the wire's insulation. Aerial cable is also very susceptible to rodent attacks, especially from squirrels.

19.1.8 Splicing Copper Cable
Splicing copper cable—joining thousands of small wires together—is a task that requires years of practice. Cable splicers use two methods to determine which wires should be spliced to one another. If the cable is old and encased in paper insulation, tone is used to identify the correct wire. If the cable is newer and insulated with plastic, the color code is used to identify the pairs. For cable that is not color coded, the individual wires are fanned out on a splice board, as diagrammed in Figure 19–28. A tone is placed on one end of the wire at the central office, cross box, or some other entrance point. The splicer in the manhole, at the pole, or at another cross point uses a tone receiver to pick out the tone being sent from the far end. The same happens in the opposite direction—a tone is placed on the wire and the splicer picks it out using a tone receiver. The two ends of the wire are then spliced together using a B-connector. The time and number of people required—two people sending tones and the splicer sitting in the middle looking for the tone—increase the cost and the complexity of maintaining the outside plant.

Figure 19–28
Splicing paper cable requires a special splice board. The individual wires are fanned out on each side of the board and are spliced together using special connectors.

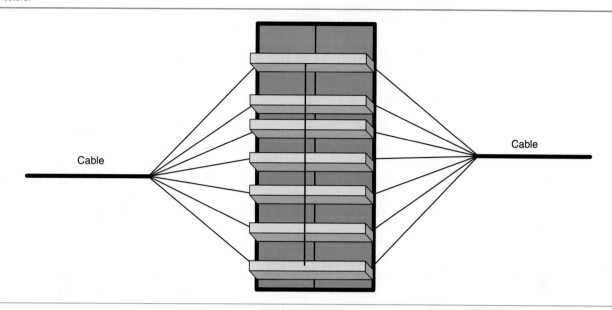

Color-coded plastic insulated cable (PIC) reduces the number of people required to splice two segments together from three to one. Color-coded cable follows a predefined color code that has been standardized across the industry. The color code consists of colors that can differentiate groups of twenty-five pairs of wires. To differentiate more than twenty-five pairs of wires, an additional color code is used to bind the twenty-five-pair groups together. Therefore, copper cable has a binder group color plus an individual wire pair color. The binder group sections are shown in Figure 19–29. Special super binder groups are used when the cable exceeds 600 pairs. These include super binder groups, associate groups, and pair colors. Pair 601 in a 3000-pair cable would be designated by:

Super binder group—red

Associate group—blue and white binder

Wire pair color—blue and white pair color

Figure 19–29
This photo provides a good illustration of the number of wires contained in a cable. The binder groups are separated out in bundles, each containing a group of wires with identifying colors. (Photo courtesy of M. A. Rosengrant)

Obviously, splicing color-coded cable is much easier than splicing paper-insulated cable. Scotchloks and the color code help reduce splice time to two hours for a 100-pair cable depending on the splicers, the weather, and the location. Splicing 100 pairs of paper cable would require a full day's work.

■ 19.2 EQUIPMENT IN THE OUTSIDE PLANT OF THE LOCAL PSTN

19.2.1 Digital Loop Carriers

One of the greatest dilemmas that the outside plant engineer has to deal with is the size of the CSA. With a larger CSA, fewer central offices are needed and therefore fewer personnel are needed to run them, less equipment is needed in the transmission room, and (most important) one less switch has to be purchased. Before the introduction of digital loop carriers and pair gain systems, the outside plant engineer's only option was to increase the diameter of the wire being used in order to serve customers that lived farther from the central office—larger diameter, lower resistance, less signal attenuation.

Another dilemma facing the outside plant engineer was the thousands of cables required to feed a densely populated section of a CSA. Every time a new housing development emerged, new cable had to be run. The ductwork underground and the telephone poles were being strained by the number and size of the cables.

The *digital loop carrier* (*DLC*) was introduced to solve both problems. A digital loop carrier interfaces directly with the subscriber on one side—the line side—and multiplexes those signals together onto T1 circuits that feed the switch in the central office. Figure 19–30 illustrates a typical DLC placed on a cement slab in a field. Hopeville, Indiana, has ninety-six homes located along Route 19. The town receives telephone service from Corn Communications, a small rural telco located in Springville, Indiana, five miles away. Corn Communications' only outside plant engineer, Hank, looks at the cable records showing the cable that feeds Hopeville and realizes that Judy Brown's request for a second line for Internet has just maxed out the 100-pair cable feeding Hopeville. Hank pulls out his calculator and figures out what it would cost to add a second 100-pair cable to Hopeville, then calculates the cost of placing a small DLC system in Hopeville. The analysis shows that the placement of a DLC in Hopeville will extend the life of the existing 100-pair cable for many years. The DLC is placed near the local post office on a cement slab. The ninety-six subscribers' lines are disconnected from the 100-pair feeder cable that connects Springville to Hopeville and are moved over to a cross box sitting next to the new DLC. Cable is placed between

digital loop carrier (DLC)
Network transmission equipment used to provide pair gain on a local loop.

Figure 19–30
Shown is a digital loop carrier next to a cross box. The DLC is used to reduce the number of copper pairs between the central office and the remote locations. T1s are used to carry multiple customer channels from the DLC to the CO. Drop lines extending to customer premises feed out of the DLC to the cross box. (Photo courtesy of M. A. Rosengrant)

the cross box and the DLC termination panel. Cross connects are made in the cross box, connecting the subscriber lines to the cable that connects to the DLC. The subscriber lines are now extended all the way to the DLC where they interface into channel cards. The DLC multiplexes the individual subscriber signals together and sends them out onto T1 circuits that connect to the Springville central office switch. Five T1 circuits are turned up between the DLC and the CO, providing ninety-six channels plus one protection T1 in case of a circuit failure. The number of wires required between the two locations using a DLC at a 1:1 concentration ratio would be five T1s plus four POTS lines, or fourteen wires. Using a 2:1 concentration ratio would require three T1s and four POTS lines, or ten wire pairs. The final number of wire pairs needed to connect Hopeville to the Springville central office is

$$4 \text{ individual POTS pairs} + (5 \text{ T1s} \times 2 \text{ pairs}) = 1 \text{ orderwire} + 1 \text{ m channel} = 16 \text{ wire pairs}$$

The savings from the reduced number of pairs between the residents and the central office more than pays for the cost of the DLC. Orderwire refers to a pair of wires reserved for the technician during turnup and troubleshooting. The orderwire pair is a dry pair, meaning there is no –48 V battery on the line. One of the technicians connects the test set and a battery source on one end of the orderwire, and the second technician connects his or her test set on the other end, completing the connection, allowing them to communicate.

Many housing complexes, strip malls, and small towns are fed by DLCs. The technology, along with the savings realized from the reduction of copper cable, has transformed the way outlying and densely populated areas are served. The original DLCs, known as AT&T's SLC-96 or Nortel's Urban, served only a small number of locations and did not provide any form of concentration on the T1 circuits. DLCs such as Lucent's FAST shelf now provide 512 lines along with concentration ratios varying from 2:1 up to 45:1. This means that if Hank placed a FAST shelf in Hopeville and chose to use a 4:1 concentration ratio, he could feed all 100 customers using two T1s or twenty-five time slots.

Two types of DLCs are now deployed: the universal DLC and the integrated DLC. The universal DLC requires that one unit (called the COT) be placed in the central office and another unit (the RT) be placed in the area being served. The subscriber pairs feed into the RT and are multiplexed onto the T1 circuits. The T1 circuits in a universal DLC connect into the COT at the switch site that is responsible for demultiplexing the individual signals and feeding them into the line side of the switch. Therefore, the signals travel as analog signals from the customer to the RT, are converted into digital signals on the T1 circuits, and are then converted back to analog signals at the COT before being fed into the switch.

The integrated DLC does not require that a COT DLC be placed in the central office. Instead, the remote DLC interfaces with the switch directly, using T1 circuits. Most DLCs are now integrated directly into the switch, as shown in Figure 19–31.

A DLC can be viewed as an extension of a switch. The switch can talk to the DLC the same way it talks to its line units. The difference is that the DLC is not able to switch calls. It must send the call back to the switch to be switched. The DLC does, however, receive signaling instructions from the switch by way of special control channels defined by the DLC standard. The two standards used to interface the switch and the DLC are TR-008 and TR-303. TR-008 is the older standard and is quickly being replaced by TR-303. The advantage of TR-303 is its high level of concentration between the switch and the DLC. Concentration provides additional savings in outside plant cables by allowing the engineer to oversubscribe the T1 circuits by a specific concentration ratio, because subscribers may share the channels on the T1. Concentration between the DLC and the switch is possible due to subscriber calling patterns. For instance, if we knew that not all of our subscribers would go off-hook at one time, we could reduce the number of outgoing channels available or, in essence, reduce the number of T1 circuits needed to connect the DLC to the switch. TR-008 allows for a

Figure 19–31
The integrated digital loop carrier
connects directly into the class 5
switch by T1 or higher rate signals.
The digital loop carrier connects to
the drop side, also known as the
subscriber side, through two wire pair
POTS lines. The two wire pairs are
connected into a cross box.

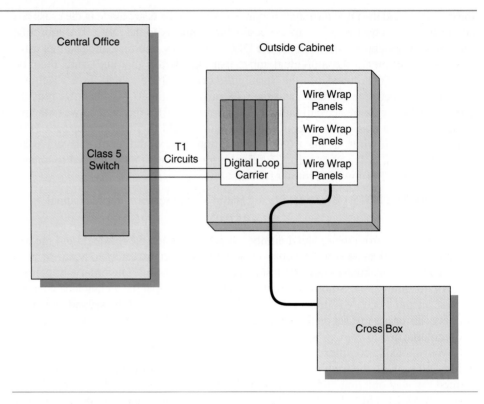

2:1 concentration ratio, meaning that we could feed ninety-six subscribers with two T1s instead of four. TR-303 allows us a concentration ratio as high as 45:1. For example, we could use a 4:1 ratio, meaning we would only need one T1 to serve ninety-six customers.

Beyond allowing us to concentrate calls, the two standards also define how the signaling information between the switch and the DLC will be handled. The signaling channels know when to provide dial tone, ring the end station, and disconnect the circuit. TR-303 defines the signaling channels as the embedded operations channel (EOC) and the transmission operations channel (TOC). Both are used to carry signaling and control information from the switch to the DLC in the field, which is one of the main reasons DLCs have become so popular with ILECs and CLECs. CLECs depend almost exclusively on DLCs when providing dial tone in an ILEC's territory. The CLEC places the DLC in the leased cage in the ILEC's CO and provides dial tone to the subscribers served out of that CO through the DLC. Figure 19–32 illustrates a typical CLEC colocation cage with a DLC mounted in a relay rack in the cage. Notice how the line-side terminations connect to a connection bay located outside the cage. The cable pair from the POT bay, as it is called, becomes the demarcation between the CLEC and the ILEC.

The physical makeup of a typical DLC consists of a chassis that holds common cards, which include redundant processors, redundant power, T1 line cards, and POTS line cards, as shown in Figure 19–33. Most DLCs today are very sophisticated pieces of transmission equipment, and the number being deployed is growing exponentially thanks to the influx of CLECs in the market. People often refer to DLCs as SLCs. The SLC is a trade name for the Lucent, formerly AT&T, digital loop carrier. Many equipment manufacturers produce DLCs, including Nortel, Alcatel, Fujitsu, and AFC. Each of their products performs the same function—it multiplexes line-side terminations into the switch.

19.2.2 Pair Gain Systems

pair gain
Electronic systems that
combine two voice signals
onto one copper pair.

A *pair gain* device allows two subscribers to share the same line. Similar to DLCs, pair gain systems help reduce the number of cable pairs needed to feed subscribers from a central office. One end of the pair gain device is placed near the subscriber's location, such as in a specially mounted terminal or in a protector on the house. The opposite end

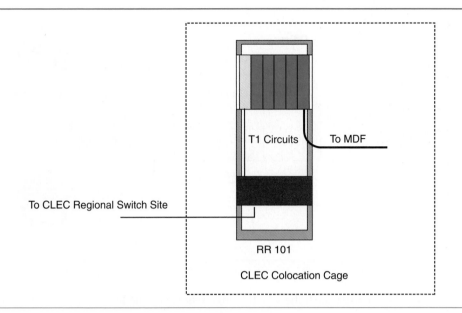

Figure 19–32
DLCs are used by CLECs to offer voice services. The DLC connects directly into the CLEC's class 5 switch via T1 or higher rate circuits. The drop side of the DLC is connected to the ILEC's MDF.

of the line is fed into a central office pair gain device that accepts multiple pair gain lines, demultiplexes the two signals on each line out, and then feeds the signals to the switch. Figure 19–34 illustrates a typical pair gain scenario showing several subscribers' lines feeding into a pair gain system located in a CO. Most local telephone companies deploy pair gain systems. The older systems used analog technologies and were not extremely effective. Current systems use digital technologies that provide a better signal and are able to offer more features. Services such as DSL and ISDN cannot work through pair gain devices so the units must be removed or a different pair assigned when those services are deployed.

19.2.3 Amplifiers

Amplifiers boost the analog signal to increase the distance that the signal is able to travel. The analog signal attenuates as it travels down the wire until, at a certain point,

amplifier
Electronic device that boosts the amplitude of the signal.

Digital Loop Carrier

High-Speed Cards	High-Speed Cards	P O T S	P O T S	P O T S	P O T S	P O T S	P O T S	P r o c e s s o r s	P r o c e s s o r s
		L i n e s	L i n e s	L i n e s	L i n e s	L i n e s	L i n e s		
High-Speed Cards	High-Speed Cards								

Figure 19–33
Digital loop carriers are used to extend the reach of the switch. A typical digital loop carrier has common cards that act as the brain of the units—high-speed interface cards that may be copper or fiber fed. POTS cards are used to feed individual subscriber cards.

Figure 19–34
Pair gain systems are deployed by
telephone companies to help relieve
congestion in the outside plant. Two
customers' signals travel on the same
wire pair from the central office to the
terminal. At the terminal, the two
signals are split apart and fed to the
correct house.

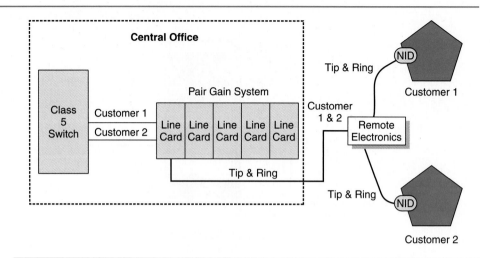

it is impossible to distinguish the signal from the noise on the line. Amplifiers receive the signal and increase the strength by amplifying it the same way you turn the volume up on your radio. Every time the signal is amplified, so is the noise. Amplifiers work as long as the noise on the line does not exceed the signal, referred to as the signal-to-noise ratio or S/N. The S/N can be calculated on a per circuit basis and is measured by using special transmission equipment.

The disadvantage of using amplifiers is that they are very sensitive to climate changes and require tuning every time the temperature shifts. Tweaking analog circuits must be done by a specially trained technician who understands all the variables that impact the signal. If the signal is out of tune, it may affect adjacent signals or radios that are located near the amplifier.

It is now rare to find an amplifier in the local network because most special service and carrier circuits carry digital signals. The introduction of T1 carrier circuits quickly replaced the old analog A-carrier circuits and DDS. In local telephone networks, however, some customers refuse to upgrade their old analog data circuits to digital.

19.2.4 Repeaters

repeater
Electronic device that repeats
or regenerates the signal.

The digital signal is not amplified—it is regenerated. The original signal leaving the central office is regenerated at the *repeater*. The repeater receives a ones pulse and generates an entirely new ones pulse of the same amplitude as the original. Repeater housings can be seen hanging from telephone poles throughout local telephone networks. The most common housing types are three-foot-high silver or white cylinders, as shown in Figure 19–35. Square white rectangular boxes are also commonly used.

Each T1 circuit is fed into the repeater (a small circuit card housed in the repeater chassis). The first repeater housing is spaced 3000 ft. from the central office, and each repeater after that is spaced every 6000 ft. The last repeater, like the first, is placed about 3000 ft. from the subscriber, as illustrated in Figure 19–36. T1 systems travel together through the outside plant, normally in groups of twenty-five circuits. Most repeater housings are built to hold twenty-five repeaters, a test line, and orderwire.

The twenty-five T1 circuits require that a fifty-pair cable be spliced into the repeater housing because a T1 circuit is a four-wire circuit—twenty-five pairs. The engineer makes sure the twenty-five-pair transmit cable—the cable traveling from the central office to the second central office—is separated from the twenty-five-pair receive cable. This is also called side 1 (S1) of the span. The second twenty-five-pair cable, called side 2 (S2) of the span, carries the signal from the second central office to the first—a signal has to travel in both directions—and because the T1 is a four-wire circuit, one pair is designated as transmit and one pair as receive. The reason that transmit and receive cables are separated by a minimum of 100 pairs is to alleviate cross talk

Figure 19–35
A T1 repeater housing (shown) containing twenty-four repeaters is used to extend the reach of a T1 signal. Repeaters are placed 6000 ft. apart, and 3000 ft. from end locations. (Photo courtesy of M. A. Rosengrant)

between the two signals. The cable splicer must be conscious of the difference between side 1 and side 2 when splicing T1 circuits into repeater housings, as shown in Figure 19–37. It is common for new repeater housings to be cut in wrong so the signal coming into the repeater housing is spliced to the cable heading back to the first central office. The transmission engineer is responsible for determining which office is designated as the S1 originating office and which as the S2 office.

Understanding how to design traditional copper T-Carrier circuits is critical for the local telephone company. The copper T1 spans, though now being replaced by fiber spans, still make up a good portion of the network that connects central offices and feeds DLCs, business developments, and high rises. The span is given a number to identify the cable route and the end locations. For example, the cable between Hopeville and the Springville CO was given a span number of 100 that included cable 700, pairs 200 to 300. The cable splicers look at the span number, the cables, and the pair counts to determine what should be spliced to what and what goes where.

ISDN lines and *HDSL* T1 lines may also need to be regenerated in order to reach the customer premises. The main difference between T1 circuits, ISDN circuits, and HDSL circuits is that the cable used to carry ISDN and HDSL rides within the cable that carries the standard POTS lines. The circuits are not separated from the rest of the

HDSL
Digital subscriber line using the 2B1Q line code.

Figure 19–36
T1 spans are built between two locations. The first section of the span is 3000 ft. from the building to the repeater as is the last section of span. The sections between repeaters average around 6000 ft.

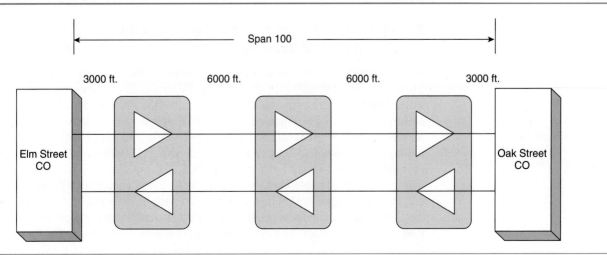

Figure 19–37
Repeaters are used extensively in the
local telephone network. The outside
plant designates the cables feeding
into the repeater housing as side 1
and side 2. The outside plant splicers
must be able to determine which wires
should be spliced to which. Many
times the transmit and receive pair of
side A are spliced together, causing
the signal to loop back to its source.

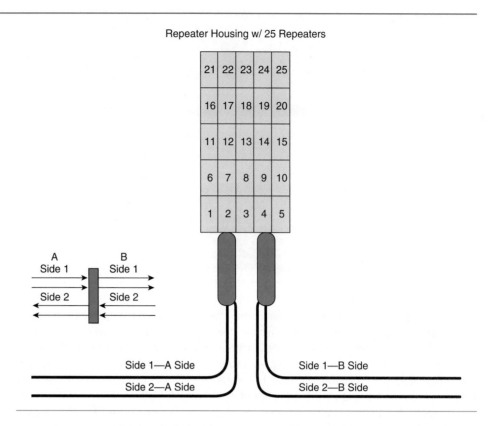

Repeater Housing w/ 25 Repeaters

wire pair in the cable. HDSL and ISDN appear on the MDF instead of on a special service block in the transmission room. An ISDN and HDSL repeater may be used to regenerate as few as one line or as many as twelve lines. Consequently, the repeater
housings are much smaller, serving between three and twelve circuits per repeater housing. The second difference between a T1 and an ISDN or HDSL repeater is the spacing
required between repeaters. ISDN and HDSL signals can travel much farther than a T1
signal. They only have to be repeated about every 12,000 ft. or twice the distance of a
T1 signal, because HDSL operates at 378 KHz compared to T1's 756 KHz. (The lower
the frequency, the farther the signal travels.) The repeater still regenerates the incoming
signal and passes it on down the line. The HDSL signals also do not require the 100-pair
spacing that a traditional T1 circuit requires. The HDSL T1 circuit's transmit and receive
signals might ride right next to one another in addition to riding along with the rest of
the POTS lines. Special groups of twenty-five cables do not have to be engineered, and
special T-screen cable does not have to be used when transmitting HDSL T1 signals.
These circuits are now being deployed faster than traditional repeated T1 spans.

■ 19.3 OUTSIDE PLANT LAYOUT FOR THE LONG DISTANCE NETWORK

19.3.1 Connecting the Long Distance Network to the Metropolitan Network

There are many different ways that a long distance carrier connects into the local network. Manhole 0 is a common meeting point, but the local carrier might also build facilities into the long distance carrier's POP. Either way, both carriers must agree to a
mutual connection point referred to as the *point of interconnection* (POI). Network op-
timization personnel are cognizant of the tariffs that define how each party will be
charged according to the location of the POI. The POI can be seen as the line of demarcation between two carriers and, as such, the place where billing begins. Figure 19–38 illustrates a POI and shows which party is responsible for which network.
The long distance carrier is responsible for connecting the equipment in its POP to the
long distance facilities. The metropolitan network used to carry traffic from the local

point of interconnection
Point at which a long
distance carrier connects into
the local network.

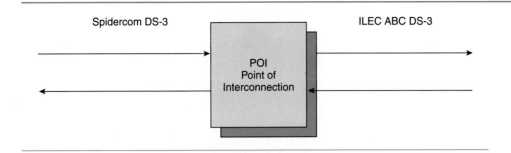

Figure 19–38
The POI is designated by both carriers as the point where they will interconnect services. The POI may be a multiplexer or a DSX panel.

area to the long-haul network is shown in Figure 19–39. The long distance company either builds fiber optic connections or leases circuits from a local provider to connect with the various local telephone providers, other long distance providers, and their private line customers. The POI defines the handoff between the local provider and the long-haul provider. Other service providers may ride on the local provider's trunks if the long distance company establishes interconnection agreements between both parties. The local ILEC then becomes a carrier for the long-haul provider's traffic, both its own and that of the other providers.

The layout of the fiber optic cable plant is similar to that of the local telephone provider's fiber network. In most cases, the fiber runs side by side through the same manholes and between the same telephone poles as the local provider's network. Specialized companies called *competitive access providers* (CAPs) build fiber in metro areas and lease the capacity to CLECs, long distance carriers, and large businesses. The last mile, as it is called, is still the most difficult piece to acquire because it is the most difficult piece to build. Buildings, streets, and old underground raceways cause headaches for construction crews, engineers, and financiers trying to establish connections around a metropolitan area. An example should help summarize: Spidercom is a long-haul provider that wants to connect its long distance network to the local ILEC. Spidercom establishes a POI at its POP in the city where the local ILEC terminates an OC-48 fiber connection. Spidercom, using the OC-48 circuit, connects into the local network through the ILEC's tandem, which switches all the local calls that are destined for the Spidercom network. The variations of this scenario are too numerous to cover. The important thing to remember is that the long-haul network and the local network must connect for traffic to be carried across both.

competitive access providers
Specialized companies that build fiber in metro areas and lease the fiber to CLECs, long distance carriers, and large businesses.

19.3.2 Microwave Links in the Long Distance Network
Microwave technology was the first method to replace copper cable in the long distance network. Since the early 1980s, microwave has been carrying long distance traffic between two points. Although most of the large, long distance providers have now migrated traffic off microwave links and onto fiber optic connections, there are still many microwave links carrying traffic across the country and between countries.

Figure 19–39
The long distance companies connect into the local telephone network through a POI located normally at the ILEC's tandem switch office.

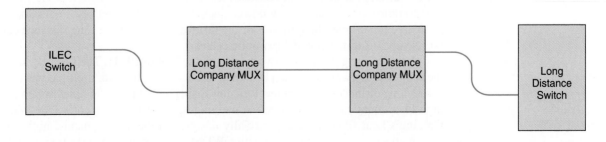

Figure 19–40
A microwave tower and antenna are used to carry information through the air to an opposite end antenna. The microwave tower is strategically placed and requires line of sight to the receiving antenna. (Photo courtesy of M. A. Rosengrant)

The advantage of microwave is that it does not require extensive builds as do fiber and copper cable. Microwave transmission requires microwave towers to be placed at a high point in an area with the antenna pointed toward the receiving antenna located on a distant hill or building. The microwave signal can travel thirty to thirty-five miles before it must be regenerated, but the towers must have a direct line of sight for the signals to be received. If a building, tree, or other tower stands between the two antennas, the site has to be changed or the tower has to be made taller.

Beyond the problems with terrain, the most daunting task the radio frequency (RF) engineer has is gaining right-of-way from the local authorities. Communities often protest the placement of an antenna in their area because of aesthetic and health concerns. In addition, the landowners of the high spot in the area demand financial incentives before allowing the tower to be placed on their property. The time it takes to work through all the paperwork and red tape is usually greater than the time it takes to build the tower.

The microwave antenna is placed on a roof or on a tower and pointed in the direction of the receiving antenna. Waveguides connect the antenna to the equipment, which is housed in a small building or, if the antenna is on a rooftop, in a small access room. Facilities connecting the tower to the local or long distance network are fed into the access building the same way they enter a central office. Manhole or pole 0 resides close to the enclosure, and that is where the handoff between the two networks takes place. Once in the building, cables are run between DSX panels that act as interconnection points between the outside network and the equipment. Microwave buildings have complete power backup and air conditioning. Figure 19–40 shows a typical microwave antenna.

19.3.3 Fiber Optic Layout of the Long Distance Network

One of the most significant occurrences in telecommunications over the past ten years has been the construction of a fiber optic network across the country. Large long distance carriers have invested billions of dollars in a fiber optic infrastructure that connects every corner of the United States. Fiber cable has been buried along thruways, next to railroad beds, and along power lines. Building the fiber optic network can be compared to building the railroads or the thruway system. Most long distance carriers now send their traffic across fiber optic cables.

A typical long distance network is shown in Figure 19–41. The fiber cable used in the long-haul network has been designed to carry high bit rate, multiple wavelengths across long distances. Special low loss, low dispersion at the 1550 nm wavelength, and high fiber count cable is deployed in the long-haul network. Lucent's True Wave and Corning's Leaf fiber cable are specially designed to be able to handle high bit rate, long-haul systems that work well with DWDMs. Long-haul networks typically deploy

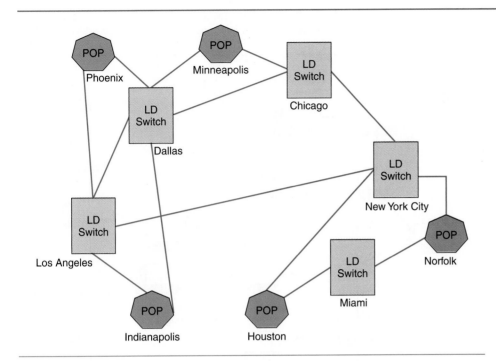

Figure 19–41
Typical long distance network showing POP sites connecting into switch sites and switch sites connecting together.

high strand count cables such as 432 or 864. They also deploy ribbon cable to take advantage of mass fusion splicing techniques. The trenching and splicing methods used are very similar to metro networks. Testing long-haul fiber is much more critical than metro fiber due to the higher bit rates being sent and the use of multiple wavelengths. Conditions such as chromatic dispersion and polarization mode dispersion require that the fiber is characterized before the network is designed and the equipment ordered. Every anomaly within the fiber affects the total long-haul system. The most significant difference between the local and long distance fiber builds is that the long-haul network travels for miles instead of blocks and requires amplifiers and regeneration sites along the way. However, the fiber cable may travel as far as sixty miles before regenerative or amplifying equipment is required and as far as 1500 miles or more before a regenerator is needed.

The introduction of erbium-doped fiber amplifiers (EDFAs) has revolutionized the long-haul network. EDFAs are small, low-powered devices that amplify each wavelength of the signal, eliminating the need to locate DWDMs every sixty miles. The cost to demultiplex, regenerate, then multiplex again each wavelength is enormous. The amplifiers are typically small and are placed in small cases along the route. In contrast, the repeaters require huts where the optical-to-electrical (O/E) equipment is mounted. Optical repeaters convert an incoming optical signal into an electrical signal, regenerate the signal, convert it back into an optical signal, and send it back out onto the fiber strand. The number of amplifiers and regenerators needed and the distance between them are changing every day. New types of fiber and amplifiers are increasing the distance a signal can travel before requiring a regenerator site. Previously, the maximum number of amplifiers allowed in a section was six, which allowed a signal to travel about 350 miles, or 50 miles per amplifier. Now that distance is much greater. Because the mileage between amplifiers varies so much among vendors, an engineer's first question to the vendor should concern the distance between amplifiers and regenerators.

As shown in Figure 19–42, the long-haul fiber network uses SONET ring technology to provide diversity. The rings are interconnected at two points to provide full redundancy. If one point fails, traffic is directed to the second entrance point on the ring. It is similar to having two doors to enter and exit a room. If one door is closed and

Figure 19–42
Most long distance communications companies depend on fully connected SONETs that carry traffic from switch site to switch site, and POP to switch site. This provides a logical view of an interconnected SONET.

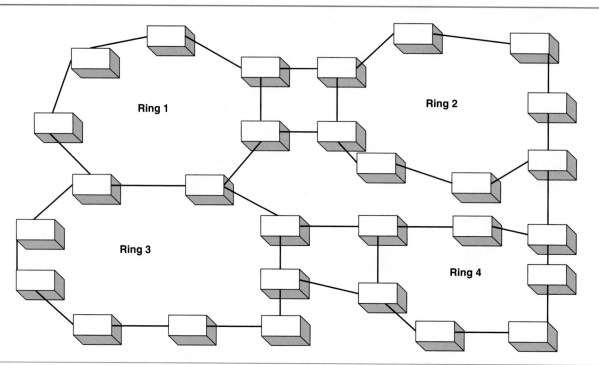

locked, the second door can be used. Each ring has a maximum of sixteen nodes, also known as fiber optic multiplexers. The cable between the nodes is strung between telephone poles or buried underground, and huts are placed along the fiber route to house regenerators or amplifiers. In addition, specially built amplifier cases are placed along the route between huts, which are small buildings that have full power backup, Global Positioning System (GPS) antennas, and air conditioning.

Maintaining the network requires technicians to travel to each of the remote huts, some of which are located in the middle of a windswept field and only accessible by a rough, rocky access road. The task of maintaining the network is formidable. Gophers often chew on fiber optic cable. When they chew through even one of the fiber strands in a cable, a technician must isolate the problem, dig up the cable, and splice it back together. The problem is multiplied by the distances that have to be traveled. If a gopher chews through a cable buried in a field in central Kansas and the closest technician is located in Denver, travel time for the technician and crew to reach the site adds to the amount of time it takes to repair the problem. The placement of long distance fiber routes depends on what right-of-way is available. Trenches are often dug next to railroad beds, as is the case of Qwest and Global Crossings networks. Large earth-digging devices run along the railroad track digging the trench as they move. The huts are strategically placed near the tracks, and the railroad companies grant access. *Ultra long haul* describes very long fiber routes. Ultra-long-haul networks include intercontinental routes that connect cities around the world. Ultra-long-haul networks depend on very detailed engineering plans to function. Fiber optic cable laid across the ocean requires specially built amplifier cases that are sunk deep under the ocean. Trials have been completed that show signals traveling at OC-768 rates and as many as eighty wavelengths covering distances of 3500 km or 2100 miles.

19.3.4 Fiber Optic Equipment in the Long Distance Network
As in all fiber optic networks, whether local, long distance, or ultra long haul, multiplexers are used to multiplex, demultiplex, and pass through signals at a particular location. The fiber optic multiplexer in the long distance network is typically an OC-48

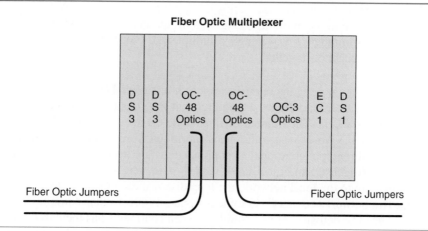

Fiber Optic Multiplexer

Figure 19–43
Long distance companies deploy many optical multiplexers.

that provides 48 DS-3s, or an OC-192 that provides 192 DS-3s of capacity. Fiber optic multiplexers may be found in the POP, in leased space, and in huts. The OC multiplexers are there to groom traffic onto and off of the large network. Returning to the highway analogy, the thruway provides high-speed lanes for vehicles bypassing a city, the exit ramps provide a way for vehicles to reach the city, and entrance ramps provide a way for vehicles to leave the city and merge onto the thruway. The fiber optic multiplexer performs the same function in the long distance network. If a signal leaving Detroit is destined for Atlanta, it must travel on the long distance network that passes by cities such as Columbus, Ohio; Lexington, Kentucky; and Nashville, Tennessee, before exiting the thruway in Atlanta. The fiber optic multiplexer also performs the protection switching required when one of the fiber strands is damaged, thus rerouting traffic in the opposite direction. Figure 19–43 shows a fiber optic MUX located in a long distance network.

A fiber optic regenerator is similar to a T1 repeater. The equipment receives the optical signal from the line, converts it to an electrical signal, converts that back to an optical signal, and transmits it out on the line. The electrical signal is of the same strength as the initial signal. As long as the regenerator is able to distinguish the difference between the ones and zeros, the outgoing signal will be the exact duplicate of the original one and zero bit stream. A signal traveling on a fiber optic cable requires regeneration about every sixty miles depending upon the strength of the lasers and the loss in the fiber. In the long distance network, a signal that is amplified can travel about 300 miles before it requires regeneration. In all fiber optic, long-haul networks, amplifiers are installed to increase the distance between regenerators.

In-line amplifiers are possibly the most important breakthroughs in fiber optic technology. They are spliced in the fiber strand along the fiber route. The cost savings of placing an amplifier compared to a regenerator is substantial. Additionally, the amplifier does not convert the optical signal into an electrical signal, thus reducing the complexity of the equipment. Instead, it simply amplifies the optical signal. In-line amplifiers are placed about every sixty miles along the fiber route, but the number used in sequence is restricted due to interference caused by amplification. Every time the signal is amplified, so is the inherent noise on the line. Five to six amplifiers can be placed in a row before the noise overtakes the signal; therefore, the long distance network has five or six amplifiers placed about every sixty miles between multiplexers and regenerators.

One of the most expensive pieces of equipment in the long distance network is the dense wave division multiplexer (DWDM). DWDMs are placed next to each optical MUX and regenerator and are responsible for providing multiple paths on one fiber optic strand. DWDM systems are now multiplexing as many as 160 wavelengths onto one transmit and one receive strand of fiber. The DWDM is installed in long distance POP sites, regenerator sites, and any other locations where the signals need to be converted into an electrical signal.

Smaller metropolitan networks rarely deploy high wavelength DWDM systems. Newer coarse wave division multiplexing systems are being deployed at a lower cost due to the fewer number of wavelengths multiplexed. CWDM typically mixes four, eight, or sixteen channels together.

19.3.5 Connecting the Long Distance Network to International Links

Connecting circuits around the world requires that long distance carriers interconnect to international carriers. The number of fiber optic cables connecting the East Coast with Europe and the West Coast with Asia is enormous. Long distance carriers connect to these cables through international meet points; the most common ones in the United States are Los Angeles, San Francisco, Miami, Providence, and New York City. Similar to connecting to the local network, the long distance carriers agree to a meet point where the two networks connect—in some cases at a T1 level and in some cases as high as an OC-192. The demand for bandwidth between continents has increased by 100% during the past few years, and analysts predict the need for bandwidth between countries will more than double in the near future. The largest long distance carriers have formed consortiums to help fund fiber builds across the oceans.

The outside plant network connecting countries differs little from the network extending across the United States. The main difference is that cable strung across the ocean (submarine cable) requires a special sturdy jacket that wards off water, debris, and shark attack. In addition to cable, underwater amplifiers and regenerators are placed in the ocean to regenerate and amplify the signals. The newer submarine regenerators are about the same size as a standard fire extinguisher. International carriers use customized ships that carry small submarines to maintain the fiber at the bottom of the ocean.

■ 19.4 CARRIERS' CARRIERS

An interesting phenomenon over the past few years is the sharing of bandwidth between carriers. The term *sharing* might be misleading because each carrier charges the other to carry its traffic. Carriers do, however, depend on each other to provide access to areas where they have not built their own network or where their network is full. For example, Spidercom's long distance network does not service Birmingham, Alabama. Spidercom leases capacity from ABC's local network. The ability of one carrier's traffic to ride on a second carrier's network is one reason long distance rates have decreased so dramatically. A ubiquitous transmission standard allows for the interconnection of networks without complicated translation equipment.

Carriers buy bandwidth in many different ways. They buy T1 circuits; T3 circuits; or OC-3, 12, or 48 capacity on the network. Carriers may also purchase dark fiber or wavelengths to help complete their networks or switchless services such as 800-number termination and calling cards. Carriers may also buy switched services. For example, a CLEC might pay a long distance carrier to handle all of its long distance traffic but brand it as its own. Long distance carriers now provide network, switching, and special services to other carriers as a product. One division in an organization may be competing with another carrier for customers while another division in the organization is trying to sell the competing carrier network capacity and other services.

Resellers also buy capacity and services from a carrier to provide services to their customers. The difference between the reseller and the wholesale service buyer is that resellers do not own their own network. They often do not even own a switch. They have an office, a brand name, and a staff that handles billing and accepts customer calls. The reseller pays the carrier to provide all of the physical network and switching. The reseller brands the service under its own name and is responsible for dealing with the customer regarding service. Resellers are found in both the long distance and local network. Therefore, the term *outside plant* literally means the physical plant that includes the poles, the repeaters, the DLCs, and so forth. Not every long distance company has an outside plant.

■ 19.5 RIGHT-OF-WAY FOR TERMINALS

Permission to place terminals, telephone poles, and even microwave towers on a property is commonly referred to as right-of-way. The local municipality may require permits before placing telephone poles along a county road, and the state may place certain restrictions against building a microwave tower on a particular piece of property. Property owners must give their permission before equipment is placed on their land. Obtaining right-of-way for a project is critical for the service provider. Engineers have had to change their initial designs due to local zoning restrictions or a property owner's refusal to lease space on his or her land. Every telephone company has a right-of-way department that is responsible for receiving permission for access, contesting regulatory rulings, maintaining permits and licenses, and keeping up to date on all regulatory issues concerning right-of-way. New fiber providers or wireless companies staff their organizations with several lawyers who negotiate with the local government and utilities for space on the poles or interduct.

SUMMARY

This chapter focused on the outside plant network, describing how the outside plant is laid out, the equipment used in the outside plant, and the different carriers' cabling methods. In focusing on the copper local plant, the layout of the copper cable was defined by the terms *feeder cable* and *distribution cable*. Splice boxes, terminals, and cross boxes were described, and digital loop carriers, amplifiers, and repeaters were defined. The fiber optic network layout was also discussed, and the terminations between the copper plant and the fiber plant were outlined. The various types of equipment used to connect the long distance fiber network together were discussed, as were regenerators, amplifiers, and DWDMs. The chapter concluded with an explanation of a carrier's carrier and the need for a right-of-way department. The purpose of the chapter was to provide an overview on what the outside plant of the local telephone network and the long distance network looked like.

CASE STUDY

In Chapter 18, you designed two central offices for Green Grass. You now need to determine the CSA for each office and lay out the outside plant around the central office. Using the map in Appendix A, determine the number of copper or fiber strands you will need to connect all the buildings in Green Grass. You will also need to run cable to the outlying ranches and movie stars. In addition to the cable, you will also be responsible for placing the termination boxes. You may wish to use an *X* as a designation for a termination box. If you wish, you can deploy a DLC to help conserve cable. Finally, it is important that you develop a naming scheme for the cables, the DLCs, and the T1 spans.

REVIEW QUESTIONS

1. What does the term *CSA* stand for? Explain why outside plant engineers design a CSA per central office.

2. What is the difference between the feeder cable, the distribution cable, and the drop cable?

3. What is meant by buried cable?

4. What is meant by aerial cable?

5. What is copper cable that runs through interduct under the streets of a city called?

6. Why is the cable color code important? List the first five pairs of the color code.

7. Termination boxes are used to splice two separate cables together. True or false? Explain.

8. What is the purpose of a cross box and what is F1 and F2 cable?

9. Why are load coils placed in the local telephone network? How are they spaced along the path of the circuit?

10. Why do load coils have to cut out when the circuit is going to carry digital traffic?

11. Why are T1 repeater housings placed at intervals along a T1 span? What is the distance between each housing?

12. Why does it take so long to splice together paper-insulated cable?

13. Explain how fiber optic cable is placed between poles along a span.

14. What type of test equipment is used to determine where the end of the fiber is in a network?

15. What is a digital loop carrier and how is it used to increase the diameter of the CSA?

16. What type of circuit connects the DLC to the central office?

17. What is the difference between an integrated DLC and a universal DLC?

18. Why are pair gain devices placed in the local telephone network?

19. What are the two most common types of transmission media used in the long distance network?

20. What is the main difference between a long-haul fiber network and a local fiber network?

TROUBLESHOOTING

You have been called out on a case of trouble along a T1 span between two COs. First you need to decide where to start troubleshooting the circuit. Assume you have already stopped by the CO and determined the trouble is outside. Using a copy of Diagram 19–1 in Appendix C, mark where you would begin troubleshooting the span. Once you get to the repeater housing, you will need to look both ways down the circuit. Using the test equipment, you will bridge onto the line using a special test repeater that is placed in the repeater slot of the circuit in trouble. The test leads are now attached to side 1 as shown in Diagram 19–1. If you see a good signal coming into the test set, what does this mean? You move your leads to S2. You now see bipolar violations coming into you from the span. Mark the next repeater housing you will check. When you get to the next housing, you will perform the same tests. You place your test leads in side 1 transmit and see errors. You place your test leads in side 1 receive but do not see any errors. What does this mean? Should you continue? If yes, what is the next step?

KEY TERMS

manhole 0 (541)

telephone pole 0 (541)

carrier serving area (541)

cleaving (548)

fusion splicing (548)

optical time domain reflectometer (548)

splice case (548)

cross box (550)

bridged tap (552)

load coil (553)

terminal box (553)

digital loop carrier (562)

pair gain (564)

amplifier (565)

repeater (566)

HDSL (567)

point of interconnection (568)

competitive access providers (569)

20

The Customer Premises

Objectives

After reading this chapter, you should be able to

- ■ Identify customer premises architecture.
- ■ Identify residential voice and data network architecture and equipment.
- ■ Identify business customer voice and data network architecture and equipment.
- ■ Identify Internet service provider network architecture and services.
- ■ Identify data communications equipment.
- ■ Identify customer premises equipment.
- ■ Identify data network topology.
- ■ Identify local area network architectures.

Outline

Introduction

20.1 Customer Premises Architecture

20.2 Internet Service Providers and Application Service Providers

20.3 Data Communications Equipment

20.4 Customer Premises Equipment

20.5 Data Network Topologies

20.6 Data Networks—LANs

■ INTRODUCTION

Today, communications networks and equipment found at the customer premises mimic the elements found in the service provider's network. The key difference is scale. A residential customer has an inside plant network that consists of copper wiring, a communications transmitter/receiver—the telephone, and a computing device—the PC. Business customers have switches that compare in structure to that of the ILEC and computing devices and networking equipment that use the same protocols deployed by the large service providers. Chapter 20 focuses on the variety of networks, equipment,

Figure 20–1
The structure of the metropolitian statistical area is divided into four distinct regions—the core; the regional network; the metropolitan core network; and numerous access networks, which cover the customer premises and building locations.

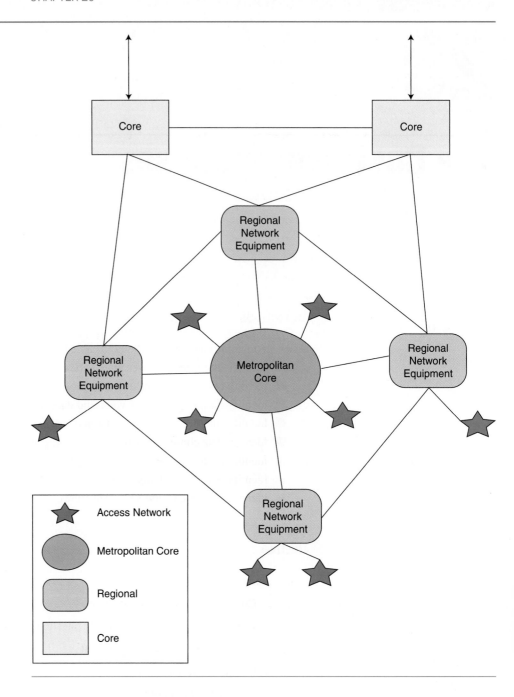

and communications methods found at the end users' premises. This includes voice and data networks and equipment used to carry the signals to the end user, all of which are considered part of the access network (see Figure 20–1).

The chapter is divided into six general areas. The first area discussed in the chapter looks at residential and business customers' communications networks. This includes defining the physical plant, that is, the equipment deployed. The second area presented in Chapter 20 looks at Internet service providers' applications and services. The third section focuses on data communications equipment used by end subscribers, and the fourth section focuses on networking equipment commonly deployed. The two final sections of the chapter look at data network topologies and data networks—LANs.

■ 20.1 CUSTOMER PREMISES ARCHITECTURE

20.1.1 Residential Customers

We all have telephones in our homes, many of us have personal computers linked to the outside world by telephone lines and modems, and some of us have fax machines. Today's residential subscriber requires many features and services in order to communicate with the rest of the world, and all these devices need to be networked and maintained. New homes are wired for home office LANs, and each room is wired for multiple telephones and TV connections. *Category 5 cable* is now being deployed in residential space the same way it is used in the business market, because it is able to carry high-frequency signals between networked home PCs. Telecommuters often have ISDN or high-speed DSLs that allow them to log into the Internet and talk on their phone simultaneously. Even customers that do not network their PCs or telecommute require multiple connections in their homes due to the increased demand for second telephone lines. Additional extensions have also complicated the communications network at the residential customer premises.

Ten years ago, we would not have discussed the residential customer's premises, because at that time the telephone was the only communications device in most homes. In the past ten years, the information age has moved into the residential market, and many residential customers now need the same communications equipment and circuits as small business customers. Residential customers' premises will continue to change as dramatically over the next ten years as it did in the past ten. They will have wireless LANs, short reach cellular phones, and intelligent notebooks. *Fiber to the home* (FTTH) may become widely available, providing an endless source of bandwidth for residential customers. Everyone may soon be able to work out of their homes without losing any of the communications features found at their offices.

One application that is creating a need for high-bandwidth pipes at residences is the introduction of gaming through peer-to-peer networks. Some games require the user to have at least a 1 Mbps access pipe in order to play an interactive strategy game. The growth in the number of higher bandwidth pipes needed by customers also increases the bandwidth needs in the metro core and regional core. DSL and cable modems are commonly installed in all areas.

20.1.2 Small Businesses
Demarcation Point and Customer Wiring

Business customers are categorized by their size—small, medium, or large. Small businesses have communications needs similar to medium-size businesses but often lack the resources to hire a telecommunications staff. A typical small business has one or more computers, several telephones, a multibutton telephone, and possibly a Web server. Small business customers have one to six telephone lines. Medium-size businesses include those with 6 to 100 telephone lines. They often contract telecommunications services from data and voice service contractors. Large business customers have many voice lines and a sophisticated LAN requiring an on-site telecommunications staff. The similarity between the different business customers is that all three depend on circuits from the local telephone company or service provider.

The service provider terminates communications circuits at the demarcation point, the place where the access network meets the enterprise network. The telephone company defines the demarcation point in tariffs that are filed with the public service commission (PSC). The PSC approves the demarcation definition, which then establishes the business customer's demarcation terminations. The tariff defines several variations of demarcations, all related to the type of building the circuit terminates at. For example, the demarcation point for a multitenant building is different from a building with a single proprietor. Customers must be well versed in the tariff descriptions in order to make the best decisions when wiring the building or ordering new services. The information that is critical to the customer is what part of the cable they will be responsible

category 5 cable
Copper cable that has more twists per foot in order to improve the transmission of data.

fiber to the home
Fiber deployment architecture in which fiber optic cables reach all the way to residential customers' homes.

Figure 20-2
The demarcation point at the customer premises is often located in the basement of the building. The cable from the outside plant is terminated onto some type of block.

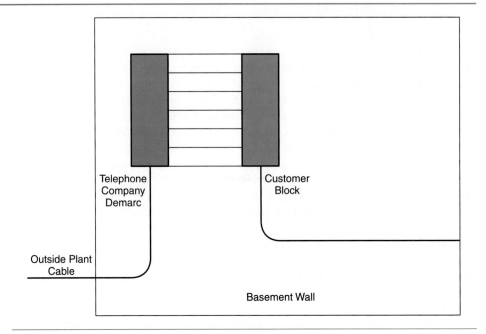

for. Unless it is specified in the contract between the customer and the telephone company, the business owner is charged for maintenance on any circuit problems that occur past the demarcation point, as shown in Figure 20–2. Many service providers require that any wiring past the demarcation point, even in multitenant buildings, is the responsibility of the customer. Constructing the inside wire point at the customer's premises is a very costly venture. As such, proper planning is critical in order to keep network costs in line.

The inside wire past the demarcation point varies depending on the building type, the customer's traffic needs, and the existing inside wire plant. The telephone company may extend the demarcation from the entrance point in the building to the customer's office but will charge a substantial fee for the service. If the building owner controls the house cable, the small business may negotiate the use of the house cable in order to extend the demarcation point to its location. In either instance, the small business owner must connect the outside cable that terminates at the demarcation point to the office wiring to complete the circuits. Figure 20–3 illustrates three common scenarios of three business locations, each with different inside wiring schemes. Figure 20–3a illustrates a demarcation point in a single proprietor building. The demarcation point terminates in the telephone closet behind the receptionist. The cable between the demarcation point in the closet and the three offices was installed by a contractor procured by the customer. The customer has established a cross connect between the 66 block designated as the demarcation point by the telephone company and the 66 block designated as the customer wiring. Note the number of circuits that are cross connected between the two blocks. The customer has three telephone lines and one T1 circuit. The telephone lines terminate into small key systems, also located in the telephone closet. The T1 line is extended to the small server room where the company's small router is located. Fireproof cable, called plenum, is used to connect the wires at the demarcation point to the wires terminated in the small server room. All of the employees are connected to the server and to shared printers via RJ-45 jacks behind their desks that allow for an Ethernet connection for each PC. The cable connecting the jack to the server in the server room is CAT5 cable and is distributed around the building in the same raceways as the twisted pair copper pairs connected to the telephones. The key system receives the three outside telephone lines and uses them to provide an outside line to the ten employees in the office. Each employee has a telephone line that can be traced back to the key system and an RJ-11 jack behind the employee's desk where the telephone connects.

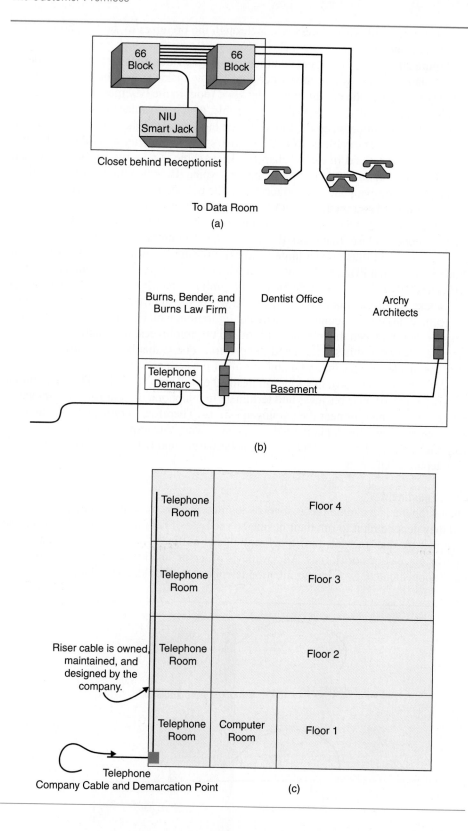

Figure 20–3
Multitenant buildings require that the telephone company demarcation point resides in one location, typically in the basement. The outside cable is connected to inside wiring and fed to the appropriate locations. (a) A small customer. (b) A multitenant building. (c) A company using the entire building.

This is just one example of a small business customer. Looking at ten different small businesses, we would find ten different wiring configurations. Some small businesses have a wiring closet with termination blocks and different types of cable—CAT3 or CAT5. Most small business customers, though, have a communications network that consists of a terminal block placed in a convenient location. From the terminal block, a conglomeration of wiring often winds through the premises to the telephones, PCs, and fax machines.

Figure 20–3b shows a larger communications provider, such as one that resides in a multitenant building where the owner controls the inside riser cable. The telephone company terminates the circuit in the telephone room in the basement, where termination blocks are mounted on the walls. Riser cable, which is similar to cable found in the outside plant, is run between the blocks in the basement and the telephone closets on each of the floors. The cable is grouped in twenty-five increments and follows the standard color code defined in earlier chapters. Riser cable is now run for both voice and data—CAT5 for data and twisted wire pair for voice. Blocks mounted on the wall by the telephone company are used to terminate the outside plant cable. A cross connect is made between a pair on these blocks and the correct pair on the riser blocks. The wire is extended to employees' desks by connecting the riser cable termination on each floor's telephone closet to a block that terminates the interhouse wiring.

Figure 20–3c illustrates a large business customer with many floors, a full telephone room with a PBX, and a data room with many servers. The telephone company terminates the outside plant cable on a frame similar to an MDF in the telephone room. The blocks are cross connected onto blocks where the interbuilding cable terminates. The technicians on staff control the wire configuration, which is similar to the previous two scenarios between the telephone room and the employees' workstations. The main difference between this scenario and the previous one is that here the customer owns all of the inside house wire and maintains all of the connections.

Most businesses have the wire run inside the walls of their suite. They are similar to those run in smaller business and terminate in telephone or data jacks, as shown in Figure 20–4, behind or near the employee's desks. Therefore, a typical connection between an employee's telephone and the outside wire pair requires several cross connections between the riser cable and the suite wiring and between the riser cable and the telephone wire pair.

Voice Equipment

The telephone is the main piece of voice equipment used by small business customers, and they depend on it to run their business. The type of telephone used depends on the customer. A one-line standard touchtone phone suffices for many small customers, such as pizza shops or florists. An attorney's office or small real estate firm may use a more sophisticated, two-line multibutton telephone with hold, conference, and transfer

Figure 20–4
Combining RJ-11 and RJ-45 modular jacks in one module.

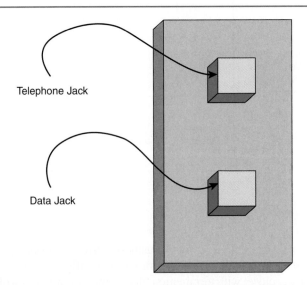

Telephone Jack

Data Jack

buttons. ISDN telephone sets serve small business customers well. They provide multiple call appearances per ISDN line, in addition to two distinct telephone channels on one telephone line. The receptionist's phone is programmed to transfer calls, place a client on hold, and send calls into the voice-mail system. Many small business owners have purchased ISDN or Centrex to provide the multitude of features that before were only possible with PBXs or *key systems*. Telephone companies now provide many products built for the small business customer, for example, Centrex provides the features of a PBX but removes the customer from having to configure and maintain the device.

key system
Phone system used to terminate lines and provide services to users at that location.

A fax machine is another piece of voice equipment often found in a small business office. It is used to receive documents electronically and is often combined with a printer. The fax machine may have its own telephone line or it may share the voice line. There are a variety of types and brands available.

Data Equipment

The type of small business and the applications and services offered by the business determine the business data requirements. A real estate office requires that each of its employees have a networked PC that can talk to one another and to the outside world. For this to happen, the PCs must be connected to some type of LAN that, in turn, must be connected to the outside world via a telephone line, an ISDN line, a DSL, or a special service line such as a T1. In most cases, the circuit connecting to the outside world connects to the Internet. The LAN normally has a file server where shared files are housed. It uses hubs to connect all of the PCs together; and, in some cases, the business has a router to interface to the outside world. The circuit types used to connect to the outside world also vary, but ISDN dial-up connections, DSL always on, and conventional dial-up modem lines are common. Any business that connects to the Internet should have a *firewall* that screens for unauthorized messages that may contain viruses or other undesirable information, as well as protects the company's information from "hackers."

firewall
Device placed on the customer's network to filter out unwanted information and attacks by hackers.

Many small businesses now have a Web server that houses the Web site as well as product information, company information, telephone numbers, e-mail addresses, and so forth. The growth of e-commerce is forcing even the smallest businesses to either install Web servers or contract an ISP or ASP to maintain one for them. Figure 20–5 shows a typical small business LAN with a firewall, a proxy server, and a Web server.

There are also small businesses that have one or two PCs that access the outside world via telephone lines and modems that are not networked together. They may share a printer and the one dial-up data telephone line. Again, small businesses vary from simple dial-up Internet connections to fully networked LANs with Web servers attached.

20.1.3 Medium-Size Businesses

Demarcation Point and Wiring

Medium-size business customers' demarcation points are similar to small businesses', in that the telephone company defines them. Medium-size customers, those with 6 to 100 lines, rarely run their own cabling. They often hire wiring contractors to handle their wiring architecture. The following is an example of a typical medium-size business customer.

This business sits on the fifth floor of a thirteen-floor office building. The building owner owns all of the house cabling, and the telephone company defines their demarcation point in the basement where the cable enters from the street. Our customer, President, Inc., has ordered three T1 circuits and twenty telephone lines that need to be terminated in its telephone room on the fifth floor. We will divide the telephone room in the basement into two sections—outside world entrance cable and house cable. Terminal blocks mounted on one of the walls by the telephone company are the demarcation

Figure 20–5
Small businesses have fully networked LANs. A typical small business has file servers, workstations, scanners, printers, and multimedia workstations. Shown here is an Ethernet LAN with a router used to access the outside network.

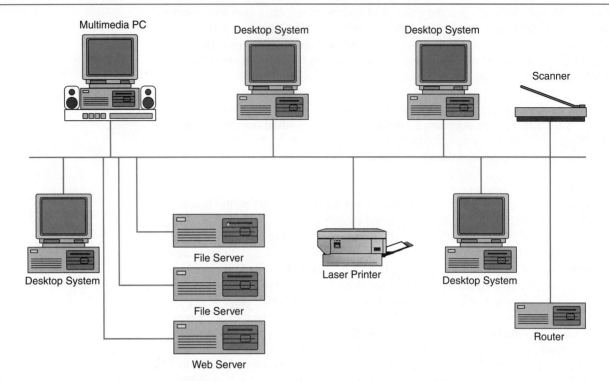

points for terminating the multiple pairs of wires that connect the high rise to the serving central office. The terminal blocks vary in size and style.

In many large business complexes and high-rises, the telephone company uses fiber optic cable and fiber optic multiplexers to terminate the large number of lines in the building. The fiber multiplexer is often located in the basement telephone room in a relay rack with backup power. The multiplexer converts the optical signal into an electrical signal that is then terminated on blocks that are mounted on the wall. The copper cable fed into the building by the telephone company is also terminated on blocks mounted on the wall. They are labeled with the cable number, the pair number, and the termination block number.

Multiple terminal blocks are mounted on the wall or in a rack. Each block is labeled with the floor number it feeds, the cable number, and the cable pair. The types of blocks used to terminate house cable vary depending on the preference of the owner, but 66 blocks are the most commonly used today. The blocks require that the cable be punched down on small metal points. The technicians do not have to strip the wire beforehand because the punch-down pins cut into the cable and make the connection. Figure 20–6 diagrams a typical 66 block, 110 block, and wire wrap block all commonly used in medium-size business locations.

Returning to our example, the telephone company tags pairs 220 to 319 of cable 805 for President, Inc., as well as four pairs for the two T1 circuits. President, Inc. has decided to pay the telephone company to extend the demarcation point to the fifth floor; and, because they negotiated with the building owner for space on the house cable, the telephone company has permission to extend the demarcation point by using the existing building cable. If the building owner would not allow the phone company to use the house cable, or if there were no additional wires available on the house cable, the phone company or customer would have to run cable from the basement to the fifth floor.

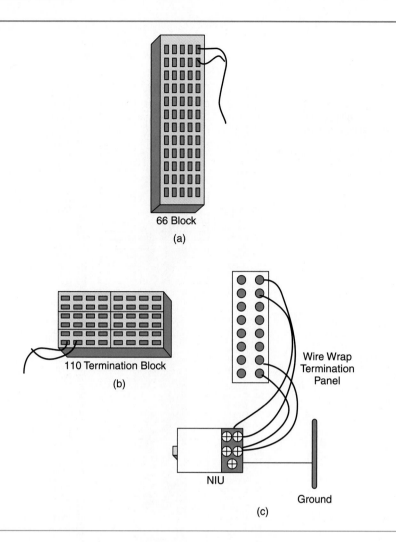

Figure 20–6
The types of termination blocks found in small business locations vary depending on the choice of the network manager. Three types of termination panels are shown, along with a Smart Jack that terminates T1 circuits.

The telephone technician runs jumper cable between the telephone demarcation point on the 66 block and the house cable that is terminated on the 66 block feeding the fifth floor. The transmit and receive pairs of the T1 circuits are separated by twenty-five pairs to avoid frequencies infiltrating adjacent pairs, causing interference on the line, and disrupting the signals. Cables 5101 and 5102 are connected to the transmit pairs of the incoming T1s, and 5150 and 5151 are connected to the receive pairs of the incoming T1s. The 100 pairs of wires are connected one for one on house cable 500 pairs 5000 to 5100.

Once the technician finishes punching down the wires on the 66 blocks, shown in Figure 20–7, he or she heads to the telephone room on the fifth floor. All tenants located on the fifth floor share the telephone room there. 66 blocks are mounted on the wall with the same cable and pair designations as in the basement. The connection between the fifth floor telephone closet and President, Inc.'s telephone room in their office space must be run physically. The technician has brought a reel of 100-pair inside wire cable. Because President, Inc. needs 100 pairs plus four pairs for the T1 circuits, the technician will run two 100-pair cables between the floor's shared closet and President's telephone closet. The technician runs the cable, punches one end down on pairs 5000 to 5100, and then goes to the telephone room in the business's office and punches down the 100 pair on a 66 block mounted there. The technician also mounts two Smart Jacks, also called *NIUs*—network interface units, on

NIU
Termination device for T1 circuits. Also called a Smart Jack.

Figure 20-7
The 66 block is used to connect two
pieces of cable at a customer's
premises.

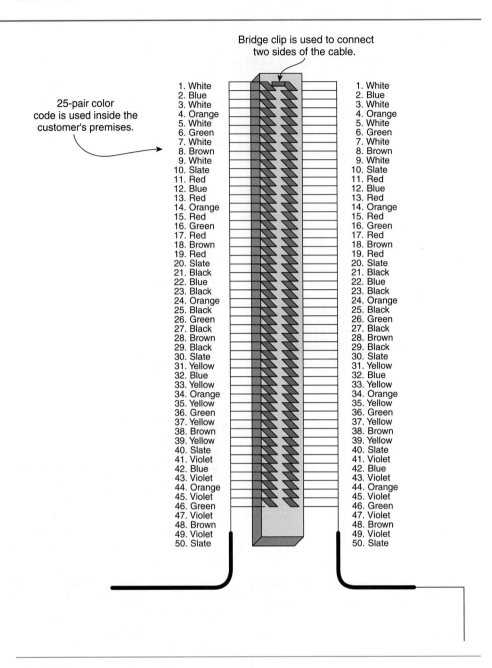

Bridge clip is used to connect
two sides of the cable.

25-pair color
code is used inside the
customer's premises.

	Left side		Right side
	1. White		1. White
	2. Blue		2. Blue
	3. White		3. White
	4. Orange		4. Orange
	5. White		5. White
	6. Green		6. Green
	7. White		7. White
	8. Brown		8. Brown
	9. White		9. White
	10. Slate		10. Slate
	11. Red		11. Red
	12. Blue		12. Blue
	13. Red		13. Red
	14. Orange		14. Orange
	15. Red		15. Red
	16. Green		16. Green
	17. Red		17. Red
	18. Brown		18. Brown
	19. Red		19. Red
	20. Slate		20. Slate
	21. Black		21. Black
	22. Blue		22. Blue
	23. Black		23. Black
	24. Orange		24. Orange
	25. Black		25. Black
	26. Green		26. Green
	27. Black		27. Black
	28. Brown		28. Brown
	29. Black		29. Black
	30. Slate		30. Slate
	31. Yellow		31. Yellow
	32. Blue		32. Blue
	33. Yellow		33. Yellow
	34. Orange		34. Orange
	35. Yellow		35. Yellow
	36. Green		36. Green
	37. Yellow		37. Yellow
	38. Brown		38. Brown
	39. Yellow		39. Yellow
	40. Slate		40. Slate
	41. Violet		41. Violet
	42. Blue		42. Blue
	43. Violet		43. Violet
	44. Orange		44. Orange
	45. Violet		45. Violet
	46. Green		46. Green
	47. Violet		47. Violet
	48. Brown		48. Brown
	49. Violet		49. Violet
	50. Slate		50. Slate

the wall in the telephone room to terminate the two T1 circuits. Figure 20–8 illustrates the connection between the NIU and the 66 block. The technician tags the pair
and leaves.

The owner of President, Inc. persuades her network savvy niece to come in and
complete the connections into the company's phone system and LAN. The office
wiring was designed to provide one telephone jack and one data jack per employee.
President, Inc.'s sixty-nine employees will all have their own telephone lines. Category 5 cabling was used to connect each of the data jacks, and CAT3 cabling was used
for each telephone jack to the telephone/LAN room. In the telephone/LAN room, 66
blocks are mounted on the wall and labeled with the telephone jack's number. Connections are made between the 66 block that is connected to the 100-pair cable connected to the telephone closet and the 66 block connected to all of the telephone jacks

Figure 20–8

The 66 termination block is used as a cross-connect point between outside wire cable, inside wire cable, and terminating equipment such as a Smart Jack, shown in the diagram.

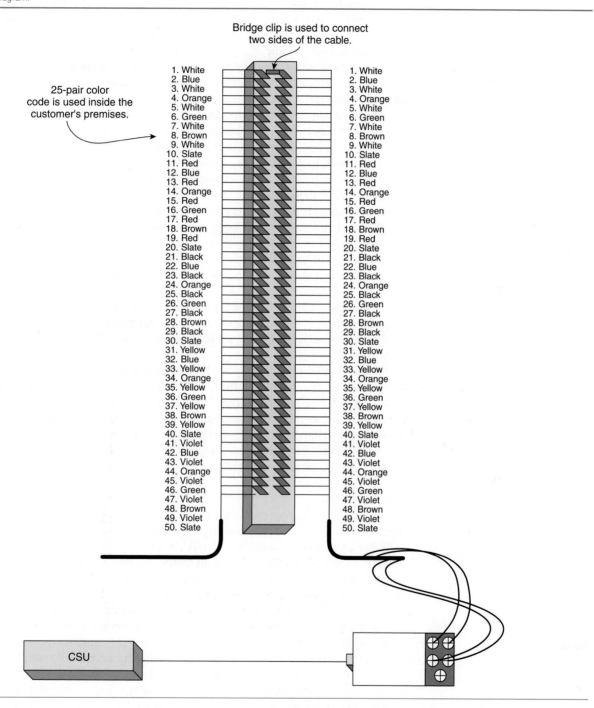

in each of the cubes. Once the connection between the two blocks has been completed, the employees can plug one end of a two-ended RJ-11 jack cord into the telephone and the other into the jack and receive dial tone.

One T1 circuit will tie President, Inc.'s office LANs together; and a second T1 will connect it to their ISP for Internet access.

Voice Equipment

The voice equipment at President, Inc. consists of single-line and dual-line telephones, two fax machines, and one multibutton receptionist set. The company determined that it was easier to purchase all of the telephone features from the phone company than it was to place a key system on site. The president and five of the top executives have caller ID and a second line. Each of the employees has voice mail, call forwarding, and remote call forwarding. Fax machines have their own telephone lines. The receptionist has a special multibutton phone that allows him or her to transfer calls and place callers on hold. It also has a special hotline into the president's office.

We have just described a medium-size business that has decided to buy many of its voice services from the telephone company. Today, phone companies provide centrex services and multiple features that reduce the need for small and medium-size business customers to purchase key systems or small PBXs, although some companies believe that it is beneficial to have control of their own services on site. A key system is a small switch that can interface with telephones and outgoing telephone lines. It has software features loaded into its processor that allow the company to provide features on all of its lines. A telephone company would charge a monthly fee for each feature placed on each line, but the key system provides the features for free. The key system also helps reduce the number of telephone lines needed to serve all the telephones. Because not everyone goes off-hook at the same time, fewer outside lines are needed to carry the calls between the central office and the business site. Company management determines the number of telephone lines needed for the number of telephone stations. The key system has trunks that connect the telephone line to the phone company, which, in turn, sells the business a block of numbers called DID or DOD numbers to assign to each telephone station.

A small PBX may also be used by medium-size customers. Larger than the key system, it is a true switch. The PBX often has an internal voice-mail system, eliminating the need for the phone company's voice-mail service. The PBX also provides trunk connections between the central office switch and the PBX, which alone often justifies the cost of installation. The PBX has sophisticated routing tables that allow call blocking, least-cost routing, and other switchlike features. Figure 20–9 illustrates a typical PBX located in the telephone room of the company.

Data Equipment

President, Inc. relies on a dedicated T1 circuit connection to tie the two LANs together. The T1 terminated at the NIU represents the demarcation point for the circuit. The company president's niece connects the NIU to a CSU interface in the company's router. She configures the router to direct all traffic heading toward the remote office LAN onto the T1 serial interface. The T1 connecting President, Inc. to its ISP, Bit Plus, is connected to the second T1 serial interface on the router. The NIU is connected to the internal CSU router serial port by an RJ-48 to RJ-48 cable, as shown in Figure 20–10. Connected to the router is an Ethernet connection feeding into a hub that connects all of the employees' PCs, the *file server*, the *Web server*, and the printers via Ethernet CAT5 cable. Figure 20–11 shows how everything connects in the telephone room.

The common data equipment used by medium-size businesses is personal PCs, file servers, Web servers, routers, hubs, gateways, bridges, proxy servers, CSUs, remote access servers, modems, printers, and a mainframe.

20.1.4 Large Businesses

Demarcation Point and Wiring

Large businesses, those with more than 100 lines, require large networks. The demarcation point in a large business is similar to that in a medium-size business. The greatest difference is the number of lines being terminated. To explain how the outside network and the internal network connect, we will again use a hypothetical example. NOTEBooks has 300 employees working at each of its three sites around the country, but we will focus on its Phoenix location. NOTEBooks has full-time voice and data

---

file server
Server that stores files on a LAN.

Web server
Server that stores Web information.

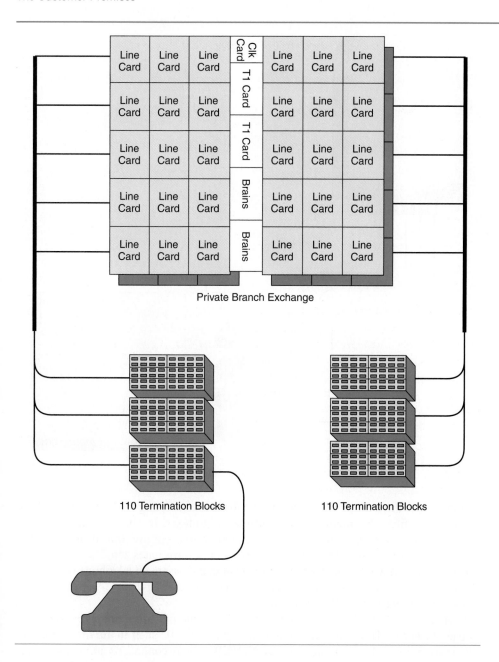

Private Branch Exchange

110 Termination Blocks   110 Termination Blocks

Figure 20–9
Small business owners often place small switches called private branch exchanges (PBXs) at their location to switch all calls within the business. Trunks are established between the PBX and the telephone company's switch. The trunks are used to switch calls leaving the premises.

staff. Alex, its network manager, oversees both. Alex knew that he needed dedicated high-speed circuits to use as WAN connections among the three NOTEBooks locations. He also realized that the building was old and would require additional wires to be run in order to support the data traffic around the building. Alex decided to design the wiring architecture himself but to hire a wiring contractor to physically pull and terminate the cables. His first task was to make wiring diagrams that illustrated where each cable would run, what type of cable would be used, and which jacks or blocks it would terminate on.

The telephone company's demarcation point was located in the basement of NOTEBooks's five-story building. Because NOTEBooks required 300 individual lines, data circuits, and a tie line, the telephone company decided to place fiber optic cable into the building. It installed an OC-3 multiplexer in a relay rack in the basement telephone room, in addition to fiber optic cable connecting the central office to the building. The fiber cable was terminated on a fiber distribution panel and, from there, optic jumpers connected the outside plant cable to the multiplexer. The phone company owns

Figure 20–10
The RJ-48 jack is used to terminate
four-wire circuits such as T1s.

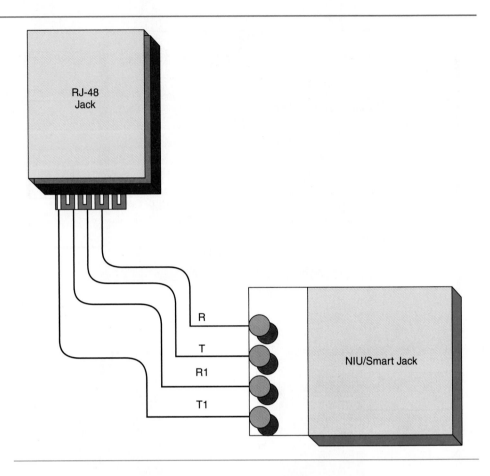

and maintains the multiplexer and is responsible for connecting the fiber optic jumpers between the fiber distribution panel and the multiplexer. In addition, the telephone company is responsible for terminating the electrical circuits that dropped out of the optical multiplexer. The telephone technician terminates ten T1 circuits onto the 110 punch-down blocks located on the wall in the basement telephone room. Each of the T1's transmit and receive pairs are separated by at least twenty-five pairs of wires in order to reduce cross talk and interference.

A cable tying the basement telephone room to the first-floor telephone room has already been run. The telephone technician installs an NIU shelf in the first-floor telephone room in a relay rack provided by NOTEBooks. A connection jumper is placed between the 110 punch-down blocks of the interbuilding cable and the NIU shelf sitting in the relay rack, as shown in Figure 20–12. Once the cables have been terminated at the NIUs, the telephone company's demarcation point, a second cable is placed between the NIUs and the CSUs. The CSUs belong to NOTEBooks and are now ready to be connected to the appropriate pieces of equipment such as the PBX, router, and so forth.

NOTEBooks has several copper telephone lines installed in its building to provide backup dial tone in case the fiber optic cable is cut, and full-power backup with an on-site generator and an uninterrupted power supply (UPS) system for the data equipment. Alex decided not to order diverse entrance fibers from the telephone company to ensure better fiber diversity, due to the additional cost.

Alex decided early on to wire each cubicle with two phone lines and one data line using CAT5 wire for data and twisted pair wire for voice. Each jack consists of two RJ-11 jacks and two RJ-45 connections. The second RJ-45 was not connected. Offices received two data-line jacks and two voice jacks. The first-floor communications room houses both the data and voice equipment and is the wiring center for the building because all of the wiring is hubbed out of the first-floor communications room.

Figure 20–11 A common data room layout. Terminals are connected to hubs. The hubs feed into a router or are fed back to file servers depending on the destination request. The VAX mainframe has an RS-232 interface that feeds into an access terminal that converts the RS-232 into Ethernet. The Ethernet interface feeds into the router. The database also requires an RS-232 conversion at the access terminal. The UPS system sitting at the corner is used to provide backup power if commercial power is lost.

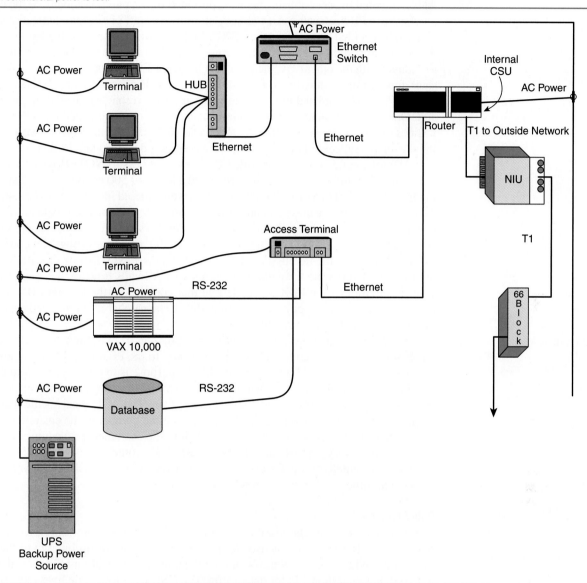

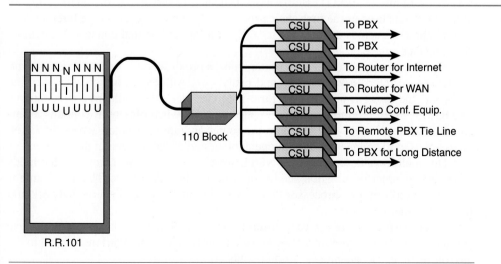

Figure 20–12
Typical business location that has multiple T1 lines. A Smart Jack shelf is placed in a relay rack in the telephone room at the business premises. The T1 provides connections to the outside world for both voice and data traffic.

Voice Equipment

Alex compared the cost of using centrex, provided by the phone company, to the cost of purchasing a PBX. Because Alex already employed a staff of telecommunications technicians, he decided to purchase the PBX, which was sized to serve 700 telephone lines. A voice-mail system was added as a feature, and direct tie-line interfaces were included to connect with the other two cities' PBXs. Of the ten T1 circuits terminating in the CSU shelf, four terminate into the PBX. A connection was made between the CSUs in the rack and the PBX DTI card, a digital T1 interface card used to connect the PBX to the T1 circuit. One T1 was designated for incoming calls, one for outgoing, and the other two for tie lines to connect to the two other NOTEBooks sites. The channels on the two T1s feeding the telephone central office switch were configured as DID and DOD trunks.

The telephone company provided DID line numbers to NOTEBooks for its telephone station numbers. NOTEBooks received a block of 500 telephone numbers ranging from 534-6000 to 534-6500. The numbers were entered into the PBX the same way they are entered into a class 5 central office switch. Restrictions were added to certain numbers not allowing long distance calls, 900 numbers, or any other restricted service. Voice mail was added to each line, as were caller ID, call forwarding, and conference calling. Thanks to the tie circuits that connect the three NOTEBooks locations, four-digit dialing is possible throughout the company. All calls traveling between the two cities are routed by the PBX to the tie line that connects to the far end PBX. NOTEBooks does not have to pay a per-minute long distance charge for any calls riding on the dedicated tie line between the sites.

NOTEBooks also has an *automatic call distributor* (ACD) to route calls to customer service representatives. The ACD interfaces to the PBX by a T1 circuit and to a database using an RS-232 interface. The ACD's purpose is to provide an automatic answering device that retrieves customer information from the customer information database and sends it to the customer service screen. The ACD receives the incoming number from the PBX, asks the database to match the number to the customer record, and then takes the customer record and feeds it to the customer service representative. ACDs are now very sophisticated and can selectively route calls to specific customer service representatives. ACDs also track employee statistics such as answer performance, call times, and other employee measurements.

Data Equipment

NOTEBooks has built a large data network that provides data connections to each employee. A fairly large router is used to interface the outside world Internet to the business office and to route LAN traffic among the three office locations. Two channels are reserved on the tie-line T1s for data traffic. The CSU used for the tie lines can drop out two 56 kbps channels on each T1. The 56 kbps channels interface into the router by way of a V.35 interface. NOTEBooks has a dedicated T1 circuit into the Internet by way of its ISP that is used as the first-choice route when employees access the Internet. The second choice is across the data circuit riding on the tie line that connects the remote sites together.

The data network consists of a router, hubs, a proxy server, a remote access server, file servers, printers, a Web server, and many PCs. All devices are connected to the LAN, and the router is used to direct traffic to the Internet or to the remote locations. The RAS is used for employees to dial into the LAN when off site. It contains multiple modem connections that feed into the router and are distributed across the LAN. Hubs are used to connect the PCs to one another and to the servers on the LAN.

One additional device found at NOTEBooks is a mainframe computer that holds all of the customer information and handles all billing transactions. The mainframe is preceded by a front-end processor that acts as the interface between the network and the mainframe.

Large businesses are not very different from medium-size businesses. Both have similar pieces of data equipment and voice equipment. The main differences are in the size of the units and the number of interconnections.

---

automatic call distributor
Device that routes incoming calls to the appropriate phone.

20.1.5 Large Corporations
Voice Network

Many large corporations have campus networks that are as large as some small telephone companies. Connecting the many buildings on a corporate campus requires fiber optic cable runs. Systems as large as OC-48s are used to provide the needed bandwidth between buildings, and entire floors or even buildings are dedicated to their communications networks. Most corporations separate their voice and data equipment into two separate areas and have two groups of employees to maintain them. The voice network must be able to handle thousands of calls, both internal and external. Voice conferencing, call centers, and international connections are some of the unique requirements of a large corporate voice network. Again, we will look at a hypothetical corporation.

Dakco Corporation is the country's largest manufacturer of chocolate-covered cherries, employing 30,000 workers at its Rochester, Minnesota, plant. The plant has twenty buildings that cover a 300-acre footprint. Each building has a need for multiple voice lines, fax lines, and data lines; and each is connected by fiber optic cable that Dakco installed three years ago. The fiber is terminated in a telephone room with relay racks, backup power, and an OC-48 multiplexer. Building 10 is used as the communications center, where floor 2 holds all of the voice equipment and floor 5 is the data communications area. Dakco has a class 5 switch located on floor 2 that is used to switch calls throughout the campus, out to the PSTN, and to Dakco's international manufacturing plants.

Floor 2 also holds numerous other pieces of voice equipment, including multiplexers, channel banks, and even an MDF. The layout of the room is the same as that of a telephone company's central office. The voice switch has line-side terminations that feed all of the telephones on the campus. It also connects to outside carriers and maintains a direct connection to two different long distance companies. A satellite antenna located on the roof of building 10 connects the Rochester plant with the other Dakco plants around the world. The satellite connection is similar to the tie lines discussed previously. Dakco avoids high international calling rates by routing all inter-company traffic over the satellite link.

Each building in the Dakco plant has telephone cable running through it to provide telephones for each cubicle, office, and workstation on the factory floor. The Dakco class 5 switch provides calling features for the employees. An ACD is attached to the switch for the customer service department's incoming calls. Essentially, Dakco's telecommunications network is more like that of a telephone company than a business customer.

Data Network

Dakco's data complex is located in building 10 on floor 5. The room holds large mainframe computers, file servers, routers, gateways, hubs, and the newest addition—a Web server. The data room has a false floor with thousands of cables running beneath the removable tiles. The walls of the data room hold many cable termination blocks where the cable that runs through the campus terminates and interconnects to the data equipment. Each building has a LAN that is connected by CAT5 cable to the other LANs in the data room in building 10. The multiple LANs form a corporate WAN that is managed by the data communications group.

Data traffic travels on the interbuilding fiber optic network to reach the LANs in the other buildings. Traffic heading overseas to Dakco's international plant is transmitted across the satellite link connecting the two sites. The international traffic is transmitted in the middle of the night to take advantage of the off-peak hours of voice traffic. Dakco connects directly to its ISP for Internet access using a full T3. Figure 20–13 shows the Dakco network.

20.1.6 University Campuses

A university is similar to a large corporation in that both have a number of buildings, telephones, and computers that need to be connected. Compared to businesses, universities have special communications needs. Because some students live on campus, the

Figure 20–13
A typical business location that has two buildings connected by fiber. Each building has an optical MUX that drops out T1 circuits.

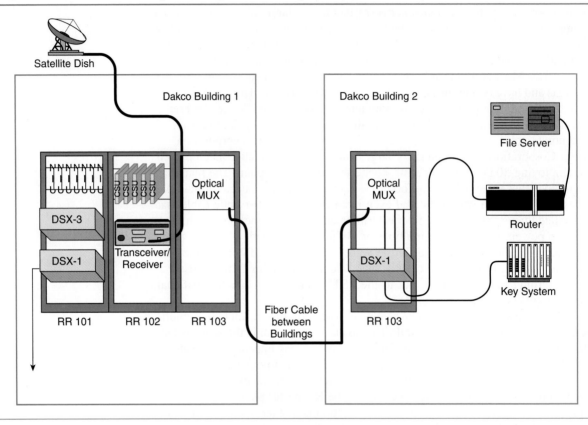

university looks more like a town than a business complex. Using the Internet for research, providing mainframes for internal computing needs, and maintaining an extensive library database force the university to provide large amounts of bandwidth throughout the campus.

We will look at a typical university campus to see how its communications needs are handled. The voice network requires two large PBXs, each capable of handling 20,000 line-side terminations. The PBX is tied to the telephone company's central office class 5 switch that provides multiple DID/DOD trunks. Each room in the residence halls has one telephone connection that interfaces into one of the PBXs. The twisted pair cable is fed through the residence halls in the same way it is fed through other high-rise buildings. Termination blocks are found on each floor to terminate the telephone lines for the rooms on that floor. In the case of the university, unlike the large corporation, fiber optic cable was only run between five of the main campus buildings. The rest of the campus depends on the existing copper cable that was installed in the early 1950s. T1 circuits have been deployed to increase the information-carrying capacity of the cabling between buildings. The greatest problem that the university faces is the need to wire the residence halls for broadband service for Internet and LAN access. The one telephone line in each room does not provide enough bandwidth for Internet use. Adding lines to each room is cost prohibitive due to the age of the building and the congestion in the ductwork.

The telecommunications committee decided to place DSLAMs in the basement of each of the residence halls and classroom buildings. The DSLAM provides each room with a 768 kbps data link with a voice circuit riding on the lower frequencies. The one copper pair already terminated in each room provides two data links and one voice line, thanks to DSL. The DSLAMs are connected to the data network by copper T1 cables that terminate in the residence halls. One hundred users are placed on one T1 circuit. The voice

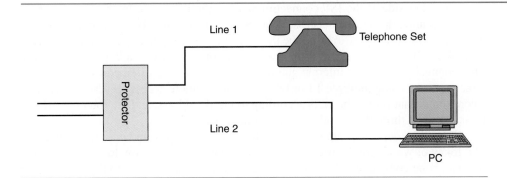

Figure 20–14
Typical residential customer that requires voice and data circuits. Residential communications networks vary dramatically due to different requirements. Voice and data require separate analog lines if the user wishes to use both simultaneously.

signals are stripped off at the DSLAM and fed into a channel bank, which is connected to the PBX by way of a T1 circuit. Voice signals enter the PBX DTI card from the T1 circuit. Data signals travel on the T1 that connects the DSLAM to the core router in the data room. Figure 20–14 illustrates how a residence hall handles both voice and data circuits.

The buildings that have fiber optic cable terminating in their basements are connected to the internal LAN by CAT5 cable feeding a router in the basement. The router has a DS-3 circuit that connects to the fiber optic multiplexer. The signal then travels into the data room where a core router is used to interface multiple campus LANs. The data room also has a mainframe, a Web server, hubs, firewalls, and multiple file servers. Large databases housed in the data center for the library have been added to the local network. Using their DSL service, students log in to the university LAN to reach the library databases, the shared file servers, the e-mail server, and the Internet.

▓ 20.2 INTERNET SERVICE PROVIDERS AND APPLICATION SERVICE PROVIDERS

ISPs and ASPs need to connect into the PSTN.

20.2.1 Internet Service Providers

An *Internet service provider* (ISP) is a company that interfaces customers to the Internet. It is the middleperson between the customer and the Internet, which is a conglomeration of networks connected together by communications circuits. ISPs provide a gateway into these networks. They vary in size from small, with one PC and a few access lines, to large fiber optic fed networks with multiple connections into the Internet and thousands of access lines.

The ISP has a point of presence (POP) similar to the long distance telephone company's POP. It may reside in a location owned and maintained by the ISP or it may be colocated in a long distance company's site. Either way, the ISP has equipment that interfaces to the subscriber's telephone line and circuits that connect into the Internet. Remote access servers are normally used to connect to the subscriber. Remote access servers terminate calls coming in from dial-up modems. They look at the access number dialed by the subscriber and connect the subscriber into a port on the network. Once the connection between the subscriber and the ISP has been established, the ISP feeds the Web page to the subscriber's PC showing each of the services available to the customer.

The most common method used by ISPs to connect to the local telephone company is the ISDN PRI circuit. It is similar to T1 circuits but has a signaling channel in the twenty-fourth channel that allows the ISP's servers to interpret the destination of the incoming call. The telephone switch receives the digits dialed by the subscriber, looks them up in the translations table, sees that they belong to the ISP, and routes them out on the ISDN PRI circuit(s) connecting to the ISP. ISDN PRI circuits have simplified the interconnection of the ISP to the local telephone company, whether it is a CLEC or an ILEC.

Internet service provider
Provides access to the Internet. Examples are AOL and EarthLink.

The other side of the ISP connection is from the ISP to the Internet interconnection point called the network access point (NAP). The ISP has to size its backbone network to the Internet by estimating the number of users it expects to dial in at any given time. A small ISP may have dial-up lines to a second ISP, while a large ISP will have fiber connections into the Internet gateway. The medium ISP often connects to the Internet gateway using multiple T1 or DS-3 circuits. Remote access servers connect into a core router that routes the calls according to the IP address, and the router interfaces to the network through a serial T1 or DS-3 interface.

The ISP is dependent upon the local telephone company and a long distance carrier, many of which are themselves ISPs. Many large ISPs are now looking at becoming telephone companies. Beyond offering network connections into the Internet, the ISP is also responsible for setting up the customer's e-mail address. This requires an e-mail server and a domain name, which, in turn, requires a domain name server. The ISP must be able to handle customer troubles, no matter how simple. The typical ISP has many servers, PCs, RASs, and routers all connected to provide services for the Internet client. Figure 20–15 illustrates the different devices found in an ISP's POP.

20.2.2 Application Service Providers

application service provider
A company that manages and houses applications and content for clients.

An *application service provider* (ASP) is a company that manages and houses applications and content for clients. A fairly new service provider, it has emerged because of the increased complexity of managing software applications. An ASP may have servers holding software programs that can be accessed through the Internet or dedicated circuits. Customers do not have to manage the software nor do they need to buy the entire software package. Instead, clients can lease the software for the amount of time they need to use it. The ASP manages the software, makes sure it is upgraded, and partitions the package according to the customer's needs. For instance, a small accounting firm would not want to purchase a software program that included manufacturing quality control features because those features would not be needed. The ASP could lease the firm only the portion of the program that was needed.

Analysts predict ASPs will grow dramatically over the next few years. The advantage for the small customer is that ASPs provide technical support. The small customer often does not have the resources to maintain software applications. In addition,

Figure 20–15
Dial-up Internet lines interface RASs before being aggregated onto circuits interfacing the switch. RASs replace modem pools. From the switch, the traffic is routed to a router that points the traffic to the Internet.

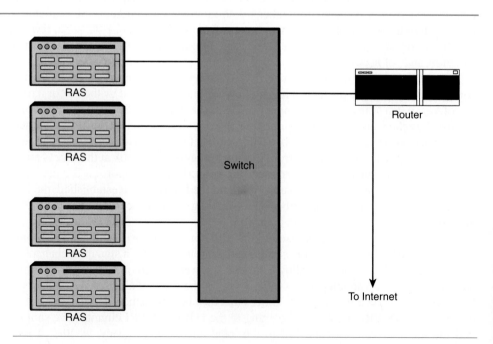

companies can lease programs for the time they are needed instead of purchasing a program that will only be used once or twice.

■ 20.3 DATA COMMUNICATIONS EQUIPMENT

20.3.1 Introduction

Data communications networks need data communications equipment. Many vendors manufacture data communications equipment. This creates an environment of stiff competition between vendors. Telephone companies, Internet service providers (ISPs), cable television companies, and cellular companies all use data communications equipment in their networks. Business users (enterprise customers) also use data communications equipment to tie their LANs and WANs together. Residential subscribers also deploy data communications equipment to provide access to the Internet and to their LANs.

20.3.2 Overview of Data Communications Equipment

The data communications network is similar to the voice network, in that devices are placed at customer premises to communicate with devices placed at other customer premises. The network consists of equipment that routes information between two devices, formats the information properly, and maintains the integrity of the content being transported. The voice network uses telephones, telephone switches, and transmission equipment. The data network uses workstations, routers, packet or cell switches, and transmission equipment.

The devices in a data communications network have been defined generically to simplify provisioning and maintenance. Figure 20–16 shows a generic data connection between two locations. Notice that the workstation has been identified as *Data Terminal Equipment* (DTE) and the modem as *Data Communications Equipment* (DCE).

A DTE can be a PC, a switch, or a mainframe. In the case of a PC that connects to a modem and then out to the Internet, the PC is where the signal ends. Therefore, the PC is the DTE. The signal may also be terminated in the mainframe or in a peripheral device containing a database, such as an SCP in the SS-7 network. The varieties of DTEs are too numerous to list, but the concept of what determines a DTE is simple and can be used when deciding whether or not a device is a DTE.

DCE describes any device that interfaces with the outside physical network, such as a dedicated T1 circuit, a POTS line, an ISDN line, and so forth. Common DCE devices are CSUs, DSUs, modems, frame relay access devices (FRADs), and NT-1s. The

Data Terminal Equipment
Computer, file server, and so forth.
Data Communications Equipment
CSU, DSU, and so forth.

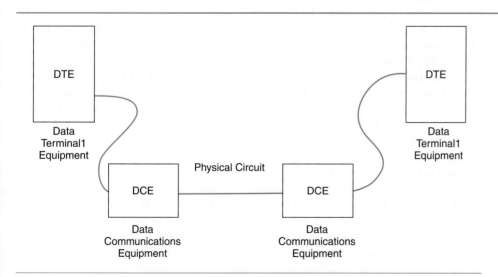

Figure 20–16
Generic depiction of a data point-to-point system.

Figure 20–17
Data circuits travel across the telephone network and are classed as either two-wire (a) or four-wire circuits (b).

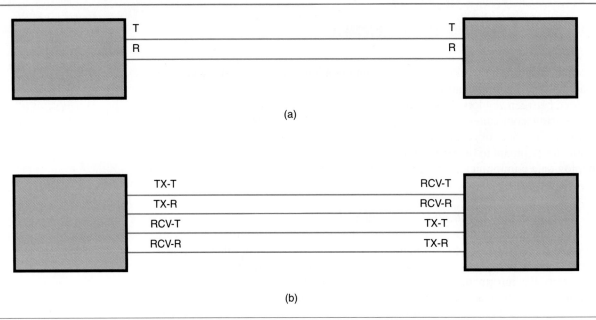

DCE is often combined with DTE as an interface pack that fits into a common chassis. Though physically the same, the distinction remains valid in that the DCE continues to function as the DCE.

Generically, data networks are also defined by the type of physical circuit used to interconnect the locations. Figure 20–17 shows the physical interface connecting two locations and defines it as the physical circuit. The physical medium may be a telephone circuit, a dedicated circuit, or anything else that connects two locations. The network between locations is often shown hidden inside a cloud. We will call the hidden equipment the network equipment. Equipment shown at the edge of the cloud will be called edge equipment, and equipment shown at the customer premises will be called customer premises equipment.

The terms *ingress* and *egress* are often used to define the way data flow into and out of the network. For instance, traffic flowing from the access device at the edge of the cloud to an ATM switch inside the cloud is flowing into the network; it is the ingress of the network. Traffic flowing from the ATM switch to the access device is traveling into the egress of the network. Unfortunately, some vendors and network designers reverse the terms, causing considerable confusion during network discussions. Figure 20–18 shows the difference between the two terms, as defined by the majority of those in telecommunications.

Data communications equipment's purpose is to transport information across a variety of networks. The size, sophistication, and cost of the equipment are dependent upon the application it will be used for. For a network that will carry mission-critical data, *carrier class* data networking equipment may be needed. Carrier class equipment has been constructed to carry large amounts of data and provide component redundancy. For a small business shipping files between two offices, the cost of a fully redundant network may be unnecessary.

The term *access* refers to the portion of the network controlled by the local telephone company. The access side of the network carries traffic from the core network to the customer demarcation. The term *network* or *backbone network* defines the portion of the network that carries all of the traffic between large network POPs or switching centers. *Core* represents the core of the backbone network. For example, an Internet provider has a long-haul backbone network that connects many POP locations together.

carrier class
Term used to define a 99.999% uptime network or equipment.

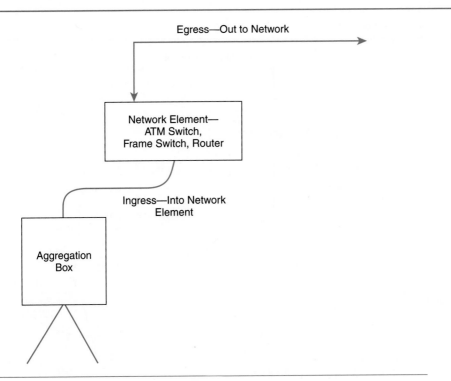

Figure 20–18
The ingress and egress of a network.

The ISP only hands traffic off to the Internet at four of the twenty locations. The four sites are connected in a mesh topology as shown in Figure 20–19 forming a core network that is used to access the Internet. Figure 20–20 shows where the access network, the backbone network, the egress, the ingress, and the cloud reside.

Although many terms are frequently used to help define communications equipment, our discussion will be limited to the few terms just defined—DTE, DCE, ingress, egress, carrier class, access side, network side, core, and cloud.

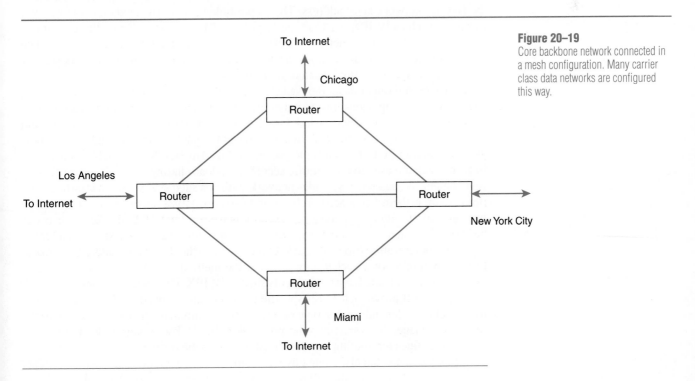

Figure 20–19
Core backbone network connected in a mesh configuration. Many carrier class data networks are configured this way.

Figure 20–20

How the backbone network and the access network connect in a region.

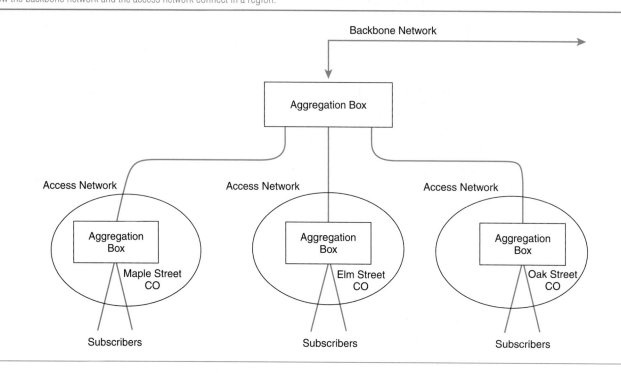

20.3.3 Packet Equipment
Routers

router

Device that routes packets using a source or destination address.

A *router* is a device that connects two or more data networks together, as shown in Figure 20–21. It can route individual data packets across an internal network, to a second network, or through an intermediary network by using Layer 2 and Layer 3 protocols. The most common use of a router is to send packets through the network using the Layer 3 network layer address. This is possible because of standard network layer protocols such as IP, IPX, DECnet, or AppleTalk. The packet may pass through multiple routers and multiple networks before reaching the terminating equipment. The Layer 3 address is used to route the packet through each device in the network step by step until it reaches the final destination router.

A router not only cracks the packet to look at the destination address, it also uses special load-sharing algorithms to balance the traffic across the links. It is also responsible for rerouting traffic when a link fails. For example, if the link between Chicago and Los Angeles fails, the router in Chicago reroutes the packets onto the link to St. Louis and the router in St. Louis sends the packets to Los Angeles. Routing tables that reside in each router are used to route traffic according to load sharing, time-of-day routing, or link outages. A router also provides network statistics such as link use and traffic statistics that are gathered on a per-port basis and used to analyze the network. Most routers provide ways to filter traffic on a per-client or per-type-of-traffic basis. Access lists can be used to screen users by looking at the originating address and the destination address.

The router's ability to route packets according to the destination address is critical to the functioning of a packet network. It is the method used by all Layer 3 devices. When routing multiple Layer 3 protocols such as IP, IPX, DECnet, and AppleTalk, each of the protocols provides features specific to the end user's needs. Which Layer 3 protocol is chosen depends on the type of terminating equipment and the preference of the network manager. The most popular protocol is the IP. The router holds the software code that defines the routing protocol used to format the traffic.

Today a router can also route Layer 2 protocols such as frame relay and ATM. It accepts a Layer 2 interface that allows it to handle Layer 2 routing protocols. The term

Figure 20–21
Routers connect data networks together and route packets to the correct destination.

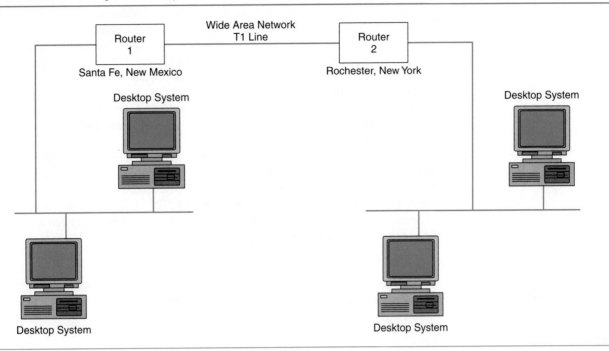

interface refers to the physical port card that is plugged into the equipment chassis. The port card has the ability to generate and terminate Layer 2 protocols. The device itself has processors that run software code, which enables the interface card to produce the desired protocol. In addition, the hardware provides the correct switch matrix or backplane bus to carry the lower layer protocol. Routers are quickly becoming the class 5 switches of the data communications network. They can no longer be viewed as just Layer 3 devices.

The size of a router varies depending on where it sits in the network, the number of interface ports that are needed, and the manufacturer. The device may be as small as a cable set-top box or as large as a small television. Routers are connected to one another by different types of circuits, depending on the amount of traffic or bandwidth required. Enterprise routers often are configured with T1, 56 kbps, and Ethernet interfaces. Access routers terminate rates from Ethernet up to OC-3. A T1 serial interface, a DS-3 serial interface, an OC-3 interface, and an Ethernet interface are commonly found in smaller network routers. Larger network routers are built to be carrier class devices able to interface circuits as small as T1 and as large as an OC-192. A router is a very complex piece of equipment that has moved from the enterprise space to the core network in less than five years. Routers must be able to handle thousands of transactions every second and must exchange routing information with adjacent routers and networks. They are the most widely deployed pieces of data communications equipment in the network today and are used by enterprise customers, local telephone companies, long distance carrier network providers, and Internet service providers.

Bridges
A *bridge* is placed in a network for two reasons—to alleviate a congested network and to connect two LANs or two data networks together. A bridge is a less-sophisticated device than a router. It routes according to the data link layer, thereby making it a Layer 2 routing device. The advantage of being a Layer 2 device is that the bridge will pass any high-level protocol transparently, thus reducing the time it takes for information to pass through. The disadvantage of using a bridge instead of a router is that the bridge does not have sophisticated routing tables or load-sharing algorithms and is thus limited in its ability to handle traffic end to end.

bridge
Device that transmits packets into a network using the MAC address.

The bridge establishes routing tables by polling adjacent devices and asking for their MAC addresses. When a bridge is first turned up, it sends out discovery packets asking each piece of equipment to send back its MAC address. The bridge uses the addresses to build a routing table that it will use to route incoming packets to their destinations. A bridge is similar to the post office. Residents have house addresses that postal workers memorize. A letter addressed to George Smith at 101 Hyland Street is given to the local postal worker named Jane. Jane has learned the directions to George's house and is therefore able to deliver the letter without having to ask for directions. The bridge, like Jane, memorizes the local addresses of the devices connected to its interface. The address is unique to a particular piece of equipment, just like a postal address is unique to a particular house.

As already noted, a bridge is a less-sophisticated piece of equipment than a router. Bridges route according to MAC addresses and therefore must be physically connected to the device they route to. The router, on the other hand, can route past adjacent devices by using routing algorithms such as BGP, OSPF, or Cisco's IGRP. A bridge is less expensive than a router and easier to configure, but the choice between a router and a bridge depends on what tasks the equipment must perform.

hub
Physical interface that interconnects multiple devices together.

Hub

A *hub* can be a passive or active unit that connects multiple devices onto one termination, such as multiple PCs onto one outgoing link. As shown in Figure 20–22, a hub

Figure 20–22
How LAN connections are made between two locations.

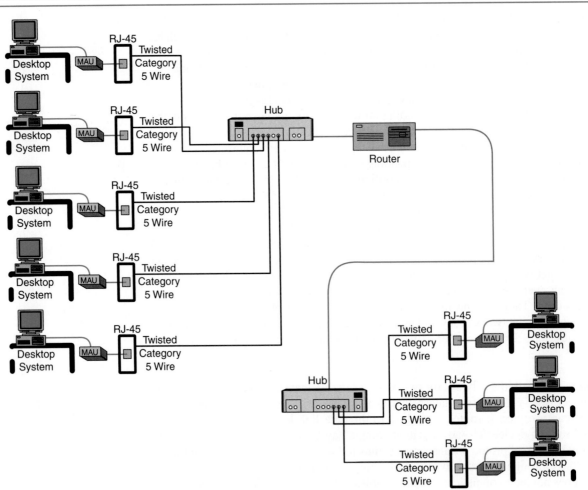

allows five PCs to share the same circuit on a LAN. Hubs are commonly used in LANs to connect multiple workstations before attaching them to a LAN backbone. Token ring LANs use hubs to form physical star networks. A hub can also be used to help manage cabling in a building by acting as a central connection point.

A hub is a small, very inexpensive device that is used in almost every business's data communications system. Residential users buy hubs to connect all PCs in the house together on one telephone line. They are deployed extensively in the enterprise space and used for internal use in carrier networks. The hub does not produce routing tables or handle any type of protocol conversion. It is not sophisticated, and it works at the physical layer, Layer 1.

Data Switches

A data switch, unlike a telephone circuit switch, is a fairly simple device. The data switch, often called an Ethernet switch, is used to switch packets between network segments. The switch relieves traffic congestion between devices and points the traffic to the destination device. If the switch were not in the network, then all of the traffic would travel down all of the links, causing lost packets.

A switch does not perform routing decisions by building routing tables and access lists. Instead, it routes packets coming into those going out by reading predefined addresses assigned to each of the ports. For instance, if a packet entering port 1 has a destination address of 2 and address 2 is assigned to port 500, the packet would be switched to port 500. The benefit of using a switch in place of a router is the time it takes the packet to traverse through the device. The switch can pass packets within 10 ms compared to a router, which passes a packet within 100 ms. The downside is that the switch is limited in its ability to route all information through the network. The switch is less sophisticated and is used primarily in small networks where the originating and terminating addresses are completely defined.

20.3.4 Frame Relay Equipment
FRADs

A frame relay network requires that data leaving a PC be assembled into a frame relay frame. The frame relay assembler/disassembler *(FRAD)* is an interface device that assembles or disassembles the frames. The FRAD is the interface between the network circuit and the local data network or device. Figure 20–23 shows a circuit with a FRAD being used to interface the outside world to the DTE device.

FRAD
Frame relay assembler/disassembler.

FRAD interfaces have been built into routers to eliminate the need for an additional box. The FRAD in the router is a circuit pack or a daughter board on the interface port that performs the frame relay encapsulation function. In a typical network, all of the data traffic from each of the data devices is fed into the FRAD, which performs the encapsulation and then ships the traffic to the CSU. The advantage of frame relay is being able to aggregate all of the data circuits into one CSU and onto one circuit. If frame relay were not used as the encapsulation method, each data circuit would require individual CSUs.

Frame Relay Switch

A frame relay switch is used to switch frame relay frames through the network. At the central locations, as shown in Figure 20–24, frame relay switches are placed with

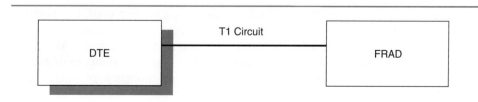

Figure 20–23
Frame relay networks require a frame relay device to interconnect to the DTE. The DTE may be a router, a file server, or a front-end processor.

Figure 20–24
Frame relay switches route traffic between points.

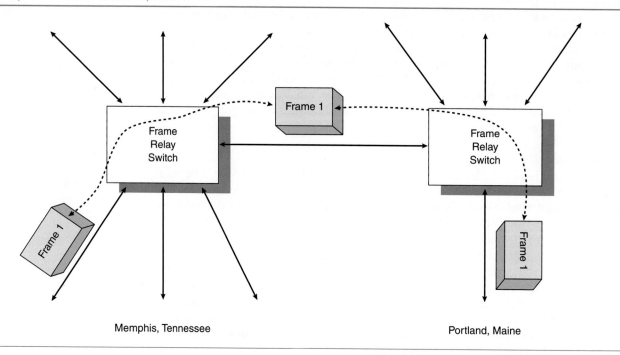

digital T1 or T3 circuits connecting the switches together. The 56 kbps circuits or T1 circuits connect the edge devices to the frame relay switch. Multiple locations interface into the frame switch that switches each of the frames through the network the same way a circuit switch switches a call through the network. A frame relay switch is a Layer 2 network device, in that it operates at the Layer 2 level. Unlike the router, the frame switch does not have to crack the IP packet to know where to switch or route the traffic. Instead, the frame switch reads the frame address and knows immediately which port to direct the traffic to. The size of a frame relay switch depends on the requirements of the network. Switches vary in size from one or two interfaces to hundreds.

DSLAM

digital subscriber line access multiplexer
DSL multiplexer that aggregates DSL signals onto a transport pipe.

A *digital subscriber line access multiplexer* (DSLAM) is a device used at the edge of the network to aggregate DSL circuits together onto one ingress circuit that connects to the central switch site. A DSL circuit carries high-speed data, up to six million bits per second, on one twisted pair. The DSLAM is made up of a processor, individual line cards, and an interface card. The line card is used to connect the DSL circuits to the subscriber, and the interface card is used to connect the DSLAM to a remote switch site. Figure 20–25 depicts a typical DSLAM and how it connects into the network.

RAS

remote access server
Used to terminate dial-up data signals.

The *remote access server* (RAS), which functions as a modem pool, is a common sight in many ISPs, telephone company offices, and large corporate data centers. The RAS provides a central point for the interconnection of dial-up circuits. Many subscribers use the same telephone number to dial the ISP. The RAS allows multiple subscribers to dial one telephone number to access the same modem pool located at the ISP's POP.

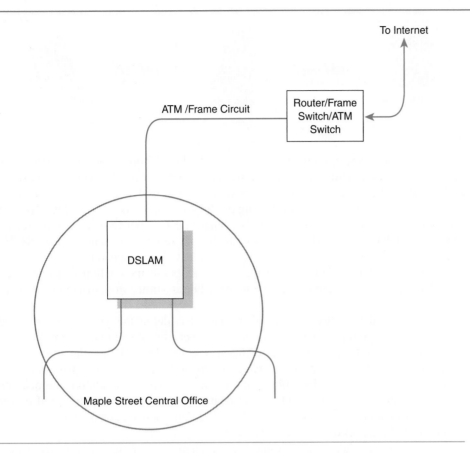

ATM /Frame Circuit

To Internet

Router/Frame
Switch/ATM
Switch

DSLAM

Maple Street Central Office

Figure 20–25
A DSLAM is located at the serving wire center and feeds subscribers on the normal cable pair. The DSLAM aggregates all of the subscribers' traffic and sends it on to the regional switch site. At the regional switch site, data are routed onto the backbone network. The data travel across the backbone network to the Internet.

■ 20.4 CUSTOMER PREMISES EQUIPMENT

20.4.1 Modems

The purpose of a modem is to convert a digital signal into an analog signal. The modem modulates a digital signal into standard telephone frequencies at the transmit end and demodulates the analog signal back into a digital signal at the receive end. Figure 20–26 illustrates a typical point-to-point network using two modems to interface the PCs to the telephone lines. Modems are now as common in homes as they are in business settings. PC manufacturers place a modem slot right in the PC chassis, allowing the customer to connect the phone line directly into the workstation. A modem is a very small device that can be embedded into the DTE or left as a stand-alone unit. The type of modem required depends on the application and the customer's budget. The cost of a 56 kbps modem, which has become the modem of choice, may be as low as $50. A stand-alone modem is the size of a paperback novel and is sold by any computer store.

A whirling sound and high-pitched connect tone occurs when a session is initiated via a modem. The modem outpulses a telephone number the same way a telephone outpulses digits using DTMF signaling. The modem waits as the terminating modem, the modem at the other end of the line, responds with a "handshake." A modem's handshake is an exchange of tones used by the two modems to set the speed, define the type of signal each will send or accept, and determine whether the two devices are compatible and willing to talk to one another. The modem on one end auto trains its signal speed to match the opposite end's modem speed. Auto training is also used to adjust the speed of the signal when the loop is out of spec, causing greater signal loss or errors at higher signal rates. Unless you live very close to your serving central office, your 56 kbps modem will probably run at 33 kbps or less due to loop length and other line characteristics that slow the signal down.

Figure 20–26
Dial-up connection between two PCs.
Both telephone lines are served out of
the same central office switch.

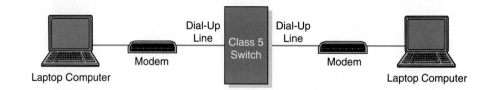

The method used to modulate the signal onto the telephone line varies depending on the manufacturer. Different modulation techniques have been devised to increase the number of bits being sent out on a line per signal change or per baud. A normal modem transmission using frequency shift keying modulation works like this: A 2600 Hz tone represents a 1. When the computer sends a binary 1 pulse, which is a +5 V pulse, the modem modulates a 2600 Hz tone out on the line. A zero is represented by a 1200 Hz tone being modulated onto the carrier signal. The newer high-speed modems combine different types of modulation techniques, such as phase modulation and amplitude modulation, to send more than one bit out on the line at a time. Modem technology continues to improve.

Table 20–1 lists modem specifications. The specs define the type of interface used to connect a computing device to modem, to connect a telephone line to a modem, and to connect power to a modem. In addition, the standards define the modulation/demodulation technique deployed, such as frequency shift keying or quadrature amplitude modulation. Beyond the physical interfaces and the modulation technique, the standards also list environmental requirements, such as heat dissipation and power drain. Overall, modem standards are fairly well defined and are considered to be some of the most ubiquitous data devices available.

DB-25/DB-9
Type of connector commonly
used for serial RS-232
connections.

The serial interface used to connect the modem to the terminating device is normally an EIA-232 interface using either a *DB-25* or *DB-9* connector. The RJ-11 standard is used as the telephone interface into the modem to connect to the telephone line. The modulation technique determines the maximum data rate possible. The parameter that concerns most consumers is the bit-per-second rate of speed of the modem. The standard also details the modem's transmission as being asynchronous or synchronous, half-duplex or full duplex, and dial-up or private line.

In the past, it would have been necessary to differentiate dial modems from private line modems. Private line modems are now rare. The sophistication of the dial-up modems and the introduction of digital private line circuits have all but replaced the need for private line modems.

Table 20–1
Common modem standards defined
by the ITU.

Modem Standard	Modulation Technique	Baud	Speed (bps)
Bell 103	Frequency shift keying	300	300
V.21	Frequency shift keying	300	300
V.22	DPSK	600	1200
V.23	Frequency shift keying	1200	1200
Bell 202	Frequency shift keying	1200	1200
Bell 212A	DPSK	600	1200
V.22bis	Quadrature amplitude modulation	600	2400
V.32bis	Trellis coding modulation	2400	14,400
V.32ter	Trellis coding modulation	2400	19,200
V.34	Trellis coding modulation	2400	28.8 k
V.34bis	Trellis coding modulation	2400	31.2 k & 33.6 k
V.42bis	Data compression algorithm increases thruput		57.6 k (varies)

20.4.2 DSL Modems

Digital subscriber line (DSL) increases the speed of the standard POTS line. It has become one of the hottest product offerings in the industry. DSL terminates into *DSL modems* at the customer premises the same way that analog line modems terminate. The key difference between the two technologies is that the DSL modem does not establish a session by going off-hook and sending out DTMF tones. The DSL modem is always on. The DSL attaches to a DSLAM located in the serving central office where it is routed to the Internet via a router or Layer 2 switch. The modem interconnects the copper loop to the customer's PC or computing device. Once the DSL is connected into the DSL port and the DSLAM's port has been configured, the connection to the terminating end is up and stays up.

The method used to connect the modem to the PC is shown in Figure 20–27. An Ethernet port on the modem is normally connected to an Ethernet port on the computing device. The telephone line terminates into a standard RJ-11 telephone jack in the modem. Beyond the Ethernet port and the RJ-11 jack, the modem is fairly ordinary. The size of a DSL modem is similar to the size of an analog modem and requires an AC power feed, as does the analog modem. DSL modems are often used by businesses for Internet connections or to connect remote sites together. In this case, the modem is connected into a hub, a bridge, or the company's router.

The real advantage DSL provides is the increased bit-per-second rate obtained on the standard copper pair. DSL is sold as a broadband Internet offering that provides an always-on connection eliminating the need to dial up every time a connection is needed. The cost of the DSL modem varies depending on the manufacturer. The cost of the loop increases as the speed of the service increases. The speed of a DSL ranges from 128 kbps to 6 Mbps.

DSL modem
Modem used to terminate a DSL signal.

20.4.3 Channel Service Units and Data Service Units

A channel service unit (CSU) or a data service unit (DSU) is used to connect digital circuits from the PSTN to the computing device defined as the DTE. A CSU, that today includes a DSU, sits between the telephone line and the DTE, which could be a router, a bridge, or a mainframe. The CSU formats the signal and readies it for transmission onto the network. A CSU often receives a unipolar signal from the DTE and must convert the unipolar format to a bipolar format. The CSU is also responsible for filtering out any voltages coming in on the circuit from the computing device or from the line.

The common interfaces used in a CSU vary depending on where the circuit is coming from and where it terminates into. The line interface from the PSTN is normally an RJ-48 connection using a modular plug, as illustrated in Figure 20–28. A V.35 connection is used to connect the CSU to the DTE. CSUs usually are built into the DTE's chassis as a circuit pack or as a daughter board on a circuit pack. CSUs are used to terminate 56 kbps DDS lines and T1 circuits.

Special codes are used to loop back a circuit at the CSU. The test equipment at the far end sends a CSU loop-back code, which causes the CSU to place a loop back on that line

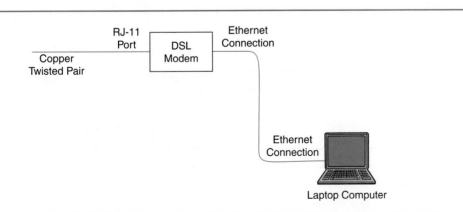

Figure 20–27
DSL modems connect to the network via RJ-11 jacks similar to analog modems and to the computing device via Ethernet.

Figure 20–28
Typical interfaces used when connecting a CSU to the terminating equipment.

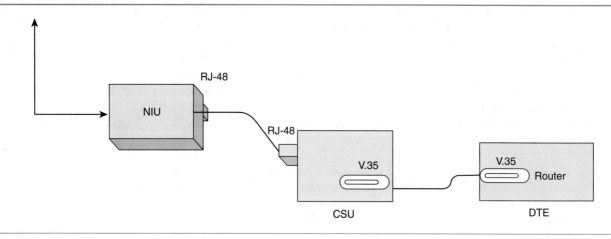

toward the device sending the code. Figure 20–29 depicts a CSU loop-back condition. When troubleshooting a circuit, it is critical to be able to sectionalize the circuit in order to identify the trouble location. The ability to loop back the CSU helps the technician determine whether the trouble is on the outside line or within the customer's equipment.

20.4.4 Routers, Bridges, and Hubs

Routers, bridges, and hubs are commonly found at customer sites. A business customer uses a router to segment traffic, determine the best route, and route it out onto the network. Bridges are also used to route traffic. Hubs are very common devices used by end customers to organize their LANs. A switch is used to point traffic to the correct port. Customers place routers, bridges, or hubs at their sites to handle network traffic internal to their business and external through the PSTN. The devices are smaller than those used by the carrier networks and often are powered by AC instead of the normal telephony −48 V power. For small networks, the redundancy is limited, reducing the resiliency of the local network and the complexity of the equipment. However, large companies often buy the same size and type of devices that the large carriers buy due to the increasing size and scope of the internal local network.

20.4.5 Multiplexers

Business customers of all sizes use multiplexers (MUXes) to aggregate their traffic. The type and size may differ from that found in the telephone network because the size and processing power needed at the customer location are much smaller than the carrier's POP. A typical end-user multiplexer mounts on the wall and has a CSU built into

Figure 20–29
A loop-back condition at the CSU. The purpose of a loop back is to help isolate trouble on the line.

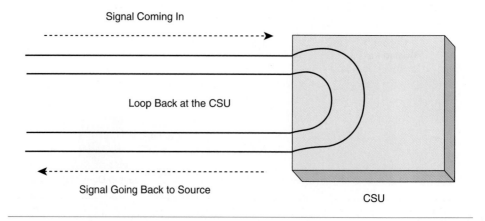

Figure 20–30
Shown is a circuit connection that routes from an outside network to a CSU, passes through a multiplexer and an FEP, then terminates into a mainframe computer. The multiplexer also drops off other circuits at the location.

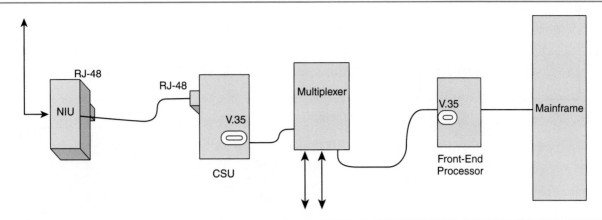

its interface pack. One side of the multiplexer interfaces with the outside world, and the other side interfaces with the individual lines running to the different workstations or other DTE. Figure 20–30 illustrates a multiplexer connecting to DTE at a customer site.

20.4.6 Concentrators
A concentrator could possibly be confused with a multiplexer. In fact, the differences are dependent upon the manufacturer. Swapping the terms would not matter a great deal. A concentrator accepts multiple slow-speed signals and concentrates them onto one high-speed signal. For instance, multiple 9.6 terminals can be concentrated into one 1.544 Mbps signal. The advantage a concentrator provides over a standard TDM (not a statistical MUX) is the dynamic use of bandwidth. The concentrator takes information from the multiple terminals only when they have information to send. If the terminal is idle, it does not eat up any of the bandwidth. The advantage this provides is the ability to oversubscribe the output high-speed circuit. A concentrator can accept 200 9.6 kbps signals and feed them onto a 1.544 Mbps circuit. The total amount of bandwidth required by the 200 terminals is 200 × 9.6 kbps = 1.92 Mbps. The total bandwidth exceeds the 1.544 Mbps line rate of the T1 circuit. Data will not be lost because it is unlikely that all of the 200 terminals will be online at once. The engineer feels comfortable oversubscribing the upstream circuit, and the concentrator sends a signal only when a signal is present.

20.4.7 Front-End Processors
A *front-end processor* (FEP) is placed in front of a mainframe or host computer to offload processing and control functions from the mainframe. The FEP was first designed by IBM and is often associated with IBM networks. It can be compared to the information officer who receives messages from lower-level officers, filters them, organizes them, controls their content, and then hands them to the general. The FEP is the intermediary between the host computing device and the communications network.

> **front-end processor**
> Device that sits between the network and the mainframe computer. Reduces the processing requirements of the mainframe.

A typical FEP has interface ports that connect to the host computer. It also has line interface ports that interface with all the terminals that need to access the host computer. A typical FEP can handle between 20 and 200 lines. It also holds sophisticated software programs that work with terminals and mainframe languages. Figure 20–31 shows a typical FEP in a communications network.

20.4.8 Mainframes
The first of the three main computing devices placed at the end of a data communications network is called the *mainframe*. The mainframe computer was the first to be developed in the late 1960s. It is large, has enormous processing power, and is expensive. Today's client-server networks have reduced the need for large mainframe processors.

> **mainframe**
> Large computer that holds files.

Figure 20–31
A circuit interfacing into a front-end processor. An FEP's circuit rates vary depending on the amount of information required by the customer.

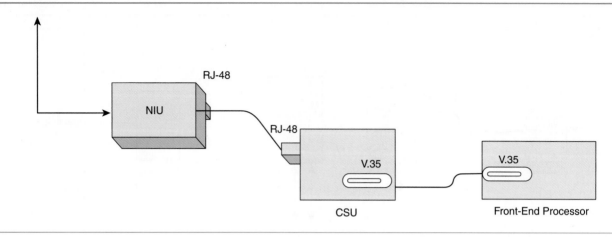

The first data communications circuits that were built interfaced into a mainframe computer and depended on its data and processing power to tell them what to do.

A mainframe computer, also called a host computer, is still used by many companies. The airline industry uses mainframes to process, store, and interface with the numerous ticket taker computers around the world. The banking industry still depends on mainframe computers to hold and process customer information and bank records. Each time you swipe your ATM card, your request travels to a mainframe for processing.

Mainframe computers can be viewed as the bigger brains that are needed by certain applications. They will never be completely replaced by PCs. Figure 20–32 shows a typical mainframe.

20.4.9 Intelligent Terminals
An intelligent terminal is nothing more than a PC workstation. It is able to process information, store information, and communicate with other devices either locally or through dial-up connections.

20.4.10 Smart Terminals
A smart terminal relies on a host computer to perform most of the processing and storage functions. It can store small amounts of data and can connect to multiple devices

Figure 20–32
A circuit connection from the outside network to the mainframe. The mainframe is preceded by an FEP that organizes the information before passing it on to the mainframe.

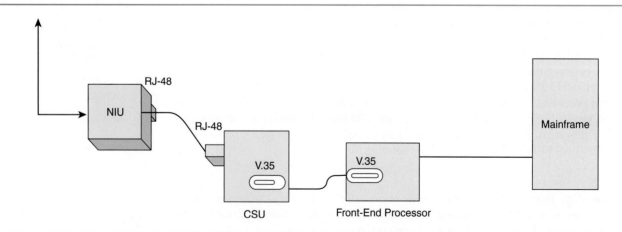

if needed. The processing power of a smart terminal is limited, as is its storage space. Smart terminals are beefed-up dumb terminals.

20.4.11 Dumb Terminals

A dumb terminal was the first type used to talk to the mainframe computer. The dumb terminal has a monitor and the ability to connect locally or remotely to an intelligent computing device. The airline industry provides a good example of using dumb terminals. The ticket agent types in a request for a reservation on the next flight to Atlanta. The request leaves the dumb terminal via an interface port that connects to a data circuit that connects to the mainframe. The mainframe processes the request and returns the information to the ticket taker. Dumb terminals communicate with communications devices. Switch technicians use a dumb terminal connected to a switch to program any changes in the switch, monitor the alarms, and so forth.

20.4.12 Physical Layer Interfaces Found at the Customer Premises

Different physical layer protocols interface with various data communications devices. Standards bodies such as the IEEE, ANSI, the ITU, and the former Bellcore have all been involved in defining the electrical parameters of each type of interface. The pinouts of the connector, the gauge of cable, the number of twists per foot, the distance the signal travels, the voltage levels, and the frequency range are some of the parameters defined by each standard. The most popular physical interfaces follow with definitions of their specific attributes.

- EIA/TIA-232 is the serial data interface standard for connections between data terminal equipment and data communications equipment. The RS-232 interface connects a computer to a modem. The standard defines the voltage, current, time, and duration of the signal. Table 20–2 lists the standard pinouts for an RS-232 connector, and Figure 20–33 and Figure 20–34 illustrate what a

Table 20–2
RS-232 Pinouts.

RS-232 Pin Number	Term	Function
1	FG	Frame ground
2	TD	Transmit data
3	RD	Receive data
4	RTS	Request to send
5	CTS	Clear to send
6	DSR	Data set ready
7	SG	Signal ground
8	DCD	Data carrier detect
9		+DC test voltage
10		−DC test voltage
11		Unassigned
12		
13		
14		
15	TC	Transmit clk
16		
17	RC	Receive clk
18		
19		
20	DTR	Data terminal ready
21		
22	RI	Ring indicator
23		
24	EXT	Ext. timing from DTE
25		

Figure 20–33
The RS-232 cable has both nine-pin male and nine-pin female connectors. RS-232 cables are commonly used to connect computing devices. (Photo courtesy of M. A. Rosengrant)

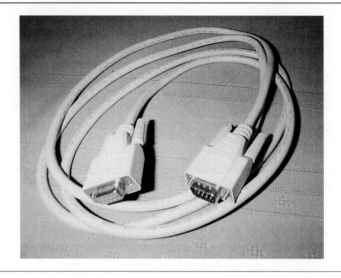

Figure 20–34
RS-232 connectors are found in either a twenty-five-pin or a nine-pin configuration.

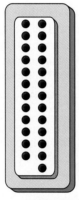

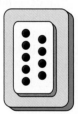

DB-9
1. Frame Ground
2. Secondary Receiver Ready
3. Secondary Send Data
4. Secondary Receive Data
5. Signal Ground
6. Receive Common
7. Secondary Request to Send
8. Secondary Clear to Send
9. Send Common

DB-25
1. Frame Ground
2. Transmit Data
3. Receive Data
4. Request to Send
5. Clear to Send
6. Data Set Ready
7. Signal Ground
8. Data Carrier Detect
9. Reserved
10. Reserved
11. Not Used
12. Secondary Received Signal Detector
13. Secondary Clear to Send
14. Secondary Transmitted Data
15. Transmitter Clock
16. Not Used
17. Receive Clock
18. Local Loop Back
19. Secondary Request to Send
20. Data Terminal Ready
21. Remote Loop Back/Signal Quality Detector
22. Ring Indicator
23. Data Signal Rate Select
24. External Timing from DTE
25. Test Mode

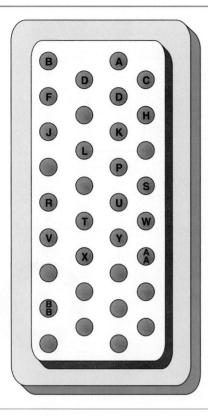

Pinout Description

A Chassis Ground
B Signal Ground
C Request to Send
D Data Set Ready
F Carrier Detect
H Data Terminal Ready
J Local Loop Back
K Test Mode
L Test Pattern
P Transmit Data – A
R Receive Data – A
S Transmit Data – B
T Receive Data – B
U Terminal Timing – A
V Receive Timing – A
W Terminal Timing – B
X Receive Timing – B
Y Transmit Timing – A
AA Transmit Timing – B
BB Remote Loop Back

Figure 20–35
The V.35 interface is commonly used for data speeds above 19.2.

common RS-232 connector looks like. The 232 interface was designed to handle speeds up to 9.6 kbps.

- EIA/TIA-449 is a serial interface that connects DTE to DCE. The standard defines the voltage, current, time, and duration of the signal, in addition to the type of connector and pin configuration. The 449 interface is not as popular as the 232 but may be used when a higher bit-per-second rate is required. It can handle speeds up to 19.2 kbps.
- V.35 is a serial interface used to connect DTE to DCE. The main difference between the V.35 and the 232 or 449 is the distance the signal can travel because of a separate ground lead. The V.35 interface is used to connect a T1 CSU/DSU to a DTE. For example, when a CSU connects to a router, the CSU represents the DCE, and the router represents the DTE. The V.35 is one of the most popular standards used in North America to interface equipment to T1 circuits. It is used for network speeds above 19.2 kbps and is shown in Figure 20–35.
- G.703 is the standard used to define the digital T1 circuit interface. G.703 started as an ITU standard defining physical characteristics of the E1 –2.03 Mbps circuit. Today G.703 includes the 1.544 Mbps T1 circuit. The T1 physical layer interface is used to connect data communications equipment to the wide area network. The name given to the G.703 is RJ-48 or registered jack 48. The standard defines the gauge of cable, the type of connector, the pinout configuration, and the distance the signal can travel before being repeated. Figure 20–36 depicts a typical G.703 interface and shows the RJ-48.
- DS-3 interface is used when a 45 Mbps signal is terminated. It requires coaxial cable with Bayonet Neill Concelman (BNC) connectors. The standard defines the gauge of the cable, the type of connector, and the distance the signal can travel before being regenerated. Figure 20–37 illustrates a typical BNC connector.
- Optical (OC) interfaces are becoming more common in the world of data, as most routers now support optical interfaces such as OC-3, OC-12, OC-48, and OC-192.

Figure 20–36
G.703 standard defines the T1 interface. The T1 circuit is a four-wire circuit.

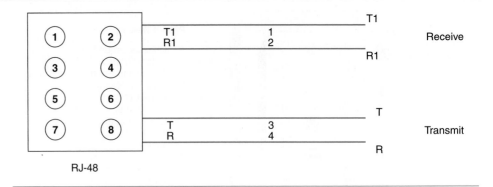

Figure 20–37
BNC connectors are used to terminate DS-3s, thin and thick Ethernet signals, and higher-rate STSs.

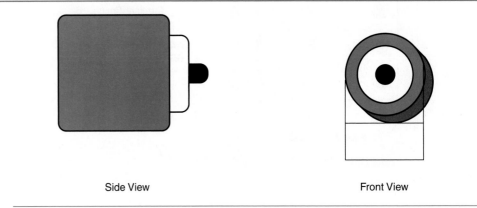

Side View   Front View

The optical interface may be defined as either single mode or multimode depending on the type of fiber deployed. The signal rate depends on the lasers used in the equipment and may range from an OC-1 to an OC-192. The type of transmitter determines the pulse level, the type of fiber, the loss budget, and the distance the signal can travel before being amplified. Figure 20–38 shows typical optical interfaces.

Figure 20–38
Typical fiber optic termination. (a) The side view of a connector; (b) front view; (c) jack the connector fits into.

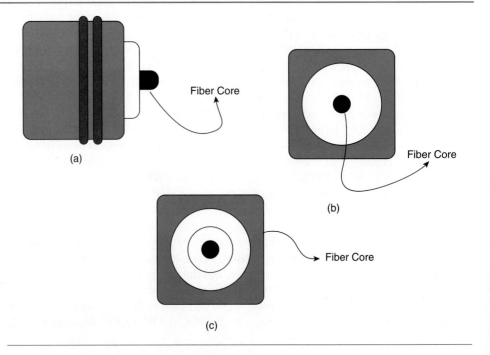

20.5 DATA NETWORK TOPOLOGIES

20.5.1 Defining Data Network Topology

The term *topology* is used in data communications to explain the network layout, such as point to point, ring, mesh, star, or bus. Network engineers take many factors into consideration when designing a data network, including the size of the network, the number of locations, the distance between the locations, the amount of traffic between each of the locations, the transport costs for link, and the types of data circuits available at that location. Different types of data applications lend themselves to specific types of data topologies. For example, the Internet is built on a mesh network topology, in which multiple locations are connected forming a spider-web-like network. A bank's network is often built with multipoint circuits allowing connectivity from one central database to multiple ATMs or branch offices. This portion of the chapter looks at the different architectures that are used to build a data network and the reasons they are used for specific applications.

20.5.2 Point-to-Point Network Topology

The *point-to-point circuit* is the simplest network design, consisting of one link connecting two end locations, such as two corporate LANs. For example, IGE Pharmaceuticals has a research campus in Burlington, Vermont, and a factory in Huntsville, Alabama, that exchange large amounts of data every day. The information being exchanged is highly sensitive and must travel on secure pipes. The network engineer knows that a dial-up circuit will not provide the needed bandwidth and is much less secure than a dedicated pipe. An ISDN line does not provide the bandwidth needed and the cost-per-minute charges would exceed the cost of a dedicated pipe. The IGE engineer decides to connect the two locations with a T1 1.544 Mbps circuit, thus creating a point-to-point network topology that provides both the needed bandwidth and the required security. Point-to-point networks are very common in this type of situation and are a viable solution when the amount of data exchanged between the two locations warrants a dedicated pipe.

point-to-point circuit
Circuit that carries information only between two points.

20.5.3 Multipoint Network Topology

A *multipoint circuit* is, as it sounds, a network tied together at multiple points, as depicted in Figure 20–39. A common example of a multipoint network is that of ATMs feeding into a central bank. The machines must talk to the bank's main database in order to exchange customer balance information. Often, dedicated 56 kbps digital data circuits are used to connect the geographically dispersed ATMs into the bank's central database. The machines do not need to talk to each other, just to the central database. The central bank, on the other hand, must be able to communicate with each of the ATMs and therefore must have a connection to each device.

multipoint circuit
Circuit that carries information between one hub and multiple locations.

The multipoint topology fits this application perfectly. Each of the machines has dedicated circuits that feed back to a central hub, usually located in the telephone company's serving central office. One circuit is placed between the hub location, as shown in Figure 20–40, and the main branch of the bank where the database resides. The dedicated 56 kbps circuit provides a secure pipe and substantial bandwidth for the transactions being made. The front-end processor in front of the database polls each of the branch locations asking whether it has data to send. If a branch site does have data to send, it sends them at that time. The database continuously queries each of the branch

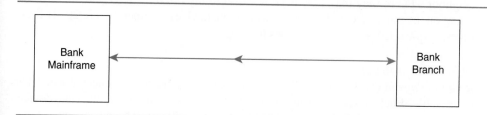

Figure 20–39
Point-to-point data circuit. The circuit speed may vary depending on the bandwidth needs. The circuit architecture is commonly used by banks in order to keep the network secure.

Figure 20–40
Multipoint data circuit with three remote ATM locations feeding into a front-end processor (FEP). From the front-end processor, the transaction is routed to the mainframe. The mainframe polls each of the sites and asks for information. The remote sites are not able to talk with each other, only to the mainframe at the main office.

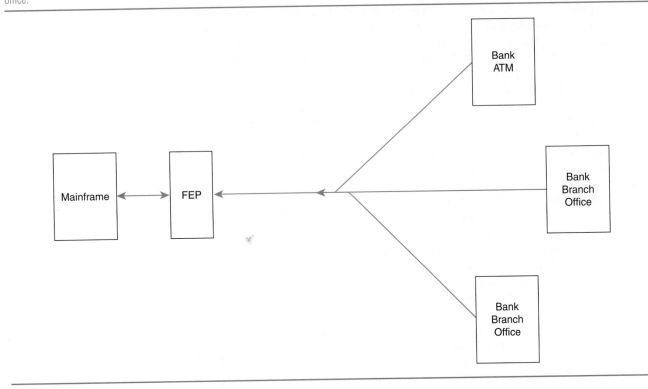

sites using a round-robin polling scheme to ensure that each site is given a chance to exchange information.

A multipoint architecture is fairly common and has been used for years in data communications. With the Internet, networks normally expect every device to be able to talk to every other, making the multipoint and point-to-point networks obsolete. Still there are many applications, such as the one mentioned, that benefit from the older architectures. Like the point-to-point topology, a multipoint network is justified if the amount of traffic to be exchanged is great enough to warrant dedicated circuits and if the application requires secure links.

20.5.4 Bus Topology

bus topology
Network topology based on a shared bus architecture.

The *bus topology* is very common in the world of local area networks (LANs). A LAN is a network that connects multiple computing devices with one another in a local area. A common architecture used to connect the devices is the bus topology. In fact, bus topology is rarely used for anything except LANs. It is similar to a local bus run in which a bus travels around the city and stops at each bus stop, picking up some passengers and letting others off. Multiple PCs are attached to the LAN. The data leaving the PC jump onto the link and travel to the desired location on the LAN, such as another PC, a file server, or a printer. The data may need to travel from my PC to my neighbor's PC or they may only need to travel to my printer. Either way the data are riding on the linear circuit that connects all of the devices together. Figure 20–41 illustrates a typical LAN connected by bus topology.

20.5.5 Ring Topology

ring topology
Closed network topology based on passing information around the ring.

Ring topology may refer to a LAN's design or to a large SONET connected with fiber optic cable. Each device on ring architecture is connected to the adjacent device, thereby forming a closed ring, as shown in Figure 20–42. The advantages of the

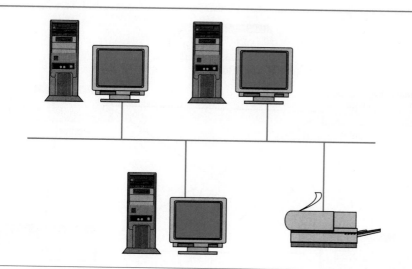

Figure 20–41
The bus network topology. The network consists of elements tied together along a shared route. Traffic from each device flows through the adjacent device as it travels down the line.

network should be apparent. If the connection between device A and device B is severed, device B is still able to communicate with device A by way of devices F, D, C, and B. The reliability of the ring architecture is the key selling point when determining the type of topology to implement. A second advantage is its ability to send traffic to any node on the ring by simply building a connection to that node. The circuit does not have to be ordered or physically built; the ring is already in place. Instead, the circuit is provisioned through software, which takes less than two minutes. Ring architecture has become one of the most popular architectures for large fiber optic networks. A typical fiber optic network carries anywhere from 32,000 channels to $80 \times 32,000$ channels and higher on one cable. The need for a fully redundant path is apparent, and ring topology provides redundancy, simple provisioning methods, and a secure network. For years, backbone data networks have been built around a mesh design, but they are now beginning to be designed in ring configurations.

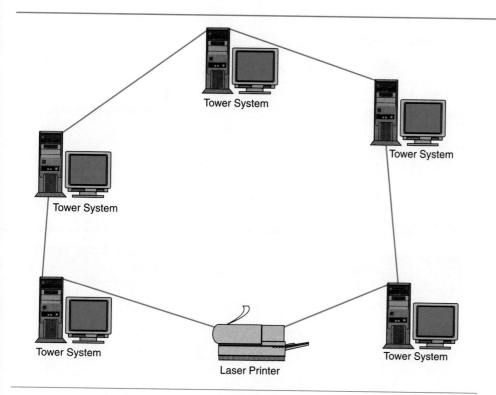

Figure 20–42
A typical ring architecture. Traffic flows around the ring passing through each node along the route.

Figure 20–43
This partial mesh network provides connectivity between at least two other nodes. Some networks are fully meshed, meaning that each node is connected to every other node in the network.

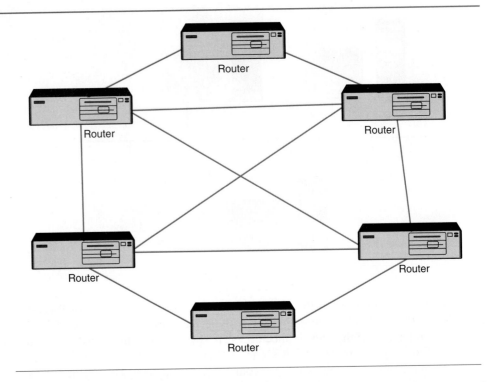

20.5.6 Mesh Topology

mesh topology
Fully connected network topology based on each device connected to two or more devices.

Mesh topology is one of the most common topologies used in data communications. Not only does it provide connections to and from every device on the network, it also provides diverse routes to every device on the network. The Internet is the best example of a mesh network. Web sites are all tied together by way of a large "spider web" of connections.

Figure 20–43 illustrates a typical mesh network in which each device has a way to reach every other device. Unlike the multipoint design discussed earlier, each device in a mesh network is able to talk to every other device. Figure 20–44 also illustrates the

Figure 20–44
Mesh networks provide redundant paths to each device on the network. For example, if the link between A and B fails, the traffic to C is rerouted through F, then to C.

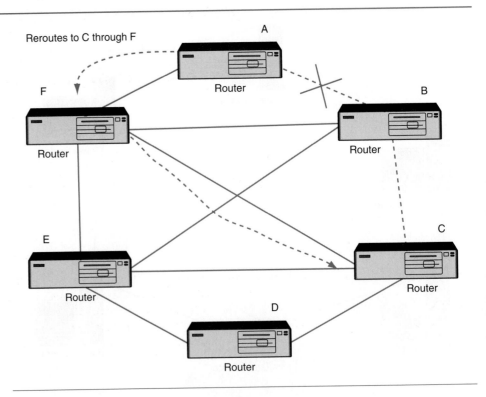

diverse routing functionality provided by a fully meshed network. If route A to route C is severed, the information is shipped from route A to route F and then to route C. IP and ATM transport protocols were built to be able to reroute packets and cells onto diverse links. Most IP and ATM networks use mesh topologies.

■ 20.6 DATA NETWORKS—LANs

The data communications network has evolved into a highly sophisticated system of protocols and devices. The overall network can be divided into two general areas—the local area network and the wide area network. Similar devices are used to carry traffic on both, and both depend on transmission interfaces and standard protocols to define how the network operates. The main difference is that one is concerned with the devices in the local area, such as the user's PCs, the company's file server, and the printers, while the other is concerned with connecting multiple diverse networks and routing traffic across the connections. Both local area networks and wide area networks will be discussed.

20.6.1 LANs—A Definition

The local area network can be defined as a group of data communications devices that are tied together allowing each to share network resources. Though the explanation is simple, the actual design and maintenance of a LAN is not. A computer that is attached to a LAN has the ability to interface with other devices attached to that LAN. Normally a LAN consists of multiple computers, file servers, printers, and an e-mail server. Users can access software programs from the file server, send e-mail, or access printers. They can share files with coworkers and print documents on a community printer. Users are able to send messages to one another via an internal e-mail system, and schools and businesses can provide one shared access circuit to the outside world from the LAN.

20.6.2 The Ethernet
Ethernet LAN
The *Ethernet LAN* provides a way for multiple devices to be connected in order to share information and resources. For example, a company with 1000 employees who all have PCs on their desks must determine how it will interface all 1000 devices onto a shared network. The solution is to implement a LAN, and a large percentage of the LANs deployed use the Ethernet standard, technically referred to as 802.3. Xerox, which later worked with DEC and ITT, developed the Ethernet standard in the late 1970s.

Ethernet LAN
LAN that uses the Ethernet Layer 2 protocol to send information out across bus topology.

Network Topology
Ethernet uses a bus topology, as shown in Figure 20–45, that provides access to each of the devices by connecting them to the same physical medium. The architecture is fairly simple to implement because adding a new device only requires extending the medium and connecting the device onto the medium at that point. The device is added to the LAN, allowing it access to all other devices on the LAN, and allowing the other devices access to it. The term *segment* defines a grouping of devices that are attached to a portion of the LAN. Multiple segments are interconnected forming the entire LAN. The purpose is twofold. Adding users is simpler if the total LAN is divided into groups, because adding or deleting the new user affects only those in the segment. Dividing LANs into multiple segments also simplifies troubleshooting and maintenance of the network. The technician is able to pinpoint a problem faster by eliminating segments one at a time instead of eliminating PCs one at a time.

Physical Interface
The physical cable used to connect Ethernet LANs varies depending on the structure of the building or buildings, the distance the signal has to travel, the number of devices connecting to the network, and the amount of traffic expected. Three types of cable are used to connect Ethernet LANs: category 5 or 6 twisted pair, thin Ethernet, and thick Ethernet. The most important criterion to use in deciding which type of medium to use is the distance between the devices.

Figure 20–45
LANs are composed of devices tied to a shared bus, forming one segment. One segment is defined by the same IP address structure.

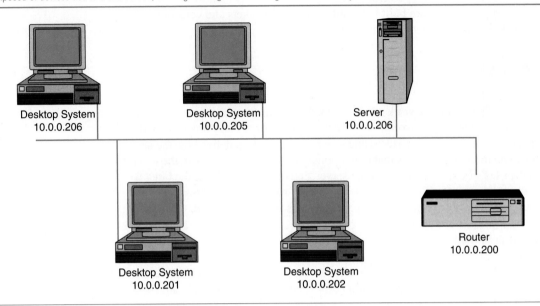

Thin Ethernet, also called 10Base2, allows a signal to travel 185 m or 500 ft. and allows thirty connections per segment. Thick Ethernet, also known as 10Base5, allows a signal to travel 500 m or 1640 ft. with a maximum of three segments and three re-generation points, as shown in Figure 20–46. The most popular and most practical physical medium is category 5 or 6 twisted pair, also called universal twisted pair (UTP), which allows the signal to travel distances as far as 3000 ft. while carrying up to 100 MHz. The advantage of using CAT5 cable is that it requires less space in the conduit, is less expensive, is easy to acquire, and is much easier to install and maintain. Figure 20–47 illustrates a typical LAN connected with copper twisted cable.

Access Method

CSMA/CD
Carrier Sense Multiple Access with Collision Detection.

The term *access method* refers to the way data are handled by the different devices in the network. Ethernet uses a *CSMA/CD* contention scheme, which stands for Carrier Sense Multiple Access with Collision Detection. Each workstation on the LAN listens to the network to see if the network is being used by another workstation before transmitting its data. Once it determines that the network is clear of other transmissions, it transmits its data out onto the network. If two workstations do send their data at the same time and the data collide, both terminate transmission and wait a predefined amount of time before transmitting again. The predefined time is different for each workstation, eliminating the chance of a second collision.

A typical CSMA/CD access method is shown in Figure 20–48. Notice that orphan data are absorbed at the end of the segment so that they do not repeat through the network. Each workstation picks up each piece of data and determines whether or not the data belong to it or whether the data should be passed on to the next entity. CSMA/CD is the most popular access method used by LANs, and it works well with Ethernet's bus topology.

Addressing

media access control sublayer
Interface between the physical layer and the upper portion of the data link layer.

Each device on the LAN must have a unique address identifier in order to receive data. The data link layer (DLL) handles addressing for the individual physical devices connected to the LAN. The DLL is divided into two sublayers, one called the *media access control* (MAC) *sublayer* and the other, the logical link control sublayer (LLC sublayer). Both are shown in Figure 20–49. The MAC layer interfaces the physical layer to the LLC layer, and the LLC layer interfaces Layer 2 with the upper-level layers. They work together when routing information through the network.

Figure 20–46 Three-segment Ethernet LAN.

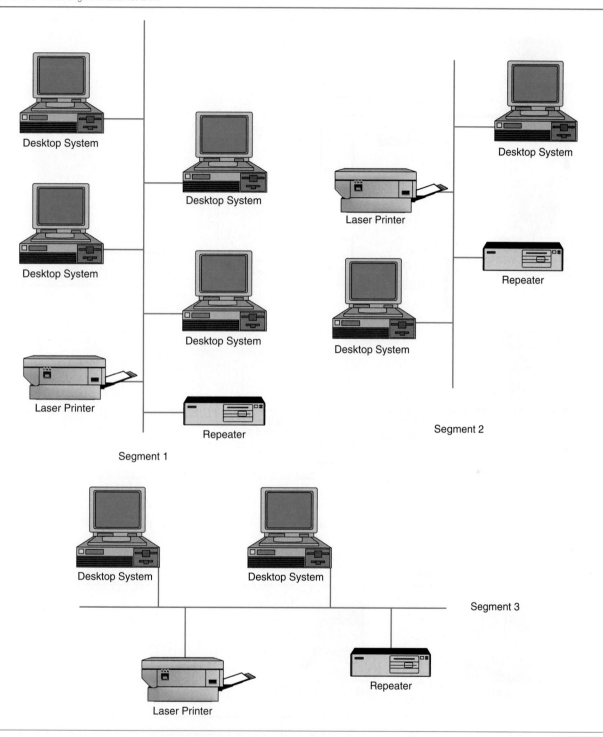

The address given to each device connected to the LAN is a MAC address. It is a unique identifier burned into the ROM of the equipment that specifies the vendor and the serial number. The MAC address is used by routing protocols as a way to direct traffic to the end device. The forty-eight-bit address is divided into two sections—the first twenty-four bits designate the vendor, and the second twenty-four bits designate the serial number unique to that equipment. The MAC address is a flat addressing scheme because it is not dependent upon an address hierarchy, as is an IP address.

Figure 20–47
LAN connections are made between multiple users' devices and a hub using twisted copper pair as the connection medium.

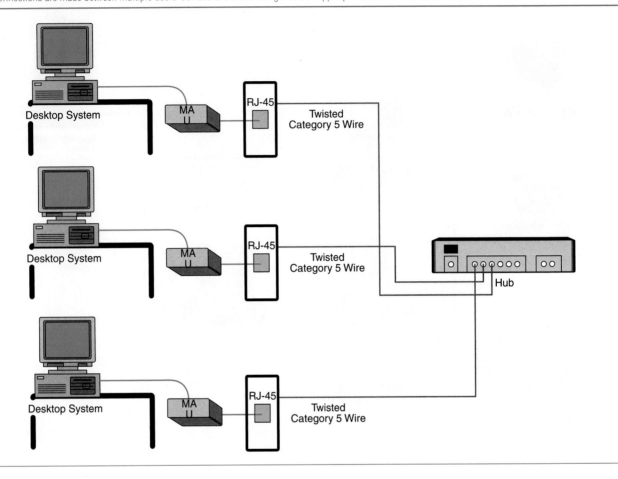

Figure 20–48
Data packets are absorbed at the end of the segment in a CSMA/CD.

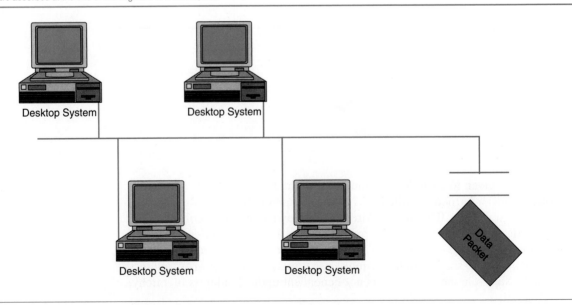

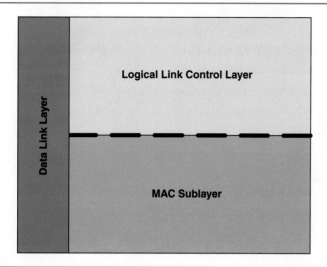

Figure 20–49
The data link layer is divided into two separate functions—the MAC and LLC layers. The MAC layer interfaces the physical hardware letting the LLC layer interface with Layer 3 functions.

The MAC address is used by data networking devices to learn what devices are hanging off the network. For instance, a router will send out a message and ask for everyone's MAC address. The equipment will automatically respond (address resolution protocol) the MAC address to the router, which will then store the address in memory and use it when routing information to that device.

The MAC address is placed in the data frame of the data packet and is viewed by each device as it works its way around the network. The workstation looks at the MAC address in the frame to see if it matches its own. If it does, it keeps the frame; if it does not, it passes it on.

20.6.3 Token Ring LAN Architecture

Token ring LAN architecture was developed by IBM in the 1970s. It is similar to the IEEE's 802.5 standard and, as with Ethernet 802.3, we will use the two standards synonymously. Token ring is the second most popular LAN technology after Ethernet. It is one of the oldest LAN technologies and, as such, continues to be used in the network.

token ring LAN
LAN architecture invented by IBM. Devices on the ring pass a token to the next device signifying whether anyone is transmitting at that time.

Network Topology

Token ring LANs are configured logically as a ring but physically as a star. Tokens are passed around the ring. Each device on the network depends on the token to notify it when it may send data or receive data, as shown in Figure 20–50. The workstation, which is attached to a central hub called a multistation access unit (MSAU) (shown in Figure 20–51), receives a signal from its nearest active upstream neighbor (NAUN). The device then repeats the signal to its nearest downstream neighbor. Workstations attached to a token ring LAN also act as repeaters, thus increasing the number of devices allowed on one segment.

Each MSAU is able to handle up to eight devices. One disadvantage associated with token ring LANs is that each time a new device is added to the LAN or taken off the LAN, service to all devices on the LAN is disrupted. In addition, only eight devices are allowed per MSAU, fewer than the thirty devices per segment on the Ethernet LAN. These are the two main reasons that token ring LANs have lost popularity.

Physical Interfaces

The common physical medium used to connect devices on a token ring LAN is twisted copper pair. Initially, IBM required shielded twisted pair (STP) cable be used, allowing a data rate of 16 Mbps. Currently, the cost of STP cable has forced network managers to use UTP cable, which provides a data rate between 4 and 16 Mbps.

Figure 20–50
Token ring LANs depend on tokens
that are passed between terminals
through the network. If one device
fails on the network, the entire
network fails.

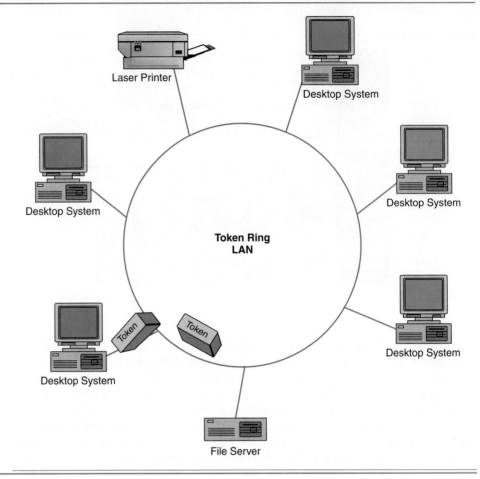

Figure 20–51
Token ring LANs are logically
configured as a ring but physically as
a star. Each device attaches to a
central aggregation device called an
MSAU.

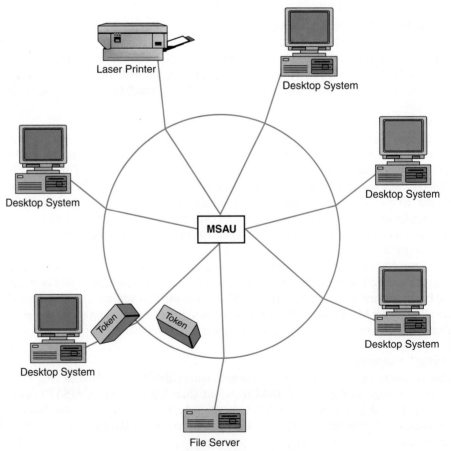

Access Method

The method used to transmit data around a token ring LAN requires that a three-octet token be passed around the ring stopping at each workstation. The workstation determines whether the priority indicator in the token is marked with a low- or high-priority bit. If the workstation's data require high priority, the token is marked as high and the information from the workstation replaces the data already in the packet with its own. The purpose of the token is to ensure that the network is fully utilized while reducing the possibility of contention between workstations.

One workstation is designated the active monitor (AM) and is responsible for overseeing the network. The AM watches for errors and congestion in the network and monitors the ring for any breaks or disruptions. The AM has the power to discard the token if congestion or ring disruption occurs. Each workstation has the capability of serving as the AM. A secondary AM is normally configured to fill in if the primary AM fails.

Addressing

The addressing method used is the same as for Ethernet.

20.6.4 Fiber Distributed Data Interface

Fiber distributed data interface (FDDI) uses fiber optic strands to carry information between devices. It is a ring architecture, and the purpose of implementing an FDDI LAN is to provide large amounts of bandwidth between devices and to extend the distance between the devices and the segments. FDDI LANs may be found on college campuses where large amounts of information must be carried between all of the buildings.

fiber distributed data interface
Fiber LAN architecture.

FDDI, similar to token ring, uses tokens to determine when data should be sent or held. The addressing method is the same as the token ring and Ethernet LAN technologies—the MAC layer addressing scheme. Though not as common, FDDI still provides a viable solution for networking a local area.

SUMMARY

Chapter 20 presented information on the types of data and voice communications network topologies, data networking equipment, and types of protocols and multiplexing schemes used by end users. The information presented in Chapter 20 illustrated how communications networks traditionally deployed by the communications carriers have expanded their reach to the end users' premises. The chapter was divided into six general areas: customer premises architecture, Internet service providers and applications service providers, data communications equipment, customer premises equipment, data network topologies, and data networks—LANs.

CASE STUDY 1

Three business customers have approached Big Bob and asked that he design their voice and data network. The first customer is Rita, from Rita's Tattoo, who needs two voice lines and one data line. The data line will be used for Internet access so Rita can download the latest tattoo stencils. Hank Alherts, who owns and operates the local senior citizen home, also approached Big Bob and asked that he design the telephone and data network for the home. The home holds 150 beds, of which 130 are occupied. Hank would like to give each of his residents his or her own phone and phone number but realizes usage would be fairly low because most people in Green Grass visit their relatives there often. Hank also wants four DSLs—two for the recreation room, one for the conference room, and one for his office.

The third business owner to approach Big Bob was Kate Duckworth from the WoodCarvings.Inc lawn ornament factory. The company has seventy-five office workers who would require telephone lines and numbers, eight floor phones that would require intercom functions, and five pod phones for conference rooms. Kate would also like a LAN to be designed for all seventy-five office workers' PCs, and an additional fifty for growth. WoodCarvings.Inc has agreed to electronically bond with its wood supplier, Mark

Woodruff, and therefore requires a VPN between itself and the Woodruff Lumber Yard. Kate has also asked for interconnection to the Internet and would like to be able to reach the Internet from any of the PCs on the LAN.

Your mission is to design the three networks and provide a solution for Big Bob.

CASE STUDY 2

Green Grass Telephone Company has received digital divide grants from the government. The grants specify that Green Grass must deploy broadband services to its subscribers. Big Bob has hired you as a data consultant and has asked that you build a big robust data infrastructure that will carry the thousands of bits of traffic and satisfy both Green Grass residents and the government officials who gave Big Bob the check.

Your first assignment is to strategically place DSLAMs in the network and establish an aggregation point for all of the digital subscriber lines that need to terminate into the Internet. The cost of the equipment needed for DSL deployment follows:

- DSLAM—$55,000 for a 300-line unit
- IP services box (SMS)—$100,000 can handle 30,000 sessions
- Router—$70,000
- ATM switch—$105,000
- Frame relay switch—$53,000

Once you have determined where you would like to place the DSLAMs and have established a means to send the information onto the Internet, you will need to develop a DSL product. Developing a telecommunications product is one of the most time-consuming tasks in the industry. Following, we have listed several of the pieces you will need to resolve in order to establish DSL as a product that can be marketed by Green Grass Telephone Company.

- Determine a price for the product. You may use pricing from your local providers.
- Establish provisioning procedures that encompass a way for the product to be billed.
- Give a name to the product.
- Define the product.

REVIEW QUESTIONS

1. List several types of voice equipment that can be found at a customer's premises.

2. List several types of data equipment that can be found at a customer's premises.

3. What is the most common type of equipment found at a customer's home or office?

4. Explain the difference between a small business customer and a medium-size business customer.

5. Explain the difference between a medium-size business customer and a large business customer.

6. Explain why business customers purchase PBXs.

7. What does a PBX do?

8. Explain the term *tie line* and list two places tie lines are used.

9. Does a PBX do least-cost routing?

10. What is the difference between a key system and a PBX?

11. Why do business customers require routers?

12. What is the term used to define the router that connects the business to the network?

13. Explain why data communications services are routed onto CAT5 cable instead of CAT3 or standard twisted pair.

14. Why does T1 cable have to be grounded on one end? Why does T1 cable have a special T-screen or metal jacket surrounding the wires?

15. Why are the transmit and receive pairs of a T1 cable run in separate cable binders?

16. Why is multimode fiber used at customer locations to connect equipment or buildings?

17. When information is carried across a wide area network, what three layers play a vital role in making sure the traffic arrives safely?

18. Define the term *data communications*.

19. The connection between a PC and a printer is a parallel connection. The connection between a modem and a PC is a

21

Cable Television As a Telecommunications Service Provider

Objectives

After reading this chapter, you should be able to

- Define CATV as a telecommunications service provider.
- Identify the DOCSIS protocol as a physical and data link layer protocol.
- Identify the logical network topology of the CATV network.
- Define the devices used in the CATV broadband network.
- Define how services are provisioned and maintained.

Outline

■ INTRODUCTION

MSO
Multiple System Operator. Main switch location for CATV communications equipment. Signals travel from the MSO to the customer's premises through a transmission network.

Over the past few years, CATV providers referred to as *MSOs*—Multiple System Operators—have entered the telecommunications market in force. The "bundle," as it is called, includes cable television, broadband Internet connectivity, and voice telephony. For the first time, local telephone companies or ILECs have a facilities-based provider competing for residential and business customers. Coaxial cable entering the customer premises gives the CATV service provider a pipe to carry the "bundle" of services to the end device—PC, television, and/or telephone. MSOs are aggressively entering the local telephone services space creating a contentious battle between ILECs and CATV providers, similar to the battle fought by Western Union and the Bell Company in the early 1900s.

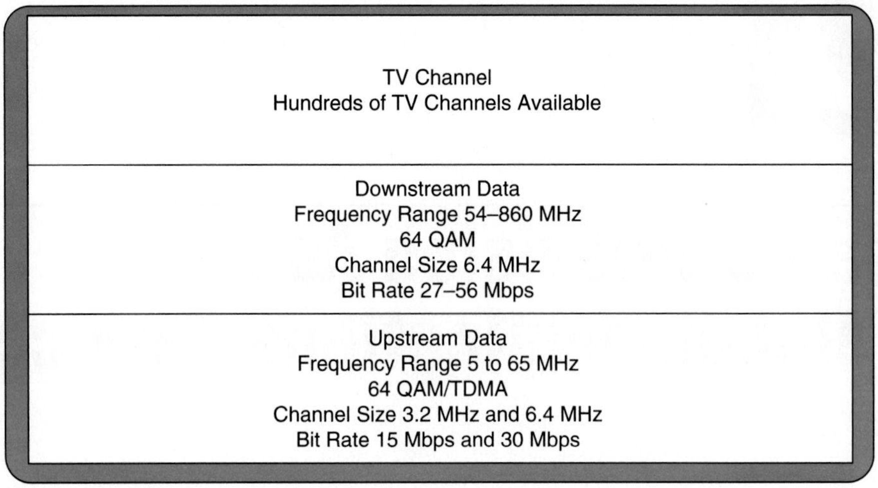

Figure 21–1
A CATV network is similar to a
roadway divided into three lanes, one
lane for each type of vehicle: car,
truck, bus.

TV Channel Hundreds of TV Channels Available
Downstream Data Frequency Range 54–860 MHz 64 QAM Channel Size 6.4 MHz Bit Rate 27–56 Mbps
Upstream Data Frequency Range 5 to 65 MHz 64 QAM/TDMA Channel Size 3.2 MHz and 6.4 MHz Bit Rate 15 Mbps and 30 Mbps

■ 21.1 CATV

CATV
Cable television. Term used
to describe the industry that
provides television and other
services to end users through
a coaxial/fiber optic network.

As mentioned earlier, MSOs have become the newest entrants in the access space. *CATV* providers offer bundled services across the coaxial connections feeding into the customers' homes. The services they offer include hundreds of television channels, asymmetrical data connections to the Internet, and VoIP—Voice over IP. Digital technology has made it possible for all three services to ride together on one pipe servicing hundreds of end points. New modulation schemes, high-speed chip sets, and hybrid fiber/coax systems are responsible for the convergence of services across the CATV network.

The signals traveling together on a CATV system are segmented into specific frequency ranges for each service. A good way to understand this concept is to relate it to a road that has been divided into three lanes, one for cars, one for trucks, and one for buses. If you are driving a car, you must drive in the lane designated for cars; if you drive a bus, you must drive in the bus lane; and so forth. The CATV network is similar in that the frequencies assigned to the system are divided into three lanes: one for television channels, one for upstream data, and one for downstream data, as shown in Figure 21–1.

■ 21.2 DOCSIS PHYSICAL AND DATA LINK LAYER PROTOCOL

DOCSIS
Data Over Cable Service
Interface Specification. The
Layer 1, 2 protocol used to
carry the communications
signals across the CATV
network.

The *DOCSIS*—Data Over Cable Service Interface Specification—is a Layer 2 protocol developed by the CableLabs® corporation and standardized by the ITU. DOCSIS 1.0, 1.1, 2.0, and eDOCSIS are the protocol versions developed over the past decade and deployed in the CATV network today. Each subsequent version of DOCSIS improved the MSO's service offerings and the network's ability to manage and maintain information flow. The DOCSIS protocol defines the architecture of the network, the modulation scheme deployed, the frequencies allocated for upstream and downstream traffic, bandwidth available in each direction, and the Layer 2 media access functions used to maintain network integrity. The following sections describe how DOCSIS defines architecture, modulation schemes, bandwidth allocation, and Layer 2 functionality.

21.2.1 Architecture

headend
Communications hub where
multiple circuits meet. Signals
are aggregated and routed
by the equipment found at
the headend.

Commonly, in the industry, CATV networks are said to resemble a tree and a leaf network topology. A CATV network consists of one or possibly two communications hubs referred to as the *headend*. The headend can be compared to the telecom service provider's POP (point of presence). The headend is used to distribute media content to the thousands of customers paying for cable television services. Today, the headend also houses all communications equipment that aggregates traffic and hands it to the Tier 1 ISP or other core network provider.

The headend connects to multiple *hybrid fiber/coax* (HFC) distribution points that are located strategically throughout the network. The HFC aggregates all of the circuits tied to the end customer and hands them on to circuits tied to the headend. The HFC locations tie back to the headend similar to a star network topology. The network architecture deployed between the HFC cabinets and the customer's premises is referred to as a tree/leaf configuration. The trunk of the tree is the cable leaving the HFC. It spurs off as it travels through the neighborhood forming branches. From the branches, leaves spur off to the customer's location—thus, the term *tree/leaf architecture.*

The headend is considered the root that services hundreds of leaves hanging off tree branches, that is, homes residing off streets and roads. The HFC distribution hubs connect to the main switching center or MSO. The branches spur from the HFC distribution points that tie back to the headend forming a star configuration. Figure 21–2 shows the combination tree/star network just described.

Channels referred to as time slots are carved out of the frequency spectrum and used to carry signals across the physical infrastructure. Each service, as explained earlier, rides in its own lane or frequency. Dividing the spectrum helps reduce signal interference between services and improves network scalability. Television channels and broadband services ride on the same physical pipe but ride on different frequencies.

The first CATV networks carried traffic in one direction, from the headend to the customer. This worked great for distributing television signals that did not need to

hybrid fiber/coax
Hybrid fiber/coax HFC refers to a network that combines fiber optic systems with coaxial cable systems in order to carry information to the end customer.

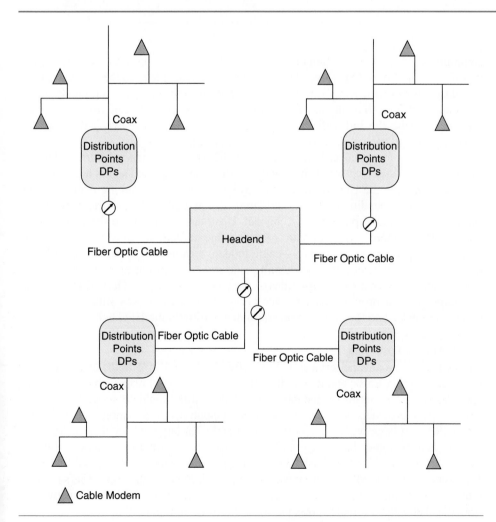

Figure 21–2
The distribution points connect to the headend using a star network topology. The cable modems form a tree network topology back to the distribution points. The term *HFC* refers to a hybrid fiber/coax distribution system.

Figure 21–3
A downstream signal refers to the channel traveling from the CMTS residing at the headend to the cable modem—root to leaf. The upstream signal refers to the channel traveling from the cable modem to the CMTS.

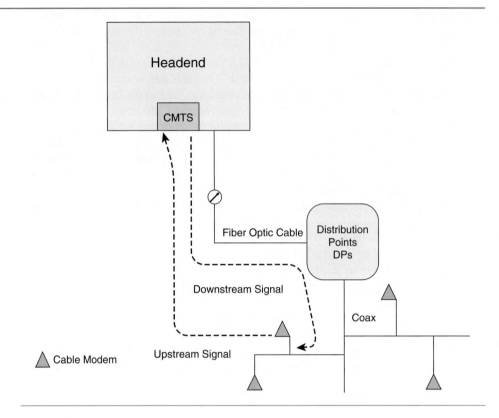

downstream flow
The path the signal travels from the communications hub to the customer's premises.

upstream flow
The path the signal travels from the customer's location to the CATV MSO or main communications hub.

cable modem
Device placed at the customer's premises that provides the interface between the end device and the broadband link that rides on the CATV network.

CMTS
Cable modem termination system. Equipment placed in a CATV network to manipulate and route the hundreds of signals to the distributed end points or customer locations. The CMTS is built on the DOCSIS protocol.

communicate with the upstream device. Today, CATV networks are designed to carry two-way traffic flows. One flow, which travels from the headend to the customer, is referred to as the *downstream flow*. The second path from the customer's premises to the headend is referred to as the *upstream flow*. The terms *upstream* and *downstream* are used to describe the bidirectional nature of the connections between the root and leaves—headend and cable subscriber, as shown in Figure 21–3.

A downstream signal as defined in DOCSIS 2.0 has a frequency range of 54 to 860 MHz capable of carrying between 27 and 56 Mbps. The upstream signal has a range of 5 up to 65 MHz in the United States with available bandwidth ranging from 320 kbps up to 30 Mbps depending on the modulation scheme deployed. A traffic burst from a *cable modem* can eat up a 3.2 MHz channel. Upstream channels are shared by all devices on the segment. Bursts of data from the cable modem are shoved into a channel and carried upstream to the *CMTS*—cable modem termination system. The CMTS is responsible for channel allocation as all cable modems on the network contend for channel space. Downstream traffic rides on the pipe to all devices. The CMTS is a piece of equipment that provides an interface between the cable boxes sitting at the customer's premises and the aggregation equipment sitting in the HFC hub.

21.2.2 Layer 3 Functions

Similar to other broadband services, CATV networks depend on IP as the underlying protocol used to carry traffic end to end. IP is the Layer 3 protocol responsible for cracking open the packet to determine what interface port the traffic should be shipped out on. The largest CATV providers are Tier 1 ISPs. They maintain peering relationships with other Tier 1 ISPs and large organizations. They also lobby for large blocks of IP addresses to service their customer base. Regional MSOs hand traffic off in the same fashion as CLECs and ILECs. They establish a POP in a region where they lease pipes into the Tier 1 ISP's network. The MSO interconnects the network router to the Tier 1 ISP's network router establishing a path to the Internet. Traffic from the ingress side of the network is aggregated and passed to the egress interface, as shown in Figure 21–4.

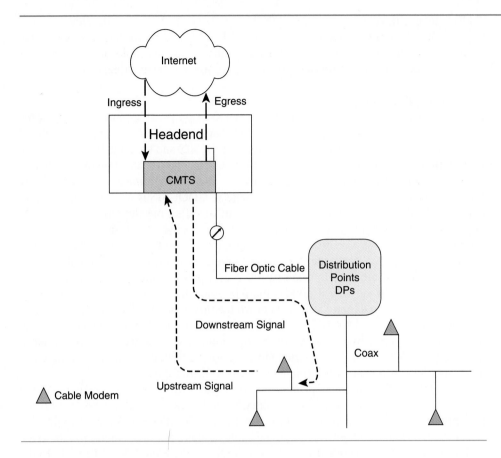

Figure 21–4
The CMTS connects to the supernet to carry traffic outside the CATV network.

The Layer 3 routing function is almost identical to that of other service providers offering broadband services. Assigning addresses, traffic shaping, enabling or disabling UDP ports, and maintaining a secure network are handled by the MSO. Typically, residential customers' addresses are provided on a dynamic basis using the Dynamic Host Configuration Protocol (DHCP), as explained in Chapter 12. DHCP is deployed in order to reduce the time spent maintaining address tables in addition to conserving IP address space. Business customers and home offices are given the option to purchase fixed addresses, though, typically, a limit is placed on the service. NAT— network address translation—functionality is also used by many of the providers in order to conserve address space. The devices hanging off the network tied to the CMTS are given private IP addresses according to the IP address implementation standards defined by the MSO. At the CMTS, public IP addresses are attached to the interface ports allowing communications to the outside world.

Large MSOs maintain domain name servers on site. Most smaller MSOs sub out DNS functionality. DNS within the network functions in the same way as other ISPs' networks. The server translates the URL address into the IP address, as explained in Chapter 5. Customers link to the desired Web site once the domain name server completes the translation and places that information inside the IP packet. The MSO's job is the same as any ISP. It must figure out how to route the traffic and establish sessions between the customer and the designation address. MSOs must also decide what UDP ports they will open and which they will close. The difficulty most network managers face is keeping the customer base happy and at the same time keeping the network secure. Robust firewalls, in addition to firewall service offerings, are maintained by most MSOs in business today. Similar to all ISPs, it is easier for the ISP to offer free firewall service than it is to clean up an attack in the network. Traffic-shaping devices are often deployed in most MSO networks. Customers requesting business services at a set data rate can be groomed out of the mix and given larger data flows than residential broadband customers.

Traffic-shaping devices are installed at the headend to rate or shape traffic flows according to either data rate—anyone paying for 1 M, 2 M, and so forth—or service level agreement—gold, silver, bronze. In either case, the equipment looks at the customer's flow and places it in the appropriate bucket—gold, silver, bronze, or paper.

21.2.3 Layer 2 Functions

DOCSIS defines how traffic flows from the headend to the cable modem sitting at the customer premises. DOCSIS compares to an Ethernet's physical structure and logical shared bus architecture. The difference between DOCSIS and Ethernet is in the way the information is placed on the wire. Ethernet frames contend for space on the wire using a CSMA/CD mechanism allowing all devices to throw their frames onto the wire any time they please as long as the pipe is not congested. DOCSIS is more sophisticated in that it establishes connections through the use of specific time slots that are in a sense tacked up for the length of the session. The devices still share the same medium; the difference is in the way the information is multiplexed onto the line. Ethernet throws it out there, while DOCSIS devices talk to each other, set up the time slot, and then place the information on the line. In one respect, it is similar to time division multiplexed (TDM) networks in the way it divides the pipe into time slots. It differs from TDM, through, in the way it dynamically tears down the time slot once the communication is complete. In summary, DOCSIS is similar to Ethernet in that it communicates to many distributed end points by talking to all the end points on the network and is similar to a TDM network in that it sets up time slots during the time the session is being used. This, of course, is a simplistic explanation on how DOCSIS works.

The DOCSIS protocol is primarily concerned with the MAC—medium access control—portion of the Layer 2 stack. The CMTS depends on MAC sessions in order to communicate and maintain connectivity with the cable modem. A parent-child relationship exists between the CMTS and the cable modem. Every child has to talk to the parent in order to talk to one of its siblings. Simply stated, the CMTS can talk to the cable modems but the cable modems only communicate with their parent, the CMTS. The CMTS has to maintain control over all of its children spread out across the entire network.

The three functions related to the Layer 2 and Layer 1 portion of the OSI stack performed by the CMTS are assigning time slots or frequencies, maintaining transmit levels, and maintaining timing. As the parent, the CMTS communicates with the cable modem frequently, making sure it behaves and follows the rules. Since the CMTS has so many children, as many as 2000, it must maintain order at all times. In the cable network, unlike real life, parents have complete authority over their children. The CMTS determines which time slot the child will use at any given time and at what transmit level. It will determine if the cable modem needs to move its clock up or back depending on the status of the network. Overall, the broadband service provided by the CATV network establishes connections with end devices using Layer 2 MAC functionality. Overall, broadband service provided by the CATV network establishes connections with end devices using Layer 2 MAC functionality built into the DOCSIS protocol and Layer 1 functionality in the way it multiplexes information into time slots established for that session.

The cable modem is told what frequency to use by its parent and when to use it. Bandwidth available on the pipe is divided into frequencies or time slots that are used to carry the information to the CMTS. As explained in the previous section, a channel has a defined bandwidth capacity depending on the DOCSIS version deployed: the downstream channel equals 6 MHz; the upstream channel equals 3.4 MHz. Expanding on the analogy referenced earlier, the cable network is similar to a road divided into different lanes for different types of traffic. The lane set aside for broadband services is further segmented into time slots that are used to carry information to and from the CMTS from the cable modem. The road is used by all cable modems in the same way all houses on the street use the same street to drive to the store, go to work, and so forth. When a cable modem has data to transmit, it shoves it into the shuttle or time slot heading to the CMTS at the headend. Each vehicle or time slot travels at a particular frequency. A cable modem residing on the same street may also ship data using a different time slot. It is the CMTS that maintains who uses what time slot and when—a very controlling parent—in order to avoid traffic jams and collisions. Therefore, one of the primary functions defined by DOCSIS at Layer 2 is allocating bandwidth to devices across the segment.

A second function defined by the standard relates to maintaining consistent signal levels across the network. Look at leaves attached to a tree branch. The distance between the leaf and the branch varies; some have shorter twigs, some longer. The same is true of cable modems tied to the cable feeding the street. The variation in the distance between the cable modem and the cable results in a difference in level as the signal reaches the network. As you may recall from earlier chapters, the greater the distance, the greater the loss. Signals that have a greater distance to travel will arrive at a lower level than signals traveling shorter distances. The result is a variation in signal strength that may disrupt traffic flow across the segment. For example, if a signal is too low, the CMTS will not see the signal and not assign it capacity on the network. The CMTS will not even know that the cable modem had data to send; it will not realize that congestion or contention occurred resulting in lost data. Therefore, DOCSIS was designed to provide the CMTS with a means to adjust the cable modem strength remotely in order to maintain the appropriate signal level. The CMTS talks to the cable modem asking it to adjust the transmit level if it is too low or too high.

The CMTS also is responsible for maintaining a reference clock for all cable modems on the network. Similar to other communications networks, a clock is required to maintain synchronization between elements and to ensure that the information traveling on the shared medium does not collide. The CMTS has the ability to adjust the cable modem clock to prevent it from selecting a time slot in use by a different cable modem. If we return to our scenario depicting the cable network as a road divided into lanes, we can easily see why a network clock is essential in maintaining order and traffic flow. A cable modem transmits its information in a time slot or, in our analogy, in a shuttle. The CMTS uses a technique called *ranging* to determine if a cable modem is marching out of step. The CMTS groups five time slots together (the number of time slots varies depending on the algorithm) and asks the cable modem to transmit in the third time slot. It then looks at the incoming data and determines if the cable modem is transmitting data too soon or waiting too long, that is, lagging behind. If out of step with the network clock, the CMTS asks the cable modem to adjust its local clock to match that of the network clock. Though this process sounds tedious, it is required in order to reduce the chance of collisions on the network.

The CMTS also connects to a provisioning registration server that ties online enrollment systems to the network, a topic discussed later in the chapter. When a new broadband customer enrolls for service, the provisioning server tells the CMTS to recognize a new cable modem on the network. The CMTS adopts the new cable modem as its own and adds it to the network, a new child in the family.

ranging
Method used to adjust the levels of the remote devices depending on the signal level.

21.2.4 Layer 1 Functions

DOCSIS, similar to Ethernet, defines both the Layer 1 and Layer 2 transmission specifications needed to format and regulate information flow across a network. As with all transmission protocols, modulation schemes are needed in order to manipulate and prepare the one and zero states for transmission. DOCSIS, unlike Ethernet, may deploy different modulation schemes on the upstream and downstream traffic flows.

Downstream traffic is modulated onto the physical medium using 64 QAM—quadrature amplitude modulation. (The modulation scheme is dependent on the version of DOCSIS deployed.) The frequency range slotted for the downstream signal ranges from 65 to 850 MHz with 6 MHz assigned per channel. The QAM method deployed assigns six symbols per second resulting in a data rate ranging between 27 and 56 Mbps on the shared medium. It is important to emphasize the term "shared" in

CATV systems. Even though the pipe can carry as much as 56 Mbps, it does not mean that each individual user will be able to send that amount of data at one time. Bandwidth is reduced as new users access the network. As traffic increases, speed decreases similar to cars on the road. When traffic increases on the highway, your speed decreases.

Upstream modulation schemes vary depending on the vendor's equipment. Today, due to the increase in peer-to-peer traffic such as gaming and file sharing, 64 QAM is being deployed on the upstream signal in addition to the downstream signal. QPSK—quadrature phase shift keying—is also a common modulation scheme used by MSOs for upstream data flows. Typically, the upstream frequency range for a DOCSIS 2.0 system is within the 5 to 65 MHz range with each channel requiring 3.2 MHz or a 6.4 MHz channel. The data rate for upstream traffic relates directly to the modulation scheme and channel size. Earlier versions of DOCSIS employed either QPSK or 16 QAM techniques resulting in 3 Mbps available bandwidth to the user. DOCSIS 2.0 has introduced 64 QAM, resulting in an increase in the data rate from 3 Mbps to 10 Mbps for upstream traffic. Again, it is important to emphasize that systems vary between providers and between equipment manufacturers. It would be impossible to say in black and white terms that the bandwidth will always be such and such and the modulation/multiplexing scheme is so and so.

time division multiple access
TDMA is a multiplexing method that divides frequencies into defined subcarriers.

code division multiple access
CDMA is a modulation/multiplexing scheme that divides data into codes that are dispersed across the medium.

Multiplexing schemes used in CATV networks vary depending on vendor specifications. Commonly, *time division multiple access* (TDMA), a technology deployed in cellular networks, is used to multiplex signals into upstream channels. *Code division multiple access* (CDMA) may also be used depending on the equipment and the MSO's design.

As a side note, it is important to understand why upstream traffic rates continue to increase with each new DOCSIS protocol version. Business customers subscribing to broadband services require symmetrical circuits. Unlike residential customers who primarily use their broadband pipe to surf the Net, business customers need to exchange information in both directions. In order to freely exchange information in both directions, it is important for the upstream line speed and the downstream line speed to be equal—thus, the term *symmetrical.* Consequently, MSOs continue to push equipment manufacturers and the standards bodies to incorporate symmetrical service capability into the standard and, in turn, the equipment.

■ 21.3 LOGICAL NETWORK TOPOLOGY

Data communications services travel across the CATV treelike network from the cable modem to the CMTS—cable modem termination system—located at the CATV headend. The CMTS is used to connect cable modems to the Internet, maintain connections between it and each cable modem, manage registration of new devices and termination of old devices, and monitor the network. The logical path between the Internet and the customer's computing device is fairly simple. A signal traveling upstream from the cable modem to the CMTS starts at the customer's computing device as an Ethernet signal. The signal travels from the Ethernet port on the PC into the Ethernet port of the cable modem. The cable modem packages the signal into the CATV standard transport protocol, DOCSIS, and sends the frames out onto the coaxial cable terminated on the cable box.

The signal rides on the coaxial cable with all the other "leaves"—customers' signals—on that segment, similar to an Ethernet. The signals arrive at the distribution point, which is nothing more than a device that formats and aggregates all the electrical signals coming in from all the coaxial buses in the area (branches), and converts them into an optical signal for transport to the headend. The HFC—hybrid fiber/coax—network is the most common network architecture deployed today in CATV systems. Once the signal arrives at the headend, it is received by the CMTS, which that strips off the DOCSIS protocol and hands the information to the core router tied to the Internet.

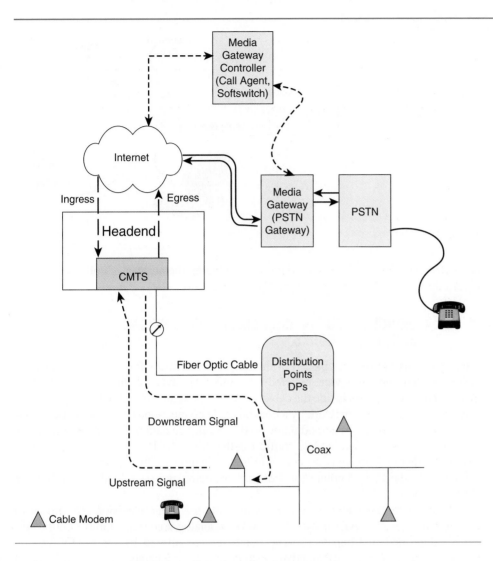

Figure 21–5
Voice over IP calls travel across the broadband network provided by the CATV service provider. A Media Gateway Controller directs the call flow to the Media Gateway where it travels on a trunk connected to a PSTN switch. From the PSTN switch, the call travels to the terminating telephone.

A signal coming in from the Internet—a downstream signal—destined for a computing device in the network travels through the CMTS where it is prepared for transport across the optical portion of the network to the distribution point. Once the signal arrives at the distribution point, it is converted into an electrical signal and pushed out onto the coaxial cable feeding the branch servicing that customer's cable modem. Though the coaxial cable connections feeding the subscribers—as explained earlier, the CATV network—physically look like a tree/leaf architecture, the actual logical network is similar to a bus. A bus topology—also explained earlier—refers to a network in which all devices share the same channel or medium.

The logical path of a voice signal is similar to that of data. The main difference is that the cable modem contains an *ATA* (analog telephone adapter) that converts the analog telephone signal coming from a traditional analog telephone into a digital signal, as described in Chapter 11. At the headend, the voice signal is often split off and handed to the PSTN for call completion. Some smaller CATV providers subcontract switching functions to a VoIP service provider. In this scenario, the signal is routed across the Internet in the same way the data signal is routed. The signal travels to the VoIP service provider where it is routed to the PSTN or termination network. Figure 21–5 shows the logical path for both data and voice services across the CATV network. In order to improve service quality, some providers have set aside frequencies specifically for voice traffic. The value to this model is that the voice signal does not need to contend with

ATA
Analog telephone adapter. A device that sits between the broadband pipe and the analog telephone. The ATA converts the VoIP digital signal into a standard analog signal.

Figure 21–6

The cable box located at the
customer's premises connects to the
outside CATV network via coaxial
cable. The television connects to the
cable box through coaxial cable. The
cable modem and ATA are integrated
into the cable box. An RJ-45 jack
connects broadband service to the
computing device, and a standard
RJ-11 jack connects to an analog
telephone set.

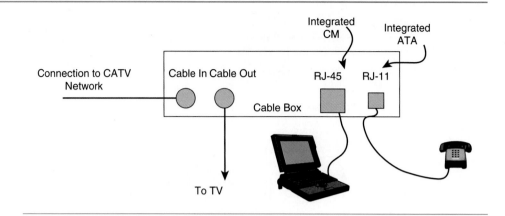

best-effort data traffic on the shared medium reducing the number of collisions, which
in turn decreases the chance of latency, jitter, and echo.

21.4 DEVICES USED FOR BROADBAND CONNECTIONS IN CATV NETWORKS

Cable modems are devices placed at the customer premises for the purpose of connecting the computing device, referred to as the host device, to the Internet. A cable
modem typically resides inside the cable television box, and the cable television box is
connected to the CATV network by a coaxial cable connection. The cable modem consists of a modulator and a demodulator used to modulate data traffic onto the network
or, in the reverse, to demodulate traffic off the network. The customer's computer,
router, or other end device is connected to the cable modem through an Ethernet cable
using the standard RJ-45 connector. Connections made on the cable television box are
shown in Figure 21–6.

Analog telephone adapters are integrated into the cable television box similar to
the cable modem. The ATA is used to convert the analog voice signal into a digital signal, as described in Chapter 11. The voice service offered by the CATV service
provider is built on the VoIP standard, also described in Chapter 11. The VoIP signal
leaving the ATA is tagged using *QoS*—quality of service—flags that prioritize the traffic across the access network. This means that if a customer places a call using the VoIP
service offered by the CATV provider, the traffic generated is given priority over data
traffic traveling on the network. The purpose of prioritizing traffic types is to reduce the
chance the voice signal will experience latency, jitter, or delay—events that degrade the
quality of the signal. CATV providers must maintain ILEC quality service in order to
compete with traditional voice service.

QoS

Quality of service. A term
used to describe a traffic flow
that has been marked with
the priority flags. Priority
traffic takes precedence over
traffic that is not tagged with
QoS flags.

21.5 PROVISIONING BROADBAND SERVICES

When you drive to work, notice the number of houses in your area that are serviced by
cable television. Multiply this by the number of streets in your area. The number of cable modems deployed in the country has been speculated to be in the millions. Now
imagine you are a customer service representative tasked to sign up new customers and
make sure they receive the equipment and are given access to the network. MSOs realized they needed to develop an automated method for provisioning and maintaining end
devices. The provisioning system deployed ties directly to the CMTS telling it when to
add or delete or modify an end point. The provisioning system sits in the headend or,
as with large MSOs, in a regionally diverse area.

The provisioning system receives information regarding the new end device and
passes that information to the CMTS. The CMTS receives the API—automatic provisioning information—from the provisioning server and sets up the new account in its

database. Once the cable modem is received and connected to the coaxial cable, the CMTS receives a message via the MAC layer telling it a new device has been added to the network—a new child has arrived. The CMTS welcomes its new family member providing it with information necessary to communicate. The television now resides in the CMTS and in the provisioning system's registration server where billing and other ancillary functions are performed. In the end, it is critical to see that service providers must maintain flexible, easy-to-manage networks in order to maintain a competitive advantage in the marketplace.

SUMMARY

The CATV network has evolved over the past few years into a multiservices network. Chapter 21 covered general functional areas associated with how CATV carries the multiservices content. DOCSIS, the CATV Layer 2 protocol, was explained—specifically, how it places information on the network. A brief explanation of the Layer 3 protocol carried on the CATV broadband pipe was discussed as were the Layer 1 protocol functions. The logical CATV network architecture was presented. Additionally, the services offered across the CATV network were discussed.

CASE STUDY

Big Bob has asked his trusted nephew Stan to design a portion of the new CATV network he is installing in Green Grass. He has asked Stan to lay out the network for Lilly Valley, a small suburb just west of Green Grass. Stan normally takes care of Big Bob's horse ranch and feels very uncomfortable with the task. On the side, Stan has employed you as a consultant to provide a network design for the CATV network. You will only need to design from the HFC hub to the customer's premises. Draw the design and note the equipment required at each end.

REVIEW QUESTIONS

1. Why are the CATV companies referred as MSOs?

2. The ILECs and CATV companies are battling for market share similar to what?

3. The early CATV networks carried traffic from the customer to the headend only.

    a. true
    b. false

4. The CATV network divides the frequency spectrum in order to

    a. reduce the chance that different services' signals will interfere with each other.
    b. provide a higher level of service to "real-time," two-way services such as VoIP.
    c. allow the upstream television signal from the customer's premises to reach the satellite transmitter.
    d. provide a guaranteed quality of service according to the type of content being shipped.
    e. a, b, and c.
    f. b, c, and d.
    g. a, b, and d.
    h. all of the above.
    i. none of the above.

5. List four functions that DOCSIS provides.

6. The tree/leaf architecture refers to a network that has multiple end points distributed around a network feeding back into a core tree trunk.

    a. true
    b. false

7. Equipment sitting in an HFC remote cabinet has an interface that terminates a(n)_____and a(n)_____.

8. The HFC network has two network topologies. They are

    a. coaxial cable and fiber feeding back to the headend used to aggregate traffic to the switching equipment.
    b. coaxial cable between the customer's premises and fiber optic cable between the HFC cabinet and the communications hub or MSO headend.
    c. fiber cable between the HFC cabinets.
    d. a and c.
    e. b and c.
    f. a and b.
    g. all of the above.

9. What regulatory agency oversees the DOCSIS standard?

10. What is the frequency range of the DOCSIS 2.0 downstream signal?

11. What is the frequency range of the DOCSIS 2.0 upstream signal?

12. What Layer 3 protocol does the CATV network typically deploy to provide broadband services to the end customer?

13. DOCSIS functions as a broadcast protocol.

    a. true   b. false

14. List the three primary Layer 2 and Layer 1 functions the CMTS performs.

15. The CMTS is responsible for maintaining consistent signal levels between end points. In order to do this, it adjusts the _____ level of that branch when it falls below the defined threshold.

    a. transmit and receive at the end point
    b. transmit at the end point and receive at the CMTS
    c. receive at the device
    d. transmit at the CMTS
    e. receive at the CMTS
    f. receive at the end point and transmit at the CMTS
    g. b and f
    h. c, d, and e
    i. all of the above

16. List two common modulation schemes deployed by DOCSIS to provide high bit-per-second range links.

17. Explain the difference between a system that deploys TDMA versus one that deploys CDMA.

18. The following represents the logical path a signal travels on a broadband CATV pipe.

    a. Ethernet interface on the computing device to Ethernet port on the cable modem; Ethernet interface on the cable modem to the HFC equipment in the cabinet
    b. Ethernet interface on the computing device to Ethernet port on the cable modem; IP interface on the cable modem to the HFC equipment in the cabinet
    c. Ethernet interface on the computing device to Ethernet port on the cable modem; GigE interface on the cable modem to the HFC equipment in the cabinet
    d. Ethernet interface on the computing device to Ethernet port on the cable modem; DOCSIS interface on the cable modem to the HFC equipment in the cabinet
    e. none of the above

19. The network from the HFC to the headend resembles a traditional metropolitan network similar to that deployed by the ILEC's PSTN.

    a. true   b. false

20. List three services a CATV provider offers the end user.

KEY TERMS

MSO (629)	hybrid fiber/coax (631)	CMTS (632)	code division multiple access (636)
CATV (630)	downstream flow (632)	ranging (635)	ATA (637)
DOCSIS (630)	upstream flow (632)	time division multiple access (636)	QoS (638)
headend (630)	cable modem (632)		

22

Cable Television Network Infrastructure

Objectives

After reading this chapter, you should be able to

- ■ Describe cable telephony infrastructure.
- ■ Define outside plant cable layout.
- ■ Define physical infrastructure.

Outline

Introduction

22.1 Cable Telephony

22.2 Direct Broadcast Data Transmission

■ INTRODUCTION

In years past, the cable television network was only allowed to carry video signals to its customers. A change in regulation has opened the door for cable television operators to offer telephone and data services. Over the past few years, several cable television operators have been building large-scale networks capable of carrying voice, data, and video to the end user. The introduction of fiber-to-curb technologies has given cable operators a cost-effective method to offer telephony services.

Many large cable operators are beginning to compete with incumbent telephone companies by offering voice and data services. The introduction of RoadRunner data services is the best example of cable carrying traditional telephone company traffic. Analysts believe that, in the future, cable companies will be the strongest competitors against ILECs in local telephone markets.

A second technology that has emerged over the past few years is that of satellite television services capable of carrying data traffic to the Internet. The dish systems can carry data through their satellite network to their Internet handoff gateway. In this chapter, we will look at both the cable television industry's telephony offerings and new satellite dish data offerings.

Figure 22–1
A typical CATV headend office that only provides cable TV to its subscribers. Video feeds are received by satellite antennas and fed into a receiver. The receiver passes the signal onto a video switch that switches various signals to modulators. The modulators modulate the signals into specific frequencies that are then fed into a combiner that combines the signals onto one trunk. It is that trunk that feeds throughout the network, eventually arriving at the subscriber's TV.

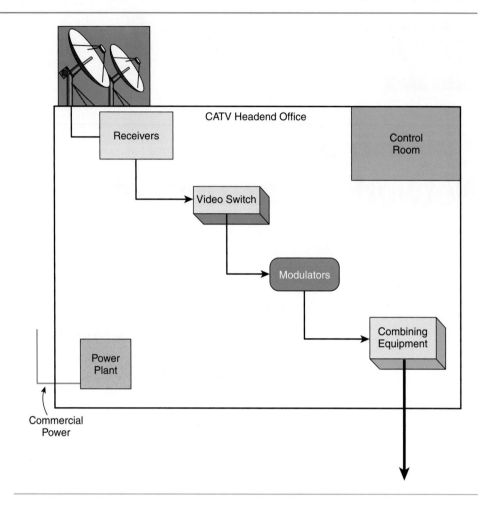

22.1 CABLE TELEPHONY

22.1.1 The Traditional Cable TV Network

headend office
Cable TV location where the video signals are fed downstream to the users.

The traditional cable network may be divided into four categories—*headend office, trunking cable, feeder cable*, and *user drops*. The headend office can be compared to a telephone company's central office. The satellite feed is located at the headend office, as shown in Figure 22–1. Inside the headend office, receivers accept the video signals coming in from off-air antennas, microwave antennas, and satellite dishes. From the receiver, the signals are passed to a video switch used to switch signals and from there they are passed on to modulators that modulate the signals onto specific frequencies. These are fed into combining equipment that combines all the signals into one pipe and transmits it out on trunk cable. The traditional cable TV network is built to carry analog signals, so amplifiers must be placed at intervals along the trunk path, as shown in Figure 22–2.

Figure 22–2
The RF signal requires that amplifiers be placed along the route in order to boost the signal and ensure proper signal strength downstream.

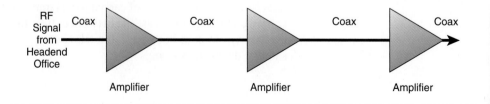

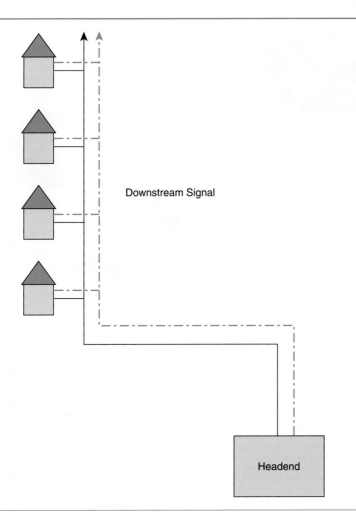

Downstream Signal

Headend

Figure 22–3
The cable signal travels from the
headend office past all subscribers.

The subscriber drop spurs off the feeder cable at a tap point. Each subscriber in the area feeds into the same feeder cable, sharing the same downstream signal. Figure 22–3 illustrates a downstream signal feeding several subscribers. Older analog systems were one-way transmission systems, downstream only, as shown in the diagram. It was not necessary to transmit signals upstream because video transmission at the time was a one-way feed to subscribers. As new features such as pay-per-view, Internet connectivity, and other two-way services were introduced, the need to convert the one-way network into a bidirectional network helped encourage the building of a fiber-optic-fed digital cable TV plant.

22.1.2 The New Cable TV Network

In addition to the one-way transmission scheme, the problems associated with analog transmission caused many cable companies to replace their older networks with digital fiber optic networks. Fiber cable replaced the coaxial trunking cable, allowing for additional bandwidth, two-way transmission, elimination of line amplifiers, and a channelized pipe that could be used to segment services such as voice, video, and data.

The new fiber optic digital *CATV* networks are called *hybrid fiber/coax (HFC) networks*. The structure of the network changed dramatically with the need to replace the old analog equipment with new high-speed digital equipment. In addition, the outside plant had to be transformed into a high-speed, digital fiber optic network. The only portion of the network that was not touched was the subscriber drop.

CATV
Cable TV.

Figure 22–4
Several changes have occurred at the
CATV headend office over the past few
years. CATV networks have been
converted into digital networks. The
modulator has been replaced by an
analog-to-digital conversion unit, and
the combiner has been replaced by a
SONET MUX. The trunk medium has
also been changed from coax to fiber.
Conversion from analog to digital and
from coax to fiber has allowed CATV
providers to increase the number of
available channels.

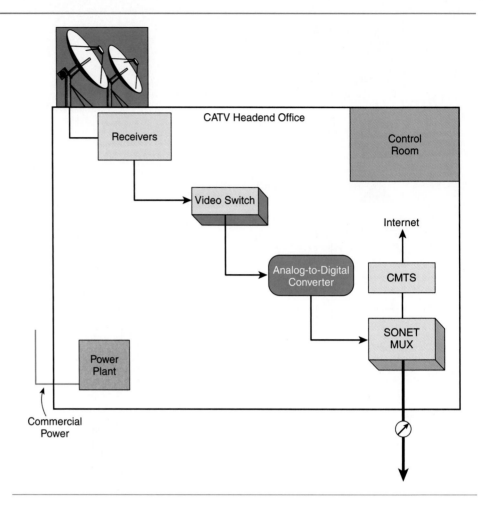

The headend office of the CATV network now has several new pieces of equipment, as shown in Figure 22–4. The site still has receivers that pick up the video signals from the satellite, off-air antennas, and microwave antennas. From the receivers, the signal enters a video-switching unit from which it travels into a digital-to-analog converter. The converter combines the signals and passes them on to a SONET multiplexer that connects to the outside fiber optic cable. Multiplexers vary in size depending on the number of subscribers.

The outside cable plant consists of fiber optic cable that travels around the area and terminates into hybrid fiber/coax termination cabinets. The fiber optic cable is terminated into a fiber-to-the-curb (FTTC) optical-to-electrical device, similar to the telephone company's digital loop carriers. An environmentally secure cabinet is placed on a cement pad close to the subscribers it needs to feed. A fiber optic multiplexer is placed inside the cabinet and is used to terminate the fiber cable running from the headend office. The equipment in the cabinets are powered by feeds from the local electric company, and fiber optic cable is terminated into the fiber optic multiplexers in the cabinet. The fiber optic multiplexer converts the optical signal into an electrical signal and sends the electrical signals out onto coaxial feeder cables that flank the street. The subscriber's drops continue to tap off the coaxial feeder cable running down the streets. Figure 22–5 shows the termination box at the curb and the cable feeding into and out of it.

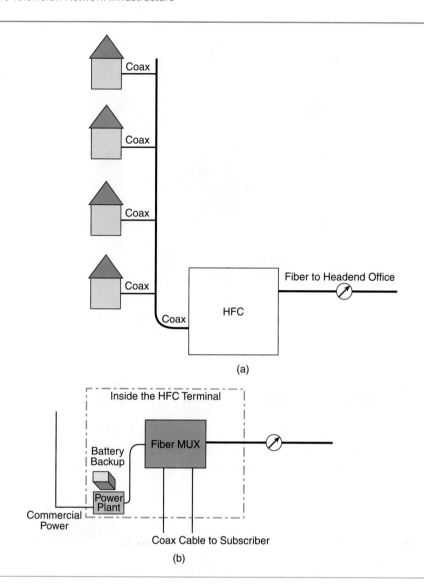

Figure 22–5
(a) The fiber from the headend office is terminated at a unit located near the curb of a residential area. (b) A pullout box shows the different components needed in the HFC terminal.

In order to understand the difference between a CATV network and a telephone company's network, it is important to understand the difference between a bus topology and a star topology, as shown in Figure 22–6. The cable telephone network is constructed of coaxial cable feeding each customer. The customer's coaxial connection taps into the main coaxial bus that runs along the street. All customers tap into the same signal that is being fed from the cable headend office where the transmission equipment resides. When two neighbors watch the same channel on television, they are watching the exact same signal. They share the signal because the signal streams through on one bus. The cable company broadcasts a signal down a pipe and everyone has access to the signal. The telephone company provides one individual copper pair per customer. It cannot be shared by anyone else. Each subscriber line hubs back to the central office where it interfaces into a switch, creating a star topology. A CATV network design is similar to an Ethernet LAN in that each cable box hangs off one shared pipe. The main difference between the two systems is that the information sent to the television is a one-way

Figure 22–6
The main difference between the CATV
network and the telephone network is
the fundamental underlying physical
network architecture. (a) The
telephone network is logically a star.
(b) The CATV network is logically a
bus.

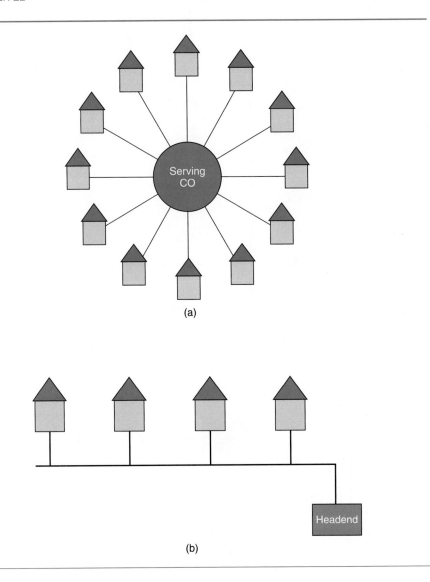

signal referred to as a downstream signal—from the CATV source at the headend
to the end device.

CATV networks depend on fiber optic systems to carry the large amounts of information between distributed locations and the main communications hub or headend.
Figure 22–7 illustrates the topological design of a CATV network that uses fiber-to-the-
curb architecture. FTTC in the CATV network relies on hybrid fiber/coax systems to
convert the optical signal into an electrical signal that is shipped out onto the coaxial
cable connected to each residence. Today, in addition to carrying the television signal
from the headend to the customer, the CATV providers sell broadband connectivity to
the subscriber. A cable modem is used to terminate the broadband connection at the
customer's site. CATV offers the same services as the telephone company: two-way
communications paths that can be used to carry voice and data.

Some CATV providers deploy class 5 switches at the headend. The switches offer
the same line-side functionality as that of a telephone commmpany's end-office switch.
The main differences found at the headend are the class 5 switch and the connections
between the ATM switch or aggregation device used to convert packets/cells into TDM

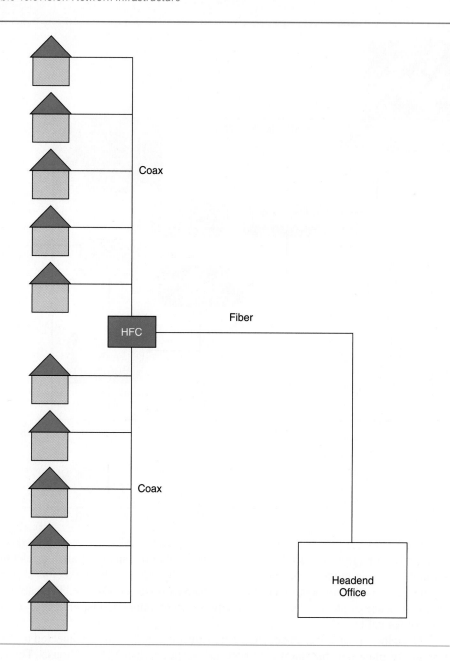

Figure 22–7
The data and video signals share the coax cable from the home to the HFC. At the HFC, the two signals are multiplexed into separate channels and transported across the fiber to the headend office.

digital signals, as shown in Figure 22–8. The CATV provider may opt for a distributed switching architecture and deploy a VoIP network. The VoIP network differs in that the connection at the headend would terminate into a router, which would direct the traffic flow accordingly.

At the customer's premises, the connection also differs from that found in a traditional telephone company network. As explained earlier, the outside plant network is a hybrid fiber/coax network, meaning that over 90% of the network is fiber optic cable and the last mile continues to use coaxial cable. This means the telephone signal travels across the coaxial signal into the customer's home. A cable modem is used to terminate and convert the digital IP signal into an analog voice signal. Most cable modems have ATAs (analog telephone adapters), a topic discussed in Chapter 11, built

Figure 22–8
The addition of data switching equipment at the headend office when cable modems are deployed in the network.

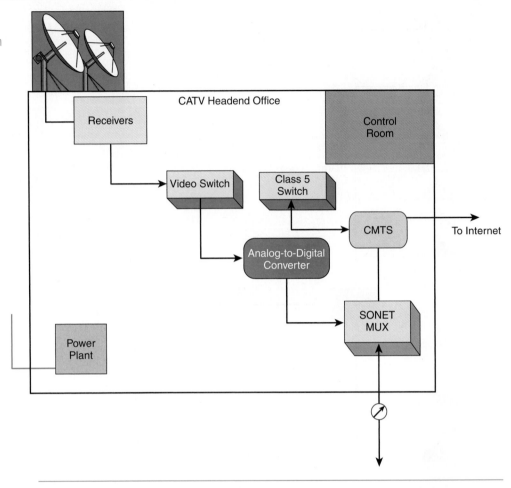

into the form factor. This allows the customer to plug the standard analog telephone set into the cable modem and receive dial tone. The primary difference, as shown in Figure 22–9, is the connection coming into the customer's residence. The cable connection is a coaxial feed, whereas the telephone interconnection is, of course, twisted pair copper cable.

Once the signal leaves the cable modem analog port, it can be connected to the inside house plant and be distributed onto all telephone jacks in the house. The term *whole house wiring* or *backlighting* is used to describe this function. As long as the jacks are tied together at one point, the ATA can be the source of dial tone throughout the house. The customer does not need to place an ATA at each jack.

■ 22.2 DIRECT BROADCAST DATA TRANSMISSION

direct broadcast satellite
Powerful signals are transmitted to users' satellites from receiving antennas on orbiting satellites.

In North America, two *direct broadcast satellite* (DBS) companies sell television programming to consumers, EchoStar and DIRECTV. Both use small eighteen- to twenty-four-inch receivers that are positioned to pick up signals from an orbiting satellite. Not all consumers can use DBS since trees, buildings, and other obstructions interfere with the satellite feed. Data services are currently provided one way from the Internet to the subscriber. To send information to the Internet, a land line is needed to carry upstream traffic. EchoStar has currently launched a new product called DishPlayer. DishPlayer turns the television into a WebTV that allows it to interact with the Internet by using the television as a monitor. If information has to be stored, an external hard drive must be attached to the system.

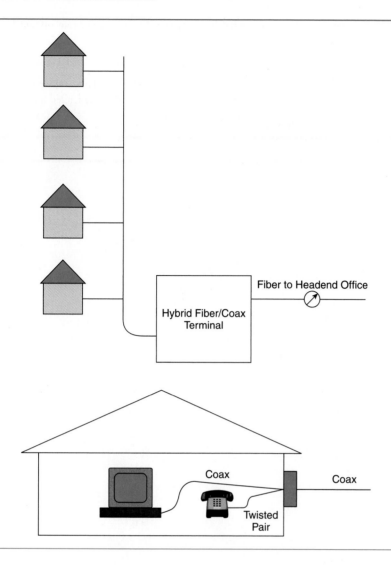

Figure 22–9
CATV companies have recently joined the growing number of competitive telephone companies in providing dial tone to consumers. The CATV providers have already deployed a network all the way to the customer's site and therefore should use the existing medium to carry other signals such as voice and data.

Satellite networks carry data transmission very efficiently because data are not as sensitive to delay as voice. Satellite dishes are now used to connect business customers' data networks together, and they do an outstanding job. The movement toward providing data connectivity to the Internet by using a satellite dish at the customer's residence will continue to grow.

SUMMARY

Chapter 22 focused on the cable television network and how the network handles voice communications. The hybrid fiber/coax network was described, as was the method used to carry voice to the subscriber. A brief explanation of the DBS network was included in this chapter.

REVIEW QUESTIONS

1. List two services offered by CATV providers.

2. Explain how a cable system is able to carry voice services.

3. Draw a typical remote terminal and explain how that terminal is connected to the cable headend office.

4. Using the drawing from question 3, diagram how a subscriber is connected to the remote terminal. Label the drop end of the CATV plant.

5. Data services deployed by the CATV network provide individual dedicated pipes to each customer. True or false? Explain.

6. What is the typical bit-per-second rate offered by a cable modem?

7. What may cause the cable modem speed to slow down?

8. Explain why you feel cable telephony will or will not become a major provider of dial tone in the United States.

KEY TERMS

headend office (642)

CATV (643)

direct broadcast
satellite (648)

23

Cellular Telephone Network

Objectives

After reading this chapter, you should be able to
- Describe the cellular telephone network.
- Define satellite telephone.

Outline

Introduction

■ INTRODUCTION

The advent of the cellular telephone network changed the landscape of communications around the world. The cellular telephone was first introduced in the early 1980s, but Bell Labs had developed cellular radio in the late 1940s. Cellular technology was not feasible until the advent of the transistor in the mid-1960s. Mobile telephone service was the predecessor of cellular radio service and was the first robust method used to provide wireless communications. The main differences between the mobile radio system and the cellular radio system are the way the networks are designed and the number of subscribers allowed in one area. Cellular radio is similar to a bee's honeycomb. Each cell within the honeycomb has a transmitter and receiver that hands off calls between cells as the user travels around the area.

The best way to understand cellular telephony is to equate it to all we have learned so far about the wire-line telephone network. Cellular telephones use DTMF dialing. Voice frequencies are the source of the signal. The network uses T1 circuits as trunks, fiber optic cable and multiplexers in their networks, and digital switches to switch calls. The difference between the two networks is that the cellular network uses the airways to carry information the last mile to the customer. In addition, the customer is mobile—no longer tied to one location. The cellular network must transmit information through the airways and follow the user around the globe. Other than these two differences, the cellular network is very similar to the land-line telephone network.

Cellular providers are now becoming stiff competitors to land-line telephone systems as additional cellular carriers per market have helped reduce the price of a cellular call. In the beginning, the FCC allotted frequencies to two carriers per *cellular geographic serving area* (CGSA). The FCC granted the local telephone company one set of frequencies and gave the second set of frequencies to a new carrier who won a bid for the air space. In 1996, the FCC opened up the spectrum, allowing six additional carriers to enter each CGSA. Most areas now have more than two cellular carriers and, consequently, are converting from an analog cellular network to a digital cellular network.

■ 23.1 THE CELLULAR TELEPHONE NETWORK

23.1.1 The Last Mile of the Cellular Network
The Base Transceiver Station

Cellular base stations—towers with small rectangular antennas perched on top—are common sights in most regions. The cellular network design is based on covering a region with hexagon-shaped cells equal in diameter. One base station is placed in each cell, preferably at the highest point in the area. A *base transceiver station* (BTS) is placed in each of the hexagon-shaped cells that are designed to mesh together similar to a bee's honeycomb, as shown in Figure 23–1. The honeycomb shape of the cellular network provides neatly interconnected cells that do not overlap but still provide sufficient coverage for the territory. The BTS consists of a cellular antenna, similar to the one shown in Figure 23–2, a base station that holds the signal generation equipment and network multiplexers, and cable from the local telephone company. Figure 23–3 shows the inside of a typical cellular base station.

The antennas are mounted according to engineering specs that define the height and position relative to the adjacent cells. Because each antenna only has to transmit the distance of the cell, the transmit power required by a cellular transceiver is less than that for the older IMT mobile radio system, an FM or AM radio station, or a local emergency radio station. The FCC allows a maximum transmit power of 100 W per cellular

cellular geographic serving area
Area that provides boundaries for cell companies' services.

cellular base station
Antenna where cellular calls are sent and received, then sent to the next station and eventually to the called party.

base transceiver station
Consists of the cellular antenna, the base station equipment, and telephone company cable.

Figure 23–1
Typical cellular networks have a design that is similar to a bee's honeycomb, with hexagon-shaped sections that are covered by particular frequencies. Duplicating patterns reuse the frequencies. The cellular architecture uses the frequency spectrum much more efficiently than the mobile phone network due to the reuse of frequency throughout the area.

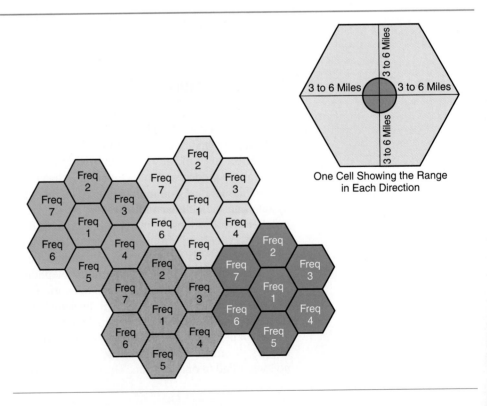

One Cell Showing the Range in Each Direction

Figure 23–2
Cellular towers are one part of the base transceiver station (BTS) and have specific placement in relation to one another. (Photo courtesy of M. A. Rosengrant)

antenna. The public's concern that cellular radio towers may cause illness to residents living close to the towers is hard to justify because the signal power is so low. As with a signal traveling down a wire, the farther the signal has to travel, the stronger the power level of the transmitter has to be. Due to community protest, many exteriors are now camouflaged to look like farm silos, trees, or water towers in hopes that the public will

Figure 23–3
The BTS houses equipment that is able to convert incoming radio frequency signals into electrical ones and zeros that are transported out to the PSTN via T1 circuits.

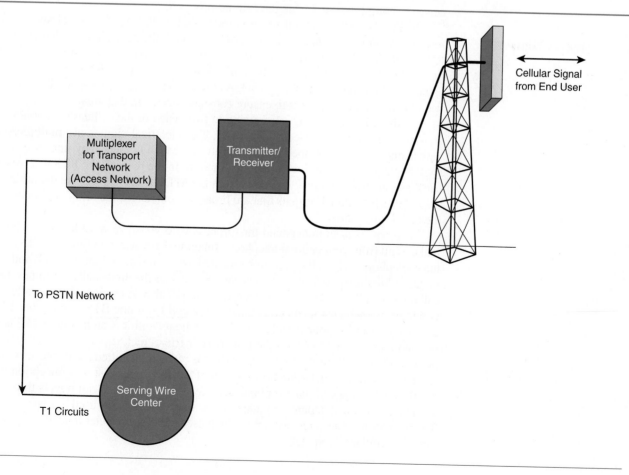

Figure 23–4
Powering cell sites becomes one of
the critical engineering tasks for a
cellular engineer. Power is normally
set up as shown. Commercial power
feeds the site during normal
operation. During a commercial power
outage, backup batteries kick in and
provide power to the equipment.
Within a few seconds, the backup
generator initiates and takes over as
the power source. The process
ensures constant uptime during a
power outage.

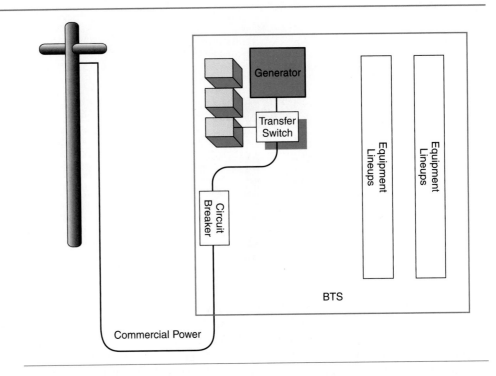

**mobile telephone switching
office**
An "office" where
mobile/cellular calls are
switched.

not protest the installation of a tower in its area. Still, RF engineers face numerous zoning boards, town committees, and angry neighbors when determining cellular tower placement. The only happy participant in the process is often the landowner who receives a monthly stipend from the cellular provider.

Inside the BTS is a cell-site controller that "talks" to the digital switch in the *mobile telephone switching office* (MTSO). The cell-site controller watches each of the RF signals traveling to the subscribers' cellular telephones and reports their status to the MTSO. Within the BTS are also multiplexers that connect the BTS to the land-line cable that then connects to the MTSO. As illustrated in Figure 23–4, a BTS has battery backup and small generators to ensure constant service in that area.

At the BTS, the RF signals coming in from each of the cellular telephones in the area are converted into electrical signals. The electrical signals are multiplexed together onto T1 circuits, then transported to the MTSO where they interface into the switch and are routed to the correct destination. In the reverse, the BTS receives electrical signals traveling on T1 circuits from the MTSO. The signals are demultiplexed, then converted into RF signals that are broadcast out across the allotted frequencies to the cellular telephones.

The best way to understand this process is to visualize what happens when you make a call from your cellular telephone. Imagine that you are calling your friend's cellular telephone, which happens to be located across town. You go off-hook and punch in the cellular telephone number. The switch receives the digits and determines that the call is a cell call to cell call within its own region. It also determines that the called party is one of its own customers and connects the call from one BTS to the other BTS. It sends a message to your friend's cellular telephone, telling it an incoming call is waiting and that it needs to sound the ringer. Your friend picks up his or her cellular telephone and goes off-hook. Your conversation is being transmitted across the airways and through a wire-line network similar to the PSTN. The biggest misconception by the public concerning cellular telephone service is that the entire call travels through the air. The cellular call depends on many of the same components used in a wire-line call. The main difference is the last mile, which uses radio frequencies to reach the end customer's cellular telephone.

Cellular Radio Frequencies

A-Band

824 to 835 MHz transmit
869 to 880 MHz receive
845 to 846 MHz transmit
890 to 892 MHz receiver

B-Band

835 to 845 MHz transmit
880 to 890 MHz receive
847 to 849 MHz transmit
892 to 894 MHz receive

PCS Frequencies

1.5 to 1.8 GHz

Table 23–1
Frequency Bands Used for Cellular Radio Systems and the Newly Allotted PCS Frequencies.

Cellular Transmission Schemes

The FCC doled out frequency spectrum to cellular providers in each of the cellular telephone service groups (CTSGs). Table 23–1 shows the frequencies used in the cellular network. Providers are now given 20 MHz to carry signals from cellular telephone sets to base stations. They divide their spectrum into thirds and give each cell 1/3 of the allotted field. In order to avoid interference between cells, adjacent cells are not allowed to use the same frequencies. Frequencies are reused in nonadjacent cells in order to increase the number of possible subscribers per region. The older IMT mobile phone system designed by AT&T in the 1960s was one of the first commercial uses of RF for telephone service. It did not allow the frequencies to be reused within an area and, consequently, limited the total number of possible subscribers in that region. The beauty of cellular architecture is its ability to reuse frequencies within the region, thereby increasing the number of subscribers on the network. The pursuit of every cellular provider in the world has been to find a way to increase the number of people it can service per cell. The solution has evolved over the past few years and, like every other technology in telecommunications, varies by country and provider. Three methods are used today to carry signals in the cellular network. The older standard is the FDM. The newer digital standards, called portable communication systems (PCSs), are TDMA and CDMA. Each uses different techniques to encode the information and transport it from cellular telephones to the BTS.

Frequency Division Multiplexing

Frequency division multiplexing (FDM) is the oldest and simplest standard used in the cellular network. It is similar to the FDM used in the wire-line network. The frequencies are divided up, and each subscriber is given a waveband that he or she uses for the duration of the call. FDM is an analog cellular technology. The analog cellular network is limited in the number of subscribers it can service in a given region. The *American mobile phone system* (AMPS) is currently the most widely used cellular standard in North America. The AMPS standard gives each subscriber 30 kHz of bandwidth. When a cellular provider is given 10 MHz of spectrum, each cell can carry 111 subscribers, each using 30 kHz. The low number of subscribers per cell has spurred designers to invent a better transmission scheme that increases the number of users per cell while reducing the power consumption of the cellular telephone and consequently the size and cost of the phone. The advent of digital cellular radio is a result of the AMPS's limited

frequency division multiplexing
The oldest and simplest standard used in the cellular network. An analog cellular technology.

American mobile phone system
Most widely used standard in North America.

number of subscribers per cell. It could provide cellular service to every person in North America—the ultimate goal of cellular providers.

Time Division Multiple Access

Time division multiple access (TDMA) is a digital cellular technology introduced a few years ago and accepted by many cellular providers. TDMA is similar to time division multiplexing in that it digitizes the voice into one and zero pulses by running the voice signal through a Codec in the cellular telephone. TDMA increases the number of conversations per frequency by digitizing the signal, using frequencies only when there is a conversation going on, and compressing the voice signal from a 64 kbps to an 8 kbps word. The result is that TDMA technology increases the number of calls that can travel across the same frequency spectrum.

Some of the TDMA standards used today are Digital American Mobile Phone System (D-AMPS), IS-136, Global System for Mobile Communications (GSM), Personal Digital Cellular (PDC), and Integrated Digital Enhanced Network (iDEN). D-AMPS is one of the most widely used standards in North America. D-AMPS frames consist of three time slots, as shown in Figure 23–5, each of which represents one communications channel. The main differences among the standards is the number of time slots, the cycle length, and the frequency width. GSM is the most popular standard in Europe. It is built around eight time-slot frames and requires a very low cycle length. The nuances between the standards have been debated continuously over the past few years. Both standards allow for a greater number of subscribers per cell, lower power requirements, and better quality transmission than analog cellular technology. PDC is also a popular TDMA standard used almost entirely in Japan. iDEN is a proprietary Motorola standard.

Code Division Multiple Access

Code division multiple access (CDMA) is one of the hottest standards to emerge on the telecommunications scene over the past few years. The technology was developed by QUALCOMM as a way to increase the number of subscribers per cell for wireless transmission. CDMA completely changes the way information is processed and transmitted in the cellular network. It is currently being deployed by many cellular providers who want to change their old analog systems to digital and by new cellular providers just building their networks. CDMA technology uses a technique referred to as spread spectrum when transmitting signals from the cellular phone to the BTS.

time division multiple access
A cellular technology that digitizes a voice signal into one and zero pulses by running the signal through a Codec. The technology increases the number of conversions per frequency.

code division multiple access
A technology developed by QUALCOMM as a way to increase the number of subscribers per call for wireless transmission.

Figure 23–5
The two types of TDMA technology widely deployed in the United States and Europe have different frame structures. (a) The D-AMPS uses a three channel per frame structure. (b) The GSM is built around an eight channel per frame structure.

(a)

(b)

CDMA works a little like a jigsaw puzzle. The information signal is chopped up into sections. A code is attached to each piece, and the information is thrown out into the spectrum—like throwing all the pieces of the puzzle in the air. The puzzle lands haphazardly. It makes little sense until the code is deciphered and pieced together to form the picture. The cellular equipment that receives the signal is able to piece together signals by reading the codes attached to the pieces and reassembling the signal in its correct order. The advantage of spread spectrum technology, spreading the information across the spectrum, is the increased number of subscribers per cell as multiple signals can now use the same frequencies. The signal is not given a frequency. It is given a code. Designers claim that PCS increases the number of subscribers per cell by twenty times that of TDMA. They expect even greater numbers per cell in the future releases.

There are several advantages to using the CDMA standard. Similar to TDMA, the signal is digitized, compressed, and shipped out with all the other digitized pieces, increasing the number of subscribers per cell. Another advantage is the security provided by spreading the information across the spectrum. Listening in on a CDMA call is relatively impossible. Several years ago, a couple picked up one of Representative Newt Gingrich's conversations on their two-way radio and recorded his cellular conversation. That was FDM analog cellular service. If Gingrich had been talking on a PCS phone, the couple would not have overheard his conversation.

Handing calls off between cells by using a soft handoff technique is another advantage of using CDMA technology. A call is connected before it is handed off, thus reducing the chance that the call will be lost during handoff and eliminating the click that is normally heard between handoffs. A final advantage is the reduced power consumption of the telephones due to the way the bits are transmitted and received. Telephones continue to decrease in size as the life of their batteries increases.

CDMA works like analog cellular technology. The cellular antenna receives the calls being placed within a cell, then hands the calls to the cellular switch site called the MTSO. Cellular calls are placed the same way a wire-line call is placed. The cellular phone has a touchtone pad, a transmitter, and a receiver. Instead of a copper wire dangling from the phone, it has an antenna. The air acts as the conduit between the cellular telephone and the base station. The base station sends the signal to the cellular phone or receives the signal from the cellular phone. Each cellular phone is given one of the 128 channels allotted per cell.

Cellular Telephones

A cellular telephone is very similar to a standard wire-line telephone. The telephone has a touchtone pad that uses DTMF signals to transmit the telephone number to the MTSO and a switch hook used to go off-hook and on-hook as needed. In addition, a cellular telephone has several other components to transmit and receive signals, including a transmitter, receiver, antenna, diplexer, frequency synthesizer, and logic board, as shown in Figure 23–6. The transmitter and receiver are bridged to an antenna that transmits and receives signals. The diplexer separates the transmit and receive signals to eliminate cross over interference. The frequency synthesizer generates a reference frequency and the logic board tells it what frequency needs to be transmitted or received. The frequency synthesizer is responsible for converting the reference frequency to the appropriate frequency for that particular call. The logic board and the frequency synthesizer work together to make the cellular telephone a tunable device, unlike older mobile telephones.

The logic board may be viewed as the "brain" of the unit. It sends and receives messages from the BTS telling it when and how to set up a call, tear down a call, and monitor the strength of the signal during a call. The logic board interprets the signal level and sends messages back up to the controller at the BTS. The controller relays this message to the MTSO, which decides whether or not the call should be handed to the next cell. Constant communication between the MTSO, the BTS, and the cellular telephone make cellular technology much more sophisticated and complicated than regular land-line technology. The process is similar to an air traffic controller talking to the

Figure 23–6
(a) A typical cell phone showing the four main components—antenna, receiver, touchtone pad, and transmitter. (b) A logical view of the components contained in a cell phone.

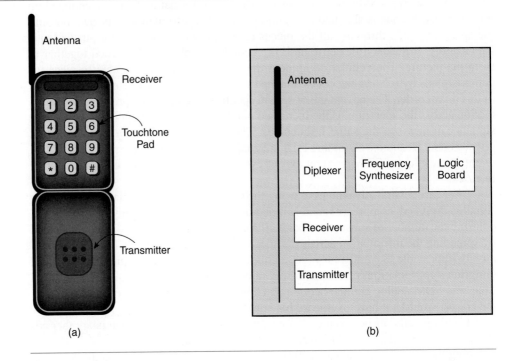

pilot of a plane. The controller and the pilot continuously exchange information about where the plane is, its status, and where it is going. The air traffic controller changes every time the plane enters a new air space, and the pilot exchanges messages with the new controller at the new air tower. The cell phone is similar to the pilot, and the cell site controller is similar to the air traffic controller. The cellular telephone must continuously exchange messages between itself and the cell site controller. Once the cellular user moves to a new cell, the phone starts to exchange messages with the new cell site controller at the new BTS.

A 32-bit code is also sent from the telephone to the base station. It is passed on to the MTSO where it is looked up in a database to confirm that the phone is one of the provider's subscribers. Unfortunately, people have been able to grab codes as they travel through the airways with special receiving black boxes. They then program stolen telephones with the new stolen codes and use the telephones to place fraudulent calls.

In the early 1990s, the power required to make a cellular telephone functional was fairly substantial. Small cellular phones, digital transmission schemes, more sophisticated microprocessors, and better cell design have contributed to a reduction in the power required by a cellular telephone. The FCC maximum power output is 7 W. A typical cellular phone requires much less. One-watt phones are now common. The battery life has also improved over the past few years. Depending upon the brand, some telephones will last hours without needing to be recharged. The prediction is that phones will continue to become more efficient and battery life will continue to improve.

Cellular phone use is being blamed for many automobile accidents. The introduction of hands-free phones has helped combat the growing concern over automobile drivers and cellular phones. Still, law enforcement and insurance groups are advocating bans on driving while talking on a cellular phone.

A second phenomenon associated with cellular technology is the introduction of stylish cell phones. Many colors, sizes, and carrying cases are available to consumers. Cellular phones, like land-line phones, are being transformed from utilitarian devices to stylish accessories.

23.1.2 The Mobile Telephone Switching Office

A typical MTSO contains multiplexers to multiplex and demultiplex circuits entering and leaving the outside world and a digital switch to switch calls both between cells and between the local and long distance PSTN. It also has a power plant similar to that of a wire-line central office. The cellular MTSO is very similar to a wire-line central office. The main difference is the lack of thousands of subscriber lines terminated on the MDF. Additionally, special software is required for each switch to communicate with the BTS and cellular calls. The MTSO is much smaller and the equipment is much newer than that in an ILEC's central office. The switch is the same type of switch used by the PSTN and the multiplexing equipment the same as that used by local and long distance carriers. The MTSO performs the same functions as the wire-line sites—it switches calls.

The MTSO switch sets up and tears down calls across the network. It is constantly talking to the BTS and asking for information on the signals and the status of the calls. For example, a person on his or her cellular phone driving across town may pass through several cells along the way. The MTSO has to be able to recognize the strength of the signal and know when the low-threshold mark is reached, thus requiring a hand-off to the next cell. It also has to recognize off-hook, on-hook, and DTMF signaling. Billing the customer for time online is a responsibility of the MTSO switch, just as it is for any telephone switch. Unlike the wire-line network, cellular calls are billed on both an incoming and outgoing basis. Users pay for every minute they are on the telephone. This strategy is being challenged by new competitors that offer special packages allowing for a certain number of free minutes a month.

The MTSO digital switch, similar to a typical class 5 end-office switch, also switches traffic to the local incumbent telephone companies, CLECs, long distance carriers, and other cellular providers. The type of interconnections vary between providers and are determined based on up-front negotiations instead of technical reasons. The MTSO accepts several types of interconnections. Connections from the BTSs come in on T1 circuits leased from the ILEC. T1s may ride on either fiber cable or copper cable depending on the size of the site and the availability of the cable. The switch also connects directly with the incumbent telephone company to be able to switch calls between its networks. The cellular provider normally interconnects with the local tandem but in certain cases may connect directly with end-office switches. Cellular providers also connect directly to the long distance carrier or have the ILEC switch the traffic to the appropriate long distance carrier. Many cellular providers are now establishing interconnection agreements with new CLECs. The CLECs are connecting directly with the cellular provider instead of paying the ILEC to handle the calls between the two networks. The method of connection is decided by performing a business analysis on the cost of circuits compared to ILEC switching costs. The cellular network has to connect with the wire-line network and vice versa in order to provide full coverage to its subscribers. Figure 23–7 shows a typical MTSO interconnection scheme.

The cellular digital switch must also interconnect to emergency agencies such as 911, operator services, and informational locations, just as a typical wire-line switch does. The cellular switch deals with 911 issues similar to the local telephone company. The cellular switch passes the call immediately to the 911 center; however, the 911 center does not receive the location of the calling party as it does with land-line calls. Over the past few years, there has been a push to introduce a *V and H coordinates* locator that would identify the location of calling party. V and H (vertical and horizontal) coordinates are often used to identify a location on Earth. In fact, every location on Earth has its own V and H coordinate. V and H are being used by other industries that require some sort of locator as defined by longitude and latitude. For example, the new On-Star network uses V and H coordinates to locate stranded motorists. The government will soon require that all cellular providers offer location information to emergency 911 centers. To accomplish this, cellular phones and the cellular network will have to be able to send V and H location information through the network to the 911 center. Because the cellular network is already using out-of-band signaling, sending additional

V and H coordinates
Latitude/longitude coordinates used to determine distance between points and to determine location on Earth.

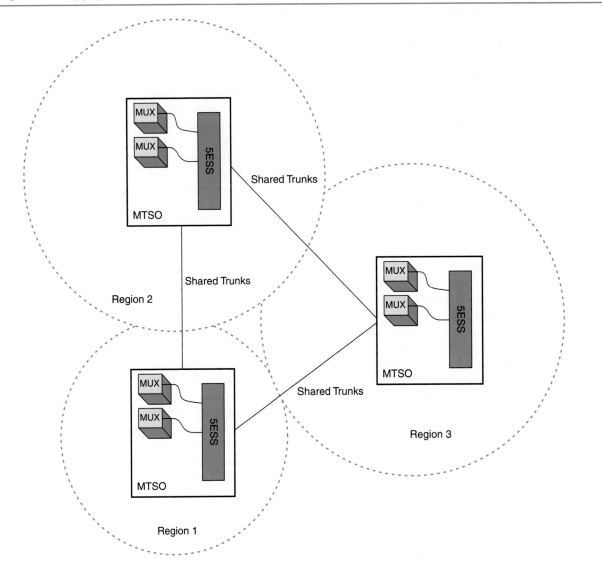

messages is not difficult. The difficulty comes from identifying the V and H coordinates of a moving target—the cellular user. The solution is to place a GPS receiver, which is fairly common, in every cellular telephone. Boaters, hikers, and mountain climbers often have handheld GPS devices that tell them where they are at any given time. The GPS satellite was placed in orbit by the U.S. military to guide military operations around the world, but the commercial market, including telecommunications, has taken advantage of GPS locator ability. A cellular phone will read its location's coordinate and transmit it to the MTSO where it will be interpreted and passed on to the 911 emergency center.

23.1.3 Connecting the Base Transceiver Station to the Mobile Telephone Switching Office

Connecting the BTS to the MTSO can be compared to connecting remote switching centers to tandem offices. Many BTS sites use T1 circuits to connect their location to the MTSO. The T1 circuits carry the many calls coming from and going to the cellular users within that area. Similar to shared trunks between switches, the circuits between the BTS and the MTSO are also shared by the different users. The biggest difference between the cellular scenario and the wire-line scenario is that the cellular users can

Figure 23–8

The MTSO is the switch center for the regional cellular network. It looks very much like a typical local central office. A class 5 switch, multiplexers, and backup power reside at the site. The main difference is that the MTSO does not have a large MDF with wire pairs feeding subscribers. The medium to the subscriber is through the air.

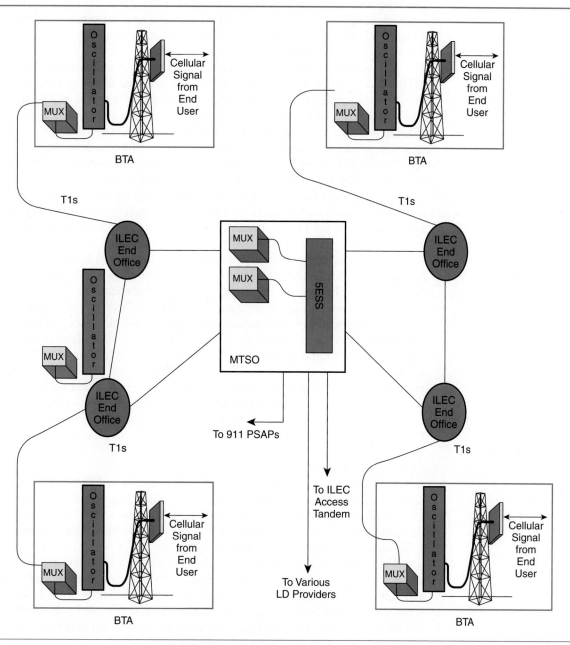

move out of the range of the BTS and discontinue using the shared T1 trunks from that particular BTS. The call then jumps to the next cell station and is placed on the shared trunks between that BTS and the MTSO. Figure 23–8 illustrates a typical cellular network including the MTSO and the BTS.

23.1.4 Signaling

Signaling in the cellular network may be divided into three categories: cellular telephone to the MTSO, the base station to the MTSO, and the MTSO to the outside world. Cellular telephone to base station is similar to the common wire-line network. The subscriber presses the switch hook and goes off-hook. The difference is that a message is sent to the switch telling it the telephone has gone off-hook. A circuit closure is not possible because there is no copper loop. In the cellular world, messages are used to transmit requests back and forth. When subscribers hear a dial tone, they can punch in the number they wish to reach. The same frequencies used for the numbers on a wireline telephone are used on a cellular telephone. DTMF is the signaling method that sends the telephone number to the MTSO switch. Once the subscriber goes off-hook, a message is sent through the air to the base station, which relays the off-hook condition to the MTSO switch.

IS-634
Signaling protocol used by cellular networks.

The second two devices that have to exchange information about the call are the MTSO and the BTS. *IS-634*, which is the standard used to connect the two sites, defines how calls will be processed and handled between the two sites. The standard was designed to work with all types of transmission schemes—AMPS, D-AMPS, CDMA, TDMA, and so forth. Not only does the BTS need to send messages to the MTSO switch regarding whether the telephone is on-hook, off-hook, or in use, it also needs to send the signal level so that the MTSO can determine whether or not the call needs to be handed to the next cell. The exchange of messages between the two sites takes place across control links that are used to carry all signaling messages. The control links are data communications channel (DCC) links. The signaling messages carry the call supervision signals used to set up calls, tear down calls, and monitor the signal level of the transmission.

The final signaling that has to happen in the network is between the MTSO and the outside world. The MTSO uses the SS-7 network to switch calls between carriers. In fact, roaming between areas is dependent on SS-7 connections. The cellular provider establishes connection agreements with local ILECs, CLECs, and long distance carriers. Calls passing between the networks depend on SS-7 messages much like local and long distance networks depend on SS-7. A call between two regional cellular providers rides on leased facilities between the two carriers. A call traveling from New York City to Ohio may travel across several carriers' facilities before it is handed to the cellular provider in Ohio. The cellular portion of the call is only between the base station and the telephone. The rest of the call depends on transmission facilities the same way other long distance calls do. SS-7 is as important to a cellular provider as it is to a long distance or local carrier.

23.1.5 Roaming between Regions

roaming
Ability to use the cellular telephone outside the cellular serving area.

Roaming becomes possible when SS-7 messages are handed between cellular providers. When the two providers have not established agreements, the user may have to enter his or her credit card number before being allowed to place a call in the area. SS-7 interconnection agreements have reduced the number of times a user has to enter a special access code or credit card number.

The greatest change within the past few years is the advent of national cellular carriers such as Sprint, AT&T, Verizon, and others. When a customer subscribes to a large national carrier, the need to roam is reduced and the handoffs are almost automatic. Roaming costs and technical difficulties cause the most complaints from cellular users. The advent of national carriers has reduced and will continue to reduce the frustration of cellular users who need to cross carrier boundaries.

23.1.6 Billing

Billing a cellular call is not as straightforward as billing a local or long distance landline call. Cellular providers must work together to bill the correct number of minutes for a call traveling between carriers. For instance, if a subscriber lives in upstate New York and subscribes to Spidercom cellular service but often travels outside Spidercom's cellular network, the customer will use airways owned by that region's cellular provider. Hopefully, Spidercom will have established an agreement with the local ILEC's cellular network or some other provider in that region. If it has, the call will go through without interruption, while behind the scenes a complex exchange of billing information is going on in order to make sure the bill shows the number of minutes of use in that territory.

Most bills show a roaming charge plus the minutes-per-use charge, which may be higher than those billed normally. The exchange of billing information is handled by special billing record exchange. As the call occurs, usage charges are placed according to location. The home location register (HLR) is a database that holds all the information about a customer's cellular provider and is used to ensure proper billing of usage. Cellular providers comply to the IS-41 standards that define how signaling messages will be used to carry billing information between carriers. The HLR database holds this information and helps levy fees between carriers.

■ 23.2 CELLULAR TELEPHONE FEATURES

Cellular telephone features are similar to land-line telephone features. The greatest differences are the new message-based features that are becoming very popular. Some of the most popular cellular telephone features are as follows.

On-hook dialing *On-hook dialing* allows the user to correct the digits of a dialed call before the call is sent. Because many users of cellular phones are moving at the same time they are dialing, the ability to correct a number before it is transmitted is not only convenient but also frugal because misdialed calls are expensive.

on-hook dialing
Feature that allows the user to correct the digits of a dialed call before it is sent.

Signal strength indicator A *signal strength indicator* shows how strong a signal will be or is at a particular time. In an area with poor cellular coverage, the signal strength indicator will show low strength, which means that the call will probably not go through or, if it does, the signal will not be good.

signal strength indicator
An indicator that shows how strong a signal is.

Call restriction *Call restriction* is a security feature that reduces the chance that an unwanted call will be made by a nonallowed caller.

call restriction
A security feature governing users.

Alphanumeric memory *Alphanumeric memory* is similar to a Rolodex. It holds names and numbers in the phone's memory.

alphanumeric memory
Memory installed that can be used to hold names and numbers of commonly called parties.

Distinctive ringing *Distinctive ringing* allows cellular phones to be set to one of several combinations of rings, music, or other unique alerting techniques. One reason different rings are used is so the user can tell it is his or her cell phone that is ringing and not someone else's. Of course, there are a limited number of different rings.

distinctive ringing
Phones can be programmed to ring in several different ways.

Hands-free *Hands-free* cellular telephones have become very popular due to the increased number of automobile accidents blamed on cellular use while driving. A brain cancer scare in 1999 also increased interest in hands-free phones. Though most research conducted so far has not linked cellular use to any type of brain cancer, the increased number of requests for a hands-free solution prompted some innovative hands-free cellular phones.

hands-free
Features that enable users to talk on a cellular phone without holding it.

Voice mail *Voice mail* in the cellular world is very similar to voice mail offered in the wire-line world. Users enter a greeting, are given access to their own voice mailbox, and are allowed to access the mailbox from any location. The cellular phone displays a message saying "voice-mail message" when a call is waiting in the voice mailbox. Voice mail will continue to be one of the most popular features provided with cellular service. One reason cellular usage costs have dropped is the addition of value-added features such as voice mail that allows cellular providers to gain income from items other than minutes of use.

voice mail
Callers are able to leave voice messages for the cellular subscriber.

Call waiting *Call waiting*, too, is similar to a call made with wire lines. Many cellular users need to be constantly connected, and call waiting provides a way to never miss a call. A call waiting beep or tone is received when the cellular user is already engaged in a conversation on the cellular telephone. The user flashes the switch hook to

call waiting
Alerts cellular subscribers if a second call comes in while he/she is using the phone.

accept a call waiting in queue. If the user wishes to ignore the call, it may be forwarded to his or her voice mail or the caller will continue to hear ringing until he or she decides to hang up.

Call return   *Call return* redials the last call dialed. For instance, if you call 567-8899 and the line is busy, you can press *send* again after a few minutes and the phone calls the number again.

Speed dialing   *Speed dialing* is the same as the speed dial we use at home. We can program the phone with several frequently called telephone numbers and dial any of these numbers by punching in two digits instead of seven or more. Because cellular telephone users are often on the move, they welcome such shortcuts when dialing, accepting calls, and listening for calls.

Message register   Cellular users can now receive data messages on their telephone screens. A good example is that of receiving stock information, such as the price at that time. The user can also request specific information such as the temperature, football scores, stock prices, and so forth. The information streams in at certain intervals, and the telephone can be programmed to alert the user when a specific piece of information arrives.

call return
Redials the last number called.

speed dialing
Allows the phone to be programmed so that a number can be dialed by dialing two numbers instead of seven or more.

message register
Allows users to receive streams of information from the Internet.

■ 23.3 CARRYING DATA ACROSS THE CELLULAR NETWORK

Carrying data over a cellular network is not as common as carrying data across a wireline network, but it can be done on either the analog or the digital cellular network. Modems connect the PC to the cellular telephone, and special protocols have been devised to carry data across the radio frequencies. The most common protocols are MNP10, MNP10EC, TXCEL, and EC2. They are found in the PC card and must also exist at the far end modem owned by the ISP. Before trying to send data traffic using your cellular telephone, you must call your ISP to find out if it supports the correct wireless protocols. Many ISPs do not.

The highest speed you can expect to reach using your cellular telephone is 14.4 kbps. The most common speeds are 9.6 kbps or 4.8 kbps, which are very slow compared to wire-line modem speeds. Facsimiles may be transmitted by using your cellular phone at rates of 4.8 kbps or less. Transmitting large bits of data across the analog wireless network is not really possible. Most uses for cellular data are short text messages. Most cellular networks have been converted into digital cellular transmission schemes that do allow for larger bits of data.

The digital cellular network is capable of carrying higher data speeds than the analog network. The three digital cellular protocols handle data transmission differently and, as expected, are able to carry varying amounts of data. Entrepreneurs are working diligently on new and powerful data transmission standards that will allow high-speed data to be carried by the cellular telephone network. The goal of the HDR—high data rate—protocol is to allow a 2.4 Mbps standard to ride on a 1.25 MHz channel. Independent design companies claim they have protocols that allow transmission of higher bit rates than HDR. The race between the different protocols will be the focus for many engineers, analysts, and designers for the next few years.

■ 23.4 SATELLITE TELEPHONE SERVICE

Satellite telephony has been one of the most-often-discussed technologies at the hundreds of telephony conferences held over the past few years. The idea of eliminating the need for extensive land-line connections intrigues the industry, and the advent of the low-orbiting satellites helped push the concept of satellite telephony into a reality.

Three large consortiums were formed—Iridium, Globalstar, and Teledesic—with wealthy investors backing the venture.

Globalstar has established a worldwide telephony network that uses forty-eight low-orbiting satellites strategically placed around the globe. Its network consists of the satellites and ground stations, called gateways, that interconnect to the local telephone networks and cellular network. Because the satellite's orbit is low, the signal strength is greater, allowing for smaller telephones. The telephones alternate between cellular telephones and satellite phones. Globalstar can carry voice, data, video, fax, paging, and any other signal you can think of.

Globalstar provides service anywhere in the world except countries that have not allowed it to interconnect into local telephone networks. The local PTTs (public telephone and telegraphs) still have control over who provides telephone service in their region. Globalstar does offer a way to provide telephone service in remote areas of the world, though, by using the existing land-line and cellular network to enhance its service and help reduce the cost of usage.

Teledesic plans to deploy low-orbiting satellites around the globe to carry all types of information—voice, data, and video. Teledesic does not plan to sell service directly to end users but instead is targeting service providers who are looking for network.

Iridium, the first to deploy satellite telephony, faced enormous financial trouble in the early part of the new century. Fortunately, Iridium received a large sum of cash that helped the company continue providing services around the world. The technology utilizes the Ka-band between satellites and between satellites and gateways. The L-band or low-orbiting band is used to connect to the base station telephones.

A few years ago, several large investors decided to launch a worldwide satellite telephone service. The intent was to provide dial tone anywhere in the world without the need for wires, cellular towers, or other land-oriented connections. The consortium paid billions of dollars to launch low-orbiting satellites strategically around the globe. The satellites would be used to relay calls to switching centers that would link into the worldwide PSTN. The object of the venture was to provide service to business travelers who often lost communications when traveling through remote parts of the world. In addition to business travelers, the consortium saw the enterprise as a way for remote villages, cities, and towns to become connected to the world communication network.

Iridium, however, is ahead of its time. Business people do not travel to remote places on Earth often enough to warrant the fairly high cost of a satellite phone ($3000) in addition to the cost per minute charged for each call, which is about $7. Remote villagers also are not able to pay for the high-priced telephone or the usage charges associated with each call. The only group embracing the satellite phone are the adventurous explorers who trek through the Himalayas, forge through snow to the South Pole, and navigate around the world in rowboats. Unfortunately for Iridium, there are not yet enough adventurous souls and their phone usage is not enough to repay the billions of dollars used to build the network.

SUMMARY

Chapter 23 focused on wireless telephony. Cellular networks and satellite telephony networks were described, and wireless analog cellular networks were discussed. Because of the need to add subscribers and improve the data-carrying capacity of the wireless network, digital cellular technology emerged. The chapter explained the difference between TDMA, PCS, and GSM and listed the advantages and disadvantages of each. The methods used to carry data across the wireless network were also explained. The chapter finished by discussing the three large satellite telephony consortiums—Iridium, Globalstar, and Teledesic.

REVIEW QUESTIONS

1. The cellular telephone network has recently converted from an analog network to a digital network. True or false? Explain.

2. Explain the main advantage digital cellular technology provides over analog cellular technology.

3. Explain why the advent of the national cellular carriers has improved the handoff of calls between regions.

4. What happens when a user passes between cellular providers' regions?

5. A cellular telephone is not able to transmit data due to the RF signal latency. True or false? Explain.

6. Define *MTSO*.

7. Draw a picture showing how an MTSO connects to a cell site.

8. Why are T1 circuits used to connect the MTSO to the cell site?

9. List three types of communications equipment found in the MTSO and explain what each is used for.

10. List three types of communications equipment in the cell site and explain what each is used for.

11. Draw a picture showing how individual cell sites mesh together.

12. Why is the cell architecture designed the way it is?

13. List the features that cellular service providers offer to cellular subscribers.

14. How does a cellular telephone send information back to the cell site?

15. Explain how calls are handed off between cells.

16. Explain why multiple calls can be made within the same cell site.

17. Is cellular usage increasing or decreasing? Why?

18. The satellite telephone system offers connectivity anywhere in the world. True or false? Explain.

19. List two of the satellite telephone consortiums formed to offer satellite telephone service.

20. Briefly explain why you think satellite telephone service will either fade away or become a means to offer dial tone in any part of the world no matter how remote.

KEY TERMS

cellular geographic serving area (652)

cellular base station (652)

base transceiver station (652)

mobile telephone switching office (654)

frequency division multiplexing (655)

American mobile phone system (655)

time division multiple access (656)

code division multiple access (656)

V and H coordinates (659)

IS-634 (662)

roaming (662)

on-hook dialing (663)

signal strength indicator (663)

call restriction (663)

alphanumeric memory (663)

distinctive ringing (663)

hands-free (663)

voice mail (663)

call waiting (663)

call return (664)

speed dialing (664)

message register (664)

24

Wireless Broadband Providers

Objectives

After reading this chapter, you should be able to

- Define broadband cellular protocols.
- Define broadband cellular network topology.
- Define Wi-Fi as a broadband service.
- Define WiMAX as a broadband service.
- Define Bluetooth wireless technology.

Outline

■ INTRODUCTION

Wireless service providers traditionally offering cell phone service to millions of subscribers are now offering high-speed untethered data connectivity. The newest wireless technologies and protocols make it possible for the cellular networks to carry broadband data rates throughout a footprint. The subscriber may surf the Web from remote locations, eliminating the need to be tied to a wired premises such as a home or office.

Over the past few years, the introduction of low-cost microwave systems, along with the opening up of additional frequencies for unlicensed services, has created a new industry lead by start-up Wi-Fi/WiMAX providers. Both the cellular and Wi-Fi carriers are contributing to the rapid expansion of broadband services over wireless media throughout the country. The following few sections look at the structure of the protocols used for broadband cellular and Wi-Fi/WiMAX networks, network infrastructure, and the service offerings.

■ 24.1 BROADBAND CELLULAR SERVICES

Randy works for a consulting company in Boston, Massachusetts. For the past month, he has been planning a mountain biking trip in Vermont. Just before he walks out the door Thursday afternoon, Randy's boss stops by, hands him a list of reference documents, and asks him to read and summarize them by Monday. Randy sighs, picks up the list of references, packs up his laptop, and heads to Vermont. Friday morning is beautiful, the sky is clear, the sun is out, and the birds are singing. Randy knows if he wants to enjoy the weekend, he will need to finish his task—reading and summarizing the reference material—as promised. He packs up his laptop, climbs a small hill behind his friend's house, finds a comfortable tree, turns on his laptop, and logs in to his broadband Internet provider's high-speed wireless network.

Randy quickly accesses the Web sites listed on his reference list and downloads the files to his laptop. After reading several paragraphs, he e-mails Sarah, a colleague at work, to find out if she has any additional information on the subject. Sarah forwards multiple files to Randy letting him know their boss has flown to Madrid to go shopping. Randy has no problem opening the large documents. He reads through the materials, summarizes the subject matter, formats the file, and ships it off to his boss. Within a millisecond, Randy's boss e-mails him from his BlackBerry saying, "Good work. Enjoy your weekend." Randy packs up his laptop and heads down the hill to meet up with his friends.

The significance of this scenario should not be overlooked. The evolution of the *cellular/PCS* protocols is responsible for Randy being able to work out of office, in a different state, in the middle of a field. Randy is a virtual office worker, meaning that his office is anywhere he has access to the Internet through his broadband wireless provider. He was able to sit next to a tree in the middle of a field and do the same work he would have done sitting in his cube in the office.

■ 24.2 BROADBAND CELLULAR PROTOCOLS

Cellular protocols evolved from defining how simple ordinary telephone service works over cellular systems, as explained in Chapter 23, to a full suite of products—voice, broadband data, instant messaging, and video. Digital cellular systems' first deployment under the 2G—2nd generation—standard in the late 1990s laid the path for sophisticated protocols such as *3G*—3rd generation—protocols that allow Randy to sit in the middle of a field on a Vermont hillside and surf the Web. Two protocols dominated the 2G rollout: CDMA and GSM. The two protocols divided the industry in the same way Pepsi and Coca-Cola divided cola drinkers. The *Global System for Mobile Communication* (*GSM*) became the most popular cellular method in Europe and much of the rest of the world. Code division multiple access (CDMA) ruled North America, South America, and Japan, in addition to several Asia markets. From this division, two standards groups emerged: the first, *3GPP*—3rd Generation Partnership Project, and the second, *3GPP2*—3rd Generation Partnership Project 2, both advocating their favorite transmission method—Pepsi user group and Coca-Cola user group. In order to guarantee deployment of 3G worldwide, the ITU produced a 3G standard that contains five terrestrial radio interfaces of which both the 3GPP and 3GPP2 favorites reside. The ITU has set specific requirements that an interface has to meet in order to be called 3G. The ITU states that the protocol has to provide 2 Mbps worth of bandwidth in fixed locations or within buildings. It must provide 384 kbps in urban environments; 144 kbps in WAM—wide area mobile; and variable data rates in remote, large geographic areas. Today, over eighty countries have 3G networks deployed with hundreds of users subscribing to the services.

The five ITU International Mobile Telecommunications–2000 (IMT-2000) interfaces, defined in Table 24–1, are the *Universal Mobile Telecommunications System*

---

cellular
Mobile wireless communications network that allows mobile computing devices such as cellular phones to accept and send calls.

PCS
Personal communication services or portable communication system. A term used to describe digital wireless mobile services. The term *cellular* is often used to describe PCS even though, technically, PCS refers specifically to digital cellular.

3G
3rd generation mobile telecommunications standard that defines the new technologies used to carry high-speed data across the mobile network.

Global System for Mobile Communication (GSM)
A mobile telephone standard used throughout the world.

3GPP
3rd Generation Partnership Project. The standards group that advocates the UMTS 3G standard.

3GPP2
3rd Generation Partnership Project 2. The standards group that advocates the CDMA2000 3G standard.

Terrestrial Radio Interface	Names
UMTS	Universal Mobile Telecommunications System
CDMA2000	Code Division Multiple Access 2000
TDD	Time division duplexing
TDMA	Time division multiple access
FDMA/TDMA	Frequency division multiple access/time division multiple access

Table 24–1
ITU IMT-2000 3G Terrestrial Radio
Interface Standards.

(*UMTS*), *CDMA2000*, *time division duplexing* (*TDD*), *time division multiple access* (*TDMA*), and *frequency division multiple access/time division multiple access* (*FDMA/TDMA*), of which three are based on CDMA. The two we will focus on are UMTS and CDMA2000, both developed on CDMA transmission methodologies. 3GP has adopted UMTS, and 3GPP2 has adopted CDMA2000 for 3G deployments. Though overlap between the two camps is becoming more common, a division continues to drive multiple protocols and standards within the wireless community.

The division between the two camps started during the deployment of 2G, the first digital cellular systems. The CDMA camp standardized on QUALCOMM's IS-95 cdmaOne protocol that used spread spectrum technology to multiplex and modulate streams of information. CDMA throws out the information similar to throwing out pieces of a puzzle, allowing the sharing of radio frequencies. In 3G systems, a special code called a pseudo-random noise (PN) code is generated, multiplying the signal for the purpose of increasing the density of the signal across the spectrum. In simple terms, this means that it reduces the amount of data needing to be sent per signaling change by using codes generated by the system that are manipulated by the data signal to condense the number of bits needed to represent the information. Though this subject is well beyond the scope of this text, it is important to point out the significance of CDMA as a multiplexing technology that helps improve the number of subscribers allowed per frequency band within a cell. Figure 24–1 shows what happens when a signal is spread across a wide band versus a narrow band.

The GSMs depended on the second dominant digital cellular method, TDMA, as the protocol for the 2G deployments. Time division multiple access establishes time slots that are used to carry a call from the time it is established to the time it is torn down. Again, as with CDMA, the details of how it works are beyond the scope of this text. The key point to grasp is that TDMA is a methodology found in the mobile network today. Going forward, the 3GPP and 3GPP2 interface standards are replacing 2G systems, based on CDMA transmission methods. TDMA-based systems are being phased out.

24.2.1 Universal Mobile Telecommunications System
The interface standard adopted by the 3GPP alliance is IMT-2000 CDMA, also called the Universal Mobile Telecommunications System (UMTS). The UMTS, often referred

Universal Mobile Telecommunications System (UMTS)
One of the ITU 3G mobile telecommunications standards used to carry high-speed data across the mobile network.

CDMA2000
Code Division Multiple Access 2000 is one of the ITU 3G mobile telecommunications standards used to carry high-speed data across the mobile network.

time division duplexing (TDD)
A 3G standard defined by the ITU to carry high-speed data across the wireless PCS network.

time division multiple access (TDMA)
A multiplexing/modulation standard used to carry information across cellular/PCS networks.

frequency division multiple access/time division multiple access (FDMA/TDMA)
Multiplexing/modulation standards used to carry information across the cellular/PCS networks.

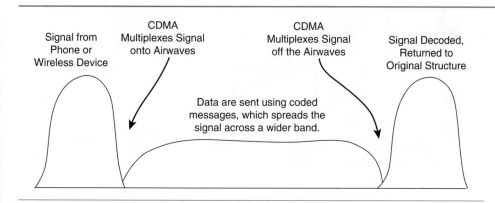

Figure 24–1
CDMA spread spectrum multiplexing schemes manipulate data across a wide frequency spectrum.

to as 3GSM, uses the Wideband Code Division Multiple Access (WCDMA) protocol and the GSM network architecture to provide broadband wireless services to the public. The UMTS is the standard defined by the ITU, and WCDMA is the protocol the standard deploys in the network to transmit information across the radio frequency (RF) spectrum. For example, when we hear the word *Coke,* we immediately think of the soft drink. The recipe used to produce the beverage is not used to define the drink. The same is true with UMTS even though the industry tends to use the names interchangeably—UMTS, WCDMA, and 3GSM.

WCDMA uses CDMA multiplexing techniques to manipulate information onto the RF carrier signal. The WCDMA protocol extends beyond CDMA as its transmission method to multiplex signals onto the wire. WCDMA is a protocol similar to other protocols that define interfaces, the multiplexing scheme, how the packets or frames are structured, and how the service works overall.

The UMTS provides close to 2 Mbps of bandwidth within the local area and 384 kbps within the WAN, though it is important to point out that newer schemes and methods continue to increase bandwidth across the network. Therefore, it is possible that systems developed by different vendors provide higher bandwidth values than those we have stated. Bandwidth rates as high as 14 Mbps have been promised by some systems.

The UMTS uses a 5 MHz channel for the upstream or uplink signal within the 1900 MHz band and a 5 MHz wide channel for the downstream or downlink in the 2100 MHz band. In comparison, narrowband CDMA deployed in 2G systems used 200 kHz channels to carry information. Obviously, sending higher amounts of data requires greater channel capacity. One drawback the UMTS faces in North America is the overlap between 2G portable communication systems (PCSs) that operate at the 1900 MHz band and satellite systems using the 2100 MHz band. Luckily, 2G GSM systems use the 900 to 1800 MHz bands, reducing the clash between the two systems.

The network architecture of the UMTS is very similar to GSM 2G deployments. The 3GPP alliance divides the network into two domains: the *User Equipment Domain* and the *Infrastructure Domain*. Each domain is segmented further and used to define interface handoffs. Figure 24–2 shows 3GPP's logical description of the UMTS network. Note that the User Equipment Domain is divided into a Universal Subscriber Identity Module (USIM) and the Mobile Equipment Domain. You may relate this portion of the network to the customer's mobile phone, the RF path, and the cellular antenna. The term *air interface* is often used to describe the handoff between the phone and the antenna. The standard defines a frequency spectrum; a multiplexing method—in this case, WCDMA; a cell size; signal strength; and receiver sensitivity. Before we move on to describe the rest of the UMTS network, we need to first

User Equipment Domain
The portion of the UMTS network that defines the user's mobile equipment used in 3G networks.

Infrastructure Domain
The portion of the UMTS network that defines the network portion of the 3G network.

Figure 24–2
Universal Mobile Telecommunications Systems are segmented into the User Equipment Domain and the Infrastructure Domain.

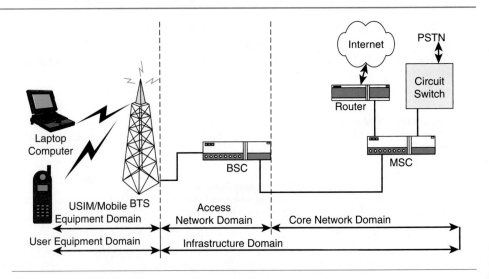

define the underlying architecture of the 2G GSM network as it provides the foundation for 3G services.

The GSM network is structured according to the size of the cell, the antenna coverage, and antenna placement. Cell size falls into four categories: *Macro, micro, pico,* and *umbrella* refer to the range the signal travels between the base station where the antenna resides and the end user. Macro cells are those where the antennas are placed so they project out over the rooftops. The signal travels a fairly substantial distance as it is not obstructed by buildings or other elements. A micro cell covers a smaller area, allowing the antenna to sit below rooftop level. A pico cell is used within a building, and an umbrella cell is used to fill in the gaps between other cells.

The GSM network structure consists of a *base transceiver station (BTS)* and a *base station controller (BSC)*, which are used to aggregate the signals from multiple towers and hand them off to the metropolitan network. The final portion of the network, referred to as the core network, varies little from any other service providers. It includes links that tie the BSC points to a location where traffic is groomed and dispersed accordingly. A significant difference—other than multiplexing schemes—between GSM and North American cellular 2G systems is the *Subscriber Identity Module (SIM)* placed in the mobile telephone. The SIM is used to hold all the information about the customer. The SIM card may be moved from handset to handset and from service provider to service provider with ease. The SIM card is used by the service provider to handle billing, assist in handoffs between providers, and maintain features per customer, not per mobile phone.

The UMTS network architecture follows the same structure as the GSM 62 network. The two most significant differences can be found within the access network and the frequencies used to carry the information from the BTS to the BSC. The frequency channel assigned per call is 5 MHz up and 5 MHz down utilizing the 1900 MHz band for up and the 2100 MHz band for down. Both the ITU and the FCC in the United States have opened up additional frequencies for 3G services. The downside is that those licenses are expensive and often difficult to obtain. The BTS is updated to include equipment capable of handling the higher data speeds. In addition, a phone built to handle the UMTS air interface does not work in a GSM network, and vice versa. Today, most phones are deployed with dual mode—GSM and UMTS—as the migration from 2G to 3G progresses.

The size of the pipe needed to connect the 3G BTS to the BSC is greater in 3G networks than in 2G networks. Connections between the BSC and the *mobile switching center (MSC)* must also be upgraded in order to carry the large volumes of broadband data connections. In many cases, depending on the area, this requires fiber optic cable builds that include the addition of fiber equipment and all the ancillary needs surrounding the link. The primary difference found at the MSC or MTSO is the addition of data networking equipment such as routers, aggregation devices, and all other piece parts required to interface the Internet.

In summary, the UMTS, though similar to the 2G GSM, requires a substantial investment by providers if they decide to enter the broadband arena. Today, several companies have launched 3G systems using UMTS and, as noted by the 3GPP, many others are planning deployments in the near future.

24.2.2 CDMA2000

CDMA2000, similar to the UMTS, is a 3G terrestrial radio interface defined by the ITU specifically to carry broadband services across the cellular network. CDMA2000 is the 3G standard that is most often deployed by service providers with 2G cdmaOne *IS-95A* networks. The migration between QUALCOMM's CDMA cellular platform to CDMA2000 is less significant than deploying an entirely new protocol such as WCDMA. Therefore, previous advocates of GSM will remain GSM-centric deploying UMTS WCDMA and advocates of IS-95A and *IS-95B* will choose CDMA2000 for their 3G deployments.

The evolutionary path of the CDMA2000 standard that started with CDMA2000 includes two offerings referred to as CDMA2000 1X and CDMA2000 1xEV.

macro
Term used to define the cell coverage area of the GSM network. Macro cells allow transmission over rooftops.

micro
Term used to define the cell coverage area of the GSM network. Micro cells cover smaller areas than macro cells, typically, below rooftops.

pico
Term used to define the cell coverage area of the GSM network. Pico cells cover space within a building.

umbrella
Term used to define the cell coverage area of a GSM network. Umbrella refers to coverage across an area not covered by other cell footprints.

base transceiver station (BTS)
Accepts the mobile phone signal and hands it to the base station controller.

base station controller (BSC)
Accepts the signal from the base transceiver station and packages it for transport onto the land-line network connected to the mobile switching center.

Subscriber Identity Module (SIM)
Card placed in the mobile device and used to define the user information. The card may be moved between devices and carriers.

mobile switching center (MSC)
Communications hub where all mobile traffic is aggregated. Interconnection point between the access network feeding the BTS/BCS sites and the core network.

IS-95A
First IS-95 2G standard that uses cdmaOne methodologies.

IS-95B
The second 2G standard that uses cdmaOne methodologies.

Figure 24–3
CDMA2000 networks include the mobile wireless network from the phone to the BTS, the access portion from the tower to the BSC, and the BSC connection to the MSC. The MSC has dedicated links to the PSTN and to the Internet.

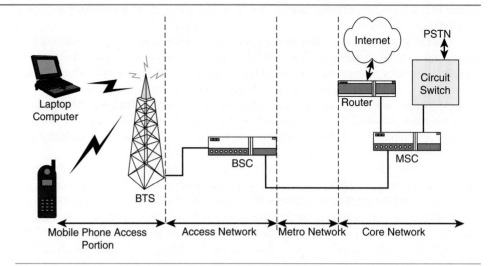

CDMA2000 1X doubles voice capacity when compared to the previous 2G technologies. It also gives the user the ability to transmit/receive 307 kbps of packet data. The second, CDMA2000 1xEV, includes the following: CDMA2000 1xEV-DO, CDMA2000 1xEV-DV, 1xEV-DO, and 1xEV-DV. All offer higher bandwidth rates than CDMA2000 1X. Currently, data rates as high as 3.1 Mbps are possible on CDMA2000 1xEV systems. Both CDMA2000 1X and the newer CDMA2000 1xEV require a 1.25 Mhz channel for upstream and downstream transmission.

Advocates for CDMA2000 note the protocol's ability to handle any type of packet data protocol, specifically IP—Internet Protocol. Additionally, the protocol was developed to handle multiple application types allowing it to service different device types such as laptops, PDAs (personal digital assistants), smart phones, and so forth.

Similar to WCDMA, the network architecture mimics that of 2G IS-95 systems. As shown in Figure 24–3, the access network can be divided to include a mobile phone access portion, which includes the end device, such as a phone, and the BTS (base transceiver station). The BTS ties to the BSC (base station controller), where traffic is aggregated and handed off to the metropolitan network. Finally, the signals are fed to an MTSO—also called the MSC—where it is groomed onto the PSTN or Internet.

One significant difference between UMTS and CDMA2000 systems is how they use the frequency spectrum. CDMA2000 jumps or hops across the 450, 800, 1700, 1900, and the 2100 MHz spectrum reducing the chance that it will conflict with existing PCS or satellite systems deployed in North America.

Though UMTS and CDMA2000 standards vary technically in how they package and transport information, both strive for the same end point. Their goal in life is to carry all types of information across the network whether that network is the PSTN or the Internet. Though, initially, the two standards resided in different parts of the globe, several companies—primarily in the United States—have opted to integrate UMTS (GSM) into their footprints. As the 3G standard evolves, where UMTS versus CDMA2000 (Coca-Cola versus Pepsi) is deployed is no longer defined geographically (Europe versus North America). Instead, the choice is being decided by the 3G carrier, not the continent.

■ 24.3 BROADBAND CELLULAR SERVICE PROVIDERS AND APPLICATIONS

Whether a network wireless service provider deploys UMTS or CDMA2000, the end results to the subscriber are higher bandwidths for data transmission; the ability to use multiple devices on the wireless network ranging from notebooks to handheld PDAs;

a variety of applications such as video, text messaging, and music downloads; and hundreds of other potential perks that more bandwidth provides. The applications available to 3G users include location positioning, video conferencing, application downloading, file sharing, video sharing, Internet connectivity, virtual local area networks (VLANs), multimedia messaging, and on and on and on.

Most important, the broadband wireless services offer the consumer the ability to access the network from almost anywhere within the provider's footprint—as Randy experienced as he sat on top of the hill in Vermont downloading files from the Internet.

■ 24.4 WI-FI: WIRELESS FIDELITY

Wi-Fi is a trademark established by the Wi-Fi Alliance, a nonprofit group that is devoted to spreading Wi-Fi technologies throughout the world. Originally, Wi-Fi was referred to as *WLAN* or wireless LAN, falling under the IEEE's *802.11* wireless Ethernet standard. Today, Wi-Fi continues to be defined by the 802.11 wireless standards and may be called WLAN, though the name is less memorable than Wi-Fi. The following section looks at Wi-Fi networks, defines where they are deployed, and explains how they interconnect to the larger world—the Internet.

24.4.1 Wi-Fi Network

Hot spot is the term used to describe an area serviced by a Wi-Fi network. A hot spot is a geographic region within which the user with Wi-Fi access can connect to the Internet. A region may be as small as a person's home network, such as from the computer sitting in the dining room to the computer sitting in the office. A region may also be defined as the area within a coffee shop such as Starbucks or a McDonald's restaurant. Paying for Internet access is handled in a variety of ways. Paying per minute is common in airports or other public locations. Cities or small towns may opt to provide wireless access free of charge at the library or other municipal buildings. College campuses deploy wireless networks throughout the campus, allowing students to access campus, databases and so forth from multiple locations. Wi-Fi is a very versatile technology that gives the user the freedom to connect easily from anywhere within the hot spot.

The distance a signal can travel within a hot spot depends on the strength of the wireless transceiver. Most indoor networks allow distances between 100 and 300 ft. Outdoor distances range as high as 2000 ft. Access points typically support fifteen to twenty users or devices. Most transceivers transmit at 1 W signal strength. The receiver must be located close enough to hear the signal. If the receiver cannot hear the signal, the connection is not established and, consequently, no data are received.

Designing a Wi-Fi network requires more than just distance measurements between the transmitter and receiver. The engineer also needs to take into consideration the surrounding environment and all the interferers, or obstacles, it contains. Other radio signals, noise from electronic devices, and so forth often corrupt the signal prior to it reaching the receiver. Therefore, though Wi-Fi networks are fairly easy to design and implement, it is important to understand that you are using the air as your medium. As such, the "concrete rules" used to design a wire-line network are not so black and white.

The components that make up a Wi-Fi network include the transceiver installed in the end device such as a wireless *PCMCIA* network interface card (NIC) or a *USB* wireless interface. The wireless cards may talk to another wireless card in a peer-to-peer network, as shown in Figure 24–4. It may also talk to a gateway or *access point*, as shown in Figure 24–5. The access point is used to interconnect to the wired network or other wireless networks—network-to-network interface. For example, you may have a wireless router in your home that connects to your PC, your printer, and your laptop. The router, in turn, connects to the DSL modem that provides broadband access to the Internet. An enterprise may set up wireless access throughout its building allowing employees to connect to the LAN from conference rooms, the break room, and so forth. Wireless and wired networks are common in most companies today. A hotel may deploy wireless access to all the rooms; an individual's laptop connects using a special

WLAN
Wireless local area network. The first name used to describe Wi-Fi LAN technologies.

802.11
Ethernet wireless LAN standard. Standard used by Wi-Fi.

hot spot
Area covered by a Wi-Fi network.

PCMCIA
The Personal Computer Memory Card International Association is a standard that defines interface cards for a personal computer.

USB
Universal Serial Bus. Interface port on a computer that provides a connection point to the device. Wireless USB adapters are built to allow Wi-Fi connections.

access point
Device used in a Wi-Fi network that aggregates traffic from multiple devices and interconnects devices with Wi-Fi interfaces such as USB wireless adapters or wireless PCMCIA cards.

Figure 24–4
Peer-to-peer wireless networks are used to connect two devices with Wi-Fi cards.

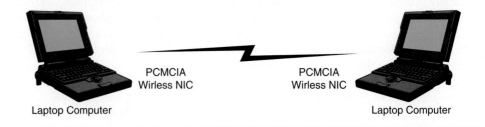

PCMCIA
Wirless NIC

PCMCIA
Wirless NIC

Laptop Computer

Laptop Computer

encryption key to ensure that security is maintained. The hotel places access points throughout the building, aggregating the signals in a communications room where the signals access the outside network through wire line, wireless broadband, or dedicated circuits such as a TI.

In summary, the Wi-Fi network is fairly simple in that it requires wireless transceivers placed in devices such as PCs that either connect to each other or to access points capable of aggregating traffic, managing traffic flows, and providing interconnection points between networks.

24.4.2 Wi-Fi Protocol

The Wi-Fi protocol 802.11 as defined by the IEEE is a wireless transmission standard used to network devices within a small geographic area. There are several variations of the protocol, as noted in Table 24–2. The most commonly deployed today is the

Figure 24–5
Wi-Fi networks allow multiple devices to interface to one access point that aggregates and routes traffic accordingly.

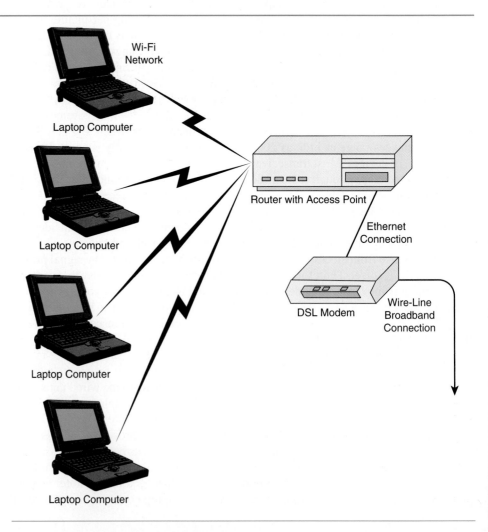

Wi-Fi
Network

Laptop Computer

Laptop Computer

Laptop Computer

Laptop Computer

Router with Access Point

Ethernet
Connection

DSL Modem

Wire-Line
Broadband
Connection

IEEE Standard	Bandwidth	Frequency Spectrum
802.11a	54 Mbps	5 GHz
802.11b	11 Mbps	2.4 GHz
802.11g	54 Mbps	5 GHz
802.11r	100 Mbps	2.4 GHz

Table 24–2
IEEE Wi-Fi Standards.

802.11b/g, which provides 11 Mbps of shared bandwidth across the 2.4 GHz unlicensed spectrum. The 802.11b/g standard functions the same as Ethernet's 802.3 standard that defines the physical layer, how the bits are put on the wire, and Layer 2 functions. The network as defined by the standard is a broadcast network, meaning that all end points receive all data transmitted by the access device. Each device looks at the frame to determine if it should grab it off the wire or ignore it. Carrier Sense Multiple Access with Collision Detection (CSMA/CD) is the method used to handle collisions on the wire—similar to Ethernet. The difference, of course, in 802.11 is that the protocol a wireless defines a wireless interface, not wired.

24.4.3 Security on the Wi-Fi Network

Security is an important topic discussed in length at conferences and user group meetings. If you have ever used Wi-Fi service, for example, at your home, you may have noticed your neighbor's network popping up in your network connections window. If your neighbor does not have wireless security keys enabled on his or her network, you would be able to access the Internet through your neighbor's broadband connection. Several methods are available to reduce the chance of the preceding scenario happening. Three wireless encryption methods are available to the user: *Wired Equivalent Privacy* (*WEP*), *Wi-Fi Protected Access* (*WPA*); and WPA2—Wi-Fi Protected Access 2. The first method used, WEP, requires the user to select an encryption key that is entered on network devices and disallows any traffic from devices not configured with the code. Newer methods expand on this by requiring authentication and more robust encryption schemes. An additional method used to assist in warding off intruders is referred to as MAC filtering. MAC filtering requires the user to enter MAC addresses of devices allowed to enter the network. *Virtual Private Networks* (*VPNs*) are implemented by corporations in order to secure the network. A VPN in the wireless world mimics that in the wired world. A VPN establishes a closed network, typically with private IP addresses, access lists, and so forth to guarantee security. Firewalls are also available to assist in securing the Wi-Fi network. Remote Authentication Dial-In User Service (RADIUS), used to authenticate users by user name and password, may also be part of the security methodology deployed by enterprise networks. Typically, the home user depends on WEP, WPA, WPA2, and MAC filtering. Other methods such as VLAN and RADIUS are deployed by businesses.

Wired Equivalent Privacy (WEP)
Method used to secure Wi-Fi networks through the use of encryption keys.
Wi-Fi Protected Access (WPA)
Protection method used in Wi-Fi networks.

Virtual Private Network (VPN)
A network established between devices using virtual paths across physical pipes. Method used to secure wireless networks.

■ 24.5 WIMAX

WiMAX is the wireless network protocol used to provide coverage across the access and metropolitan networks. The IEEE 802.16 standard defines broadband wireless access across unlicensed microwave spectrum. The term *wireless metropolitan area network* (*WMAN*) is also used to describe this network. WiMAX is predominantly deployed in areas not covered by wire-line media or in areas in which wire-line costs are substantial.

24.5.1 WiMAX Network

The WiMAX network may reach as far as thirty-one miles depending on the geography of the region. Most WiMAX networks are *LOS*—line of sight—meaning that the antennas must have a clear path in order to transmit and receive signals from each other.

wireless metropolitan area network (WMAN)
Network that covers the metropolitan network. Used to describe the area a WiMAX network covers in a region.
LOS
Line of sight. Term used to describe microwave technologies that require a line-of-sight path between the transmitter and the receiver.

Figure 24–6
Wi-Fi networks often tie into WiMAX networks that carry traffic across the metropolitan region.

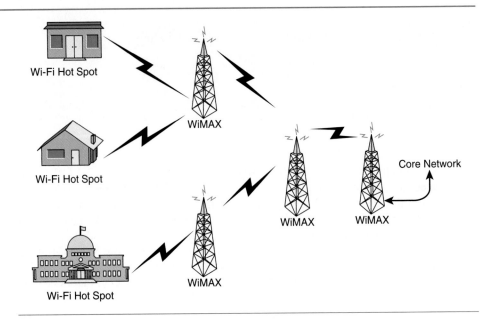

NLOS
Near line of sight. Term used to describe microwave technologies that work with a near-line-of-sight path between the transmitter and the receiver.

Some systems are said to be *NLOS*—near line of sight—meaning that the signal can travel between antennas even though trees or other objects may obstruct the path. Therefore, most providers state the network is LOS and NLOS.

The frequency spectrum used by WiMAX networks ranges from 2 to 11 GHz. NLOS networks work with the 5 to 6 GHz range. Most networks deployed today are point to multipoint, referred to as PTMP networks. Point-to-point networks are also common, particularly in rural areas.

WiMAX networks often are used as aggregation points for Wi-Fi hot spots, as shown in Figure 24–6. A WiMAX network may provide connections for multiple business or residential customers depending on the need. Customers or devices share a 70 Mbps pipe.

24.5.2 WiMAX Protocol

A WiMAX network uses the 802.16 IEEE standard. The standard defines how the traffic is placed on the wire and how Layer 2 connectivity is handled. The significant differences between the Wi-Fi and WiMAX protocol are the way WiMAX handles bandwidth allocation and the distance or coverage area WiMAX provides. As mentioned in the previous section, Wi-Fi is a broadcast CSMA/CD transmission method. Traffic shares the bus, using a collision/contention methodology to ensure that no one device hogs the line. WiMAX uses a scheduling method that requires a device to schedule a time slot for transmission. All devices compete for the time slot prior to transmission. Obviously, larger networks, such as a WiMAX network aggregating multiple end points, need a more robust means of managing traffic on the wire. The ability to pre-assign time slots allows the protocol to avoid congestion situations that result in lost data.

24.5.3 Multiplexing/Modulation Scheme

orthogonal frequency division multiplexing (OFDM)
Multiplexing/modulation technique used in WiMAX technologies.

WiMAX networks deploy *orthogonal frequency division multiplexing (OFDM)* schemes to carry traffic end to end. OFDM uses the discrete multitone (DMT) modulation scheme to impose data onto individual carrier frequencies. OFDM improves network efficiency when compared to time division multiplexing (TDM) technologies deployed in traditional DS-3 microwave systems. OFDM divides the frequency spectrum into multiple tones or frequencies, each of which is used to carry traffic. If, for example, a signal is experiencing interference, say from an overlapping wireless signal, OFDM adjusts the frequencies to compensate for the overlap, that is, it does not send traffic on that frequency. Therefore, system bandwidth relates directly to signal level and interferers or the signal-to-noise ratio of the link.

The frequency spectrum used by WiMAX services ranges from 10 to 66 GHz under 802.16 and from 2 to 11 GHz under 802.16a. Most WiMAX deployments take advantage of the unlicensed frequency space that has been allocated internationally.

■ 24.6 WI-FI AND WIMAX NETWORK SCENARIOS

Commonly, WiMAX networks are used to provide broadband access to small rural towns and cities. Service providers offering wire-line and even broadband 3G cellular networks bypass rural areas due to the high cost of network builds and low return on the investment—not enough subscribers to warrant a network build. Mayors and town boards have opted to "wire" their towns themselves and offer both residential and business customers access to the digital world. Thanks to WiMAX, the cost to deploy such networks is very reasonable.

WiMAX networks are also used to combine Wi-Fi networks together such as in the situation in which a small ISP decides to bypass the local telephone company and offer hot spots and broadband connections using the free air space. As mentioned in the previous section, a Wi-Fi network covers a minimal distance. Combining the two networking methods provides an end-to-end solution across the last mile through the metropolitan network to the POP (point of presence) where traffic is handed off to the Internet.

■ 24.7 BLUETOOTH

The *Bluetooth* protocol is often misrepresented as a Wi-Fi or WiMAX network protocol. Bluetooth, unlike Wi-Fi and WiMAX, was developed by electronics manufacturers wanting to build a standard wireless connection for their devices. Bluetooth is used to create wireless connections between a PC and a keyboard, between a headset and a PDA used as a mobile phone, and between a mouse and a PC. Bluetooth has a coverage area of around 30 ft or 10 m. Unlike infrared technology deployed on remote control systems—television remote and so forth—the Bluetooth standard uses the unlicensed frequency spectrum of 2.45 GHz.

Bluetooth
Wireless protocol developed by equipment manufacturers to create pico networks between devices and ancillary units such as between a PDA and an earpiece.

The standard uses spread spectrum technology to build small wireless network connections referred to as *piconets*—also called *personal area network* (PAN)—between devices or between one device and multiple end points. The networks hop between seventy-nine radio frequencies using a spread spectrum multiplexing scheme similar to CDMA.

A Bluetooth transmitter is very weak, around 1 mW transmit power. Devices connect to each other automatically without needing end user setup or teardown. For example, a wireless mouse tying to a PC establishes a connection by exchanging addresses that tell each device they belong to each other. The connection may be half or full duplex depending on the need. Typically, the piconet is capable of carrying around 1 Mbps with a 20% overhead penalty, reducing the pipe to around 700 kbps. Regardless of the bandwidth available, the main feature Bluetooth offers is ease of use to the consumer. It also eliminates the annoying wires used to connect individual devices to each other.

■ 24.8 CONNECTING THE DOTS

We have discussed how 3G cellular networks have allowed the cellular service providers to enter the broadband space. We have also discussed how Wi-Fi and WiMAX networks are being used to carry traffic across the last mile and metropolitan space. We now need to explain how each either complements or competes with the other. In most situations today, the cellular 3G networks compete with WiMAX for broadband wireless customers. Certain 3G providers have incorporated WiMAX technologies to extend their footprint and offer coverage to areas without 3G services. The two technologies differ primarily in the way they pass traffic between areas, how they handle billing between providers when a customer roams, and the cost of deploying the network.

The most significant difference between the two technologies is Wi-Fi/WiMAX's inability to roam across networks on a global scale. Cellular networks through the years have developed signaling protocols and roaming techniques that allow customers to move freely between service regions without losing service—or at least in most cases not losing service. Wi-Fi/WiMAX providers are currently working diligently on solutions that will allow for mobility between locations globally, not just within their regional footprint. Ironically, it is likely that providers in the two camps—3G and Wi-Fi/WiMAX—will need to work together to create a solution to this problem.

A second issue cited is the difficulty Wi-Fi/WiMAX providers have exchanging billing information between multiple network providers. 3G providers also face this issue if they have not integrated billing transfers between themselves and other 3G providers. The main difference is that the 3G providers have built extensive networks that cover a good portion of metropolitan areas. Again, Wi-Fi/WiMAX providers and 3G providers will more than likely work together to resolve this issue.

A third and final issue often cited by critics is the cost to deploy a full 3G network that covers every area that cellular telephone service covers today. A 3G-capable cell site requires fairly substantial interconnection pipes between it and the MSC. In order to meet the bandwidth needs, costly fiber builds are required. In addition, equipment has to be added to each cell site and new methods for tracking and billing are needed. In the end, the concern many analysts still have is whether a 3G network meets the return on investment standard that most companies demand.

As with all discussions in telecommunications, there are instances in which cellular 3G providers depend on Wi-Fi/WiMAX networks to extend their service footprint. In other instances, there are situations in which Wi-Fi/WiMAX providers will interconnect with 3G providers to offer global connectivity. Therefore, at this time, it is difficult to predict whether one or the other or both will dominate the broadband wireless service provider space.

SUMMARY

Wireless broadband technologies are the newest entrants into the broadband services space. Terms such as 3G, Wi-Fi, and WiMAX are understood by lay people using the services today to carry their information traffic.

The two most common technologies deployed to carry broadband services across the wireless space are 3G and Wi-Fi/WiMAX. The 3G technologies can be further divided into UMTS and CDMA2000. UMTS evolved from the GSM cellular and is supported by the 3GPP user group. CDMA2000 evolved from the CDMA side of the industry and is supported by the 3GPP2 user group. Both camps support broadband wireless services using spread spectrum transport techniques.

Wi-Fi is a wireless Ethernet standard used to connect devices to a network. Wi-Fi is used in homes, businesses, across campuses, and so forth. The Wi-Fi standard, referred to as 802.11, has become extremely popular over the past few years. The term is used frequently by even the least technologically savvy individuals.

WiMAX, similar to Wi-Fi, is a networking technology used to carry information across the access and metropolitan networks. The WiMAX standard, also called WMAN, is a line-of-sight or near-line-of-sight transmission method that uses the unlicensed microwave space to carry traffic. WiMAX service providers are springing up all across the country. Often, municipal governments are building their own WiMAX networks in order to offer broadband services to their constituents.

Bluetooth is a pico network technology that builds connections between a device such as a PDA and an ancillary device such as a Bluetooth earpiece. Bluetooth was developed by electronics manufacturers to assist consumers—by having fewer wires wrapping around the ends of the chair and so forth.

Wireless services that used to carry broadband signals have evolved in the early 2000s to provide an alternative to DSL and cable modem services. The wireless technology today is proving many futurists' prediction of a virtual office space with people working, communicating, and living anywhere in the world.

The Green Grass council has asked you to design a Wi-Fi/WiMAX network that connects all county offices, the library, the high school, the grade school, the three outlying one-room schools, and three of the council members' homes to the Internet using a broadband pipe.

Using the map in Appendix A, draw up the design for the network and note where you would place access points, antenna sites, and so forth. You are free to use any technology you like.

REVIEW QUESTIONS

1. What two protocols dominated the 2G cellular industry?

2. What countries have adopted the GSM cellular protocol as the foundation for their cellular networks?

3. What countries and/or regions have adopted the CDMA protocol as the foundation for their cellular networks?

4. List the five ITU IMT-2000 interfaces and circle the two that dominate the industry today.

5. The 3GPP and the 3GPP2 organizations have adopted the same set of cellular standards.

   a. true   b. false

6. TDMA multiplexing methods used in cellular networks establish time slots between the customer's cell phone or PC and the network equipment. The connections are nailed up for the duration of the session.

   a. true   b. false

7. CDMA is the portion of the cellular protocol, similar to collision detection on an Ethernet link, that watches the link for collisions and congested links.

   a. true   b. false

8. List two acronyms used to describe UMTSs.

9. List the uplink and downlink frequency bands used by UMTSs.

10. The 3GPP alliance divides the UMTS into the _____ and the _____.

11. The USIM Domain and the Mobile Equipment Domain relate to the _____, the _____, and the _____.

12. The term *air interface* relates to the handoff between the BTS and the cell tower.

    a. true   b. false

13. The four antenna sizes defined by the GSM standard are

    a. macro, micro, pico, and nano.
    b. macro, micro, pico, and umbrella.
    c. macro, micro, pico, and domain.
    d. none of the above.

14. The SIM card

    a. allows the user to change phones without having to notify the service provider.

    b. allows users to change service providers and keep their phone number, phone lists, and other phone-related functions.
    c. is used by service providers for administrative purposes such as billing, adds, moves, changes, and other account-related functions.
    d. converts GSM to CDMA.
    e. a, b, and c.
    f. b, c, and d.
    g. all of the above.
    h. none of the above.

15. IS-95 and IS-95B are offshoots of the GSM protocol that was adapted by CDMA 2G advocates.

    a. true   b. false

16. List the CDMA2000 1xEV offerings.

17. CDMA2000 1xEV and CDMA2000 1X require a _____ Hz channel for both upstream and downstream.

18. CDMA2000 1X quadruples the voice capacity of the network.

    a. true   b. false

19. The CDMA2000 1X is divided into the _____ network and the _____.

20. CDMA2000 may conflict with existing PCS or satellite networks in North America.

    a. true   b. false

21. List three applications that are offered over 3G networks.

22. The Wi-Fi network was built around the IEEE 802.3 protocol.

    a. true   b. false

23. When you surf the Web while sitting at Starbucks, you are in an Internet _____.

24. List three components used to establish a Wi-Fi connection.

25. A Wi-Fi network handles security using the following:

    a. WEP
    b. WPA2
    c. MAC filtering
    d. WPA
    e. RTP port lists
    f. a, b, d, and e
    g. a, b, c, and d
    h. all of the above

26. WiMAX networks are built around the IEEE 802.16 protocol.

  a. true   b. false

27. WiMAX has the capability to cover _____ km and offer _____ Mbps of bandwidth.

28. What modulation scheme does WiMAX deploy?

TROUBLESHOOTING

You have received reports from your supervisor that multiple customers have complained that the cellular service in the Oak Hill region is terrible. Customers are complaining of slow upload and download speeds to the Internet. You have been asked to determine where in the network the traffic is experiencing congestion.

Draw up the network from the tower to the MSC and note at which points you plan on monitoring traffic.

KEY TERMS

cellular (668)

PCS (668)

3G (668)

Global System for Mobile Communication (GSM) (668)

3GPP (668)

3GPP2 (668)

Universal Mobile Telecommunications System (UMTS) (669)

CDMA2000 (669)

time division duplexing (TDD) (669)

time division multiple access (TDMA) (669)

frequency division multiple access/time division multiple access (FDMA/TDMA) (669)

User Equipment Domain (670)

Infrastructure Domain (670)

macro (671)

micro (671)

pico (671)

umbrella (671)

base transceiver station (BTS) (671)

base station controller (BSC) (671)

Subscriber Identity Module (SIM) (671)

mobile switching center (MSC) (671)

IS-95A (671)

IS-95B (671)

WLAN (673)

802.11 (673)

hot spot (673)

PCMCIA (673)

USB (673)

access point (673)

Wired Equivalent Privacy (WEP) (675)

Wi-Fi Protected Access (WPA) (675)

Virtual Private Network (VPN) (675)

wireless metropolitan area network (WMAN) (675)

LOS (675)

NLOS (676)

orthogonal frequency division multiplexing (OFDM) (676)

Bluetooth (677)

Telecommunications Applications

Technician working on a main distribution frame in a
telephone office.
Photo courtesy of the Library of Congress

25

Services Offered to Residential Customers by the PSTN

Objectives

After reading this chapter, you should be able to

- ■ Discuss residential telephone services.
- ■ Describe call features.
- ■ Describe residential Internet services.

Outline

Introduction

■ INTRODUCTION

In years past, the telephone was just a telephone. Subscribers called across the street, across town, or across the country. Today, the calling features offered to residential telephone subscribers have changed all that. The network has turned the simple telephone set into a sophisticated communications device. From your home, you may forward your calls, interrupt a call when a call-waiting tone comes in, see the number and often the name of an incoming caller, have a call conference with friends, and much more. Residential telephone subscribers can dial their neighbor across the street or their pen pal living halfway around the world. Through direct distance dialing for long distance and international calling, the global telephone network provides connectivity around the world for any callers with a telephone line connected to their local telephone company.

Chapter 25 describes the services available to the residential telephone subscriber and the most popular call features. It also discusses the global calling network and how calls can travel around the world.

■ 25.1 THE GLOBAL CALLING NETWORK

The global calling network refers to the connections between continents, countries, cities, and towns around the world. The structure of the telecommunications network has made it possible for the residential customer to pick up the phone and dial anyone in the world. This interconnection of networks between countries has been made possible by international direct dialing (IDD) and by standards organizations, such as the ITU, that define how networks connect.

Seven-digit dialing is used to call within a local calling area; *ten-digit dialing* is used to call outside a local calling area, and (at least) sixteen-digit dialing is used for international calls. If Louise decides to call her friend Erma who lives two streets away, she picks up her telephone and dials Erma's telephone number—454-6789. After speaking with Erma, she decides to call Hal who lives three states away. Louise picks up her phone and dials 716-555-3456. In a very short time, she hears the phone ring. After Hal hangs up, Louise decides to call her good friend Karen living in Madrid. She picks up the phone, dials 011-34-1-678932, and waits about 45 s while the call travels between her local phone company and Karen's local phone company. The call leaves Louise's local phone company, travels through her long distance phone company, and connects into an international carrier that then connects to Spain's Public Telephone and Telegraph (PTT) in Madrid. If the country had not upgraded its network to handle IDD, Louise would have had to dial 01 and ask the international operator to connect her to her friend Karen's telephone in Madrid. The time it takes to build a connection to a distant country depends on several variables, such as the international carrier, the country's local telephone network, and other factors. Thanks to the expansive fiber optic cables spanning the oceans, international calling has become much more precise, reliable, and less expensive.

seven-digit dialing
A seven-digit number used to route a call.

ten-digit dialing
A ten-digit number used to route a call.

■ 25.2 SUBSCRIBER CALLING FEATURES

25.2.1 Local Telephone Company Subscriber Features

The number of subscriber features has increased dramatically during the past ten years. The introduction of the digital switch allowed telephone companies to increase their revenues by offering special calling features to their customers. The number and types of features offered are fairly consistent between local telephone companies. The prices charged for each feature are approved by the state's PSU or PUC. Following are the most common features offered by local telephone companies.

Call Waiting

When a second person calls during a phone conversation, the switch sends a *call-waiting* signal to notify you that another caller is trying to reach you. You may hear a tone, see a light on the telephone, or see a message scroll past on your telephone LCD screen. At this point, you may wish to place the person you are talking to on hold and pick up the incoming caller. A phone with a hold button is not necessary. All you need to do is flash, or click, the switch hook and the central office switch will recognize your wish to answer the second incoming call. The switch places the first caller in que while you speak with the second caller. When you finish speaking with the second caller, flashing the switch hook again notifies the switch that you wish to disconnect from the second caller and talk to the first caller. The first caller then comes back on line and you continue your conversation. You may also bounce between both callers by flashing the switch hook; or you might want to flash back to the first caller, finish your conversation, say good-bye, and flash back to the second caller.

call waiting
Feature offered by the switch that notifies the user that a second call is waiting.

The telephone subscriber does not have to accept the incoming call. The tone will sound two or three times before the switch stops sending the call-waiting beep. The second caller will hear ringing and assume that you are not there or that you are not willing to interrupt your first call. The switch can also forward calls to your voice mailbox or return a busy signal. Depending on the telephone company, you may be required to pay an extra fee for directing unanswered calls to a voice mailbox. The caller's name and number will show up on your caller ID if you subscribe to the caller ID feature with name. Most phone companies vary the tone depending on whether the call is coming in from outside the calling area or whether the call is from a cellular telephone. However, some phone companies do not distinguish between the two. The switch can be programmed differently according to the desires of the telephone company.

A drawback to call waiting occurs when you dial into the Internet or other dial-up data connection using a modem. Modems are very sensitive to the incoming call-waiting beep; and, in fact, the beep causes the modem to disconnect the data session. The simplest way to solve this problem is to turn off call waiting by dialing *70 before you dial the access number. This can be done by placing *70 in your dial-up number on your PC dial screen. When the number is dialed, you will hear dial tone, the three-digit tones representing *70, then two short bursts of dial tone to signify that call waiting was disabled. When you hang up, or disconnect, call waiting is automatically enabled for the next call.

Distinctive Ringing

Distinctive ringing is a feature that is popular with parents of teenagers. Residential telephone subscribers may have as many as four telephone numbers assigned to their telephone line. Each number has its own ringing pattern. For instance, a family with two teenage daughters may wish to have three telephone numbers, each with its own ringing pattern. When the older daughter's telephone number is dialed, the phone rings with two short bursts. The rest of the family knows that the call is for her and they do not bother to pick it up.

Special adapters can also be placed on the line to route the different ringing patterns to different telephones. Many residential customers are willing to pay for distinctive ringing to lessen the times they have to jump up to answer the phone.

Call Forwarding

Call forwarding is one of the most popular features offered by telephone companies. They have made millions of dollars on this feature alone. The call-forwarding feature allows the subscriber to forward all incoming calls to a different telephone number. For example, if you were going to spend the evening at a neighbor's house but did not want to miss an important call, you could forward all your calls to the neighbor's telephone number. The method used to set up call forwarding on a telephone in the United States is as follows:

1. Determine which number you wish to forward the calls to.
2. Dial *72 on a touchtone phone or 1172 on a rotary dial telephone or your local access code if it is different.
3. When you hear a second dial tone, enter the telephone number you want all your calls to be forwarded to.
4. Someone must answer the telephone of the number you dialed for the calls to route to that number.
5. If the telephone is busy or if no one answers, you will need to hang up, wait two minutes, then dial the access code and telephone number again. You will receive a confirmation tone telling you that the feature is activated.
6. To deactivate call forwarding, punch in *73 on a touchtone phone or 1173 on a rotary dial telephone or your local code, if different.

Call forwarding can be used several ways, including call forwarding on busy, call forwarding unanswered, and call forwarding all calls. Call forwarding on busy and unanswered is used with voice-mail systems to forward calls to your voice mailbox when the phone is in use and when no one answers the incoming call.

A final type of call forwarding is call forwarding remote access, which is used to activate call forwarding from a remote site. Business customers use this service more frequently than residential subscribers.

Three-Way Calling

Three-way calling is another very popular feature offered by most telephone companies. It allows telephone subscribers to add a third person to the conversation. For instance, if Mary Jane and Annie wanted to conference in their friend, Bridgett, to discuss their homework assignment, Mary Jane would need to flash the switch hook, enter *33

---

distinctive ringing
Feature offered by the switch that provides distinctive ringing on the same telephone line.

call forwarding
Feature offered by the switch that forwards calls to a predefined alternative number.

three-way calling
Feature offered by the switch that allows three callers to communicate at one time.

(or some other code), wait for a confirmation tone, and enter Bridgett's telephone number. Once Bridgett's phone starts to ring, Mary Jane can again flash the switch hook and bring Annie back on line. If Bridgett does not answer, Mary Jane can flash the switch hook again and disconnect the call to Bridgett. If Bridgett does answer, the three girls can carry on a conversation until one of them hangs up. However, if Mary Jane hangs up, the connection between Bridgett and Annie is also disconnected, because she initiated the conference.

Most telephone companies allow conference calling to be set up either on a per-call basis or as an always-on feature. The telephone subscriber has the choice of purchasing the feature as a monthly recurring charge or a per-call charge. The following steps illustrate how a conference call is set up when the feature is always on the line:

1. Dial the telephone number of one of the people you wish to call.
2. When the call is connected, flash the switch hook or press the conference button on your telephone or the call-waiting button on the caller ID box.
3. When you hear dial tone, dial the third party's telephone number.
4. When the third party picks up the phone, flash the switch hook, press the conference button on the phone, or press the call-waiting button on the caller ID box.
5. All three parties are now conferenced together.
6. To disconnect the second caller from the conference, the flash hook, the conference button on the phone, or the call-waiting button on the caller ID box must be pressed to disconnect the second caller.

Subscribers who do not have the conference call feature turned on must use the per-call activation method to conference a third party on the line. The method used is as follows:

1. Before dialing the first participant's telephone number, you must dial *33 or your local code to activate the feature. Then follow the steps already listed to make the three-way call.
2. When the call is complete and the phone is hung up, the conference call feature is deactivated in the switch on your line. To place a second conference call, you will need to activate the feature again by pressing the code on the touchtone pad.

Call Transfer

Call transfer is very similar to three-way conference calling. Pressing the flash hook, then dialing the telephone number of the person you wish to transfer the call to activates the feature. The main difference between the two features is that the originator of the third-party call may hang up the phone once the first and second callers are connected. Most phone companies call this *enhanced* three-way calling. Call transfer is also used to transfer callers to a voice mailbox. The steps used to initiate a call transfer are as follows:

1. Dial the telephone number of one of the people you wish to call.
2. When the call is connected, flash the switch hook, press the conference button on your telephone, or press the call-waiting button on the caller ID box.
3. When you hear dial tone, dial the third party's telephone number.
4. When the third party picks up the phone, flash the switch hook, press the conference button on the phone, or press the call-waiting button on the caller ID box.
5. As soon as you hear the third party's telephone ring, you may hang up the telephone. The transfer is complete.

Speed Dial

Speed dial is a very popular feature that allows customers to dial a telephone number using two digits instead of seven. The advantage of speed dial is obvious—it takes less time to dial a friend, family member, or frequently called location. A two-digit number is easier to remember. The feature is packaged differently by the various telephone companies, but the most common method is to sell a specific number of speed dial numbers for a set price. For example, a telephone company may have an eight-number package, a fifteen-number package, and a thirty-number package. If you only have eight numbers you call frequently, you can subscribe to the eight-number package, meaning you can program a two-digit number to replace the seven-digit telephone numbers for those eight locations. Emergency numbers are often programmed as two-digit speed dial numbers. The steps needed to program the speed dial numbers are as follows:

1. Dial the setup code for the number of speed codes you have purchased.
2. When you hear the second dial tone, dial the speed code number you want to use to represent the seven-digit number and then the telephone number.
3. A confirmation tone will come from the switch.
4. To use the service, you will dial the speed code followed by a #.

Call Blocking

Call blocking is enabled when the telephone subscriber wishes to block certain types of calls, such as all long distance or 900 number calls, from being completed from that telephone line. For example, if 900 number calls are blocked and a resident dials a 900 number, an intercept recording tells the customer that the call is not valid and will not be completed. The switch translations department at the telephone company has placed call blocking on that line for all 900 numbers. Subscribers must request that the numbers be blocked, and telephone companies do not charge to block 900 numbers.

Voice Mail

Voice mail is a feature that provides a value-added service to the telephone subscriber from the telephone company. The term *value-added service* is used by all telephone companies to define additional calling features that they hope differentiate them from their competitors. The voice-mail platform is attached to the central office switch. Calls are forwarded to the customer's voice mailbox when the called party does not answer or when the called party's telephone is busy. The voice-mail system provides all of the features of a telephone answering machine, but the customer does not have to change tapes in the machine or maintain the answering machine; the telephone company maintains the voice mailbox. The voice mailbox is also easy to access from a remote location. A telephone access number is dialed and a pass code entered allowing the subscriber to listen to the messages from anywhere in the world. Newer answering machines also allow customers to access their messages remotely, though not as easily, as with voice-mail systems.

Voice-mail systems allow customers to configure the mailbox from their telephone. Subscribers may customize the recording with a personal greeting that can be updated or changed at any time. A caller's message may be saved for later reference, deleted, or forwarded to someone else's mailbox. When the called party is on the telephone or not at home, the calls coming in are forwarded to the subscriber's voice mailbox. Customers who subscribe to voice mail must also purchase the call-forwarding feature to be able to route calls to their voice mailbox, but most telephone companies bundle the cost of both services together to help sell voice mail to the frugal subscriber. The popularity of the answering machine pushed the telephone companies into building voice-mail services. Today, voice mail is a common and very profitable feature for telephone companies.

25.2.2 Long Distance Telephone Company Features

Long distance telephone companies also offer residential calling features, though fewer than those offered by the local telephone company. Long distance companies are mainly concerned that the residential subscriber use their long distance network when placing long distance calls, because its revenue is dependent on the minutes of usage charged to the subscriber. The two most common residential features or value-added services are the personal 800 number and calling cards.

Personal 800 Number

The *personal 800 number* service has become a very popular long distance residential offering in the past few years. The most common example is that of the college student who frequently calls home for advice, financial help, and support. The student may call on the 800 number and not incur long distance charges because the parent pays for the incoming call. The charge per minute for the incoming call is less than that charged for a normal outgoing long distance call; however, the subscriber does pay a monthly fee for the 800 number service. The main advantage is peace of mind for the parent because the student can call anytime without worrying about having enough change at the pay phone, remembering a calling card number, or placing a high-priced collect call.

personal 800 number
Subscriber 800 number used to reduce the cost of incoming LD calls.

Calling Card

Calling card service is offered by most long distance carriers to residential telephone subscribers. A calling card is used when the subscriber is away from home and wishes to make a call. Each subscriber is given an 800 access number that connects him or her to the long distance calling card platform. Callers enter a unique calling card number followed by the number they wish to call. The long distance telephone company then connects the caller to the called party number. The customer is billed at the rate used for calling card calls by the long distance carrier. Calling card costs are normally more expensive than direct distance dialed calls but less than collect calls.

calling card
Card with a special pin used for long distance calling.

Customers do not have to use the same long distance company for their calling card service as for their long distance provider. Rates for calling card service vary among the different companies. There are now companies that only offer calling card services. The prepaid calling card industry has boomed in the past few years. Each prepaid card has a specific number of minutes assigned to it and is only good for that number of minutes.

25.2.3 Class Features

Local telephone companies offer *customized local area signaling service* (CLASS) features as defined by Bellcore. CLASS features are dependent upon the SS-7 network and are only offered if the switch is SS-7 capable. CLASS features can travel between service provider networks as long as interconnection agreements have been initiated between the two companies. In other words, the two companies must agree to accept and carry the other company's signaling messages. Numerous CLASS features are now offered by most telephone companies, and more features are being developed by switch manufacturers at the request of service providers. Service providers are very interested in adding calling features because features bring revenue.

customized local area signaling service
Features dependent upon the SS-7 network and offered only if the switch is SS-7 capable.

Caller ID

Caller ID allows the incoming caller's telephone number to appear on a caller ID screen. The subscriber has to purchase the feature and place a caller ID box on the end of the telephone line. Many new telephones have caller ID screens built into the telephone. *Calling line identification* (CLID) is the term used to describe the caller ID feature. Calls coming in from outside the local area sometimes show the incoming telephone number on the caller ID screen, but sometimes the screen shows "out of area." Interconnection agreements between telephone companies allow for calling numbers to pass between their networks, but some companies are not capable of passing caller ID information outside their network. The goal in the industry is to have all companies' calling information pass between networks.

caller ID
Feature offered by the switch that delivers the calling party's number to the called party.

The calling number restrict option allows the subscriber the choice of blocking the transport of his or her telephone number to the person he or she is calling. The second option provides a per-call restrict feature that requires the caller to enter a special code before dialing the number. The code tells the switch to block the transport of the calling party's telephone number for that call only. The telephone company does not charge for the permanent call-block feature but may charge for the per-call block feature.

Caller Name

caller name
Feature offered by the switch that delivers the name of the calling party to the called party.

Caller name (CNAME) is similar to caller ID in that it works with a caller ID box or special feature phone. If you subscribe to caller ID and caller name, you will see both the telephone number and the caller's name on the caller ID screen. For the caller name to appear, the SS-7 signal must carry the caller name and address through the network. When a call comes in from someone using a different service provider, the caller name feature will only work if both service providers have agreed to pass on the information. In order to pass the information between their networks, both must place specific network information in their SS-7 network's translations. If the two companies have not established interconnection agreements allowing CNAME information to pass, the subscriber will receive the calling number without caller name and address information. Caller name has also caused controversy between telephone companies and consumer groups. The consumer groups believe that passing the name with the telephone number is an invasion of privacy and also provides too much information to the called party. Call centers like the information because they are able to capture the incoming information, place it in a database, and use it as a calling list for telemarketers.

Call Restriction

call restriction
Feature offered by the switch that allows the called party to reject incoming calls based on the telephone number.

Call restriction is a free feature given to all telephone subscribers. When call restriction is turned on, the caller's telephone number and name are not transmitted to the called party's telephone. Call restriction can be turned on as a per-call or always-on feature. Telephone companies must offer call restriction if they sell caller ID and caller name. The method used to activate and deactivate call restrict follows:

1. To activate call restrict, dial *77 or your local code, if different. Callers will hear an announcement stating that the subscriber does not accept anonymous calls.
2. To deactivate, dial *87 or your local code.
3. The caller may deactivate his or her caller ID block by dialing *82 or your local code.

Call Return

call return
Feature offered by the switch that allows the called party to ring back the last caller by dialing *69 or another local code.

Call return allows subscribers to ring back the last caller by dialing *69 or another local code. Once the subscriber dials the code, the switch determines who the last caller to that subscriber's telephone line was and automatically places a call using that caller's telephone number. Thanks to SS-7, the switch has access to both the called party's number and the calling party's number. In most cases, calls will not be returned between telephone companies. For the feature to work between networks, the telephone companies must be SS-7 capable and have interconnection agreements providing translation information that can be placed in the switching and signaling devices. Call return is currently one of the most popular features offered by the local telephone company. The telephone company charges the customer every time he or she uses the call-return feature. The method used to activate call return is outlined here:

1. Dial *69 on a touchtone phone or 1169 on a rotary dial phone or your local code, if different.
2. The switch will provide the number of the last caller and the date and time of the last call.
3. By pressing 1, you will automatically reach the last number.
4. If the telephone is busy, you may hang up and wait until you hear three short rings that indicate the line is free.

Call Trace

call trace
Feature offered by the switch that allows the called party to initiate a trace on an incoming call.

Call trace also depends on the signaling network to work. It is used in conjunction with the telephone company to find out who has called. When the call-trace feature is activated, the subscriber is given the date and time of activation and is then required to contact the security bureau at the telephone company and notify the bureau of the trace. The telephone company is able to look in the switch and determine where the call came from. The phone company charges the subscriber for each call trace. Call trace is activated as follows:

1. Dial *64 on a touchtone phone, 1164 on a rotary phone, or your local access code, if different.
2. You will receive a message telling you whether the call was within the calling area or outside the calling area.
3. Contact the phone company's security department and have the department complete the trace.

Ring Again

The *ring again* feature is useful when calling a number that is busy. When the caller gets a busy signal, he or she activates the feature and hangs up the phone. The telephone switches watch the line and when the subscriber hangs up they send a ring-back message to the caller who activated the feature. The caller hears the tone and knows the person's line is now free. When the caller picks up the phone, the call automatically goes through to the called party's telephone, and his or her telephone rings. The following outlines how the feature works:

1. When you hear the busy signal, hang up the phone.
2. Dial *66 on a touchtone phone, 1166 on a rotary phone, or your local code, if different.
3. If the phone is still busy, you will hear a confirmation tone.
4. When the called party hangs up, the switch notifies you with three short rings.
5. When you pick up the phone, the call is completed and the called party's phone rings.

ring again
Feature offered by the switch that alerts the caller that the called party is free.

Selective Call Acceptance

Selective call acceptance is a feature that allows subscribers to enter a list of numbers that they will accept. All other numbers are intercepted and an announcement is played stating that the subscriber does not accept the call. The subscriber's telephone rings only when a calling number matches one of the numbers in the selective number table in the switch. Selective call acceptance is not as popular as other features but does allow subscribers some control over whom they will receive calls from.

selective call acceptance
Feature offered by the switch that allows subscribers to enter a list of numbers that they will accept.

Selective Call Forwarding

Selective call forwarding is a feature that defines where particular incoming calls should be directed. For example, the number 567-8900 may be selectively forwarded to voice mail. The subscriber can use this feature to direct certain calls to specific telephones, voice mailboxes, or attendants.

selective call forwarding
Feature offered by the switch that defines where particular incoming calls should be forwarded.

Selective Call Rejection

Selective call rejection allows customers to reject some incoming telephone numbers with an announcement telling the caller that his or her call is not being accepted. This is the opposite of selective call acceptance. Selective call rejection allows the subscriber to restrict incoming calls on a per-number basis. For example, if a vendor continues to call even though you have asked him not to, you can place his telephone number in the selective call rejection table in the switch and his call will never reach your telephone. He will hear a recording denying him access to your telephone.

selective call rejection
Feature offered by the switch that allows customers to reject some incoming telephone numbers with an announcement telling the caller that his or her call is not being accepted.

■ 25.3 RESIDENTIAL INTERNET SERVICE

25.3.1 Dial-Up Internet Access

Residential subscribers can use their telephone lines, a modem, and a personal computer to dial into the Internet. The Internet browser loaded on the subscriber's PC is accessed by double-clicking the icon. A screen pops up asking for the access number that the subscriber wishes to call. The access number is the number that rings to the customer's ISP, which has banks of modems that are shared by all of its subscribers. Once the log-in button on the pop-up screen is clicked, the modem goes off-hook, dial tone comes on line, and the digits are dialed. The modems on each end perform their handshake and are able to communicate. The ISP responds to the subscriber's request with its home page.

If the access number is busy, the subscriber may select a different access number to connect to the ISP. The speed at which the subscriber connects varies depending on modem speed. A typical page from the Internet has more than 79,000 bits of information. Therefore, the faster the modem, the faster the subscriber's PC screen will paint (fill the screen). A typical modem is 28,000, 33,000, or 56,000 bps. The second factor that affects the connection speed is the quality of the phone line. The distance between the subscriber and the central office and the condition of the copper wire connecting them dictate the speed at which the circuit can sync up. The final factors that determine how many bits per second will travel between the customer and the ISP are the gauge of the cable, the number of bridge taps, the quality of the terminations, and the quality of the splices. Each of these variables contributes to the signal attenuation. The greater the signal attenuation, the slower the connection.

Beyond the customer's telephone line and connection into the ISP, the speed at which the Internet responds to the subscriber's request depends on the backbone between the ISP and the Internet. Figure 25–1 illustrates a typical Internet connection from the residential telephone subscriber to the Internet. *Oversubscription* is the ratio at which the ISP concentrates customers on the backbone. A common oversubscription ratio for residential subscribers is as high as 100 to 1 meaning 100 subscribers for every T1 circuit. The oversubscription ratio varies depending upon the ISP.

The number of ISPs in the market grows daily. Subscribers must sift through the many ISP offerings to determine which fits their needs. Telephone companies are often ISPs, and they may bundle the service in with the other services.

25.3.2 Residential Digital Subscriber Line Services
The newest data service offering to residential customers is digital subscriber line (DSL) service that allows customers to connect into the Internet at very high speeds.

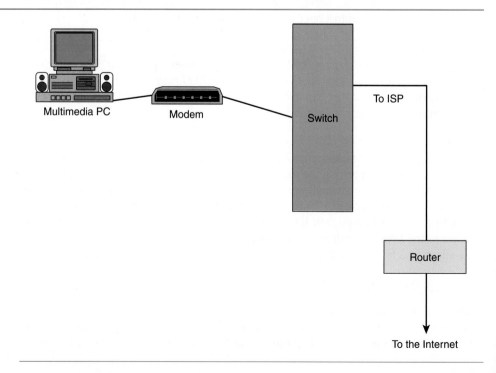

Figure 25–1
A typical dial-up modem circuit architecture. The user dials the telephone number provided by the ISP. The switch then routes traffic to the ISP where the router aggregates all of the traffic and sends it out across the backbone network to the Internet.

DSL is a new technology that runs on regular copper pairs that connect the subscriber to a DSLAM in the central office. Asymmetrical digital subscriber line (ADSL) is the most common type of DSL sold to the residential subscriber. The ADSL provides a 6 Mbps speed downstream from the Internet to the subscriber, and a 256 kbps speed upstream from the subscriber to the Internet.

DSL does not require the subscriber to dial into the Internet. The DSL modem is instead connected to the end user's PC with an Ethernet cable. Figure 25–2 shows a typical DSL deployment at a residential customer's home. DSL modems are always on, unlike dial-up circuits that require a connection to be made before connecting to the Internet. DSL is one of the fastest growing broadband services for Internet connectivity. The cable modem is its only competitor.

A residential customer purchases a DSL from his or her ISP, ILEC, or CLEC/ICP (integrated communications provider). The ILEC or CLEC installs the DSL on the customer premises. The circuit consists of one unloaded copper pair. It may be an old telephone line, as long as there are no load coils on the line and the distance does not exceed the signal strength. The amount of bandwidth available to the customer is dependent upon the distance the cable runs, the gauge of the cable, the number of bridged taps, and the quality of the termination at the customer site. DSL provides a high-speed link into the Internet for the residential customer and is one of the fastest growing telecommunications technologies in the residential market space.

25.3.3 Broadband Services

CATV and wireless service providers offer residential broadband access to the Internet. Cable modem dominates the residential broadband space offering customers asymmetrical high-speed Internet access through a device incorporated into the cable box, as described in Chapter 21. In addition to Internet access, CATV is offering a bundled product that includes Internet access, cable television, and voice services.

Cellular providers also have a broadband Internet residential access product offering. Cellular service providers provide residential customers with a broadband service that can be used from any location served by their network. For example, if I subscribe to broadband through my cellular provider, I can receive high-speed access from my home in Florida and from my friend's home in Wisconsin—if the carrier has built network in both locations.

The final access providers offering broadband services are the Wi-Fi service providers that have built hot spots around an area to provide last-mile service to the residential customer. Often, local government have opted to build wireless access networks in their areas in order to "wire" the town and provide high-speed Internet access

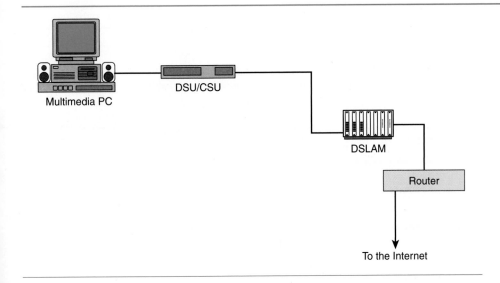

Figure 25–2
Comparing this figure to Figure 25–1 shows the difference between a dial-up data scenario and an always-on data circuit using DSL. The DSL does not interface into the switch, and the user does not have to initiate a session by dialing a number. The session is always on and is routed directly into a router, then to the backbone network and out to the Internet.

to the residents. Typically, small rural cities and towns that have a hard time persuading the larger service providers to build out network in their areas opt for government owned and run Wi-Fi networks.

The four broadband services providers—DSL, cable modem, cellular wireless, and Wi-Fi—offer enhanced services to residential customers, the most common being Web space, e-mail services with additional e-mail address options, firewall services, and file backup. Web space allows the customer to place his or her Web site on the provider's Web server. E-mail is one of the most popular services offered by the providers. Most providers offer the customer several e-mail accounts and give the customer the ability to purchase additional e-mail addresses. Firewall services benefit both the customer and the service provider. Often, the feature is given to the customer with no charge. Protecting the customer also protects the service provider's network from attack from within. The final offering, file backup, refers to giving the customer the ability to backup files on a server sitting in a secure server farm. The residential consumer may opt to set up an automatic backup of his or her files to the service provider's file backup system. Corrupted files and system crashes and be restored easily through the backup system.

■ 25.4 APPLICATION SERVICE PROVIDERS

New application service providers (ASPs) have begun to offer special applications such as gaming, special user groups, and other peer-to-peer types of services to residential subscribers. The user pays a monthly fee to gain access to the application server, which stores games and functions as a gateway to other players. Highly sophisticated games allow thousands of players to interact in a make-believe kingdom. The game continues indefinitely as players vie for position and power.

Analysts predict that gaming and peer-to-peer resource sharing will boom over the next few years as people gain interest and as the software becomes more realistic and usable. Presently, though, controversy over sharing copywritten information may stifle the growth of shared resources. The technology is ready to handle all of the applications, yet the legal climate may thwart its growth.

SUMMARY

Chapter 25 focused on the services offered to residential telephone subscribers. The importance of the global telephone network was addressed, and residential call features were described in detail. The CLASS features were defined and the importance of interconnection agreements between carriers was explained. The chapter detailed the two types of Internet connectivity available to the residential subscriber—dial-up Internet and digital subscriber line (DSL).

CASE STUDY

1. Develop a residential product offering that includes special features bundled into specific packages.
2. Draw the residential POTS line from the customer location to the switch. Label each component in the network.

REVIEW QUESTIONS

1. Define the term *residential subscriber* as it pertains to the telecommunications industry.

2. Define the term *residential POTS line* as it pertains to the telecommunications industry.

3. List and explain five call features commonly offered by local telephone companies to residential telephone subscribers.

4. What communications equipment can be found at a residential telephone subscriber's home?

5. Define *rate center* as it pertains to local telephone service.

6. Who is responsible for determining the rate center boundaries for a calling region?

7. Explain the difference between the *rate center* and the *wire center* as they pertain to the local telephone serving area.

8. Explain what service is offered over an ISDN BRI circuit to residential telephone subscribers.

9. How many voice channels are available on one ISDN BRI circuit?

10. Can an ISDN line carry data information and voice information simultaneously? If yes, what bit-per-second rate is available for the data transmission?

11. When commercial power is lost at the resident's home, the ISDN line continues to function. True or false? Explain.

12. What is the difference between a *dial-up data line* and a *voice grade POTS line?*

13. A dial-up data line's bit-per-second rate is directly related to _____ and _____ of the copper loop.

14. Define two CLASS features and explain the difference between a CLASS feature and a typical call feature.

15. Why does a voice mailbox require that the customer have call forwarding optioned on his or her line along with voice-mail service?

16. How is call waiting disabled? Why should you disable call waiting before dialing into the Internet?

TROUBLESHOOTING

Problem 1: Mrs. Able has called into the repair bureau and is complaining that there is static on the line. List the tests you would perform to determine the cause of the static. List all of the possible causes of static on the line.

Problem 2: Mr. Jenkins is complaining that he hears his neighbor's voice every time he talks on his phone. List the tests you would per-

form to determine the cause of the voice on Mr. Jenkin's line. List all of the possible causes of phantom voices on the line.

Problem 3: Mrs. Avery has asked for a second line into her home. Explain the process that has to be followed from the time the order is taken to the time the second line is turned up.

KEY TERMS

seven-digit dialing (683)

ten-digit dialing (683)

call waiting (683)

distinctive ringing (684)

call forwarding (684)

three-way calling (684)

personal 800 number (687)

calling card (687)

customized local area
  signaling service (687)

caller ID (687)

caller name (688)

call restriction (688)

call return (688)

call trace (688)

ring again (689)

selective call acceptance (689)

selective call forwarding (689)

selective call rejection (689)

26

Business Services—Voice and Data

Objectives

After reading this chapter, you should be able to

- Describe switched business services.
- Describe dedicated business services.

Outline

Introduction

26.1 Switched Voice Services

26.2 Dedicated Voice Services

26.3 Switched Data Services

26.4 Dedicated Data Services

26.5 High-Speed Digital Connections

■ INTRODUCTION

Business customers' communications needs continue to change and grow as the world becomes a boundaryless information exchange. The purpose of Chapter 26 is to discuss the various services offered by telephone companies to business customers. The chapter is divided into two general areas—switched voice services and dedicated services.

   Business customers' networks can be as complicated as telephone companies' networks. Multiple technologies that handle the different types of traffic are essential. Telephone companies have subsidized the residential telephone user by charging higher rates to business clients, but the introduction of CLECs to the local market is lowering the costs to the business customer. They will continue to decrease at the same time that new services are introduced. Understanding the many services and technologies used by business subscribers is essential for the telecommunications student because the industry is built around providing advanced services to business customers.

■ 26.1 SWITCHED VOICE SERVICES

26.1.1 Local Measured Service

Though many telephone companies do not charge residential customers a per-minute charge for telephone usage, they do charge business customers a per-minute usage charge for all local calls. The charge varies depending upon the tariff filed by the telephone company, but it ranges between 4 cents and 8 cents per minute depending on the region and service provider.

Local measured service (LMS) is charged on a per local call basis. For example, if the owner of a florist shop calls his accountant who lives two blocks away, he pays 4 cents for every minute they stay on the telephone. But, if the accountant calls the florist, the florist is not charged for that call. Therefore, LMS is only charged on outgoing calls. Currently, the competition between ILECs and CLECs is causing LMS charges to drop, much like the price of long distance dropped after the 1982 Modified Final Judgment.

In certain instances, the telephone company charges *local measured unit* (LMU) instead of LMS. LMU refers to charging per call instead of per minute. The LMU method of charging is less popular than LMS. One of the highest costs to do business for the business customer is the telephone cost charged for each call. One of the main drivers of being able to call across the Internet is to reduce or completely eliminate LMS charges for local calls.

26.1.2 IntraLATA Voice Services

IntraLATA voice services consist of charges billed on calls that are rated as long distance within the LATA. The *rate center* defines what a local call is and what a long distance call is within the LATA. Business customers select which communications company will carry their intraLATA voice calls, just like they choose who will carry their long distance traffic. The ILEC, CLEC, or long distance carrier offers intraLATA long distance service. In most cases, the long distance carriers or CLECs pay the ILEC to carry the intraLATA long distance call on their network but brand the bill with their own logo. The long distance carriers and CLECs are able to bundle long distance interLATA service with intraLATA long distance service and thus compete with the ILEC.

A rate center is a local calling area defined by a geographic area surrounding one or many central offices. ILECs decide the geographic boundaries for the rate center. These boundaries determine what calls within a LATA are considered intraLATA toll and which are nontoll calls. Figure 26–1 shows a region with four rate centers. PUCs and telephone companies sometimes send out questionnaires to subscribers asking which areas they would like to be able to call toll free. The PUC reviews the ILEC's request for rate center changes and agrees or disagrees with the proposed rate center boundaries. CLECs must now determine whether they mirror the ILEC's rate centers. Most CLECs elect to copy the rate center boundaries established by the ILEC due to billing issues between the two carriers.

26.1.3 InterLATA Voice Services

Long distance companies and CLECs offer *interLATA voice services*. Business customers must review call plans from the carriers to determine who offers the lowest cost-per-minute price. Comparing the many long distance packages becomes confusing because one package may offer a lower rate for calls on the East Coast while other packages offer lower rates on calls after 9 p.m. but before 6 a.m. The long distance carriers continuously reduce their rates, bundle services such as 800 number or local calling into one package, or eliminate charges in order to gain new customers. Long distance service is now a commodity, not a fixed service as it was in years past.

Business customers should perform "traffic studies" before choosing a long distance package. The number of calls, the busiest time of day, the calling patterns, where the calls are going, and expected future call volumes should be analyzed before selecting a long distance carrier and a long distance calling package.

In the late 1980s, long distance and local phone companies offered *wide area telephone service* (WATS) to help reduce the cost of long distance services for business customers. WATS refers to established bands of geographic calling areas where customers may call for a reduced rate. For instance, if a florist frequently needs to call a distributor who lives in Chicago, the florist might purchase WATS for that area. Instead of

local measured service
Charging a per-minute cost for telephone service.

local measured unit
Charging per call instead of per minute.

intraLATA voice services
Calls traveling within the LATA and charged as long distance within the LATA.

rate center
The LATA is divided into regions called rate centers. A rate center is defined by the telephone company and approved by the public utilities commission.

interLATA voice services
Calls traveling between LATAs and charged as interLATA calls.

wide area telephone service (WATS)
Established bands of geographic calling areas where customers may call for a reduced rate.

Figure 26–1
The local telephone company is divided into areas called rate centers. Rate centers determine what calls will be charged and what calls will be free of charge. Most CLECs mirror the rate centers in order to eliminate confusion.

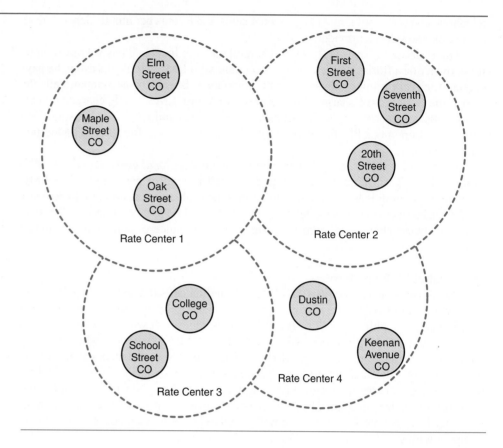

paying 8 cents per minute, she would pay 5 cents for all calls going to Chicago. WATS bands are shown in Figure 26–2. WATS is no longer as popular thanks to the increased competition between carriers. Rates have been reduced across the country and are often cheaper than purchasing WATS. WATS was initially defined as incoming, 800 number, or outgoing and needed its own circuits. Today, 800 number is called 800 service, and outgoing WATS is still purchased depending on traffic patterns and call volumes.

26.1.4 800 and 900 Number Services
800 Number Services

Inward wide area telephone service (INWATS), better known as 800 service, was first offered in the late 1960s. Once AT&T was divested and competitive carriers moved into the market, 800 service became one of the most popular services marketed to business customers. The benefit offered by 800 service is that the caller is not charged for placing the call. Instead, the called party pays for the call. The area code 800 is used to designate a call that is free for the calling party.

Basic 800 Number Features   Before 800 number portability, each carrier was given its own NXX that followed the 800 number area code. AT&T used 542, while MCI used 999. Therefore, when a customer called 800-999-1111, it meant that the call was routed through the MCI network and that the business customer using the 800 number had purchased its service from MCI. If that customer wanted to switch 800 number providers, it could not take its 800 number along. In 1993, 800 number portability, which allowed customers to keep their 800 numbers even when they changed 800 number service providers, was introduced.

800 number portability depended on the SS-7 network that routed the 800 number to a database where it was translated into a regular POTS number. The POTS number was then routed to the correct carrier and the call was completed.

The popularity of 800 numbers has depleted the supply of numbers and caused the industry to add new area codes for free calling. The new area codes—888, 887, and

Figure 26–2
Long distance companies determine the cost per long distance call by where the call is terminating. Some business customers subscribe to WATS to help reduce their long distance charges. The map depicts possible WATS segments. If most calls being placed by a business are on the East Coast, that business should purchase a Band 1 rate.

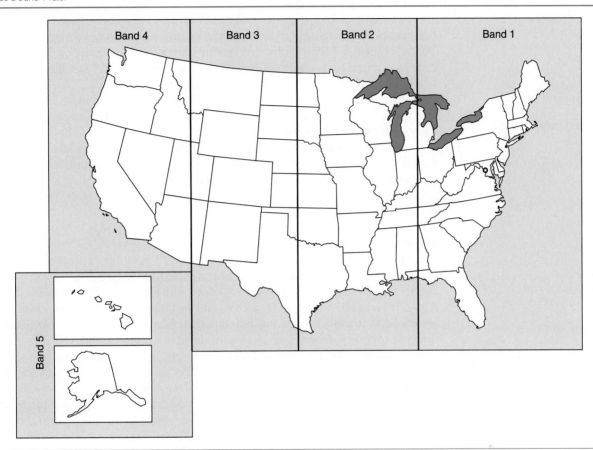

866—function the same way as the 800 number area code. Today, buying 800 numbers that spell out words or names is possible thanks to 800 number portability. For example, 1-888-call-edd represents 1-888-225-5333.

Additional 800 Number Features   The 800 number system in North America provides more than just free inbound calling for the calling party. The service also provides special routing features that help the business provider handle incoming calls. 800 number routing provides several pieces of information that can be used by the 800 number service provider, or by the customer's equipment, to direct incoming calls to a particular location. The information is available because of the SS-7 network and 800 number portability and includes originating telephone number, terminating telephone number, and time of day, week, and month.

The customer may request that 800 number queries be routed to two different call centers based on the time of day. This is called *time-of-day routing*. An example of a customer that may request time-of-day routing is a catalog company with two call centers—one in California and one in Maine. Because of the time zones, the catalog company may request that all calls received before 9:00 a.m. be routed to its Maine call center because California is three hours behind Maine.

Call blockage is used to block calls coming in from specific areas in the country. A small honey factory may want to ship only to customers in the Midwest. It can ask its 800 service provider to block all calls coming from areas other than the Midwest.

Call-attempt profile provides the customer with a report on the number of call attempts made on a particular 800 number. Call-attempt profile is a value-added service that most carriers offer to their customers.

time-of-day routing
Routing long distance calls according to the time of day.

call blockage
Ability of the subscriber to block specific calls.

call-attempt profile
A report profiling the number of calls made to a particular 800 number.

automatic number identifier
Provides the telephone operator with the caller's telephone number.

Automatic number identifier (ANI) provides the telephone operator with the caller's telephone number. The customer receives the number and uses it to retrieve information from his or her database. One of the biggest advantages of 800 number service is its ability to pass ANI.

point-of-call routing
Routes calls from specific area codes to specific locations.

Point-of-call-routing routes calls from specific area codes to specific locations. For example, if a catalog company has four locations—one in Los Angeles, one in New York City, one in Minneapolis, and one in Tampa—calls originating in the 716 area code have a first route choice of New York City. Area code 404 has Tampa designated as the first route choice.

dialed-number identification service
Allows one trunk or circuit to be used to carry multiple 800 numbers.

Dialed-number identification service (DNIS) is one of the most popular services offered by the 800 number provider. DNIS allows one trunk or circuit to be used to carry multiple 800 numbers. For example, an insurance company offers auto, life, and homeowner's insurance. The insurance company has three different 800 numbers—one for auto, one for life, and one for homeowner's insurance. Incoming calls all travel on the same circuit, and the customer's equipment reads the incoming 800 number and routes it to the correct department. The advantage DNIS offers is the reduced number of circuits required for multiple applications.

900 Number Calling

900 number
The charges levied on 900 number calls are set by the company offering the service, not by the telephone company.

The introduction of 900 service was viewed differently by different people. Entrepreneurs immediately realized the profits that could be made from having a *900 number*, and consumers soon learned of the problems associated with the abuse of 900 numbers. The technical differences between the 900 number and the 800 number are minimal. 900 numbers are routed in the same manner as 800 numbers. Many of the 800 number features are available for 900 numbers, such as time-of-day routing, DNIS, and ANI. The major difference between 900 numbers and 800 numbers is that customers must pay when they initiate a 900 number call, not the business. The second issue is that the company that owns the 900 number receives a large portion of the money paid for the 900 number call. The telephone company bills the caller for the 900 call by charging a per-minute charge, a billing charge, and a miscellaneous charge. The owner of the 900 number receives the money for the call from the telephone company. Telephone companies have become very sensitive to the problems associated with 900 numbers, and most of the large carriers have placed rules and restrictions on what they will allow businesses to offer.

26.1.5 Centrex

centrex
Special calling packages that provide numerous features and special routing capabilities such as four-digit routing from the telephone company.

There are two types of *centrex*—analog centrex and digital centrex. Analog centrex uses regular telephone lines with special switch features configured on each line. ISDN lines deliver digital centrex. Digital centrex offers many more features than analog centrex but bears an increased cost to the business customer. The purpose of a centrex line is to provide special features similar to those supported on a PBX or key system, thus eliminating the need for a PBX or key system at the customer location.

Business customers should analyze three main areas when comparing the cost difference between centrex and owning their own PBX or key system: cost, maintenance requirements, and technology obsolescence. Cost becomes difficult to compare because centrex has a monthly recurring cost while a PBX or key system is a one-time equipment cost. The telephone company pricing for centrex, the number of lines the customer has, and the required feature set contribute to the final cost for each solution. Beyond cost, business customers must recognize the increased need for provisioning and maintenance personnel required for PBX or key systems. The telephone company takes full responsibility for maintaining the central office switch and all the centrex features residing in the switch; business customers do not have to. With a centrex system, the telephone company is responsible for maintaining and upgrading equipment as technology changes.

Centrex Features

Switch manufacturers are continuously adding features to their centrex feature set. The following are the most popular features currently being used by business customers.

Four-digit dial is one of the most important and popular centrex features. Employees of a firm can call any employee in the centrex group by dialing the employee's four-digit exchange rather than the full seven-digit telephone number. The main reason centrex has been able to compete with PBX and key systems is the four-digit dial feature.

Class-of-service restrictions restrict outgoing call usage on a per-telephone basis. Toll numbers, 900 numbers, and radio station call-in numbers can all be blocked on employee telephones.

Automatic identification of outward dialing monitors call usage on a call-per-call basis for each employee's telephone. A call usage report is compiled showing what calls were made, the length of the calls, and the telephone numbers the calls originated from. Business customers require detailed call records to determine their networking needs, watch for telephone abuse, and control costs.

Call pickup allows a user to pick up a call designated for a different telephone. Call pickup is used to distribute calls around the office. If one employee is busy or out of the office, his or her telephone number is dealt with by another employee.

Call transfer allows a telephone user to transfer a call to a different telephone station in the centrex group. The benefits related to call transfer are obvious; the caller does not have to call back using the new station number.

Call forward is similar to the call-forward feature for residences. Several types of call-forward features are available. Call forward on busy forwards an incoming call to a different line or voice mailbox when the telephone is already in use.

Call hold allows telephone users to place a call on hold and accept an incoming call while the first caller remains on hold. Analog centrex does not allow the user to pick up a second call, but ISDN centrex does.

Conference calling allows customers to hold a conference call with three to six users. Conference calling has become one of the most popular features offered to businesses, and centrex provides a more cost-effective way to place the feature on employees' telephone lines.

Distinctive ringing is offered on centrex lines just as on regular POTS lines. Several ring types are available to the customers to differentiate their stations.

Advantages and Disadvantages of Centrex Services

The biggest advantage centrex offers is the reduction of maintenance costs. The telephone company handles all switch problems, feature issues, and staffing needs. Customers receive a telephone line similar to a residential customer. The added features provided by centrex allow customers to improve the efficiency of their business by improving the way they communicate. Four-digit dialing is a second key advantage offered by centrex. Employees no longer have to dial all seven digits when calling another employee. A third and significant benefit of centrex is the insurance against purchasing obsolete equipment. Technology changes at an extremely fast pace. A PBX purchased today will more than likely be out of date within two years. In the past, the cost comparison between centrex and customer-owned equipment allowed for a ten-year depreciation schedule for all PBX or key systems. A final advantage of centrex is the customer's ability to provision his or her own circuits. Similar to a PBX, the customer may access the telephone company's switch and provision the features, change numbers, and add numbers on each of his or her lines. The customer no longer has to depend on a telephone company representative to make simple adds, moves, or changes.

The disadvantage of using centrex is the monthly fee charged by the telephone company. Business customers are also often tied into a term contract that prohibits them from changing providers. The PBX, on the other hand, is a one-time purchase that may be capitalized and may offer special features that the local centrex package is unable to provide. Whether a business customer should choose centrex over a business-owned PBX or key system is a decision that must be made on a case-by-case basis.

---

four-digit dial
Allows employees to dial another employee's four-digit exchange rather than the seven-digit number.

class-of-service restrictions
Restrict outgoing call usage on a per-telephone basis.

automatic identification of outward dialing
Monitors each employee's call usage.

call pickup
Allows a user to pick up a call for a different phone.

call transfer
Allows users to transfer calls.

call forward
Forwards incoming calls.

call hold
Allows users to place a call on hold.

conference calling
Allows users to hold three to six user conference calls.

distinctive ringing
Users can choose from several different ring types.

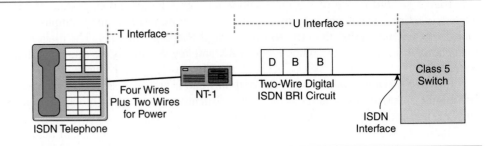

Figure 26–3
ISDN BRI circuits have two 64 kbps channels and one 16 kbps data channel. The T interface refers to the portion of the network from the NT-1 to the telephone. Today, most ISDN telephones have NT-1s. The U interface refers to the two-wire pair that connects the customer to the class 5 switch at the serving central office.

ISDN Basic Rate Interface (ISDN BRI)
Telephone circuit that provides two 64 kbps channels plus one data channel.

26.1.6 ISDN Services
ISDN Basic Rate Interface Services

Integrated services digital network (ISDN) is a service that offers a digital circuit all the way to the customer site. *ISDN Basic Rate Interface* (ISDN BRI) is the digital equivalent to the analog POTS line. There are several differences between the two, but they both provide voice services to the customer. ISDN BRI gives the customer two voice channels and a data channel on one wire pair.

ISDN is a very popular service in Europe and Asia; but in North America, it had a rough start and, consequently, did not establish a strong following. The cost of the ISDN telephone along with the RBOCs hesitancy to upgrade existing switches to be ISDN capable contributed to the low numbers of ISDN BRI services in the United States. ISDN BRI is now starting to receive attention due to the increased desire for high-speed Internet access.

ISDN BRI gives the customer two distinct telephone lines and a data channel that can handle four 9.6 kbps packetized data signals. A customer may have two employees share the same telephone line and attach their PCs to the ISDN data channel. The two 64 kbps channels are referred to as B channels or bearer channels. The data channel is the D channel. Figure 26–3 illustrates an ISDN BRI circuit. The two B channels serve two separate telephones, and the D channel carries signaling information for the two 64 kbps B channels. The signaling channel carries messages from the central office switch that tells the telephones to ring, provides dial tone, delivers caller ID information, and provides other signaling messages. The structure of the D channel is shown in Figure 26–4. The D channel also carries customer data information. The ISDN line is capable of carrying 144 kbps of information that can be a combination of customer voice traffic, customer data traffic, and signaling messages.

The D channel is why ISDN is able to offer multiple services. It has the advantage of carrying out-of-band signaling the same way SS-7 does between PSTN switches. Out-of-band signaling increases the services and features available to the end customer. Essentially, ISDN extends the signaling network all the way to the customer's premises. Some of the special features that ISDN provides follow.

fast-connect times
ISDN feature that connects calls much faster than analog lines.

Fast-connect times are possible because of the out-of-band signaling method. ISDN connects calls in about 1 s compared with the 20 s call setup times found with analog lines.

Figure 26–4
The D channel consists of four 9.6 kbps subchannels plus a 16 kbps signaling channel.

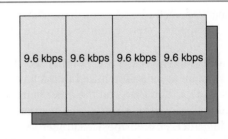

D Channel

Call appearances are additional telephone numbers but not additional telephone lines. Each ISDN telephone may have multiple call appearances, and each employee may be given a pseudo telephone number that appears on the receptionist's phone. For example, if a business has ten employees but wishes to purchase only one telephone line, it may ask the telephone company for ten call appearances. When a call comes in, everyone knows who the call is for because the call appearance belongs to the employee. In this case, only two employees can go off-hook at any one time, but all ten employees can have their own telephone number.

Call appearances are also used to distinguish different types of calls. Employees may have ten call appearances on their telephones, each representing a different product.

Call transfer, detailed call reporting, and *distinctive ringing* are all features available from ISDN service. The user may pick up one line, place the person on hold, and pick up the second incoming call without dropping the first. The user may also conference both calls together. The ISDN line provides two distinct call paths and therefore allows two calls to use the same telephone line.

Video conferencing is possible because of high-speed digital ISDN links. Users might wish to see the person they are talking to or join a video conference held thousands of miles away.

Desktop conferencing allows conference attendees to talk on their telephones while they all watch the same applications on their computer screens. For example, the attendees may work on the same spreadsheet in real time during the conference. Often, each participant is given a different color to use to signify who is making the changes. The value of the conference call is increased by the addition of the visual graphics.

Voice and data integration allows the user to talk on the telephone and work on line whether connected to the Internet or to a remote database. ISDN provides this service through both B and D channels. The customer may use the 9.6 kbps data channel and one of the B channels to integrate voice and data communications. The customer may also use one B channel for voice and the second for a dial-up Internet connection. Again, the versatility of ISDN is one of the reasons it is a viable solution for business customers.

High-speed Internet access is possible by bonding the two B channels together to form one 128 kbps data channel. Many residential customers purchase ISDN specifically to increase their access speed to the Internet.

ISDN Primary Rate Interface Services

ISDN Primary Rate Interface (ISDN PRI) service provides a 1.544 Mbps digital circuit to the customer. The difference between it and T1 service is that ISDN uses the twenty-fourth channel for signaling. ISDN PRI circuits give customers twenty-three 64 kbps channels to use any way they wish. They may fill the circuit with DID/DOD trunks or with twenty-three data connections. The advantage ISDN PRI provides is similar to ISDN BRI—the signaling is out of band and thus improves the efficiency of the circuit along with increasing the number of features available to the user.

One of the most common uses of ISDN PRI is to connect an ISP and the telephone company. An ISDN PRI circuit connection establishes multiple trunks between the ISP's PBX and the telephone company's central office switch. The ISP's customers may dial one access number that directs them onto the ISDN PRI link connecting into their remote access server (RAS). The ISDN PRI link is an efficient way to connect ISPs to telephone company switches.

ISDN PRI is also a popular solution for call centers. These circuits connect long distance telephone companies to call centers that take advantage of the out-of-band signaling feature of ISDN PRI. The call center is able to route calls according to incoming information in addition to increasing the connection times for incoming and outgoing calls. The efficiencies, features, and out-of-band signaling methods make ISDN PRI a popular solution for certain types of applications.

Figure 26–5 depicts a typical ISDN PRI solution showing a connection between a call center and an ISP.

call appearances
Additional telephone numbers without additional lines.

video conferencing
Similar to teleconferencing but using video cameras so all parties can see each other.

desktop conferencing
All parties can view the same things on their computer screens while talking on the phone.

voice and data integration
Allows users to talk on the phone and work online at the same time.

ISDN Primary Rate Interface (ISDN PRI)
Four-wire circuit that provides 1.544 Mbps of bandwidth.

Figure 26–5
The ISDN PRI circuit provides twenty-three channels that carry information and one D channel used for signaling. A common use of PRI circuits is between a PBX and the central office class 5 switch or LD switch. The channels are similar to trunks between two LD switches. The signaling, similar to the out-of-band SS-7, rides outside the call path within the D channel.

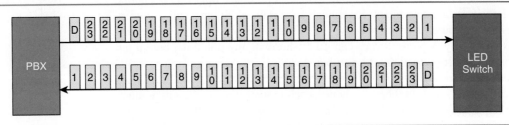

26.1.7 Direct Inward Dialing/Direct Outward Dialing Services

Direct inward dialing (DID) and direct outward dialing (DOD) are services offered to business customers that have some sort of voice switch or voice routing device. The telephone company switch is connected to a PBX using DID/DOD trunks. The service gives the customer the advantage of establishing trunks between its equipment and the central office switch the same way the phone company connects its switches to each other using trunks. The business customer uses DID/DOD trunks to reduce the number of connections between the two switches.

Direct inward dialing (DID) trunks carry traffic coming from the central office switch to the business customer site. Calls originating off site are routed onto the DID trunks that are built between the telephone company switch and the PBX. The telephone company sells DID service along with blocks of telephone numbers that may be provisioned as one-way or two-way trunks. Business customers must decide how many telephone numbers they wish to purchase. A customer with 500 employees may purchase a block of 500 numbers, but it does not purchase 500 DID trunks. Instead, the business customer must determine the concentration ratio between the number of employees and the number of available circuits. For example, if the business customer determines that there will be only ninety-six customers on line at one time, it will purchase four T1 circuits but 500 telephone numbers. Ninety-six trunks will be available between the customer's PBX and the central office switch, but each of the 500 employees will have his or her own telephone number.

Direct outward dialing (DOD) trunks are the opposite of DID trunks. They carry traffic from the customer site to the telephone company. They also normally ride on T1 digital circuits and are provisioned by the telephone company and the business customer. Similar to DID trunks, DOD trunks require blocks of telephone numbers from the telephone company, and the service may be provisioned on one-way or two-way trunks.

For example, WordPlus has 500 employees at its corporate headquarters in Burlington, Vermont. The telecommunications manager decided to place DID/DOD trunks between the company's internal PBX switch and its new service provider—Spidercom. The ratio of trunks to employees was determined to be ten employees to one trunk. The DID trunks were built using the 10:1 ratio. Because there are 480 employees, WordPlus ordered forty-eight trunks from Spidercom, who turned up two T1 circuits between WordPlus's PBX and Spidercom's class 5 switch. The trunks were designated as two-way DID/DOD trunks. Spidercom sold WordPlus 600 telephone numbers for its employee telephone number pool, and both parties agreed that the main billing number would be 678-3455. Signaling on the DID/DOD circuits would be ground start with reverse battery. The central office switch would send a wink to cause the PBX to reverse the battery on the line and to notify the central office switch to initiate billing.

The advantage of DID/DOD to business customers is obvious. WordPlus has saved an enormous amount in monthly recurring fees from the telephone company by reducing the number of telephone lines from 500 to two T1 circuits and receiving a block of telephone numbers from Spidercom. The DID/DOD service gives the business customer a way to connect to the local telephone provider using standard digital circuits configured as trunks. The connection proves to be more efficient, less expensive, and easier to provision and maintain.

26.1.8 Analog Trunks

Analog trunks are not as popular as they once were. They are used mainly by small business customers who own older, small key systems or PBXs with analog trunk terminations. Some PBXs or key systems do not have a digital interface and consequently cannot accept a digital trunk. They can, however, accept analog trunks from the serving central office switch.

Similar to the digital DID/DOD service, analog trunks are used to carry incoming and outgoing calls between the customer site and the serving central office and are also referred to as DID or DOD. The main difference between the two services is that one is analog and one is digital.

26.1.9 Features

Features used by business customers are basically the same as those used by residential subscribers. Descriptions of these features follow. Slight changes have been made in the descriptions to explain how the business customer uses the features compared with the residential customer.

Hunting describes the way an incoming call seizes an idle circuit. For example, if a business has three lines and the first line is being used, an incoming call will seize the first idle line, which in this case is the second line. There are two popular hunt types—terminated and circular, or round robin. Terminated hunting always starts from the top and proceeds down, so every incoming call looks to see if the first line if available. If it is not, it hunts to the second line, then the third line, and so on. With circular hunting, an incoming call hunts to the line down from the last line seized, as shown in Figure 26–6. In a terminal hunting group, the first line is seized more often than it is in circular hunting, which starts wherever the last call stopped.

hunting
Feature offered by the switch that hunts down the lines until it finds an idle circuit.

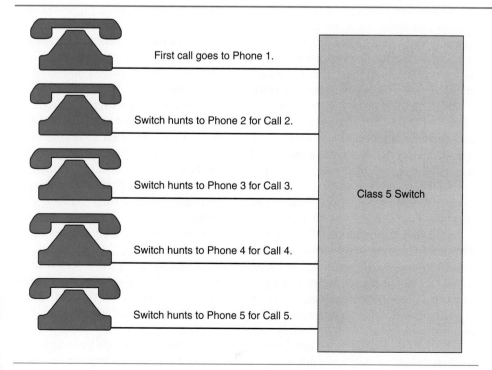

First call goes to Phone 1.

Switch hunts to Phone 2 for Call 2.

Switch hunts to Phone 3 for Call 3.

Class 5 Switch

Switch hunts to Phone 4 for Call 4.

Switch hunts to Phone 5 for Call 5.

Figure 26–6
Many business customers require that incoming calls be answered; therefore, they use hunt groups and hunt patterns. This is referred to as the round robin because it moves from 1 through 5, then back to 1.

Call waiting allows a business customer to pick up a second incoming call while on line with someone else. The call-waiting feature is the same as the one placed on the residential customer's line. When a beep interrupts a phone conversation, it means that another call has arrived and is waiting in que. The switch sends a signal to notify you that a second caller is trying to reach you. It may be a tone, a light on the telephone set, or a message scrolling past on the telephone's LCD screen. To place the person you are talking to on hold and pick up the incoming caller, simply flash the switch hook and the central office switch will recognize that you want to answer the second incoming call. The switch places the first caller in queue while you speak with the second caller. When you finish speaking with the second caller, flashing the switch hook again notifies the switch that you want to disconnect from the second caller and go back to the first caller. The first caller comes back on line and you continue your conversation. You may also bounce between both callers by flashing the switch hook; or you may flash back to the first caller, finish your conversation, say good-bye, and flash back to the second caller.

The user does not have to accept the incoming call. The beep will sound two or three times before it stops. The second caller will hear ringing and assume that you are not there or that you are not willing to interrupt your first call. The switch can also forward calls to your voice mailbox or return a busy signal. Depending on the telephone company, you may be required to pay an extra fee for directing unanswered calls to a voice mailbox. The caller's name and number will show up on your caller ID if you subscribe to the caller ID feature with name. Most phone companies vary the tone depending on whether the call is coming in from outside the calling area or whether it is from a cellular telephone; however, some do not distinguish between the two. The switch can be programmed differently according to the desires of the telephone company.

A drawback of call waiting has to do with dialing into the Internet or other dial-up data connection using a modem, which is very sensitive to the incoming call-waiting tone. In fact, the tone causes the modem to disconnect the data session. The simplest way to solve this problem is to turn off call waiting by dialing *70 or your local access code, if different, before you dial the access number. The switch is set to turn off call waiting when it receives the tones associated with the access codes, which are normally listed in the front of the telephone directory. Once you hang up, call waiting is automatically enabled for the next call.

Distinctive ringing is a popular feature with small business owners. Several employees can share the same telephone line. Different telephone numbers can be assigned to that line with a different ring pattern for each of the numbers. For instance, if two professors share an office with only one telephone line, distinctive ringing allows each to know whom the call is for. Special adapters can also be placed on the line to route the different ringing patterns to different telephones.

Call forwarding is very popular with business customers. This feature allows the subscriber to forward all incoming calls destined for one telephone number to a different telephone number. This is useful for the business user who often leaves the office and spends the day at a remote location. Call forwarding to a cellular phone is an important feature, especially for salespeople who rarely sit at their desks in the office.

Call forwarding can be used several ways. Call forwarding on busy and unanswered is often used with voice-mail systems to forward calls to a voice mailbox when the phone is in use and no one answers the incoming call.

Call forwarding automatically forwards all incoming calls. Call forwarding remote access activates call forwarding from a remote site.

Three-way calling allows the telephone subscriber to add a third person to the conversation. Business customers frequently subscribe to three-way calling because they often need to communicate with more than one other person at a time. Most telephone companies allow conference calling to be set up either on a per-call basis or as an always-on feature with a monthly recurring charge.

Call transfer, also called *enhanced three-way calling*, is very similar to three-way conference calling. The main difference is that, unlike three-way calling, the originator of the third-party call may hang up the phone when the other two parties are connected. Call transfer is also used to transfer callers to a voice mailbox.

Speed dial saves time and offers business customers a way to simplify their daily tasks. Programming the telephone for speed dial is fairly simple, and the obvious advantage is that a phone number is dialed using two digits instead of seven. The feature is packaged differently by different telephone companies, but the most common method is to sell a specific number of speed dial numbers for a set price. For example, a telephone company may have an eight-number package, a fifteen-number package, and a thirty-number package. If there are only eight numbers that you call frequently, you may only subscribe to the eight-number package and so forth. Emergency numbers are often programmed as two-digit speed dial numbers.

Call blocking ensures that employees do not abuse the telephone. A good example is placing a call block on the telephone line for 900 number calls. When an employee dials the 900 number, an intercept recording tells the employee that the call is not valid and will not be completed. The subscriber must request that the numbers be blocked, but the telephone company does not charge to block 900 numbers.

Voice mail is a value-added service, a term used to describe additional calling features that phone companies hope will differentiate them from their competitors. Business customers are the most frequent users of voice mail. The voice-mail platform is attached to the central office switch, and calls are forwarded to the customer's voice mailbox when the called party does not answer or when the called party's telephone is busy. The voice-mail system provides all the features of a telephone answering machine, but the customer does not have to change tapes in the machine or maintain it. Voice mail is easily accessed from a remote location by dialing an access number and a pass code. This allows subscribers to listen to their messages from anywhere in the world. Newer answering machines also allow customers to access their messages remotely though not always as simply as with the telephone voice-mail system.

The customer can configure the mailbox from his or her telephone and can customize the recording with a personal greeting. A caller's message may be saved for later reference, deleted, or forwarded to someone else's mailbox. When the called party is on the telephone or not available, calls coming in are forwarded to the subscriber's voice mailbox. Customers who subscribe to voice mail must also purchase the call-forwarding feature in order to route calls to their voice mailbox, but most telephone companies bundle the cost of the two services to make them more affordable.

26.1.10 Long Distance Telephone Company Features

Long distance telephone company features for business customers are also very similar to those for residential customers. The main difference is that business customers are more likely to subscribe to them. Long distance telephone companies bundle many services, reduce the cost-per-minute usage fees, and offer additional features to enhance their calling packages. The three most common business features or value-added services are 800 numbers, calling cards, and conference calling.

Business 800 numbers have always been a popular feature. Many businesses depend on 800 numbers to pull in customers. For example, if a person who wants to purchase a computer finds two companies that sell computers, one with an 800 number and one without, the person would probably call the company with the 800 number.

Calling-card service is offered by most long distance carriers to business telephone subscribers. It is used when the subscriber is away from the office and wishes to make a call. Subscribers are given access numbers that connect them to the long distance calling-card platform. Callers enter their unique calling-card PIN followed by the number they wish to call. Long distance telephone companies connect the caller to the called-party number, and the customer is billed at the rate used for calling-card calls, which is normally higher than direct distance dialed calls.

Business customers buy calling-card service that they hand out to their employees. Each employee has his or her own PIN. The PIN is used to track usage for each employee. Customers do not have to use the same long distance company for their calling-card service as for their long distance provider. Rates for calling-card services vary among the different companies.

Some calling-card companies now offer only calling-card services. Prepaid cards have a specific number of minutes assigned to them and are only good for the specified number of minutes assigned.

26.1.11 CLASS Features

Local telephone companies offer customized local area signaling service (CLASS) features that are dependent on the SS-7 network and are offered only if the switch is SS-7 capable. CLASS features can travel between different service provider networks if interconnection agreements have been initiated between the two companies.

Caller ID, or calling line identification (CLID), allows an incoming caller's telephone number to appear on a caller ID screen. The subscriber has to purchase the caller ID feature and place a caller ID box on the end of the telephone line, although many newer telephone sets have caller ID built into the telephone.

If "out of area" appears on the caller ID, it indicates that the caller's phone company is not capable of passing caller ID information outside its network. The goal in the industry is for all companies to pass calling information between networks so caller ID can be carried between regions.

Caller ID is seen as an invasion of privacy by some. One of the main concerns is the ability of call centers to use an incoming telephone number to pull up a customer's records. Telephone companies were forced to offer a second feature—calling-number restrict—in conjunction with caller ID.

Caller name (CNAME) is similar to caller ID, but it supplies both the telephone number and the caller's name on the caller ID screen. When a call comes in from someone using a different service provider, the caller-name feature will only work if both service providers agree to pass the information. If the two companies do not establish interconnection agreements allowing for CNAME information to pass, the subscriber will receive the calling number without the caller name and address information. Caller name, like caller ID, has caused controversy between telephone companies and consumer groups. Consumer groups believe that it is an invasion of privacy and provides too much information to the called party. Call centers, on the other hand, like it. They can capture the incoming information, place it in a database, and use it for telemarketing.

Call restrict is free to all telephone subscribers. When it is turned on, the caller's telephone number and name are not transmitted to the called party's telephone. Call restriction can be used as a per-call feature or an always-on feature. Telephone companies must offer call restriction if they sell caller ID and caller name.

Call return lets the subscriber ring back the last caller by dialing *69 or some other code number. When the subscriber dials the code, the switch determines who the last caller was and automatically places a call using that telephone number. The calling party's number is resident in the SS-7 messages after the call and is used to return the missed call. In most cases, call return will not return calls between telephone companies. The telephone company charges the customer every time the call-return feature is used.

Call trace is used to find out who called and also depends on the signaling network to work. When the call-trace feature is activated, the subscriber is given the date and time of activation, then contacts the security bureau at the phone company and notifies it of the trace. The telephone company is able to look in the switch and determine where the call came from. The subscriber is charged for each call trace.

Ring again is useful when calling a number that is busy. The caller activates the feature, hangs up the phone, and waits for the ring-back message indicating that the called party has hung up. When the caller picks up the phone, the call automatically goes through to the called party's telephone.

Selective call acceptance lets a subscriber enter a list of numbers that he or she will accept. All other calls are intercepted and an announcement is played stating that the subscriber does not accept the call. The subscriber's telephone rings only when a calling number matches one of the numbers in the selective number table in the switch. Selective call acceptance is not used as frequently as other features.

Selective call forwarding defines where particular incoming calls should be directed. Subscribers use this feature to direct certain calls to specific telephones, voice mailboxes, or attendants.

Selective call rejection lets customers reject specific incoming phone numbers with an announcement telling the caller that his or her call is not being accepted. It is the opposite of selective call acceptance.

■ 26.2 DEDICATED VOICE SERVICES

26.2.1 Tie Lines

For years, telephone companies have provided special service circuits to connect two distant locations. A *tie line* is a circuit that ties together two remote locations' PBXs or other terminating equipment. Tie lines are similar to circuits that connect class 5 switches together and are used to carry shared trunks between two customer switches. The signaling method normally used is E&M. The advantage of using tie lines, which are offered by the local or long distance provider, is the reduction in LMS or long distance charges for the business customer.

tie line
Dedicated circuit used to connect two locations.

The tie line is a four-wire copper circuit that interfaces into the PBX on a trunk port. For example, Typo Company has a tie line connecting the PBX located in its Cincinnati corporate office and the PBX located in its Denver corporate office. When Luther, a Cincinnati employee, calls Horatio, a Denver employee, Luther only has to dial Horatio's four-digit extension. The call does not incur LMS or long distance charges, but Typo Company pays a monthly recurring fee for the tie-line circuit. Employees can use the circuit in any way they wish. Proving in the flat rate monthly recurring fee is fairly easy because the cost of per minute charges adds up quickly. Figure 26–7 illustrates a typical tie line connecting two switches at different offices.

Figure 26–7
Many larger companies with two locations tie their PBXs together to allow their employees to dial just four digits anywhere in the company. The tie line saves long distance charges because the circuit is leased by the company for any use it wishes.

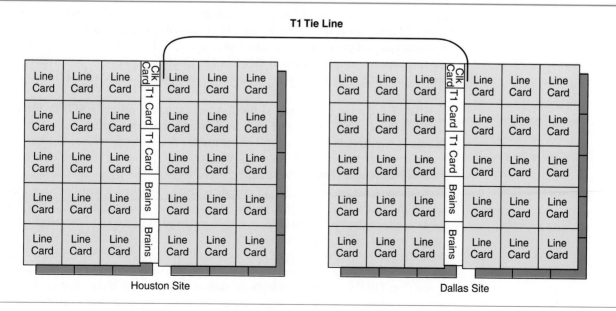

Figure 26–8
FX circuits are used to provide dial tone from one CO to a subscriber served by a different CO. Call forwarding has replaced FX service in many areas.

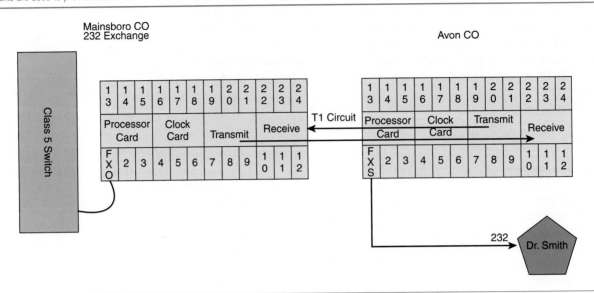

26.2.2 Foreign Exchange Circuits

Foreign exchange (FX) circuits allow customers to use their telephone number at a remote location. The best way to explain FX circuits is with an example. Dr. Smith has an office in Mainsboro. The office is served by the Mainsboro central office. He would like his office telephone number to also ring at his home, but his home is served by the Elmwood central office. A foreign exchange circuit physically connects Dr. Smith's house to the Mainsboro central office, thus allowing his office number to ring at his home. The method used to establish the FX circuit requires a connection between the Mainsboro switch and the Elmwood central office that serves the doctor's house. The connection from Dr. Smith's home to the Elmwood central office is a two-wire copper pair that is cross connected into a T1 channel bank located at the central office. A dedicated T1 circuit with three available channels links the two central offices together. Dr. Smith's FX channel is multiplexed into one of the three available DS-0 channels at the Elmwood central office. Once at the Mainsboro central office, the DS-0 carrying Dr. Smith's FX circuit is demultiplexed and cross connected to a switch port. The switch port has control over the telephone sitting on the hall stand in Dr. Smith's home. The Mainsboro switch sends dial tone, ringing, and so forth to the telephone. Figure 26–8 illustrates the connection.

The FX circuit is also called the foreign exchange office (FXO) or foreign exchange station (FXS). The value of the FX circuits has been reduced since the advent of call forwarding, but they are still being used.

26.2.3 Off-Premises Exchange Circuits

The *off-premises exchange (OPX) circuit* is used to extend a station off a PBX, as shown in Figure 26–9. An OPX circuit is a copper tip and ring pair that interfaces on the line side of the PBX in an OPX interface card and extends out to the telephone company's central office. At the central office, the OPX feeds into an OPX card that connects the pair to a second copper pair feeding the off-site premises. A good example of an OPX circuit is one that connects a car sales office to its body shop located across the street. The OPX line ties the telephone station sitting in the body shop to the PBX in the telephone room at the car sales office. The telephone company's central office ties the two legs together.

foreign exchange (FX) circuit
Allows calls to a customer of one central office to ring at a location served by another central office.

off-premises exchange (OPX) circuit
Used to extend the station off a PBX.

Figure 26–9

OPX circuits are used to extend the reach of the line from the switch. Here the bus garage is fed out of the PBX by an OPX circuit card in the PBX.

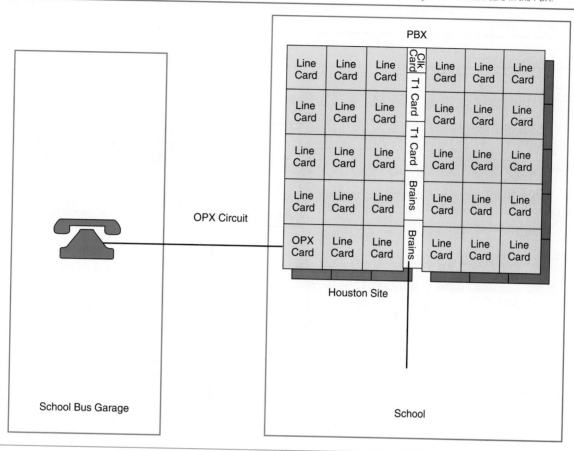

OPX circuits are also commonly used to connect a school with its administrative offices, bus garage, and athletic buildings. Many organizations and businesses still use OPX circuits.

26.2.4 Coin Phones

The telephone company sells coin phone circuits to businesses such as restaurants or stores. There are many types of coin phones, and small start-up businesses have emerged to compete with the local telephone company to provide the service. Consumers need to be careful when placing long distance calls from a coin phone. The lowest bidder usually wins the coin phone long distance business and often charges ridiculously high per-minute charges to anyone using the phone.

26.2.5 Private Voice Networks

Private voice networks are similar to tie lines in that they are circuits purchased by the business customer from the telephone company for a monthly recurring cost. Business customers use multiple circuits to connect their offices and form a private voice network. The types of circuits vary depending upon the amount of traffic the circuit has to carry and the medium used, which also depends on several factors. Many private networks are built using T1 circuits, tie lines, FX lines, and analog trunks. Some companies have built satellite links using small virtual satellite terminals. Others purchase special service circuits from the telephone companies, such as T1s, tie lines, and FX circuits, and therefore depend on the medium used by the telephone company. Figure 26–10 illustrates a typical private voice network deployed by a large corporation.

Figure 26–10
Many large corporations establish private circuits between their locations to carry their voice traffic. This is a common private voice network between four cities. All of the long distance calls from each site are routed back to the main PBX in Springfield. Off-net calls travel on the T1 circuit between the PBX and the company's LD carrier, Spidercom.

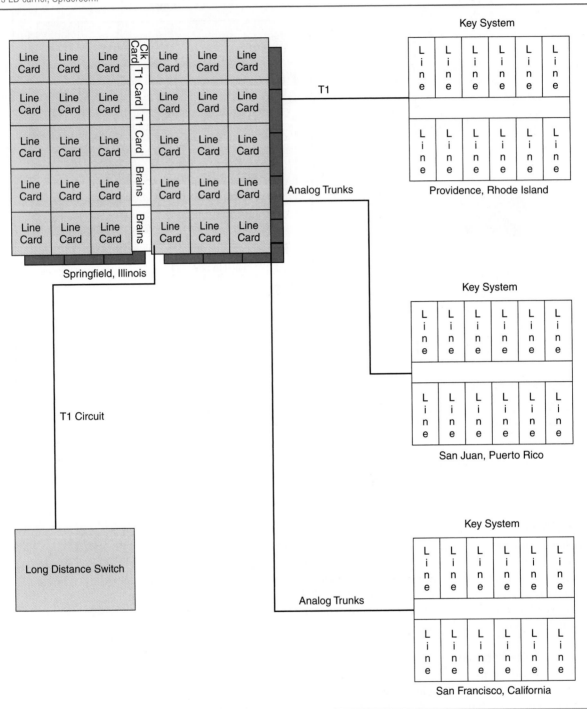

■ 26.3 SWITCHED DATA SERVICES

Like residential subscribers, business subscribers need a way to transport data across the PSTN. Four main types of switched data circuits are used in the United States to carry data information across the public network: dial-up data connections, ISDN BRI data connections, ISDN PRI connections, and switched-56 connections.

A *dial-up data* line is simply a phone line with a modem on the end. A business customer dials the number he or she wants to connect to, the modems perform their

dial-up data
A phone line with a modem on the end.

handshake, and the connection is made. The business customer transfers files, surfs the Web, or accesses a database using the established connection. The speeds available for dial-up data vary from a slow 9.6 kbps to 56 kbps. The speed depends on the distance between the customer's location and the serving central office and the type of modem being used on each end.

Small business customers commonly use dial-up data lines for their data transmission needs. The monthly recurring cost is the same as for a dial-up voice line—minimal. The greatest cost incurred with a dial-up data line is the per-minute usage charge for LMS or long distance. The small business customer that does not stay on line for long periods of time may find that a dial-up data line is the most cost-effective way to transport its data information. A small business that has to stay on line for long periods of time will easily prove in a nonswitched data service.

ISDN Basic Rate Interface (ISDN BRI) is similar to a dial-up data connection in that a telephone number is used to establish the connection between the two ends. The difference is that the ISDN BRI line consists of two 64 kbps bearer channels that may be used individually or bonded together to form a 128 kbps data path. The increased data speed, along with the reliability of a digital circuit, makes ISDN BRI an attractive solution for the small business customer who needs to transmit data.

The cost of an ISDN BRI line varies depending on the service provider. In some areas, the service is sold on a flat-rate basis. ISPs often sell ISDN BRI through the local service provider, allowing the customer to dial in and not pay a per-minute charge. A common use of ISDN BRI lines for data transmission is redundant links between routers or other data devices. If the T1 circuit fails, the router automatically dials up the far end and reroutes all traffic onto the ISDN circuit. Figure 26–11 illustrates several ISDN BRI data scenarios. In Europe and Asia, ISDN circuits are very common and are accepted as a way to carry both voice and data across a switched network.

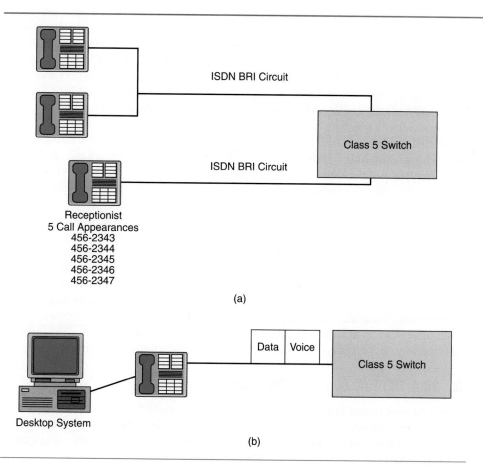

Figure 26–11
(a) A small business customer with two ISDN lines. The receptionist has five call appearances that are used to designate different types of calls coming into the business. For example, the first call appearance, 456-2343, is used for the sales department. The second number, 456-2344, is assigned for repair calls. The customer only purchases one line but can assign multiple numbers.
(b) An ISDN line is used to carry a small business customer's voice and data information.

Figure 26–12
Switched-56 circuits have been
deployed for years. They provide a
special service 56 kbps circuit that is
not nailed up end to end.
Unfortunately, switched 56 is not
offered across the country and has
been replaced by ISDN lines.

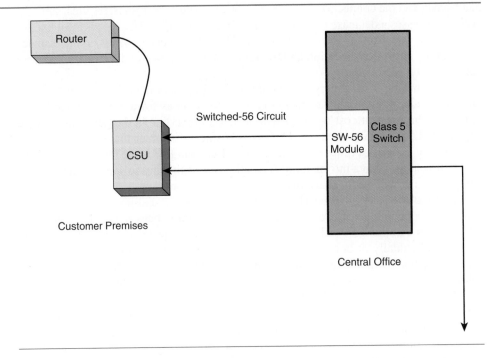

ISDN Primary Rate Interface (ISDN PRI) is a circuit used by many large companies to carry their data information. The advantages ISDN PRI offers are the number
of bearer channels allowed per circuit and the signaling channel used to handle all signaling events. For example, a company may connect into the ISP using an ISDN PRI
circuit that allows twenty-three dial-up connections at a rate of 64 kbps that may be
shared by all employees. ISDN PRI eliminates a one-for-one nailed-up channel per
user. When the employee accesses the Internet, a channel is seized for the duration of
the session, then released once the user logs off—freeing up the channel for a different
user. The employees do not have to use their voice circuits to connect into the Internet.
ISDN PRI data circuits are also used for circuit backup links between routers and, most
commonly, for connecting the ISP to the local telephone provider. ISDN PRI gives the
ISP the advantages of out-of-band signaling methods and shared trunks.

Switched 56 was first offered in the mid-1980s, but ISDN BRI lines and higher
speed modems have now all but replaced the service. Switched 56 gives business customers a transmission path that will carry 56 kbps of data. The circuit travels through
the switch but is not a dial-up circuit. The switch must have a special module that is
used to switch the 56 kbps channels to other switched-56 locations. Switched-56 circuits can only talk to other 56 kbps circuits, not dial-up modems.

switched 56
Gives business customers a
transmission path that will
carry 56 kbps of data.

The purpose of the switched-56 circuit was to give business customers a way to
connect to multiple locations using one circuit. It was also one of the first digital services offered to the business customer and, at the time, was seen as a major breakthrough
in telecommunications. The main problem associated with the service is the need for
telephone companies to install special equipment in order to connect the two ends of the
circuit. Equipment that interfaces with the incoming digital circuit must be placed at the
customer site. The DSU is used as the DCE for switched-56 service. In addition, the customer must place special equipment beyond the DSU in order to handle the 56 kbps connections. Figure 26–12 illustrates a common switched-56 service offering.

■ 26.4 DEDICATED DATA SERVICES

Telephone companies have offered dedicated data services for the past three decades.
The first services involved switched dial-up services. The selection improved as the
technology advanced to include dedicated analog data circuits, dedicated data circuits,
and high-speed optical connections. It is important to distinguish between a switched

Figure 26–13

A point-to-point analog data circuit is designed between two customer locations. The analog line feeds into a channel bank at the central office. The channel bank multiplexes the analog line with other signals and transports them to the terminating central office.

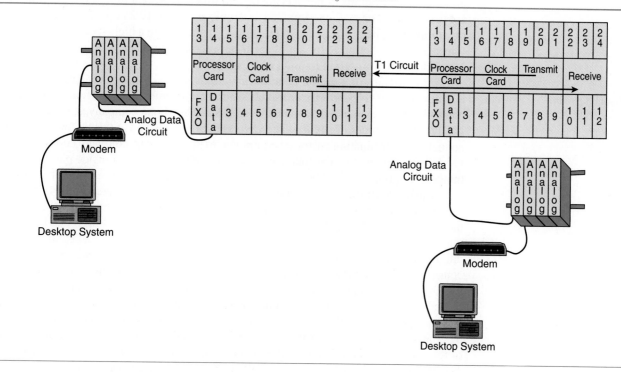

business service and a dedicated business service. A switched service depends on the switch to connect the two points together; the circuit is only established after the switch completes the connection. Dedicated circuits are always connected. Customers purchase a circuit between locations and the circuit is built, turned up, and left up. Dedicated circuits do not go through a switch.

Analog dedicated data circuits have been installed in the PSTN since the introduction of the first communications circuits. An analog data circuit is a dedicated wire connection between two business locations. Analog circuits were introduced primarily to improve the speed of the data exchange and to ensure the security of the information being transferred. The analog circuit was one of the first services offered as a special service by the telephone company.

Figure 26–13 depicts a typical analog data circuit connecting two locations. A device similar to a modem or DSU is placed at the customer site, and the copper pair (or pairs, depending on whether it is a two- or four-wire dedicated circuit) is terminated in an analog line unit in the transmission room in the central office. The signals may be amplified if needed and transmitted to the terminating location. The path of the circuit depends on the two end locations. The greatest problems with using analog circuits to carry high-speed data are the need to tune the frequencies between the two end points several times each year and the limitation on the distance the signal can travel before it needs to be amplified. Analog data circuits are still maintained but are rarely added to a network. The newer dedicated digital circuits have replaced the need for dedicated analog circuits; but many business customers continue to use their analog equipment, refusing to upgrade because of the high capital investment it would require. Their analog circuits provide ample bandwidth for their needs, and the telephone company incurs the maintenance costs, not them.

Digital dataphone was the first dedicated digital circuit offered by AT&T before divestiture. DDS™ (digital data service) is a trademark of AT&T's digital service, and AT&T does not allow other telephone companies to use the name to describe their services. However, most industry professionals use the term DDS when describing *dedicated 56 kbps services*. As with dedicated analog circuits, digital 56 kbps dedicated

analog dedicated data circuit
Dedicated wire connection between two business locations.

dedicated 56 kbps service
A special service circuit ranging in speed from 9.6 to 56 kbps.

service is categorized as a special service circuit and is therefore priced, maintained, and sold as one. Speed ranges from 9.6 to 56 kbps. A customer can live no farther than about 5000 ft. from his or her serving central office to order a full 56 kbps digital data service circuit. The customer may wish to only pay for a 19.2 kbps data connection if the information he or she needs to transfer does not require a high-speed line. The customer pays a higher price for a higher bit-per-second rate. There are two types of dedicated data circuits—point to point and multipoint.

point-to-point circuit
A dedicated 56 kbps circuit.

The structure of a digital data service circuit is very similar to analog dedicated circuits. A *point-to-point circuit* travels from the customer site on a four-wire copper pair that interfaces into a customer-owned DSU. The phone company will also lease a DSU to the customer, if needed. The DSU is the DCE located at the customer site and connects the terminating customer equipment to the telephone company copper circuit. From the customer site, the four-wire circuit travels through the outside plant with the thousands of other lines. The four-wire circuit arrives on the vertical side of the MDF (main distribution frame) and is connected to a frame block mounted on the horizontal side of the frame using a frame jumper. This is cabled to the transmission room, not to the switch room. The digital data circuit, therefore, travels from the MDF into the transmission room on the interbuilding cable and is terminated into a channel bank. The digital data signal is multiplexed into a T1 signal in the channel bank, and the T1 circuit connects to the second central office. The signal travels to this distant central office on the T1 circuit. At the remote central office, the 56 kbps dedicated circuit is demultiplexed out of the T1 circuit. The four-wire digital data circuit leaves the channel bank and travels up to the MDF, similar to the first central office, where it is terminated onto a frame termination block. A frame jumper connects the frame block to the vertical side of the frame. From there, the circuit is connected all the way out to the customer site. The 56 kbps circuit leaves the central office on a four-wire copper circuit that connects to the customer's other location. Figure 26–14 traces the dedicated digital circuit from Point A to Point B—a point-to-point circuit.

multipoint dedicated digital circuit
A circuit used to connect multiple points to one central server.

A common example of a multipoint circuit is a bank interconnecting its remote ATMs with its mainframe. Figure 26–15 shows a common *multipoint dedicated digital circuit* used to connect Big Bank's mainframe to five Big Bank branches. As shown in the figure, the four-wire copper circuit travels from the main branch of the bank to its serving central office. The circuit, as described earlier, terminates on the MDF.

Figure 26–14
A point-to-point DDS circuit. The 56 kbps circuit is used to connect two data networks together. As in Figure 26–13, channel banks are used to multiplex multiple circuits between the two central offices.

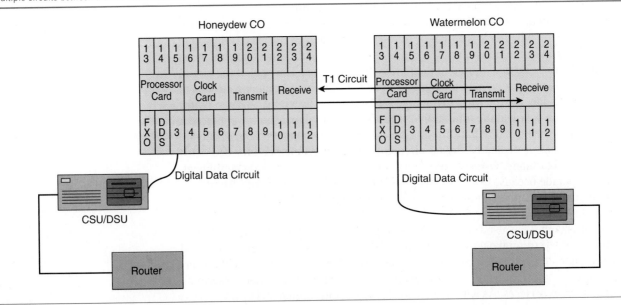

Figure 26–15
The typical route for a special service circuit. The circuit arrives at the vertical side of the MDF and is connected to the horizontal side with a frame jumper. The block on the frame is connected to a channel bank. A digital circuit card receives the signal, then multiplexes it into a T1 circuit that interfaces a DCS (digital cross-connect switch).

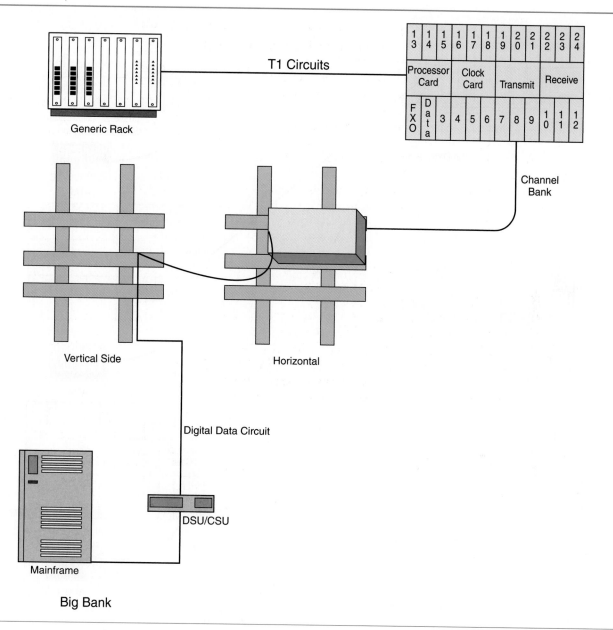

From the MDF, the circuit interfaces into a channel bank located in the transmission room. The T1 circuit feeds into a 1:0 DACS (digital access cross-connect switch) where the one dedicated digital service is bridged onto multiple channels, each riding in different T1 circuits. The T1 circuit fans out, feeds the different central offices, and interfaces into channel banks. The individual dedicated circuits are demultiplexed out of the channel banks and ride up from the transmission room to the MDF. From there, a frame jumper is used to connect the frame block to the pair on the vertical side of the MDF. The signal travels out onto the four-wire copper pair that terminates at Big Bank's branch office(s). Figure 26–16 shows the Big Bank data multipoint service.

Multipoint dedicated digital service is one of the most common ways to carry data between multiple locations from one main site. Each branch location is allowed to talk to the Big Bank's main site, but they cannot talk to one another. There are also multipoint services that do allow the branch sites to talk to one another without having to

Figure 26–16
The multipoint circuit is commonly deployed between banks and their branches. Each branch talks to the main branch. The main branch poles each site asking if it has data to send. The branches respond with their information.

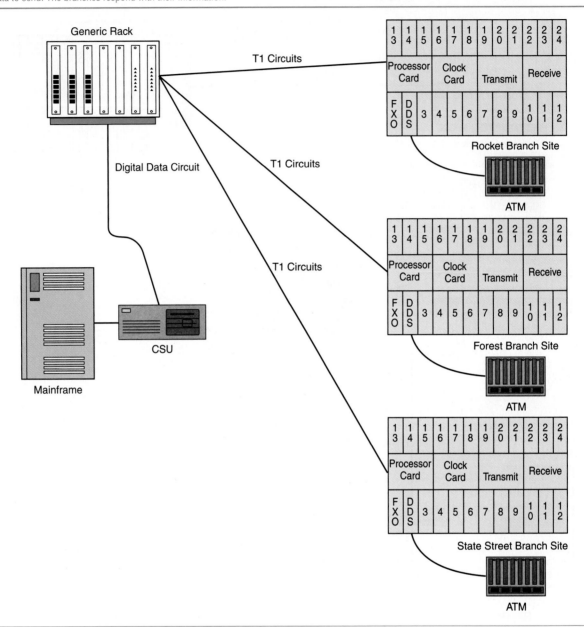

travel through the main site first. How the services are deployed depends on the telephone company. Both local and long distance phone companies offer dedicated digital service. Grooming the data signals onto T1 circuits is a common way for long distance companies to carry the data channels across long distances.

■ 26.5 HIGH-SPEED DIGITAL CONNECTIONS

Many smaller business customers cannot afford to pay for a full T1 circuit to carry their traffic. They have compensated by using dial-up data circuits, dedicated 56 kbps circuits, telephone circuits, and other individual services. The introduction of *fractional T1 circuits* was well received by the small business customer who needed a smaller pipe than a full T1 circuit. The fractional circuit, which allows the customer to buy between two and eight DS-0s per circuit, normally requires a four-wire circuit between the central office and the customer premises. The difference between a T1 circuit and a

fractional T1 circuits
Allow customers to buy between two and eight DS-0s per circuit.

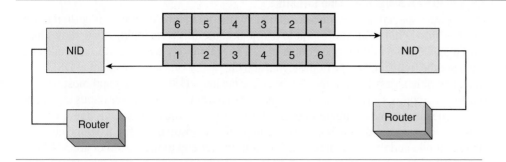

Figure 26–17
Fractional T1 circuits are point-to-point circuits used to carry voice and data between two locations. Customers that do not have enough information to warrant a full T1 circuit may choose to deploy fractional T1 circuits. The circuits function the same as a T1 circuit. Fractional T1s are offered in various speeds—normally 128, 256, 384, and 512 kbps.

fractional T1 circuit is more evident within the network than it is between the customer site and serving central office.

26.5.1 Fractional T1 Services

Fractional T1s are often used to connect two locations' LANs together, as shown in Figure 26–17. There, a network interface device (NID) is located at each location and is used to terminate the four-wire circuit from the telephone company. From the NID, the circuit interfaces into the customer's CSU; and from the CSU, the circuit is terminated into the end piece of equipment—a mainframe, PBX, router, bridge, or switch.

The fractional T1 circuit may ride with regular POTS wires or within its own group of cable pairs that pass through repeaters every mile. The circuit enters the central office into the transmission room where it interfaces into a digital cross-connect device or into a special channel bank. From the DCS or channel bank, the fractional T1 channels are groomed together with other individual channels and travel to the distant terminating central office—whether in the local calling area or across the country. Figure 26–18 illustrates how the channels are groomed into one T1 circuit. At the terminating office, the fractional T1 channels are again groomed onto their very own four-wire copper circuit that connects the central office to the customer location. The four-wire circuit is demarcated into another NID and then wired into the customer's CSU. The CSU interfaces into the customer's DTE equipment, which may be a mainframe, PBX, router, bridge, or switch.

Similar to dedicated data service, fractional T1 circuits are offered by both local and long distance providers. The benefit of fractional T1 circuits is the lower cost when compared to a full T1 circuit. Not all telephone companies offer fractional T1 service because channel grooming requires special network equipment.

Figure 26–18
Each channel is multiplexed onto a T1 circuit. Eight bits are taken from each channel, placed in a frame, and shipped out across the line.

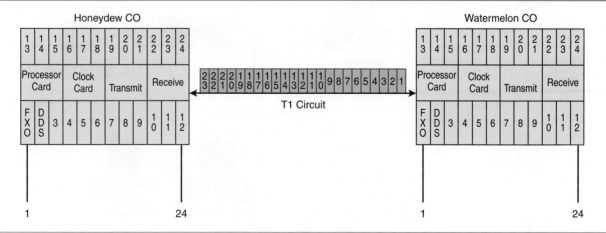

26.5.2 Digital Subscriber Line Circuits

digital subscriber line
Line that allows high-speed signals to travel on one copper pair.

The newest special service circuit offered by the telephone company is a *digital subscriber line* (DSL). DSL is difficult to define because it is a technology that allows high-speed signals to travel on one copper pair. The technology uses special line codes that increase the frequency range of the signal and bit-per-second rate. Residential DSL is different from business DSL in that the price for business DSL is higher and most business customers require synchronous speeds instead of asynchronous. Synchronous in the DSL world means that the upstream line speed is the same as the downstream line speed. If a business customer orders a 384 kbps circuit, both the upstream traffic from the customer to the far end and the downstream traffic from the far end to the customer are 384 kbps.

DSL is one of the newest transmission methods offered by the telephone company. It is mainly used for two types of services—Internet access and VPNs. Currently, it is primarily being used to carry data.

DSL Internet access gives the customer an always-on connection into the Internet, along with a higher bit-per-second rate across the access circuit. The path of a DSL Internet access circuit is shown in Figure 26–19. The circuit is terminated at the customer

Figure 26–19
Business customers use DSLs to access the Internet, create virtual private networks, and provide other applications normally provided by special service circuits.

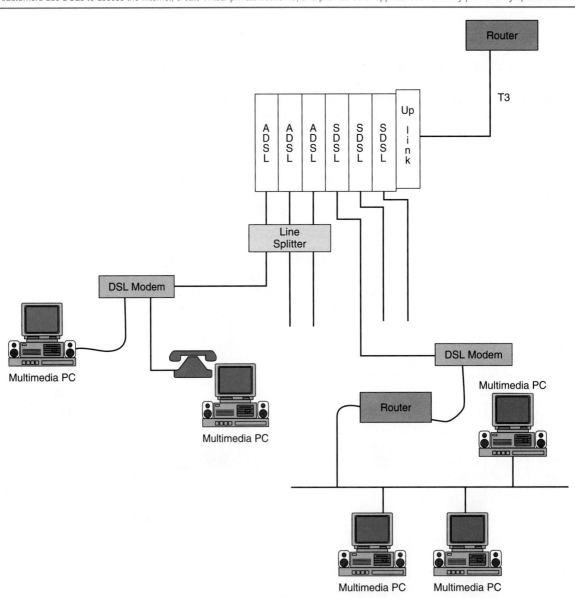

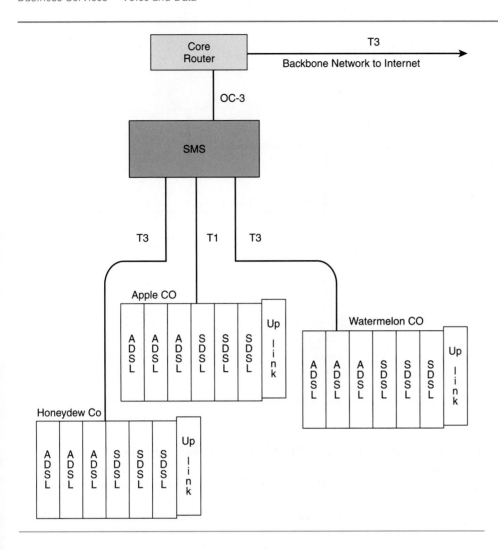

Figure 26–20
The backbone network that carries DSL traffic. Many COs are fed into an SMS aggregator that hands the information to a core router. The core router directs the traffic to the Internet.

site into a DSL modem that may or may not be the demarcation point for the telephone company. The copper pair terminates into the DSL modem using a common RJ-11 jack, similar to a standard telephone connection. The modem then connects into various types of equipment—the customer's PC (using an Ethernet connection, a router, or a bridge) or an Ethernet switch (using an Ethernet connection). Some DSL modems connect into a CSU using a V.35 connector (similar to a T1 termination). The opposite end of the circuit interfaces into a DSLAM at the serving central office. From the DSLAM, a higher-speed circuit, T1 or T3, connects into the big backbone network that eventually connects into the Internet. Trace the path of the DSL circuit through the network in Figure 26–20.

26.5.3 T1 Circuits

We should, at this point, be familiar with the T1 circuit. It is the backbone circuit of the PSTN and the customer's network. The uses of T1 circuits continue to grow as new end devices are invented and the price for a T1 circuit drops. T1s were first offered as a special service product in the early 1980s, usually to connect switch sites. Equipment manufacturers were able to discount the price on channel banks, interface cards into the PBX, and other T1 interfaces due to the high volumes of T1 circuits deployed by the PSTN. The end customer found that purchasing a T1 circuit to interface into its own terminating equipment proved to be cost-effective because the customer was not charged a per-minute usage charge for any traffic traveling on the circuit. The telephone companies priced the T1 circuit as if it were a dedicated pipe paid for by the customer. Therefore, the customer

Figure 26–21
Tie cable is used to connect separate
floors together in a building.

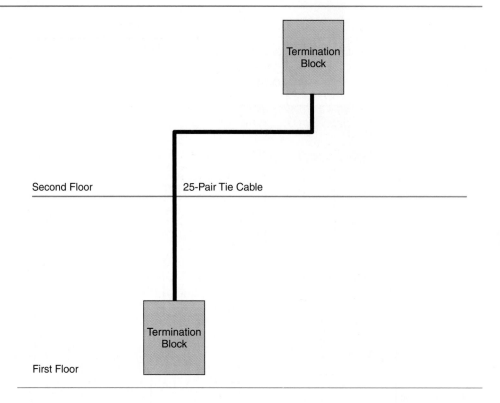

was allowed to place whatever traffic it liked on the circuit and pay only a monthly recurring fee for the pipe. T1 prices varied depending on the telephone company tariff, the
distance the circuit had to travel, and the bundled discount allowed for large volumes. T1
circuits are one of the most successful special service offerings the telephone companies
have introduced. Today, its cost is dropping rapidly, thanks to the deployment of DSLs.

T1 circuits can carry both voice and data, or only voice, or only data. The four most
common uses for T1 circuits are PBX to PBX, PBX to class 5 switch, LAN to LAN,
and channel bank to channel bank.

PBX to PBX

T1 circuits are commonly used as tie lines between PBXs. Figure 26–21 illustrates a
typical T1 tie cable. The telephone company's demarcation point at the customer site
is the network interface unit (NIU). From the NIU, the circuit interfaces into a channel
service unit (CSU) and from there the connection extends and terminates at the PBX
digital T1 card. The T1 travels through the network in the same fashion as all T1 circuits and may be multiplexed into a higher rate signal between central offices or switching centers until it reaches the end customer's serving central office. The central office
extends the T1 circuit to the customer's second location in the same way it terminated
at the opposite end. The T1 circuit now connects two PBXs located across town, across
the country, or somewhere in between.

PBX to Local Class 5 Switch

T1s are commonly used to connect a customer's PBX to the local telephone provider's
class 5 switch. DID/DOD circuits are configured to run on the T1 circuit. The path the
T1 signal travels is shown in Figure 26–22. The T1 leaves the central office on a four-
wire copper pair that either rides with the rest of the POTS copper pairs or separately
in a twenty-five-pair group of T1 circuits. The T1 circuits interface into repeater housings spaced about one mile apart. If the T1 rides on an HDSL circuit, the signal is repeated every 10,000 ft.

Figure 26–22
T1 feeding a customer's premises from the serving central office.

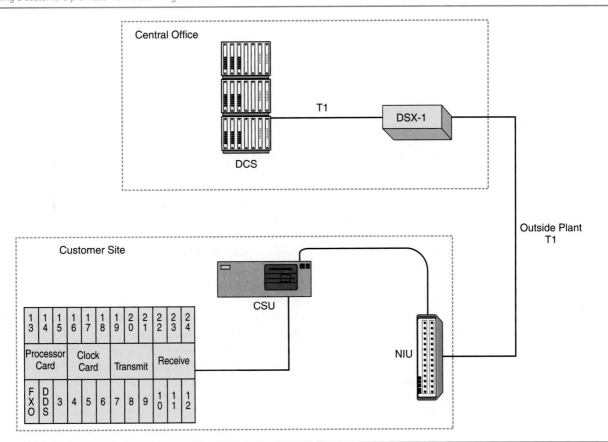

The copper pair enters the customer site and is demarcated into a Smart Jack called an NIU. The customer connects the NIU to its CSU, which interfaces into the digital TI card in the PBX. Some PBXs have CSUs built into the digital T1 interface card. The opposite end of the circuit terminates a DS-1 interface into the class 5 switch.

T1 LAN to LAN
A wide area network (WAN) contains multiple LANs connected together, often with T1 circuits. The physical structure of the connection is very similar to the tie line connecting a PBX. The business customer buys a 1.544 Mbps pipe from the telephone company, who is responsible for providing a physical circuit, whether fiber, copper, or RF. The circuit terminates at the customer premises into an NIU and from there connects into a CSU, which may be located in the router port or reside as an external unit. The T1 travels through the telephone network as a T1. It is normally multiplexed into higher order pipes such as DS-3s or OC-12s, then carried between switch sites. The greatest advantage that the T1 gives the customer is a dedicated digital pipe that carries 1.544 Mbps of information. Figure 26–23 depicts T1 circuits connecting remote LANs.

Channel Bank to Channel Bank
A business may require that part of the T1 circuit carry voice services and part carry data services. Channel banks are commonly placed at each end of the circuit to multiplex and demultiplex the individual DS-0s that, in turn, interface into the routers, PBXs, mainframes, and so forth. Figure 26–24 illustrates two channel banks connected with a T1 circuit. The copper circuit is demarced at the customer site into an NIU, the NIU is connected to a customer-owned CSU, and the CSU is connected to

Figure 26–23
Wide area networks are commonly deployed using T1 circuits. The routers direct the packets to the correct destination by reading the destination address.

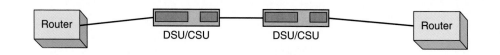

the channel bank where the individual twenty-four DS-0s are demultiplexed and fed out the twenty-four copper wires to a termination block. From the termination block, the signals are fed into their respective terminating equipment—a PBX, a mainframe, or a router. The same terminations happen at the opposite end of the T1 circuit. The customer has back-to-back channel banks to multiplex and demultiplex individual circuits.

The telephone company carries the T1 circuit intact through its network. The phone company does not groom or demultiplex any of the DS-0 channels inside the T1 circuit, but it may multiplex the T1 circuit into a higher bit rate signal such as a DS-3 or OC rate signal. The channels within the DS-1 are not touched. The customer purchases a point-to-point T1 circuit.

26.5.4 T3 Circuits

Large business customers, government agencies, and universities subscribe to T3 services from the telephone company. The T3 circuit is a point-to-point connection that carries large volumes of traffic between two points. A typical T3 connection is shown in Figure 26–25. Birdy, Incorporated has just ordered a dedicated T3 circuit from the telephone company to connect its two birdhouse factories located at opposite sides of town. Birdy, Inc. will use the T3 circuit to carry all of the inventory information, scalability, customer orders, video conferencing, and voice traffic between the two sites.

The local telephone company begins by building a fiber entrance facility into both of Birdy's sites. It places a small OC-3 multiplexer that it owns at each site with a DS-3 interface card. The telephone company completes the installation of the demarcation by connecting the DS-3 interface card to the DS-3 DSX panel in the telephone room and labeling the module as the demarcation point. Birdy cross connects the circuit at the demarcation point to its DS-3 multiplexer where the DS-3 is demultiplexed into T1 circuits that are cross connected to the appropriate locations. For example, three T1s are cross connected to the PBX for voice, five T1s are connected to a router to extend the LAN between the sites, and three T1s are cross connected into the video conferencing equipment. The T3 runs between the two sites on the telephone company's fiber as shown in Figure 26–25.

Figure 26–24
Business locations often use channel banks to multiplex the various types of circuits needed at each location.

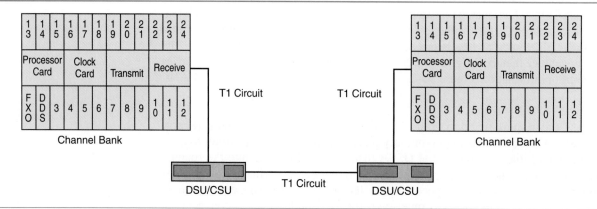

Figure 26–25
DS-3 circuits are multiplexed into optical transport cable.

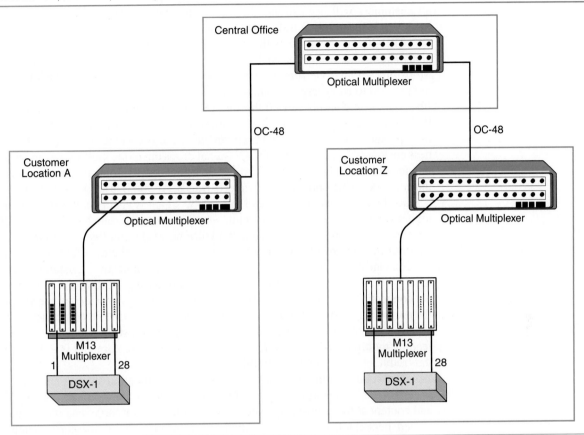

Fiber Optic Cable

The T3 circuit rides inside a higher-order circuit such as an OC-3, OC-12, or even an OC-48 signal. The fiber optic cable is terminated into an OC multiplexer where the DS-3 is dropped out on two coaxial cables. The cable is terminated onto a DS-3 DSX panel. For the individual T1 circuits to be terminated out of the T3 circuit, a second multiplexer is used to terminate the T3 circuit. An M13 multiplexer takes the T3 circuit and demultiplexes twenty-eight T1 circuits that are wired onto a DS-1 DSX panel and cross connected to the correct terminating equipment. The same process occurs at the opposite end of the T3 circuit. The customer must have fiber optic cable terminating in its building from the telephone company or other optical provider.

Satellite Termination

Some customers, such as large department store chains, have their own microwave frequencies that they use to send store data to a main central processing site. Antennas transmit out a T3 signal that is carried through the air to a satellite where it is pointed toward the antenna sitting on top of the main processing center's roof. The T3 carries all interstore information.

26.5.5 Optical Carrier Rates

One of the newest offerings by telephone companies is optical interconnection at an OC rate. The most common rate deployed to business customers is the OC-3c rate used to carry packetized or cell traffic such as IP or ATM. The large-size customer interfaces into the PSTN's ATM or IP network with a full OC-3c pipe. The advantage of using a full OC rate pipe is its ability to carry traffic using dynamic bandwidth allocation. The pipe is used only when data are being transmitted; therefore, it can be oversubscribed at an oversubscription ratio to be determined by the customer. For instance, to over-subscribe by a 10:1 ratio, ten customers would be placed on one transmission pipe. Not all customers will need to use the pipe at the same time, so the customer can save money by concentrating traffic on the circuit.

The OC-3c circuit gives the customer 155 Mbps of bandwidth. The *C* stands for *concatenate* and means that each of the DS-3s is viewed as one pipe instead of three 52 Mbps pipes. Only one framing header is used for all three STS-1s. OC-3cs are commonly used to carry large amounts of packetized or cell traffic. Traffic always rides on fiber optic cable and is carried through the network intact from point A to Point B. A typical scenario shows an optical multiplexer placed at the customer's location. The multiplexer may be owned and maintained by the telephone company or by the business customer. Either way, the OC-3c signal drops out of the multiplexer and terminates into an ATM switch or a router. The OC-3c signal travels from customer Site A through the network's fiber optic SONET rings to Site B. At Site B, a second optical multiplexer is placed like the one at site A and the OC-3c signal is again demultiplexed out of the higher-order signal and terminated into an ATM switch or router. OC-3cs and OC-12cs are becoming more popular as data and voice networks converge. Similar to the applications offered to residential customers, new services are being offered to business customers. One that is growing in popularity is Ethernet connection. Customers can extend their LANs between two remote locations by purchasing a 100 Mbps or 1 Gbps Ethernet connection. The advantages that Ethernet connections provide are the simplicity of the protocol and low cost due to volume production of Ethernet components. The customer does not have to have a data networking specialist to manage IP-based WANs feeding multiple circuit types. The current employee base normally is Ethernet-trained, thus reducing labor costs and training time.

A second offering to business customers is application sharing and data storage. Data-storage companies are allowing customers to locate their servers, applications, and content at the data-storage site, reducing the need for large data rooms and highly skilled personnel. Large storage devices are used to hold customer content. High-bandwidth circuits are used to connect the customer to the storage facility.

SUMMARY

The goal of this chapter was to describe some of the numerous business services offered by the telephone company. New services are constantly being developed and offered by the PSTN to its business customers. The chapter was divided into several sections that defined the services as switched or dedicated and then further defined them as voice or data. The importance of understanding that a dedicated service does not interface into the telephone switch was emphasized in this chapter. The different ways a T1 circuit is used by the business customer were discussed, and the DSL as a business service fractional T1 circuit was introduced. The chapter concluded by introducing OC-3c circuits that are now being used to carry packetized or cell data. OC-3c circuits are the first to allow the convergence of voice and data across the PSTN.

CASE STUDY

1. Select five small business customers from the list in Appendix A and provide a voice and data services solution. Determine the type of circuit, the type of customer premises equipment (CPE) needed, and the number of circuits needed.
2. Select one of the factories from the list in Appendix A and provide a voice and data services solution. Determine the type of switch, features, circuits, type of CPE, and the number of circuits needed.
3. Design the school system communications network for the elementary school, high school, and community college.

4. Design voice and data services for the municipal offices located in the courthouse.
5. Select one other business and provide a detailed network diagram, including the type of voice and data circuits you will use, and the reason you have selected the design.

REVIEW QUESTIONS

1. Explain what the term *business subscriber* means as it pertains to the telecommunications industry.

2. Business clients are often charged a per-minute usage fee when calling within the local call area. What term is used to describe this charge?

3. List five common call features used by business subscribers.

4. Explain why business customers purchase PBX or key systems.

5. What is the difference between a PBX and a key system?

6. What does the term *four-digit dialing* refer to?

7. Explain the purpose of a tie line and draw a picture showing how a point-to-point tie line is deployed between two business locations.

8. The term *special services* is used to describe features provided by the switch primarily for business customers such as call return or call detail recordings. True or false? Explain.

9. Provide an example of where an FX circuit is used. Draw the end-to-end path of an FX circuit and explain why a business customer would require one.

10. What does the term *OPX* stand for?

11. Provide an example of where an OPX circuit would be placed. Draw the end-to-end path of the OPX circuit and explain why it is used.

12. A special service analog data circuit may be a two- or four-wire circuit. True or false? Explain.

13. List and explain two reasons why analog data circuits are not commonly used in today's network.

14. Who developed and first deployed digital data services lines?

15. Are digital data circuits two- or four-wire circuits?

16. Think of a situation in which a digital data circuit is deployed. Draw the end-to-end circuit path of a typical digital data point-to-point circuit. Draw the end-to-end circuit path of a typical digital data multipoint circuit.

17. Explain why a business customer would use an ISDN line for its receptionist's telephone line.

18. Provide a scenario that shows a use for an ISDN PRI circuit in a business setting.

19. Why do many businesses purchase T1 digital circuits?

20. List three scenarios showing uses of a T1 digital circuit in a business setting.

21. List the digital circuit types offered to business customers by telephone providers.

KEY TERMS

local measured service (695)
local measured unit (695)
intraLATA voice services (695)
rate center (695)
interLATA voice services (695)
wide area telephone service (WATs) (695)
time-of-day routing (697)
call blockage (697)
call-attempt profile (697)
automatic number identifier (698)
point-of-call routing (698)

dialed-number identification service (698)
900 number (698)
centrex (698)
four-digit dial (699)
class-of-service restrictions (699)
automatic identification of outward dialing (699)
call pickup (699)
call transfer (699)
call forward (699)
call hold (699)
conference calling (699)

distinctive ringing (699)
ISDN Basic Rate Interface (ISDN BRI) (700)
fast-connect times (700)
call appearances (701)
video conferencing (701)
desktop conferencing (701)
voice and data integration (701)
ISDN Primary Rate Interface (ISDN PRI) (701)
hunting (703)
tie line (707)

foreign exchange (FX) circuit (708)
off-premises exchange (OPX) circuit (708)
dial-up data (710)
switched 56 (712)
analog dedicated data circuit (713)
dedicated 56 kbps service (713)
point-to-point circuit (714)
multipoint dedicated digital circuit (714)
fractional T1 circuits (716)
digital subscriber line (718)

The Internet

Objectives

After reading this chapter, you should be able to
- Describe the Internet.
- Describe Internet services.

Outline

Introduction

27.1 The Internet

27.2 The Internet Network

27.3 Internet Services Data Centers, Equipment, Content, and Applications

■ INTRODUCTION

The Internet was the phenomenon of the twentieth century. It changed the way the world performs business, searches for information, exchanges ideas, finds new friends all over the world, and of course communicates. The purpose of this chapter is to explain how the Internet works, to discuss what services are available on the Internet, and to discuss who provides those services. The chapter begins by defining the Internet and what makes it work. Services such as e-commerce, browsing the Web, and chat rooms are explained. The chapter also looks at ISPs, ASPs, ICPs, and network providers.

■ 27.1 THE INTERNET

27.1.1 The Internet Structure
History of the Internet
The history of the Internet involves university professors, the military, and researchers looking for information. The *Advanced Research Projects Agency Network* (*ARPANET*) started it all. The ARPANET was funded by the federal government to tie universities and military installations together on a network designed to survive catastrophic events such as nuclear war, multiple earthquakes, and other catastrophe-related disasters that might threaten world peace. The ARPANET was where packet-switching networks evolved and data communications were perfected. Initially, only the computer-literate

Advanced Research Projects Agency Network (ARPANET)
A Department of Defense data network.

university or government personnel accessed the ARPANET due to the complexity of the networking protocols. Clicking on an icon was still years away. In 1983, the government formed the *DARPANET* from ARPANET and MILNET; in 1990, the DARPANET was phased out and control over the Internet was given to the public. The *Internet Engineering Task Force (IETF)*, formed in 1986, consisted of engineers and programmers who became the keepers of the Internet. They still oversee Internet standards and help resolve issues relating to the Internet.

Corporate America soon saw the value of the interconnected network as a way for its research and development departments to exchange information. As the number of sites grew on the network, managing and maintaining all of the addresses and connections became almost impossible. The advent of the domain name and domain name resolvers helped to increase the scalability of the network, while the introduction of the HyperText Transfer Protocol (HTTP) improved the usability of the network and consequently increased the number of lay users on the Net. Browsers, search engines, and simplified networking methods all helped to simplify Web surfing. The Internet was gradually evolving into a very usable entity that provided information to the masses. Presently, millions of Web sites are connected to the Internet and there are well over 1 billion users.

Today, very young children access the Web, as do great-grandmothers who send e-mail to their grandchildren. The fastest growing area on the Internet, however, is e-commerce. Some critics fear that the Internet will soon be nothing more than a conglomeration of marketing hype and items to buy. The altruistic intention of the founders of the Internet is slowly being eroded by the millions of dollars being made from it.

Routing Packets

The Internet cannot be defined easily. A topographical view of the Internet shows a mishmash of networks all connected like a spiderweb built by a disoriented spider. An analogy that best defines the Internet is that of our road system. Each of us lives on a road that connects to larger roads that in turn connect to even larger roads or to smaller roads or even driveways. You could view the Internet like roads meandering through the country, connecting small towns to big towns to big cities. When a house is built, a driveway provides a path to the road, allowing the people in the house to travel wherever they wish. The same holds true with the Internet. When a new Internet host is attached to the network, it is like a house being built in a neighborhood; the connection between the host and the network provides a pathway to the new host site. This pathway allows all users on the Internet to access the new site and allows the new host site to access anyone on the Internet.

The address of the new home is noted in the directory at the post office, and the post office uses the address label on the envelope to route the mail to the recipient. The router, similar to the post office, uses the Internet Protocol (IP) address to route information to the recipient. The address used to direct mail to the new home owner is hierarchical and allows for each layer of the postal network to route according to a portion of the address. Each post office looks at the ZIP Code first and routes to the correct city's end post office. The home post office then looks at the house number and street address and delivers the mail to that address. Once the mail reaches the house, the name on the envelope is used to distribute the letter to the correct person in the house.

Routing information through the Internet is very similar to the snail-mail analogy. The IP address is used first to determine to which network the information wants to go. The network portion of the IP address is used in the same way the ZIP Code is used to direct the letter to the correct town or city. Once the packet of information reaches the correct network router, the next portion of the IP address is used to route the packet to the correct host. The host may be a single PC, a router, a Web server, or another terminating device. The host device receives the information and, if necessary, looks at the remaining portion of the address to determine to which device in the subnetwork the information needs to go. The *uniform resource locator (URL)* is the term used to define the entire HTTP address. It is important to understand that the IP address is part of

DARPANET
Defense ARPANET.

Internet Engineering Task Force (IETF)
A task force that sets the standards that run the Internet.

uniform resource locator (URL)
An Internet address.

Table 27–1
Methods Used to Access Web Servers on the Net.

Type	Address	Description
Harvard	Responds with multiple listings	Web search "worm" goes out and retrieves sites with the word *Harvard* in them.
Henry@Spidercom.com	Sends the message through Spidercom's network to Henry's workstation	e-mail address
http://www.Harvard.com/http/science/	Responds with the site information from the science department	Specific Web site going all the way to the science department's Web server

the URL and is used to route information through the network all the way to the gateway host of the subnetwork. The remaining portion of the URL address, as shown in Table 27–1, routes the packets to the correct destination device, also called a host. The destination device may be a Web server, an e-mail server, a supercomputer, or something else. You may view this portion of the address as being similar to the home owner's name above the street address. The home recipient looks at the address on the envelope to see to which person the letter should go. The postmaster does not come into the house and hand each individual person in the house the mail destined for him or her. Instead, the postmaster depends on the home owner to distribute it to the appropriate recipient.

The routers in the network are able to read the IP address and know where to send the packets. The Internet is able to route traffic through the networks because of the hierarchical addressing scheme that uses source and destination addresses. Hosts are attached to networks, networks are attached to other networks, and routers are the post offices in the networks.

Network Connections

The Internet consists of thousands of smaller networks connecting together to form pathways to every host on every network. For all of these networks to interconnect, defined Internet hubs or gateways have been established to act as central routing points for all users on the network. Hubbing points are large central sites where all of the ISPs, network providers, and other large network carriers can interconnect. The two large hubbing points are MAE East and MAE West. *Metropolitan area exchange (MAE)* was built to provide a hub point for the thousands of interconnecting networks on the Internet. MCI WorldCom (Metropolitan Fiber Systems [MFS]) maintains MAE East in Vienna, Virginia, and MAE West in San Jose, California. ISPs that want to connect into MAE East and MAE West must pay a connection fee to MCI WorldCom (MFS). When additional space was needed for both MAE East and MAE West, additional MAE sites were built in the same vicinity as the originals. The new sites are called MAE West, PlusPlus and MAE East PlusPlus. They contain thousands of wire connections, large switches, routers, and hubs—all used to interconnect all of the networks, as shown in Figure 27–1.

Additional interconnection points have been organized on the East Coast and West Coast to help alleviate some of the congestion at the MAE locations. *Network access points (NAPs)* were built in San Francisco, Chicago, and New Jersey to provide additional interconnection points for ISPs and other networks. Additional NAPs are still being constructed. NAPs either consolidate the traffic before connecting into the MAE sites or route traffic to other NAP interconnection partners. When you dial into the Internet, you do not access the MAE sites directly. You first hit your ISP, then travel with

metropolitan area exchange (MAE)
A point where ISPs interconnect to exchange traffic at the national backbone level.

network access point (NAP)
Software within a switch capable of recognizing a call that requires processing by Advanced Intelligent Network (AIN) logic, which, upon recognizing such a call, routes the call to a service switching point (SSP) or ASC switch.

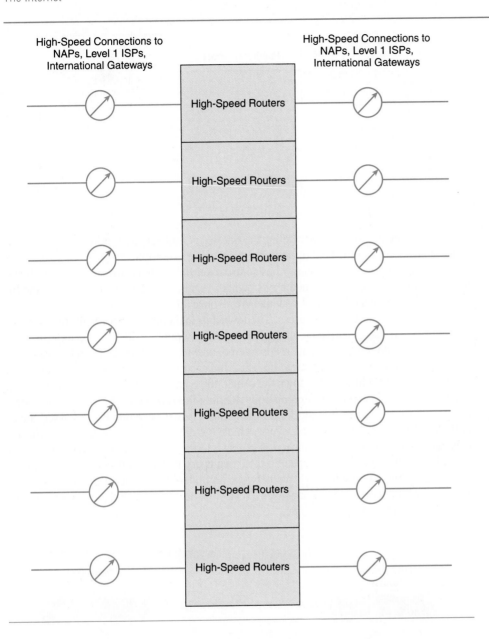

High-Speed Connections to
NAPs, Level 1 ISPs,
International Gateways

High-Speed Connections to
NAPs, Level 1 ISPs,
International Gateways

High-Speed Routers

High-Speed Routers

High-Speed Routers

High-Speed Routers

High-Speed Routers

High-Speed Routers

High-Speed Routers

Figure 27–1
MAE sites may be seen as the hubs where the millions of packets of information traveling around the Internet pass through on their way to the Web site. The MAE site houses lots of equipment that routes all of those packets.

hundreds of other people's traffic to an NAP. At the NAP, you are combined with thousands of others' traffic and routed to one of the MAEs. If you are accessing a Web site located on a network served by a different ISP, your request will travel from your ISP to an NAP, to one of the MAEs, back to an NAP, to the other ISP, and finally to the host site you have requested. The router selects the paths according the source and destination addresses attached to the packet. The route may vary depending on the network, the geographic location, the time of day, and the ISP being used. The Internet depends on the connectionless nature of the IP and makes connections as the packet travels along the path from router to router.

It is not always necessary to interconnect to a different network through one of the MAE sites. Peering relationships allow traffic to pass between one network provider to a second network provider as long as they have a peering point to route traffic through, as shown in Figure 27–2. NAP locations are used as peering points, as are other private peering locations. Peering with multiple providers helps to reduce the number of connections needed into MAE East and MAE West. Figure 27–3 shows a typical peering relationship among three big carriers.

Figure 27–2
Peering points are interconnection
points between two Internet providers.

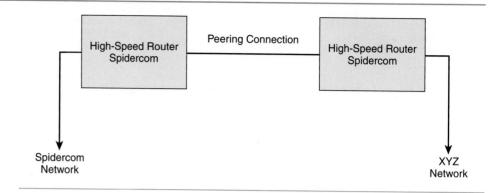

It would be nice to be able to define the Internet completely and show each of the possible connections, but the task of defining the Internet is complicated by the fact that not one company or government agency has control over the entire network. It is a worldwide network of networks, each defined by its owners. Again, the best way to visualize the Internet is to compare it to our road systems. We all have different types of driveways—some are paved, some are dirt, some grass. The township maintains the township roads in our community and makes sure they interconnect with the county roads. The county roads interconnect to state roads, and state roads interconnect to federal roads. Each township has its own rules and regulations, such as how fast you can travel, how wide the shoulders will be, if there will be a line drawn down the center, the type of road signs used, and so forth. Each county also has rules and regulations defining the road systems, as does each state and each country. The important thing to remember is that everyone in the United States drives on the right side of the road; every vehicle has an accelerator, a brake, and turn signals; and every state has interconnection points into the federal highway system. The Internet is no different. There are specific rules that every ISP and network provider have to follow for traffic to flow through the network unhindered. But each ISP's network may vary in its traffic speed, the protocol used to carry the information, and the types of circuits used to interconnect to the customer.

Figure 27–3
Three carriers peering using Border
Gateway Protocol (BGP) routing.

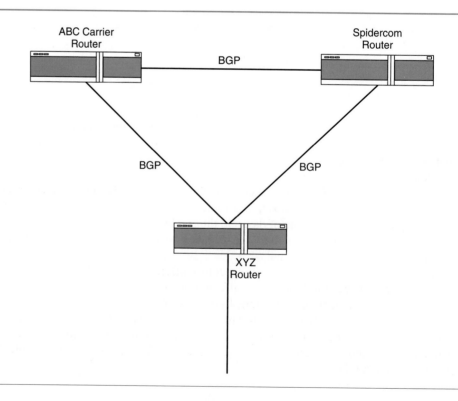

Continuing with our analogy, if you begin a journey in California and want to reach New York, you have the option of traveling many different routes, including federal interstates or smaller secondary roads. The Internet is no different. The packets have many choices of paths to reach their final destination. In addition, even though there are multiple Internet providers, there are specific rules that each must follow, such as using TCP/IP to route traffic. HTTP is the protocol used to route Web information. A *domain name server* (DNS) is the server used to resolve naming derivations between the IP address and the domain name. ISPs have certain nuances they can adopt to differentiate themselves from competitors, but the overall structure of each varies little among providers.

domain name server
A server used to resolve naming derivations between the IP address and the domain name.

27.1.2 Layer 3 Internet Network—Routing Traffic
IP Addresses—Host and Network
In Chapter 12, the IP address structure was defined. We will summarize IP addressing here; but, for a full review, return to Chapter 12.

A thirty-two-bit binary number is used to define one IP address, making 2,294,967,296 addresses available for network devices. The IP address is divided into a host and a network portion, and each network interface requires an IP address. Multiple devices may hang off a router, such as a Web server or internal routers, each of which needs an IP address.

The IP addressing structure is a five-class architecture—Class A, Class B, Class C, Class D, and Class E. The Class A address uses the first seven bits to represent the network and the remaining twenty-four bits to represent the host device. Class A addresses provide 128 network addresses and 16,780,000 host addresses for each network address. Only large organizations and countries are given Class A addresses.

Class B addresses designate the first fourteen bits as the network address and the remaining sixteen bits as the host addresses. This provides 16,384 network addresses and 65,536 host addresses per network address, which are given to large corporations that require thousands of host addresses and multiple network addresses.

Class C addresses use the first twenty-one bits for the network address and eight bits for the host address. This makes 2,000,000 network addresses and 256 host addresses per network address. These are given to small corporations and businesses. Table 27–2 lists the Class A to Class E address structure.

Figure 27–4 illustrates a typical network showing a router connected to two networks and four devices. Most Class C addresses do not provide enough address space for many networks. Therefore, the two-level address—network and host—was increased to a three-level address—network, subnetwork, and host. The eight-character host address was subdivided into a subnetwork and a host address. The advantage of adding a subnetwork address is twofold. Besides increasing the number of networks and, in turn, hosts allowed per Class C address, it reduces the number of entries required in the router's routing tables and, consequently, the processing power needed per router.

Class	Address Structure	Number of Networks	Number of Hosts
A	Byte 1 = network byte 2, 3, 4 = hosts	128	16.78 million hosts
B	Byte 1, 2 = network Byte 3, 4 = hosts	16,384	65,536 million hosts
C	Byte 1, 2, 3 = network Byte 4 = host	2,000,000	256 hosts
D	1110 header remaining bytes = multicast address	268,000,000 multicast addresses	NA
E	1111 header; remaining bytes = experimental addresses	268,000,000 addresses	NA

Table 27–2
Class A through Class E Address Structures.

Figure 27–4
Four devices connected to two network devices. The routing is performed by cracking the packet and reading the source and destination addresses.

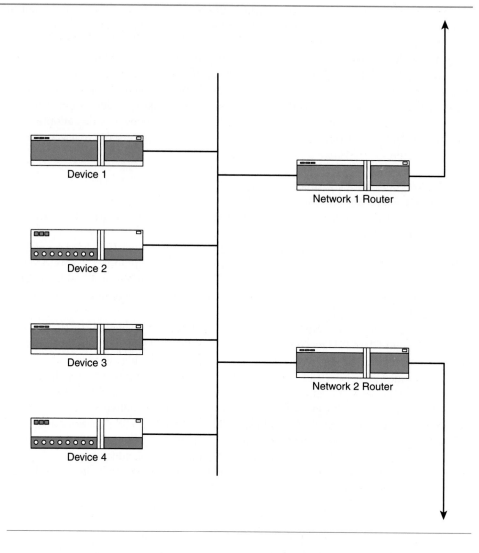

Device 1

Network 1 Router

Device 2

Device 3

Network 2 Router

Device 4

The network and host addresses just described are used to direct traffic through the Internet. For example, if you ask for information from http://www.Chocolate.com, your request is directed to the network address associated with Chocolate.com. Domain name resolution is discussed later in the chapter, but for now understand that Chocolate.com has an equivalent IP address such as 172.20.10.10. The request travels to your ISP where it is routed to a peering point, such as an NAP, to determine whether the destination address belongs to one of the peering providers or needs to be passed on to one of the MAEs. If one of the peers has Chocolate.com on its network, the request is routed through that provider's network. Once the provider receives the request, it passes it on to the correct destination network. In our case, the destination network is a second ISP that buys address space from the larger ISP. The smaller ISP looks at the network and host address to determine where to send the request and sees that the address belongs to a client connecting into its network. The request is then shipped to the client's router, which looks at the host address and sees that it belongs to the Web server that hangs off the router. The router passes the packet on to the Web server where it interprets the request and returns the Web page from Chocolate.com to my PC.

Domain names are used to steer the Internet user to the site he or she wishes to reach. A domain name has an IP address hidden behind it. The IP address is used to

Table 27–3
Description of Domain Name
Governing Bodies.

Acronym	Full Name	Region or Responsibility
InterNIC	Internet's Network Information Center	Overall responsibility of the Internet Registry. Oversees top-level domains and regional registries.
ARIN	American Registry for Internet Numbers	Americas and southern region for DNS allocation
APNIC	Asia Pacific Network Information Centre	Asia/Pacific region for DNS allocation
RIPE NCC	Réseaux IP Européens Network Coordination Centre	Europe and surrounding areas for DNS allocation

route the packets of information through the network, while the domain name is used because it is much easier for the user to remember than the IP address.

The organization responsible for assigning, monitoring, and policing domain names and other Internet-related protocol issues is the Internet Assigned Numbers Authority or IANA. IANA oversees the InterNIC and the regional registries that allocate and deal with Internet addressing issues for their geographic areas. The InterNIC deals with top-level domains and other higher-order issues relating to the Internet. The regional registries report into the InterNIC and are responsible for allocating DNS names for their regions. Table 27–3 lists the organizations and their regions. Because the government does not regulate Internet standards, each of the Internet organizations must adhere to the agreed-upon rules for assigning domain names and Internet numbers in order to avoid the consequences of an unruly Internet.

Initially, seven *top level domains* (TLDs) were established by the InterNIC. Due to the rapid growth of Internet users, seven additional TLD domains have been added to the Internet. Table 27–4 lists the fourteen domains in use today and Figure 27–5 shows a graphical representation of all fourteen domains. The country code top-level domains are used to route information to the correct country. The suffix assigned is a

top-level domains
Domains established to route information through the Net—.com, .net, and so on.

Table 27–4
Fourteen Internet Domains.

Suffix	Definition
.edu	Educational institutions, specifically, colleges and graduate schools
.gov	Governmental agencies such as IRS, FCC
.net	Networking companies such as UUNET, AOL, AT&T, Spring, Global Crossing
.com	Commercial businesses
.mil	Military institutions such as Marines, Army, Navy, Air Force, Coast Guard, National Guard
.int	International treaties and international treaty databases
.org	Organizations that do not fit under commercial or government such as nonprofits and user groups
.biz	Businesses similar to commercial; established to alleviate demand on .com
.museum	Museums such as the Smithsonian
.name	Individual names
.info	Unrestricted use; helps free up addresses for businesses
.pro	Lawyers, doctors, accountants, and other professionals
.coop	Cooperatives like farmers coop, food coops
.aero	Airports and transportation industry

Figure 27–5
Top-level domains (TLDs) established
to organize the Internet into
categories. The darker circles
represent older standard TLDs; the
lighter circles represent the newly
designated domains.

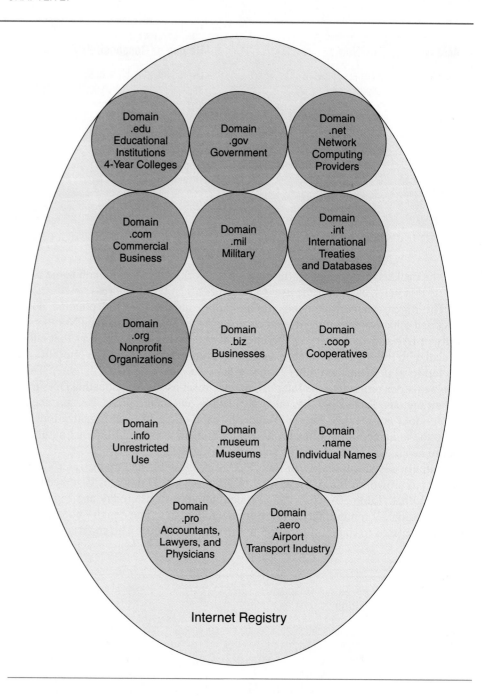

two-letter suffix and is kept in the Internet Registry, which the InterNIC oversees. Table 27–5 lists each country suffix as defined by the IANA's InterNIC branch. An individual country may further segregate the TLDs internally to accommodate that country's network structure.

The IANA also assigns autonomous system (AS) numbers to ISPs through regional registries. Service providers apply to ARIN for AS numbers and pay a nominal processing fee. The AS numbers may be compared to the carrier identification codes (CICs) in the voice world. Carrier identification codes 10-10-220, as advertised on television, are used as dial-around numbers that define a particular service provider's network. AS numbers are used in a similar way to help route millions of packets across multiple service providers' networks more efficiently. Sophisticated routing protocols, such as BGP, depend on AS numbers when exchanging routing information between service providers' networks. No two networks have the same AS number, and no two long distance telephone companies have the same CIC.

Table 27–5
Country Codes.

Country	Suffix	Country	Suffix
Ascension Island	.ac	Djibouti	.dj
Andorra	.ad	Denmark	.dk
United Arab Emirates	.ae	Dominica	.dm
Afghanistan	.af	Dominican Republic	.do
Antigua and Barbuda	.ag	Algeria	.dz
Anguilla	.ai	Ecuador	.ec
Albania	.al	Estonia	.ee
Armenia	.am	Egypt	.eg
Netherlands Antilles	.an	Western Sahara	.eh
Angola	.ao	Eritrea	.er
Antartica	.aq	Spain	.es
Argentina	.ar	Ethiopia	.et
American Samoa	.as	European Union	.eu
Austria	.at	Finland	.fi
Australia	.au	Fiji	.fj
Aruba	.aw	Falkland Islands (Malvinas)	.fk
Aland Islands	.ax	Micronesia, Federal State of	.fm
Azerbaijan	.az	Faroe Islands	.fo
Bosnia and Herzegovina	.ba	France	.fr
Barbados	.bb	Gabon	.ga
Bangladesh	.bd	United Kingdom	.gb
Belgium	.be	Grenada	.gd
Burkina Faso	.bf	Georgia	.ge
Bulgaria	.bg	French Guiana	.gf
Bahrain	.bh	Guernsey	.gg
Burundi	.bi	Ghana	.gh
Benin	.bj	Gibraltar	.gi
Bermuda	.bm	Greenland	.gl
Brunei Darussalam	.bn	Gambia	.gm
Bolivia	.bo	Guinea	.gn
Brazil	.br	Guadeloupe	.gp
Bahamas	.bs	Equatorial Guinea	.gq
Bhutan	.bt	Greece	.gr
Bouvet Island	.bv	South Georgia and the South	
Botswana	.bw	Sandwich Islands	.gs
Belarus	.by	Guatemala	.gt
Belize	.bz	Guam	.gu
Canada	.ca	Guinea-Bissau	.gw
Cocos (Keeling) Islands	.cc	Guyana	.gy
Congo, the Democratic		Hong Kong	.hk
Republic of the	.cd	Heard and McDonald Islands	.hm
Central African Republic	.cf	Honduras	.hn
Congo, Republic of	.cg	Croatia/Hrvatska	.hr
Switzerland	.ch	Haiti	.ht
Cote d'Ivoire	.ci	Hungary	.hu
Cook Islands	.ck	Indonesia	.id
Chile	.cl	Ireland	.ie
Cameroon	.cm	Israel	.il
China	.cn	Isle of Man	.im
Colombia	.co	India	.in
Costa Rica	.cr	British Indian Ocean	
Cuba	.cu	Territory	.io
Cape Verde	.cv	Iraq	.iq
Christmas Island	.cx	Iran, Islamic Republic of	.ir
Cyprus	.cy	Iceland	.is
Czech Republic	.cz	Italy	.it
Germany	.de		*(Continued)*

Table 27–5
Continued

Country	Suffix	Country	Suffix
Jersey	.je	Netherlands	.nl
Jamaica	.jm	Norway	.no
Jordan	.jo	Nepal	.np
Japan	.jp	Nauru	.nr
Kenya	.ke	Niue	.nu
Kyrgyzstan	.kg	New Zealand	.nz
Cambodia	.kh	Oman	.om
Kiribati	.ki	Panama	.pa
Comoros	.km	Peru	.pe
Saint Kitts and Nevis	.kn	French Polynesia	.pf
Korea, Democratic		Papua New Guinea	.pg
People's Republic	.kp	Philippines	.ph
Korea, Republic of	.kr	Pakistan	.pk
Kuwait	.kw	Poland	.pl
Cayman Islands	.ky	Saint Pierre and Miquelon	.pm
Kazakhstan	.kz	Pitcairn Island	.pn
Lao People's Democratic		Puerto Rico	.pr
Republic	.la	Palestinian Territories	.ps
Lebanon	.lb	Portugal	.pt
Saint Lucia	.lc	Palau	.pw
Liechtenstein	.li	Paraguay	.py
Sri Lanka	.lk	Qatar	.qa
Liberia	.lr	Reunion Island	.re
Lesotho	.ls	Romania	.ro
Lithuania	.lt	Russian Federation	.ru
Luxembourg	.lu	Rwanda	.rw
Latvia	.lv	Saudi Arabia	.sa
Libyan Arab Jamahiriya	.ly	Solomon Islands	.sb
Morocco	.ma	Seychelles	.sc
Monaco	.mc	Sudan	.sd
Moldova, Republic of	.md	Sweden	.se
Madagascar	.mg	Singapore	.sg
Marshall Islands	.mh	St. Helena	.sh
Macedonia, of Former		Slovenia	.si
Yugoslav Republic	.mk	Svalbard and Jan Mayen Islands	.sj
Mali	.ml	Slovak Republic	.sk
Myanmar	.mm	Sierra Leone	.sl
Mongolia	.mn	San Marino	.sm
Macau	.mo	Senegal	.sn
Northern Mariana Islands	.mp	Somalia	.so
Martinique	.mq	Suriname	.sr
Mauritania	.mr	Sao Tome and Principe	.st
Montserrat	.ms	El Salvador	.sv
Malta	.ms	Syrian Arabic Republic	.sy
Mauritius	.mu	Swaziland	.sz
Maldives	.mv	Turks & Ciacos Islands	.tc
Malawi	.mw	Chad	.td
Mexico	.mx	French Southern Territories	.tf
Malaysia	.my	Togo	.tg
Mozambique	.mz	Thailand	.th
Namibia	.na	Tajikistan	.tj
New Caledonia	.nc	Tokelau	.tk
Niger	.ne	Timor-Leste	.tl
Norfolk Island	.nf	Turkmenistan	.tm
Nigeria	.ng	Tunisia	.tn
Nicaragua	.ni	Tonga	.to

Table 27–5
Continued

Country	Suffix	Country	Suffix
East Timor	.tp	Saint Vincent and the Grenadines	.vc
Turkey	.tr	Venezuela	.ve
Trinidad and Tobago	.tt	Virgin Islands, British	.vg
Tuvalu	.tv	Virgin Islands, U.S.A.	.vi
Taiwan	.tw	Vietnam	.vn
Tanzania	.tz	Vanuatu	.vu
Ukraine	.ua	Wallis and Futuna Islands	.wf
Uganda	.ug	Western Samoa	.ws
United Kingdom	.uk	Yemen	.ye
U.S. Minor Outlying Islands	.um	Mayotte	.yt
United States	.us	Yugoslavia	.yu
Uruguay	.uy	South Africa	.za
Uzbekistan	.uz	Zambia	.zm
Holy See (Vatican City State)	.va	Zimbabwe	.zw

Subdomains branch off each TLD like leaves on a tree. An example of a subdomain is Harvard.edu, with Harvard as a subdomain of the .edu top-level domain. Harvard may also have subdomains hanging off its address. The structure of the domain address is similar to a file name. Table 27–6 illustrates how the domain address is structured.

In the early days of the Internet, designers realized in order for the general population to be able to utilize the Web a method was needed to simplify the naming schema used to designate a host or end device. DNS was designed to resolve or simply match up the IP address associated with the device to the friendly URL name. The domain name server's function in life is to resolve (translate) the user-friendly Web address to the cumbersome, hard-to-remember IP address. Early on, the network designers decided that domain name servers must be distributed throughout the network and be maintained by the ISPs using the services. Therefore, network providers are responsible for having their own DNS database, where inquiries enter and are translated. The databases, called name servers, sit on service providers' networks. For example, if Spidercom is my friend's ISP, it is responsible for adding her chosen Internet name to its DNS. All traffic destined for her host (her PC or router) that hangs off Spidercom's network is first directed to Spidercom's DNS by using the Spidercom domain name. When I send e-mail to my friend Caroline at Caroline@Spidercom.net, my message travels from my ISP, where the address is translated to reach the next point on the network, in

Table 27–6
The Internet Address.

Web Address	Host Address	Subdomain	TLD or ccTLD
Harvard.edu	NA	Harvard	edu
Chocolate.com	NA	Chocolate	com
marine.mil	NA	marine	mil
aol.net	NA	aol	net
fcc.gov	NA	fcc	gov
iana.org	NA	iana	org
dulles.aero	NA	dulles	aero
cambridge.gb	NA	cambridge	gb
nepal.np	NA	nepal	np
farmers.coop	NA	farmers	coop
Henry@AOL.net	Henry	AOL	net
Lisa@Earthlink.com	Lisa	Earthlink	com
Jsmith@Spidercom.net	Jsmith	Spidercom	net
Birdiebear@att.net	Birdiebear	att	net

Figure 27–6
An Internet address is divided into
several sections. The first portion of
the address routes the packet to the
network by using the domain address.
Once the packet enters the domain,
the address further routes the packet
to a subdomain. Finally, the packet is
routed to the correct zone by reading
the zone address.

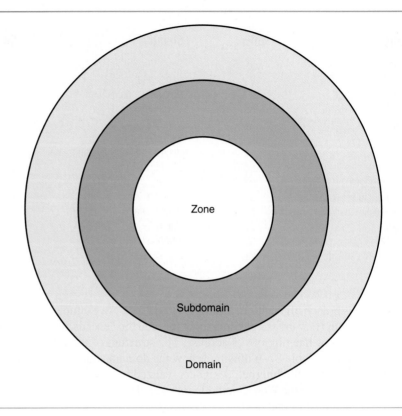

our case to a name server on the .net network. From .net, the message is pointed toward the Spidercom network using the Spidercom portion of the Internet address. When it gets to Spidercom's network, it is translated further within its name server into Caroline's host IP address. Spidercom uses the host portion of the IP address and routes my message all the way to Caroline's PC.

This is a very simplified view of how a DNS functions, but it does provide a good overview of how the Internet routes traffic through the mesh of networks. It is very similar to the U.S. Postal Service. The hierarchical address routes the information from the source post office through the large postal network to the state distribution site and, finally, to the local post office. Spidercom may be seen as the local post office and Caroline's computer as the house being served by the town's post office.

The DNS administrator is responsible for adding and deleting names in the database. A portion of a network within a domain is called a *zone*. A zone handles the administration for that portion of the domain. Figure 27–6 illustrates how the domain space is parsed out as domains, subdomains, and zones. There may be multiple zones within a domain or subdomain. In our previous example, Harvard is a subdomain hanging off the .edu top-level domain. Spurring off the Harvard subdomain are multiple siblings such as the science department at Harvard, which may control its own domain addresses. Therefore, science is a zone living in the Harvard subdomain. The science center has its own name servers that add zone files when a host within the zone is added or changes. Similar to the post office adding or deleting a recipient on its route, host addresses are added to or deleted out of the DNS database. A host can be anything from a simple PC to a full-blown Web server. They are all seen as hosts, and each one requires some sort of address. Figure 27–7 demonstrates the entire picture from the domain space down to the multiple zones.

Zone files contain the host address. When Caroline subscribed to Spidercom's Internet product, her Internet name "Caroline" was added to the database through a zone file. I could not have e-mailed her at Caroline@Spidercom.net until the zone file was transferred into Spidercom's name server.

zone
A portion of a network within
a domain.

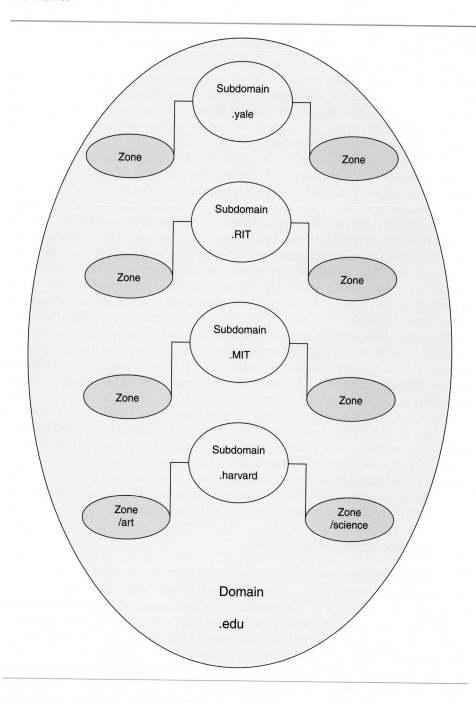

Figure 27–7
Domains contain subdomains, which
contain zones.

TCP/IP Routing through the Network

TCP/IP is the protocol used to carry traffic through the conglomeration of networks and is most responsible for our ability to access Web servers, send e-mail, and transfer files anywhere in the world. The best way to explain how a message travels through the network is to follow a transaction as it travels the Internet. We begin by defining the network elements we need to travel through for this example and then will explain how each element handles the incoming request and how it forwards it on to its destination.

Alfred loves Vidalia onions and each year calls the Onion House in Savannah, Georgia, and orders two bushels of them. This year, he has learned that he can go on the Internet to order his two bushels of onions. The transaction begins when Alfred logs on to the Internet. Because he uses Spidercom as his ISP, Spidercom receives the log-in request and immediately connects Alfred to its Web page. Alfred types in the Onion

House Internet address: *http://www.onionhouse.com/ect:orders.html*. Alfred presses the Enter button on his computer keyboard and waits for Onion House's Web page to respond. The name server looks in its tables to determine if it has information on the network address. Onion House uses Sprint as its Internet provider. Because Spidercom and Sprint have established a peering relationship with each other, Sprint's network IP address is housed in Spidercom's database. If Sprint's address did not reside there, the request would travel to the .com domain, where the address would be translated and Sprint's IP address returned to Spidercom.

Spidercom's router looks at the IP address and knows that it has to send it to the peering point that was agreed upon. The request from Alfred reaches Sprint's gateway router where it is sent to Sprint's DNS to determine which subnetwork owns the Onion-House domain name. The remaining portion of the address is translated into Onion House's host IP address. The request then travels the links established between Onion House and Sprint. The Onion House host, which happens to be a router connecting to the various departments in the company, receives the request and looks at the incoming address. The router sends the request to the Onion House Web server that houses the order forms for ordering onions. The Web server receives the request from Alfred and responds with the Onion House Orders Web page and sends it back to the Onion House router. The Onion House router looks at the destination address, which is Alfred's domain address—Alfred@Spidercom.net—and ships the response out on the physical link connecting it to the Sprint network. When the information hits the Sprint network, its gateway router decides which path the information should follow by looking at the destination address. The router determines that the transaction needs to go to Spidercom's network and, to do so, must be sent to the Sprint router, which peers with the Spidercom network.

Spidercom receives the information and again translates the domain name into Alfred's host IP address that was given to his computer. The information flows down to Alfred's computer where the Onion House Web site paints across his screen. Alfred fills in the order form on line, then clicks the Send button. The information travels through the same interconnection points that the first request did. However, because TCP/IP is a connectionless service, the physical route that the packets of information travel may vary because it is up to the routers to determine the best path to take.

■ 27.2 THE INTERNET NETWORK

27.2.1 Physical Connections

The physical connections used as interconnection circuits for the Internet vary depending on who is connecting to whom and where they are connecting. Earlier, we explained how the Internet began and described the use of MAE East and MAE West as connection points for Internet traffic. An ISP that connects into MAE East or MAE West determines the speed of that link by performing a traffic analysis. T1s, T3s, OC-3s, OC-12s, OC-48s, and even OC-192s may be used to connect nodes on the Internet. Large ISPs have huge pipes between their gateway locations and may require a large OC-48 pipe. Smaller ISPs, such as our imaginary Spidercom, may require connections as low as T1s. The physical circuits used to connect networks to networks vary dramatically throughout the industry.

Figure 27–8 shows a typical backbone network of a medium-size ISP. DS-3s connect smaller cities and OC-3s connect larger cities. The network has core routers used to interface with other ISPs or networks. The medium-size ISP does not connect directly into a NAP site or MAE site. Instead, it has an interconnection agreement with one of the handful of large Internet service providers. As shown in the figure, the medium-size ISP depends on the larger ISP's network to carry its traffic through the Internet.

Figure 27–9 illustrates typical peering relationships a larger ISP has with some of its larger customers. The more *peering points* a provider has, the more efficiently the traffic will flow through the network. For instance, if I run a large corporate network, I may choose to establish a peering relationship with my ISP, meaning that the number of hops through the network will be reduced due to the reduced number of networks the traffic has to travel through. The term *hop* describes an instance when the traffic

peering point
Point where two Internet providers exchange information.

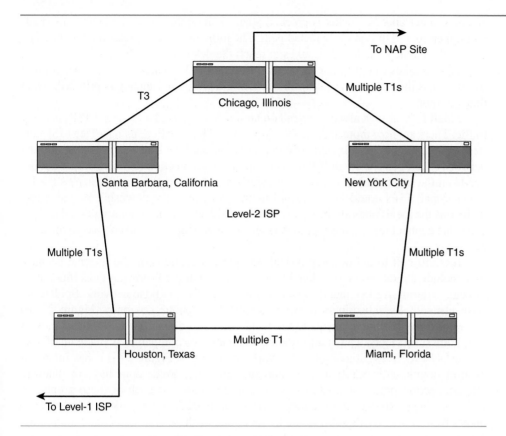

Figure 27–8
Medium-size ISPs are often referred to as level-2 carriers. Typically, they connect into NAPs and level-1 ISPs.

To NAP Site

Multiple T1s

T3

Chicago, Illinois

Santa Barbara, California

New York City

Level-2 ISP

Multiple T1s

Multiple T1s

Multiple T1

Houston, Texas

Miami, Florida

To Level-1 ISP

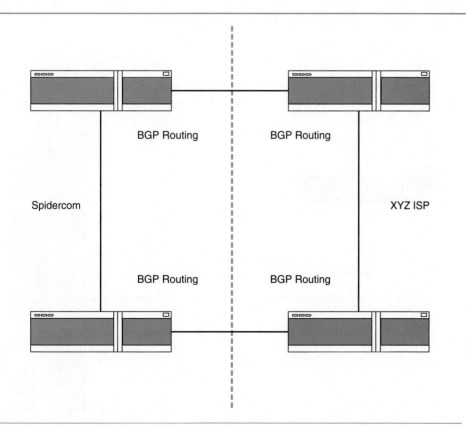

Figure 27–9
Peering points between two large ISPs. Two connections are established between the two networks for reliability. BGP routing is used to exchange routing tables between the two networks. Peering points are established between the two to reduce the number of hops the packets need to travel and, thus, reduce the time it takes the information to travel the Net.

BGP Routing

BGP Routing

Spidercom

XYZ ISP

BGP Routing

BGP Routing

enters a router and the packet is cracked open to show the source and destination address, then routed out on the selected path. The more hops, the greater the latency. The greater the latency, the longer it takes to reach the destination address and, ultimately, the longer it takes to return the requested information. ISPs are constantly performing traces across the network when troubleshooting a problem. Latency is primarily what they monitor.

Small ISPs almost always depend on larger ISPs referred to as Tier 1 ISPs to carry traffic. There are now thousands of ISPs in service. The benefit of using a Tier 1 ISP carrier is the resiliency of a larger network, the additional peering relationships, and the network coverage. The small ISP provides the local connection to the customer but depends on the larger ISP or network provider's backbone to carry traffic across the Internet. Small ISPs could be compared to the township. The township provides and maintains the local roads all the way out to the residents but does not maintain the highways that connect the regions together. However, township roads do connect to the large highways.

It is difficult to define the physical network connections on the Internet because they include almost every speed and type available, but the following does illustrate a typical medium-size ISP that depends on a larger ISP's backbone network. Spidercom connects with its subscribers through the local telephone company. Figure 27–10 shows how Spidercom connects to the local ILEC to gain access to the thousands of telephone subscribers. The connection between the two companies is made up of several ISDN PRI lines. Spidercom's network, as shown in Figure 27–11, has interconnection points with XYZ, a large network provider. Spidercom has six gateway interconnection points located across the United States. At each of these points, an OC-3 has been purchased to connect Spidercom to XYZ's network. All Internet requests from Spidercom's subscribers travel to one of these six interconnection points. The network provider is responsible for routing the requests through the Internet. Depending on the request, the traffic may flow across several backbone networks before it reaches its destination host. For instance, traffic heading overseas may travel from Spidercom to XYZ, from XYZ to an international Internet carrier, from there to the country's PTT Internet network, and finally to the end site.

Figure 27–10
ISPs often establish connections with ILECs or CLECs using ISDN PRI circuits.

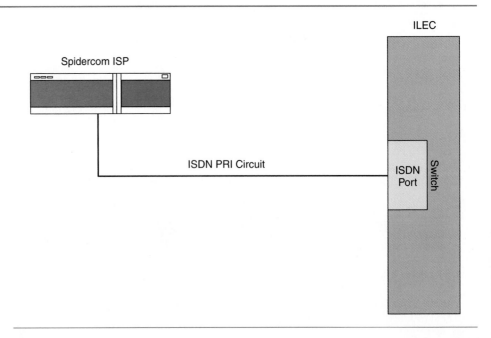

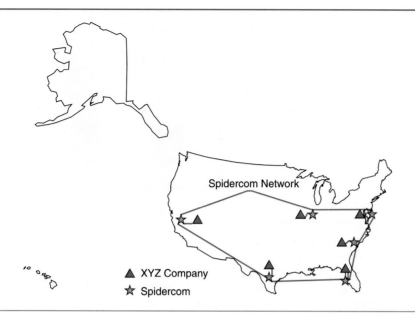

Figure 27–11
Spidercom connects with XYZ at six different interconnection points. The purpose of interconnecting is to ensure routes between the two networks even during outages.

27.2.2 Network Equipment

Equipment found on the Internet may be divided into two general categories—network and service. The network requires several pieces of networking equipment and transport equipment. The equipment used for Internet services depends on the different types of servers used to store and compile the information.

Network equipment exists to carry traffic across the network. The type of equipment needed depends on several factors—the size of the pipe, the number of hosts it will serve, and the type of backbone established.

Remote Access Servers

A remote access server (RAS) aggregates multiple dial-up connections from subscribers. It may be thought of as a modem pool that establishes a connection with the subscriber and also with the network. The information between the subscriber and the Internet flows through the RAS.

Subscriber Management Systems

Subscriber management systems (SMSs) are the newest aggregation devices used by CLECs, ICPs, and cable companies. The SMS is similar to the RAS in that its main function is to aggregate subscriber connections before handing them off to the network. The difference between the two is that the RAS aggregates dial-up lines while the SMS aggregates broadband connections and cable modems. The SMS is able to terminate thousands of subscriber sessions and aggregate the traffic before handing it on to the routed network. Vendors have carved a nice niche in the networking market providing a way to help off-load overloaded routers by terminating subscriber sessions. In the future, SMSs will offer many value-added services that will improve networking for everyone who uses the Internet. Currently, any provider that offers broadband Internet services, DSL, or cable modems has to have an SMS in its network. Physically, the SMS resides close to the edge of the network where the subscriber's lines terminate. Figure 27–12 illustrates a typical configuration in which an SMS is used as an aggregation device for broadband services.

Ethernet Switches

Ethernet switches may be used as load-balancing devices between the SMS or router and the cache, radius, or other servers. The Ethernet switch helps reduce the chance of failure between devices. If one server failed and a backup server was available, the Ethernet switch would direct traffic away from the failed server and to the

Figure 27–12
SMS is used to terminate circuits
carrying DSL signals. The SMS maps
the signals onto larger pipes out to
the network. It is a Layer 3 device that
routes the packets using a Layer 2
protocol.

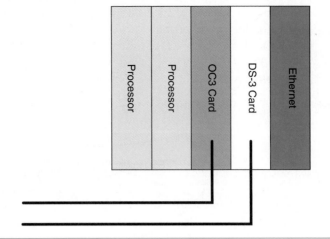

backup server. Figure 27–13 shows where an Ethernet switch may reside in an Internet network.

Routers

The router in the ISP is more important than that used by the enterprise customer. Routers are why the Internet is so versatile and resilient. Routers are Layer 3 switching devices that take in packetized information and distribute it through the network by using IP addresses. ISPs and service providers use routers in their core backbones and their access

Figure 27–13
An Ethernet switch switches Ethernet
traffic. The Ethernet LAN feeds into the
switch where the packets are routed to
the correct interface port. Many
DSLAMs output Ethernet signals that
are fed into an Ethernet switch.

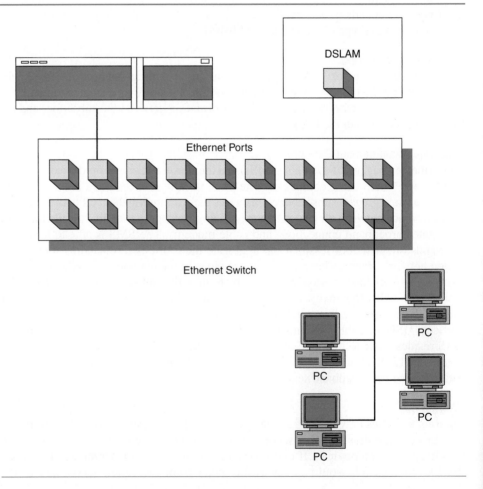

networks to move traffic around the network. The two types of routers in the network are core routers, which sit closest to the Internet, and access routers, which feed into the core routers. For instance, a core router would hand traffic from Spidercom's network into XYZ's network. As shown in Figure 27–14, there are always fewer core routers than access routers. Today, many networks depend on SMSs to handle access routing. They feed into the core routers and depend on them to hand traffic to the Internet.

Domain Name Servers
Domain name servers may be considered network equipment because they are essential pieces of the backbone network. Without the DNS, traffic would come to a standstill. The DNS connects into the network through a port on the router, which sends DNS requests out the port connected to the server and receives messages back from the DNS in return. Some ISPs pay other ISPs to handle their domain name service. Whether the provider has DNS in-house or uses a second provider, the address resolution must take place before traffic can flow.

Multiplexer
An M13 multiplexer is commonly found in an ISP location. It is used to demultiplex DS-3s into DS-1s and to multiplex DS-1s into DS-3s. Depending on the network connections, an M13 multiplexer may or may not be required. Understand that the M13 multiplexer receives a DS-3 from the network and outputs twenty-eight DS-1 circuits that are distributed to the routers and other devices at the site.

Fiber Optic Multiplexer
Again, fiber optic multiplexers may be used by the ISP to connect into the backbone network. The type of connection determines whether it is required. A fiber optic multiplexer accepts lower speed circuits and multiplexes them into a higher-speed output. For example, eighty-four DS-1s may be fed into an OC-3 multiplexer and fed out at an

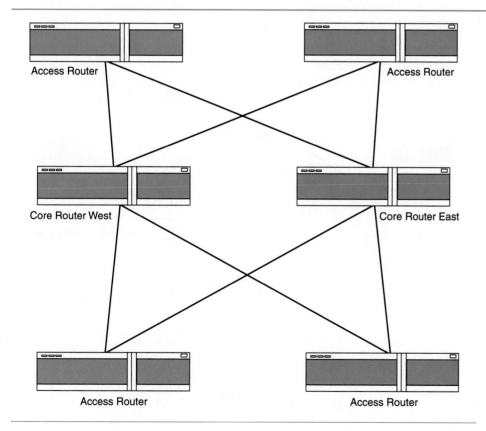

Figure 27–14
The two core routers are used to interconnect to the outside world in the same way the gateway router connects host routers to the outside world. The access routers feed into the core routers that are responsible for handling all higher-level routing protocols. The purpose of reducing the number of interconnecting routers in the network is to reduce the complexity of the network.

OC-3 pipe. Larger ISPs often use OC-48s and even OC-192s as interconnection multiplexers to tie their backbones together.

27.2.3 Protection Switching Schemes on the Network

Protection switching refers to switching traffic over to an alternative path on the network. Protection switching is handled several ways on a data network. We call these a Layer 1 switching approach, a Layer 2 switching approach, and a Layer 3 switching approach. Layer 1 refers, of course, to the physical layer of the Open System Interconnection (OSI) stack, the actual circuit that is connected to the equipment. When a network has Layer 1 protection switching, it means that there is Layer 1 equipment, such as a fiber optic multiplexer, that can switch paths when one path fails. The SONETs provide full Layer 1 protection switching and are the most common methods used to provide Layer 1 protection. The time it takes for SONET to switch to the backup path is less than 50 ms. Data may receive an error, but connectivity will not be dropped.

Layer 2 protection switching refers to networks built around a Layer 2 architecture, such as ATM or frame relay. The Layer 2 protection switching methodology depends on the actual ATM or frame relay switch to handle moving traffic from one route onto a different route. The permanent virtual circuits (PVCs) may be dual homed, meaning there are two routes to the same location. When one circuit fails, traffic is automatically switched to the alternative PVC route. The switchover time for ATM circuits varies depending on the distance of the routes, the type of ATM switch, and other variables that affect latency. When ATMs use switched virtual circuits (SVCs) instead of PVCs, the time to switch routes is even less. Layer 2 switchover time is about three to four seconds. The best way to differentiate Layer 1 protection and Layer 2 switching is to view Layer 2 switching as alternate virtual circuits and Layer 1 as alternate physical circuits. The Layer 2 alternate virtual circuit must ride on an alternate physical circuit, but it does not depend on protection switching at the physical level. Figure 27–15 shows a network that depends on the ATM switch to handle all redundancy issues.

Figure 27–15
ATM switches are used to reroute traffic during failures.

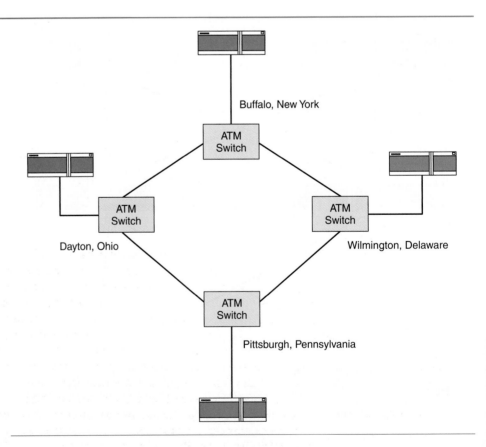

Layer 3 routing depends on the type of routing protocol used in the network. Routers are responsible for rerouting traffic onto a second route. The biggest difference between a Layer 3 and a Layer 2 method is that the router looks at each individual packet before determining which circuit the packet should be routed onto. ATM nails up logical paths and ships all traffic heading to the same destination over the PVC or backup PVC. Routers are often used to load balance circuits; therefore, they constantly monitor the ins and outs of the network before transmitting information. The time required to reroute traffic using Layer 3 depends on several factors: the type of protocol being used, the number of redundant circuits, the path of the redundant circuits, the number of new hops the traffic has to travel through on the new route, and the load on the router. Therefore, it is difficult to define the time needed to reroute traffic from a failed circuit to a backup when using Layer 3 protection switching methods. Layer 3 protection switching is not instantaneous as with Layer 1, and it is not as simple as with Layer 2.

The best way to design a network is to consider all three layers of redundancy—Layers 1, 2, and 3—when all three layers exist. If a network only has Layers 1 and 3, the design should include both protection switching schemes for both layers and intentionally have both work together to form a "five nines" network in which uptime = 99.999%. The voice world was built around the "five nines" uptime philosophy. Data networks today are striving for the same reliable, always up, redundant network.

27.3 INTERNET SERVICES DATA CENTERS, EQUIPMENT, CONTENT, AND APPLICATIONS

27.3.1 Data Centers

A data center is a site where content is developed, stored, and distributed to the Internet. The many servers it contains hold content and interface with the network. Every time you log on to the Internet and surf the Web, you access someone's file server in a data center. The structure of the data center is similar to a switch site. Backup power is required, and the site must have air conditioning to keep the servers cool. The site has alarms and secured access to serve as a fortress against cyber attacks. Many data centers, unlike switch rooms, have raised floors that allow cabling to be run underneath the equipment. The relay racks are more like cabinets, and the equipment is smaller than typical transport and switching equipment. Data centers, like switch sites, are kept clean. Dust filtration is a must to preserve the computing equipment.

In the United States, there are several large content providers that operate huge data centers. Exodus, Global Center, UUNET, and AOL are just a few that store information for clients. Large data centers can be viewed as large content centers. The world today revolves around content, so everyone wants to be a content provider, not just a network provider. A data center is the first step toward being a full-fledged content provider.

In addition to the large data centers, small Web site houses set up Web servers, build Web pages for their clients, and offer Web hosting on the side. Most large- and even medium-size businesses have Web servers that hold their Web pages. As with the network, many options are available for developing, storing, and distributing content. A data center may be a small site run by one or two people or a large international corporation.

27.3.2 Data Center Equipment

A data center may be a large site with multiple servers humming away or a small garage with a couple of Web servers sitting on crates in the corner. The equipment is the same; its purpose is to handle content. The following section describes a typical corporate data center.

A *mail server* is a computing device with software that allows it to receive and forward mail requests. Small LANs have internal mail servers that house employee e-mail accounts. ISPs such as CompuServ, AOL, and other providers have large mail servers that handle all of their subscribers' e-mail transactions. The mail server is technically just a server with mail software loaded—the post office of the messaging network. *Messaging* is the term used for e-mail transactions between hosts.

mail server
Computing device with software that allows it to receive and forward mail requests.

mail gateway
Connects mail servers together.

The *mail gateway* connects mail servers together. It can connect two mail servers located on different networks, even when the mail servers are dissimilar. The mail gateway works with the mail server to ensure that transactions are able to travel from any service provider, country, or city to any other service provider, country, or city. Imagine the number and types of mail servers stationed throughout the globe. A mail server feeding the country of Bhutan is able to talk to a mail server sitting in the garage of a small start-up ISP in California. Gateways form the bridge between the two worlds.

Web server
A computer responsible for retaining content for the Internet.

A *Web server* is a computer that sits on the Internet and holds content. The information may be documents, graphics, video clips, audio clips, text documents, or something else. It must be very powerful, because it holds large files and must continuously access the files as requests are received. The Web server may sit at an ISP, an ASP (application service provider), or an ICP (integrated communications provider) site. It can also be found in a corporate data room. A corporation places all corporate information on the Web server that may serve the corporation's Intranet, its external Internet, or both.

video server
A computer responsible for handling video streams for Internet content.

A *video server* is similar to a Web server and is often called a Web server because it sits on the Internet. It holds video content and is built specifically to handle video streams. Video streaming is one of the hottest new applications being touted by Internet enthusiasts. The video server is a high-powered computer that can handle multiple video streams and hold large amounts of data files.

audio server
A server responsible for handling audio content for the Internet.

An *audio server* is similar to a video server and a Web server. It sits on the Internet and holds audio content. Various types of protocols used to format audio content are housed in the audio server.

cache unit
Buffer for Web server.

A *cache unit* does not hold Web pages and is not used to develop content, but it does solve two significant problems. The first is to save the server's processing power. For instance, when a Web site is bombarded with millions of hits, the Web servers holding the content may crash from the volume of requests at one time. A cache server is placed in front of the Web server to buffer the server from being inundated by requests. The advantage of the cache is that it is a much less expensive device than the server and can parse content right from the cache back to the requester without talking to the server.

The second use of a cache is to place it on the ISP's network close to its clients. The cache receives a request from a client, sends out a request to the Internet, receives a response from the Web site, and caches the content in addition to passing it on to the requester. The next time a request comes in for the same Web site, the cache engine feeds the customer the Web content from its own cache. Portions of Web pages are noncacheable and still need to be pulled directly from the Web site, but a good portion of the Web pages are cacheable and can sit in the cache engine instead of at the Web site, as shown in Figure 27–16. The benefits of caching are clear. The amount of traffic traveling across the backbone network from the ISP to the Web site is reduced because the cache engine sits at the ISP's site. In addition, the speed at which the customer receives the content is increased due to the fewer number of hops the information must make. The cache is becoming an important piece of network equipment, both at the data center and at the ISP's site.

27.3.3 World Wide Web Content
HTML Standards

HyperText Markup Language (HTML)
Language used to create World Wide Web pages.

HyperText Markup Language (*HTML*) is the programming language used to create World Wide Web pages. It uses tags to define different types of content, and the creator of the Web page uses HTML to explain how the content should be presented. It divides content into three general areas: tags used to define styles, comments used by the Web page creator but not displayed, and the actual content presented. Tags are also used to mark the content as cacheable or noncacheable and may be used to provide hyperlinks to another site on the Web site or to a distant Web site on the Internet. HTML was designed to simplify the delivery of content across the Web. Its ease of use is one of the main reasons the Internet is so popular. Other standards are constantly being introduced.

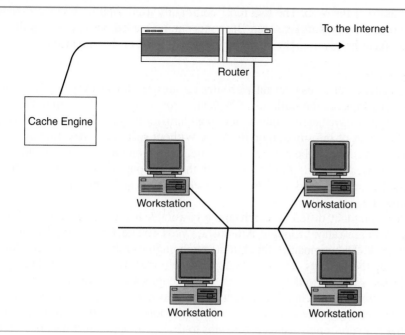

Figure 27–16
Cache engines are used to store Web site information closer to the end user.

HTTP Standards

HyperText Transfer Protocol (*HTTP*) is the protocol or language that the Web server and the browser use to carry information around the Internet. It is the supreme communicator for HTML Web pages. When you click on your Web browser, an HTTP session starts up and the path between Web sites and hosts are established using this upper-layer protocol. HTTP appears before the Web site name when browsing the Internet. It is the glue that sticks the IP address and the TCP address together allowing an end-to-end connection through hyperlinks. A hyperlink is one that allows the user to point and click on an item and be taken to a site that is attached via HTTP to that item.

HyperText Transfer Protocol (HTTP)
The protocol that Web servers use to carry information on the Internet.

Defining URLs

A uniform resource locator (URL) is an address on the Internet. The DNS discussed earlier in the chapter translates URL strings into IP addresses. The structure of the address is commonly seen when accessing a Web site on the Internet. The URL standard structure is:

method//host_spec{port}{path}{file}{misc}

An example of a URL is as follows:

http://www.appleorchard.com/macintosh

In the preceding example:

> *http* is the *protocol* used to link the Web site to the address;
>
> *www* stands for the World Wide Web. The request is being sent out onto the *Internet;*
>
> *appleorchard* is the name of the *Web server;*
>
> *.com* is the *domain* where the site resides; and
>
> */macintosh* is a *subdomain* or *zone* on the appleorchard Web site.

Web Browsers

A *Web browser* helps users navigate around the Web by using HTTP to set up a communications path between the Web server and the client's computer. The browser functions as a guide that simplifies surfing the Internet and consequently increases the

Web browser
Helps users navigate around the Web by setting up a communications path between the Web server and the client's computer.

functionality of the Web. The two most commonly used Web browsers are Netscape Navigator and Microsoft Explorer. Browsers can be loaded onto the user's PC by the PC manufacturer, given to the client by the client's ISP, or downloaded off the Net.

Search Engines

A *search engine* retrieves relevant Web sites for the user. It takes key words typed in by the user and looks at the millions of Web sites for information matching the request. Search engines have become much more popular over the past few years due to the increased amount of information on the Web. Without search engines, finding information on the Web would be almost impossible. Search engines use a program called a *spider* that weaves itself through each of the Web pages looking for content that matches the request typed in by the user. Web crawlers and robots are also search engine programs.

There are many different search engines available to the Internet user. A few of the most popular are Yahoo!, Google, Alta Vista, Lycos, Excite, Go2, AOL.NetFind, Alexa, and Infoseek. The companies that build search engines rely on advertising dollars to compensate their efforts. The search engines are portals to the Internet. They provide thousands of *eyeballs* to the advertisers. Search engines are becoming more and more intelligent and are thus helping to reduce the information overload that overwhelms many Internet users. The newest types of search tools are intelligent retrieval applications that run very sophisticated algorithms to find information that matches a request.

Portals

Portal is a term being used frequently by Internet entrepreneurs who wish to cash in on the growing number of advertising dollars being spent through the Internet. A portal is simply a door that leads the user to the content.

Packaging Audio for the Web

Audio on the Web, or *Voice over IP* (*VoIP*), depends on the H.323 protocol. The protocol packages voice and places a priority bit in the frame that notifies the upstream routers that the information being carried is time sensitive and needs to be given priority over data traffic.

Packaging Video for the Web

Video information crossing the Internet is commonplace. Today, video streaming, online video downloads, and peer-to-peer gaming have increased the need for robust video encapsulation methods on the Net. Video information is formatted and compressed before being transported across the network. The most common type of compression is via the *Moving Pictures Experts Group* (*MPEG*). The MPEG forum introduced the standard as a way to compress full motion video and get it ready for transport across a network. Today there are several versions of MPEG: MPEG 1, 2, 3, and 4. Each is used to compress video frames into smaller files before they are transmitted onto the backbone network. MPEG 1 compresses a file from 7.7 Mbps to 150 kbps. This represents a 52:1 compression ratio.

Video is a time-sensitive medium requiring special formatting of the frames to ensure that the information arrives in a timely fashion. Depending on the transport protocol, priority bits are set to make sure traffic is not discarded along the way. For ATM, the video signal goes through AAL 2 adaptation and is marked as either VBR-nrt (variable bit rate–non real time) or VBR-rt (variable bit rate–real time) QoS standard. The newest method for shipping video over the Net is IPVideo. Video over IP, similar to Voice over IP, requires fairly sophisticated protocols in addition to fairly substantial line rates end to end.

Packaging Graphics and Text for the Web

A common way to package graphics, texts, and pictures before sending them on to the Web is to place the files into Adobe Acrobat .pdf files. Acrobat allows any user to view

search engine
Searches for relevant Web sites for users. It matches key words to Web sites that may be of interest.

spider
A program that prowls the Internet for publicly accessible resources.

eyeball
A viewing audience.

portal
An entrance to another Web site.

Voice over IP (VoIP)
Gives voice streams priority over data.

Moving Pictures Experts Group (MPEG)
Video compression algorithm.

any software file and see all of the colors, pictures, correct fonts, and so forth, even if they do not have the creator's software on their PC. Adobe Acrobat is used extensively to help carry files across the Internet.

27.3.4 Applications

This section focuses on the applications available on the Internet. The Internet is a huge resource for information in addition to being a network that allows anyone to talk to anyone else logged on to the Net. Following are a few of the many applications being deployed on the Web.

Web Content

This is, as it sounds, information "sitting" on the Web. *Web content* is one of the most important Internet applications provided by the Internet.

Web content
Information available on the Internet.

Streaming Video—Video on Demand

Streaming video is a continuous video stream. You may have logged on to a site that has a camera focused on a particular location, such as the capitol building in Kansas City or a storm brewing in Texas. The camera is feeding the video stream into the video server that, in turn, sends the video stream across the Net to your PC. New algorithms and methods for packaging on-demand video graphics have allowed content providers to offer new "pay-per-view"-type services from the Internet. The increased number of broadband connections into subscribers' homes has provided a large enough pipe to handle high bandwidth video files. Today, video off the Net is good quality and fairly robust. The question that will be answered over the next few years is, Will the technology evolve to the point at which content providers will no longer be tied to sending real-time video signals over satellite or CATV connections?

streaming video
A continuous video stream.

E-Commerce

E-commerce is one of the most overused buzzwords of this era. It refers to providing online catalogs where the consumer may go in and click on a product, provide credit card information, press the send button, and buy the product. Today, e-commerce is no longer the new kid on the block. A good example of how Net sales have become a standard in the commercial space is the ever-increasing number of Internet sales raked up every Christmas. Every year, the percentage of purchases from the Net increases substantially. One interesting phenomenon that has evolved thanks to e-commerce is the advent of e-commerce home businesses. Entrepreneurs no longer have to purchase expensive window space in a busy traffic area sacrificing large chunks of capital for physical space. Instead, they set up a Web site, set up a supply chain, and start to push their product while sitting at home in front of their computer.

e-commerce
Commercial use of the Internet.

Information Sources

The Internet provides a great avenue for researchers to find documentation and *information sources*. Libraries, governmental agencies, and corporate research centers place their files on line for research. No longer is it necessary to physically drive to the university library to find information for a term paper. One of the most altruistic applications on the Internet is the ability to find information.

information sources
Sources of information available on the Internet.

FTP Files

Files being transferred between two users' computers often use the file transfer protocol, which is an efficient way to send files across the Internet. Many users have large database logs that need to be sent to a coworker across the country or just across town. FTP is a great way to send these files, just like e-mail is a great way to send messages.

FTP files
Files being transfered via the file transfer protocol.

Audio

One of the most popular applications on the Web today is downloading audio files. Companies such as Napster pushed the envelope to the point at which the music industry had to move quickly to accommodate the high demand for downloadable music sites. Devices such as the iPod and cell phones with Web access and storage built in significantly changed the way music is distributed and sold to the public. Additionally, similar to e-commerce, small, unknown artists now have a chance to distribute their music to large populations of people tuned into the Web. The change has been rapid and, amazingly, pushed primarily by the actual user, not the marketing types on Madison Avenue. Beyond downloading the newest pop hit before it has been released in the stores, other audio applications—such as downloading a segment from a popular radio broadcast—are common. The possibilities are endless. The unresolved issue surrounding this application is the misuse of copyrighted material. Record companies, artists, and Internet enthusiasts are trying to set standards ensuring that artists and the record companies receive their money while the Internet user enjoys the convenience of buying songs online.

VoIP

Voice over IP is another hot application being touted by ISPs. The ability to use a PC as a telephone and the Internet as the telephone connection is being seen as quite a boon to many users, especially those with family or friends living overseas. The Net, at this time, does not charge a per-minute usage fee. However, there are two drawbacks. VoIP across a public network is not yet stable. The nature of a connectionless architecture such as IP does not provide the quality of service parameters needed to ensure toll-quality voice communications. The second issue is that, while there are no usage charges associated with VoIP calls yet, the very high cost of network equipment, network facilities, and network maintenance will probably not allow that to continue. The question at this point is whether VoIP will take over as the method used to carry voice calls through the network or whether it will only be used by a small portion of the population, mainly for expensive international calls.

Chat Rooms

chat room
A virtual room where people from anywhere can "meet" and "chat."

A *chat room* is a place to chat—something of a conference bridge in the voice world. A person who wants to talk about how to grow persimmons can enter a chat room and talk live with other people who are also interested in growing persimmons. Chat rooms are very popular today, especially with teenagers.

Newsgroups

newsgroup
An online discussion group.

A *newsgroup* discusses topics online. The difference between a newsgroup and a chat room is that the newsgroup is not normally live. Users submit questions, articles, information, and comments to a particular newsgroup and then receive responses from other members of the group. If you subscribe to a newsgroup, you may be overwhelmed by the amount of mail you receive. Newsgroups are great places to gather information, meet experts in the field, and exchange new ideas.

Virtual Private Networks

Virtual Private Network (VPN)
Used to connect two or more remote locations.

With *Virtual Private Networks* (*VPNs*), corporations are using the Internet to connect two or more remote locations. In the past, corporations have leased dedicated circuits or depended on dial-up circuits to connect their locations. Today, connecting through the Internet allows users more flexibility and the corporations a lower network cost. Figure 27–17 illustrates a typical VPN across the Internet.

Internet gaming is becoming one of the fastest Internet applications. Players pay monthly fees to a gaming provider and are allowed to interact with other players. The sophisticated graphics and complexity of the games require high-speed Internet connections to the end users. Thanks to the high bandwidth requirements of the gaming applications, cable modems and DSLs are being deployed extensively by ILECs, DLECs (data local exchange carriers), and CLECs.

Figure 27–17
VPNs allow remote locations to connect through the Internet without establishing point-to-point dedicated circuits.

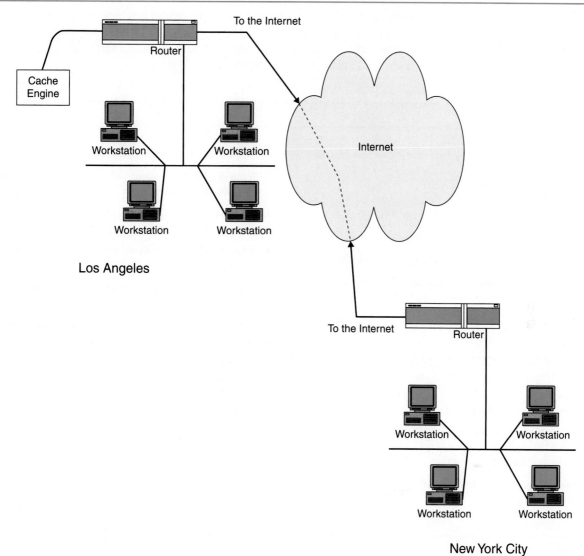

SUMMARY

The Internet was one of the most significant phenomena of the twentieth century. Many historians see the advent of the Internet as being as significant as the invention of the automobile, the gas engine, the electric lightbulb, and, of course, the telephone. The structure of the Internet is often difficult to visualize because there are no clean network boundaries and there are numerous types of equipment and applications. The purpose of this chapter was to clarify what the Internet is and how it functions by explaining the structure of the Internet network, the methods used to fetch information, and the applications available to Internet users.

CASE STUDY

Joe Cosmos wants to become a millionaire, so he has decided to become an ISP in Green Grass. The name of his ISP is Green Grass Internet and he plans to tap into the newly built Green Grass telephone network infrastructure to carry all his data. Joe has asked that you help develop his ISP. Include the network design, the servers, the product offerings, and the logo.

REVIEW QUESTIONS

1. What agency initially developed the network we now call the Internet?

2. What is the Internet?

3. What is meant by the terms *MAE East* and *MAE West?*

4. Where is MAE East located and who maintains the site?

5. Where is MAE West located and who maintains the site?

6. What is an NAP and what is its function in the world of the Internet?

7. What does the term *peering* refer to as it pertains to the Internet?

8. What does the term *private peering* refer to as it pertains to the Internet?

9. Explain what a DNS does and why it is critical to the transfer of information across the Internet.

10. The Internet cloud contains devices that route packets of information through the mesh of networks to the correct Web site. What are these devices called?

11. If I lived in Iceland, what Internet suffix would be attached to my Internet address?

12. What is the difference between .com and .net?

13. Explain each portion of the following address: *http://www.oatmeal.com*

14. Why is IP addressing such a critical issue now in the world of internetworking?

15. Explain how to search for information on the Internet with the correct URL.

16. Explain how to search for information on the Internet without the specific URL.

17. List and define each of the suffixes used to categorize the different organizations on the Internet.

18. Explain what the term *domain* means as it relates to the Internet.

19. What devices store content that can be accessed by the World Wide Web?

20. What is a chat room?

21. What is a newsgroup?

KEY TERMS

Advanced Research Projects Agency Network (ARPANET) (726)

DARPANET (727)

Internet Engineering Task Force (IETF) (727)

uniform resource locator (URL) (727)

metropolitan area exchange (MAE) (728)

network access point (NAP) (728)

domain name server (731)

top-level domains (733)

zone (738)

peering point (740)

mail server (747)

mail gateway (748)

Web server (748)

video server (748)

audio server (748)

cache unit (748)

HyperText Markup Language (HTML) (748)

HyperText Transfer Protocol (HTTP) (749)

Web browser (749)

search engine (750)

spider (750)

eyeball (750)

portal (750)

Voice over IP (VoIP) (750)

Moving Pictures Expert Group (MPEG) (750)

Web content (751)

streaming video (751)

e-commerce (751)

information sources (751)

FTP files (751)

chat room (752)

newsgroup (752)

Virtual Private Network (VPN) (752)

PART SEVEN

Emerging Technologies

Chapter 28: Emerging Technologies

Alexander Graham Bell
Photo courtesy of AT&T Archives and History Center,
Warren, New Jersey

28

Emerging Technologies

Objectives

After reading this chapter, you should be able to

- Describe next generation switch architecture.
- Describe the innovations in information transmission.
- Describe the new trends in the access network.
- Describe new services offered to the end user.
- Describe wireless network services.

Outline

Introduction

28.1 Next Generation Switching

28.2 Transmission Network Innovations

28.3 Access Network

28.4 Wireless Technologies

■ INTRODUCTION

The telecommunications landscape changes continuously. Service providers, equipment vendors, and software companies are constantly introducing new products that improve or change the current telecommunications infrastructure. Chapter 28 looks at the direction the industry is heading with a focus on several key areas. The chapter describes next generation distributed switches that will be able to handle all types of traffic: voice, video, and data. It also describes the innovations made in fiber optic transmission equipment such as an increased number of wavelengths per DWDM system, wavelength routing, and the combination packet and time division multiplexers. The chapter looks at the future wireless network that will be able to carry broadband data signals. The direction the access network is heading will be discussed as will new service offerings such as VoIP.

■ 28.1 NEXT GENERATION SWITCHING

28.1.1 Next Generation Switching Architecture

At this time, one of the most significant changes occurring in the industry is the introduction of next generation switching architectures. Several companies, including

Lucent, Nortel, Telcordia, Cisco, Marconi, CopperCom, and Alcatel, are introducing innovative switching architectures that will replace current network circuit switches. Within the next two years, the digital circuit switch will begin to be replaced by the new distributed packet/cell-based distributed switch architecture. The advantage of the packet-based architecture is the efficiency gained by dynamic bandwidth allocation, as defined in chapter 12. In addition to saving bandwidth, the new distributed models are much more scalable than the current architecture and are less expensive on a per-port basis. The change from the traditional circuit switch to the next generation architecture is as significant as was the change from analog switching to digital switching.

The structure of *next generation switches* is shown in Figure 28–1. The switch is comprised of a *call server* that is used to hold customer information, define services, and interface with switching gateways. A call server may be placed in one central location and feed multiple regions. A signaling gateway converts the SS-7 messages into a new signaling standard called *Megaco protocol* or *MGCP* (Media Gateway Control Protocol) that signals the distributed devices. SS-7 is used in the circuit switched world,

next generation switch
New distributed network switch developed around IP or ATM transport and advanced signaling protocols.

call server
Server that holds all routing information in the next generation network.

Megaco protocol or MGCP
Signaling protocol used with next generation switching architectures.

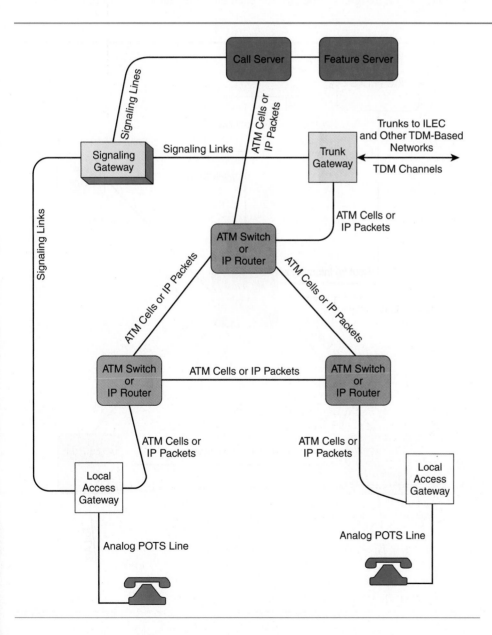

Figure 28–1
A next generation switch is built around a distributed architecture. The switch controller holds all routing tables, customer information, and signaling intelligence. The signaling gateway is responsible for all signaling conversions, such as SS-7 to Megaco. The trunk access gateway (TAG) converts the ATM or IP packets into the traditional TDM protocol. The local access gateway (LAG) accepts all customer analog and digital circuits, converting them into ATM or IP packets for transport.

and Megaco or the MGCP is used in the packet/cell world. Both perform the same functions—they set up calls, tear down calls, and carry AIN (Advanced Intelligent Network) messages for features. The third component of the distributed next generation switch is the local gateway. The local gateways interface to the customer's lines and are similar to the digital loop carriers commonly used today. The gateways take in the analog or digital signal from the customer. If necessary, the signal is first converted into a digital signal, which will then be converted into a cell or packet.

Connections between the gateway and the outside world vary depending on where the call is heading. If the call is staying within that carrier's network, it rides on an ATM or IP transport network that connects the originating gateway to the terminating gateway, as shown in Figure 28–2. The Megaco signaling protocol sets up the connection the same way the SS-7 messages set up routes between switches. The main difference is that local gateways sit closer to the customer and the call no longer needs to go back to one class 5 switch in order to be switched through the network. The call server, in conjunction with the signaling gateway, builds paths between the two local gateways. A gateway may be a device sitting at the customer location. The second advantage to this architecture is the use of packet or cell transport methods such as ATM or IP. Unlike TDM, which assigns a time slot for each call, the statistical nature of the cell or

Figure 28–2
Data and voice information is separated in the transport network by placing voice information on one virtual circuit (VC) and the data on a second. (TAG = truck access gateway; LAG = local access gateway.)

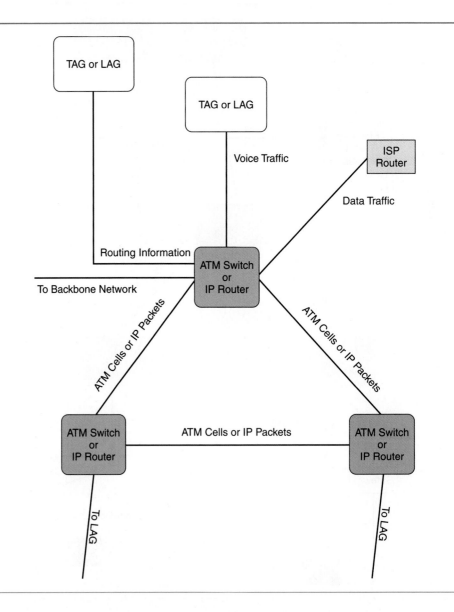

packet network allows more calls to travel across the same number of facilities. In certain instances, a local call gateway may be placed at the customer premises to allow analog lines to travel just the distance of the in-house wiring. Figure 28–3 shows the entire network from the LAG (local access gateway) through the ATM network to the TAG (trunk access gateway).

A third advantage to this architecture is that data can ride on the same network as the voice in a similar format, either packet or cell. Data travel in one of two formats on the local cell or packet network. The first is through the local gateway. Some vendors are building local gateways with broadband data interfaces such as DSL. The customer's data traffic arrives at the local gateway on a digital subscriber line. The gateway converts the incoming signal into a packet or cell and transports it through the network to a data handoff, such as an ISP. If the data are dial-up analog data, the network will be able to separate them from the voice traffic at the local gateway, as shown in Figure 28–4.

Calls traveling outside the carrier's network, such as from a CLEC to an ILEC, or from CLEC to a long distance carrier, must be routed through a trunk gateway normally

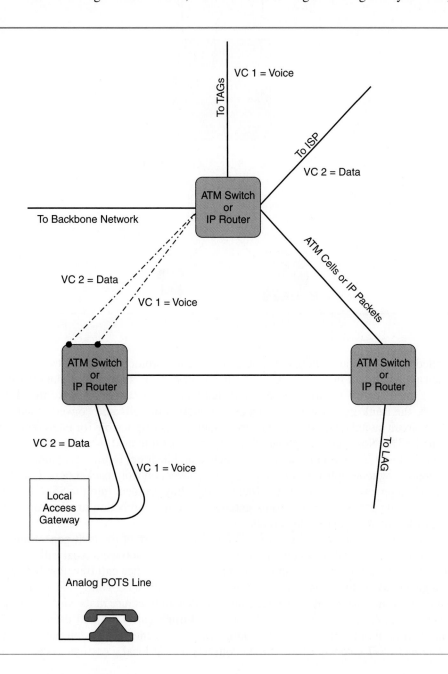

Figure 28–3
The ATM network or IP network used to connect the TAGs, LAGs, and call server is similar to the currently deployed ATM or IP networks. The purpose of building a cell or packet network versus the traditional TDM network is to separate voice and data efficiently before hitting the switch. In addition, bandwidth savings are realized on statistically multiplexed networks such as ATM and IP. (VC = virtual circuit.)

Figure 28–4
The LAG is able to separate the data
and voice signals onto their own
logical path. The switch does not deal
with data.

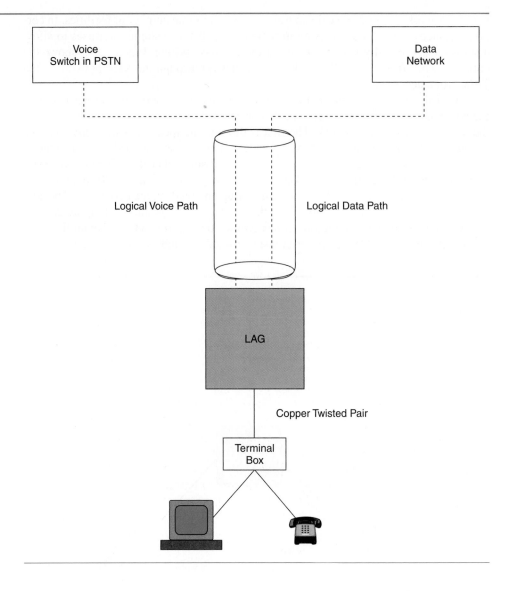

located at the central switch site. The SS-7/Megaco protocols also control the trunk
gateway. As shown in Figure 28–5, the trunk gateway is similar to a class 5 switch
where interconnecting circuits are terminated from other carriers. Unlike the class 5
switch, the trunk gateway accepts incoming packets or cells and converts them into
TDM circuit switched time slots. The trunk gateway is responsible for interconnecting
to the PSTN. Not every telephone company will institute next generation switching.
Some will rely on their current circuit switched architecture. In this situation, the next
generation switches must be able to perform a conversion from packet or cell to TDM.
The signaling gateway converts the Megaco signaling protocol into the SS-7 protocol.
The signaling gateway and the trunk gateway work in conjunction to allow the traffic
to flow between the two architectures.

One of the best ways to understand the new structure of the next generation network is to follow a call from beginning to end. We will first trace a local call that travels within the carrier network and then follow a long distance call traveling between a
next generation network and the traditional circuit switched network. Spidercom Communications has deployed a next generation switch architecture in its Burlington, Vermont, office. At its regional site, it has installed trunk gateways, and it has signaling
and control circuits connecting to its large switching center in Boston where its call
server and SS-7 gateway reside. Spidercom has placed local gateways in each of the

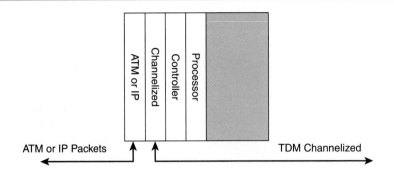

Figure 28–5
The TAG accepts traffic in both directions—from the ILEC or other TDM network and from the next generation network. The TAG converts the various protocols such as ATM to TDM or TDM to ATM to interface both types of networks.

ILEC's central offices. Along with the gateways, it has deployed an ATM access network. The access network connects each of the central office sites to one another and to the regional switch site, as shown in Figure 28–6. An ATM backbone network connects the Burlington, Vermont, site to the Boston, Massachusetts, control site.

Sarah is a new Spidercom customer who is served out of the Maple Street central office. She decides to call her cousin Abby who lives on the other side of town. Sarah picks up her telephone handset, and the local access gateway sees the loop closure the same way the class 5 switch does and instantly places dial tone on the line. Sarah hears dial tone and dials Abby's telephone number on the touch pad on her touchtone telephone. The digits are received by the local access gateway and via signaling links carried to the call server in Boston. The call server pulls up its translation tables to determine how the call should be routed and what features should be attached to the path. It looks at the seven-digit number and determines that the call is within its own local network and that the destination number resides out of the Woodburne central office. The signaling gateway works in conjunction with the call server and sends a message to the Woodburne local access gateway to see if Abby's phone is available to accept a call. The Woodburne local access gateway responds that the phone is idle and ready to accept an incoming call. A path is built between the two central offices' local access gateways on the ATM access network. In Spidercom's case, switched virtual circuits (SVCs) are automatically built using signaling information, thus creating a virtual logical circuit path between the Maple Street and the Woodburne Street offices' local access gateways. Abby's telephone rings, she picks it up, and the call path is completed. Abby and Sarah begin their conversation and continue to talk for more than an hour. One of the beneficial features of carrying the girls' conversation across an ATM network is the use of silence suppression and voice concentration. The ATM equipment senses silent periods and instead of holding an open channel uses that bandwidth space for other traffic. The path between the two girls is bidirectional from Sarah to Abby and back again. Though both girls may talk at the same time once in a while, normal human conversation is one person talking at a time. The ATM equipment recognizes this and takes advantage of the silent periods. Therefore, the logical path between the two girls is used only when they are talking. The advantage of silence suppression is that it frees up bandwidth and allows more conversations to share the same pipe than if a time slot were nailed up as it is with TDM.

A second feature that helps make the ATM or IP transport more efficient is the ability of the equipment to compress the voice signal into smaller chunks than 64 kbps; 8, 16, or 32 kbps are now commonly used to carry voice signals. The advantage of using 8 kbps per voice circuit compared with using 64 kbps per voice circuit, as with TDM, is obvious. The network engineer responsible for capacity management on the access network loves this feature, just like transmission engineers loved it when shared trunks were first introduced. Reducing the cost of bandwidth helps justify changing the traditional circuit switched network to next generation switching architectures. The savings come from the use of the statistical nature of the packet network and the efficient use of the bandwidth. Silence suppression and voice compression are just two of

Figure 28–6
The dial tone Sarah hears when she picks up her phone is generated by the LAG that resides in her serving central office—Maple Street.

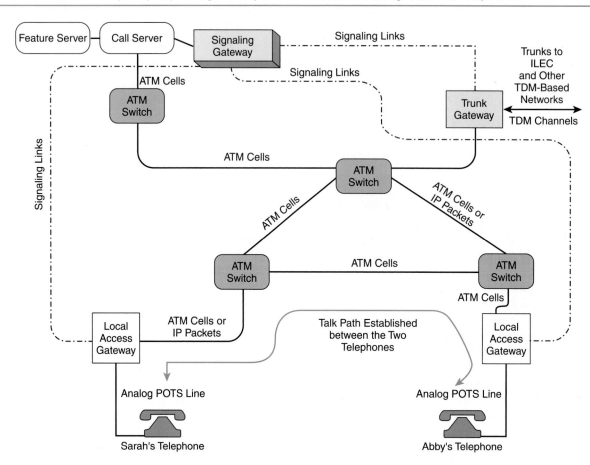

1. Dial tone is generated onto Sarah's line once Sarah picks up the telephone receiver.
2. Sarah dials Abby's telephone number.
3. The LAG sitting in Maple Street issues a request for information via the signaling links to the Boston call server.
4. The call server works on the Maple Street LAG's request by looking up Abby's telephone number in its routing tables.
5. The call server queries the Woodburne CO to see if Abby's line is idle. The Woodburne LAG responds with an idle condition.
6. The call server then returns the routing information to the Maple Street LAG and has the Woodburne CO LAG initiate ringing Abby's telephone.
7. The Maple Street LAG sends the request for circuit seizure to Woodburne CO LAG through the ATM access network.
8. The Maple Street CO establishes a connection between the two central offices through the ATM network by sending information out on the defined VC.
9. Abby picks up the telephone and the call is established end to end.

the advantages of the next generation architecture that help justify the millions of dollars required for the conversion from circuit switching.

Once Sarah and Abby finish their conversation and hang up their telephones, the access gateways see the loop open and relay this information through the signaling links to the signaling gateway in Boston. The virtual paths are torn down between the ATM switches and between the ATM switches and the local access gateways—one in the Maple Street central office and one in the Woodburne Street central office. Sarah looks at her watch and realizes she needs to call her friend Jamie who lives in Illinois to find out if he is going to visit this summer. She picks up the handset of her telephone again and punches in a 1, then Jamie's telephone number. The local access gateway sends the digits out on the signaling links that connect to the signaling gateway. The signaling gateway asks the call server what should be done with the called number. The call server looks at the number and determines that it must be sent out across XYZ's long distance network because Sarah has chosen XYZ as her primary carrier. The call server realizes that the trunk gateways located in the Burlington regional office have direct connections to XYZ's network. The call server, in conjunction with the signaling gateway, sends requests for a call path between Spidercom's network and XYZ's long distance network. A path between XYZ's network and Jamie's local telephone provider, which happens to be the ILEC, must also be established. The SS-7 messages run ahead to see if there is a path available and to see if Jamie's telephone is idle, the same as if setting up a call between two circuit switched networks, as explained in Chapter 9. XYZ's network is connected to the trunk gateway, which has channelized TDM circuit interfaces. The message saying that a trunk is available between the two networks and that Jamie's telephone is idle returns to the signaling gateway.

The signaling gateway instantly sends a request to the local access gateway to build an SVC between itself and the trunk gateway and at the same time tells the trunk gateway to establish a path between itself and XYZ's network. A virtual path is built between the two gateways. The trunk gateway establishes a TDM trunk between the two switches, thus connecting the trunk gateway and XYZ's long distance switch via shared TDM trunks. The trunk gateway can convert the ATM signal into a TDM signal and, conversely, a TDM signal into an ATM signal. A TDM-to-ATM conversion is performed in the trunk access gateway on the traffic coming in from XYZ and going to Sarah's local access gateway. The scenario just explained illustrates an example of a call leaving the local next generation network and traveling across a traditional circuit switched TDM network. When next generation networks begin to be implemented, this scenario will become very common.

The example described how the next generation network handles voice transmission. Now we will explain how the same network handles data transmission. The final scenario is that of a data session being established between the user on the next generation network and the ISP. One of the greatest advantages touted by next generation enthusiasts is the network's ability to handle data and voice in a ubiquitous manner—they are both seen as ones and zeros. Abby decides to log on to the Internet to do some research for a term paper that she is writing for biology class about gladiolas. She sits down at the computer and brings up her Internet browser. Abby's computer is connected to a DSL modem that interfaces the local access gateway's DSL port. DSL is an always-on service, meaning that it is not necessary to log on to the Internet—the link is always up. Abby clicks on her search engine and types in the word *gladiolas*. The request travels across her 384 kbps DSL local loop and enters the DSL port on the local access gateway. The traffic is encapsulated into ATM cells and shipped out on a switched virtual circuit that connects the local access gateway to an IP services aggregation box located in the regional switch site in Burlington. The IP services box terminates thousands of broadband data sessions similar to Abby's. ATM virtual circuits are built between the two boxes and used to carry all data information across the ATM infrastructure. The IP services box receives Abby's request to search the Internet. It first terminates the ATM session or simply discards the ATM cell uncovering the IP packet inside. Using the source and destination address, the IP services box routes the request to the correct output port. In this example, Abby has chosen Spidercom as her ISP. The packets are routed out onto Spidercom's IP backbone that travels to their Internet peering point in Boston. The information traveling downstream to Abby's computer is first received by the IP services box and routed to the appropriate output port, where it is encapsulated into an ATM cell and shipped out to the local access gateway. The local access gateway terminates the ATM session and transports the information out onto the DSL. The DSL connects to Abby's DSL modem, which connects to Abby's computer. Figure 28–7 illustrates how the signal flows from source to destination through the next generation network.

Next generation switching will truly integrate the voice and data world into one network. The advantages of using a highly distributed architecture are evident when tracing calls through the network. Large, locally controlled switches are not the most efficient means to switch traffic through a network. The circuit switched network worked when a large percentage of traffic was voice. The growth of data is forcing the industry to redesign the network to carry voice, video, and data efficiently.

Figure 28–7

The advantage of using a next generation network to carry both voice and data, often referred to as a converged network. The TAG and call server are not involved when data traffic flows through the network. Dial-up data calls still must be routed via the routing tables housed in the call server.

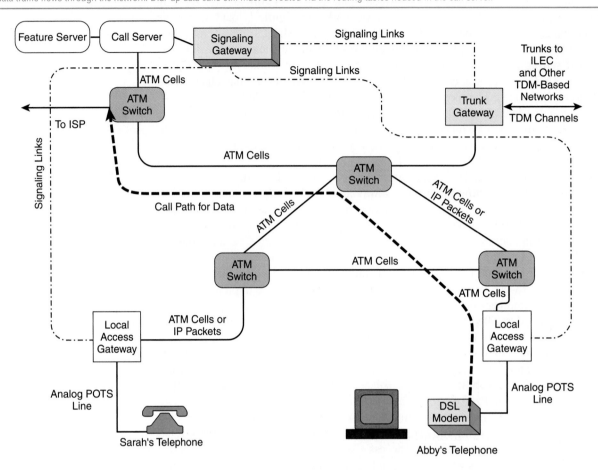

1. Abby's always-on DSL service initiates a session when Abby presses Enter on her keyboard.
2. The Woodburne LAG routes the packet onto a VC provisioned specifically for data.
3. The packet of data moves through the ATM network on already established VCs to the Boston ATM switch.
4. The ATM switch in Boston routes the packets onto the Internet.

28.1.2 Soft Switches

soft switch
Switch that handles all off-net and long distance calls, relieving switch trunks.

Soft switches have been introduced to help eliminate the need to use circuit switches for all calls, such as dial-up Internet calls, long distance calls, and other off-net calls. The soft switch is similar to the next generation architecture in that it is based on a packet network design and its purpose is to combine voice and data routing in one device. The difference is that the soft switch was not built to replace the class 5 switch. Instead it was designed to complement the class 5 switch by offloading non–locally switched traffic. Several vendors have introduced soft switches, including Lucent, Nortel, CopperCom, and Cisco. We first walk through a call that passes through the soft switch to illustrate how it fits in the network.

Spidercom's Denver office has decided to wait to replace its class 5 switch with a next generation switch and instead is adding a soft switch in its central regional switching center. The soft switch is connected directly to the Internet by OC-3 pipes and to the access network via circuits from the IP services box. The soft switch may be viewed as a controller more than a true line aggregation box. It depends on devices such as the IP services box, VoDSL gateways, routers, and possibly ATM switches. The soft switch has translation tables loaded into its memory and uses signaling links to tell connecting equipment what to do with the calls coming in and going out. As shown in

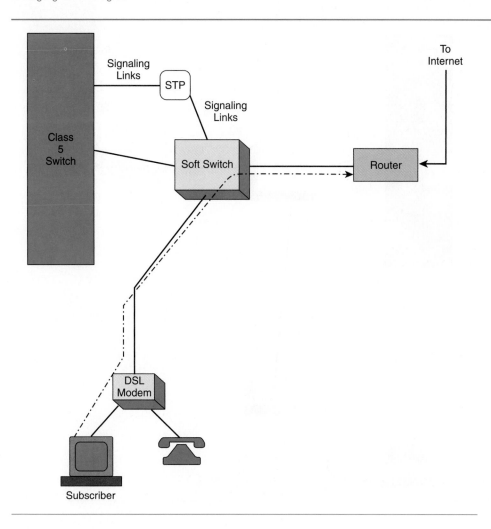

Figure 28–8
The soft switch is used to route data
traffic off the voice network.

Figure 28–8, the soft switch sits at the regional switch site and connects the access network to the backbone network.

Our first example shows a VoIP call traveling from the end user's site to the Internet and finally to the recipient at the far end. Our caller in this scenario is Curtis who is very excited about the VoIP software he just downloaded from the Net. He decides he will call his friend Sandeep in Bombay. Curtis logs on to the Internet as usual using his analog voice line. The call request travels to Spidercom's Denver regional switch site where the signaling links in the soft switch recognize that the number Curtis dialed belongs to his ISP. The soft switch knows that the call does not have to travel through the circuit switch because it is destined for the ISP. The call path is established between Curtis's modem and his ISP via Spidercom's transport network, as shown in Figure 28–9. The advantage to Spidercom is saving switch processing power, switch ports, and the switching matrix. Because Internet users stay online for hours instead of three or four minutes, they are forcing service providers to add expensive switch ports and processing power to their class 5 switches. Soft switches provide a much more economical way to route Internet traffic through the network by bypassing the class 4 and 5 switches.

A second use of the soft switch is to route long distance calls from the access network directly to the long distance carrier, again bypassing the class 5 switch. When Curtis says good-bye to Sandeep, he decides to call his mother who is living in Charlotte, South Carolina. He picks up his telephone handset and punches in a 1 plus his mother's telephone number. The soft switch recognizes that the call needs to be routed to the long distance network and directed across shared trunks that have been built between Spidercom's local access network and Spidercom's long distance network. The

Figure 28–9
The soft switch directs calls coming in
from traditional analog POTS line and
a modem.

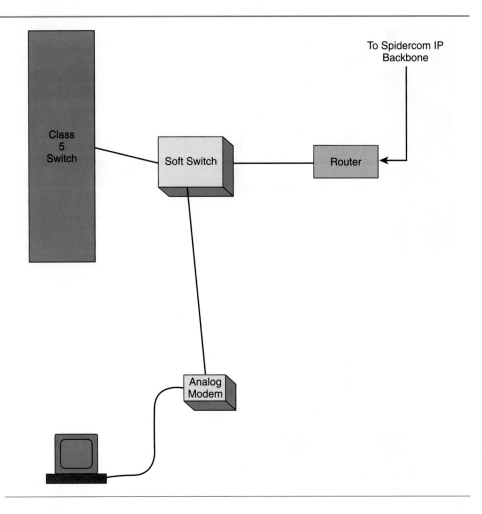

call does not have to be routed through the class 5 switch before it is handed off to the long distance provider. There are two reasons why this is better than continuing to use the class 5 switch to route long distance traffic. A line port on a class 5 switch is very expensive, as is a trunk port on the egress side of a class 5 switch. The soft switch ports tend to be priced at a quarter of the cost of a switch port. For example, a soft switch port costs about $25, while a class 5 switch port costs about $100. A second advantage is the reduced latency experienced if the call remains in the packet world, as with the first example. When the Internet call has to travel through a TDM switch like the class 5 switch, latency is attached and becomes a real hazard when transporting packetized voice. The longer the traffic is carried in the packet format and the fewer TDM-to-packet conversions, the better. The soft switch eliminates at least one conversion, in addition to avoiding the inherent latency of a class 5 switch.

■ 28.2 TRANSMISSION NETWORK INNOVATIONS

The transmission network, like the switching network, is evolving at an incredible pace. The greatest areas of innovation are in the packet/cell transport equipment and the optical networking devices. They are working hand in hand to create a very significant change in the way traffic is transported across the network. The first innovation is improvement that is being made within the dense wave division multiplexing (DWDM) designs as the number of wavelengths per fiber strand continue to increase. The second new technology being touted by optical manufacturers is combining routing, TDM, and ATM into one multiplexer. The multiplexer will be able to route traffic within the box, along with separating different kinds of traffic, such as ATM or TDM. A third optical innovation is that of cross connecting and switch optical wavelengths. In addition, the efficiencies of the ATM protocol are being refined and improved to help reduce the negative effects caused by the large cell tax inherent with ATM. The biggest breakthrough in the world of IP is *IP Class of Service (IP CoS)*. This section focuses on these new technologies and techniques and explains how they will change the current transport infrastructure.

IP Class of Service (IP CoS)
Quality service classes
defined within the Internet
Protocol.

28.2.1 Optical Networking

The optical network has evolved from point-to-point asynchronous multiplexers to SONET ring architectures to actual routing multiplexers. Today, optical multiplexers are being introduced that are able to route IP traffic the same way a router is able to route IP packets. The advantage of the optical multiplexer over the router is that the signal does not have to be converted into an electrical signal before the packet is cracked and the source and destination addresses are revealed and used to route. Optical multiplexers are also being built to carry ATM traffic directly and, similar to IP, to route ATM switching or concentration within the optical signal. Again, the advantage is obvious—the signal does not have to be converted into an electrical signal in order to be switched. The efficiency of an optical multiplexer that can carry and route cells or packets, in addition to being able to keep TDM traffic intact, is a reality. Next generation SONET will be much more sophisticated and will eliminate the need for external routers and ATM switches. Some believe ATM and IP boxes will replace the optical multiplexers. The point is that one box will be used to perform all of the transmission functions required by the network.

A recent introduction by many SONET multiplexer providers is a multiple-ring-capable multiplexer. The newer multiplexers allow more than one OC-48 ring to interface into the box. The advantage of this feature is that it allows dual rings to terminate into the same multiplexer, as shown in Figure 28–10. In the past, two very expensive SONET OC-48 multiplexers were needed to provide dual-ring internet working, also referred to as *matched nodes*. Additionally, the cost of a SONET multiplexer has been reduced to about one third of the cost of the initial device. As chips become more compact and less expensive, the cost of optical networking also becomes less expensive and more powerful.

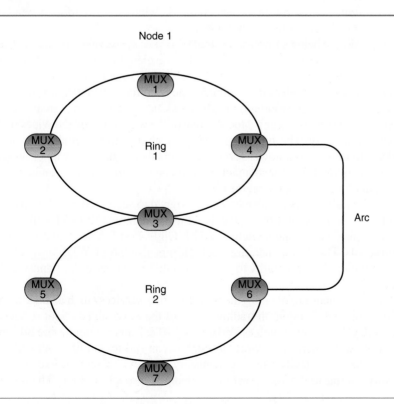

Figure 28–10
Connecting two SONET fiber optic rings together only requires one SONET multiplexer at each of the interconnection points. New methods for connecting a node riding on a ring using an arc connection between the two are becoming very popular.

The area of fiber optic transport has emerged as one of the most innovative and fastest-growing sectors of the industry. The introduction of wave division multiplexing technology, optical-to-electrical-to-optical cross connects, wavelength routers, and optical-to-optical switches is transforming the way information is carried.

Wave division multiplexing (WDM) has been refined to include wide wave division multiplexing (WWDM), coarse wave division multiplexing (CWDM), and dense wave division multiplexing (DWDM). WWDM is also called *Fat WDM*. The multiplexers are built to combine two to four wavelengths that have very wide spaces between them. WWDM is an inexpensive way to increase the number of signals on a fiber strand while remaining within a budget.

CWDM also has wide divisions between each wavelength but not as wide as in WWDM. CWDM is able to multiplex eight to sixteen wavelengths per fiber for a reasonable cost. The final WDM technology, DWDM is the most expensive of the three due to the close wavelength spacing requiring very precise lasers and filters. DWDM multiplexes as many as 140 wavelengths onto one fiber.

The introduction of OC-768 transport technology is also changing how companies establish networks. OC-768 is equivalent to 768 DS-3s, each able to carry 672 64 kbps channels. Thus, the total number of 64 kbps channels carried on an OC-768 signal is:

$$768 \times 672 = 516,096 \text{ DS-0 channels}$$

Today, lab trials have shown systems running 140 wavelengths of OC-192 signals on one fiber strand. The number of DS-0s would equal:

$$(192 \times 672) \times 140 = 18,063,360 \text{ DS-0s}$$

Because of WDM technologies, new optical cross-connect and switching equipment is being developed. The optical cross connect is a device that accepts optical wavelengths from DWDM systems and grooms and cross connects wavelengths to other wavelengths. Optical cross connects are now optical-to-electrical-to-optical (OEO) devices but will in the future be all optical—OOO. Optical switches, similar to optical cross connects, will dynamically switch wavelengths according to signaling protocols such as MPLambdaS or GMPLS. Agilent's optical switch uses bubbles to redirect the light waves through the switch matrix while other vendors depend on gratings and other light-directing technologies.

Another device being showcased at trade shows is the wavelength router. The wavelength router uses special tags attached to the wavelength, similar to IP addresses, to route the wavelength in real time. Lucent's wavelength router is an example of this new technology. Routing wavelengths eliminates the need for static cross-connect or switch ports, which require provisioning time when making changes to the circuit destination.

digital wrapper
Protocol used to encapsulate all protocols onto an optical pipe, eliminating the need to access the electrical signal.

Using the *digital wrapper* protocol is one of the ways all optical devices will maintain network integrity. The only monitoring capability now available for all optical networks is to monitor the power level of the signal. The digital wrapper protocol encapsulates the optical signal and attaches overhead information that can be used to detect errors and other problems on the circuit path.

SONET multiplexers are also being redesigned to accommodate the various transport protocols. Next generation, also called third generation, SONET multiplexers are able to multiplex, route, and switch IP, ATM, GigE, Fast Ethernet, and the traditional TDM protocols. The third generation multiplexers also have CWDM systems built in, allowing additional wavelengths to be provisioned to increase the bandwidth on the network.

All these devices are opening the door for other services, such as bandwidth trading, to be deployed. Bandwidth trading is one of the most talked about services in the industry today. Companies such as Verizon and AT&T and are deploying billion-dollar fiber optic networks across the world to establish an end-to-end fiber network that will allow them to dynamically exchange capacity. Bandwidth will be exchanged like a commodity similar to the way power is exchanged through brokers. The brokers will sit at terminals and buy and sell bandwidth by punching in information, causing flow-through provisioning systems to activate a circuit on the network. Analysts predict that once bandwidth trading begins, the price per bit will drop dramatically.

28.2.2 Packet/Cell Networking

The use of an all-IP or an all-ATM network is becoming a reality. IP and ATM transport methods will be the next generation transport technologies. The equipment vendors have been building larger, more robust IP networking boxes that are able to forward packets at incredibly high speeds. In the past, the IP networks, which are synonymous with the Internet, were built using enterprise-type routers that lacked processor and input/output card redundancy. Carriers are now demanding carrier-class devices that have fail-over power, processor, and interface module redundancy. In addition to building carrier-class routing devices, the industry is looking to combine both IP and ATM capability in one large box. The ability of the IP router is no longer just IP routing. The router must either be able to provide protection switching when facilities are lost or be able to handle routing protocols without stressing the processors. The new robust IP routers, such as those produced by Foundry Networks, Juniper, Cisco, and others, are setting the bar for others to reach.

IP CoS is an innovation similar to ATM QoS. It is a way to prioritize packets by tagging them with priority bits similar to the classes of service offered by ATM—CBR, VBR, and UBR. The Internet Engineering Task Force (IETF) has created DiffServ to perform traffic management on IP traffic, making IP a ubiquitous protocol. The majority of end devices use IP as the transmission protocol. The Internet is everywhere, and because IP is the protocol used to access the Internet, IP is everywhere. Data are less temperamental than time-sensitive traffic such as voice and video. If a packet is lost out of a data stream, the end device asks that it be retransmitted. If, on the other hand, there is congestion in the network and the voice has been converted into an IP packet, the person listening at the terminating end may hear interruptions in the conversation. Voice is very sensitive to delay and therefore requires that the packet be given priority over the data traffic riding on the same pipe.

DiffServ is the solution presented by the IETF forum and the reason some analysts are saying IP will win out as the transport technology of the future, beating out ATM technologies. Analysts do have a valid argument if DiffServ proves to be able to provide the same secure, reliable transport as ATM. If DiffServ cannot provide defined classes of service, IP will continue to be used to carry data. IP and ATM advocates are still warring over whose technology is best. Both will continue to be used. The questions are, *Where will they be used?* and *What type of traffic will they carry?*

ATM switches are also becoming much more robust as they add features and services to their processors. The biggest change in the ATM world is moving toward a public SVC network that allows switched VCs between carriers. Today many large ATM networks depend on SVCs within their own network but must build PVCs between external networks. The advantage of using SVC technology is the simplification of provisioning circuits between networks and reducing bandwidth needs because the SVC is built only when a session needs to be established.

In addition to moving toward a fully public SVC network, SVCs are moving closer to the end points—customer locations. A few equipment vendors, mainly those building devices for VoDSL, are now deploying SVCs to the end point. Accelerated Networks is building its product to be able to support SVCs to the end point, mainly because the provisioning requirements for ATM to the customer site are cumbersome and often reduce the profits that may be made on data services. The ATM vendors realize that SVCs eliminate the need to provision the PVC every time a new service is added or a new customer is placed on line.

The quality of service parameters built for ATM are also being improved and refined to help improve transmission and bandwidth savings. The introduction of silence suppression and AAL2 voice are both providing improved bandwidth savings because of the way the cells are structured.

DiffServ
Newly defined portion of the IP packet used to prioritize traffic, such as time-sensitive voice and video, to ensure timely delivery of the information.

Figure 28–11
MPLS is becoming one of the most talked about protocols in the transport industry. MPLS provides a way to lessen the latency on IP routed networks. A tag is attached to each packet, similar to attaching a ZIP code to an address. The tag is used to route the packet through the network, eliminating the need for the router to crack open the IP frame and read the source and destination address. MPLS is one of the most promising protocols today.

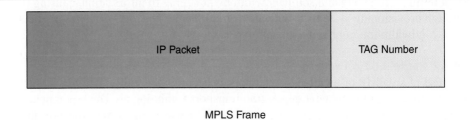

MPLS Frame

The telecommunications network is evolving into one that uses statistically multiplexed packets or cells to carry traffic across the network. The merging of data, voice, and video onto one network is the goal of the industry. The battle between ATM and IP supporters will not deter the growth of this new packet/cell network. Yes, the old world of circuit switched services will remain in place for some time to come. The innovations being made in optical networking will allow all three transport technologies to live in the same space.

28.2.3 MultiProtocol Label Switching

The *multiprotocol label switching* (MPLS) has recently been deployed in large backbone networks. It is a new protocol standard that improves the reroute capability of packetized information. A label that defines the destination address of the packet is attached to each packet, similar to adding a ZIP Code to a letter, so the packet no longer has to be cracked at every router in the network. Instead, the label is used to switch the packet through the network. Figure 28–11 depicts the MPLS frame structure.

multiprotocol label switching
New protocol used to simplify routing packets through the network.

■ 28.3 ACCESS NETWORK

28.3.1 Broadband Technologies

Digital subscriber line (DSL) is fast becoming one of the most talked about services in the Internet world. It offers high-speed Internet service to the customer and the reuse of the copper outside plant for the ILEC. DSL is transport technology that uses different line codes to send more bits per second than on a typical analog line. The most common types of DSL available today are as follows:

- SDSL—With synchronous digital subscriber line (SDSL), the upstream and downstream speeds are the same. For example, a 1.544 Mbps DSL from Spidercom is able to carry 1.544 Mbps upstream and, at the same time, 1.544 Mbps downstream, thus the term *synchronous*.
- ADSL—With asynchronous digital subscriber line (ADSL), the upstream speed is less than that of the downstream speed. ADSL is the line code offered to residential customers who mainly use the service to pull information from the Internet and consequently require more bandwidth on the downstream side of the circuit. Common line rates are 6 Mbps downstream, 978 kbps upstream.
- G.Lite—This is a light version of ADSL for customers who do not want to pay for the higher speeds of ADSL but want to have a DSL. G.Lite is also the standard that the industry is hoping will spread to every residential subscriber. Eventually G.Lite modems may be placed right in the PCs, similar to 56 kbps analog modems.
- RADSL—Rate adaptive digital subscriber line (RADSL) is similar to ADSL but allows the speed to rate-adapt when the physical makeup of the circuit changes. For example, copper cable tends to increase in resistance as the temperature increases. When the resistance increases, the speed of the circuit decreases. RADSL allows the circuit to automatically decelerate without dropping the connection.
- MVL—This is a proprietary line code developed by Paradyne Corporation. The line code falls within the acceptable frequency spectrum and, thus, has passed all FCC requirements. The code allows for a very long loop length of the circuit.

- IDSL—ISDN digital subscriber line (IDSL) is used when a customer's loop is fed out of a digital loop carrier (DLC). The only DSL method able to work through a DLC is IDSL. The IDSL code allows a maximum of 128 kbps circuit speed.
- HDSL—High bit-rate digital subscriber line (HDSL) comes in two versions—one that rides on a four-wire circuit and one that rides on a two-wire circuit. The four-wire HDSL circuit has been deployed by carriers for years and has been used to carry standard T1 carrier circuits through the network. An HDSL signal does not have to be regenerated as often as a traditional T1 circuit. The HDSL circuit is able to ride on the standard POTS wiring, eliminating the need to build special carrier spans. HDSL2 wire provides the advantage of using just one pair instead of two.
- HDSL2—High bit-rate digital subscriber line 2 (HDSL2) is one of the newest standards and is being touted as T1 replacement. The HDSL2 standard allows 1.544 Mbps to travel on one pair of wire. The difference between HDSL and HDSL2 wire is the line code used by HDSL2, which allows it to travel farther than the 2B1Q line code used by HDSL wire.

Each of these categories may be further defined by the line code used to carry the information. Line code is the method used to format the bit stream. In other words, as shown in Figure 28–12, the line code determines how many bits are transmitted every cycle and how they are modulated, similar to quadrature amplitude modulation. The three most common line codes currently used are carrierless amplitude phase (CAP), discrete multitone (DMT), and 2B1Q. The industry forum that standardized DSL still

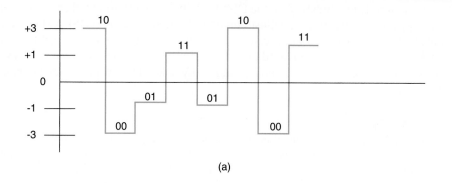

(a)

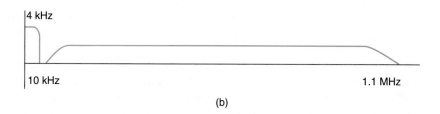

(b)

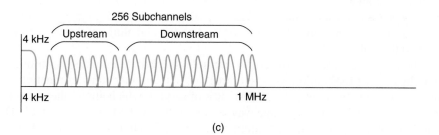

(c)

Figure 28–12
The three most popular line codes used to transport information across digital subscriber lines:
(a) 2B1Q line code
(b) CAP line code
(C) DMT line code

Figure 28–13
A DSL modem attached to a laptop computer. A CAP line code that provides an asymmetrical DSL service—higher downstream speed than upstream speed—is being used.

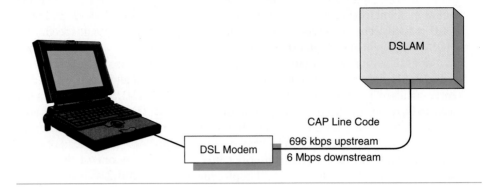

argues about which line code should become the industry standard. DMT has been named the defacto standard by the DSL Forum, though many vendors still deploy the CAP line code. Again, both line codes provide a means to carry high-speed, bit-per-second rates across an ordinary copper loop—one pair. The advantages are obvious. The customer may now use this high-speed loop to connect into the Internet. DSL is able to carry information both upstream and downstream, similar to the cable modem. Unlike cable modems that have their customers share the same bus, DSL users each have their own circuit, as explained in the previous chapter. The major advantage of using broadband DSL compared with cable modems is the added security offered by the individual circuits that hub back to the DSLAM. DSL is a technology that uses the existing copper plant. Figure 28–13 illustrates a typical DSL service. A DSLAM is placed in the serving central office and is connected to the regional switch site via T1, T3, or optical interfaces. The protocol used to connect the DSLAM to the regional switch site varies depending on the vendor. Today, thanks to VoDSL, DSLAMs are quickly being converted into ATM capable devices.

28.3.2 Voice Over Digital Subscriber Line

Voice over DSL (VoDSL) is one of the hottest products being deployed today. The technology is popular because it allows multiple telephone terminations on one digital subscriber line, in addition to allowing data to travel on that same local loop. VoDSL was first developed after the advent of DSL service. Large-scale deployment of VoDSL is limited, but analysts see the service as growing exponentially over the next few years. The way the technology works is fairly simple. The customer is given a device similar to a DSL modem called an *integrated access device* (IAD). The IAD has several RJ-11 jacks that may be used to connect multiple telephones. It also has a port where data devices may be terminated. At the DSLAM, the digital subscriber line connects as it does in a normal DSL deployment. From the DSLAM back to the regional switch site, the voice traffic from the customer is routed onto one virtual circuit and the data traffic is routed on a second. Once at the switch site, the voice PVC is groomed toward a VoDSL gateway, as shown in Figure 28–14. The data traffic is groomed toward a router or some other IP device. The VoDSL gateway accepts the ATM signal from the access network and converts the signal into a traditional TDM signal. The VoDSL gateway and the class 5 switch are connected by multiple trunks that use the GR-303 protocol to handle the signaling needs at the customer site. The GR-303 protocol carries the call supervisory information inside the pipe. The GR-303 carries the signaling information from the switch to the IAD.

The advantage VoDSL provides is multiple telephone terminations on one telephone line. Most IADs allow between three and sixteen telephone terminations. The voice is encapsulated in the AAL2 ATM class of service, which allows the signal to be statistically multiplexed in the same way as data traffic. VoDSL gateways and IAD are being built by many new start-up equipment vendors such as CopperCom, Jetstream Communications, TollBridge, Accelerated Networks, and others.

integrated access device
Device placed at the customer premises in order to terminate VoDSL or VoIP signals.

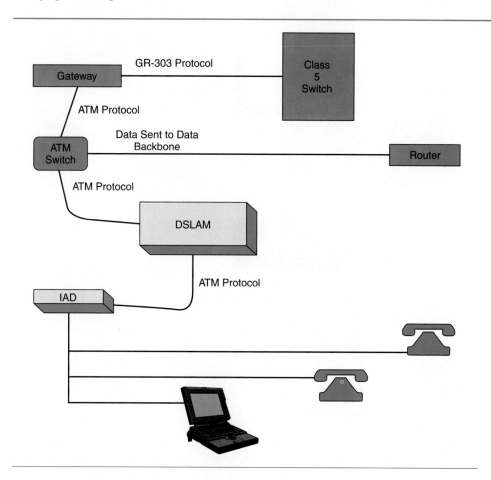

Figure 28–14
The deployment of DSLAMs in the local network produced alternative ways to provide dial tone to the business customer. One digital subscriber line provides numerous voice lines and a data connection.

28.3.3 Voice Over Internet Protocol Services

Similar to VoDSL, Voice over IP travels through the access network from the customer site to the central office, regional switch site, or ISP. The VoIP traffic is fed into a VoIP gateway. The gateway is responsible for either converting the signal into a TDM signal or routing the IP packets through the IP network to a gateway at the distant end. VoIP is going to be one of the fastest growing services due to the zero cost for usage. VoIP networks are built differently depending on the design engineers. The soft switch may be used to switch the IP packets through the gateways, or the gateway may convert the IP traffic into GR-303 protocol in the same way the VoDSL gateways convert the ATM cells into TDM GR-303.

28.3.4 Internet Protocol Services Systems

With broadband services, cable modems, and DSL, a new piece of equipment has been added to the access network. An IP services system is used to aggregate broadband circuits and apply special value-added services. The services range from firewall service to directing the user to specific Web pages. The IP services systems, also called Subscriber Management Systems (SMSs), are capable of terminating thousands of broadband digital subscriber lines. They have multiple virtual routers that can direct traffic to different terminating locations. The SMS has many access protocols used to connect to the DSLAM, as shown in Figure 28–15. The most common protocol used between the two pieces of equipment is ATM. A virtual path (VP) or multiple VCs are built between the two devices, and individual subscribers' traffic rides in the VCs. The SMS is being touted as the box that provides IP centrex. The introduction of the SMS device shows that the intelligence in the network is moving farther away from the customer's end office and into the access network. The SMS will perform routing for customers, eliminating the need for them to own and

Figure 28–15
IP services boxes are placed in the
network and used to aggregate DSL
traffic and route the information to the
correct network. In addition to
aggregating digital subscriber lines,
the IP services box may offer special
services such as firewalls.

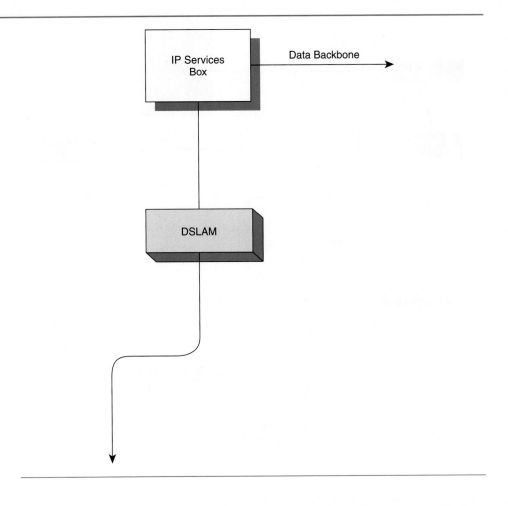

maintain their own routers. The SMS will also provide firewall protection, eliminating the need for a firewall at the customer site. The SMS connects the end customer to a remote location and to the ISP.

28.3.5 Customer Premises

The changes happening at the customer premises revolve around Internet service. Customers today have many options available to them for both Internet and telephone service. New technologies such as DSL and cable modems are providing high-speed access into the Internet, and VoIP is offering voice services across the Internet so the user can speak with friends who are continents away for free. Special value-added services offered by service providers reduce the customer's network responsibilities. The goal of the service provider is to make the customer's equipment as dumb as the telephone. They want to provide and maintain all the extra services the customer currently has to control.

One of the most interesting technologies being deployed is wireless LAN (WLAN) service. A wireless LAN carries LAN traffic between the user's PC and the server without needing to install expensive and hard-to-maintain inside wiring. Wireless LANs use infrared frequencies to carry traffic through the building and around the floor.

**Broadband over Power Line
(BPL)**
Access technology that uses
the power line infrastructure
to carry a broadband signal
into the customer's premises.

28.3.6 Broadband Over Power Line

Over the past few years, a new entrant into the broadband access space has emerged—the power company. *Broadband over Power Line (BPL)* refers to using the power lines to carry an RF signal capable of transporting broadband services to the end customer. As discussed in Chapter 17, the primary means of access to a subscriber's home is through cable television coaxial cable drops, telephone company telephone drop wire, fiber to the home or business feeds, and Wi-Fi/WiMAX wireless hotspots. Engineers, physicists, and lots of other very smart people realized that one other cable entered a person's home—the power wire. From this realization, BPL emerged.

HomePlug Standard

Broadband over Power Line network topologies vary somewhat between service providers. The majority, though, use the *HomePlug standard* as the Layer 1 and Layer 2 protocols. HomePlug was first developed as a means to carry communications signals across inside electrical wires. You could plug a HomePlug modem into an electrical outlet anywhere in the house and plug your PC into the modem using standard Ethernet cable. Though this technology was adopted by some techie types, it did not take off as hoped.

HomePlug standard
Protocol used to package and transport the BPL signal across the power line.

Layer 2

Broadband over Power Line today uses the HomePlug standard to package and transport the frames of data across the power lines. It does this in a manner similar to that of a traditional Ethernet. It broadcasts the packets to all devices on the segment, employs a collision/contention CDMA/CD scheme to handle congestion and allows new devices to be added to the network through the use of MAC and IP address.

Layer 1

The HomePlug standard uses orthogonal frequency division multiplexing (OFDM) to multiplex the data onto the wire. As explained earlier, it divides the frequency spectrum into discrete bands or parts. Discrete multitone (DMT) is the modulation method employed by OFDM to divide the spectrum and manipulate the data. Tone maps are created by the transmitting device. Each tone map carries data depending on the signal-to-noise ratio (S/N) for that frequency. In fact, if conditions are poor and the S/N is weak, the standard will adjust the bandwidth for that connection to compensate for the low levels. The HomePlug standard uses the unlicensed frequencies defined by the FCC and used today or Wi-Fi, WiMAX, and other wireless technologies. Most BPL equipment manufacturers carve out frequencies used by ham radio operators in order to reduce the chance of bleed through and interference.

Network Topology

The BPL network topology is often referred to as a hybrid fiber BPL network. A BPL network has two parts: (1) the link that carries the signal from the power line to the POP for access to the outside world and (2) the portion of the network feeding into the customer's premises. In order to build either of these sections of the network, equipment has to be placed on the poles. In our example, we will look at a network that uses both the primary and the secondary power feed to service the customer.

In order to couple the signal onto the primary portion of the power line, a BPL device is mounted on the electrical pole and through a passive conductor tied to the primary power line. The BPL equipment also ties to a fiber optic cable that is connected to a distribution point in the field. The BPL equipment has two jobs: (1) it has to convert the optical signal coming from the fiber optic cable into an electrical signal, and (2) it has to transmit an RF signal out onto the primary power line. In addition, the BPL equipment has to handle the HomePlug Layer 1 and 2 functions.

The second half of the network contains a second BPL device that is placed on the pole where the transformer feeding a neighborhood resides. The BPL equipment must be able to receive the incoming RF signal off the primary line, bypass the transformer, and inject the signal onto the secondary line that feeds into the customer's home.

The signal rides on the secondary power feed into the customer's home through the electrical meter and throughout the electrical wiring in the house. This means the customer may connect his or her PC to any outlet in the home and receive broadband connectivity.

Figure 28–16
Broadband over Power Line network
topology takes advantage of the power
lines that feed a neighborhood. As
shown in the figure, the network
topology is a combination of fiber
optic cables and power lines. The
term *hybrid fiber/BPL* is often used to
describe the BPL network.

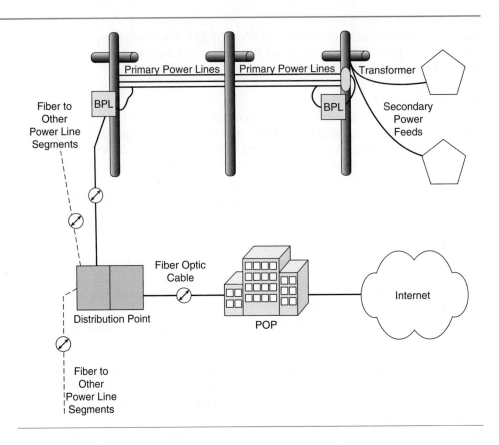

The fiber optic network required in a BPL design is similar to the cable television company's "tree and leaf" design. Multiple strands of fiber feed BPL equipment sitting on electrical poles. The fiber optic cables feed back to a regional location referred to as a *distribution point* where signals coming from the hundreds of BPL devices are aggregated and shipped out onto a fiber connection tied to the point of presence (POP) as shown in Figure 28–16.

At this time, most BPL networks are deployed across aerial power feeds. Currently, the goal is to push signals across underground power networks as well. Broadband over Power Line network topology varies depending on the service provider's design. For example, some BPL providers do not deploy fiber; they opt for wireless transport methods from the BPL equipment on the pole to the POP or distribution point. The only drawback to using wireless is when trees and so forth obstruct the signal to an unacceptable level, forcing the provider to implement an alternate solution.

The reality of this technology is often unclear as there are numerous ways to provide the same service. In most situations, the power line portion of the network—across the primary and secondary power lines—is minimal. Note that the signals are not traveling across the high power lines, or grid, as it is called. It is a last-mile technology that is valuable because it is able to eliminate the need to build another pipe into the customer premises.

**power line interface device
(PLID)**
Modem used to convert the
HomePlug signal from the
power line to an Ethernet
interface that can be
connected to the end
computing device such as
the PC, X-box, and so forth.

Customer Premises

For customers to access their broadband connection, they need a *power line interface device (PLID)* or simply a power line modem. The modem is a small cubelike device that plugs directly into the AC outlet. The PLID has standard Ethernet interface that is used to connect an Ethernet cable to the host device, as shown in Figure 28–17. The PLID can be plugged into any outlet in the house and used to access the broadband signal riding on the electrical wiring.

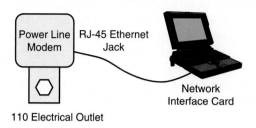

Figure 28–17
A power line interface device (PLID),
also called a power line modem, is
plugged into the AC outlet. The laptop
or other computing device is
connected via Ethernet cable to the
PLID. The HomePlug signal rides on
frequencies well above AC power
signals, eliminating interference from
the power feed.

The PLID uses the HomePlug standard to format and manage traffic between the BPL network and the customer's end device. The PLID received Ethernet frames from the customer's PC, formats the frames with the HomePlug frame, attaches overhead, and ships them out across the tone map.

Enhanced Power Distribution Services

One of the most promising service offerings HomePlug provides is *Enhanced Power Distribution Services* (*EPDS*). This service offering provides power companies the ability to remotely read power meters, remotely monitor transformer operation, and remotely monitor other devices on the power line network. Some analysts see the broadband connection to the subscriber as a gray area. The real meat of the BPL network is its ability to help the power company maintain its network.

EPDS uses the HomePlug standard to carry messages, for example, from the transformer to a central network operations center. The messages tell the network operations technician that the transformer is above its allowed temperature or the transformer is in jeopardy. The HomePlug standard PLID is built right into the power meter hung on the outside of the consumer's home. The service technician, now sitting in a central location, can read the meter at any time of the day or night. In fact, the readings may be gathered over a period of time to assess trends in power usage. Many analysts predict that appliances in the home will have HomePlug chips inside that regulate when they run, how much energy they use, and on and on and on. In summary, the technology behind the BPL access network is the most exciting technology to emerge over the past few years.

Enhanced Power Distribution Services (EPDS)
The BPL network carries information from the customer's power meter or from the utility's transformers. The information is used to provide the power company with the status of the device, the power usage of the end point, and so forth.

28.4 WIRELESS TECHNOLOGIES

A newer type of wireless technology uses lasers to transmit signals through a building or between buildings. Campus settings may use new laser technology to transmit information between buildings and thus eliminate the need to run wires. The lasers are placed on top of the roof and are aimed toward the receiver located on an opposite rooftop. The advantage of using laser technology is the greater amount of bandwidth available and, most important, the free spectrum available.

Fourth Generation Wireless (4G) is being discussed at conferences, in boardrooms, across dinner tables, and in any other location that telecom professionals haunt. 4G systems being tested today by NTT DoCoMo in Japan show speeds of 100 Mbps while mobile and 1 Gbps when stopped. Video services across *4G systems* are the focus of the wireless community. The second directive being pushed by the industry is to make 4G a ubiquitous standard that allows end users to access any type of wireless network regardless of protocol. The term *smart radio* is being used to describe a device that can connect to anything, anywhere. For example, if a subscriber is sitting in Phoenix, Arizona, he or she can connect to the UMTS 4G system. If that same user travels to Green Grass, Wyoming, he or she can connect to the locally run WiMAX network. In summary, 4G is in its infancy. The ability to provide smart radio and/or video able networks is predicted to occur in about the 2010 time frame.

4G systems
4th Generation cellular systems that will use 3G technology and newer, more robust communications protocols capable of carrying higher amounts of data, specifically video.

Wireless technology is slowly improving the amount of data that can be transmitted in the cellular network. Equipment manufacturers are working many hours to prove that high-speed data can be transported across a cellular network. The race to produce a reliable high-speed wireless service is helping push the service as far as it can go.

Another wireless service being touted is that of palm devices that are now capable of transmitting and receiving information. Palm wireless protocols have been developed to allow information to flow in from the Internet.

SUMMARY

The purpose of this chapter was to introduce a few of the many new technologies being deployed in the telecommunications network. The chapter discussed the evolution to the next generation switch and the importance of the change from a circuit switch to a packet-based distributed switch. The soft switch was described as a controller that is able to offload traffic from the circuit switch and direct traffic to the Internet or long distance network.

The focus on the access network was DSL, the new IP services device, and ATM in the local access network. The chapter looked at the customer premises and emphasized the direction change from intelligence at the end point to intelligence at the edge: the switch site.

Finally, the chapter briefly described the new wireless data networks that are able to carry high-speed data across the cellular network. The wireless LAN environment was mentioned, as was the new laser wireless service that does not depend on the expensive microwave or cellular frequencies.

REVIEW QUESTIONS

1. Explain what changes are occurring in the area of circuit switching.

2. What is meant by the term *distributed architecture* when discussing switching technologies?

3. List and explain two emerging technologies that will dramatically change the way we communicate.

4. How will wavelength switching change the way networks carry voice and data information?

5. A soft switch is used to switch calls between international gateway switches. True or false? Explain.

6. What is the difference between ADSL and SDSL technologies?

7. Define the term *BPL*.

8. Explain why BPL is considered an access network technology.

9. The signal on a BPL network rides on the power line infrastructure.

   a. true   b. false

10. HomePlug is the protocol used to define how the BPL signal is packaged and transported across the access network.

    a. true   b. false

11. Why do you think a power company would be interested in an EPDS system?

12. What application will 4G cellular radio systems provide?

KEY TERMS

next generation
  switch (757)

call server (757)

Megaco protocol or
  MGCP (757)

soft switch (764)

IP Class of Service
  (IPCoS) (767)

digital wrapper (768)

DiffServ (769)

Multiprotocol label
  switching (770)

integrated access device (772)

Broadband over Power Line
  (BPL) (774)

HomePlug standard (775)

power line interface device
  (PLID) (776)

Enhanced Power Distribution
  Services (EPDS) (777)

4G systems (777)

Green Grass Case Study

BEGINNING

Setting the Stage

Green Grass, Wyoming, is a small town nestled in the hills of Wyoming that, until recently, saw no need for a telephone company. The residents of Green Grass were content to talk with each other at the local coffee shop, post office, feed mill, church, and the Grange. The change happened rather suddenly on the day of the big Green Grass centennial celebration. Horace Michaels, mayor of Green Grass, announced that a new company, Flipper.com, was interested in locating in Green Grass. The residents of Green Grass listened intently as the mayor talked about the many amenities a big .com company would bring—money for a new community hall; a traffic light; a new snowplow; and, most important, a memorial to the great grasshopper invasion of 1873. The residents of Green Grass felt a rush of pride as the mayor spoke about the memorial and the significance it had for all Green Grass residents. Green Grass was the only town to have survived the grasshopper invasion and, as such, should be recognized. The memorial would remind all of the grit and tenacity of their forefathers and mothers in the time of hardship.

The mayor then introduced Big Bob Buchman, the town's entrepreneur and funeral director. The mayor explained to the residents that Big Bob had received financing from some progressive-thinking venture capitalists. Big Bob, in turn, had engaged a group of telecommunications consultants to design the Green Grass telecommunications network. Big Bob expressed his great joy at being able to bring communications to Green Grass and explained that he had already hired a telecommunications firm that was going to design the best network in the state.

Throughout the text, you are acting as the communications consultant hired by Big Bob to design the town's telephone network. Your mission is to build solutions for the different communications needs required by the Green Grass population. At the end of several chapters, you are asked to design a particular portion of the network. The case study begins with Chapter 4 and concludes in Chapter 22. By the time you finish the case study in Chapter 22, you have developed a foundation in telecommunications networks.

Background Information

The following pages contain information about the demographic makeup of Green Grass, along with a detailed description of each location that requires special telecommunications services.

- There are 5000 people living in Green Grass according to the last census report. The breakdown of homes is as follows: 2000 households reside within the city limits, 1000 residents and 250 households live within a 10 mi. radius around Green Grass.
- Map A–1 (page 786) shows the layout of Green Grass including the street names and topological layout. Map A–2 (page 787) shows the layout of the surrounding township. Map A–3 (page 788) indicates where the businesses, schools, senior citizens' home, county buildings, and churches are located. Map A–4 (page 789) shows where the ranches, homes, and movie stars' residences are located.
- Table A–1 lists each business, school, senior citizens' home, county building, and church along with its reference number.

The following descriptions describe each of the businesses and their telecommunications requirements. You will need to use this information when completing the case studies in each chapter.

- *Gertrude Stone's Hardware & Lawn Mower Repair.* Five employees work at Gertrude's hardware. Chet and Jane work in the hardware and the lawn mower repair

Table A–1
Locations in Green Grass.

Site	Map Reference No.
Gertrude Stone's Hardware & Lawn Mower Repair	1
Straight or Curled Hair Salon	2
BB's Barber Shop	3
Louden's Department Store	4
Cowboy Clothes	5
Chaser, McDonald, and Rhetorick, Attorneys at Law	6
Mayfair's Five & Dime	7
BookWorm's BookStore	8
Angel's Diner	9
Blessings Insurance Plus	10
Buchman Funeral Parlor	11
Buchman Real Estate	12
Ed's Tattoos & Rug Cleaning	13
Stuff and Stiff Taxidermy	14
Travel Much Travel Agency	15
Kelly's Green Grass Florist	16
Green Grass Grocery	17
Dan's Auto Parts and Car Repair	18
Radishes Ice Cream	19
Pumpkin Stop Mini Marts	20
Five Star Restaurant & Hotel	21
Green Grass Oil & Gas	22
Green Grass Bank	23
Green Grass Municipal Offices	24
Green Grass Municipal Garages	25
Green Grass Library	26
Green Grass Elementary School	27
Green Grass High School	28
Green Grass Community College	29
Green Grass Senior Citizens' Home	30
Green Grass Hospital	31
Wyoming & Western Railway	32
Woodland's Wooden Lawn Ornaments	33
Western Ware Plus	34
Large ranches	35
Small ranches	36
Movie star estates	37

service departments. Larry is responsible for stocking shelves and cleaning. Clarence repairs lawn mowers and helps occasionally in the store. Brenda, Gertrude's daughter, works on accounting, manages the employees, and purchases all of the stock. Purchasing stock is one of Brenda's most time-consuming jobs. Every time a product is purchased, Brenda has to manually enter the transaction, mail the inventory request to the distributor, and then track the product through the system.

- *Straight or Curled Hair Salon*. Violet Tang owns and operates Straight or Curled Hair Salon. At this time, Violet employs two hair stylists and one floor girl—Jessie, Janice, and Jennifer. Violet has longed for Internet access that would help her research the newest hair trends and keep an eye on her stock market investments.

- *BB's Barber Shop*. Bill Barnes has been running BB's for fifty years by himself. His grandson Bernard

recently joined him in running the business. Bernard hopes to improve the efficiency of the shop by adding automatic scheduling.

- *Louden's Department Store*. Established in 1899, Louden's carries all types of items, ranging from clothes to furniture. It is still owned and managed by the Louden family, including Walden Louden, president; his daughter Carol, vice president of apparel and dry goods; and Walden Jr., vice president of furniture, marketing, and finance. Louden's employs fifty people. Ten work in the office handling mailings, customer complaints, and inventory database. Two of those ten handle the books. Thirty employees serve customers. Five employees work in the warehouse and shipping department, and the remaining five act as supervisors for floor, office, and shipping personnel. Everyone at Louden's is very excited about the new telecommunications network in Green Grass. Currently, information within the store is handled by an old pneumatic system that was installed in 1900. Sales slips, messages, and cash are inserted in a canister that is placed in the enclosed air system that shoots it through to the office.

- *Cowboy Clothes*. Todd opened Cowboy Clothes three years ago. He employs two salespeople and one assistant manager. The store carries locally made cowboy boots, cowboy hats, and other Western apparel. Todd hopes to begin e-commerce and take advantage of the world's fascination with Western wear.
- *Chaser, McDonald, and Rhetorick, Attorney's at Law*. Bill Chase, Sarah McDonald, and Ernest Rhetorick are equal partners in the only law office in Green Grass. In addition to the three attorneys, there are two paralegals, John and Megan, and one legal secretary, Bart.
- *Mayfair's Five & Dime*. Mayfair's has been running in Green Grass since 1956, when Abby and Brandon Mayfair moved in from New York City. The store carries everything from knickknacks to fabric. A small soda fountain located on one side of the store is run by Laura Jones. Abby and Brandon still run the store every day and employ five floor workers and one stock clerk.
- *BookWorm's BookStore*. Marian Page started the bookstore within the past year. Her goal was to provide Green Grass with a place to come in, relax, browse, and purchase books. Marian does not have any employees at this time. She would like to add an Internet kiosk once the communications network is up.
- *Angel's Diner*. The diner has been running in Green Grass since 1929. Angel started the diner hoping to save her failing ranch. The diner is now open from 6 a.m. until 3 p.m. and serves breakfast and lunch. Angel works every day of the year except for the five that she spends in Las Vegas with Henry Blessing, the local insurance agent. Erma Mudge, Val Littleton, and Henrietta Sprata wait tables; and Chris Henley cooks.
- *Blessings Insurance Plus*. Barbara Blessing runs the only insurance business in Green Grass. She sells auto, home, and life insurance. Barbara employs two agents and one secretary.
- *Buchman Funeral Parlor*. Big Bob Buchman owns and operates both funeral parlors in Green Grass. Big Bob also owns a funeral parlor in Prairie Corner, a town four miles from Green Grass. He employs three funeral directors and one bookkeeper.
- *Buchman Real Estate*. Big Bob also owns and runs the only real estate agency in Green Grass. Buchman Real Estate covers all of Green Grass and the surrounding areas. Big Bob was the person responsible for enticing the movie stars to buy estates outside of Green Grass. He employs two real estate agents, Deb Kettle and Horace Buchman, his less-ambitious brother. Big Bob has decided that the first circuit to be turned up will be an Internet connection to his real estate office. He has high hopes for setting up a Web site to advertise all the beautiful properties for sale in Green Grass.

- *Ed's Tattoos & Rug Cleaning*. Ed recently attended a seminar in Denver on how to apply permanent tattoos. On his return, he quickly added the tattoo business to his established rug cleaning business. He has only applied two tattoos, both on the mayor's prize heifers. Ed employs the Jones brothers as rug cleaners. The Jones brothers run a ranch south of town but often need spare cash to keep up with their tractor-pulling hobby.
- *Stuff and Stiff Taxidermy*. Stanley Stanowski owns and operates the taxidermy shop. He teaches calculus at the local high school and runs the taxidermy shop during hunting season in the fall. The store also sells antiques and is run by Stan's wife, Arbuta. Arbuta is looking forward to eBay.
- *Travel Much Travel Agency*. Kelly Jones runs the travel agency that is currently limited to booking rooms at the local motel or bed and breakfast. Sally is the only person running the business but hopes to expand once the network is in.
- *Kelly's Green Grass Florist*. Kelly also operates the Green Grass Florist shop out of the same building as the travel agency.
- *Green Grass Grocery*. Green Grass Grocery is owned and operated by George Hatch. George has two stores—one on the east side of town and one on the west. He employs thirty people at the east side store and twenty-five at the west side store. George is hoping to link the two stores together and keep a running inventory of all items. He is also planning to place ATMs in each store, along with credit card validation machines.
- *Dan's Auto Parts and Car Repair*. Dan runs an auto parts and car repair shop. He employs three mechanics, two salespeople, and one stocker.
- *Radishes Ice Cream*. Radishes is open for the summer season and is owned and operated by Todd Tillman, the Green Grass grade school principal, and his wife Mary. Todd and Mary hire several counter people to serve customers.
- *Pumpkin Stop Mini Marts*. Paul Snyder owns four Pumpkin Stop Mini Marts in Green Grass—one on the east, one on the west, one on the north, and one on the south. The stores located in the north and south are open twenty-four hours a day while the stores located in the east and west are open only sixteen. Paul always has two counter workers per store, per shift, and one pump person for the diesel pumps.
- *Five Star Restaurant & Hotel*. Five Star seats 400 people for banquets, conferences, and meetings. The total number of employees working at Five Star is twenty to forty servers, five janitors, two events planners, five managers, six hotel clerks, two bookkeepers, one gardener, two executive chefs, five prep cooks, and four dishwashers.
- *Green Grass Oil & Gas*. Paul Snyder also owns the local oil company. The company offices house all the oil employees and Paul's office. Paul has ten employees—four office workers, one accountant, three full-time truck

drivers, and two part-time truck drivers. Paul services all of Green Grass and the surrounding territory. He started by supplying gas to ranchers, then expanded to gas and oil for consumers.

- *Green Grass Bank*. Green Grass Bank, established in 1870, survived grasshopper invasions, the Depression, the oil crisis, and the invasion of easterners. The bank now has one main office and two branch offices. It is run by Karl Gordan III. He employs four clerks, one loan officer, and one administrator at the main branch. The two branch offices each have two clerks to serve customers. Karl has been nagging the city council for years for a communications network. He sees the value of tapping into the ATM market and the chance to link their banks into Grass Hoppers Bank, also owned by Karl.

- *Green Grass Municipal Offices*. The courthouse houses all county employees. The mayor has two offices in the courthouse—one for himself and one for his secretary, Karen Jackson. Also located in the courthouse are Judge Gary Keenan, district attorney Steve Stephenson, business planner Bob Wright, and Sheriff Bob French. In addition to the county officers, the courthouse has four clerks in the county records office, two clerks in the justice department, four clerks in the department of transportation office, and two clerks and four deputy sheriffs in the sheriff's office.

- *Green Grass Municipal Garages*. The municipal garage houses the road crew and road equipment such as graders, snowplows, and mowers. The road crew consists of five road men and one supervisor. A clerk and purchasing agent also work at the garage.

- *Green Grass Library*. The library was constructed in 1905 from money obtained from a private foundation. Abby and Sarah Keating have run the library for forty-five years. Additional community volunteers help to maintain the 5000 books. Abby and Sarah are hoping to add Internet connectivity once the communications network is up.

- *Green Grass Elementary School*. The elementary school is used for kindergarten through sixth grade. Each grade has two classes, each with one teacher and one teacher's aide. The elementary school has a principal, vice principal, counselor, school nurse, and three office workers. The bus garage for the entire Green Grass system is located across from the elementary school, which has 400 students.

- *Green Grass High School*. The high school holds grades 7 through 12. Each grade has two classes with one teacher and one aide. The school also has a principal, vice principal, business director, counselor, school nurse, and four office workers. The administrative office where the superintendent Hal Hooking works is located opposite the high school, which holds 350 students.

- *Green Grass Community College*. The community college has three separate buildings, located on the same campus, and one satellite office at the local high school. The college employs 130 people—50 professors, 30 clerical workers, twenty administrative personnel, 15 food service workers, 10 janitors, and 5 security guards.

- *Green Grass Senior Citizens' Home*. The senior citizen home has 100 residents and a staff of six nurses, 10 health aides, and six cafeteria workers. Arliss Abbott is the director of the home.

- *Green Grass Hospital*. Green Grass Hospital has fifty beds, an emergency room, doctors' offices, an X-ray division, and administrative offices. The hospital employs 150 people—50 nurses, 70 aides, five doctors, five administrators, 10 food service workers, five security officers, and five clerical employees.

- *Wyoming & Western Railway*. Wyoming and Western has a depot north of Green Grass near the feed mill. The depot houses two employees who maintain the line between Green Grass and Grass Hopper. Wyoming & Western is interested in selling right-of-way along the track between Green Grass and Grass Hopper.

- *Woodland's Wooden Lawn Ornaments*. Woodland's factory is located on the south side of Green Grass. The factory employs 500 people; 350 of the employees work on the factory line building the ornaments, 50 are designers, 20 are clerical workers, 10 are buildings people, 30 are managers, and 40 are salespersons.

- *Western Ware Plus*. Western Ware Plus designs and manufactures Western apparel and sells its products worldwide to major retailers. Western Ware Plus employs 300 people—200 seamstresses; 50 packers, cutters, and gofers; 30 clerical workers; 10 designers; and 10 managers.

Locations surrounding Green Grass city limits include the following:

- Jeff Jopkers, million-dollar movie star, lives at Rancho Jopkers located west of the city.
- Mabel Currie, half-million-dollar movie star, lives on a ranch also west of Green Grass.
- Hal Avery, famous talk show host, has an estate south of Green Grass.
- Emery Luftinkin, owner of three NFL teams, owns 1000 acres east of the city along Green Grass lake.
- Green Grass lake has seventy-five homes built along the shoreline as shown on the map.
- Buchman Estates is a new housing development located east of the city. At this time, ten homes have been completed in the development.
- Rancho Lopez, located five miles from Green Grass, consists of a 5000-acre ranch with five homes and a fifty-bed bunkhouse.

Equipment/Product	Cost
20,000 port class 5 switch	$1,200,000
Fiber optic OC-48 multiplexer	$60,000
Fiber optic OC-12 multiplexer	$45,000
Fiber optic OC-3 multiplexer	$30,000
M13 DS-3 multiplexer	$3700
Digital loop carrier—500 POTS lines to feed customers and twenty T1s for backhaul	$55,000
Digital cross connect—3:1, 256 DS-3s	$200,000
DSX DS-1 eighty-four-position panel	$2000
DSX DS-3 sixteen-position panel	$2000
66 block—100 position	$25
Voice-mail system	$79,000
Fiber optic cable—twenty-four strand	$1/foot
Fiber optic cable—144 count	$6/foot
Fiber optic cable—288 count	$12/foot
Fiber optic distribution panel—100	$1,500
Fiber optic distribution panel—300 port	$2,000
Copper cable—drop wire—two pairs	$300/per run per cable
Copper cable—100 pairs	$4/foot
Copper cable—1000 pairs	$41/foot
Copper cable—3000 pairs	$125/foot
Inside 100-pair copper cable	$4/foot
Category 5 cable	$7/foot
Telephone pole	$500
Terminal—twenty-five position	$300
Terminal—ten position	$200
Splice case	$1,000
Media Gateway	$25,000
Soft switch	$50,000
SIP proxy	$50,000
Ethernet 100-port patch panel	$1000
Twenty-five-port Ethernet hub	$300
Twenty-five-port Ethernet switch	$2000
Router	$3000
CAT5 cable 1000 ft.	$200
HFC remote cabinet setup (includes all piece parts—CMTS, etc.)	$200,000
Headend system components (includes all piece parts)	$200,000
Cable modems	$50.00 Wholesale
WiMAX remote package (includes antenna, transmitter/receiver)	$50,000
Twenty-five-port Ethernet fiber switch (WIMAX aggregation point equipment)	$25,000
Wi-Fi package (antenna, access, point, etc.)	$3000

Table A–2
Costs for Equipment and Cable, and Cost of Building Cable per Mile.

- Montana Ranch is a 3000-acre ranch with three homes and six ranchettes.
- Waterford Ranch is a 7000-acre ranch with eight homes, a seventy-bed bunkhouse, a restaurant, and a gift shop.
- There are also twenty smaller rancheros scattered around Green Grass. Each of these is shown in Map A–4.

To complete the case studies, you will also need to reference cost and DS-2 line code speed. Table A–2 lists generic costs for equipment and cable. Table A–3 shows DSL speeds. Table A–4 shows Tier 1 ISP parameters (Chapter 15). Table A–5 lists services required at specific sites (Chapter 16). Table A–6 provides a list of access services (Chapter 17). Table A–7 provides CATV subdivision information and refers to Map A–5 (Chapter 21). Refer to Map A–1 through Map A–4 to complete the case study for Wi-Fi/WiMAX design information (Chapter 22).

Table A–3
The Standard Digital Subscriber Line
Code, DMT, Speed versus Distance.

Wire Gauge	Line Code	Speed	kft.
26 g	DMT	5440 k	9.0
26 g	DMT	1720 k	13.5
26 g	DMT	256 k	17.5
26 g	DMT	640 k	9.0
26 g	DMT	176 k	13.5
26 g	DMT	96 k	17.7
26 g	HDSL2	1.544 M	9.0

Table A–4
Tier 1 ISP.

Parameter	XYZ Tier 1 ISP	ABC Tier 1 ISP
No. of peering partners	15	25
NAP connections	All	All
Interconnection interfaces	DS-3, OC-3, OC-12, FastE	DS-3, OC-3, OC-12, OC-48, GigE
Fully redundant interconnection	Yes—fiber working path; wireless protection path	Yes—fiber primary working and protection paths
Cost per OC-3 interconnection	$3500/month; $1000 one-time installation cost	$4500/month; $1000 one-time installation cost

Table A–5
Required Services at Selected Sites.

Site	Services Required
Central office 1	Trunks to central office 2; trunks to cellular site; OC-3 to CATV headend; ISDN PRI trunks to ISP; OC-3 to remote DLC
Central office 2	Trunks to central office 2; trunks to cellular site; OC-3 to CATV headend; ISDN PRI trunks to ISP; OC-3 to Tier ISP; OC-3 to remote DLC
CATV headend	DS-3 to Tier 1 ISP
VoIP switch center	FastE to Tier 1 ISP
Cellular switch site	Trunks to central office 1; trunks to central office 2
Local ISP	ISDN PRI trunks to central office 1 and central office 2

Table A–6
List of Access Services.

Access network services
Broadband pipe—DSL & cable modem
Traditional voice lines
VoIP services
CATV
Wi-Fi hotspots
DS-1
Fast Ethernet
OC-3

Table A–7
CATV Subdivision Information.

Fifty homes in the subdivision
Refer to Map A–5 for street layout information
Equipment list
- CMTS
- Ethernet optical switch
- Router
- Cable modems

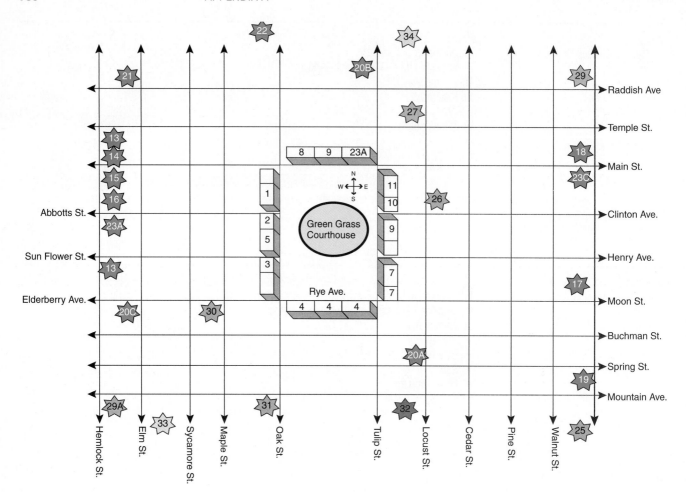

Map A–1
Green Grass town layout.

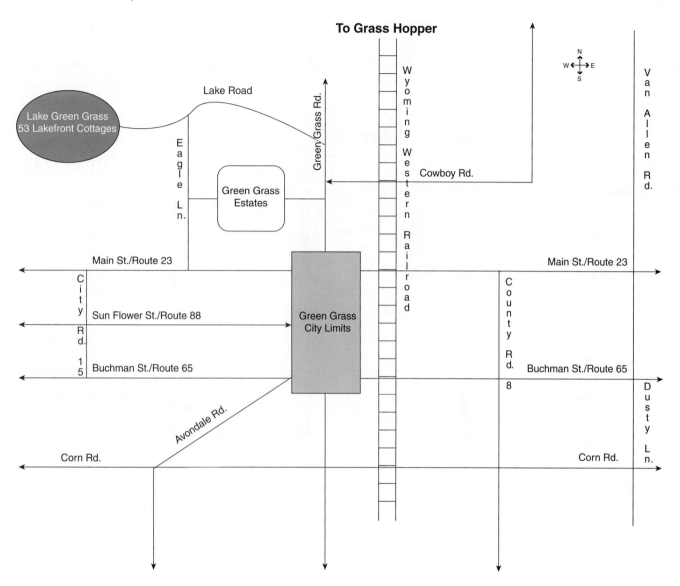

Map A–2
Green Grass rural layout.

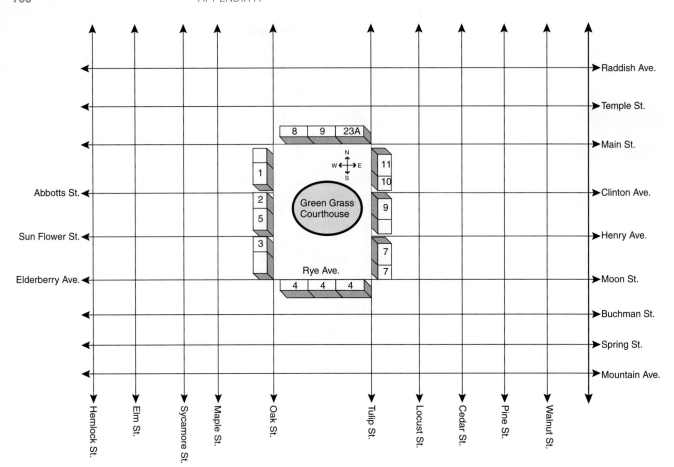

Map A–3
Green Grass town layout.

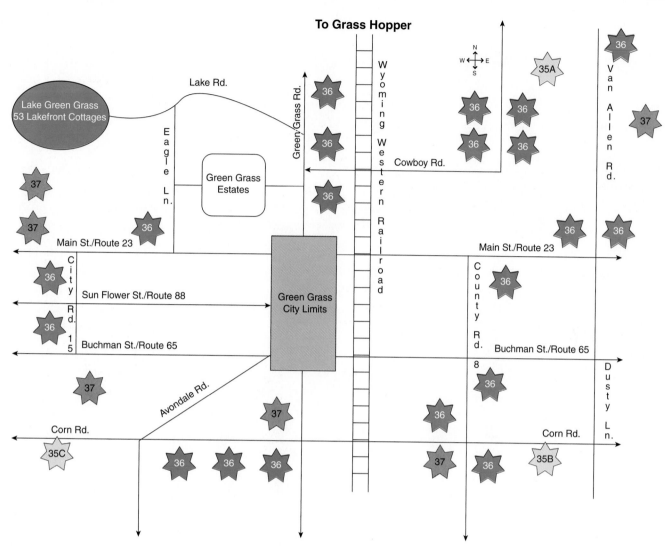

Map A–4
Green Grass rural layout.

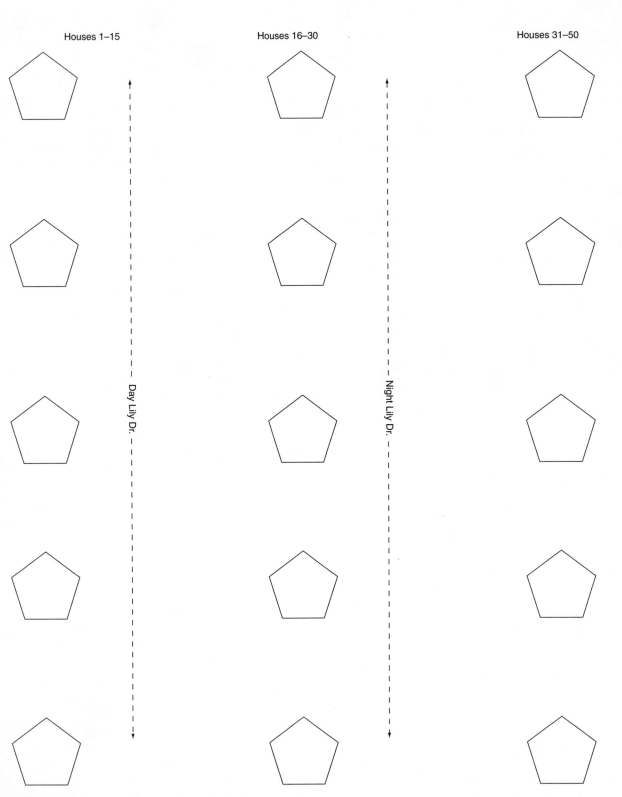

HFC Space

Houses 1–15

Houses 16–30

Houses 31–50

Day Lily Dr.

Night Lily Dr.

Spring Lily Dr.

Map A–5
Lily Valley subdivision.

Troubleshooting Guide

COPPER CABLE TESTS
POTS Lines
I. Power measurement
1. Voltage measurement
Using a volt/ohm meter (see Figure B–1), measure voltage by
Measuring between tip and ring = –48 to –52 VDC

Measuring between ring and ground = –48 to –52 VDC
Note: –48 to –52 VDC source is applied to the ring conductor; tip conductor's job is to be an insulated ground.
2. Current measurement
Using volt/ohm meter (see Figure B–2), measure current by

Figure B–1
Measuring between tip and ring.

Volt/Ohm Meter

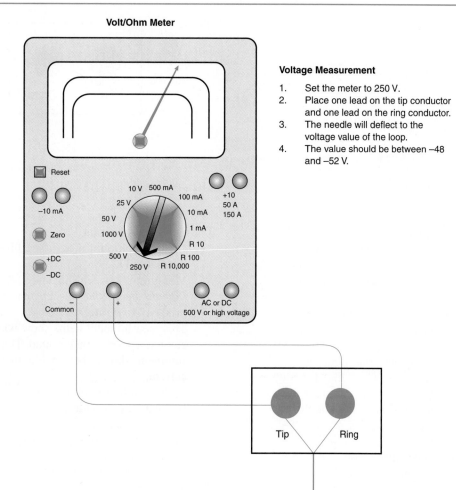

Voltage Measurement
1. Set the meter to 250 V.
2. Place one lead on the tip conductor and one lead on the ring conductor.
3. The needle will deflect to the voltage value of the loop.
4. The value should be between –48 and –52 V.

Figure B–2
Loop current measurement.

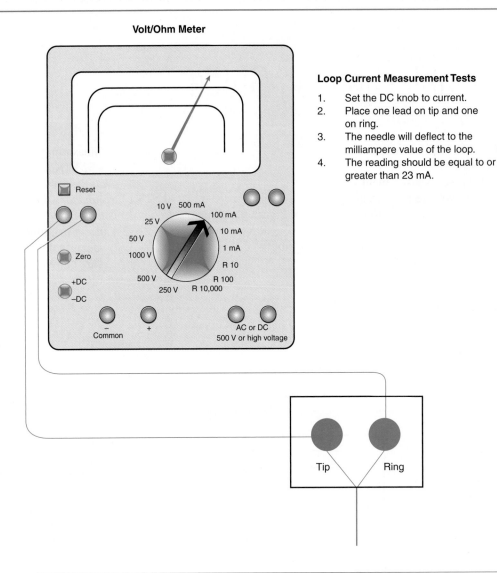

Volt/Ohm Meter

Loop Current Measurement Tests

1. Set the DC knob to current.
2. Place one lead on tip and one on ring.
3. The needle will deflect to the milliampere value of the loop.
4. The reading should be equal to or greater than 23 mA.

Good measurement

Measuring between tip and ring = 26 mA @ 52 VDC

(Varies with loop length due to resistance. Should not be less than 23 mA.)

Measuring between ring and ground = 42 mA

(Ground has 0 resistance.)

Problem found

Measuring between tip and ring <23 mA

Other than very long lines, which may function with as little as 18 or 19 mA, a current measurement less than 23 mA is telling you there is a problem on the line.

Resolution

a. Place loop extenders if the loop is too long.

b. Check the loop characteristics—balance, resistance, ground.

3. Ground measurement

Using a volt/ohm meter (see Figure B–3), set the dial to DC.

Measuring between ring and ground = 42 mA

(Should be double the current measured between tip and ring.)

This measurement helps show whether the circuit has a good or bad ground. The current measurement should be double that of the line current.

4. Loop continuity measurement

Using a volt/ohm meter (see Figure B–4), measure resistance by

Idle circuit—Handset in place

Good measurement

Measuring between tip and ring = Over 3.5 MOhms

Figure B–3 Ground measurement.

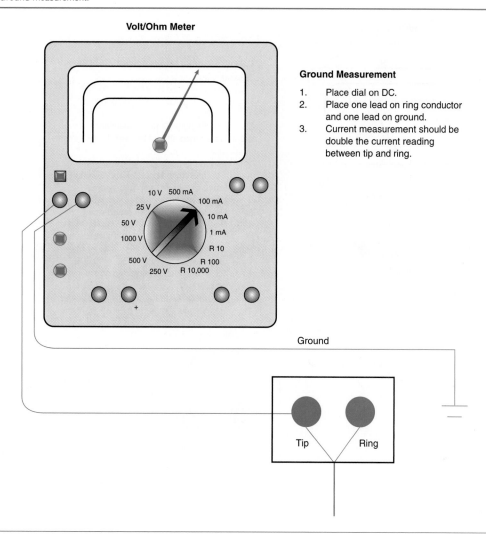

Volt/Ohm Meter

Ground Measurement

1.   Place dial on DC.
2.   Place one lead on ring conductor and one lead on ground.
3.   Current measurement should be double the current reading between tip and ring.

10 V   500 mA
25 V   100 mA
50 V   10 mA
1000 V   1 mA
                R 10
500 V   R 100
250 V   R 10,000

Ground

Tip   Ring

Measuring between tip and ground = Over 3.5 MOhms
Measuring between ring and ground = Over 3.5 MOhms
   Problem found if
Measuring between tip and ring = Less than 3.5 MOhms
   Telling you there is a short on the line.
Measuring between tip and ground = Less than 3.5 MOhms
   Telling you a ground exists.
Measuring between ring and ground = Less than 3.5 MOhms
   Telling you a ground exists.
   Resolution

If resistance is less than 5000 ohms, the fault is called a hard short. You may determine the distance to the short by
a. Calculating the resistance value to feet by knowing the cable gauge
b. Using a TDR (time domain reflectometer) or a fault locator to locate the short
   If the resistance is greater than 5000 ohms, the fault is a highly resistive fault. You may determine the distance to the short by
   • Using a TDR or fault locator to locate the trouble

Figure B–4 Loop continuity measurement.

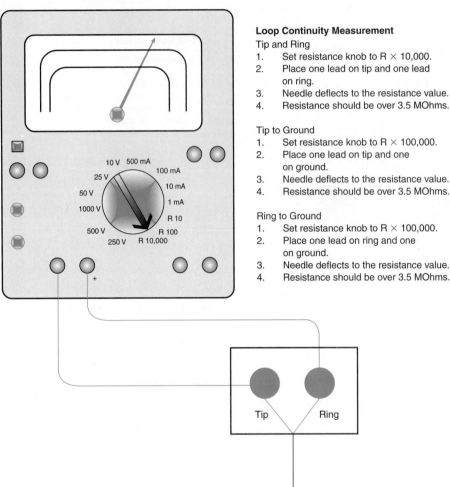

Volt/Ohm Meter

Loop Continuity Measurement

Tip and Ring
1. Set resistance knob to R × 10,000.
2. Place one lead on tip and one lead on ring.
3. Needle deflects to the resistance value.
4. Resistance should be over 3.5 MOhms.

Tip to Ground
1. Set resistance knob to R × 100,000.
2. Place one lead on tip and one on ground.
3. Needle deflects to the resistance value.
4. Resistance should be over 3.5 MOhms.

Ring to Ground
1. Set resistance knob to R × 100,000.
2. Place one lead on ring and one on ground.
3. Needle deflects to the resistance value.
4. Resistance should be over 3.5 MOhms.

5. Balance measurement
Using a volt/ohm meter (see Figure B–5)
a. Place the dial on the resistance scale to R × 10,000.
b. Touch one lead to tip and the second lead to ground.
c. Move the AC volts dial from –DC to +DC.
d. Note how far the needle kicks each time you toggle between the +/–.
e. Perform the test on each conductor noting the kick value.
f. The kick values for the tip and the ring conductor should be the same.
Problem found if
One conductor has a much higher or lower kick value than other

Resolution
a. One wire—tip or ring may be coming open.
b. Additional wire attached to one wire may cause higher resistance value and thus cause an unbalance between the two conductors.
c. Use a TDR to check the line for impairments.
II. Noise measurements for voice and data circuits riding on copper medium
Using a transmission test set referred to as a TIMS to perform the following noise tests
1. Loss
a. Connect the circuit to the test set.
b. The loss test involves sending a 1004 Hz tone across the circuit. First you will measure for the AML (actual measured loss). Once

Volt/Ohm Meter

Figure B–5
Balance measurement.

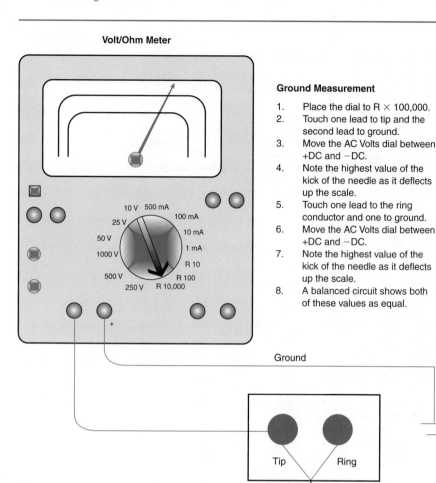

Ground Measurement

1. Place the dial to R × 100,000.
2. Touch one lead to tip and the second lead to ground.
3. Move the AC Volts dial between +DC and −DC.
4. Note the highest value of the kick of the needle as it deflects up the scale.
5. Touch one lead to the ring conductor and one to ground.
6. Move the AC Volts dial between +DC and −DC.
7. Note the highest value of the kick of the needle as it deflects up the scale.
8. A balanced circuit shows both of these values as equal.

complete, the AML will be compared to the EML (expected or engineered measured loss) to determine whether the circuit is good or not good.

c. Measure the dB loss at each point in the circuit as shown in the example in Figure B–6 that defines each point that requires testing.

d. Circuit loss for voice circuits is referred to as TLP (transmission level point). The TLP is a method used to reference loss at defined points in the circuit.

e. Circuit loss for data circuits is referred to as DLP (data level point). The DLP, similar to the TLP, is used to reference loss at defined points in the circuit.

2. 3-tone slope
   a. 3-tone slope uses frequencies other than 1004 Hz to uncover problems with circuit balance.
   b. Using a transmission test set, connect the circuit to the test set.
   c. Send 404 Hz tone down the circuit and note the loss.
   d. Send 2804 Hz tone down the circuit and note the loss.
   e. The loss should range from −2 to +7.5 with an AML reading of 16.7 dBm as shown in Figure B–7.

3. C-message noise
   a. C-message noise test measures the white noise level of a voice frequency within a digital circuit.

Figure B–6 Loss test.

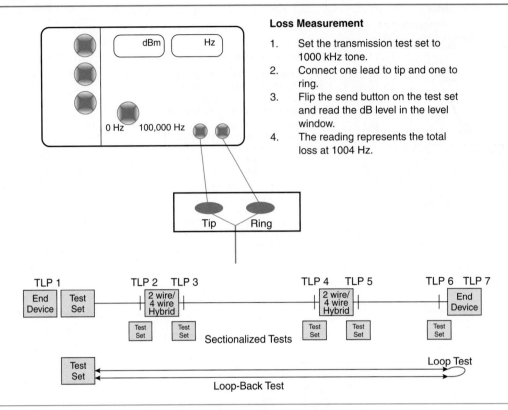

Loss Measurement

1. Set the transmission test set to 1000 kHz tone.
2. Connect one lead to tip and one to ring.
3. Flip the send button on the test set and read the dB level in the level window.
4. The reading represents the total loss at 1004 Hz.

Figure B–7 3-tone slope test.

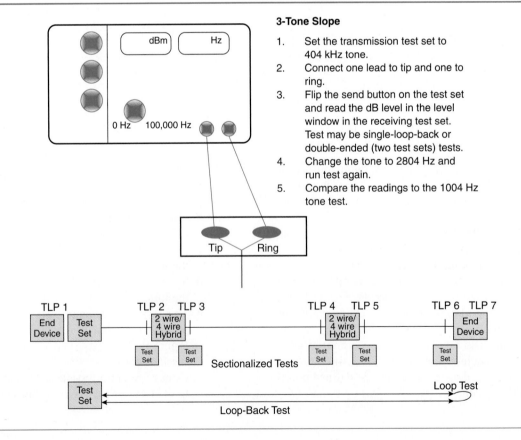

3-Tone Slope

1. Set the transmission test set to 404 kHz tone.
2. Connect one lead to tip and one to ring.
3. Flip the send button on the test set and read the dB level in the level window in the receiving test set. Test may be single-loop-back or double-ended (two test sets) tests.
4. Change the tone to 2804 Hz and run test again.
5. Compare the readings to the 1004 Hz tone test.

b. The far end of the circuit must be terminated—600 ohm termination in order to conduct this test. Using a TIMs test set, test the circuit by terminating each end, then measuring using the message filter that eliminates all frequencies below 250 Hz and above 300 Hz, looking for noise only within this window.

4. C-notched noise
   a. Measures the noise caused by the data signal.
   b. Using a TIMS test set, send out a 1004 Hz holding tone from the distant end at a –13 dBm0 level.
   c. The receiving end notches out the 1004 Hz tone allowing the remaining harmonics to be measured.
   d. A measurement of 45 dBrnc0 TLP shows the signal is good.
   e. Circuit impairments found by C-notched noise test may be caused by the analog-to-digital converter or analog amplifiers.

5. Signal to noise
   a. S/N ratio test is used to determine whether the loss, slope, and c-notched noise performed properly.
   b. The three tests should be performed and compared as shown in Figure B–8.
   c. The higher the value, the better.

6. Impulse noise
   a. Impulse noise tests look for spikes in the signal such as those caused by lightning or electrical interference.
   b. The first step before testing impulse noise is to set a signal threshold to measure against. A 1004 Hz tone may be used to perform this test.
   c. Monitor the signal for spikes occurring above the defined threshold.

T1 Testing

I. Signal monitoring
   1. Setting up the test set
      a. Setting up the test set at a DSX—Refer to Figure B–9.
      b. Setting up the test set at a termination block—Refer to Figure B–10.
      c. Setting up the test set at a repeater housing—Refer to Figure B–11.
   2. Read and interpret the output on the test set.
      a. The test set should indicate a live signal; frame sync; and, if applicable, pattern sync. Each value has an associated LED that lights green when it likes what it sees and red when it doesn't. Look first at the LEDs to make sure you are receiving a signal. If the signal LED is not green, you have a real problem.
      b. Look for BPV errors—Bipolar violations are caused by ones of the same polarity arriving consecutively. BPVs may indicate there is a problem at a repeater, there is a cable problem, or a piece of equipment is ready to fail. M13, fiber MUXes, and microwave equipment correct BPVs, thus making this test useless on fiber and microwave links.
      c. Look for CRC errors—Cyclic redundancy errors may only be seen on T1 circuits optioned for ESF framing. SF framing does not provide CRC error checks. The CRC error is a robust error-checking algorithm that has an accuracy rate of 98.4%. CRC is one of the most reliable monitoring tests available.
      d. Look for frame errors—Frame errors occur once every 193rd bit. The information bits are not checked.

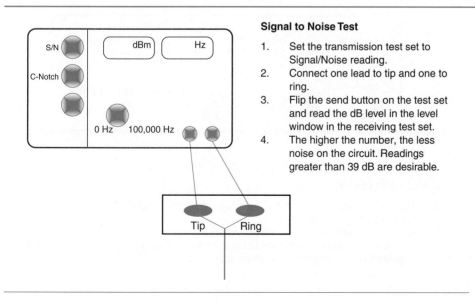

Signal to Noise Test

1. Set the transmission test set to Signal/Noise reading.
2. Connect one lead to tip and one to ring.
3. Flip the send button on the test set and read the dB level in the level window in the receiving test set.
4. The higher the number, the less noise on the circuit. Readings greater than 39 dB are desirable.

Figure B–8
Signal to noise test.

Figure B–9
Monitoring T1 circuit at a DSX panel.

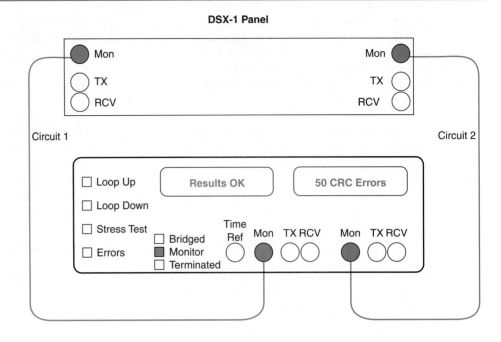

DSX-1 Panel

Monitoring T1 Circuit

1. Connect the monitor jack of the DSX to the monitor jack of the test set as shown.
2. Set the test set to monitor.
3. Review the results in the associated window.
4. The illustration above shows two circuits being monitored by one test set. Circuit one shows "Results OK" meaning the signal looks good. Circuit two shows CRC errors meaning the signal has errors.

Figure B–10
Monitoring T1 circuit at a termination block.

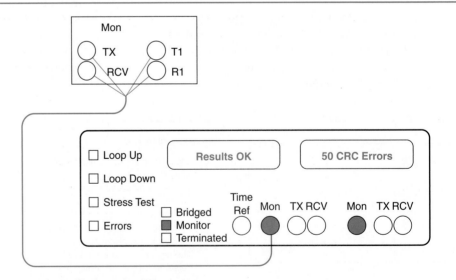

Monitoring T1 Circuit

1. Connect the leads of the termination screws to the monitor jack of the test set as shown.
2. Set the test set to bridged.
3. Review the results in the associated window.
4. The illustration shows two circuits being monitored by one test set. Circuit one shows "Results OK" meaning the signal looks good.

Figure B–11
Monitoring T1 circuit at a repeater housing.

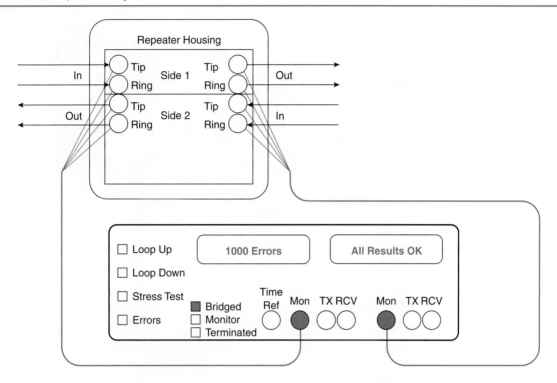

Monitoring T1 Circuit

1. Connect the leads to the pins on the repeater housing and to the monitor jack of the test set as shown.
2. Set the test set to bridged.
3. Review the results in the associated window.
4. The illustration shows two circuits being monitored by one test set. Circuit one shows
    "Errors" showing a problem coming in on side 1or going out on side 2. Further isolation is required.

3. Timing test
   a. This test may be performed only if you have a stable timing source such as a T1 signal coming from a switch or piece of equipment being referenced back to a Stratum 1 clock.
   b. Place the timing reference cord into the monitor jack of the T1 timing source. Set up the test set to show the timing slips. The indicator on the test set should show 0 timing slips. Even slowly recurring slips—1 every 2 or 3 minutes—indicate a timing problem in the network.
   c. Common cause of timing problems
      • Equipment optioned incorrectly. For example, if a channel bank or PBX is optioned as the master clock and the opposite end channel bank or PBX is also optioned as the master clock, timing problems occur. One end of the circuit must be optioned as the slave.
      • Equipment on the network timed to different clock references. For example, if one multiplexer's timing source references back to a

Stratum 1 and the second multiplexer is using the internal Stratum 3 timing source, timing slips will occur.

4. Loss of signal indicators
   a. Red alarm is telling you that there is a network problem at the near end of the circuit. The incoming pattern loses frame sync and the test set shows an all ones keep alive signal.
   b. Yellow alarm is telling you that the far end equipment is out of service and is shipping you a yellow alarm to let you know there is a problem on the circuit. The test set will show an all ones keep alive signal.
   c. All ones pattern indicates that the framed signal is gone and an all ones signal is being sent to continue synchronization.

5. Monitoring the signaling bits—A, B bits
   a. Monitoring the A, B bits in a T1 bit stream can be done by using a T1 test set with drop channel capability. Table B–1 shows the different states of the A, B bits according to their state.

Table B–1
CAS bits are read by T1 test equipment for the purpose of troubleshooting CAS problems.

Ground Start	Digital Loop Carrier				Foreign Exchange			
	Station		CO		Station		CO	
	A	B	A	B	A	B	A	B
Station Called by CO								
On-hook	0	0	0	0	0	1	1	1
Off-hook	0	0	0	1	0	1	0	1
Ringing from CO	0	0	1	1/0	0	0	0	1
Off-hook talking	1	0	0	1/0	0	1	1	1
On-hook	0	0	0	1/0	0	1	0	1
CO disconnects	0	0	0	0	0	1	1	1
CO Being Called								
On-hook	0	0	0	0	0	1	1	1
Off-hook	0	1	0	0	0	0	1	1
Dial tone	1	0	0	1/0	1	1	0	1
Dialing	1	0	0	1/0	1	1	0	1
Talking	1	0	0	1/0	1	1	0	1
On-hook	0	0	0	1/0	0	1	0	1
CO disconnects	0	0	0	0	0	1	0	1
Loop Start								
CO Being Called								
On-hook	0	0	1	1	0	1	0	1
Off-hook	1	0	1	1	1	1	0	1
Dial tone	1	0	1	1	1	1	0	1
Dialing	1	0	1	1	1	1	0	1
Talking	1	0	1	1	1	1	0	1
On-hook	0	0	1	1	0	1	0	1

b. Problems associated with signaling may be caused by
- Incorrectly optioned equipment such as robbed-bit signaling not set for that channel.
- Mismatch signaling type. For example, the channel card in the channel bank may be optioned for ground start signaling and the switch for loop start.

II. Intrusive tests
1. Setting up the test set
a. Setting up the test set at a DSX—Refer to Figure B–12.
b. Setting up the test set at a termination block—Refer to Figure B–13.
c. Setting up the test set at a repeater housing—Refer to Figure B–14.
d. Setting up two test sets at each end of the circuit—Refer to Figure B–15.
2. Stress pattern—Stress patterns are used to stress the circuit. Table B–2 lists and defines the stress patterns used to test a T1 circuit. Each tests a particular parameter, such as ones density or excessive zeros.
3. Loop-back tests
a. Two types of loop backs may be used in a loopback test. The first is a hard loop, which stands for a physically hard loop placed between the transmit and receive jacks at a DSX panel or in the field. The second is a soft loop back initiated by a loop-back code sent by the test set to the far end equipment. The equipment bridges the incoming signal to the outgoing port creating a closed loop for the signal to travel around. The benefit of the loop-back test is to quickly determine whether the circuit has continuity end to end. Once this is determined, stress patterns can be sent in order to stress the line.
b. The first test that should be performed once the loop back is established is to send a bit error from the test set and watch for its return in the

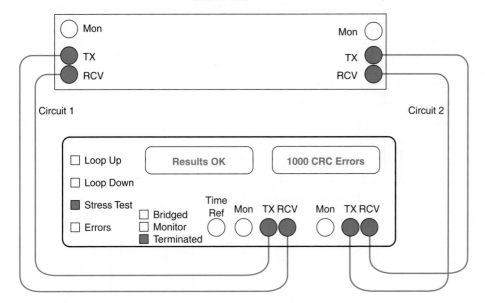

DSX-1 Panel

Figure B–12
Intrusive testing of a T1 circuit at a DSX panel.

Intrusive T1 Circuit Test

1. Connect the TX and RCV jacks of the DSX to the TX and RCV jacks of the test set as shown.
2. Set the test set to terminated.
3. Select a stress pattern such as 1:7, 3:24, QRSS, multilevel test.
4. A loop-back or second test set must be placed at the far end of the circuit.
5. Circuit one shows "Results OK" meaning the signal looks good.
   Circuit two shows CRC errors meaning the signal has errors.

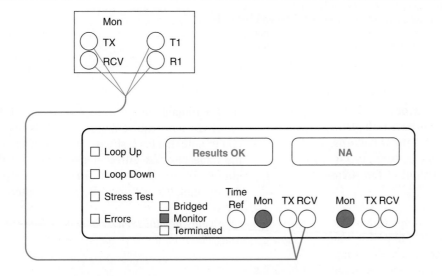

Figure B–13
Intrusive test at a termination block.

Intrusive T1 Circuit Test

1. Connect the leads to the termination screws and to the TX and RCV jacks of the test set as shown.
2. Set the test set to terminated.
3. Select a stress pattern such as 1:7, 3:24, QRSS, multilevel test.
4. A loop-back or second test set must be placed at the far end of the circuit.
5. Review the results in the associated window.
6. Circuit one shows "Results OK" meaning the signal looks good.

Figure B–14
Intrusive test of a T1 circuit at a repeater housing.

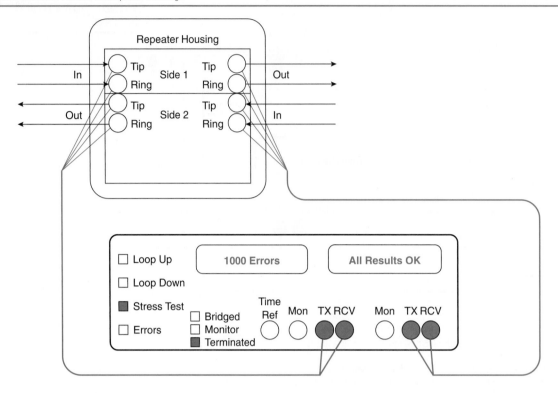

Intrusive T1 Circuit Test

1. Connect the TX & RCV jacks of the test set to the TX and RCV pins on the repeater housing and the TX & RCV jacks of the test set as shown.
2. Set the test set to terminated.
3. Select a stress pattern such as 1:7, 3:24, QRSS, multilevel test.
4. A loop-back or second test set must be placed at the far end of the circuit.
5. Circuit one shows "Results OK" meaning the signal looks good.
   Circuit two shows CRC errors meaning the signal has errors.

display of the test set. The bit error returning to the test set ensures that the loop back is in place.

   c. Once the loop back is verified, stress patterns, as shown in Table B–2, should be run. The circuit should run clean through all of the stress patterns.

4. End-to-end test

   a. Once the loop-back test is complete and the loop taken down, you may wish to perform an end-to-end test. At the time of circuit turnup, the loop-back test is performed to ensure circuit continuity. Once the loop-back test proves good, an end-to-end test is initiated to truly stress the circuit. An end-to-end test requires that two people physically be at each end of the circuit with test sets.

   b. Similar to the loop-back test, stress patterns should be sent between the two test sets and the signal monitored for errors. The stress patterns

should remain on the line for at least 1 hr. Many carriers require 24 hr. tests. It is important to note that each stress pattern as shown in the table stresses different parameters of the circuit. Troubleshooting can be made simpler if you understand what the stress pattern is stressing.

Note: The errors may be interpreted the same whether they show up during a loop-back test or an intrusive test.

DS-3 Testing

  I. Nonintrusive tests

    1. Setting up the test set

      a. Setting up the test set at a DSX—Refer to Figure B–16.

    2. The three common nonintrusive tests performed on a DS-3 are a pulse shape test, signal level, and a BPV test. All three help to indicate trouble on the line.

Figure B–15
Double-ended test of a T1 circuit.

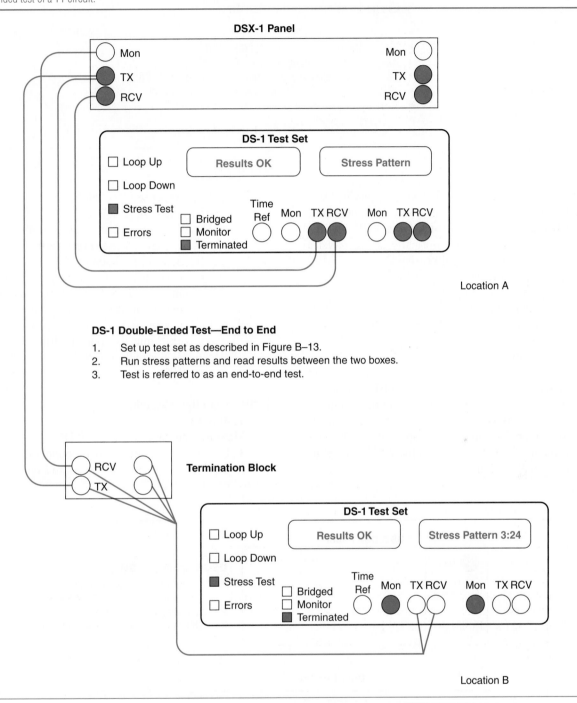

DS-1 Double-Ended Test—End to End

1. Set up test set as described in Figure B–13.
2. Run stress patterns and read results between the two boxes.
3. Test is referred to as an end-to-end test.

3. Pulse shape test and signal level are performed by
   a. Setting the display on the test set to pulse shape and referencing the value displayed as shown in the following example.
      - Pulse shape = pass (or fail)
      - Pulse width = 11.0 ns
      - Rise time = 8.0 ns
      - Fall time = 11.0 ns
      - Pulse mask = 93 ANSI
      - Power = 17.5 dBm
      - Level = 0.045 V
   b. Cause of the problem
      - Cable length issue
      - LBO set wrong for the cable length
      - Bad crimps

Table B–2
Common stress patterns used to stress a T1 link.

Stress patterns used on T1 links are:

- QRSS—Pseudorandom sequence generates every combination of patterns to stress circuit.
- 3:24—Stresses ones density and maximum zeros.
- 1:7—Stresses the timing recovery of repeaters.
- 2:6—Used to determine if equipment mis-optioned for B8ZS.
- All Ones—Stresses for DC problems, repeater problems.
- All Zeros—Stresses for all zeros rule (clear channel).
- 55 Octets—Stress repeaters.

4. BPV tests can be performed on an in-service DS-3.
5. Reference the display on the test set to determine whether there are BPVs on the line. BPVs indicate the signal is not good normally at the physical level since this is a nonintrusive DS-3 test.
   a. Cause of the problem may be
      - Unterminated plugs
      - Taps on the line
      - Equipment grounded incorrectly
      - Static charges

II. Intrusive tests
1. Two intrusive tests may be performed on the out-of-service DS-3—jitter test and a BERT test.
2. A loop back may be placed at one of the ends of the circuit, or two test sets one on each end may be used to perform this test. Loop backs on DS-3 circuits are normally placed at the DSX-3 panel using a coaxial jumper.

a. The test set sends a pattern out onto the line to stress the system.
b. The BERT test calculates the number of bit errors experienced, then converts them into a specific BER value such as $1 \times 10 - 7$. If the circuit complies with this standard, the circuit is deemed good. If the BER exceeds this standard, the circuit is considered bad.
c. Jitter is a second test performed to ensure that network timing is good. If jitter is found, the cause is normally misoptioned equipment or a poor timing source somewhere in the network.

Testing Fiber Optic Systems

A. Testing fiber for loss
   Measurement taken with an OTDR
   1. Using a fiber optic jumper, connect the OTDR to the fiber to be tested as shown in Figure B–17.

Figure B–16 DS-3 test setup.

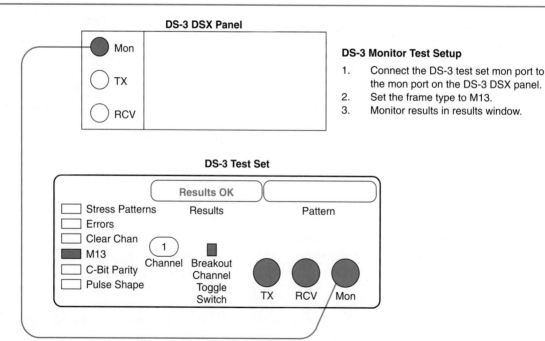

DS-3 DSX Panel

Mon

TX

RCV

DS-3 Monitor Test Setup

1. Connect the DS-3 test set mon port to the mon port on the DS-3 DSX panel.
2. Set the frame type to M13.
3. Monitor results in results window.

DS-3 Test Set

Results OK

Stress Patterns
Errors
Clear Chan
M13
C-Bit Parity
Pulse Shape

Results

Pattern

1
Channel Breakout
Channel
Toggle
Switch

TX   RCV   Mon

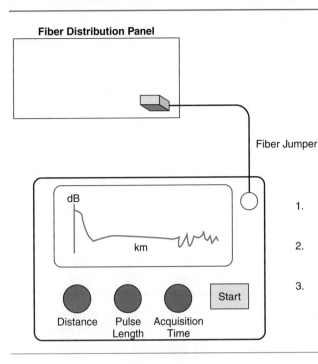

Fiber Distribution Panel

Fiber Jumper

dB

km

Distance   Pulse   Acquisition
           Length   Time

Start

OTDR Fiber Test

1. Connect the port on the fiber distribution panel to the optical output of the OTDR.
2. Set the distance, pulse length, and time on the meter, or press auto for auto configuration.
3. Read the trace in the window for loss, distance, splices, and so forth.

2. Set up the OTDR to send 1310 nm wavelength.
3. Interpret the test result.
   a. Look for end-to-end fiber distance.
   b. Look for problems on the line such as a bad splice, fiber anomaly.
   c. Reference the loss of the span.
   d. Repeat the preceding test using the 1550 nm wavelength.

Measurement taken using a power meter
1. This test requires one person with a stable light source at one end of the fiber span and a second person with a power meter at the opposite end of the fiber span.
2. Connect the power meter to the fiber to be tested.
3. Set stable light source to send 1310 nm wavelength.
4. Read the results on the power meter.
   a. No signal indicates the power coming in is too low to read.
   b. Low signal (-25.0 to -55 dB)
      • Fibers may need to be cleaned or polished.
      • Connectors are not seated correctly.
      • May have fault in the fiber span.
   c. High signal (-10.0 to -15.0 dB)
      • May need to place a pad (attenuator) to reduce signal strength.

5. Repeat the test using the 1550 nm wavelength.
Measure the return loss.
1. Most OTDRs now have stable power sources plus test for return loss, therefore set up the OTDR to test for return loss measurement.
2. Interpret the test results.
   a. Return loss low (> 30 dB)
      • Fiber may be partially open at opposite end.
      • Fibers may be dirty or need polishing.
      • Attenuate the reflection by using a mandrel wrap on the fiber. This entails wrapping the fiber around a small object four or five times. If this increases the return loss, the problem is at the opposite end. If the value does not increase, the problem is at your end. Cleaning, polishing, or, if necessary, replacing the connector should solve the problem.
Measure the PMD of the signal.
1. PMD (polarization mode dispersion) tests require a PMD tester.

Diagrams

Diagram 8–1

Using a copy of the diagram, draw connections showing a loop-back test at the far end location and connections sending a stress pattern from the near end location using the test set.

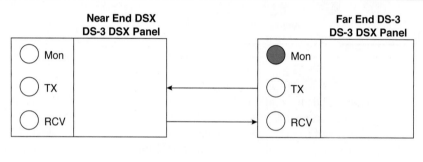

Diagram 18–1 Using a copy of the diagram, show the cross connect required to connect circuit 20 to circuit 56. Designate T1, R1, RCV, and GND.

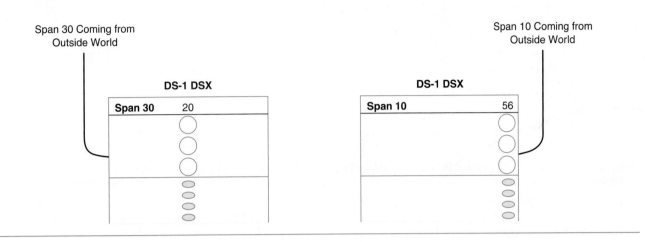

Diagram 19–1
Determine the next regenerator location to be checked for errors.

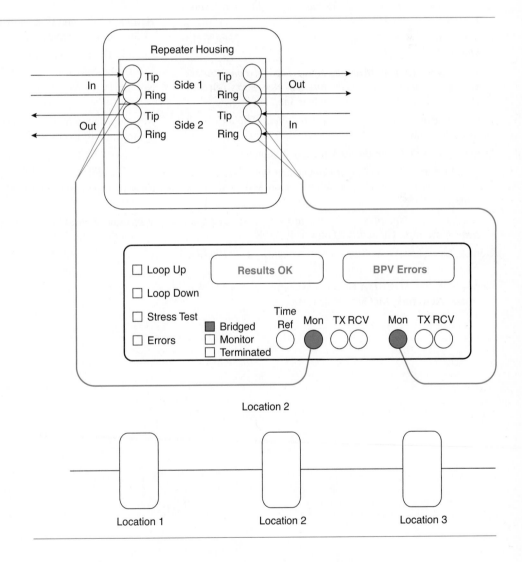

Adler, Irving. *The Wonders of Physics.* New York: Golden Press, 1966.

Ali, Syed. *Digital Switching Systems.* New York: McGraw-Hill, 1998.

Bellcore. *BOC Notes on the LEC Networks.* 1994.

Blyth, John, and Mary M. Blyth. *Telecommunications Concepts, Development, and Management.* Mission Hills: Glencoe/McGraw-Hill, 1990.

Brooks, John. *Telephone: The First Hundred Years.* New York: Harper & Row, 1976.

Brosnan, Michael and Messina, John. *The Telecom Professional's Complete Guide to the Internet.* New York: Telecom Books, A Division of Miller Freeman Inc., 1998.

Chomycz, Bob. *Fiber Optic Installer's Field Manual.* New York: McGraw-Hill, 2000.

Dempsey, Jack. *Telecom Basics.* Chicago: Telephony, 1988.

Fitzgerald, Jerry. *Business Data Communications: Basic Concepts, Security, and Design,* 4th ed. John Wiley & Sons, 1993.

Frenzel, Louis. *Principles of Electronic Communication Systems.* New York: Glencoe/McGraw-Hill, 1998.

Gonick, Larry, and Art Huffman. *The Cartoon Guide to Physics.* New York: Harper Perennial, 1991.

Goralski, Walter. *SONET: A Guide to Synchronous Optical Networks.* New York: McGraw-Hill, 1997.

Green, James. *The Irwin Handbook of Telecommunications.* 3rd ed. Chicago: Irwin Professional Pub., 1997.

Gurrie, Michael, and Patrick O'Connor. *Voice/Data Telecommunications Systems: An Introduction to Technology.* Englewood Cliffs, NJ: Prentice Hall, 1986.

Held, Gilbert. *Voice and Data Internetworking.* New York: McGraw-Hill, 2000.

Herrick, Clyde, and C. L. McKim, *Telecommunications Wiring,* 2nd ed. Upper Saddle River, NJ: Prentice Hall, 1998.

Minoli, Daniel. *Telecommunications Technology Handbook.* Boston: Artech House, 1991.

Newton, Harry. *Newton's Telecom Dictionary.* New York: Flatiron, 1997.

Newton, Harry. *Newton's Telecom Dictionary,* 16th ed. New York: CMP Books, 2000.

Rappaport, S. Theodore. *Wireless Communications Principles and Practice.* New York: Pearson Education, Inc, 2002.

Russell, Travis. *Signaling System #7.* New York: McGraw-Hill, 1995.

Van Name, F. W., Jr. *Elementary Physics.* Englewood Cliffs, NJ: Prentice Hall, 1966.

Wegner, J. D., and Rockwell, Robert. *IP Addressing and Subnetting.* Rockland, MA: Syngress Media Inc., 2000.

INDEX

Note: Italicized page numbers indicate illustrations.